LIBRARY OF CURVES: TRANSCENDENTAL FUNCTIONS

Exponential Function

$y = b^x$

($y = 2^x$ is shown)

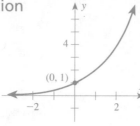

Logarithmic Function

$y = \log_b x$

($y = \log_2 x$ is shown)

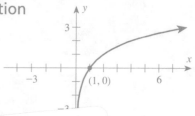

Trigonometric Functions

Cosine $y = \cos x$

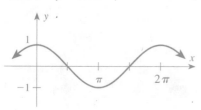

Sine $y = \sin x$

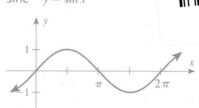

Secant $y = \sec x$

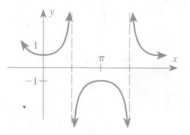

Cosecant $y = \csc x$

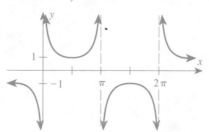

Cotangent $y = \cot x$

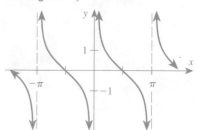

Arccosine $y = \cos^{-1} x$

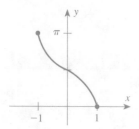

Arcsine $y = \sin^{-1} x$

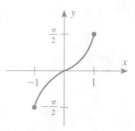

Arctangent $y = \tan^{-1} x$

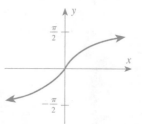

Arcsecant $y = \sec^{-1} x$

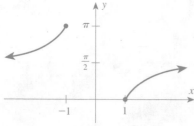

Arccosecant $y = \csc^{-1} x$

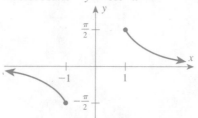

Arccotangent $y = \cot^{-1} x$

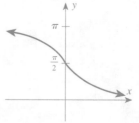

CONIC SECTIONS

$y = mx + b$ is the **line** with slope m and y-intercept b.

$y - k = m(x - h)$ is the **line** with slope m passing through (h, k).

$y - k = a(x - h)^2$ is the **parabola** with vertex (h, k) opening upward if $a > 0$ and downward if $a < 0$.

$x - h = a(y - k)^2$ is the **parabola** with vertex (h, k) opening to the right if $a > 0$ and to the left if $a < 0$.

$(x - h)^2 + (y - k)^2 = r^2$ is the **circle** with center (h, k) and radius r. This is the special case of an ellipse, where $a = b = r$.

The following standard conic sections can be translated:

$\dfrac{x^2}{a^2} + \dfrac{y^2}{b^2} = 1$ is the **ellipse** with center at $(0, 0)$, x-intercepts $\pm a$, y-intercepts $\pm b$, and foci $(\pm c, 0)$ with constant sum $2a$; $(a^2 - b^2 = c^2)$.

$\dfrac{y^2}{a^2} + \dfrac{x^2}{b^2} = 1$ is the **ellipse** with center at $(0, 0)$, x-intercepts $\pm b$, y-intercepts $\pm a$, and foci $(0, \pm c)$ with constant sum $2a$; $(a^2 - b^2 = c^2)$.

$\dfrac{x^2}{a^2} - \dfrac{y^2}{b^2} = 1$ is the **hyperbola** with center at $(0, 0)$, x-intercepts $\pm a$, foci $(\pm c, 0)$, and constant difference $2a$; $(a^2 + b^2 = c^2)$.

$\dfrac{y^2}{a^2} - \dfrac{x^2}{b^2} = 1$ is the **hyperbola** with center at $(0, 0)$, y-intercepts $\pm a$, foci $(0, \pm c)$, and constant difference $2a$; $(a^2 + b^2 = c^2)$.

$y = \pm \dfrac{b}{a} x$ are the **asymptotes** for the hyperbola.

Translation substitution:
$$x' = x - h$$
$$y' = y - k$$

Rotation substitution:
$$x = x' \cos \theta - y' \sin \theta$$
$$y = x' \sin \theta + y' \cos \theta$$
where $\cot 2\theta = (A - C)/B$

GRAPHING PROCEDURE

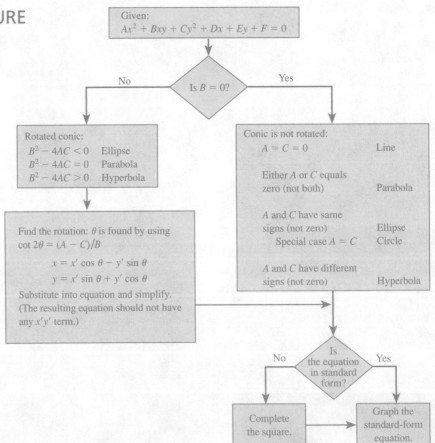

PRECALCULUS
WITH GRAPHING AND PROBLEM SOLVING

Karl Smith has gained a reputation as a master teacher and guest lecturer. He has been a department chairperson and is past president of the American Mathematical Association of Two-Year Colleges. He has served on numerous boards, including the Conference Board of Mathematical Sciences (which serves as a liaison among the National Academy of Sciences, the federal government, and the mathematical community in the United States), the Council of Scientific Society Presidents, the Board of the California Mathematics Council for Community Colleges, and was invited to participate in the National Summit of Mathematics Education. He received his B.A. and M.A. from the University of California at Los Angeles, and his Ph.D. from Southeastern University. In his spare time he enjoys running and swimming.

One of the most important goals of this book is to make mathematics real and alive for students. For that reason, personal notes of many mathematicians are included in the book.

The Precalculus Series
by Karl J. Smith

This book is part of a series of books designed to prepare students for the study of calculus. The other books in this series are:

Trigonometry for College Students
Sixth Edition

College Algebra

The Smith Business Series

Finite Mathematics
Third Edition

Calculus with Applications
Second Edition

College Mathematics
Second Edition

The Brooks/Cole One-Unit Series
by Karl J. Smith

Analytic Geometry
Second Edition

Symbolic Logic
Second Edition

Problem Solving

Geometry

Other Brooks/Cole Titles
by Karl J. Smith

The Nature of Mathematics
Sixth Edition

Mathematics: Its Power and Utility
Third Edition

Essentials of Trigonometry
Second Edition

Beginning Algebra for College Students
Fourth Edition
(with Patrick J. Boyle)

College Algebra
Fourth Edition
(with Patrick J. Boyle)

Primer for College Algebra
(with Patrick J. Boyle)

Study Guide for Algebra
(with Patrick J. Boyle)

5th EDITION

PRECALCULUS
WITH GRAPHING AND PROBLEM SOLVING

▌ **Karl J. Smith, Ph.D**

Brooks/Cole Publishing Company
Pacific Grove, California

Brooks/Cole Publishing Company
A Division of Wadsworth, Inc.
© 1993, 1990, 1986, 1983, 1979 by Wadsworth, Inc., Belmont, California 94002.

Printed in the United States of America

10 9 8 8 7 6 5 4 3 2

Library of Congress Cataloging-in-Publication Data

Smith, Karl J.
Precalculus with graphing and problem solving / Karl J. Smith.—5th ed.
   p.   cm.
Updated ed. of: Precalculus mathematics. 4th ed. c1990.
Includes index.
ISBN 0-534-16782-9
1. Functions. I. Smith, Karl J. Precalculus mathematics. II. Title.
QA331.S6173 1993   92-29752
512′.1—dc20   CIP

Sponsoring Editor: Paula-Christy Heighton
Editorial Assistant: Carol Benedict
Production Services Manager: Joan Marsh
Production Editor: Susan L. Reiland
Permissions Editor: Carline Haga
Interior Design: Kathi Townes, TECH*arts*
Cover Design: Kelly Shoemaker
Interior Illustration: TECH*arts*
Typesetting: TECHSET Composition Limited
Cover Printing: The Lehigh Press
Printing and Binding: R. R. Donnelley & Sons Co., Crawfordsville, Indiana

PREFACE

Preparation for Calculus

The selection of topics in this book is, of course, designed to prepare the student for calculus. Precalculus is a difficult subject, and there is no magic key to success. You will need to read the book, work the examples in the book using your own pencil and paper, and you will need to make a commitment to do your mathematics homework on a daily basis. I have written this book so that it will be easy for you to know what is important:

▌ Important terms are presented in **boldface type**.

▌ Important ideas are enclosed in ⎣ boxes ⎦.

▌ Common pitfalls, helpful hints, and explanations are shown in *italics*.

▌ ⊗ Warnings are given to call your attention to common mistakes.

▌ Color is used in a functional way to help you "see" what to do next.

In addition to these pedagogical aids, I have included entire sections to provide built-in redundancy and help you review for exams:

▌ The goals of each section are listed as objectives at the end of each chapter.

▌ Review problems are provided for each chapter objective.

▌ Answers to the odd-numbered problems are provided to let you know whether you are on the right track; many solutions and hints are also provided and more than 350 graphs are presented in the answer section alone to provide additional examples.

▌ Problems are graded and presented in sets so that you can work some of the simpler problems before progressing to the more challenging ones.

▌ Cumulative reviews are given as practice tests.

You will notice that this book is directly oriented to your success in calculus. Many of the topics (such as factoring) are topics that are not new, but when you look at the examples and problems presented, you will find them to be different from those you found in algebra. The reason is that I have included many of the problems *just as you will see them in calculus*. In addition, at the beginning of each chapter I have included not only a chapter overview but also a chapter perspective. This perspective points out how the techniques learned in that chapter will be used in calculus. The formulas summarized on the inside front and back covers are those ideas presented in this book that will be needed in calculus.

The main concept of this book is the one that is most needed for the study of calculus—namely, the **notion of a function**. This concept is defined and discussed in Chapter 2 and is then used as the unifying idea for the rest of the material. Although the presentation of material assumes high-school algebra, the central ideas from algebra (factoring, solving equations, simplifying algebraic expressions, the laws of exponents) are integrated into the text where appropriate since these topics are often the very ones that cause difficulty for the beginning calculus student. It is not necessary for students to have had trigonometry, since an entire trigonometry course is presented in Chapters 6, 7, and 8. I do not believe that the right-triangle approach is the best way to introduce trigonometry to students who are about to take calculus. Therefore, I use reference angles and the ratio definition for the trigonometric functions, and quickly introduce radian measure of angles.

In addition to the usual precalculus material, I have included problems that are not typical in precalculus courses, but that will be particularly useful in the study of calculus.

Techniques for Success

Success in this course is a joint effort by the student, the instructor, and the author. The student must be willing to attend class and devote time to the course *on a daily basis*. There is no substitute for working problems in mathematics. The problem sets in this book are divided into A, B, and C problems according to the level of difficulty. A word of warning is necessary here. The C problems are generally extraordinary problems and should be considered difficult and beyond the scope of the usual precalculus course. The *Student's Solution Manual* shows the complete solution to all of the odd-numbered problems.

I do not believe that students completely learn the material by working a particular type of problem only once. However, as a college instructor, I am well aware of time constraints and the amount of material that must be covered. I have therefore developed the idea of uniform problem sets for this book, which means that the instructor can make "standard assignments" consisting of the same problems being assigned from section to section. This makes it easy for the instructor and student to work problems from more than one section at a time. For example, a typical spiral assignment would be

First day:	5–60 multiples of 5 from Problem Set x
Second day:	5–60 multiples of 5 from Problem Set $(x + 1)$
	7–28 multiples of 7 from Problem Set x
Third day:	5–60 multiples of 5 from Problem Set $(x + 2)$
	7–28 multiples of 7 from Problem Set $(x + 1)$
	6–24 multiples of 6 from Problem Set x

This means that the standard assignment would be 5–60 multiples of 5, multiples of 7 from the previous set, and multiples of 6 from the assignment two class meetings before. On the other hand, if you do not wish to work spiral assignments, you can still have the same assignment from section to section:

Standard 1–2 hour assignment: Problems 5–50, multiples of 5

Standard 2–3 hour assignment: Problems 3–60, multiples of 3

If you make any of these standard assignments, or spiral assignments, you will practice all of the important ideas in this book. These standard assignments do not apply to the review problems at the end of each chapter.

In addition to a large number of drill problems, there are extensive groups of application problems, including applications from agriculture, archaeology, architecture, astronomy, aviation, ballistics, business, chemistry, earth science, economics, engineering, medicine, navigation, physics, police science, psychology, social science, space science, sports, surveying, as well as consumer and general-interest applications.

Finally, at the end of each part of the book I have included an extended application introduced by a news clipping that examines a topic in more than the usual depth—population growth, solar power, and planetary orbits.

Accuracy of the Material

It is extremely important to present the material accurately. Not only has this book been reviewed by the usual number of persons, but all of the problems have been checked and rechecked by the following persons: Gary Gislason of the University of Alaska, Audrey Rose of Portland Community College, Karen Sharp of Mott Community College, Donna Szott of Allegheny Community College, Stephen P. Simonds of Portland Community College, Mary Ellen White of Portland Community College, and Jean Woody of Tulsa Junior College. Finally, just prior to publication, the following persons checked the book's accuracy: Nancy Angle, Michael Ecker, Eunice Everett, Michael Friedberg, Diana Gerardi, Mark Greenhalgh, Scott Holm, Diane Koenig, Katherine McClain, and Janis Phillip. If you find *any* errors in this book, I want you to call me at home. My phone number is (707) 829-0606.

Calculator Usage

We are now seeing a new generation of calculators that are able to do symbolic manipulation and graphing. These calculators are characterized by the Texas Instruments TI-81, Hewlett Packard 28S, Sharp EL-5200, and Casio fx 8000G. Even though many students still do not have access to these calculators, they are beginning to revolutionize the way precalculus and calculus are taught. Nearly every change I made in this edition was the direct result of the graphing calculator. For example, solving linear and quadratic equations and inequalities can be enhanced by the use of a graphing calculator. For this reason, I have introduced two-dimensional coordinate systems as early as possible (Section 1.3). With these calculators, students can now see a *graphical* description of a process as well as an algebraic one. Consider the problem of optimization (which is considered in Section 3.3). We can not only illustrate how to find the maximum or minimum by completing the square, but can also *show* how it is the *turning point on a grap*h. Although I do not assume that students (or instructors) using this book have a graphing calculator, I have included many references to graphing calculators throughout the book. There are CALCULATOR COMMENT boxes to show how I used my graphing calculator to do a graph or enhance the concept at hand.

In the general text of this book, I assume only that the student has an inexpensive scientific calculator. Procedures for using inexpensive calculators are presented as they occur in a natural way without any special designation. This is in recognition of the assumption that most students will have such calculators. For example, the computation aspects of the logarithmic function in Chapter 5 have been minimized, and the introduction to the evaluation of the trigonometric functions in Chapter 6 assumes that each student has a calculator. In writing this book, I used a Sharp EL-531A (for which I paid $9.95) and a Texas Instruments TI-81 for graphing and matrix applications (for which I paid $125). You should always consult the owner's manual for the calculator you own. In the previous editions of this book I showed keystrokes for both algebraic and RPN logic. However, with the new generation of calculators, I now see much more diversity in possible keystrokes for a particular process. For that reason, I have removed specific keystrokes and instead have included only general remarks about process or procedure.

For the Instructor

A great deal of flexibility is possible in the instructor's selection of topics presented in this book. I have written the book so that each section represents approximately one day of class material. I have also provided more material than can be used in a single semester or quarter so that you can select material appropriate for your school or class. Some of the material that I have included may not be considered typical. Included is a chapter on graphing techniques to allow a consideration of graphing, not only of polynomial and rational functions, but also radical and absolute functions; a complete discussion of the ambiguous case in solving triangles; De Moivre's theorem; the Fundamental Theorem of Algebra; linear programming; partial fractions; translations and rotations of the conic sections; three-dimensional coordinate systems; vectors (in two and three dimensions); parametric equations; polar coordinates; and intersection of polar-form curves.

The following changes were made in the Fifth Edition:

▌ Added emphasis in illustrating each idea graphically, as well as algebraically. This will make it easy for those who have access to a graphing calculator to use it with this book.

▌ New sections on algebraic expressions, problem solving, operations of functions, cubic and quadratic functions, real roots of rational and radical equations, and inverse matrices

▌ Rewrote and expanded the section on inverse functions

▌ Split the chapter on polynomial and rational functions into two chapters; included new material on radical functions

▌ Deleted specific calculator keystrokes and added generic directions. We have provided a calculator supplement to discuss specific calculator keystrokes.

Acknowledgments

I would like to thank the reviewers of the previous editions of this book:

James Bailey, *University of Toledo*
Roy Bergstrom, *University of the Pacific*
Virginia Buchanan, *Southwest Texas State University*
David Bush, *Shasta College*
Lee R. Clancy, *Golden West College*
John Cross, *University of Northern Iowa*
Joel Cunningham, *Susquehanna University*
Daniel Drucker, *Wayne State University*
Vivian Fielder, *Fisk University*
Marjorie S. Freeman, *University of Houston, Downtown College*

Gary A. Gislason, *University of Alaska*
Sheldon Gordon, *Suffolk Community College*
Thomas Green, *Contra Costa College*
Steve Hinthorne, *Central Washington University*
James Householder, *California State University, Humboldt*
Herbert E. Kasube, *Bradley University*
Robert Keicher, *Delta College*
John Kuisti, *Michigan Technological University at South Range*
Laurence P. Maher, Jr., *North Texas State University*
Joan Mahmud, *Bergen Community College*
Raymond McGivney, *University of Hartford*
James McMurdo, *Shasta College*
Gordon D. Mock, *Western Illinois University*
Bill Orr, *San Bernardino Valley College*
Bill Pelletier, *Wayne State University*
Anna Penk, *Western Oregon State University*
Hubert J. Pollick, *Killgore College*
Richard Redner, *University of Tulsa*
Kenneth M. Shiskowski, *Eastern Michigan University*
Thomas Strommer, *Oxford College of Emery University*
David Tabor, *University of Texas*
Donald Taranto, *California State University, Sacramento*
John R. Unbehaun, *University of Wisconsin, La Crosse*
Robert Webber, *Longwood College*
Howard L. Wilson, *Oregon State University*
Terry Wilson, *San Jacinto College*

I would also like to thank the reviewers of the fifth edition:

Carl E . Cuneo, *Essex Community College*
Carol Harrison, *Susquehanna University*
Patricia Jones, *Methodist College*
Paula Kemp, *Southwest Missouri State University*
Lee McCormick, *Pasadena City College*
Stuart E. Mills, *Louisiana State University, Shreveport*
Gale Nash, *Western State College*
Froylán Tiscareño, *Mt. San Antonio College*

Finally, I would like to thank those who worked on the production of this book: Paula-Christy Heighton, Susan Reiland, Kathi Townes, and Joan Marsh.
By the time a book reaches a fifth edition, it seems inadequate to simply thank my wife, Linda, and to acknowledge the lost family time while I was working on this book. I must thank her for a lifetime of love and support.

Karl J. Smith
Sebastopol, California

CONTENTS

Karl Gauss

*Complex numbers can be omitted from this text. All parts of the book requiring the use of complex numbers will be marked as optional.

Amalie (Emmy) Noether

John Napier

Hipparchus

Copernicus

Benjamin Banneker

Gottfried Leibniz

Isaac Newton

Charlotte Scott

ADVANCED ALBEGRA TOPICS AND ANALYTIC GEOMETRY

APPENDIXES

PRECALCULUS
WITH GRAPHING AND PROBLEM SOLVING

**J. J. Sylvester
(1814–1897)**

There are three ruling ideas,
three so to say, spheres of
thought, which pervade the
whole body of mathematical
science, to some one or other of
which, or to two or all three of
them combined, every
mathematical truth admits of
being referred; these are the
three cardinal notions, of
Number, Space, and Order.
Arithmetic has for its object the
properties of number in the
abstract. In algebra, viewed as a
science of operations, order is the
predominating idea. The business
of geometry is with the evolution
of the properties of space, or of
bodies viewed as existing in space.

J. J. SYLVESTER
Philosophical Magazine, vol. 24
(1844), p. 285

It is appropriate to begin our study
of precalculus with a quotation from
J. J. Sylvester, one of the most colorful
men in the history of mathematics. This
quotation of Sylvester ties together the
ideas of number, space, and order. As we
look at the topics of precalculus, we see
that those very notions form the basis
not only for this chapter, but also for the
entire course.

Sylvester's writings are flowery and
eloquent. He was able to make the dullest
subject bright, fresh, and interesting. His
enthusiasm and zest for life were evident
in every line of his writings. He was,
however, a perfect fit for the stereotype
of an absent-minded mathematics
professor.

Born and educated in England, he
accepted an appointment as Professor
of Mathematics at the University of
Virginia in 1841. After only 3 months, he
resigned because of an altercation with a
student. He returned to England
penniless, but in 1876 he came back to
America to accept a position at Johns
Hopkins University. The 7 years he spent
at Johns Hopkins were the happiest and
most productive in his life. During his
stay there, he founded the *American
Journal of Mathematics* (in 1878).

It is in the spirit of Sylvester that I
write this book for you, the student. I
have tried to deal with your frustration
and concerns by writing a book that is
bright, fresh, and interesting. I have tried
to make the material easy to read and
understand. If I have overlooked anything, I hope that you will not dismiss
this as the work of a typical absent-
minded professor; please take time to
write me a letter or a postcard to
communicate your thoughts.

As you begin your study of precalculus, I want to remind you that learning
mathematics requires systematic and
regular study. Do not expect it to come
to you in a flash, and do not expect to
cram just before an examination. Take it
in small steps, one day at a time, and
you will find success at your doorstep.

1

FUNDAMENTAL CONCEPTS

Contents

Preview

This chapter reviews many of the preliminary ideas from previous courses and sets the foundation for the remainder of this course. Numbers, linear and quadratic equations and inequalities, as well as a two-dimensional system are introduced in this chapter. This chapter has 29 objectives, which are listed at the end of the chapter on pages 62–64.

Perspective

Calculus is divided into two parts: differential and integral calculus. Both the definition of a derivative and that of an integral are dependent on a notion defined in calculus called the *limit*. The formal definition of a limit involves both an absolute value and an inequality,

$$0 < |t - c| < \delta$$

as you can see in the portion of a page from a calculus book shown below. In the middle of the page you can also see that intervals are used. We introduce interval notation and discuss absolute value inequalities in Section 1.4.

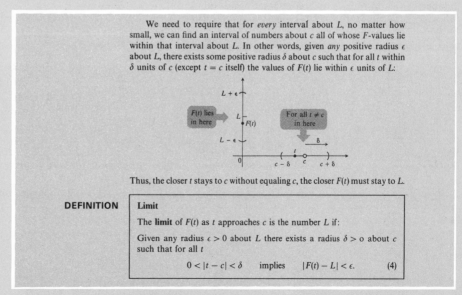

We need to require that for *every* interval about L, no matter how small, we can find an interval of numbers about c all of whose F-values lie within that interval about L. In other words, given *any* positive radius ϵ about L, there exists some positive radius δ about c such that for all t within δ units of c (except $t = c$ itself) the values of $F(t)$ lie within ϵ units of L:

$F(t)$ lies in here

For all $t \neq c$ in here

Thus, the closer t stays to c without equaling c, the closer $F(t)$ must stay to L.

DEFINITION

Limit

The **limit** of $F(t)$ as t approaches c is the number L if:

Given any radius $\epsilon > 0$ about L there exists a radius $\delta > o$ about c such that for all t

$$0 < |t - c| < \delta \qquad \text{implies} \qquad |F(t) - L| < \epsilon. \qquad (4)$$

From George Thomas and Ross Finney, *Calculus and Analytic Geometry*, 7th ed. (Reading, Mass.: Addison-Wesley), pp. 57–58.

1.1 Real Numbers

Sets of Numbers

You are, no doubt, familiar with various sets of numbers, as well as with certain properties of those numbers. Table 1.1 gives a brief summary of some of the more common subsets of the set of **real numbers.**

Table 1.1 Sets of numbers

NAME	SYMBOL	SET	EXAMPLES
Counting numbers or natural numbers	$\mathbb{N}$	$\{1, 2, 3, 4, \ldots\}$	$86; \; 1{,}986{,}412; \; \sqrt{16}; \; \sqrt{1}; \; \sqrt{100}; \ldots$ $\sqrt{a}$ is the nonnegative real number b so that $b^2 = a$. It is called the *principal square root* of a. Some square roots are natural numbers.
Whole numbers	$\mathbb{W}$	$\{0, 1, 2, 3, 4, \ldots\}$	$0; \; 86; \; 49; \; \frac{8}{4}; \; \frac{16{,}425}{25}; \ldots$ $\dfrac{a}{b}$ means $a \div b$. It is called the *quotient* of a and b. Some quotients are whole numbers.
Integers	$\mathbb{Z}$	$\{\ldots, -2, -1, 0, 1, 2, 3, \ldots\}$	$-8; \; 0; \; \frac{-16}{4}; \; 43{,}812; \; -96; \; -\sqrt{25}; \ldots$
Rational numbers	$\mathbb{Q}$	Numbers that can be written in the form p/q, where p and q are integers with $q \neq 0$; they are characterized as numbers whose decimal representations either terminate or repeat.	$\frac{2}{3}; \; \frac{4}{17}; \; .863214; \; .866\ldots; \; 5; \; 0; \; \frac{-19}{10}; \; -16; \; \sqrt{\frac{1}{4}};$ $3.1416; \; 8.\bar{6}; \; .1\overline{56}; \ldots$ An overbar indicates repeating decimals. Some square roots are also rational.
Irrational numbers	$\mathbb{Q}'$	Numbers whose decimal representations do not terminate and do not repeat	$4.1234567891011\ldots; \; 6.31331333133331\ldots;$ $\sqrt{2}; \; \sqrt{3}; \; \sqrt{5}; \; \pi; \; \dfrac{\pi}{2}; \ldots$ π is the ratio of the circumference of a circle to its diameter. It is sometimes approximated by 3.1416, but this is a rational approximation of an irrational number. We write $\pi \approx 3.1416$ to indicate that the numbers are *approximately* equal. Square roots of most numbers are irrational.
Real numbers	$\mathbb{R}$	Numbers that are either rational or irrational	All examples listed above are real numbers. Not all numbers are real numbers, however. Some of these will be considered in Section 1.5; they are called *complex numbers*.

One-Dimensional Coordinate System

The real numbers can most easily be visualized by using a **one-dimensional coordinate system** called a **real number line** (Figure 1.1).

FIGURE 1.1 A real number line

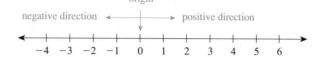

A **one-to-one correspondence** is established between all real numbers and all points on a real number line:

1. Every point on the line corresponds to precisely one real number.
2. Every real number corresponds to precisely one point.

A point associated with a particular number is called the **graph** of that number. Numbers associated with points to the right of the **origin** are called **positive real numbers** and those to the left are called **negative real numbers.** Numbers are called **opposites** if they are plotted an equal distance from the origin. The **opposite of a real number** *a* is denoted by −*a*. Notice that if *a* is positive, then −*a* is negative; and if *a* is negative, then −*a* is positive. It is a common error to think of −*a* as a negative number; it might be negative, but it might also be positive. If, say, $a = -5$, then $-a = -(-5) = 5$, which is positive.

EXAMPLE 1   Graph the following numbers on a real number line:

$$5; \quad -2; \quad 2.5; \quad 1.31331133311\ldots; \quad \tfrac{2}{3}; \quad \pi; \quad -\sqrt{2}$$

SOLUTION   When graphing, the exact positions of the points are usually approximated.   ∎

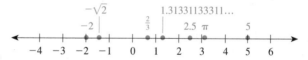

Absolute Value and Distance on a Number Line

A very important idea in mathematics involves the notion of **absolute value.** If *a* is a real number, then its graph on a number line is some point; call it *A*. The distance between *A* and the origin is the geometric interpretation of the **absolute value of *a*.**

ABSOLUTE VALUE

The absolute value of a real number *a* is denoted by $|a|$ and is defined by

$$|a| = a \quad \text{if } a \geq 0$$
$$|a| = -a \quad \text{if } a < 0$$

Thus $|5| = 5$, $|-5| = 5$, $|0| = 0$, and $|-\pi| = \pi$. Notice that $|a|$ is nonnegative for all values of *a*.

EXAMPLE 2 $|\pi - 3| = \boldsymbol{\pi - 3}$ since $\pi - 3 > 0$; use the first part of the definition of absolute value   ∎

EXAMPLE 3 $|\pi - 4| = -(\pi - 4)$ since $\pi - 4 < 0$; use the second part of the definition of
$ = \boldsymbol{4 - \pi}$ absolute value   ∎

EXAMPLE 4 $|w^2 + 1| = w^2 + 1$ since $w^2 + 1$ is positive for all real values of w

EXAMPLE 5 $|-4 - t^2| = -(-4 - t^2)$ since $-4 - t^2$ is negative for all real values of t
$$= t^2 + 4$$

Since $|a|$ can be interpreted as the distance between the point A whose coordinate is a and the origin, it is a straightforward derivation to show that the distance between any two points on a number line can be expressed as an absolute value.

DISTANCE ON A NUMBER LINE

Let x_1 and x_2 be the coordinates of two points P_1 and P_2, respectively, on a number line. The distance d between P_1 and P_2 is
$$d = |x_2 - x_1|$$

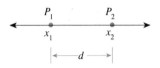

Notice that if a is the coordinate of P_1 and 0 is the coordinate of P_2, then $d = |0 - a| = |-a|$; and if 0 is the coordinate of P_1 and a is the coordinate of P_2, then $d = |a - 0| = |a|$. Since d is the same distance for both these calculations,
$$|a| = |-a|$$
It also follows that $d = |x_2 - x_1| = |x_1 - x_2|$ and that $|x^2| = |x|^2$.

EXAMPLE 6 Find the distance between the points whose coordinates are given.

a. (-8) and (2): $d = |2 - (-8)| = |10| = \mathbf{10}$
b. (108) and (-34): $d = |-34 - 108| = |-142| = \mathbf{142}$
c. $(-\pi)$ and $(-\sqrt{2})$: $d = |-\sqrt{2} + \pi| = \mathbf{\pi - \sqrt{2}}$
d. $(-\pi)$ and $(-\sqrt{10})$: $d = |-\sqrt{10} + \pi| = -(-\sqrt{10} + \pi) = \mathbf{\sqrt{10} - \pi}$

Properties of Inequality

There are certain relationships between real numbers with which you should be familiar:

LESS THAN $a < b$, read "*a is less than b*," means that $b - a$ is positive. On a number line, the graph of a is to the left of the graph of b.

GREATER THAN $a > b$, read "*a is greater than b*," means that $b < a$ or $a - b$ is positive. On a number line, the graph of a is to the right of the graph of b.

EQUAL TO $a = b$, read "*a is equal to b*," means that a and b represent the same real number. On a number line, the graphs of both a and b are the same point.

If a and b represent any real numbers, then they are related by a property called **Trichotomy** or **Property of Comparison.**

PROPERTY OF COMPARISON

Given any two real numbers a and b, exactly one of the following holds:
1. $a = b$ 2. $a > b$ 3. $a < b$

EXAMPLE 7   Illustrate the Property of Comparison by replacing the $\square$ by $=$, $>$, or $<$.

a. $-3 \square -7$ **b.** $\frac{2}{5} \square .4$ **c.** $\pi \square 3.1416$ **d.** $\sqrt{2} \square 1.4142$

SOLUTION   **a.** $-3 > -7$ Think of a number line. The graph of -3 is to the right of the graph of -7; we see that -3 is greater than -7 (so "$>$" is the answer).

b. $\frac{2}{5} = .4$ $.4 = \frac{4}{10} = \frac{2}{5}$, so the numbers are equal.

c. $\pi < 3.1416$ $\pi \approx 3.1415926$, so use "$<$," less than.

d. $\sqrt{2} > 1.4142$ $(1.4142)^2 = 1.99996164$, which is less than 2. ∎

The Property of Comparison establishes the order on a real number line. If you let $b = 0$, for example, then $a = 0$, $a < 0$, or $a > 0$. Using the definition of less than and greater than, it follows that

$a > 0$ if and only if a is positive.

$a < 0$ if and only if a is negative.

Properties of Equality

In addition to the Property of Comparison, four properties of equality are used in mathematics:

PROPERTIES OF EQUALITY

Let a, b, and c be real numbers.

REFLEXIVE PROPERTY: 1. $a = a$

SYMMETRIC PROPERTY: 2. If $a = b$, then $b = a$.

TRANSITIVE PROPERTY: 3. If $a = b$ and $b = c$, then $a = c$.

SUBSTITUTION PROPERTY: 4. If $a = b$, then a may be replaced by b (or b by a) throughout any statement without changing the truth or falsity of the statement.

EXAMPLE 8   Identify the property of equality illustrated by each:

a. $(a + b)(c + d) = (a + b)(c + d)$

b. If $(a + b)(c + d) = (a + b)c + (a + b)d$ and $(a + b)c + (a + b)d = ac + bc + ad + bd$, then $(a + b)(c + d) = ac + bc + ad + bd$.

c. If $a(b + c) = ab + ac$, then $ab + ac = a(b + c)$.

d. If $a = 3$ and $(a + 2)(a + 5) = 40$, then $(3 + 2)(3 + 5) = 40$.

SOLUTION   **a.** Reflexive   **b.** Transitive   **c.** Symmetric   **d.** Substitution ∎

Properties of Real Numbers

Finally, consider some properties of the real numbers. When we are adding real numbers, the result is called the **sum** and the numbers added are called **terms.** When we are multiplying real numbers, the result is called the **product** and the numbers multiplied are called the **factors.** The result from subtraction is called the **difference**; the result from division is the **quotient.** The real numbers, together with the relation of equality and the operations of addition and multiplication, satisfy what are called **field properties.**

Let a, b, and c be real numbers.

	Addition Properties	*Multiplication Properties*
CLOSURE:	$a + b$ is a unique real number.	ab is a unique real number.
COMMUTATIVE:	$a + b = b + a$	$ab = ba$
ASSOCIATIVE:	$(a + b) + c = a + (b + c)$	$(ab)c = a(bc)$
IDENTITY:	There exists a unique real number zero, denoted by 0, such that	There exists a unique real number one, denoted by 1, such that
	$$a + 0 = 0 + a = a$$	$$a \cdot 1 = 1 \cdot a = a$$
INVERSE:	For each real number a, there is a unique real number $-a$ such that	For each *nonzero* real number a, there is a unique real number $1/a$ such that
	$$a + (-a) = (-a) + a = 0$$	$$a\left(\frac{1}{a}\right) = \left(\frac{1}{a}\right)a = 1$$
DISTRIBUTIVE:	$a(b + c) = ab + ac$	

A set is said to be **closed** with respect to a particular operation if it satisfies the closure property, as shown in Example 9.

EXAMPLE 9 The set $\{-1, 0, 1\}$ is closed for multiplication because

$$(-1)(-1) = 1 \quad (0)(1) = 0$$
$$(-1)(0) = 0 \quad (0)(0) = 0$$
$$(-1)(1) = -1 \quad (1)(1) = 1$$

But it is not closed for addition because

$$1 + 1 = 2$$

which is not in the set.

EXAMPLE 10 Identify the field property illustrated.

  a. $4 \cdot 3$ is a real number. **b.** $\frac{1}{4}(4 \cdot 3) = (\frac{1}{4} \cdot 4)3$
  c. $(\frac{1}{4} \cdot 4)3 = 1 \cdot 3$ **d.** $1 \cdot 3 = 3$
  e. $4 + (3 + 5) = (3 + 5) + 4$ **f.** $4 + (3 + 5) = (4 + 3) + 5$

SOLUTION **a.** Closure property for multiplication
  b. Associative property for multiplication
  c. Inverse for multiplication
  d. Identity for multiplication
  e. Commutative property for addition
  f. Associative property for addition

1.1 Problem Set

A *Classify each example in Problems 1–3 as a natural number, whole number, integer, rational number, irrational number, real number, or none of these. Notice from Table 1.1 that each number listed may be in more than one of these sets.*

1. a. -5 **b.** $\dfrac{13}{2}$ **c.** $\sqrt{25}$

    d. $\sqrt{20}$ **e.** 2π **f.** $\sqrt{\dfrac{1}{8}}$

2. a. 9 **b.** $\sqrt{16}$ **c.** $\dfrac{0}{5}$

    d. $\dfrac{5}{0}$ **e.** $.\overline{3}$ **f.** $\sqrt{\dfrac{1}{9}}$

3. a. $.\overline{5}$ **b.** $.4\overline{63}$ **c.** $\sqrt{1{,}000}$

    d. $.281$ **e.** $\dfrac{\pi}{6}$ **f.** $.5656$

WHAT IS WRONG, *if anything, with each of the statements in Problems 4–6? Explain your reasoning.*

4. $\sqrt{2} = 1.414213562$

5. $\pi = 3.141592654$

6. $-n$ is a negative number

Graph each number given in Problems 7–12 on a real number line.

7. $-3;\quad -\sqrt{3};\quad \dfrac{5}{4};\quad 2;\quad .1234567891011\ldots$

8. $-1;\quad \dfrac{4}{5};\quad -\sqrt{5};\quad 2\tfrac{1}{8};\quad 1.03469217$

9. $-4;\quad -\dfrac{5}{3};\quad 0;\quad \sqrt{2};\quad -1.343443444\ldots$

10. $0;\quad \dfrac{\pi}{6};\quad \dfrac{\pi}{-3};\quad \dfrac{\pi}{2};\quad \pi$

11. $-\pi;\quad -\dfrac{\pi}{2};\quad \dfrac{\pi}{3};\quad 0;\quad .5$

12. $-\pi;\quad .25;\quad 0;\quad \dfrac{\pi}{4};\quad 1$

Illustrate the Property of Comparison in Problems 13–18 by replacing the $\square$ by $=$, $>$, or $<$.

13. a. $-10\,\square\,-3$ **b.** $6-9\,\square\,9-6$

14. a. $-4\,\square\,-8$ **b.** $\dfrac{5}{4}\,\square\,1.2$

15. a. $-\dfrac{8}{3}\,\square\,-2.66$ **b.** $\dfrac{35}{21}\,\square\,\dfrac{-30}{-18}$

16. a. $\sqrt{3}\,\square\,1.7$ **b.** $\dfrac{1}{3}\,\square\,.33333333$

17. a. $\dfrac{3}{4}+\dfrac{4}{5}\,\square\,\dfrac{31}{20}$ **b.** $\pi\,\square\,\dfrac{22}{7}$

18. a. $\dfrac{2}{3}+\dfrac{5}{2}\,\square\,\dfrac{19}{6}$ **b.** $\dfrac{2}{3}\,\square\,.66666667$

Write each expression in Problems 19–21 without using the symbol for absolute value.

19. a. $|\pi - 2|$ **b.** $|\pi - 5|$
    c. $|2\pi - 6|$ **d.** $|2\pi - 7|$

20. a. $|x^2|$ **b.** $|x^2 + 4|$
    c. $|-3 - x^2|$ **d.** $|-2 + \sqrt{5}|$

21. a. $|\sqrt{2} - 1|$ **b.** $|\sqrt{2} - 2|$
    c. $\left|\dfrac{\pi}{6} - 1\right|$ **d.** $\left|\dfrac{2\pi}{3} - 1\right|$

Find the distance between the points whose coordinates are given in Problems 22–24.

22. a. (8) and (15) **b.** (-103) and (6)
    c. (-143) and (-120)

23. a. $(-\pi)$ and (-3) **b.** (23) and (-96)
    c. $(-\sqrt{5})$ and (-3)

24. a. (π) and (-4) **b.** $(\sqrt{3})$ and $(-\pi)$
    c. $(-\pi)$ and $(-\sqrt{2})$

Identify the property of equality or field property illustrated by Problems 25–42.

25. $14x + 8x = 14x + 8x$

26. $10y + 4y = 10y + 4y$

27. $14x + 8x = (14 + 8)x$

28. $10y + 4y = (10 + 4)y$

29. $(14 + 8)x = 22x$

30. $(10 + 4)y = 14y$

31. If $14x + 8x = (14 + 8)x$ and $(14 + 8)x = 22x$, then $14x + 8x = 22x$.

32. If $10y + 4y = (10 + 4)y$ and $(10 + 4)y = 14y$, then $10y + 4y = 14y$.

33. If $14x + 8y = 22$ and $y = b$, then $14x + 8b = 22$.

34. If $22 = 14x + 8y$, then $14x + 8y = 22$.

35. 4π is a real number.

36. $\dfrac{1}{6}$ is a real number.

37. $\pi \cdot \dfrac{1}{\pi} = 1$

38. $\dfrac{1}{\frac{1}{6}}$ is a real number.

39. $5x(a + b) = 5xa + 5xb$

40. $(a + b)5x = 5x(a + b)$

41. $\dfrac{\sqrt{3} + 2}{\sqrt{5} + 1} = \dfrac{\sqrt{3} + 2}{\sqrt{5} + 1} \cdot \dfrac{\sqrt{5} - 1}{\sqrt{5} - 1}$

42. $\dfrac{4}{5} = \dfrac{4}{5} \cdot \dfrac{3}{3}$

___**B**___

43. Is the set $\{0, 1\}$ closed for addition?

44. Is the set $\{0, 1\}$ closed for multiplication?

45. Is the set $\{-1, 0, 1\}$ closed for subtraction?

46. Is the set $\{-1, 0, 1\}$ closed for nonzero division?

47. Is the set $\{0, 3, 6, 9, 12, \ldots\}$ closed for addition?

48. Is the set $\{0, 3, 6, 9, 12, \ldots\}$ closed for multiplication?

49. Is the set $\{0, 3, 6, 9, 12, \ldots\}$ closed for nonzero division?

50. Find an example showing that the operation of subtraction on $\mathbb{R}$ is not commutative.

51. Find an example showing that the operation of division on the set of nonzero real numbers is not commutative.

52. Is the operation of subtraction on $\mathbb{R}$ associative?

53. Is the operation of division on the set of nonzero real numbers associative?

___**C**___

54. What is the additive inverse of the real number $2 + \sqrt{3}$?

55. What is the additive inverse of the real number $\frac{\pi}{3} + 1$?

56. What is the multiplicative inverse of the real number $2 + \sqrt{3}$?

57. What is the multiplicative inverse of the real number $\frac{\pi}{3} + 1$?

58. What is the multiplicative inverse of the real number $\sqrt{5} + 1$?

59. Which of the field properties are satisfied by the set of natural numbers?

60. Which of the field properties are satisfied by the set of whole numbers?

61. Which of the field properties are satisfied by the set of integers?

62. Which of the field properties are satisfied by the set of rational numbers?

63. Which of the field properties are satisfied by the set of irrational numbers?

64. Which of the field properties are satisfied by the set of real numbers?

65. Which of the field properties are satisfied by the set $H = \{1, -1, i, -i\}$, where i is a number such that $i^2 = -1$?

66. Which of the field properties are satisfied by the set of nonnegative multiples of 5, namely $F = \{0, 5, 10, 15, 20, 25, \ldots\}$?

1.2 **Algebraic Expressions**

Terminology

An **algebraic expression** is a grouping of constants and variables obtained by applying a finite number of operations (such as addition, subtraction, multiplication, nonzero division, extraction of roots, or raising to integral or fractional powers). In your previous courses you have used the laws of exponents extensively, and they must also be fully understood for the study of calculus.

DEFINITION OF EXPONENTS (RATIONAL EXPONENTS)

If b is any real number and n is any natural number, then
$$b^n = \underbrace{b \cdot b \cdot b \cdots b}_{n \text{ factors}}$$

If $b \neq 0$, then $b^0 = 1$ and $b^{-n} = \dfrac{1}{b^n}$.

If $b > 0$, then $b^{1/n} = \sqrt[n]{b}$.

If m is also a natural number where m/n is reduced, then
$$b^{m/n} = (b^{1/n})^m = \sqrt[n]{b^m} = (\sqrt[n]{b})^m$$

The number b is called the **base**, n is called the **exponent**, and b^n is called the **nth power** of b. The five laws of exponents can now be summarized.

LAWS OF EXPONENTS

Let a and b be real numbers and let m and n be any integers. Then the five rules listed below govern the use of exponents except that the form 0^0 and division by zero are excluded.

FIRST LAW: $\quad b^m \cdot b^n = b^{m+n}$ FOURTH LAW: $\quad (ab)^m = a^m b^m$

SECOND LAW: $\quad \dfrac{b^m}{b^n} = b^{m-n}$ FIFTH LAW: $\quad \left(\dfrac{a}{b}\right)^m = \dfrac{a^m}{b^m}$

THIRD LAW: $\quad (b^n)^m = b^{mn}$

If an algebraic expression is a finite sum of terms with *whole-number exponents* on the variables, then it is called a **polynomial.** For example,

$$5x, \quad 3x^2 y + y^3, \quad \tfrac{1}{2}x + 4, \quad 0, \quad 8.42, \quad \text{and} \quad \frac{x+1}{5}$$

are all polynomials, but

$$\frac{1}{x}, \quad 2 + \sqrt{x}, \quad \frac{6\sqrt{x+2}}{5}, \quad \text{and} \quad x^{2/3}$$

are not polynomials.

nth DEGREE POLYNOMIAL IN x

$$a_n x^n + a_{n-1} x^{n-1} + \cdots + a_2 x^2 + a_1 x + a_0 \qquad a_n \neq 0$$

In an expression of the form $a_n x^n$, a_n is called the **coefficient** of x^n. And if a_n is the coefficient of the highest power of x, it is called the **leading coefficient** of the polynomial. If a_n and x are nonzero, the polynomial function is said to have **degree** n. Notice that if $n = 2$ (degree 2), then the polynomial is called **quadratic;** if $n = 1$ (degree 1), then the polynomial is said to be **linear;** and if $n = 0$ (degree 0), then it is called a **constant.**

Each $a_k x^k$ of a polynomial is called a **term.** Polynomials are frequently classified according to the number of terms. A **monomial** is a polynomial with one term, a **binomial** has two terms, and a **trinomial** has three terms. It is customary to arrange the terms of a polynomial in order of decreasing powers of the variable. **Similar terms** are terms with the same variable and the same degree. Two polynomials are **equal** if and only if coefficients of similar terms are the same. If some of the numberical coefficients are negative, they are usually written as subtractions of terms. Thus,

$$6 + (-3)x + x^3 + (-2)x^2$$

would customarily be written as

$$x^3 - 2x^2 - 3x + 6$$

Operations with Polynomials

A polynomial is said to be **simplified** if it is written with all similar terms combined and is expressed in the form of a polynomial, as given in the preceding box.

EXAMPLE 1 Find $(5x^2 + 2x + 1) + (3x^3 - 4x^2 + 3x - 2)$.

SOLUTION Use the distributive law to combine similar terms:

$$(5x^2 + 2x + 1) + (3x^3 - 4x^2 + 3x - 2)$$
$$= 3x^3 + \underbrace{5x^2 + (-4)x^2}_{\text{similar terms}} + \underbrace{2x + 3x}_{\text{similar terms}} + \underbrace{1 + (-2)}_{\text{similar terms}}$$

This step is usually done mentally.

$$= \mathbf{3x^3 + x^2 + 5x - 1}$$

EXAMPLE 2 Find $(5x^2 + 2x + 1) - (3x^3 - 4x^2 + 3x - 2)$.

SOLUTION Be careful when subtracting polynomials; to subtract, you must subtract *each* term. Thus

$$(5x^2 + 2x + 1) - (3x^3 - 4x^2 + 3x - 2)$$
$$= 5x^2 + 2x + 1 + (-3x^3) + 4x^2 + (-3x) + 2$$
$$= \mathbf{-3x^3 + 9x^2 - x + 3}$$

EXAMPLE 3 Find $(4x - 5)(3x^3 - 4x^2 + 3x - 2)$.

SOLUTION Use the distributive property:

$$(4x - 5)(3x^3 - 4x^2 + 3x - 2)$$
$$= (4x - 5)(3x^3) + (4x - 5)(-4x^2) + (4x - 5)(3x) + (4x - 5)(-2)$$
$$= 12x^4 - 15x^3 - 16x^3 + 20x^2 + 12x^2 - 15x - 8x + 10$$
$$= \mathbf{12x^4 - 31x^3 + 32x^2 - 23x + 10}$$

Factoring Polynomials

If a polynomial is written as the product of other polynomials, each polynomial in the product is called a **factor** of the original polynomial. **Factoring** a polynomial means to break up the polynomial into the product of individual factors. The

Table 1.2 Factoring types

TYPE	FORM	COMMENTS
1. Common factors	$ax + ay + az = a(x + y + z)$	This simply uses the distributive property. It can be applied with any number of terms.
2. Difference of squares	$x^2 - y^2 = (x - y)(x + y)$	The *sum* of two squares cannot be factored in the set of real numbers.
3. Difference of cubes	$x^3 - y^3 = (x - y)(x^2 + xy + y^2)$	This is similar to the difference of squares and can be proved by multiplying the factors.
4. Sum of cubes	$x^3 + y^3 = (x + y)(x^2 - xy + y^2)$	Unlike the sum of squares, the sum of cubes can be factored.
5. Perfect square	$x^2 + 2xy + y^2 = (x + y)^2$ $x^2 - 2xy + y^2 = (x - y)^2$	The middle term is twice the product xy.
6. Perfect cube	$x^3 + 3x^2y + 3xy^2 + y^3 = (x + y)^3$ $x^3 - 3x^2y + 3xy^2 - y^3 = (x - y)^3$	The numerical coefficients of the terms are: 1 3 3 1 or 1 −3 3 −1
7. Binomial product	$acx^2 + (ad + bc)xy + bdy^2$ $\qquad = (ax + by)(cx + dy)$ See Examples 11–13 in the text.	This procedure is used to factor a trinomial into binomial factors. It should be used after types 1–6 have been checked.
8. Grouping	See Examples 15 and 16 in the text.	After types 1–7 have been checked, you can factor some expressions by grouping.
9. Irreducible (cannot be factored over the set of integers)	Examples arise in every factoring situation. Expressions such as $x^2 + 4$, $x^2 + xy + y^2$, and $x^2 + y^2$ cannot be factored over the set of integers.	When factoring, you are not through until all the factors are irreducible.

types of factoring problems usually encountered are summarized in Table 1.2. For example, the expression $6x^2 + 3x - 18$ is properly factored by *first* finding the common factor (type 1) and then noting that types 2–6 do not apply. Finally, use type 7 (binominal factors) to write

$$6x^2 + 3x - 18 = 3(2x^2 + x - 6)$$
$$= 3(2x - 3)(x + 2)$$

We usually factor **over the set of integers,** which means that all the numerical coefficients are integers. If the original polynomial has fractional coefficients, you should factor out the fractional part first, as shown in Example 4.

EXAMPLE 4

Factor: $\dfrac{1}{36}x^2 - 9y^4 = \dfrac{1}{36}(x^2 - 6^2 \cdot 3^2 y^4)$

$$= \dfrac{1}{36}[x^2 - (18y^2)^2]$$

$$= \dfrac{1}{36}(x - 18y^2)(x + 18y^2)$$

Factoring over the set of integers rules out factoring

$$x^2 - 3 = (x - \sqrt{3})(x + \sqrt{3})$$

⊗ Note the agreement about factoring in this book. ⊗

since the factors do not have integer coefficients. In this book, a polynomial is **completely factored** if all fractions are eliminated by common factoring (as shown in Example 4) and if no further factoring is possible *over the set of integers*. If, after fractional common factors have been removed, a polynomial cannot be factored further, then it is said to be **irreducible.** Examples 5–16 review various factoring techniques.

EXAMPLE 5 $5x + 7xy = x(5 + 7y)$ Common factor x

EXAMPLE 6 $7 - 35x = 7(1 - 5x)$ Common factor 7 (note the 1)

EXAMPLE 7 $5x(3a - 2b) + y(3a - 2b) = (5x + y)(3a - 2b)$ Common factor $(3a - 2b)$

EXAMPLE 8 $3x^2 - 75 = 3(x^2 - 25)$ Common factor first
$$= 3(x - 5)(x + 5)$$ Difference of squares

EXAMPLE 9 $\dfrac{9a^2}{b^2} - (a + 3b)^2 = \dfrac{1}{b^2}[9a^2 - b^2(a + 3b)^2]$ Common factor to eliminate fractions

$$= \dfrac{1}{b^2}\{(3a)^2 - [b(a + 3b)]^2\}$$ This step can be done mentally.

$$= \dfrac{1}{b^2}[3a - b(a + 3b)][3a + b(a + 3b)]$$ Difference of squares

$$= \dfrac{1}{b^2}(3a - ab - 3b^2)(3a + ab + 3b^2)$$ Simplify

EXAMPLE 10 $(x + 3y)^3 + 8 = [(x + 3y) + 2][(x + 3y)^2 - (x + 3y)(2) + (2)^2]$ Sum of cubes
$$= (x + 3y + 2)(x^2 + 6xy + 9y^2 - 2x - 6y + 4)$$

EXAMPLE 11 $x^2 - 8x + 15 = (x - 5)(x - 3)$ Binomial product is sometimes called FOIL; you may have to try several possibilities before you find the correct binomial factors.

EXAMPLE 12 $6x^2 + x - 12 = (2x + 3)(3x - 4)$

EXAMPLE 13 $6x^2 - 9x - 15 = 3(2x^2 - 3x - 5)$ Remember to factor completely; several types may be combined in
$$= 3(2x - 5)(x + 1)$$ one problem.

EXAMPLE 14 $4x^4 - 13x^2y^2 + 9y^4 = (x^2 - y^2)(4x^2 - 9y^2)$
$$= (x - y)(x + y)(2x - 3y)(2x + 3y)$$

EXAMPLE 15 $9x^3 + 18x^2 - x - 2 = (9x^3 + 18x^2) - (x + 2)$ None of the types 1–7 from Table 1.2 seem to apply. Thus try grouping the
$$= 9x^2(x + 2) - (x + 2)$$ terms. Some groupings may lead to a
$$= (9x^2 - 1)(x + 2)$$ factorable form; others may not.
$$= (3x - 1)(3x + 1)(x + 2)$$

In calculus, one of the most commonly used types of factoring is common factoring. The following simplification uses common factoring and is required in calculus.

EXAMPLE 16 Simplify $\dfrac{x^{2/3} \cdot 2(x^2 - 9) \cdot 2x - \frac{2}{3}x^{-1/3}(x^2 - 9)^2}{(x^{2/3})^2}$.

SOLUTION You need to recognize the numerator as two terms separated by a minus sign:

$$\boxed{x^{2/3} \cdot 2(x^2 - 9) \cdot 2x} \quad - \quad \boxed{\tfrac{2}{3}x^{-1/3}(x^2 - 9)^2}$$

Looking at these terms (enclosed in boxes), we see that the factors 2, x, and $(x^2 - 9)$ are in both boxes. We choose as the common factor each of these base numbers with an exponent equal to the smaller of the exponents of these factors that appear in the boxes—namely, 2, $x^{-1/3}$, and $(x^2 - 9)^1$. In addition, since there is a fraction in the second box, we also factor out $\frac{1}{3}$:

$$\frac{x^{2/3} \cdot 2(x^2 - 9) \cdot 2x - \frac{2}{3}x^{-1/3}(x^2 - 9)^2}{(x^{2/3})^2} = \frac{\frac{2}{3}x^{-1/3}(x^2 - 9)[3x \cdot 2x - (x^2 - 9)]}{(x^{2/3})^2}$$

$$= \frac{\frac{2}{3}x^{-1/3}(x^2 - 9)(5x^2 + 9)}{(x^{2/3})^2}$$

$$= \tfrac{2}{3}x^{-5/3}(x^2 - 9)(5x^2 + 9)$$

$$= \tfrac{2}{3}x^{-5/3}(x - 3)(x + 3)(5x^2 + 9) \qquad \blacksquare$$

1.2 Problem Set

A *Write the expressions in Problems 1–6 in simplified polynomial form.*

1. a. $(x + 2)(x + 1)$ **b.** $(y - 2)(y + 3)$
   c. $(x + 1)(x - 2)$ **d.** $(y - 3)(y + 2)$

2. a. $(a - 5)(a - 3)$ **b.** $(b + 3)(b - 4)$
   c. $(c + 1)(c - 7)$ **d.** $(z - 3)(z + 5)$

3. a. $(2x + 1)(x - 1)$ **b.** $(2x - 3)(x - 1)$
   c. $(x + 1)(3x + 1)$ **d.** $(x + 1)(3x + 2)$

4. a. $(x + y)(x - y)$ **b.** $(a + b)(a - b)$
   c. $(5x - 4)(5x + 4)$ **d.** $(3y - 2)(3y + 2)$

5. a. $(a + 2)^2$ **b.** $(b - 2)^2$
   c. $(x + 4)^2$ **d.** $(y - 3)^2$

6. a. $(u + v)^3$ **b.** $(s - 2t)^3$
   c. $(1 - 3n)^3$ **d.** $(2x - y)^3$

Factor completely, if possible, the expressions in Problems 7–12.

7. a. $me + mi + my$ **b.** $a^2 - b^2$
   c. $a^2 + b^2$ **d.** $a^3 - b^3$

8. a. $a^3 + b^3$ **b.** $s^2 + 2st + t^2$
   c. $m^2 - 2mn + n^2$ **d.** $u^2 + 2uv + v^2$

9. a. $a^3 + 3a^2b + 3ab^2 + b^3$
   b. $p^3 - 3p^2q + 3pq^2 - q^3$
   c. $-c^3 + 3c^2d - 3cd^2 + d^3$
   d. $x^2y + xy^2$

10. a. $(a + b)x + (a + b)y$ **b.** $(4x - 1)x + (4x - 1)3$
   c. $x^2 - 2x - 35$ **d.** $2x^2 + 7x - 15$

11. a. $3x^2 - 5x - 2$ **b.** $6y^2 - 7y + 2$
   c. $8a^2b + 10ab - 3b$ **d.** $2s^2 - 10s - 48$

12. a. $4y^3 + y^2 - 21y$ **b.** $12m^2 - 7m - 12$
   c. $12p^4 + 11p^3 - 15p^2$ **d.** $9x^2y + 15xy - 14y$

B *Write the expressions in Problems 13–18 in simplified polynomial form.*

13. a. $(3x - 1)(x^2 + 3x - 2)$
   b. $(2x + 1)(x^2 + 2x - 5)$

14. a. $(5x + 1)(x^3 - 2x^2 + 3x - 5)$
   b. $(4x - 1)(x^3 + 3x^2 - 2x - 4)$

15. a. $(x + 1)(x - 3)(2x + 1)$
   b. $(2x - 1)(x + 3)(3x + 1)$

16. a. $(x - 2)(x + 3)(x - 4)$
   b. $(x + 1)(2x - 3)(2x + 3)$

17. a. $(x - 2)^2(x + 1)$ **b.** $(x - 2)(x + 1)^2$

18. a. $(x - 5)^2(x + 2)$ **b.** $(x - 3)^2(3x + 2)$

Factor completely, if possible, the expressions in Problems 19–44.

19. $(x - y)^2 - 1$

20. $(2x + 3)^2 - 1$

21. $(5a - 2)^2 - 9$

22. $(3p - 2)^2 - 16$

23. $\frac{4}{25}x^2 - (x + 2)^2$

24. $\frac{4x^2}{9} - (x + y)^2$

25. $\frac{x^6}{y^8} - 169$

26. $\frac{9x^{10}}{4} - 144$

27. $(a + b)^2 - (x + y)^2$

28. $(m - 2)^2 - (m + 1)^2$

29. $2x^2 + x - 6$

30. $3x^2 - 11x - 4$

31. $6x^2 + 47x - 8$

32. $6x^2 - 47x - 8$

33. $6x^2 + 49x + 8$

34. $6x^2 - 49x + 8$

35. $4x^2 + 13x - 12$

36. $9x^2 - 43x - 10$

37. $9x^2 - 56x + 12$

38. $12x^2 + 12x - 25$

39. $4x^4 - 17x^2 + 4$

40. $4x^4 - 45x^2 + 81$

41. $x^6 + 9x^3 + 8$

42. $x^6 - 6x^3 - 16$

43. $(x^2 - \frac{1}{4})(x^2 - \frac{1}{9})$

44. $(x^2 - \frac{1}{9})(x^2 - \frac{1}{16})$

WHAT IS WRONG, *if anything, with each of the statements in Problems 45–50? Explain your reasoning.*

45. $(x + y)^3 = x^3 + y^3$

46. $(x^2 + y^2) = (x + y)^2$

47. $\sqrt{x^2 + y^2} = x + y$

48. $\sqrt{x^2} = x$

49. $\sqrt{x^3} = x\sqrt{x}$

50. $(x^{-1} + y^{-1})^{-1} = x + y$

C *Many processes in calculus require considerable factoring ability. Factor Problems 51–56, which were all taken from a calculus book.*

51. $-2(2x^2 - 5x)^{-3}(4x - 5)$

52. $-(x^4 + 3x^3)^{-2}(4x^3 + 9x^2)$

53. $(2x + 1)^2[3(3x + 2)^2(3)] + (3x + 2)^3[2(2x + 1)(2)]$

54. $(5x + 1)^2[-2(4x + 3)^{-2}(4)] + (4x + 3)^{-1}[2(5x + 1)(5)]$

55. $\dfrac{(x^2 + 3)(3)(7x + 11)^2(7) - (7x + 11)^3(2x)}{(x^2 + 3)^2}$

56. $\dfrac{(2x - 5)^2(4)(x + 5)^3 - (x + 5)^4(4)(2x - 5)}{(2x - 5)^4}$

Write the expressions in Problems 57–62 in simplified polynomial form.

57. $(2x - 1)^2(3x^4 - 2x^3 + 3x^2 - 5x + 12)$

58. $(x - 3)^3(2x^3 - 5x^2 + 4x - 7)$

59. $(2x + 1)^2(5x^4 - 6x^3 - 3x^2 + 4x - 5)$

60. $(3x^2 + 4x - 3)(2x^2 - 3x + 4)$

61. $(2x^3 + 3x^2 - 2x + 4)^3$

62. $(x^3 - 2x^2 + x - 5)^3$

Factor completely, if possible, the expressions in Problems 63–78.

63. $x^{2n} - y^{2n}$

64. $x^{3n} - y^{3n}$

65. $x^{3n} + y^{3n}$

66. $x^{2n} - 2x^n y^n + y^{2n}$

67. $(x - 2)^2 - (x - 2) - 6$

68. $(x + 3)^2 - (x + 3) - 6$

69. $z^5 - 8z^2 - 4z^3 + 32$

70. $x^5 + 8x^2 - x^3 - 8$

71. $x^2 - 2xy + y^2 - a^2 - 2ab - b^2$

72. $x^2 + 2xy + y^2 - a^2 - 2ab - b^2$

73. $x^2 + y^2 - a^2 - b^2 - 2xy + 2ab$

74. $x^3 + 3x^2 y + 3xy^2 + y^3 + a^3 + 3a^2 b + 3ab^2 + b^3$

75. $(x + y + 2z)^2 - (x - y + 2z)^2$

76. $(x^2 - 3x - 6)^2 - 4$

77. $2(x + y)^2 - 5(x + y)(a + b) - 3(a + b)^2$

78. $2(s + t)^2 + 3(s + t)(s + 2t) - 2(s + 2t)^2$

The following problems require common factoring and are the types of problems you will encounter in calculus.

79. $3(x + 1)^2(x - 2)^4 + 4(x + 1)^3(x - 2)^3$

80. $4(x - 5)^3(x + 3)^2 + 2(x - 5)^4(x + 3)$

81. $3(2x - 1)^2(2)(3x + 2)^2 + 2(2x - 1)^3(3x + 2)(3)$

82. $(2x - 3)^3(3)(1 - x)^2(-1) + 3(2x - 3)^2(1 - x)^3(2)$

83. $4(x + 5)^3(x^2 - 2)^3 + (x + 5)^4(3)(x^2 - 2)^2(2x)$

84. $5(x - 2)^4(x^2 + 1)^3 + (x - 2)^5(3)(x^2 + 1)^2(2x)$

1.3 **Two-Dimensional Coordinate System and Graphs**

Two-Dimensional Coordinate System

There is a one-to-one correspondence between the real numbers and points on a real number line, called a one-dimensional coordinate system. A **two-dimensional coordinate system** can be introduced by considering two perpendicular coordinate

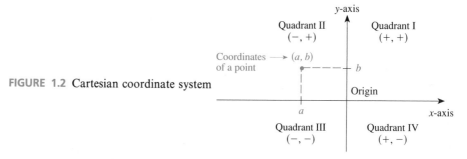

FIGURE 1.2 Cartesian coordinate system

lines in a plane. Usually one of the coordinate lines is horizontal with the positive direction to the right; the other is vertical with the positive direction upward. These coordinate lines are called **coordinate axes** and the point of intersection is called the **origin**. Notice from Figure 1.2 that the axes divide the plane into four parts called the **first, second, third,** and **fourth quadrants.** This two-dimensional coordinate system is also called a **Cartesian coordinate system** in honor of René Descartes, who was the first to describe such a coordinate system in mathematical detail.

Points of a plane are denoted by ordered pairs. The term **ordered pair** refers to two real numbers represented by (a, b), where a is called the **first component** and b is the **second component.** The order in which the components are listed is important, since $(a, b) \neq (b, a)$ if $a \neq b$.

EQUALITY OF ORDERED PAIRS | $(a, b) = (c, d)$ if and only if $a = c$ and $b = d$.

The horizontal number line is called the **x-axis** (sometimes called the *axis of abscissas*), and x represents the first component of the ordered pair. The vertical number line is called the **y-axis** (sometimes called the *axis of ordinates*), and y represents the second component of the ordered pair. The plane determined by the x and y axes is called a **coordinate plane, Cartesian plane,** or **xy-plane.** When we refer to a point (x, y), we are referring to a point in the coordinate plane whose abscissa is x and whose ordinate is y. To **plot a point** (x, y) means to locate the point with coordinates (x, y) in the plane and represent its location by a dot.

Distance and Midpoint Formulas

Finding the distance between points in two dimensions requires the Pythagorean theorem.

PYTHAGOREAN THEOREM | A triangle with sides a and b, and hypotenuse c, is a right triangle if and only if

$$a^2 + b^2 = c^2$$

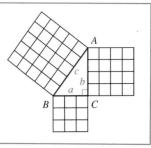

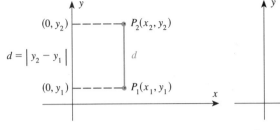

FIGURE 1.3 Vertical and horizontal line segments

a. Vertical segment, $x_1 = x_2$

b. Horizontal segment, $y_1 = y_2$

Let $P_1(x_1, y_1)$ and $P_2(x_2, y_2)$ be any two distinct points in a plane. If $x_1 = x_2$, then P_1P_2 is a *vertical line segment*; if $y_1 = y_2$, then P_1P_2 is a *horizontal line segment* as shown in Figure 1.3.

With these special cases, it is easy to find the distance d between P_1 and P_2 because these distances correspond directly to distances on a one-dimensional coordinate system, as discussed in Section 1.1. Study Figure 1.3 and see Problems 68 and 69 of Problem Set 1.3 for the details.

Since these special cases are considered in the problem set, we will focus our attention on the general case in which P_1 and P_2 do not lie on the same horizontal or vertical line. Draw a line through P_1 parallel to the x-axis and another through P_2 parallel to the y-axis. These lines intersect at a point Q with coordinates (x_2, y_1), as shown in Figure 1.4.

The distance P_1Q is $|x_2 - x_1|$; the distance QP_2 is $|y_2 - y_1|$. By the Pythagorean theorem,

$$d^2 = |x_2 - x_1|^2 + |y_2 - y_1|^2$$

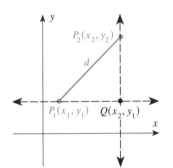

FIGURE 1.4 Distance between points

Thus, since d is nonnegative and $|a|^2 = a^2$ for every real number a, the distance from P_1 to P_2 is given by the following formula.

DISTANCE FORMULA

If $P_1(x_1, y_1)$ and $P_2(x_2, y_2)$ are any two points, then the distance d from P_1 to P_2 is

$$d = \sqrt{(x_2 - x_1)^2 + (y_2 - y_1)^2}$$

EXAMPLE 1 Find the distance between $(-3, 2)$ and $(-1, -6)$.

SOLUTION
$$d = \sqrt{[(-1) - (-3)]^2 + (-6 - 2)^2}$$
$$= \sqrt{4 + 64}$$
$$= 2\sqrt{17}$$

To find the **midpoint** M of a segment $P_1(x_1, y_1)$ and $P_2(x_2, y_2)$, you simply average the coordinates of the two endpoints.

MIDPOINT FORMULA

The midpoint M between point $P_1(x_1, y_1)$ and $P_2(x_2, y_2)$ is

$$M = \left(\frac{x_1 + x_2}{2}, \frac{y_1 + y_2}{2} \right)$$

EXAMPLE 2 Find the midpoint of the segment connecting $(-3, 2)$ and $(-1, -6)$.

SOLUTION
$$M = \left(\frac{(-3) + (-1)}{2}, \frac{(2) + (-6)}{2}\right) = (-2, -2)$$

Graph of Equation

One of the main topics of this course is the graphing of certain relations. Consider the formula for the volume V of a right circular cone whose radius is one-half its height h:

$$V = \frac{\pi h^3}{12}$$

A table of values showing the volumes for different heights is given here:

Height	1	2	3	4	5	6
Volume	$\dfrac{\pi}{12}$	$\dfrac{2\pi}{3}$	$\dfrac{9\pi}{4}$	$\dfrac{16\pi}{3}$	$\dfrac{125\pi}{12}$	18π

Do you see how these values were obtained? If $h = 1$, then

$$V = \frac{\pi(1)^3}{12} = \frac{\pi}{12}$$

If $h = 2$, then $V = \dfrac{\pi(2)^3}{12} = \dfrac{8\pi}{12} = \dfrac{2\pi}{3}$.

And so on. This table can be written as a set of ordered pairs: $(1, \pi/12)$, $(2, 2\pi/3)$, $(3, 9\pi/4)$, ... A set of ordered pairs is called a **relation.** In this example, the ordered pairs are (h, V) so that $V = \pi h^3/12$. If an ordered pair has components that, when substituted for their corresponding variables, yield a true equation, we say that the ordered pair **satisfies** the equation. For example, $(h, V) = (1, \pi/12)$ satisfies the equation $V = \pi h^3/12$. By plotting the points whose coordinates are shown in the table and then connecting them with a smooth curve, you arrive at the **graph** of this relation, a portion of which is shown in Figure 1.5.

FIGURE 1.5 Graph of $V = \dfrac{\pi h^3}{12}$

for $0 \leq h \leq 6$

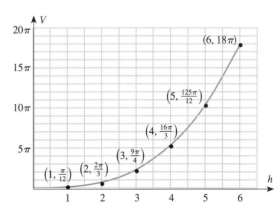

It is important that you clearly understand the relationship between an equation and its graph. This relationship is very explicit: There is a one-to-one correspondence between ordered pairs satisfying an equation and coordinates of points on the graph.

RELATIONSHIP BETWEEN AN EQUATION AND ITS GRAPH

The **graph of an equation,** or the **equation of a graph,** means:

1. Every point on the graph has coordinates that satisfy the equation.
2. Every ordered pair satisfying the equation has coordinates that lie on the graph.

This fundamental relationship between an equation and its graph tells you that to graph an equation (or check to see if a graph is correct) you can find ordered pairs satisfying an equation, and then plot those ordered pairs. After plotting many points, you can see the graph "take shape."

EXAMPLE 3   Graph $2x + 3y + 6 = 0$ by plotting points.

SOLUTION   One of the most efficient methods for graphing a curve by plotting points is to solve the equation for y and then substitute values for x, arranging the ordered pairs in table form.

$$2x + 3y + 6 = 0$$

x	y
0	-2
3	-4
-3	0
-6	2

$$3y = -2x - 6$$

$$y = -\tfrac{2}{3}x - 2$$

You choose the x values.

Do you see why we chose 3 and not 1 or 2?

This table entry represents the point $(-6, 2)$.

Plot the points $(0, -2)$, $(3, -4)$, ...
and draw a smooth curve connecting them:

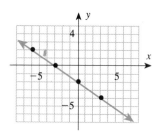

EXAMPLE 4   Graph $y = x^2$ by plotting points.

SOLUTION

x	y
0	0
1	1
2	4
-1	1
-2	4

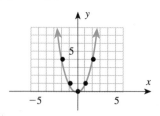

EXAMPLE 5 Graph $y = |x|$ by plotting points.

SOLUTION

x	y
0	0
1	1
-1	1
2	2
-2	2

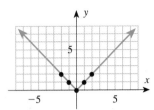

The method of drawing the graph of a relation as shown here—that of plotting points—is very primitive, and one of the primary purposes of this book is to develop more efficient methods of graphing curves than simply plotting points.

CALCULATOR COMMENT

Graphing calculators are easy to use. For example, we will show how to graph the equations in Examples 3–5:

Example 3	**Example 4**	**Example 5**
Input the value of y:	Input the value of y:	Input the value of y:
$Y = -(2/3)X - 2$	$Y = X^2$	$Y = \text{abs}(X)$

The output for each is shown below:

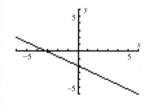 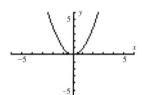 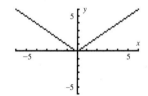

Equation of a Circle

There are several curves that you may recognize from your previous work. Example 3 shows the graph of a line; Example 4, a parabola; and Example 5, an absolute value curve. We will discuss each of these graphs at length later in the book.

The equation of a circle can be found by using the definition of a circle, along with the distance formula. By definition, a circle is the set of all points $P(x, y)$ whose distance from a fixed point, called the **center** (h, k), is a constant r, called the **radius,** as shown in Figure 1.6.

From the distance formula,

$$\sqrt{(x - h)^2 + (y - k)^2} = r$$

we obtain the desired equation by squaring both sides:

$$(x - h)^2 + (y - k)^2 = r^2$$

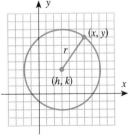

FIGURE 1.6 Circle with center (h, k) and radius r

EQUATION OF A CIRCLE

The equation of a circle with center (h, k) and radius r is

$$(x - h)^2 + (y - k)^2 = r^2$$

The **unit circle** is the circle with center $(0, 0)$ and radius 1:

$$x^2 + y^2 = 1$$

EXAMPLE 6 Write the equation of a circle whose diameter has endpoints $(-5, 3)$ and $(1, 5)$.

SOLUTION The radius is found using the distance formula to first find the length of the diameter:

$$\sqrt{[1 - (-5)]^2 + (5 - 3)^2} = \sqrt{40} = 2\sqrt{10}$$

The radius is one-half the length of this diameter, namely $\sqrt{10}$. The center is the midpoint of the segment:

$$\left(\frac{-5 + 1}{2}, \frac{3 + 5}{2}\right) = (-2, 4)$$

The equation is found using the equation of a circle where $(h, k) = (-2, 4)$ and $r = \sqrt{10}$:

$$(x + 2)^2 + (y - 4)^2 = 10$$

Symmetry

The first property of a graph we will consider is called **symmetry.** The *idea* of symmetry is the *idea* of mirror images. A graph or curve is **symmetric with respect to a line,** for example, if the graph is the same on both sides of that line as shown in Figures 1.7 and 1.8.

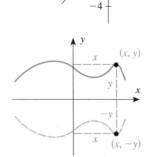

FIGURE 1.7 Symmetry with respect to a given line

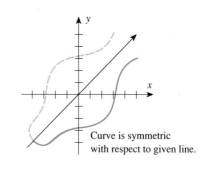

Curve is symmetric with respect to given line.

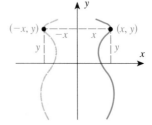

FIGURE 1.8 Symmetry with respect to the x-axis and the y-axis

a. Symmetry with respect to the x-axis   **b.** Symmetry with respect to the y-axis

EXAMPLE 7 Draw the reflection of the given curve as it would appear in the mirror.

SOLUTION The answer is shown as a dashed curve. Your paper would look like the graph on the right.

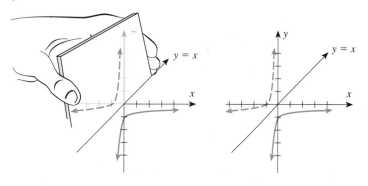

SYMMETRY WITH RESPECT TO
A COORDINATE AXIS

The graph of a relation is **symmetric with respect to the x-axis** if substitution of $-y$ for y does not change the set of coordinates satisfying the relation. The graph of a relation is **symmetric with respect to the y-axis** if substitution of $-x$ for x does not change the set of coordinates satisfying the relation.

EXAMPLE 8 Draw the given curve so that it is: **(a)** symmetric with respect to the x-axis, and **(b)** symmetric with respect to the y-axis.

SOLUTION **a.**

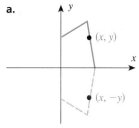

b.

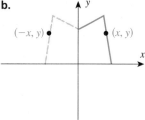

EXAMPLE 9 Discuss the symmetry of the curve $y = |x| - 5$ with respect to
a. the x-axis   **b.** the y-axis

SOLUTION **a.** Replace y by $-y$: $-y = |x| - 5$; this equation is different from the original equation so the curve is not symmetric with respect to the x-axis.

b. Replace x by $-x$: $y = |-x| - 5 = |x| - 5$; this equation is unchanged, so it is symmetric with respect to the y-axis.

CALCULATOR COMMENT
_____

We will use symmetry to sketch curves in Chapter 3, and we will use these symmetry tests at that time, but for now we can look at this curve on a graphing calculator. The graph of $\boxed{Y = \text{abs}(X) - 5}$ is shown:

The symmetry found algebraically agrees with the graph.

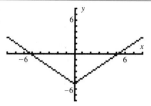

Some graphs possess another type of symmetry called **symmetry with respect to the origin** (Figure 1.9).

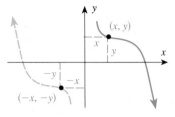

FIGURE 1.9 Symmetry with respect to the origin

The graph of a relation is **symmetric with respect to the origin** if the simultaneous substitution of $-x$ for x and $-y$ for y does not change the set of coordinates satisfying the relation.

EXAMPLE 10 Draw the given curve so that it is symmetric with respect to the origin.

SOLUTION

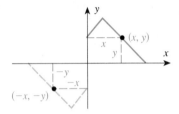

If a curve has any two of the three types of symmetry, it will also have the third. (You are asked to show this in Problem 70.) If you find that a curve has one type of symmetry but not a second, then it cannot have the third type either.

EXAMPLE 11 Test for symmetry with respect to the x-axis, y-axis, and origin.

a. $y = \dfrac{3x^2 + 1}{x^4}$ **b.** $xy = 2$ **c.** $9x^2 - 16y^2 = 144$

SOLUTION **a.** **x-axis:** $-y = \dfrac{3x^2 + 1}{x^4}$ Substitute $-y$ for y. This is a different equation.

   y-axis: $y = \dfrac{3(-x)^2 + 1}{(-x)^4}$ Substitute $-x$ for x.

   $= \dfrac{3x^2 + 1}{x^4}$ This is the same as the original equation.

Because one symmetry holds and the other does not, the graph will not have the third type of symmetry. Therefore, this curve is **symmetric with respect to the y-axis.**

b. **x-axis:** $x(-y) = 2$ This is a different equation.
   y-axis: $(-x)y = 2$ This is a different equation.
   origin: $(-x)(-y) = 2$ Substitute $-x$ for x and $-y$ for y.
   $xy = 2$ This is the same equation.

This curve is **symmetric with respect to the origin.**

c. x-axis: $9x^2 - 16(-y)^2 = 144$ This is the same equation.
y-axis: $9(-x)^2 - 16y^2 = 144$ This is the same equation.
origin: Since two symmetries hold, the third symmetry also must hold.

This curve is **symmetric with respect to the x-axis, the y-axis, and the origin.** ▮

1.3 Problem Set

A *Find the distance between the points whose coordinates are given in Problems 1–6.*

1. a. (5, 1) and (8, 5)   **b.** (1, 4) and (13, 9)
2. a. (−2, 4) and (0, 0)   **b.** (0, 0) and (5, −2)
3. a. (4, 5) and (3, −1)   **b.** (−2, 1) and (−1, −5)
4. (7x, 5x) and (3x, 2x), $x < 0$
5. (x, 5x) and (−3x, 2x), $x < 0$
6. (x, 5x) and (−3x, 2x), $x > 0$

Find the midpoint of the segment connecting the points whose coordinates are given in Problems 7–12.

7. a. (5, 1) and (8, 5)   **b.** (1, 4) and (13, 9)
8. a. (−2, 4) and (0, 0)   **b.** (0, 0) and (5, −2)
9. a. (4, 5) and (3, −1)   **b.** (−2, 1) and (−1, −5)
10. (x, 5x) and (3x, 2x), $x < 0$
11. (x, 5x) and (−3x, 2x), $x < 0$
12. (x, 5x) and (−3x, 2x), $x > 0$

Draw a reflection of each curve given in Problems 13–16. To do this, draw coordinate axes on your paper. Next draw the line x = y and the curve as shown. Finally imagine a mirror as shown here and draw the curve as it would look in this mirror.

13.
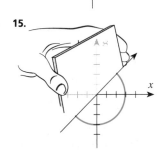

14.
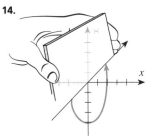

15.
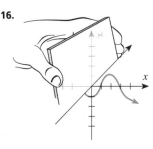

16.

Graph each relation in Problems 17–28 by plotting points.

17. $x + y + 3 = 0$ **18.** $3x - y + 1 = 0$
19. $3x + 2y + 6 = 0$ **20.** $4x - 5y + 2 = 0$
21. $y = -x^2$ **22.** $2x^2 + y = 0$
23. $x^2 + 3y = 0$ **24.** $x^2 - 2y = 0$
25. $y = -|x|$ **26.** $y = -|x + 2|$
27. $y = 2|x|$ **28.** $y = |x| + 2$

In Problems 29–34 draw a curve so that it is symmetric to the given curve with respect to: (a) the x-axis; (b) the y-axis; and (c) the origin.

29.

30.

31.
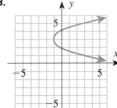

32.
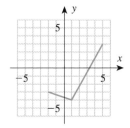

33.
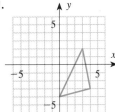

34.
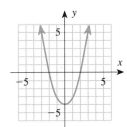
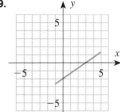
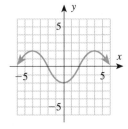

B *In Problems 35–46 test for symmetry with respect to both axes and the origin.*

35. $y = x^2$ **36.** $y = x^3$

37. $y = \dfrac{x^3 - 1}{x - 1}$ **38.** $y = \dfrac{x^3 + 8}{x + 2}$

39. $y = -\dfrac{\sqrt{x + 3}}{x - 1}$ **40.** $y = -\dfrac{\sqrt{9 - x^2}}{x + 3}$

41. $2|x| - |y| = 5$ **42.** $|y| = 5 - 3|x|$

43. $xy = 1$ **44.** $xy + 6 = 0$

45. $x^2 + 2xy + y^2 = 4$ **46.** $x^{1/2} + y^{1/2} = 4$

Find an equation of the circle that satisfies the conditions given in Problems 47–52.

47. Center $(5, -1)$, radius 4

48. Center $(-1, 3)$, radius π

49. Endpoints of a diameter $(1, -4)$ and $(3, 6)$

50. Endpoints of diameter $(\sqrt{2}, \pi)$ and $(3\sqrt{2}, 5\pi)$

51. Center (a, b), tangent to the x-axis

52. Center (c, d), tangent to the y-axis

The sides of a triangle satisfy $a^2 + b^2 = c^2$ if and only if it is a right triangle. Which of the triangles whose vertices are given in Problems 53–58 are right triangles?

53. $(1, 3), (7, 1), (7, 10)$ **54.** $(0, 0), (6, 4), (2, 10)$

55. $(0, 0), (4, 3), (-3, 8)$ **56.** $(-6, 0), (-5, -6), (0, -5)$

57. $(3, 2), (7, -4), (4, -6)$ **58.** $(8, -1), (14, 2), (6, 3)$

59. Find a formula for the set of points (x, y) for which the distance from (x, y) to $(2, 3)$ is 7.

60. Find a formula for the set of points (x, y) for which the distance from (x, y) to $(-3, -5)$ is 5.

61. Find a formula for the set of points (x, y) for which the distance from (x, y) to $(-4, 1)$ is 3.

62. Find all points on the x-axis that are 8 units from the point $(2, 4)$.

63. Find all points on the y-axis that are 8 units from the point $(2, 4)$.

64. Find all points on the y-axis that are 5 units from the point $(-3, -1)$.

65. Draw the graph of the relation $A = \pi r^2$.

66. Draw the graph of the relation $V = \frac{2}{3}\pi R^2$.

67. Draw the graph of the relation $h = \dfrac{a}{2}\sqrt{3}$.

C

68. Let $P_1(x_1, y_1)$ and $P_2(x_2, y_2)$ be two points such that $y_1 = y_2$. Show that the distance from P_1 to P_2 is $|x_2 - x_1|$.

69. Let $P_1(x_1, y_1)$ and $P_2(x_2, y_2)$ be two points such that $x_1 = x_2$. Show that the distance from P_1 to P_2 is $|y_2 - y_1|$.

70. Show that if a curve has any two of the three types of symmetry discussed in this section, then it will also have the third.

71. Show that if a curve has one type of symmetry discussed in this section but not a second, then it cannot have the third type either.

72. Find a formula for the set of points (x, y) for which the distance from (x, y) to $(3, 0)$ plus the distance from (x, y) to $(-3, 0)$ equals 10.

73. Find a formula for the set of points (x, y) for which the distance from (x, y) to $(4, 0)$ plus the distance from (x, y) to $(-4, 0)$ equals 10.

74. Find a formula for the set of points (x, y) for which the distance from (x, y) to $(5, 0)$ minus the distance from (x, y) to $(-5, 0)$, in absolute value, is 6.

75. Find a formula for the set of points (x, y) for which the distance from (x, y) to $(0, 5)$ minus the distance from (x, y) to $(0, -5)$, in absolute value, is 8.

1.4   Intervals, Inequalities, and Absolute Value

Linear Equations in One Variable

A **linear equation in one variable** is an equation that can be written in the form

$$ax + b = 0 \qquad (a \neq 0)$$

where x is a **variable** and a and b are any real numbers. An **open** or **conditional equation** is an equation containing a variable that may be either true or false, depending on the replacement for the variable. A **root** or a **solution** is a replacement for the variable that makes the equation true. We also say that the root **satisfies** the equation. The **solution set** of an open equation is the set of all solutions of the equation. To **solve an equation** means to find its solution set. If there are no values for the variable that satisfy an equation, then the solution set is said to be **empty** and is denoted by $\varnothing$. If every replacement of the variable makes the equation true, then the equation is called an **identity.** It is assumed that you know how to solve linear equations in one variable.

Linear Inequalities in One Variable

You will need to know how to solve linear inequalities involving variables. There are other inequality relationships besides less than and greater than. For example:

LESS THAN OR EQUAL TO $a \leq b$, read "*a is less than or equal to b*," means that either $a < b$ or $a = b$ (but not both).

GREATER THAN OR EQUAL TO $a \geq b$, read "*a is greater than or equal to b*," means that either $a > b$ or $a = b$ (but not both).

BETWEEN $b < a < c$, read "*a is between b and c*," means *both* $b < a$ and $a < c$. Additional "between" relationships are also used:

$$b \leq a \leq c \quad \text{means} \quad b \leq a \text{ and } a \leq c$$
$$b \leq a < c \quad \text{means} \quad b \leq a \text{ and } a < c$$
$$b < a \leq c \quad \text{means} \quad b < a \text{ and } a \leq c$$

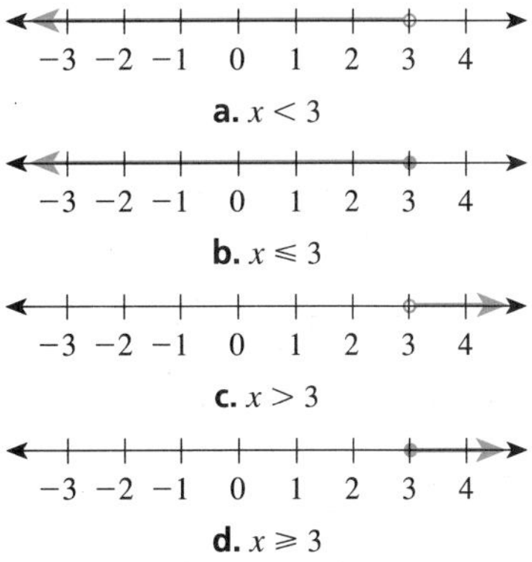

a. $x < 3$

b. $x \leq 3$

c. $x > 3$

d. $x \geq 3$

FIGURE 1.10 Graph of inequality statements

In this book, when we use the word *between* we mean strictly between—namely, $b < a < c$.

Graphs of linear inequality statements with a single variable are drawn on a one-dimensional coordinate system. For example, $x < 3$ denotes the interval shown in Figure 1.10a. Notice that $x \neq 3$, and this fact is shown by an open circle at the endpoint of the ray. Compare this with $x \leq 3$ (Figure 1.10b) in which the endpoint $x = 3$ is included. Notice that to sketch the graph you darken (or color) the appropriate portion.

There is a very useful notation to describe the graphs and intervals shown in Figure 1.10. This notation, called **interval notation,** uses brackets and parentheses to denote intervals algebraically. A parenthesis is used for an open point and a bracket for a closed point. The infinity symbol, ∞, is used to denote a ray or the entire number line. This notation is summarized in Table 1.3.

Table 1.3 Graph, inequality, and interval notation

ONE-DIMENSIONAL GRAPH	INEQUALITY NOTATION	INTERVAL NOTATION
	$a < x < b$	(a, b)
	$a \leq x < b$	$[a, b)$
	$a < x \leq b$	$(a, b]$
	$a \leq x \leq b$	$[a, b]$
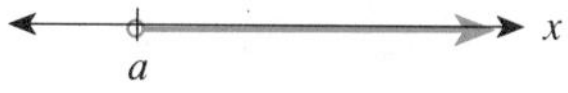	$x > a$	(a, ∞)
	$x \geq a$	$[a, \infty)$
	$x < b$	$(-\infty, b)$
	$x \leq b$	$(-\infty, b]$
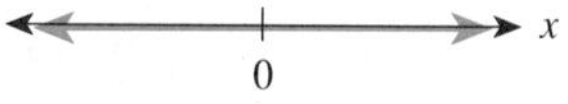	all real x values	$(-\infty, \infty)$

EXAMPLE 1 Write interval notation for each given inequality statement. Also graph each of the given inequalities.

a. $-6 \leq x < 5$; $[-6, 5)$

b. $7 < x < 8$; $(7, 8)$

c. $-9 \leq x \leq -2$; $[-9, -2]$

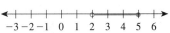

d. $-5 < x \leq 0$; $(-5, 0]$

e. $x > 2$; $(2, \infty)$

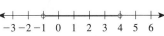

EXAMPLE 2 Graph each of the given intervals.

a. $[1, 3]$

b. $(2, 5]$

c. $[0, 4)$

d. $(-1, 4)$

e. $(2, 5] \cup [8, 10)$

The symbol $\cup$ means *union*, and is used for disjoint segments.

A **linear inequality in one variable** is an inequality that can be written in one of the following forms:

$$ax + b < 0 \qquad ax + b \leq 0 \qquad ax + b > 0 \qquad ax + b \geq 0$$

where x is a variable, a is a nonzero real number, and b is a real number. Linear inequalities are solved by using the following principles stated for $<$, but they are also true for $\leq$, $>$, and $\geq$. These symbols are referred to as the **order** of the inequality.

PROPERTIES OF INEQUALITY

TRANSITIVITY:
If $a < b$ and $b < c$, then $a < c$ for any real numbers a, b, and c.

ADDITION PRINCIPLE:
If $a < b$, then $a + c < b + c$ for any real numbers a, b, and c.

POSITIVE MULTIPLICATION PRINCIPLE:
If $a < b$, and c is a positive real number, then $ac < bc$.

NEGATIVE MULTIPLICATION PRINCIPLE:
If $a < b$, and c is a negative real number, then $ac > bc$.

PROCEDURE FOR SOLVING LINEAR INEQUALITIES

The procedure and terminology for solving linear inequalities are identical to the procedure and terminology for solving linear equations except for one fact: If you multiply or divide by a negative number, the order of the inequality is reversed. And remember, from the definition of greater than, that if you reverse the left and right sides of an inequality, the order is also reversed.

EXAMPLE 3 Solve the inequality $5x - 3 \geq 7$ and state the solution set using interval notation.

SOLUTION
$$5x - 3 \geq 7$$
$$5x - 3 + 3 \geq 7 + 3$$
$$5x \geq 10$$
$$x \geq 2$$

Solution: $[2, \infty)$

EXAMPLE 4 Solve the given inequality and write the solution set using interval notation.

SOLUTION

$2(4 - 3t) < 5t - 14$ Eliminate parentheses.

$8 - 6t < 5t - 14$ Isolate the variable on one side.

$-6t < 5t - 22$

$-11t < -22$

$t > 2$ Reverse the order of the inequality when you divide by a negative number.

Solution: $(2, \infty)$

Sometimes two inequality statements are combined, as with

$$-3 < 2x + 1 < 5$$

This means that $-3 < 2x + 1$ *and* $2x + 1 < 5$. It also means that $2x + 1$ is between -3 and 5. When solving an inequality of this type, you can frequently work both inequalities simultaneously, as shown in Examples 5 and 6.

EXAMPLE 5

$$-3 < 2x + 1 < 5$$

$$-3 - 1 < 2x + 1 - 1 < 5 - 1$$ Your goal is now to isolate the variable in the middle of the "between" statement. Each inequality statement is equivalent to the preceding one. Whatever you do to one part, you do to all three parts.

$$-4 < 2x < 4$$

$$\frac{-4}{2} < \frac{2x}{2} < \frac{4}{2}$$

$$-2 < x < 2$$

Solution: $(-2, 2)$

EXAMPLE 6

$$-5 \leq 1 - 3x < 10$$

$$-6 \leq -3x < 9$$

$$\frac{-6}{-3} \geq \frac{-3x}{-3} > \frac{9}{-3}$$ Inequality reverses when you multiply or divide by a negative.

$$2 \geq x > -3$$

However, $2 \geq x > -3$ is not convenient for converting to interval notation, so it is rewritten as

$$-3 < x \leq 2$$

Solution: $(-3, 2]$

Relationship Between One- and Two-Dimensional Coordinate Systems

We have been discussing linear equations, linear inequalities, and absolute value equations in one variable. If we show the solutions graphically, we would plot the solution points or intervals on a real number line, which in Section 1.1 was called a one-dimensional coordinate system. With the advent of graphing calculators, and the power of these calculators in graphing in two dimensions, it is worthwhile (even if you do not have such a calculator) to consider the relationship between one- and two-dimensional coordinate systems. In two dimensions we have two variables, usually x and y, and we can interpret the solution to an equation in one variable as a subset of a system of equations with the second variable equal to 0. We illustrate by taking a second look at some of the previous examples of this section.

EXAMPLE 7 (*Example 1 revisited in two dimensions*)   We were asked to graph $-6 \leq x < 5$ and found the graph to be:

In two dimensions the graph is shown in Figure 1.11. Notice that the solution for Example 1a is the intersection of the two-dimensional regions $-6 \leq x < 5$ and the line $y = 0$ (shown in color).

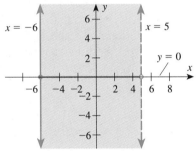

FIGURE 1.11 Graph of $-6 \leq x < 5$ shown in two dimensions

Why would we want to reconsider Example 1 in two dimensions? How can a two-dimensional solution be easier to consider when the one-dimensional solution is so easy? The answer is tied to the power of graphing calculators. A **trace key** gives coordinates of a point on a curve. That is, you move a cursor along a curve and the calculator shows the coordinates of the cursor at the bottom of the screen. Consider Example 3 on a calculator.

EXAMPLE 8 (*Example 3 reconsidered in two dimensions*)   Solve $5x - 3 \geq 7$ graphically (in two dimensions).

SOLUTION Let $y_1 = 5x - 3$ and $y_2 = 7$. The graph is shown in Figure 1.12. We are interested in finding the x values for which $y_1 \geq y_2$. The appropriate region is shown in color and the desired solution for x on the x-axis is shown in color. The endpoint is found algebraically by solving the system

$$\begin{cases} y_1 = 5x - 3, \quad y_2 = 7 \\ y_1 = y_2 \end{cases}$$

By substitution, $5x - 3 = 7$ has solution $x = 2$, so the solution is **[2, ∞)**.

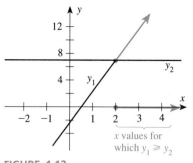

FIGURE 1.12

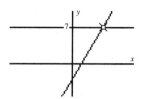

CALCULATOR COMMENT

On a graphing calculator, input

$$\boxed{Y1 = 5X - 3} \text{ and } \boxed{Y2 = 7} \text{ and } \boxed{GRAPH}.$$

With the $\boxed{TRACE}$ move the cursor to the point of intersection:

$$\boxed{X = 2 \quad Y = 7}$$

to find the solution $[2, \infty)$.

We are not suggesting that you disregard the algebraic solution shown in Example 3. We are simply suggesting that if you have a graphing calculator available, you might wish to *reinforce* the plausibility of the answer you obtained by looking at a graph.

EXAMPLE 9 (*Example 6 revisited in two dimensions*)   Solve $-5 \le 1 - 3x < 10$.

SOLUTION Graph $y_1 = -5$, $y_2 = 1 - 3x$, and $y_3 = 10$. The graph is shown in Figure 1.13.

The solution (in terms of x) is highlighted. Using a calculator $\boxed{TRACE}$, we find the upper point of intersection:

$$\boxed{X = -3.052632 \quad Y = 10.157895}$$

and the lower point of intersection:

$$\boxed{X = 2 \quad Y = -5}$$

This reinforces our work in Example 6, which gives the solution $(-3, 2]$.

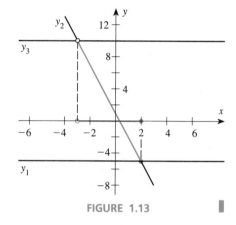

FIGURE 1.13

Absolute Value Equations

Since $|x - a|$ can be interpreted as the distance between x and a on the number line, an equation of the form

$$|x - a| = b$$

has two values of x that are a given distance from a when represented on a number line. For example, $|x - 5| = 3$ states that x is 3 units from 5 on a number line. Thus x is either 2 or 8:

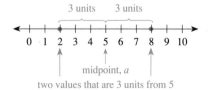

You can verify this conclusion by proving the following property.

ABSOLUTE VALUE EQUATIONS

> $|a| = b$ where $b \geq 0$ if and only if $a = b$ or $a = -b$.

PROOF   There are two parts to a proof using the words "if and only if." **(1)** If $|a| = b$, then $a = b$ or $a = -b$. **(2)** If $a = b$ or $a = -b$, then $|a| = b$. We will prove the first part here and leave the second part as a problem. That is, suppose $|a| = b$; we now wish to show that $a = b$ or $a = -b$. Begin with the Property of Comparison to compare the real number a with the real number zero:

$$a > 0 \qquad a = 0 \qquad \text{or} \qquad a < 0$$

If $a > 0$ or if $a = 0$, then $|a| = a$ and

$$|a| = b \qquad \text{Given}$$
$$a = b \qquad \text{Substitution of } a \text{ for } |a|$$

If $a < 0$, then $|a| = -a$ and

$$|a| = b \qquad \text{Given}$$
$$-a = b \qquad \text{Substitution of } -a \text{ for } |a|$$
$$a = -b \qquad \text{Multiplication of both sides by } -1$$

EXAMPLE 10   Solve $|x + 5| = 2$.

SOLUTION $x + 5 = 2 \qquad \text{or} \qquad x + 5 = -2$
$x = -3 \quad \text{or} \qquad x = -7$

Solution: $\{-3, -7\}$

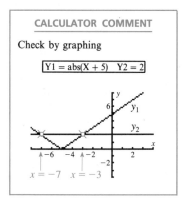

CALCULATOR COMMENT

Check by graphing

$\boxed{Y1 = \text{abs}(X + 5) \quad Y2 = 2}$

EXAMPLE 11   Solve $|x + 5| = -2$.

SOLUTION   The absolute value of every real number is nonnegative, so the solution set is **empty**.

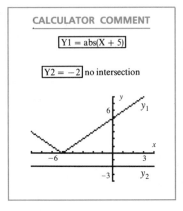

CALCULATOR COMMENT

$\boxed{Y1 = \text{abs}(X + 5)}$

$\boxed{Y2 = -2}$ no intersection

If you are solving an absolute value equation with absolute values on both sides of the equation, the following property may be useful.

| **PROCEDURE FOR SOLVING ABSOLUTE VALUE EQUATIONS** | $\|a\| = \|b\|$ if and only if $a = b$ or $a = -b$. |

You are asked to prove this in Problem 70 of Problem Set 1.4.

EXAMPLE 12 Solve $\|x + 5\| = \|3x - 4\|$.

SOLUTION Use the given property of absolute value.

$$x + 5 = 3x - 4$$
$$-2x = -9$$
$$x = \frac{9}{2}$$

or

$$x + 5 = -(3x - 4)$$
$$x + 5 = -3x + 4$$
$$4x = -1$$
$$x = -\frac{1}{4}$$

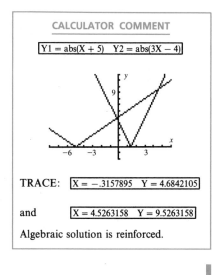

CALCULATOR COMMENT

Y1 = abs(X + 5)   Y2 = abs(3X − 4)

TRACE: $\boxed{X = -.3157895 \quad Y = 4.6842105}$

and $\boxed{X = 4.5263158 \quad Y = 9.5263158}$

Algebraic solution is reinforced.

Solution: $\left\{ \frac{9}{2}, -\frac{1}{4} \right\}$ ▮

Absolute Value Inequalities

Since $\|x - 5\| = 3$ states that x is 3 units from 5, the inequality $\|x - 5\| < 3$ states that x is any number less than 3 units from 5:

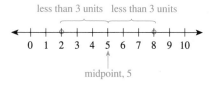

On the other hand, $\|x - 5\| > 3$ states that x is any number greater than 3 units from 5:

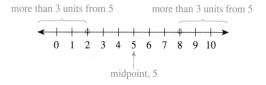

These properties are summarized by four absolute value inequality properties, where a and b are real numbers and $b > 0$.

<table>
<tr><td style="text-align:right">ABSOLUTE VALUE
INEQUALITIES</td><td>$|a| < b$ if and only if $-b < a < b$
$|a| \leq b$ if and only if $-b \leq a \leq b$
$|a| > b$ if and only if $a > b$ or $a < -b$
$|a| \geq b$ if and only if $a \geq b$ or $a \leq -b$</td></tr>
</table>

PROOF *Prove*: If $|a| < b$, then $-b < a < b$. By definition:

$$|a| = a \quad \text{if } a \geq 0$$
$$|a| = -a \quad \text{if } a < 0$$

Case i: If $a \geq 0$

$\quad |a| < b$ ⎯ Given

$\quad\quad a < b$ ⎯ Substitute a for $|a|$.

Case ii: If $a < 0$

$\quad |a| < b$ ⎯ Given

$\quad\quad -a < b$ ⎯ Substitute $-a$ for $|a|$.

$\quad\quad 0 < a + b$ ⎯ Add a to both sides.

$\quad\quad -b < a$ ⎯ Subtract b from both sides.

Therefore, $-b < a$ and $a < b$ as seen from cases *i* and *ii*. This is the same as saying $-b < a < b$. The proofs of the other parts are similar and are left for the exercises. ⬜⬜⬜

EXAMPLE 13 Solve $|2x - 3| \leq 4$.

SOLUTION $-4 \leq 2x - 3 \leq 4$. Now solve the "between" relationship:

$$-4 + 3 \leq 2x - 3 + 3 \leq 4 + 3$$

$$-1 \leq 2x \leq 7$$

$$\frac{-1}{2} \leq \frac{2x}{2} \leq \frac{7}{2}$$

$$-\frac{1}{2} \leq x \leq \frac{7}{2}$$

Solution: $\left[-\frac{1}{2}, \frac{7}{2}\right]$

CALCULATOR COMMENT

Example 13 on a TI-81 graphing calculator:

Y1 = abs(2X − 3)

Y2 = 4   GRAPH

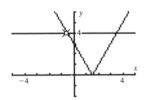

Interpretation of graph:
We are looking for $y_1 \leq y_2$. This is shown in color:

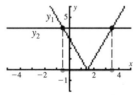

The range of x values for which $y_1 \leq y_2$ is highlighted.

To find the endpoints we use the TRACE key. For the first intersection point, the display might be:

X = −.5263158   Y = 4

For the second intersection point:

X = 3.4736842   Y = 4

If you wish greater accuracy you can use the ZOOM key to approximate the coordinates to any desired degree of accuracy.

EXAMPLE 14 Solve $|4 - x| < -3$.

SOLUTION Absolute values must be nonnegative, so the solution set is empty.

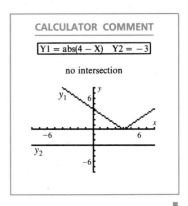

EXAMPLE 15 Solve $|x + 3| > 4$.

SOLUTION
$$x + 3 > 4 \quad \text{or} \quad x + 3 < -4$$
$$x > 1 \quad \text{or} \quad x < -7$$

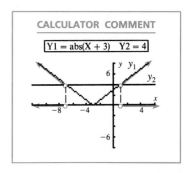

Remember that the union symbol $\cup$ is used to include more than one segment. The solution is

$$(-\infty, -7) \cup (1, \infty)$$

In calculus, it is common to use absolute value equations and inequalities to specify distances. In particular, the set of all points x within a specified distance of δ units from a real number c is summarized in Table 1.4. (The symbol δ is the lowercase Greek letter delta.)

Table 1.4 Absolute value as a distance

SYMBOL	DESCRIPTION	INTERVAL	GRAPH
$|x - c| < \delta$	All points within δ units of c	$(c - \delta, c + \delta)$	
$|x - c| = \delta$	All points exactly δ units from c	Two points (not an interval) $x = c + \delta, x = c - \delta$	
$|x - c| > \delta$	All points more than δ units from c	$(-\infty, c - \delta) \cup (c + \delta, \infty)$	

1.4 Problem Set

A *Write interval notation and graph each of the inequalities in Problems 1–4.*

1. a. $3 < x < 7$ **b.** $-4 < x < -1$
     c. $-2 \leq x \leq 6$ **d.** $-3 < x \leq 0$

2. a. $-6 \leq x \leq 0$ **b.** $-3 < x \leq 3$
     c. $-2 \leq x \leq 2$ **d.** $-5 \leq x < 1$

3. a. $x \leq -3$ **b.** $x \geq -2$
     c. $x < 0$ **d.** $2 < x$

4. a. $3 < x \leq 6$ **b.** $5 < x$
     c. $10 > x$ **d.** $-.01 \leq x \leq .01$

Graph each of the intervals in Problems 5–8.

5. a. $[-3, 2]$ **b.** $(-2, 2)$
     c. $(-\infty, 3]$ **d.** $[-2, \infty)$

6. a. $(-3, 4]$ **b.** $[-3, 4)$
     c. $[-4, 3)$ **d.** $(2, \infty)$

7. a. $(-2, 0] \cup (3, 5)$ **b.** $[-8, 3) \cup [5, \infty)$
     c. $(-\infty, -3] \cup (0, 3]$ **d.** $[-5, -1) \cup (0, 5]$

8. a. $(-\infty, 0) \cup (0, \infty)$ **b.** $(-\infty, 2) \cup (2, \infty)$
     c. $(-\infty, 1) \cup (1, \infty)$ **d.** $(-\infty, 6) \cup (6, \infty)$

Write each of the intervals in Problems 9–12 as an inequality. Use x as the variable.

9. a. $[-4, 2]$ **b.** $[-1, 2]$
     c. $(0, 8)$ **d.** $(-5, 3]$

10. a. $(-3, -1)$ **b.** $[-5, 2)$
     c. $[5, 9]$ **d.** $(-6, -2)$

11. a. $(-\infty, 2)$ **b.** $(6, \infty)$
     c. $(-1, \infty)$ **d.** $(-\infty, 3]$

12. a. $(-\infty, -6]$ **b.** $(-\infty, 0)$
     c. $[5, \infty)$ **d.** $(-1, \infty)$

WHAT IS WRONG, *if anything, with each of the statements in Problems 13–18? Explain your reasoning.*

13. If $5 > x > 1$, then $5 > x$ and $x > 1$, so $x > 5$.

14. If $1 \leq x \leq 5$, then $-1 \leq -x \leq -5$.

15. If $\dfrac{x}{y} \leq 2$, then $x \leq 2y$.

16. If $|x - 5| \leq 2$, then x is more than 2 units from 5.

17. If $|x - 3| > 4$, then x is within 3 units of 4.

18. If $|x - 2| = 3$, then x is 3 units from 2.

Solve the equations in Problems 19–30.

19. $|x| = 5$ **20.** $|x| = 10$

21. $|x| = -1$ **22.** $|x| = -4$

23. $|x - 3| = 4$ **24.** $|x + 9| = 3$

25. $|x - 9| = 15$ **26.** $|2x + 4| = -8$

27. $|2x + 4| = -12$ **28.** $|3 - 2x| = 7$

29. $|5x + 4| = 6$ **30.** $|5 - 3x| = 14$

B *Solve the inequalities in Problems 31–54 and write each solution in interval notation.*

31. $3x - 9 \geq 12$ **32.** $-x > -36$

33. $-3x \geq 123$ **34.** $3(2 - 4x) \leq 0$

35. $5(3 - x) > 3x - 1$ **36.** $-5 \leq 5x \leq 25$

37. $-8 < 5x < 0$ **38.** $3 \leq -x < 8$

39. $-3 \leq -x < -1$ **40.** $-5 < 3x + 2 \leq 5$

41. $-7 \leq 2x + 1 < 5$ **42.** $-4 < 1 - 5x \leq 11$

43. $-5 \leq 3 - 2x < 18$ **44.** $|x - 5| \leq 1$

45. $|x - 3| \leq 5$ **46.** $|x - 8| \leq .01$

47. $|x - 3| < .001$ **48.** $|3x + 2| < -3$

49. $|2x + 1| < -1$ **50.** $|3 - x| < 5$

51. $|4 - 3x| < 3$ **52.** $|x + 1| > 3$

53. $|2x + 7| > 5$ **54.** $|3 - 2x| > 5$

55. BUSINESS   If Amex Automobile Rental charges $35 per day and 45¢ per mile, how many miles can you drive and keep the cost under $125 per day?

56. BUSINESS   A saleswoman is paid a salary of $300 plus a 40% commission on sales. How much does she need to sell in order to have an income of at least $2,000?

57. CHEMISTRY   A certain experiment requires that the temperature be between 20° and 30° C. If Fahrenheit and Celsius degrees are related by the formula $C = \frac{5}{9}(F - 32)$, what are the permissible temperatures in Fahrenheit?

58. ECONOMICS   An economist estimates that the Consumer Price Index will grow by 14%, give or take 1%. Let c represent the growth rate of the Consumer Price Index.
a. Write the condition of this problem as an inequality.
b. Write the condition of this problem as an absolute value statement.

59. BUSINESS   The management of a certain company needs to monitor the activities of salespersons whose sales are not the usual $15,000 per week. That is, if the sales of a certain employee are less than $10,000 or greater than $20,000, the company needs to monitor the amount of time spent on the job. Let s represent the amount of weekly sales.
a. Write the condition of this problem as an inequality.
b. Write the condition of this problem as an absolute value statement.

C *Solve the inequalities in Problems 60–65 and write the solution in interval notation.*

60. $\left| \dfrac{5 - x}{2} \right| > 1$ **61.** $\left| \dfrac{3 - x}{2} \right| > 3$

62. $\dfrac{5}{|x-2|} \le 1$

63. $\dfrac{3}{|x+1|} \ge 3$

64. $|2x-1| + |x| - 3 = 0$

65. $|3x+1| + |x| - 5 = 0$

66. Prove that if b is a nonnegative real number, and if $a = b$ or $a = -b$, then $|a| = b$.

67. If $|a| \le b$, show that $-b \le a \le b$.

68. If $|a| > b$, show that either $a > b$ or $a < -b$.

69. If $|a| \ge b$, show that either $a \ge b$ or $a \le -b$.

70. Prove that $|a| = |b|$ if and only if $a = b$ or $a = -b$.

1.5 Complex Numbers

To find the roots of certain equations, you must sometimes consider the square roots of negative numbers. Since the set of real numbers does not allow for such a possibility, a number that is *not a real number* is defined. This number is denoted by the symbol i.

THE IMAGINARY UNIT

> The number i, called the **imaginary unit,** is defined as a number with the following properties:
>
> $$i^2 = -1 \quad \text{and} \quad \sqrt{-a} = i\sqrt{a} \quad (a > 0)$$

With this number you can write the square root of any negative number as the product of a real number and the number i. Thus $\sqrt{-9} = i\sqrt{9} = 3i$. For any positive real number b, $\sqrt{-b} = i\sqrt{b}$. We now consider a new set of numbers.

COMPLEX NUMBERS

> The set of all numbers of the form
>
> $$a + bi$$
>
> with real numbers a and b, and i the imaginary unit, is called the set of **complex numbers.**

If $b = 0$, then $a + bi = a + 0i = a$, which is a **real number;** thus the real numbers form a subset of the complex numbers. If $a \ne 0$ and $b \ne 0$, then $a + bi$ is called an **imaginary number;** if $a = 0$ and $b \ne 0$, then $a + bi = 0 + bi = bi$ is called a **pure imaginary number.** If $a = 0$ with $b = 1$, then $a + bi = 0 + 1 \cdot i = i$, which is the imaginary unit.

A complex number is **simplified** if it is written in the form $a + bi$, where a and b are simplified real numbers and i is the imaginary unit. In order to work with complex numbers, you will need definitions of equality along with the usual arithmetic operations. Let $a + bi$ and $c + di$ be complex numbers. Values that could cause division by zero are excluded.

OPERATIONS WITH COMPLEX NUMBERS

> EQUALITY: $\quad a + bi = c + di \quad$ if and only if $\quad a = c$ and $b = d$
>
> ADDITION: $\quad (a + bi) + (c + di) = (a + c) + (b + d)i$
>
> SUBTRACTION: $\quad (a + bi) - (c + di) = (a - c) + (b - d)i$
>
> MULTIPLICATION: $\quad (a + bi)(c + di) = (ac - bd) + (ad + bc)i$
>
> DIVISION: $\quad \dfrac{a + bi}{c + di} = \dfrac{(ac + bd) + (bc - ad)i}{c^2 + d^2} = \dfrac{ac + bd}{c^2 + d^2} + \dfrac{bc - ad}{c^2 + d^2}\, i$

It is not necessary to memorize these definitions, since you can deal with two complex numbers as you would any binomials, as long as you remember that $i^2 = -1$. Also notice $i^3 = i \cdot i^2 = -i$ and $i^4 = i^2 \cdot i^2 = 1$.

EXAMPLE 1 Simplify each expression.

a. $(4 + 5i) + (3 + 4i) = \mathbf{7 + 9i}$ b. $(2 - i) - (3 - 5i) = \mathbf{-1 + 4i}$

c. $(5 - 2i) + (3 + 2i) = \mathbf{8}$ or $8 + 0i$ d. $(4 + 3i) - (4 + 2i) = \mathbf{i}$ or $0 + i$

e. $(2 + 3i)(4 + 2i) = 8 + 16i + 6i^2$ f. $i^{94} = i^{4 \cdot 23 + 2}$

$\qquad\qquad\qquad\quad = 8 + 16i - 6$ $= (i^4)^{23}(i^2)$

$\qquad\qquad\qquad\quad = \mathbf{2 + 16i}$ $= 1 \cdot i^2$

$\qquad\qquad\qquad\qquad\qquad\qquad\qquad\qquad\qquad\qquad = \mathbf{-1}$

g. $i^{125} = i^{4 \cdot 31 + 1}$ h. $i^{1,883} = i^{4 \cdot 470 + 3}$

$\qquad\quad = (i^4)^{31}(i^1)$ $= (i^4)^{470}(i^3)$

$\qquad\quad = \mathbf{i}$ $= i^3$

$\qquad\qquad\qquad\qquad\qquad\qquad\qquad\qquad\qquad\quad = \mathbf{-i}$

EXAMPLE 2 Verify that multiplying $(a + bi)(c + di)$ in the usual algebraic way gives the same result as that shown in the definition of multiplication of complex numbers.

SOLUTION
$$(a + bi)(c + di) = ac + adi + bci + bdi^2$$
$$= ac + (ad + bc)i - bd$$
$$= (ac - bd) + (ad + bc)i$$

EXAMPLE 3 Simplify $(4 - 3i)(4 + 3i)$.

SOLUTION
$$(4 - 3i)(4 + 3i) = 16 - 9i^2 = 16 + 9 = \mathbf{25}$$

Example 3 gives a clue for dividing complex numbers. The definition would be difficult to remember, so instead we use the idea of *conjugates*. The complex numbers $a + bi$ and $a - bi$ are called **complex conjugates,** and each is the conjugate of the other:
$$(a + bi)(a - bi) = a^2 - b^2i^2 = a^2 + b^2$$

which is a real number. Thus you can simplify a quotient by using the conjugate of the denominator, as illustrated by Examples 4–6.

EXAMPLE 4 Simplify $\dfrac{15 - 5i}{2 - i}$.

SOLUTION
                Multiply by 1

$$\frac{15 - 5i}{2 - i} = \frac{15 - 5i}{2 - i} \cdot \frac{2 + i}{2 + i}$$

              conjugates

$$= \frac{30 + 5i - 5i^2}{4 - i^2}$$

$$= \frac{35 + 5i}{5}$$

$$= \mathbf{7 + i}$$

You can check this by multiplying $(2 - i)(7 + i)$:

$$(2 - i)(7 + i) = 14 - 5i - i^2 = 15 - 5i$$

EXAMPLE 5

$$\frac{6 + 5i}{2 + 3i} = \frac{6 + 5i}{2 + 3i} \cdot \frac{2 - 3i}{2 - 3i}$$

$$= \frac{12 - 8i - 15i^2}{4 - 9i^2}$$

$$= \frac{27}{13} - \frac{8}{13}i$$

EXAMPLE 6   Verify that the conjugate method gives the same result as the definition of division of complex numbers.

SOLUTION

$$\frac{a + bi}{c + di} = \frac{a + bi}{c + di} \cdot \frac{c - di}{c - di}$$

$$= \frac{ac - adi + bci - bdi^2}{c^2 - d^2 i^2}$$

$$= \frac{(ac + bd) + (bc - ad)i}{c^2 + d^2}$$

$$= \frac{ac + bd}{c^2 + d^2} + \frac{bc - ad}{c^2 + d^2}i$$

1.5 Problem Set

A *Simplify the expressions in Problems 1–36.*

1. $\sqrt{-36}$

2. $\sqrt{-100}$

3. $\sqrt{-49}$

4. $\sqrt{-8}$

5. $\sqrt{-20}$

6. $\sqrt{-24}$

7. $(3 + 3i) + (5 + 4i)$

8. $(6 - 2i) + (5 + 3i)$

9. $(5 - 3i) - (5 + 2i)$

10. $(3 + 4i) - (7 + 4i)$

11. $(4 - 2i) - (3 + 4i)$

12. $5i - (5 + 5i)$

13. $5 - (2 - 3i)$

14. $-2(-4 + 5i)$

15. $6(3 + 2i) + 4(-2 - 3i)$

16. $4(2 - i) - 3(-1 - i)$

17. $i(5 - 2i)$

18. $i(2 + 3i)$

19. $(3 - i)(2 + i)$

20. $(4 - i)(2 + i)$

21. $(5 - 2i)(5 + 2i)$

22. $(8 - 5i)(8 + 5i)$

23. $(3 - 5i)(3 + 5i)$

24. $(7 - 9i)(7 + 9i)$

25. $-i^2$ **26.** $-i^3$ **27.** i^3 **28.** i^4

29. $-i^4$ **30.** $-i^5$ **31.** $-i^6$ **32.** i^6

33. i^{11} **34.** i^{236} **35.** $-i^{1,992}$ **36.** $i^{1,994}$

B *Simplify the expressions in Problems 37–57.*

37. $(6 - 2i)^2$

38. $(3 + 3i)^2$

39. $(4 + 5i)^2$

40. $(1 + i)^3$

41. $(3 - 5i)^3$

42. $(2 - 3i)^3$

43. $\dfrac{-3}{1 + i}$

44. $\dfrac{5}{4 - i}$

45. $\dfrac{2}{1 - i}$

46. $\dfrac{5}{i}$

47. $\dfrac{2}{i}$

48. $\dfrac{3}{-i}$

49. $\dfrac{-2i}{3 + i}$

50. $\dfrac{3i}{5 - 2i}$

51. $\dfrac{-i}{2 - i}$

52. $\dfrac{1 - 6i}{1 + 6i}$

53. $\dfrac{4 - 2i}{3 + i}$

54. $\dfrac{5 + 3i}{4 - i}$

55. $\dfrac{1 + 3i}{1 - 2i}$

56. $\dfrac{3 - 2i}{5 + i}$

57. $\dfrac{2 + 7i}{2 - 7i}$

58. Simplify $(1.9319 + .5176i)(2.5981 + 1.5i)$.

59. Simplify $\dfrac{-3.2253 + 8.4022i}{3.4985 + 1.9392i}$.

Let $z_1 = (1 + \sqrt{3}) + (2 + \sqrt{3})i$ and $z_2 = (2 + \sqrt{12}) + \sqrt{3}i$.
Perform the indicated operations in Problems 60–65.

60. $z_1 + z_2$

61. $z_1 - z_2$

62. $z_1 z_2$

63. z_1/z_2

64. $(z_1)^2$

65. $(z_2)^2$

For each of Problems 66–68 let $z_1 = a + bi$, $z_2 = c + di$, and $z_3 = e + fi$.

66. Prove the commutative laws for complex numbers. That is, prove that:

$$z_1 + z_2 = z_2 + z_1$$
$$z_1 z_2 = z_2 z_1$$

67. Prove the associative laws for complex numbers. That is, prove that:

$$z_1 + (z_2 + z_3) = (z_1 + z_2) + z_3$$
$$z_1(z_2 z_3) = (z_1 z_2)z_3$$

68. Prove the distributive law for complex numbers. That is, prove that:

$$z_1(z_2 + z_3) = z_1 z_2 + z_1 z_3$$

1.6 Equations for Calculus

The most common types of equations you need to solve in calculus are linear and quadratic. However, there are other types of equations that you will occasionally encounter, and we review these types, as well as quadratic equations, in this section.

Quadratic Equations

A **quadratic equation in one variable** is an equation that can be written in the form

$$ax^2 + bx + c = 0 \qquad (a \neq 0)$$

where x is a variable and a, b, and c are real numbers. Quadratic equations can be solved by several methods. The simplest method can be used if the quadratic expression $ax^2 + bx + c$ is factorable over the integers. The solution depends on the following property of zero.

PROPERTY OF ZERO | $AB = 0$ if and only if $A = 0$ or $B = 0$ (or both).

Thus, if the product of two factors is zero, then at least one of the factors is zero. If a quadratic is factorable, this property provides a method of solution.

EXAMPLE 1 Solve $x^2 = 2x + 15$.

SOLUTION
$$x^2 - 2x - 15 = 0$$
$$(x + 3)(x - 5) = 0$$
$$x + 3 = 0 \quad \text{or} \quad x - 5 = 0$$
$$x = -3 \quad \text{or} \quad x = 5$$

Solution: $\{-3, 5\}$

> **CALCULATOR COMMENT**
>
> Graph verifies solution.
>
> $y_1 = x^2$ $y_2 = 2x + 15$

SOLUTION OF QUADRATIC EQUATIONS BY FACTORING

To solve a quadratic equation that can be expressed as a product of linear factors:

1. Rewrite all nonzero terms on one side of the equation.
2. Factor the expression.
3. Set each of the factors equal to zero.
4. Solve each of the linear equations.
5. Write the solution set, which is the union of the solution sets of the linear equations.

Completing the Square

When the quadratic is not factorable, other methods must be employed. One such method depends on the square-root property.

SQUARE-ROOT PROPERTY

If $P^2 = Q$, then $P = \pm\sqrt{Q}$.

The equation $x^2 = 4$ could be rewritten as $x^2 - 4 = 0$, factored, and solved. However, the square-root property can be used, as illustrated below.

Using the square-root property:

$$x^2 = 4$$
$$x = \pm\sqrt{4}$$
$$x = \pm 2$$

Using factoring:

$$x^2 - 4 = 0$$
$$(x + 2)(x - 2) = 0$$
$$x = -2 \quad \text{or} \quad x = 2$$

The square-root property can be derived by using the following property of square roots and an absolute value equation.

For all real numbers x,

$$\sqrt{x^2} = |x|$$

This means that if $x \geq 0$, then $\sqrt{x^2} = x$ and if $x < 0$, then $\sqrt{x^2} = -x$. For example, $\sqrt{2^2} = |2| = 2$ and $\sqrt{(-2)^2} = |-2| = 2$. The importance of this property, however, is that it can be applied to any quadratic! This is because every quadratic may be expressed in the form $P^2 = Q$ by isolating the variable terms and completing the square, as illustrated in the following examples.

EXAMPLE 2 Solve $x^2 = 2x + 15$.

SOLUTION

$$x^2 - 2x = 15$$ Isolate the variable.

$$x^2 - 2x + 1 = 15 + 1$$ Complete the square by adding the square of half the coefficient of x to both sides.

$$(x - 1)^2 = 16$$ Factor.

$$x - 1 = \pm 4$$ Use the square-root property. Alternatively, you can write $\sqrt{(x-1)^2} = \sqrt{16}$

$$x = 1 \pm 4$$ Solve for x. $|x - 1| = 4$

$$x = 1 + 4 \quad \text{or} \quad x = 1 - 4$$ $x - 1 = \pm 4$

$$x = 5 \quad \text{or} \quad x = -3$$

Solution: $\{5, -3\}$

CALCULATOR COMMENT

You may have noticed that Examples 1 and 2 are the same. We solved Example 1 by factoring and Example 2 by completing the square. We verified Example 1 graphically, by drawing two curves, namely $y_1 = x^2$ (the left side) and $y_2 = 2x + 15$ (the right side). The points of intersection are at $x = -3$ and $x = 5$, which verifies the solution. In this example, let $y = y_1 - y_2$ and then verify the solution by drawing a single graph, namely $y = x^2 - 2x - 15$. In this example, we look for the x values for which $y = 0$ (the x-axis). We see that this curve passes through the x-axis at the points $x = -3$ and $x = 5$. Notice that this graph is not the same as the one shown in Example 1, but the solution is the same (which it must be since it is the same problem).

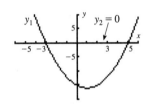

EXAMPLE 3 Solve $4x^2 - 4x - 7 = 0$.

SOLUTION

$$4x^2 - 4x = 7$$

$$x^2 - x = \frac{7}{4}$$

$$x^2 - x + \left(\frac{1}{2}\right)^2 = \left(\frac{1}{2}\right)^2 + \frac{7}{4}$$

$$\left(x - \frac{1}{2}\right)^2 = 2$$

$$x - \frac{1}{2} = \pm\sqrt{2}$$

$$x = \frac{1 \pm 2\sqrt{2}}{2}$$

Solution: $\left\{\dfrac{1 + 2\sqrt{2}}{2}, \dfrac{1 - 2\sqrt{2}}{2}\right\}$

CALCULATOR COMMENT

Let $\boxed{Y = 4X^2 - 4X - 7}$

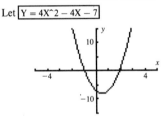

$\boxed{\text{TRACE}}$: $\boxed{X = -.9473684 \quad Y = .37950139}$

also using $\boxed{\text{ZOOM}}$:

$\boxed{X = 1.9578947 \quad Y = .50182825}$

Compare with algebraic solution:

$$x = \frac{1 + 2\sqrt{2}}{2} \approx 1.914213562$$

and $$x = \frac{1 - 2\sqrt{2}}{2} \approx -.9142135624$$

Solution approximated; $\boxed{\text{ZOOM}}$ would provide greater accuracy.

Quadratic Formula

The process of completing the square is often cumbersome. However, if *any* quadratic $ax^2 + bx + c = 0$, $a \neq 0$, is considered, completing the square can be used to derive a formula for x in terms of the coefficients a, b, and c. The formula can then be used to solve all quadratics, even those that are nonfactorable.

$$ax^2 + bx + c = 0$$

$$ax^2 + bx = -c \qquad \text{Isolate the variable.}$$

$$x^2 + \frac{b}{a}x = -\frac{c}{a} \qquad \text{Divide by } a.$$

$$x^2 + \frac{b}{a}x + \left(\frac{b}{2a}\right)^2 = \left(\frac{b^2}{4a^2}\right) - \frac{c}{a} \qquad \text{Complete the square; } \frac{1}{2} \text{ of } \frac{b}{a} \text{ is } \frac{b}{2a}.$$

$$\left(x + \frac{b}{2a}\right)^2 = \frac{b^2 - 4ac}{4a^2} \qquad \text{Factor.}$$

$$x + \frac{b}{2a} = \pm\sqrt{\frac{b^2 - 4ac}{4a^2}} \qquad \text{Use the square-root property.}$$

$$x + \frac{b}{2a} = \pm\frac{\sqrt{b^2 - 4ac}}{2a}$$

$$x = -\frac{b}{2a} \pm \frac{\sqrt{b^2 - 4ac}}{2a} \qquad \text{Solve for } x.$$

$$x = \frac{-b \pm \sqrt{b^2 - 4ac}}{2a}$$

QUADRATIC FORMULA

If $ax^2 + bx + c = 0$, $a \neq 0$, then

$$x = \frac{-b \pm \sqrt{b^2 - 4ac}}{2a}$$

EXAMPLE 4   Solve $5x^2 + 2x - 2 = 0$.

SOLUTION

$$x = \frac{-2 \pm \sqrt{2^2 - 4(5)(-2)}}{2(5)}$$

$$= \frac{-2 \pm 2\sqrt{1 + 10}}{2(5)}$$

$$= \frac{-1 \pm \sqrt{11}}{5}$$

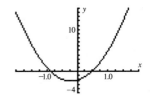

Solution: $\left\{\dfrac{-1 \pm \sqrt{11}}{5}\right\}$

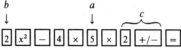

EXAMPLE 5   Solve for x: $5x^2 + 2x - (w + 4) = 0.$

SOLUTION
$$x = \frac{-2 \pm \sqrt{4 - 4(5)(-w - 4)}}{2(5)}$$

$$= \frac{-2 \pm 2\sqrt{5w + 21}}{2(5)}$$

There are some steps not shown here. Can you fill in the details?

$$= \frac{-1 \pm \sqrt{5w + 21}}{5}$$

When solving equations, it is not necessary to recopy the answer using solution set notation. If you leave your answer as shown in this example, it will be understood that the solution set is

$$\left\{ \frac{-1 \pm \sqrt{5w + 21}}{5} \right\}$$

EXAMPLE 6   Solve $5x^2 + 2x + 2 = 0.$

SOLUTION
$$x = \frac{-2 \pm \sqrt{4 - 4(5)(2)}}{2(5)}$$

$$= \frac{-2 \pm \sqrt{-36}}{10}$$

CALCULATOR COMMENT

$\boxed{Y = 5X^2 + 2X + 2}$

Since the square root of a negative number is not a real number, the solution set is **empty over the reals.**

Notice that the graph shown in the Calculator Comment box does not intersect the real number line $y = 0$. If you are using the domain consisting of the complex numbers (Section 1.5), then you can solve this equation:

$$x = \frac{-2 \pm 2(3i)}{10} = \frac{2(-1 \pm 3i)}{10}$$

$$= \frac{-1 \pm 3i}{5} = -\tfrac{1}{5} \pm \tfrac{3}{5}i$$

Since the quadratic formula contains a radical, the sign of the radicand will determine whether the roots will be real or nonreal. This radicand is called the **discriminant** of the quadratic, and its properties are summarized in the box.

THE DISCRIMINANT OF THE QUADRATIC

If $ax^2 + bx + c = 0$, $a \neq 0$, then $b^2 - 4ac$ is called the **discriminant.**
If $b^2 - 4ac < 0$, there are *no real solutions*.
If $b^2 - 4ac = 0$, there is *one real solution*.
If $b^2 - 4ac > 0$, there are *two real solutions*.

Equations in Quadratic Form

Many equations in calculus are not quadratic, but can be solved as if they were quadratic. We illustrate with two examples.

EXAMPLE 7 Solve $4x^4 + 3 = 13x^2$.

SOLUTION
$$4(x^2)^2 - 13(x^2) + 3 = 0$$
$$(4x^2 - 1)(x^2 - 3) = 0$$
$$4x^2 - 1 = 0 \quad \text{or} \quad x^2 - 3 = 0$$
$$x^2 = \tfrac{1}{4} \qquad\qquad x^2 = 3$$
$$x = \pm\tfrac{1}{2} \qquad\qquad x = \pm\sqrt{3}$$

The solution is $\{-\tfrac{1}{2}, \tfrac{1}{2}, \sqrt{3}, -\sqrt{3}\}$

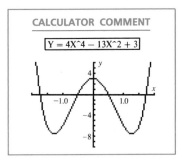

CALCULATOR COMMENT

Y = 4X^4 − 13X^2 + 3

EXAMPLE 8 Solve $x^4 + 6x^2 - 3 = 0$ over $\mathbb{R}$.

SOLUTION $x^4 + 6x^2 - 3 = 0$

$$x^2 = \frac{-6 \pm \sqrt{36 - 4(1)(-3)}}{2}$$

$$x^2 = \frac{-6 \pm 4\sqrt{3}}{2}$$

$$x^2 = -3 \pm 2\sqrt{3}$$

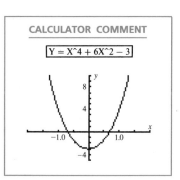

CALCULATOR COMMENT

Y = X^4 + 6X^2 − 3

Thus, $x = \pm\sqrt{-3 \pm 2\sqrt{3}}$; since $-3 - 2\sqrt{3} < 0$, the only real roots are

$$x = \pm\sqrt{-3 + 2\sqrt{3}} \approx \pm.6812500386$$

which are verified by the graph.

1.6 Problem Set

A *Solve each equation in Problems 1–12 by factoring.*

1. $x^2 + 2x - 15 = 0$ **2.** $x^2 - 8x + 12 = 0$

3. $x^2 + 7x - 18 = 0$ **4.** $2x^2 + 5x - 12 = 0$

5. $6x^2 = 5x$ **6.** $6x^2 = 12x$

7. $10x^2 - 3x - 4 = 0$ **8.** $6x^2 + 7x - 10 = 0$

9. $9x^2 - 34x - 8 = 0$ **10.** $4x^2 + 12x + 9 = 0$

11. $x^3 + 2x^2 + x = 0$ **12.** $x^4 - 25x^2 + 144 = 0$

Solve each equation in Problems 13–24 by completing the square.

13. $x^2 + 4x - 5 = 0$ **14.** $x^2 - x - 6 = 0$

15. $x^2 + 2x - 8 = 0$ **16.** $x^2 + x - 6 = 0$

17. $x^2 + 7x + 12 = 0$ **18.** $x^2 + 5x + 6 = 0$

19. $x^2 - 10x - 2 = 0$ **20.** $x^2 + 2x - 15 = 0$

21. $x^2 - 3x = 1$ **22.** $x^2 - 4x = 2$

23. $6x^2 = x + 2$ **24.** $x^2 = 4x + 2$

Solve each equation in Problems 25–48 over the set of real numbers.*

25. $x^2 + 5x - 6 = 0$ **26.** $x^2 + 5x + 6 = 0$

27. $x^2 - 10x + 25 = 0$ **28.** $x^2 + 6x + 9 = 0$

29. $12x^2 + 5x - 2 = 0$ **30.** $2x^2 - 6x + 5 = 0$

31. $5x^2 - 4x + 1 = 0$ **32.** $2x^2 + x - 15 = 0$

33. $4x^2 - 5 = 0$ **34.** $3x^2 - 1 = 0$

35. $3x^2 = 7x$ **36.** $7x^2 = 3$

37. $3x^2 = 5x + 2$ **38.** $3x^2 - 2 = 5x$

39. $5x = 3 - 4x^2$ **40.** $4x^2 = 12x - 9$

41. $3x = 1 - 2x^2$ **42.** $5x^2 = 3x - 4$

43. $\sqrt{5} - 4x^2 = 3x$ **44.** $x = \sqrt{2} - 2x^2$

45. $3x^2 - 4x = \sqrt{5}$ **46.** $5x^2 - 2x - \sqrt{3} = 0$

47. $x^4 + 6x^2 = 25$ **48.** $x^4 = 4x^3 + x^2$

B *Solve the equations in Problems 49–60 for x in terms of the other variable.*

49. $2x^2 + x - w = 0$ **50.** $2x^2 + wx + 5 = 0$

51. $3x^2 + 2x + (y + 2) = 0$ **52.** $3x^2 + 5x + (4 - y) = 0$

53. $4x^2 - 4x + (1 - t^2) = 0$

54. $x^2 - 6x + y^2 - 4y + 9 = 0$

55. $2x^2 + 3x + 4 - y = 0$

56. $y = 2x^2 + x + 6$

57. $4x^2 - (3t + 10)x + (6t + 4) = 0, \quad t > 2$

58. $(x - 3)^2 + (y - 2)^2 = 4$

59. $(x + 2)^2 - (y + 1)^2 = 9$

60. $3(x + 1)^2 + 4(y - 2)^2 = 144$

C

61. If $ax^2 + bx + c = 0$, show that the following are roots:

$$r_1 = \frac{2c}{-b + \sqrt{b^2 - 4ac}} \quad \text{and} \quad r_2 = \frac{2c}{-b - \sqrt{b^2 - 4ac}}$$

---

*If you have covered Section 1.5, you can solve these over the set of complex numbers.

1.7 Inequalities for Calculus

Quadratic Inequalities

A **quadratic inequality in one variable** is an inequality that can be written in the form

$$ax^2 + bx + c < 0 \qquad (a \neq 0)$$

where x is a variable and a, b, and c are any real numbers. The symbol $<$ can be replaced by $\leq$, $>$, or $\geq$ and it is still called a quadratic inequality in one variable.

The procedure for solving a quadratic inequality is similar to that for solving a quadratic equality. First, use the properties of inequality to obtain a zero on one side of the inequality. The next step is to factor, if possible, the quadratic expression on the left. For example:

$$x^2 - 4 \geq 0$$

$$(x - 2)(x + 2) \geq 0$$

Find the values of x that make the inequality valid. A value for which a factor is zero is called a **critical value** of x. The critical values for this example are 2 and -2. For *every other value of* x, the left side of the inequality is either positive or negative. Next, plot the critical values on a number line as in Figure 1.14a.

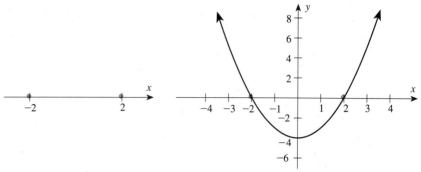

FIGURE 1.14 First step in solving $x^2 - 4 \geq 0$

a. Number line with critical values   **b.** Graph of $y = x^2 - 4$

These critical values divide the number line into three intervals. Choose a sample value from each interval. Evaluate each factor to determine the sign only—it is not necessary to complete the arithmetic to find its sign.

SAMPLE VALUE	FACTOR	SIGN OF FACTOR	SIGN OF PRODUCT
This is *your* choice	This is done mentally		
$x = -100$	$x - 2 = -100 - 2 \dashrightarrow -$ $x + 2 = -100 + 2 \dashrightarrow -$		positive product
$x = 0$	$x - 2 = 0 - 2 \dashrightarrow -$ $x + 2 = 0 + 2 \dashrightarrow +$		negative product
$x = 100$	$x - 2 = 100 - 2 \dashrightarrow +$ $x + 2 = 100 + 2 \dashrightarrow +$		positive product

This procedure can be summarized as in Figure 1.15a.

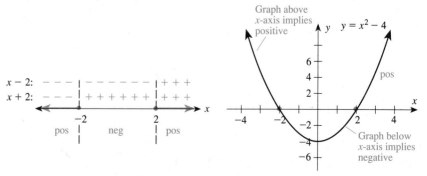

FIGURE 1.15 Procedure for solving $x^2 - 4 \geq 0$

a. One-dimensional solution   **b.** Two-dimensional solution

You want $(x - 2)(x + 2) \geq 0$. Since this is positive or zero, you pick out the parts of Figure 1.15a labeled positive and also include those that are zero (namely, the critical values). If the given inequality is of the form $\leq$ or $\geq$, the endpoints are included; if it has the form $<$ or $>$, the endpoints are excluded. The solution set is $x \leq -2$ or $x \geq 2$ or, using interval notation, $(-\infty, -2] \cup [2, \infty)$.

You may have observed a relationship between the one-dimensional solution shown above in Figures 1.14a and 1.15a and a two-dimensional solution. Notice that the positive and negative values for the one-dimensional inequality correspond to the values of x where the graph is above or below the x-axis. Also notice that the critical values are the places where the graph crosses the x-axis. If you have a graphing calculator, the two-dimensional solution shown in Figures 1.14b and 1.15b might be more efficient.

EXAMPLE 1   Solve the inequality $2x^2 < 5 - 9x$.

SOLUTION $2x^2 + 9x - 5 < 0$

$(2x - 1)(x + 5) < 0$

$$
\begin{array}{ll}
2x - 1: & - - - \mid - - - \mid + + + \\
x + 5: & - - - \mid + + + \mid + + + \\
(2x - 1)(x + 5): & \text{positive} \mid \text{negative} \mid \text{positive}
\end{array}
$$

(number line with critical values -5 and $\frac{1}{2}$)

Solution: $-5 < x < \frac{1}{2}$ or $\left(-5, \frac{1}{2}\right)$

EXAMPLE 2   Solve the inequality $x^2 + 2x - 4 < 0$.

SOLUTION   The term on the left is in simplified form and cannot be easily factored. Therefore proceed by considering $x^2 + 2x - 4$ as a single factor. To find the critical values, find the values for which the factor $x^2 + 2x - 4$ is zero:

$$x^2 + 2x - 4 = 0$$

$$x = \frac{-2 \pm \sqrt{4 - 4(1)(-4)}}{2} = \frac{-2 \pm 2\sqrt{5}}{2} = -1 \pm \sqrt{5}$$

Use the quadratic formula (as shown here) whenever you cannot factor the quadratic expression.

Plot the critical values and check the sign of the factor in each interval:

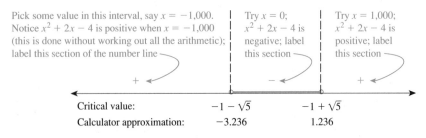

Pick some value in this interval, say $x = -1,000$. Notice $x^2 + 2x - 4$ is positive when $x = -1,000$ (this is done without working out all the arithmetic); label this section of the number line

Try $x = 0$; $x^2 + 2x - 4$ is negative; label this section

Try $x = 1,000$; $x^2 + 2x - 4$ is positive; label this section

Critical value: $-1 - \sqrt{5}$ $-1 + \sqrt{5}$

Calculator approximation: -3.236 1.236

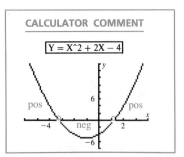

The solution is $-1 - \sqrt{5} < x < -1 + \sqrt{5}$ or, using interval notation,

$$\left(-1 - \sqrt{5}, \, -1 + \sqrt{5}\right)$$

EXAMPLE 3 Solve the inequality $x^2 + 5x + 12 > 0$.

SOLUTION The quadratic does not factor, so we look for a critical value by solving $x^2 + 5x + 12 = 0$ by the quadratic formula:

$$x = \frac{-5 \pm \sqrt{25 - 4(1)(12)}}{2}$$

which has no real roots. Since there are no critical values, the answer is either all real values or no real values. Let $x = 0$ be a test value; we see that

$$0^2 + 5(0) + 12 > 0$$

is true. Thus, all real values satisfy the inequality. The solution is $(-\infty, \infty)$. ▮

CALCULATOR COMMENT

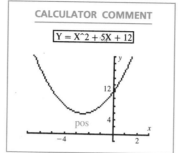

$Y = X^2 + 5X + 12$

Inequalities in Factored Form

The factoring procedure used in this section for solving quadratic inequalities can be used for other inequalities that are in factored form, even though they may not be quadratic.

EXAMPLE 4 Solve the inequality $(x - 5)(2 - x)(2 - 3x) > 0$.

SOLUTION

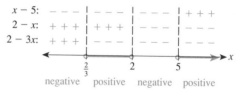

Solution: $\left(\frac{2}{3}, 2\right) \cup (5, \infty)$

CALCULATOR COMMENT

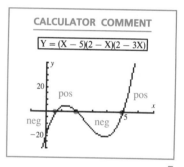

$Y = (X - 5)(2 - X)(2 - 3X)$

EXAMPLE 5 Solve

$$\frac{3x(x + 1)(x - 2)}{(x - 1)(x + 3)} \geq 0$$

SOLUTION Plot the critical values of 0, -1, 2, 1, and -3. These are included (because it is $\geq$), but *values that cause division by zero* ($x = 1$, $x = -3$) need to be *excluded*.

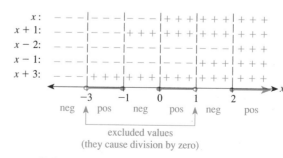

CALCULATOR COMMENT

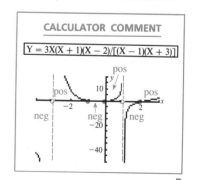

$Y = 3X(X + 1)(X - 2)/[(X - 1)(X + 3)]$

Solution: $(-3, -1] \cup [0, 1) \cup [2, \infty)$ ▮

⊘ Be careful about following the procedure described above. It is tempting to attempt these solutions without testing intervals on a number line. ⊘

You must also take care not to divide both sides by a variable. For example, if

$$x^2 > x$$

and you divide both sides by x to obtain $x > 1$, you will have made a very common mistake, which can be shown by a counterexample. Suppose $x = -2$. Then

$$x^2 > x$$
$$(-2)^2 > -2$$
$$4 > -2$$

which is true. But if you divide both sides by $x = -2$, you obtain

$$\frac{4}{-2} > \frac{-2}{-2}$$
$$-2 > 1 \qquad \text{False!}$$

Using numbers, you can easily see the mistake; the order of inequality was not reversed as it should have been.

EXAMPLE 6   Solve $\dfrac{5}{x+3} > \dfrac{-2}{x-2}$.

SOLUTION   Remember, in order to solve an inequality with degree greater than 1, we wish to obtain a 0 on one side.

$$\frac{5}{x+3} > \frac{-2}{x-2} \qquad \text{Note: } x \neq -3, x \neq 2$$

$$\frac{5}{x+3} + \frac{2}{x-2} > 0$$

$$\frac{5(x-2) + 2(x+3)}{(x+3)(x-2)} > 0$$

$$\frac{7x-4}{(x+3)(x-2)} > 0$$

⊘ Do not multiply both sides by the common denominator of $(x+3)(x-2)$. We do not know whether this product is positive or negative, so we do not know whether to reverse the inequality. ⊘

$7x - 4$: $- - - \; | - - - \; | + \; | + + +$
$x + 3$: $- - - \; | + + + \; | + \; | + + +$
$x - 2$: $- - - \; | - - - \; | - \; | + + +$

$-3 \qquad \frac{4}{7} \quad 2$

neg   pos   neg   pos

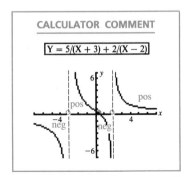

CALCULATOR COMMENT

$Y = 5/(X + 3) + 2/(X - 2)$

Solution: $(-3, \tfrac{4}{7}] \cup (2, \infty)$.

1.7 Problem Set

A

WHAT IS WRONG, *if anything, with each of the statements in Problems 1–6? Explain your reasoning.*

1. If $x(x + 5) < 0$, then $x < 0$ and $x + 5 < 0$.

2. If $x(x + 6) \geq 0$, then $x + 6 \geq 0$ (divide both sides by x).

3. If $5x^2 = 125x$, then $x = 25$ (divide both sides by $5x$).

4. If $(x + 3)(x - 2) = 4$, then $x + 3 = 4$ and $x - 2 = 4$.

5. If $bx^2 + ax + c = 0$, $b \neq 0$, then

$$x = \frac{-b \pm \sqrt{b^2 - 4ac}}{2a}$$

6. Solve $2x^2 < 5 - 9x$.

$$2x^2 + 9x - 5 < 0$$
$$(2x - 1)(x + 5) < 0$$
$$2x - 1 < 0 \quad \text{or} \quad x + 5 < 0$$
$$2x < 1 \qquad\qquad x < -5$$
$$x < \tfrac{1}{2}$$

Solve the inequalities in Problems 7–60 and write each answer using interval notation.

7. $x(x + 3) < 0$

8. $x(x - 3) \geq 0$

9. $(x - 6)(x - 2) \geq 0$

10. $(y + 2)(y - 8) \leq 0$

11. $(x - 8)(x + 7) < 0$

12. $(x + 1)(x + 6) > 0$

13. $(x + 2)(2x - 1) \leq 0$

14. $(x - 2)(2x + 1) \leq 0$

15. $(3x + 2)(x - 3) > 0$

16. $(3x - 2)(x + 2) > 0$

17. $(x + 2)(8 - x) \leq 0$

18. $(2 - x)(x + 8) \geq 0$

19. $(1 - 3x)(x - 4) < 0$

20. $(2x + 1)(3 - x) > 0$

21. $x(x - 3)(x + 4) \leq 0$

22. $x(x + 3)(x - 4) \geq 0$

23. $(x - 2)(x + 3)(x - 4) \geq 0$

24. $(x + 1)(x - 2)(x + 3) \leq 0$

25. $(x + 1)(2x + 5)(7 - 3x) > 0$

26. $(x - 2)(3x + 2)(3 - 2x) < 0$

27. $\dfrac{x + 2}{x} < 0$

28. $\dfrac{x}{x + 3} < 0$

29. $\dfrac{x}{x - 8} > 0$

30. $\dfrac{x - 8}{x} > 0$

31. $\dfrac{x - 2}{x + 5} \leq 0$

32. $\dfrac{x + 2}{4 - x} \geq 0$

B

33. $\dfrac{x(2x - 1)}{5 - x} > 0$

34. $\dfrac{x}{(2x + 3)(x - 2)} < 0$

35. $\dfrac{1}{x(x - 3)(x + 2)} \leq 0$

36. $\dfrac{1}{x(x + 1)(5 - x)} \geq 0$

37. $x^2 \geq 9$

38. $x^2 \geq 4$

39. $x^2 + 9 \geq 0$

40. $x^2 + 2x - 3 < 0$

41. $x^2 - x - 6 > 0$

42. $x^2 - 7x + 12 > 0$

43. $5x - 6 \geq x^2$

44. $4 \geq x^2 + 3x$

45. $5 - 4x \geq x^2$

46. $x^2 + 2x - 1 < 0$

47. $x^2 - 2x - 2 < 0$

48. $x^2 - 8x + 13 > 0$

49. $2x^2 + 4x + 5 \geq 0$

50. $x^2 - 2x - 6 \leq 0$

51. $x^2 + 3x - 7 \geq 0$

52. $\dfrac{(x - 3)(x + 1)}{(x - 2)(x - 1)} \leq 0$

53. $\dfrac{x(x + 5)(x - 3)}{(x + 3)(x - 4)} \geq 0$

54. $\dfrac{x(x - 6)(2 - x)}{(x - 4)(3 - x)} \leq 0$

C

55. $\dfrac{2}{x - 2} \leq \dfrac{3}{x + 3}$

56. $\dfrac{1}{x + 1} \geq \dfrac{2}{x - 1}$

57. $\dfrac{x - 3}{3x - 1} \geq \dfrac{x + 3}{2x + 1}$

58. $\dfrac{x - 7}{x^2 - 4x - 21} < 0$

59. $\dfrac{x - 4}{x^2 - 5x + 6} \leq 0$

60. $\dfrac{x^2 + 4x + 3}{x + 2} \geq 0$

1.8 Problem Solving*

You must approach word problems with the realization that algebra will not solve them—*you will*. If an equation or inequality can be found to represent accurately and fully the relationships presented in the problem, then that equation or inequality can be solved, and the solutions can be interpreted as an answer to the problem. However, the direct application of algebra to the words of the problem will not yield an equation; that is your contribution. This step can be accomplished only if you understand the problem well enough to state its relationships algebraically. You must, therefore, thoroughly *understand* the problem *before* you can solve it. The following suggestions give a strategy that is helpful in attacking such problems:

STRATEGY FOR SOLVING APPLIED PROBLEMS

1. Read the problem. Determine what it is all about. Be certain you know what information is given and what must be found. You must understand a problem if you are to solve it.

2. Analyze the problem. Focus on processes rather than on numbers and answers. Recall what you know about the subject of the problem. Display this information to help clarify the facts and relationships. In mathematical modeling it is sometimes necessary to discard information that has no bearing on the solution, or to research or experiment to find essential information that is needed for the solution, but not given as part of the original problem.†

3. Identify the variable and write the equation or inequality that evolves from an analysis of the problem after defining a variable. The discovery of this equation will be the result of your understanding of the problem, not your algebraic skill.

4. Solve the equation or inequality. Check the result in the original problem to make certain that you discovered the right equation. That is, your solution should make sense.

5. State the answer or answers to the problem. Pay attention to units of measure and other details of the problem.

Current studies of learning theories support such a strategy. The following is quoted from a recent address at a national convocation called by the National Academy of Science and the National Academy of Engineering: *Good problem-solvers do not rush in to apply a formula or an equation. Instead, they try to understand the problem situation; they consider alternative representations and relations among variables. Only when satisfied that they understand the situation and*

---

*For a more complete discussion of this topic, see the book in the Brooks/Cole One-Unit Series in Precalculus Mathematics entitled *Problem Solving*, written by the same author as this text.

†In a classroom setting it is not practical to give too little information in a problem (unless it is designated as a research problem). However, in the real world it is common to need additional information to find a solution. Since problem solving involves a long learning process, we will only occasionally give you "extra" information. Our goal in this book is to build your problem solving skills to bring you to a level of competency that will allow you to know when you have too much or too little information.

all the variables in a qualitative way do they start to apply the quantification."* You are well advised to begin to develop a strategy with which you are comfortable. If the strategy makes sense, your confidence will provide greater success.

Applied problems are often concerned with length, area, and volume. Since these are largely geometric ideas, begin whenever possible with a sketch. Watch carefully in the following examples as the strategy is applied to help solve the problems.

EXAMPLE 1   Two rectangles have the same width, but one is 40 ft² larger in area. The larger rectangle is 6 ft longer than it is wide. The other is only 1 ft longer than its width. What are the dimensions of the larger rectangle?

ANALYSIS   The *areas* differ by 40, and each area is the product of length and width:

$$\left(\begin{matrix}\text{AREA OF}\\\text{LARGER}\end{matrix}\right) - \left(\begin{matrix}\text{AREA OF}\\\text{SMALLER}\end{matrix}\right) = 40$$

$$\left(\begin{matrix}\text{LENGTH}\\\text{OF LARGER}\end{matrix}\right)(\text{WIDTH}) - \left(\begin{matrix}\text{LENGTH}\\\text{OF SMALLER}\end{matrix}\right)(\text{WIDTH}) = 40$$

The width is the same in each case, but is related differently in each figure to its length:

$$(\text{WIDTH} + 6)(\text{WIDTH}) - (\text{WIDTH} + 1)(\text{WIDTH}) = 40$$

SOLUTION   Let W be the width of each rectangle. Then

$$(W + 6)(W) - (W + 1)(W) = 40$$
$$W^2 + 6W - W^2 - W = 40$$
$$5W = 40$$
$$W = 8 \quad \text{and} \quad W + 6 = 14$$

The larger rectangle is 8 ft wide by 14 ft long.

ALTERNATE ANALYSIS   A sketch will frequently simplify a problem. In this case, since the widths are the same, the difference in area can be seen in the larger rectangle:

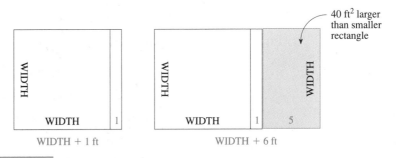

---

*An address by Lauren B. Resnick, National Convocation on Precollege Education in Mathematics, National Academy of Science and National Academy of Engineering, Washington, D.C., May 1982. (Quoted in *Science*, vol. 220, no. 4596, April 29, 1983, p. 29.)

The difference is shown as the shaded region. The area of this region is stated in the problem:

$$\begin{pmatrix} \text{DIFFERENCE} \\ \text{IN AREA} \end{pmatrix} = 40$$

$$(5)(\text{WIDTH}) = 40$$

ALTERNATE SOLUTION Let W be the width of each rectangle. Then

$$5W = 40$$

$$W = 8 \quad \text{and} \quad W + 6 = 14$$

The larger rectangle is 8 ft wide by 14 ft long. ▮

A dictionary definition of *rate* is "the quantity of a thing in relation to the units of something else." This definition is quite general, yet rate is too often limited to rate–time–distance relationships. A typing rate is the number of words typed per unit of time, and most prices are based on cost per unit. The following example deals with the rate of interest earned as a percentage of the principal invested.

EXAMPLE 2 A total of \$1,400 is invested for a year, part at 5% and the rest at $6\frac{1}{2}\%$. If \$76 is collected in interest, how much is invested at 5%?

ANALYSIS First, interest is earned from two separate investments, and since interest is the product of the principal and the rate, we have

$$\begin{pmatrix} \text{INTEREST} \\ \text{AT } 5\% \end{pmatrix} + \begin{pmatrix} \text{INTEREST} \\ \text{AT } 6\frac{1}{2}\% \end{pmatrix} = \begin{pmatrix} \text{TOTAL} \\ \text{INTEREST} \end{pmatrix}$$

$$\begin{pmatrix} \text{PRINCIPAL} \\ \text{AT } 5\% \end{pmatrix}(5\% \text{ RATE}) + \begin{pmatrix} \text{PRINCIPAL} \\ \text{AT } 6\frac{1}{2}\% \end{pmatrix}(6\frac{1}{2}\% \text{ RATE}) = \begin{pmatrix} \text{TOTAL} \\ \text{INTEREST} \end{pmatrix}$$

Now some of the known quantities may be replaced by their values. Since the total investment is known, one principal may be written in terms of the other:

$$\begin{pmatrix} \text{PRINCIPAL} \\ \text{AT } 5\% \end{pmatrix}(.05) + \left(1,400 - \begin{array}{c} \text{PRINCIPAL} \\ \text{AT } 5\% \end{array} \right)(.065) = 76$$

SOLUTION Let P be the principal invested at 5%. Then

$$.05P + .065(1,400 - P) = 76$$

$$.05P + 91 - .065P = 76$$

$$91 - .015P = 76$$

$$-.015P = -15$$

$$P = 1,000$$

The amount invested at 5% is \$1,000. ▮

EXAMPLE 3 An office worker takes 55 minutes to return from the job each day. This person rides a bus that averages 30 mph and walks the rest of the way at 4 mph. If the total distance is 21 miles from office to home, what is the distance walked each day?

ANALYSIS The total trip is composed of two distinct distances. These distances are the product of rate and time, so once again *rate* is an important ingredient of the analysis.

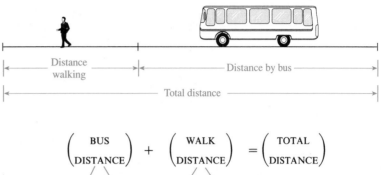

$$\begin{pmatrix} \text{BUS} \\ \text{DISTANCE} \end{pmatrix} + \begin{pmatrix} \text{WALK} \\ \text{DISTANCE} \end{pmatrix} = \begin{pmatrix} \text{TOTAL} \\ \text{DISTANCE} \end{pmatrix}$$

$$\begin{pmatrix} \text{BUS} \\ \text{RATE} \end{pmatrix}\begin{pmatrix} \text{BUS} \\ \text{TIME} \end{pmatrix} + \begin{pmatrix} \text{WALK} \\ \text{RATE} \end{pmatrix}\begin{pmatrix} \text{WALK} \\ \text{TIME} \end{pmatrix} = \begin{pmatrix} \text{TOTAL} \\ \text{DISTANCE} \end{pmatrix}$$

The total time is known; therefore, the bus time may be expressed in terms of the walk time. Be careful to express the time in *hours* since the rate is given in miles per hour:

$$(30)\left(\frac{55}{60} \; \begin{matrix} \text{WALK} \\ \text{TIME} \end{matrix} \right) + (4)\left(\begin{matrix} \text{WALK} \\ \text{TIME} \end{matrix} \right) = 21$$

SOLUTION Let W be the time walked, in hours. Then

$$30(\tfrac{55}{60} - W) + 4W = 21$$
$$\tfrac{55}{2} - 30W + 4W = 21$$
$$27.5 - 26W = 21$$
$$-26W = -6.5$$
$$W = .25$$

The worker walks .25 hours, but the problem asks for the distance, so returning to our original analysis, we find

$$\begin{pmatrix} \text{WALK} \\ \text{DISTANCE} \end{pmatrix} = \begin{pmatrix} \text{WALK} \\ \text{RATE} \end{pmatrix}\begin{pmatrix} \text{WALK} \\ \text{TIME} \end{pmatrix}$$

Be certain that the solution answers the problem posed.

$$= 4(.25)$$
$$= 1$$

The worker walks 1 mile daily.

EXAMPLE 4 Milk containing 10% butterfat and cream with 80% butterfat are mixed to produce half-and-half, which is 50% butterfat. How many gallons of each must be mixed to produce 140 gallons of half-and-half?

ANALYSIS In mixture problems of this sort, focus on the *amount* of each substance:

$$\begin{pmatrix} \text{AMT OF} \\ \text{BUTTERFAT} \\ \text{IN MILK} \end{pmatrix} + \begin{pmatrix} \text{AMT OF} \\ \text{BUTTERFAT} \\ \text{IN CREAM} \end{pmatrix} = \begin{pmatrix} \text{AMT OF} \\ \text{BUTTERFAT IN} \\ \text{HALF-AND-HALF} \end{pmatrix}$$

Each amount is given as a percentage (or rate) of its volume, so we may write the amounts as products:

$$\begin{pmatrix} \text{PCT IN} \\ \text{MILK} \end{pmatrix}\begin{pmatrix} \text{VOL OF} \\ \text{MILK} \end{pmatrix} + \begin{pmatrix} \text{PCT IN} \\ \text{CREAM} \end{pmatrix}\begin{pmatrix} \text{VOL OF} \\ \text{CREAM} \end{pmatrix} = \begin{pmatrix} \text{PCT IN} \\ \text{HALF-AND-HALF} \end{pmatrix}\begin{pmatrix} \text{VOL OF} \\ \text{HALF-AND-HALF} \end{pmatrix}$$

$$.10\begin{pmatrix} \text{VOL OF} \\ \text{MILK} \end{pmatrix} + .80\left(140 - \begin{matrix} \text{VOL OF} \\ \text{MILK} \end{matrix} \right) = .50(140)$$

SOLUTION Let M be the volume of milk used, in gallons. Then

$$.10M + .80(140 - M) = .50(140)$$
$$.10M + 112 - .80M = 70$$
$$-.70M = -42$$
$$M = 60 \quad \text{and} \quad 140 - M = 80$$

The half-and-half contains 60 gal of milk and 80 gal of cream. ∎

EXAMPLE 5 A 1-mile length of pipeline connects two pumping stations. Special joints must be used along the line to provide for expansion and contraction due to changes in temperature. However, if the pipeline were actually one continuous length of pipe fixed at each end by the stations, then expansion would cause the pipe to bow. Approximately how high would the middle of the pipe rise if the expansion was just 1 inch over the mile?

ANALYSIS Assume that the pipe bows in a circular arc, as shown in the figure. A triangle would produce a reasonable approximation since the distance x should be quite small compared to the total length. Since a right triangle is used to model the situation, the Pythagorean theorem may be employed.

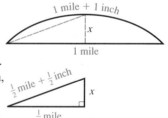

SOLUTION Let x be the height of the middle of the pipe. Then

$$(\tfrac{1}{2}\text{ mile} + \tfrac{1}{2}\text{ inch})^2 = x^2 + (\tfrac{1}{2}\text{ mile})^2$$
$$(31{,}680.5)^2 = x^2 + (31{,}680)^2$$
$$x^2 = (31{,}680.5)^2 - (31{,}680)^2$$
$$x = \sqrt{(31{,}680.5)^2 - (31{,}680)^2}$$
$$x \approx \mathbf{177.99}$$

1 mi = 5,280 ft
= (5,280)(12 in.)

The positive value is taken since x is a distance.

The solution 177.99 in. is approximately 14.8 ft. This is an extraordinary result if you consider what your estimate would have been before you worked the problem.

The pipe would bow approximately 14.8 ft at the middle. ∎

EXAMPLE 6 A container is to be constructed from an 11 in. by 16 in. sheet of cardboard. Squares will be cut from the corners of the sheet and discarded as waste. The area of the base of the container should exceed the area of wasted cardboard. Find the size of the container with the greatest possible volume if the dimensions must be integers.

SOLUTION First understand the problem. Let s be the length of a side of the square.

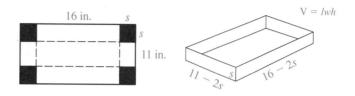

Domain: All sides must be nonnegative. That is,

$$s \geq 0 \qquad 16 - 2s \geq 0 \qquad 11 - 2s \geq 0$$
$$s \leq 8 \qquad\qquad s \leq 5.5$$

The only possible integer values are 0, 1, 2, 3, 4, and 5.

TABULAR SOLUTION Since there are very few values for s, we can actually calculate the volume of the six possible boxes. This is particularly easy to do if you use a computer.

VALUE OF s	LENGTH	WIDTH	HEIGHT	VOLUME
0	16	11	0	0
1	14	9	1	126
2	12	7	2	168
3	10	5	3	150
4	8	3	4	96
5	6	1	5	30

The maximum volume is 168 cubic inches, which occurs for a box with dimensions 12 in. × 7 in. × 2 in.

ALGEBRAIC SOLUTION

$$\text{AREA OF BASE} \geq \text{AREA OF WASTE}$$
$$(\text{LENGTH OF BASE})(\text{WIDTH OF BASE}) \geq 4(\text{AREA OF CORNER})$$
$$(16 - 2s)(11 - 2s) \geq 4s^2$$
$$176 - 54s + 4s^2 \geq 4s^2$$
$$-54s \geq -176$$
$$s \leq 3.25925$$

The possible integer values are $s = 0, 1, 2,$ and 3. Testing each of these, we find the maximum volume is 168 cubic inches, which results from a box with dimensions 12 in. × 7 in. × 2 in., formed by cutting out a 2-in. square from each corner.

The following problems are similar to related rate problems you will encounter in a calculus course.

EXAMPLE 7   A person 6 feet tall is standing 7 feet from the base of a streetlight. If the light is 20 feet above ground, how long is the person's shadow due to the streetlight?

SOLUTION   Let x denote the length (in feet) of the person's shadow as shown.

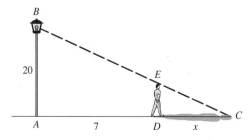

Since triangles $\triangle ABC$ and $\triangle DEC$ are similar, we have

$$\frac{6}{x} = \frac{20}{7 + x} \qquad x \neq 0, -7$$

$$6(7 + x) = 20x$$

$$42 + 6x = 20x$$

$$3 = x$$

The shadow's length is 3 feet.

EXAMPLE 8   A water tank is in the shape of an inverted cone 20 ft high with a circular base whose radius is 5 ft. How much water is in the tank when the water is 8 ft deep?

SOLUTION   The volume of a cone is $V = \frac{1}{3}\pi r^2 h$. We are given $h = 8$, and need to find r. Once again, we use similar triangles as shown:

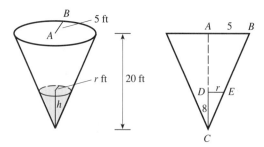

Since triangles $\triangle ABC$ and $\triangle DEC$ are similar, we have

$$\frac{5}{20} = \frac{r}{8}$$

$$40 = 20r$$

$$2 = r$$

The desired volume is $V = \frac{1}{3}\pi(2)^2(8) = \frac{32}{3}\pi$ **ft³.**

1.8 Problem Set

A *For Problems 1–60 carefully analyze the problem, develop and solve an equation, and state an answer. Watch for rates involved in the problems.*

1. Find two consecutive numbers whose sum is 125.

2. Find three consecutive odd integers whose sum is 57.

3. Find the first of three consecutive odd integers if their sum is 99.

4. What is the largest of three consecutive integers if their sum is 84?

5. The product of two consecutive integers is 62 less than the product of the next two integers. What is the second of the four integers?

6. The product of two consecutive odd integers is 208 more than the product of the preceding two odd integers. What is the third of these odd integers?

7. Two rectangles have the same width, but one is 20 ft² larger in area. The larger rectangle is 6 ft longer than it is wide. The other is only 2 ft longer than its width. What are the dimensions of the larger rectangle?

8. Two rectangles have the same length, but one is 24 in.² larger in area. The larger rectangle is only 4 in. shorter than it is long. The other is 6 in. shorter than its length. What are the dimensions of the larger rectangle?

9. Two triangles have the same height. The base of the larger one is 3 cm greater than its height. The base of the other is 1 cm greater than its height. If the areas differ by 3 cm², find the dimensions of the smaller figure.

10. Two triangles have the same base. The height of the larger one is 2 yd shorter than its base. The height of the other is 4 yd shorter than its base. If the areas differ by 6 yd², find the dimensions of the smaller figure.

B

11. A rectangle is 2 ft longer than it is wide. If you increase the length by a foot and reduce the width the same, the area is reduced by 3 ft². Find the width of the new figure.

12. The length of a rectangle is 3 m more than its width. The width is increased by 2 m, and the length is shortened by a meter. If the two figures have the same perimeter, what is it?

13. BUSINESS A total of $1,000 was invested for one year, part of it at 10% and the rest at 12.5%. If $110 interest was earned, how much was invested at 10%?

14. BUSINESS A total of $1,200 was invested for one year, part of it at 10% and the rest at 12%. If $130 interest was earned, how much was invested at 12%?

15. BUSINESS A total of $8,000 is invested for one year, part at 9% and the remainder at 8%. If $665 interest is earned, how much is invested at 9%?

16. BUSINESS A total of $12,000 is invested for one year, part at $9\frac{1}{2}$% and the remainder at 11%. If $1,275 interest is earned, how much is invested at 11%?

17. BUSINESS Part of $1,500 is invested at $9\frac{1}{2}$% and the remainder is invested at 14%. The combined investments will yield $183 interest the first year. How much is invested at each rate?

18. BUSINESS Part of $9,000 is invested at 7.5% and the remainder is invested at 11.5%. The combined investments will yield $803 interest the first year. How much is invested at each rate?

19. BUSINESS One part of $20,000 is invested at $10\frac{1}{4}$% and the rest at $11\frac{1}{4}$%. A total of $2,165 will be earned from the first year's interest. How much is invested at the lower rate?

20. BUSINESS One part of $12,400 is invested at 9.75% and the rest at 10.75%. A total of $1,256 will be earned from the first year's interest. How much is invested at the higher rate?

21. A trip is made by train and bus. The train averages 72 mph and the bus only 39 mph. The total trip of 405 mi takes 7 hr. What distance is traveled by train?

22. A trip is made by ship and train. The ship averages 32 mph and the train 52 mph. The total trip of 660 mi takes 20 hr. How many hours are spent on the ship?

23. A businesswoman logs time in an airliner and a rental car to reach her destination. The total trip is 1,100 mi, the plane averaging 600 mph and the car 50 mph. How long is spent in the automobile if the trip took a total of $5\frac{1}{2}$ hr?

24. Barry hitchhikes back to campus from home, which is 82 mi away. He makes 4 mph walking, until he gets a ride. In the car, he makes 48 mph. If the trip took 4 hr, how far did Barry walk?

25. A commuter takes an hour to get to work each day. She takes rapid transit that averages 65 mph and a shuttle bus the rest of the way at 25 mph. If the total distance is 61 mi from home to work, how far does she ride the bus to get to work each day?

26. A commuter rides a train into the city and catches a taxi to the office, traveling 50 mi to work. The train averages 60 mph and the cab 20 mph. If his trip takes a total of 1 hr, how much time is spent on the train each morning?

27. Two joggers set out at the same time from their homes 21 mi apart. They agree to meet at a point in between in an hour and a half. If the rate of one is 2 mph faster than the other, find the rate of each.

28. Two joggers set out at the same time but in opposite directions. If they were to maintain their normal rates for 4 hr, they could be 68 mi apart. If the rate of one is 1.5 mph faster than the other, find the rate of each.

29. Melissa commutes 30 mi to work each day, partly on a highway and partly in city traffic. On the highway she doubles her city speed for just 15 minutes. If the entire trip takes an hour, how fast is she able to average in city traffic?

30. Shannon drives 56 mi to his job in the city every day, part on the highway and part on city streets. Off the highway, traffic crawls along at a third of his highway speed for 10 minutes to complete the journey. If the trip takes an hour, how fast is Shannon able to average in the city?

31. Milk containing 12% butterfat and cream containing 69% butterfat are blended to produce half-and-half, which is 50% butterfat. How many gallons of each must be mixed to make 150 gallons of half-and-half?

32. Milk containing 8% butterfat and cream with 64% butterfat are mixed to produce half-and-half, which is 50% butterfat. How many gallons of each must be mixed to make 200 gallons of half-and-half?

33. CHEMISTRY A chemist has two solutions of sulfuric acid. One is a 72% solution, and the other is a 45% solution. How many liters of each does the chemist mix to get 4.5 liters of a 60% solution?

34. CHEMISTRY A chemist has two solutions of hydrochloric acid. One is a 50% solution, and the other is a 32% solution. How many milliliters of each must be mixed to get 75 ml of 44% solution?

35. CHEMISTRY How much water must be added to a 35% acid solution to obtain 100 cc of a 21% solution?

36. METALLURGY How much pure silver must be alloyed with a 36% silver alloy to obtain 100 g of a 52% alloy?

37. An aftershave lotion is 50% alcohol. If you have 8 fluid ounces of the lotion, how much water must be added to reduce the mixture to 20% alcohol?

38. If you have 8 fluid ounces of an aftershave lotion that is 20% alcohol, how much alcohol must be added to enhance the mixture to 50% alcohol?

39. You have 8 oz of an aftershave lotion that is 50% alcohol. How much of a 15% alcohol lotion must be added to reduce the mixture to 25% alcohol?

40. If 8 oz of an aftershave lotion is 15% alcohol, how much of a 50% alcohol lotion must be added to produce a 30% alcohol lotion?

41. The product of two numbers is at least 340. One of the numbers is three less than the other. What are the possible values of the larger number?

42. The product of two numbers is no larger than 300. One number is five larger than the other. What are the possible values of the smaller number?

43. The quotient of two numbers is positive. The divisor is three larger than the dividend. What are the possibilities for the smaller number?

44. Two numbers have a negative quotient. What are the possibilities for the dividend if it is five larger than the divisor?

45. A rectangular area is to be fenced. If the space is twice as long as it is wide, for what dimensions is the area numerically greater than the perimeter?

46. A rectangular area three times as long as it is wide is to be fenced. For what dimensions is the perimeter numerically greater than the area?

47. BUSINESS A small manufacturer of citizens' band radios determines that the price of each item is related to the number of items produced per day. The manufacturer knows that (a) the maximum number that can be produced is 10 items; (b) the price should be $400 - 25x$ dollars; (c) the overhead (the cost of producing x items) is $5x^2 + 40x + 600$ dollars; and (d) the daily profit is then found by subtracting the overhead from the revenue:

$$\begin{aligned}
\text{Profit} &= \text{Revenue} - \text{Cost} \\
&= (\text{Number of items})(\text{Price per item}) - \text{Cost} \\
&= x(400 - 25x) - (5x^2 + 40x + 600) \\
&= 400x - 25x^2 - 5x^2 - 40x - 600 \\
&= -30x^2 + 360x - 600
\end{aligned}$$

What is the number of radios produced if the profit is zero?

48. PHYSICS Suppose you throw a rock at 48 ft/sec from the top of the Sears Tower in Chicago and the height in feet, h, from the ground after t sec is given by

$$h = -16t^2 + 48t + 1{,}454$$

a. What is the height of the Sears Tower?

b. How long will it take (to the nearest tenth of a second) for the rock to hit the ground?

49. PHYSICS If an object is shot up from the ground with an initial velocity of 256 ft/sec, its distance in feet above the ground at the end of t sec is given by $d = 256t - 16t^2$ (neglecting air resistance). Find the length of time for which $d \geq 240$.

50. PHYSICS Find the length of time the projectile described in Problem 49 will be in the air.

C

51. A radiator contains 8 quarts of a 40% antifreeze mixture. Some of the mixture must be drained and replaced by pure antifreeze to increase the amount of antifreeze in the mixture to 60% antifreeze. How much pure antifreeze must be added?

52. A radiator contains 8 quarts of a 75% antifreeze solution. Some of the solution is to be drained for another purpose and replaced by water to decrease the solution to 60% antifreeze. How much of the original solution should be replaced by water?

53. CONSUMER ISSUE An electric company charges $40 for the first 1,000 kilowatt-hours or less, 6¢ per kilowatt-hour for the next 1,000 kilowatt-hours, and 8¢ per kilowatt-hour for anything beyond. If the charge is $119.20, how many kilowatt-hours were used?

54. Current postal regulations do not permit a package to be mailed if the combined length, width, and height exceed 72 in. What are the dimensions of the largest permissible package with length twice the length of its square end?

55. Two brothers leave home at the same time and walk in opposite directions, one at 3.5 mph and the other at 4 mph. How long will it be before the boys are 15 mi apart? How far must each boy walk?

56. Two trains leave towns 84 mi apart at the same time and travel toward one another. If one travels at 25 mph and the other at 31 mph, in how many hours will they meet? How far must each train travel?

57. A girl riding a motorcycle at 35 mph left home $2\frac{1}{2}$ hours after her brother, who was riding a bicycle at 10 mph. How long will it take the girl on the motorcycle to overtake her brother if they are traveling in the same direction?

58. An airplane flying 120 mph leaves Newport 4 hours after a yacht has sailed and overtakes it in 1 hour. What is the speed of the yacht, and how far had it sailed before being overtaken?

59. ENGINEERING Many materials, such as brick, steel, aluminum, and concrete, expand due to increases in temperature. This is why fillers are placed between the cement slabs in sidewalks. Suppose you have a 100-ft roof truss securely fastened at both ends, and assume that the buckle is linear. (It is not, but this assumption will serve as a worthwhile approximation.) Let the height of the buckle be x ft. If the percentage of swelling is y, then, for each half of the truss,

New length = Old length + Change in length

$$= 50 + (\text{Percentage})(\text{Length})$$

$$= 50 + \left(\frac{y}{100}\right)50$$

$$= 50 + \frac{y}{2}$$

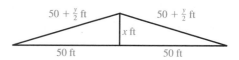

These relationships are shown in the figure. Then, by the Pythagorean theorem,

$$x^2 + 50^2 = \left(50 + \frac{y}{2}\right)^2$$

$$x^2 + 50^2 = \frac{(100 + y)^2}{4}$$

$$4x^2 + 4 \cdot 50^2 = 100^2 + 200y + y^2$$

$$4x^2 - y^2 - 200y = 0$$

Solve this equation for x and then calculate the amount of buckling (to the nearest inch) for the following materials (the percentage of swelling is given for each):

a. Brick; $y = .03$ **b.** Steel; $y = .06$
c. Aluminum; $y = .12$ **d.** Concrete; $y = .05$

60. SPACE SCIENCE Suppose a model rocket weighs $\frac{1}{4}$ lb. Its engine propels it vertically to a height of 52 ft and a speed of 120 ft/sec at burnout. If its parachute fails to open, determine the approximate time to fall to earth according to the following equation for free fall in a vacuum:

$$h = h_0 + v_0 t - \tfrac{1}{2}gt^2$$

where h is the height (in feet) at time t, h_0 and v_0 are the height (in feet) and velocity (in feet per second) at the time selected at $t = 0$, and g is approximately 32 ft/sec^2. (For this problem, $h_0 = 52$ ft and $v_0 = 120$ ft/sec.)

The material of this chapter is reviewed in the following list of objectives. After each objective there are some practice questions. For a sample test, select the first question of each set and check your answers with the answer section. For a

sample test without answers, use the second question of each set. Additional practice is given by the other questions in each set. If you are having trouble with a particular type of problem, look back to that section for extra help.

1.1 Real Numbers

OBJECTIVE 1 *Be familiar with the counting numbers, whole numbers, integers, rational numbers, irrational numbers, and real numbers.* Classify each of the following numbers using the above listed sets.

1. $\dfrac{14}{7}, \quad \sqrt{144}, \quad 6.\overline{2}, \quad \pi$

2. $\sqrt{2.25}, \quad \sqrt{30}, \quad 6.545454\ldots, \quad \dfrac{\pi}{6}$

3. $3.\overline{1}, \quad \dfrac{5\pi}{6}, \quad \dfrac{22}{7}, \quad \sqrt{10}$

4. $3.1416, \quad 4.513, \quad \sqrt{1.69}, \quad \sqrt{12}$

OBJECTIVE 2 *Graph numbers on a number line.* Graph each of the following sets of numbers on the same real number line.

5. $\dfrac{14}{7}, \quad \sqrt{144}, \quad 6.\overline{2}, \quad \pi$

6. $\sqrt{2.25}, \quad \sqrt{30}, \quad 6.545454\ldots, \quad \dfrac{\pi}{6}$

7. $3.\overline{1}, \quad \dfrac{5\pi}{6}, \quad \dfrac{22}{7}, \quad \sqrt{10}$

8. $3.1416, \quad 4.513, \quad \sqrt{1.69}, \quad \sqrt{12}$

OBJECTIVE 3 *Use $<$, $>$, and $=$ relationships.* Illustrate the Property of Comparison by replacing $\square$ by $=$, $<$, or $>$.

9. $\dfrac{5}{8} \square .625$

10. $\dfrac{5}{9} \square .555$

11. $\dfrac{5}{7} \square \dfrac{8}{11}$

12. $\sqrt{2} \square 1.414$

OBJECTIVE 4 *Know the definition of absolute value.* Write expressions without using absolute value notation.

13. $|-\sqrt{11}|$

14. $|4 - \sqrt{11}|$

15. $|3 - \sqrt{11}|$

16. $|2\pi - 9|$

OBJECTIVE 5 *Find the distance between points on a number line.*

17. (3) and (-5)

18. (-5) and (-1)

19. $(-\pi)$ and (2)

20. (4) and $(-\sqrt{5})$

OBJECTIVE 6 *Know the reflexive, symmetric, transitive, and substitution properties.* Complete the given statement so that the requested property of equality is demonstrated.

21. $a(b + c) =$ _____ (reflexive property)

22. If $a(b + c) = ab + ac$, then _____ (symmetric property)

23. If $a(b + c) = ab + ac$ and $ab + ac = 5$, then _____ (transitive property)

24. If $a(b + c) = 5$ and $a = 3$, then _____ (substitution property)

OBJECTIVE 7 *Be familiar with the real number properties.* Complete the given statement so that the requested field property is demonstrated.

25. $a(b + c) =$ _____ (commutative property for multiplication)

26. $a(b + c) =$ _____ (commutative property for addition)

27. $a(b + c) =$ _____ (distributive property)

28. $a(b + c) =$ _____ (identity property for multiplication)

1.2 Algebraic Expressions

OBJECTIVE 8 *Be familiar with the terminology of polynomials, including the definition and laws of exponents.* Fill in the blanks.

29. An nth-degree polynomial in x is _____

30. If b is any real number and n is _____, then $b^n =$ _____

31. $b^m \cdot b^n =$ _____

32. $(ab)^m =$ _____

OBJECTIVE 9 *Simplify algebraic expressions.*

33. $(3x + 1)(3x^3 + 4x^2 - 35x - 12)$

34. $(3x + 1) + (3x^3 + 4x^2 - 35x - 12)$

35. $(3x + 1) - (3x^3 + 4x^2 - 35x - 12)$

36. $(3x + 1)^3$

OBJECTIVE 10 *Factor polynomials.*

37. $\dfrac{4x^2}{y^2} - (2x + y)^2$

38. $x^4 - 26x^2 + 25$

39. $(x^3 - \frac{1}{8})(8x^3 + 8)$

40. $4x^3 + 8x^2 - x - 2$

1.3 **Two-Dimensional Coordinate System and Graphs**

OBJECTIVE 11 *Plot points on a Cartesian coordinate system.*

41. $\left(\dfrac{\pi}{2}, 0\right)$

42. $\left(\dfrac{2\pi}{3}, -\dfrac{1}{2}\right)$

43. $\left(\dfrac{\pi}{4}, \dfrac{\sqrt{2}}{2}\right)$

44. $\left(\dfrac{5\pi}{6}, \dfrac{-\sqrt{3}}{2}\right)$

OBJECTIVE 12 *Know the Pythagorean theorem and the distance formula. Find the distance between points in a plane.* Find the distance between each pair of points.

45. (α, β) and (γ, δ)

46. (x, x) and $(5x, 4x)$, where $x > 0$

47. (x, x) and $(5x, 4x)$, where $x < 0$

48. $(-3, -2)$ and $(1, -4)$

OBJECTIVE 13 *Find the midpoint of a segment.* Find the midpoint of the segment connecting the given points.

49. (α, β) and (γ, δ)

50. (x, x) and $(5x, 4x)$, where $x > 0$

51. (x, x) and $(5x. 4x)$, where $x < 0$

52. $(-3, -2)$ and $(1, -4)$

OBJECTIVE 14 *Know the definition of a relation and the terminology of graphing in two dimensions.* Fill in the blanks.

53. A relation is _____

54. By a graph of a relation we mean _____

55. If an ordered pair has components that, when substituted for their corresponding variables, yield a true equation, we say that the ordered pair _____ the equation.

56. Draw a Cartesian coordinate system, label the axes, origin, and quadrants by number.

OBJECTIVE 15 *Graph relations specified by an equation by plotting points.*

57. $2x - y + 3 = 0$

58. $y = -\frac{2}{3}x$

59. $y = -\frac{2}{3}x^2$

60. $y = -\frac{2}{3}|x|$

OBJECTIVE 16 *Draw curves showing symmetry with respect to a line; the x-axis; the y-axis; and the origin.* Given the curve shown below, draw a curve so that it has the indicated symmetry.

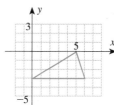

61. x-axis

62. y-axis

63. Origin

64. The line $x = y$

1.4 **Intervals, Inequalities, and Absolute Value**

OBJECTIVE 17 *Write interval notation.*

65. $-4 < x < 2$

66. $-12 \le x < -8$

67. $x > -3$

68. $4 \ge x$

OBJECTIVE 18 *Graph inequalities given interval notation.*

69. $[-5, -2]$

70. $[-3, 0)$

71. $[-1, \infty)$

72. $(-\infty, 4) \cup (4, \infty)$

OBJECTIVE 19 *Write interval notation using inequality notation.* Write each of the intervals as an inequality. Use x as the variable.

73. $[-8, -5)$

74. $[-2, 0)$

75. $(-\infty, 3)$

76. $[3, \infty)$

OBJECTIVE 20 *Solve linear inequalities.* Leave your answer in interval notation.

77. $3x + 2 \le 14$

78. $-9 \le -x$

79. $5 \le 1 - x < 9$

80. $-.001 \le x + 2 \le .001$

81. $|x| = 8$ **82.** $|x| = -5$

83. $|2x + 3| = 8$ **84.** $|2 - 3x| = 11$

85. $|x - 4| < 5$ **86.** $|3x + 1| \le 5$

87. $|5 - 2x| \le 25$ **88.** $|x - 5| \ge 3$

1.5   Complex Numbers

OBJECTIVE 23   *Define a complex number. Simplify expressions involving complex numbers.*

89. $-i^7$ **90.** $(2 - 3i) - (5 + 6i)$ **91.** $(2 + 5i)(2 - 5i)$ **92.** $\dfrac{1 - 8i}{3 + 2i}$

1.6   Equations for Calculus

OBJECTIVE 24   *Solve quadratic equations by factoring.*

93. $x^2 - x - 12 = 0$ **94.** $x^2 - 10x + 24 = 0$

95. $(3 - x)(5 + 2x) = 0$ **96.** $x^2 - 100 = 0$

OBJECTIVE 25   *Solve quadratic equations by completing the square.*

97. $x^2 - 2x - 15 = 0$ **98.** $x^2 + 6x + 8 = 0$

99. $x^2 + 9x + 20 = 0$ **100.** $x^2 - 3x + 1 = 0$

OBJECTIVE 26   *Know the quadratic formula. Solve quadratic equations over the set of real numbers.**

101. $x^2 - 5x + 3 = 0$ **102.** $2x^2 - 5x - 3 = 0$

103. $x^2 + 2x - 5 = 0$ **104.** $3x^2 + 2x + 1 = 0$

1.7   Inequalities for Calculus

OBJECTIVE 27   *Solve quadratic inequalities. Write your answer using interval notation.*

105. $3x^2 - 2x - 1 < 0$

106. $3 + 5x \ge 2x^2$

107. $x^2 + 2x + 1 \ge 0$

108. $x^2 - x - 1 \le 0$

OBJECTIVE 28   *Solve inequalities in factored form. Write your answer using interval notation.*

109. $x(3 - x)(x + 1) < 0$ **110.** $\dfrac{x + 5}{x - 9} \le 0$

111. $x(x - 2)^2(x + 1) > 0$ **112.** $\dfrac{(2x - 1)(x - 2)}{(x + 1)(x + 2)} \ge 0$

1.8   Problem Solving

OBJECTIVE 29   *Solve applied problems.*

113. Two rectangles have the same width, but one is 48 sq ft larger in area. The larger rectangle is 8 ft longer than it is wide. The other is only 4 ft longer than it is wide. What are the dimensions of the larger rectangle?

114. A total of $1,600 is invested for a year, part at 8% and the rest at 9.5%. If $134 is collected in interest, how much is invested at 8%?

115. Milk containing 10% butterfat and cream with 70% butterfat are mixed to produce half-and-half, which is 50% butterfat. How many gallons of each must be mixed in order to produce 15 gallons of half-and-half?

116. A traveler takes 9 hours to reach her destination. She takes a car that averages 65 mph and a bus the rest of the way at 36 mph. If the total distance is 498 miles, what is the distance traveled on the bus?

*Find the solution over the set of complex numbers if you covered Section 1.5.

**Leonhard Euler
(1707–1783)**

Nature herself exhibits to us measurable and observable quantities in definite mathematical dependence; the conception of a function is suggested by all the processes of nature where we observe natural phenomena varying according to distance or to time. Nearly all the "known" functions have presented themselves in an attempt to solve geometrical, mechanical, or physical problems.

J. T. MERTZ
History of European Thought in the Nineteenth Century

Euler calculated without any apparent effort, just as men breathe and as eagles sustain themselves in the air.

F. ARAGO IN HOWARD EVES,
In Mathematical Circles, p. 47

The word *function* was used as early as 1694 by the universal genius of the seventeenth century, Gottfried Wilhelm von Leibniz (1646–1716), to denote any quantity connected with a curve. The notion was generalized and modified by Johann Bernoulli (1667–1748) and by Leonhard Euler, the most prolific mathematical writer in history. Throughout his life, Euler was both a frequent contributor to research journals and a superb textbook writer who was widely known for his clarity, detail, and completeness. Euler's work with functions was later expanded by the mathematician P. G. Lejeune-Dirichlet (1805–1859). Around 1815, functions were being considered that were not "nice" and it was Dirichlet who, in 1837, suggested a very broad definition of a function—the one, in fact, that leads to the definition used in this chapter.

An interesting story about Euler is found in a book by Howard Eves:*

"In 1735, the year after his wedding and when he was in Russia, Euler received a problem in celestial mechanics from the French Academy. Though other mathematicians had required several months to solve this problem, Euler, using improved methods of his own and by devoting intense concentration to it, solved it in three days and the better part of the two intervening nights. The strain of the effort induced a fever from which Euler finally recovered, but with the loss of the sight of his right eye. Stoically accepting the misfortune, he commented, 'Now I will have less distraction.' Thirty-one years later, in 1766, when he was again in Russia, Euler developed a cataract in his remaining eye and went completely blind. Now blindness would seem to be an insurmountable barrier to a mathematician, but, like Beethoven's loss of hearing, Euler's loss of sight in no way impaired his amazing productivity. He continued his creative work by dictating to a secretary and by writing formulas in chalk on a large slate for his secretary to copy down. In 1771, after five years in darkness, Euler underwent an operation to remove the cataract from his left eye, and for a brief period he was able to see again. But within a few weeks a very painful infection set in, and when it was over Euler was once again totally blind—so to continue for the remaining twelve years of his life."

---

*From Howard Eves, *In Mathematical Circles* (Boston: Prindle, Weber, & Schmidt), p. 48.

FUNCTIONS

Contents

Preview

The central idea for this course is the notation of a function, and you are introduced to this concept here in Chapter 2. We present essential properties of functions in general that will be used throughout the course. This chapter has 17 objectives, which are listed on pages 114–115.

Perspective

The derivative in calculus involves, among other things, the evaluation of a function. In this section you will learn the definition of a function, notation for a function, and the evaluation of a function. The definition of derivative involves finding the limit of this expression:

$$\frac{f(x + h) - f(x)}{h}$$

Functional notation is discussed in Section 2.1, and you might want to compare Example 7 and Problems 35–44 in that section with what you see on the page below, which is reproduced from a leading calculus book. You will not only need to thoroughly understand the functional concept, but also need to be at ease with functional notation in order to succeed in calculus.

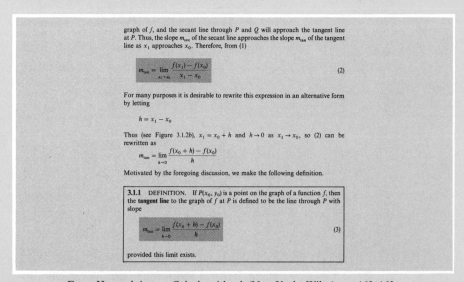

graph of f, and the secant line through P and Q will approach the tangent line at P. Thus, the slope m_{sec} of the secant line approaches the slope m_{tan} of the tangent line as x_1 approaches x_0. Therefore, from (1)

$$m_{tan} = \lim_{x_1 \to x_0} \frac{f(x_1) - f(x_0)}{x_1 - x_0} \tag{2}$$

For many purposes it is desirable to rewrite this expression in an alternative form by letting

$$h = x_1 - x_0$$

Thus (see Figure 3.1.2b), $x_1 = x_0 + h$ and $h \to 0$ as $x_1 \to x_0$, so (2) can be rewritten as

$$m_{tan} = \lim_{h \to 0} \frac{f(x_0 + h) - f(x_0)}{h}$$

Motivated by the foregoing discussion, we make the following definition.

3.1.1 DEFINITION. If $P(x_0, y_0)$ is a point on the graph of a function f, then the **tangent line** to the graph of f at P is defined to be the line through P with slope

$$m_{tan} = \lim_{h \to 0} \frac{f(x_0 + h) - f(x_0)}{h} \tag{3}$$

provided this limit exists.

From Howard Anton, *Calculus*, 4th ed. (New York: Wiley), pp. 162–163.

2.1 Introduction to Functions

The material presented in this book is designed to prepare you for the study of calculus. As the title suggests, the main thread that will lead you through this book is the concept of a *function*. You have, no doubt, been introduced to this idea before, probably in algebra. The notion of a correspondence between sets is a common idea. The price of a stock, for example, can be determined by looking at the daily quote in the newspaper; the height of a bridge can be determined by dropping a rock and measuring the time it takes to hit the bottom; and the surface area of a balloon can be determined if you know its radius. All of these are everyday examples of the use of functions.

The Meaning of Function

FUNCTION

> A **function** f is a rule that assigns to each element x of a set X *exactly* one element y of a set Y.

The set X is called the **domain** of the function. The element y is called the **image** of x under f and is denoted by $f(x)$. The set of all images of elements of X is called the **range** of the function. Let us consider some examples of rules that are functions.

Let $X = \{1, 2, 3, 4\}$ be the domain of a function called f. Think of the function f as a machine (a function machine) that accepts an input x from X and produces an output $f(x)$, pronounced "f of x."

$\gtrless$ $f(x)$ is a single symbol and does not mean f times x. $\lessgtr$

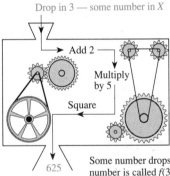

Drop in 3 — some number in X

625

Some number drops out; in this case we drop in a 3 and 625 drops out; this number is called $f(3)$. That is, $f(3) = 625$. The output for the other members of the domain is shown:

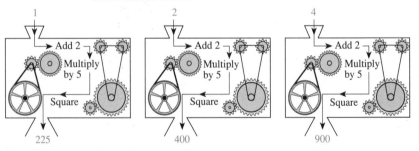

The function machine description has the advantage of being easy to understand, but it is awkward to use. We might also describe this function in a variety of ways, as shown in Example 1.

EXAMPLE 1 The function machine on the previous page describes a function. In this example, we describe this function in several different common ways.

a. Rule

For each input value, add two, then multiply by five, and finally square the result to find the output. The function in this example would probably not be defined by a verbal rule, but nevertheless a verbal rule is often the best way we have to describe a function. For example, a verbal rule would be quite useful for the function T that is the time of daylight on a particular day.

b. Table

x	$f(x)$
1	225
2	400
3	625
4	900

c. Equation

$$f(x) = [5(x + 2)]^2$$

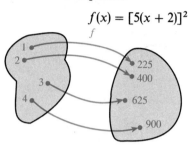

d. Graph

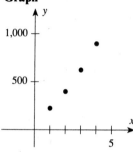

e. Mapping

f. Set of ordered pairs $\{(1, 225), (2, 400), (3, 625), (4, 900)\}$

EXAMPLE 2 For each example, name the domain, the range, and the outputs $f(x)$ for each input x in the domain.

a.

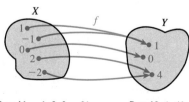

$X = \{a, b, c, d\}$ $Y = \{1, 2, 3, 4, 5, 6\}$
Domain is $\{a, b, c, d\}$ Range is $\{1, 2, 3, 4\}$

$f(a) = 1$
$f(b) = 2$
$f(c) = 3$
$f(d) = 4$

The range consists only of those elements of Y that are actually used as outputs. On the other hand, the domain and X must be the same.

b.

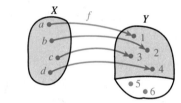

$X = \{a, b, c\}$ $Y = \{3, 7, 9\}$
Domain is $\{a, b, c\}$ Range is $\{3, 7, 9\}$

$f(a) = 3$
$f(b) = 7$
$f(c) = 9$

c.

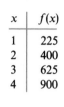

$D = \{1, -1, 0, 2, -2\}$ $R = \{0, 1, 4\}$

$f(0) = 0$ $f(2) = 4$
$f(1) = 1$ $f(-2) = 4$
$f(-1) = 1$

It is possible that different elements in the domain have the same image.

Notice that Example 2c showed some repeated outputs. The ultimate example of repeated outputs is a function $f(x) = c$ for *all* values of x. Such a function is called a **constant function.** If the outputs are always different (that is, if there are no repeated outputs), then the function is called *one-to-one.*

ONE-TO-ONE FUNCTION

> If f maps X into Y so that for any distinct elements x_1 and x_2 of X, $f(x_1) \neq f(x_2)$, then f is a **one-to-one function** of X into Y.

Example 3 illustrates a mapping that is not a function.

EXAMPLE 3

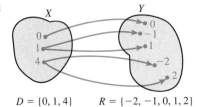

$D = \{0, 1, 4\}$ $R = \{-2, -1, 0, 1, 2\}$

Do not use $f(x)$ notation unless f is a function.

This is not a function because 1 is associated with more than one image (so is 4).

Horizontal and Vertical Line Tests

There are two tests that involve sweeping a line across a graph. The first tells us whether a graph represents a function, and the second tells us whether a graph represents a one-to-one function.

> VERTICAL LINE TEST   Every vertical line passes through the graph of a function in at most one point. This means if you sweep a vertical line across a graph and it intersects the curve at more than one point at the same time, then the curve is not the graph of a function.
>
> HORIZONTAL LINE TEST   Every horizontal line passes through the graph of a one-to-one function in at most one point. This means if you sweep a horizontal line across the graph of a function and it intersects the curve at more than one point at the same time, then the curve is not the graph of a one-to-one function.

EXAMPLE 4   Use the vertical line test to determine whether the given curve is the graph of a function, and if it is the graph of a function, use the horizontal line test to determine whether it is one-to-one.

a.

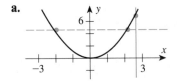

Passes vertical line test; **it is a function**
Does not pass horizontal line test; **it is not one-to-one**

b.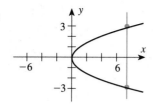

Does not pass vertical line test; **it is not a function**

c.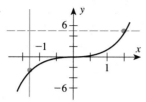

Passes vertical line test; **it is a function**
Passes horizontal line test; **it is one-to-one**

d.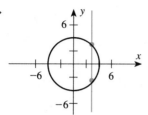

Does not pass vertical line test; **it is not a function**

e.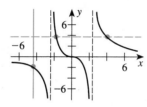

Passes vertical line test; **it is a function**
Does not pass horizontal line test; **it is not one-to-one**

Functional Notation

One of the most useful inventions in all the history of mathematics is the $f(x)$, or functional, notation:

The function is denoted by f

x is a member of the domain

$f(x)$

$f(x)$ is a member of the range

The *function* is denoted by f; $f(x)$ is the *number* associated with x. Sometimes functions are defined by expressions such as

$$f(x) = 3x + 2 \quad \text{or} \quad g(x) = x^2 + 4x + 3$$

To emphasize the difference between f and $f(x)$, some books use $f: x \to 3x + 2$ to define functions, but this book simply uses the notation $f(x) = 3x + 2$ to mean the set of all ordered pairs (x, y) such that $y = 3x + 2$.

EXAMPLE 5 Given f and g defined by $f(x) = 3x + 2$ and $g(x) = x^2 + 4x + 3$, find the indicated values: **a.** $f(1)$; **b.** $g(2)$; **c.** $g(-3)$; **d.** $f(-3)$; **e.** $f[g(2)]$

SOLUTION **a.** The symbol $f(1)$ represents the second component of the ordered pair of the function f with first component 1. Replace x by 1 in the expression

$$f(x) = 3x + 2$$
$$\uparrow \qquad \uparrow$$
$$f(1) = 3(1) + 2 = \mathbf{5}$$

b. $g(2)$: $g(x) = x^2 + 4x + 3$

$$\qquad\qquad \downarrow \quad \downarrow \qquad \downarrow$$
$$g(2) = (2)^2 + 4(2) + 3$$
$$= 4 \quad + 8 \quad + 3$$
$$= \mathbf{15}$$

c. $g(-3) = (-3)^2 + 4(-3) + 3$
$$\qquad = 9 \qquad - 12 \qquad + 3$$
$$\qquad = \mathbf{0}$$

d. $f(-3) = 3(-3) + 2$
$$\qquad = -9 \quad + 2$$
$$\qquad = \mathbf{-7}$$

e. $f[g(2)] = f(15)$ From part b
$$\qquad = 3(15) + 2$$
$$\qquad = \mathbf{47}$$ ∎

The members of the domain of a function may also be represented by variables, as shown in Example 6.

EXAMPLE 6 Let F and G be defined by $F(x) = x^2 + 1$ and $G(x) = (x + 1)^2$. Find

a. $F(w)$ **b.** $G(t)$ **c.** $F(w + 3)$ **d.** $G(x - 2)$ **e.** $F(w + h)$

SOLUTION **a.** $F(w) = w^2 + 1$

b. $G(t) = (t + 1)^2$
$$\qquad = t^2 + 2t + 1$$

c. $F(w + 3) = (w + 3)^2 + 1$
$$\qquad = w^2 + 6w + 9 + 1$$
$$\qquad = w^2 + 6w + 10$$

d. $G(x - 2) = [(x - 2) + 1]^2$
$$\qquad = (x - 1)^2$$
$$\qquad = x^2 - 2x + 1$$

e. $F(w + h) = (w + h)^2 + 1$
$$\qquad = w^2 + 2wh + h^2 + 1$$ ∎

In calculus, functional notation is used to carry out manipulations such as those shown in Example 7.

EXAMPLE 7 Find $\dfrac{f(x + h) - f(x)}{h}$ for each function.

a. $f(x) = x^2$, where $x = 5$

$$\frac{f(5 + h) - f(5)}{h} = \frac{(5 + h)^2 - 5^2}{h}$$
$$= \frac{25 + 10h + h^2 - 25}{h}$$
$$= \frac{(10 + h)h}{h}$$
$$= \mathbf{10 + h}$$

b. $f(x) = 2x^2 + 1$,   where $x = 1$

$$\frac{f(1 + h) - f(1)}{h} = \frac{[2(1 + h)^2 + 1] - [2(1)^2 + 1]}{h}$$

$$= \frac{[2(1 + 2h + h^2) + 1] - (2 + 1)}{h}$$

$$= \frac{2h^2 + 4h + 3 - 3}{h}$$

$$= \mathbf{2h + 4}$$

c. $f(x) = x^2 + 3x - 2$

$$\frac{f(x + h) - f(x)}{h} = \frac{[(x + h)^2 + 3(x + h) - 2] - [x^2 + 3x - 2]}{h}$$

$$= \frac{x^2 + 2xh + h^2 + 3x + 3h - 2 - x^2 - 3x + 2}{h}$$

$$= \frac{2xh + h^2 + 3h}{h}$$

$$= \mathbf{2x + 3 + h}$$

Functional notation can be used to work a wide variety of applied problems, as shown by Example 8 and again in the problem set.

EXAMPLE 8 If an object is dropped from a certain height, it is known that it will fall a distance of s ft in t sec according to the formula

$$s = 16t^2$$

This formula can be represented by $f(t) = 16t^2$.

a. How far will the object fall in the first second?

$$f(1) = 16 \cdot 1^2$$

$$= 16 \quad \text{or} \quad \mathbf{16 \ ft}$$

b. How far will it fall in the *next* 2 sec?

$$f(1 + 2) = 16 \cdot 3^2$$

$$= 144 \quad \text{or} \quad 144 \text{ ft in 3 sec}$$

So the answer to the question is

$$f(3) - f(1) = 144 - 16$$

$$= 128 \quad \text{or} \quad \mathbf{128 \ ft}$$

c. How far will it fall during the time $t = 1$ sec to $t = 1 + h$ sec?

$$f(1 + h) - f(1) = 16(1 + h)^2 - 16$$

$$= 16 + 32h + 16h^2 - 16$$

$$= \mathbf{(32h + 16h^2) \ ft}$$

d. What is the average rate of change of distance (in feet per second, fps) during the time $t = 1$ sec to $t = 3$ sec?

$$\frac{f(3) - f(1)}{3 - 1} = \frac{128}{2} = \textbf{64 fps}$$

e. What is the average rate of change of distance during the time $t = 1$ sec to $t = 1 + h$ sec?

$$\frac{f(1 + h) - f(1)}{h} = \frac{32h + 16h^2}{h}$$

$$= \textbf{(32 + 16\textit{h}) fps}$$

f. What is the average rate of change of distance during the time $t = x$ sec to $t = x + h$ sec?

$$\frac{f(x + h) - f(x)}{(x + h) - x} = \frac{f(x + h) - f(x)}{h} \qquad \text{Does this look familiar?}$$

$$= \frac{16(x + h)^2 - 16x^2}{h}$$

$$= \frac{16x^2 + 32xh + 16h^2 - 16x^2}{h}$$

$$= \textbf{(32\textit{x} + 16\textit{h}) fps}$$

The variable t in Example 8 represents an arbitrary number from the domain of f and is often called the **independent variable**. The variable s, which represents a number from the range of f, is called a **dependent variable** since its value depends on the value assigned to t.

2.1 Problem Set

A *State whether each set in Problems 1–12 is or is not a function. If it is a function, determine whether it is one-to-one.*

1. $\{(8, 2), (7, 1), (6, 3), (5,0)\}$

2. $\{(5, 2), (7, 3), (1, 6), (7, 4)\}$

3. $\{1, 2, 3, 4\}$

4. $\{6, 9, 12, 15\}$

5. $\{(x, y)|\, y = 4x + 3\}$*

6. $\{(x, y)|\, y \leq 4x + 3\}$

7.

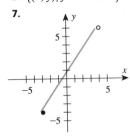

8.

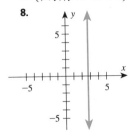

9.

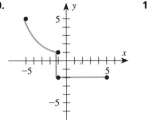

10.

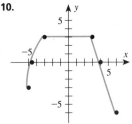

11.

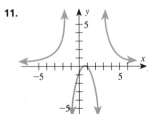

12.

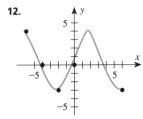

---

*This is called set-builder notation and is read as "the set of all ordered pairs (x, y) such that $y = 4x + 3$."

WHAT IS WRONG, *if anything, with each of the statements in Problems* 13–16? *Explain your reasoning.*

13. If $f(x) = 2x + 3$, then $x = 2x + 3$ so $x = -3$.

14. $f(x)$ is a function.

15. If $f(x) = x^2$, then
$$\frac{f(x + h) - f(x)}{h} = \frac{x^2 + h^2 - x^2}{h} = h$$

16. If $f(x) = 5$, then
$$\frac{f(x + h) - f(x)}{h} = \frac{5 + h - 5}{h} = 1$$

In Problems 17–28 *let* $f(x) = 2x + 1$ *and* $g(x) = 2x^2 - 1$. *Find the requested values.*

17. a. $f(0)$ **b.** $f(2)$ **c.** $f(-3)$
   d. $f(\sqrt{5})$ **e.** $f(\pi)$

18. a. $f(1)$ **b.** $g(1)$ **c.** $f(\sqrt{3})$
   d. $g(\sqrt{3})$ **e.** $g(\pi)$

19. a. $f(w)$ **b.** $g(w)$ **c.** $g(t)$
   d. $g(v)$ **e.** $f(m)$

20. a. $f(t)$ **b.** $f(p)$ **c.** $f(t + 1)$
   d. $g(t + 1)$ **e.** $f(t^2)$

21. a. $f(1 + \sqrt{2})$ **b.** $g(1 + \sqrt{2})$ **c.** $g(t + 3)$
   d. $f(t^2 + 2t + 1)$ **e.** $g(m - 1)$

22. a. $f(x + 2)$ **b.** $g(x + 2)$ **c.** $f(t + h)$
   d. $g(t + h)$ **e.** $f(x + h)$

23. $\dfrac{f(t + 3) - f(t)}{3}$ **24.** $\dfrac{f(t + h) - f(t)}{h}$

25. $\dfrac{f(x + h) - f(x)}{h}$ **26.** $\dfrac{g(t + 2) - g(t)}{2}$

27. $\dfrac{g(t + h) - g(t)}{h}$ **28.** $\dfrac{g(x + h) - g(x)}{h}$

In Problems 29–34 *compute the given value where* $f(x) = x^2 - 1$ *and* $g(x) = 2x + 5$.

29. a. $f(w)$ **b.** $f(h)$
   c. $f(w + h)$ **d.** $f(w) + f(h)$

30. a. $g(s)$ **b.** $g(t)$
   c. $g(s + t)$ **d.** $g(s) + g(t)$

31. a. $f(x^2)$ **b.** $f(\sqrt{x})$
   c. $f(x + h)$ **d.** $f(-x)$

32. a. $g(x^2)$ **b.** $g(\pi)$
   c. $g(x + \pi)$ **d.** $g(-x)$

33. $\dfrac{g(x + h) - g(x)}{h}$ **34.** $\dfrac{f(x + h) - f(x)}{h}$

B *In Problems* 35–44, *find* $\dfrac{f(x + h) - f(x)}{h}$ *for the given function f.*

35. $f(x) = 9x + 3$ **36.** $f(x) = 5 - 2x$

37. $f(x) = |x|$ **38.** $f(x) = |2x + 1|$

39. $f(x) = 5x^2$ **40.** $f(x) = 3x^2 + 2x$

41. $f(x) = 2x^2 + 3x - 4$ **42.** $f(x) = \dfrac{1}{x}$

43. $f(x) = \dfrac{x + 1}{x - 1}$ **44.** $f(x) = \dfrac{x^2 + x - 6}{x + 3}$

For Problems 45–54, *use the accompanying table, which reflects the purchasing power of the dollar from October* 1944 *to October* 1984. (Source: *Bureau of Labor Statistics, Consumer Division.) Let x represent the year, let the domain be the set* $\{1944, 1954, 1964, 1974, 1984\}$, *and let*

$r(x) =$ *Price of 1 lb of round steak*

$s(x) =$ *Price of a 5-lb bag of sugar*

$b(x) =$ *Price of a loaf of bread*

$c(x) =$ *Price of 1 lb of coffee*

$e(x) =$ *Price of a dozen eggs*

$m(x) =$ *Price of $\frac{1}{2}$ gal of milk*

$g(x) =$ *Price of 1 gal of gasoline*

| | YEAR | | | | |
	1944	1954	1964	1974	1984
Round steak (1 lb)	$.45	$.92	$1.07	$1.78	$2.15
Sugar (5-lb bag)	.34	.52	.59	2.08	1.49
Bread (loaf)	.09	.17	.21	.36	1.29
Coffee (1 lb)	.30	1.10	.82	1.31	2.69
Eggs (1 dozen)	.64	.60	.57	.84	1.15
Milk ($\frac{1}{2}$ gal)	.29	.45	.48	.78	1.08
Gasoline (1 gal)	.21	.29	.30	.53	1.10

45. Find: **a.** $r(1954)$ **b.** $m(1954)$

46. Find: **a.** $g(1944)$ **b.** $c(1984)$

47. Find $s(1984) - s(1944)$.   **48.** Find $b(1984) - b(1944)$.

49. a. Find the change in the price of eggs from 1944 to 1984.
   b. Write the change in the price of eggs using functional notation.

50. a. Find the change in the price of round steak from 1944 to 1984.
   b. Write the change in the price of round steak using functional notation.

51. a. Find $\dfrac{g(1944 + 40) - g(1944)}{40}$.
   b. In words, attach some meaning to the expression given in part **a.**

52. a. Find $\dfrac{m(1944 + 40) - m(1944)}{40}$.
   b. In words, attach some meaning to the expression given in part **a.**

53. a. What is the average increase in the price of sugar per year from 1944 to 1954? Write this in functional notation.
   b. What is the average increase in the price of sugar per year from 1944 to 1964? Write this in functional notation.
   c. What is the average increase in the price of sugar per year from 1944 to 1974? Write this in functional notation.
   d. What is the average increase in the price of sugar per year from 1944 to 1984? Write this in functional notation.
   e. What is the average increase in the price of sugar per year from 1944 to 1944 + h, where h is an unspecified number of years? Write this in functional notation.

54. Repeat Problem 53 for coffee instead of sugar.

55. According to the U.S. Public Health Service, the number of marriages in the United States was about 2,421,000 in 1987 and about 2,495,000 in 1982. Let $M(x)$ represent the number of marriages in year x.

   a. Find $\dfrac{M(1987) - M(1982)}{5}$.

   b. Give a verbal description for the following functional expression:

   $$\dfrac{M(1982 + h) - M(1982)}{h}$$

56. According to the U.S. Public Health Service, the number of divorces in the United States was about 1,157,000 in 1987 and about 1,180,000 in 1982. Let $D(x)$ represent the number of divorces in year x.

   a. Find $\dfrac{D(1987) - D(1982)}{5}$.

   b. Give a verbal description for the following functional expression:

   $$\dfrac{D(1982 + h) - D(1982)}{h}$$

57. BUSINESS A firm determines that the total cost C (in dollars) of producing x units of a certain product is given by

   $$C(x) = -.02x^2 + 4x + 500 \quad (0 \le x \le 150)$$

   Find $C(50)$ and $C(100)$.

58. BUSINESS What is the average cost per unit in Problem 57 if 50 units are produced? Repeat for 100 units. What is the per-unit change in cost for the increase from 50 to 100 units?

59. BUSINESS What is the per-unit change in cost in Problem 57 for an increase from 50 units to 51 units? Compare this answer with the answer to Problem 58.

60. BUSINESS What is the per-unit change in cost in Problem 57 for an h-unit increase in production above a level of x units?

61. PHYSICS Let d be a function that represents the distance an object falls (neglecting air resistance) in t sec. It can be shown that $d(t) = 16t^2$. Find the average rate that the object falls for the intervals of time given:
   a. From $t = 2$ to $t = 6$ **b.** From $t = 2$ to $t = 4$
   c. From $t = 2$ to $t = 3$ **d.** From $t = 2$ to $t = 2 + h$
   e. From $t = x$ to $t = x + h$

62. PHYSICS In Problem 61, give a physical interpretation for

   $$\dfrac{d(x + h) - d(x)}{h}$$

C

63. If $f(x) = x^2$, then

   $$f\left(\dfrac{1}{x}\right) = \left(\dfrac{1}{x}\right)^2 = \dfrac{1}{x^2} = \dfrac{1}{f(x)}$$

   Give an example of a function for which

   $$f\left(\dfrac{1}{x}\right) \ne \dfrac{1}{f(x)}$$

64. If $f(x) = x$, then $f(x^2) = [f(x)]^2$. Give an example of a function for which $f(x^2) \ne [f(x)]^2$.

2.2 Graph of a Function

Even if you do not now have access to a graphing calculator or computer software that does graphing, you will no doubt be using this technology in the future. Many have a misconception that if they only had this technology, they would not need to study graphing in a mathematics course. Quite the contrary is true. Even the best software will often show a blank screen when an equation or a curve is input. Most graphing calculators require input in the form $\boxed{Y=}$, which means that you are expected to input equations that are functions. In this chapter we focus on the nature of a function, in general, and on particular graphing techniques. In the next chapter, we graph specific functions.

Domain and Range

⊗ In this book, unless otherwise specified, the domain is the set of real numbers for which the given function is meaningful. ⊗

A valuable tool in curve sketching is the ability to find the domain and range. We begin by assuming that the domain is the set of all real numbers, and then focus on finding **excluded values** and **excluded regions**. These are points or regions in the plane in which a given curve cannot lie. These points and regions are found by noting where either x or y does not exist in the set of real numbers. Exclusion will occur in the following situations:

1. Division by zero
2. Negative under a radical with an even index
3. Even-powered variables equal to negative numbers

EXAMPLE 1 Find the domain for the given functions:

a. $f(x) = 2x - 1$

b. $g(x) = \dfrac{(2x - 1)(x + 3)}{x + 3}$

c. $G(x) = 2x - 1, \quad x \neq -3$

d. $h(x) = \sqrt{x + 1}$

e. $F(x) = \sqrt{2 - 3x - 2x^2}$

f. $r(x) = 2 - \dfrac{x}{x}$

SOLUTION **a.** All real numbers; $(-\infty, \infty)$

b. All real numbers except $x = -3$ because if $x = -3$, then the expression is meaningless. That is, set the denominator $(x + 3)$ equal to zero $(x + 3 = 0)$, and solve $(x = -3)$. The domain is $(-\infty, -3) \cup (-3, \infty)$. The domain is usually denoted simply by $x \neq -3$.

c. The domain has $x = -3$ explicitly eliminated. This means that the domain is $(-\infty, -3) \cup (-3, \infty)$ or simply $x \neq -3$.

CALCULATOR COMMENT

We will discuss the graphing of particular functions in the next chapter. However, if you have a graphing calculator you can use it as an effective tool in finding not only the domain but also the range. For some problems, you can draw the graph on a calculator to find the domain and the range. For other problems, you will need first to find the domain and range to be able to obtain a graph that you can see in the window or on the screen. For example, on the TI-81 the graphs for part **a** and part **b** look the same. However, as we will see in the next chapter, the correct graph for Examples 1b and 1c will display an open circle at $x = -3$ to reflect the correct domain—namely, $x \neq -3$.

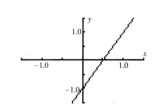

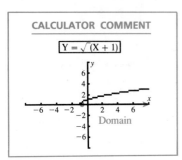
d. Here h has meaning if $x + 1$ is nonnegative. That is,

$$x + 1 \geq 0$$
$$x \geq -1$$

This domain can be described by writing **[−1, ∞)**.

e. F has meaning if $2 - 3x - 2x^2$ is nonnegative:

$$2 - 3x - 2x^2 \geq 0$$
$$(1 - 2x)(2 + x) \geq 0$$

Domain: **[−2, ½]**

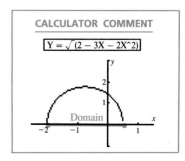
f. $r(0)$ is meaningless, but $r(x) = 1$ for $x \neq 0$, so the domain is **all real numbers except 0.**

EQUALITY OF FUNCTIONS

Two functions f and g are **equal** if and only if
1. f and g have the same domain.
2. $f(x) = g(x)$ for all x in the domain.

Compare Examples 1a and 1b, where $f(x) = 2x - 1$ and $g(x) = \dfrac{(2x - 1)(x + 3)}{x + 3}$.

In algebra you wrote

$$\frac{(2x - 1)(x + 3)}{x + 3} = 2x - 1$$

but it is *not* true that $f = g$, because their domains are not the same. However, $g = G$ (from Example 1c, $G(x) = 2x - 1$, $x \neq -3$) because both conditions of the definition are satisfied.

EXAMPLE 2 Determine whether the pairs of functions are equal.

a. $f(x) = \dfrac{(x-3)(x+5)}{x+5}$; $\quad F(x) = x - 3$

b. $g(x) = \dfrac{(2x-5)(x+1)}{x+1}$; $\quad G(x) = 2x - 5,\ x \neq -1$

SOLUTION
a. $f \neq F$ since the domain of f is all reals except -5 and the domain of g is all real numbers.

b. $g = G$ since the domain of both f and g is all reals except $x = -1$, and $f(x) = g(x)$ for all x in the domain.

Even though the usual graphing procedure is to find the domain, draw the graph, and then use the graph to determine the range, it is sometimes necessary to find both the domain and the range. We summarize these procedures:

To find the domain, D: *Solve for y* and look for exclusions for x.

To find the range, R: *Solve for x* and look for exclusions for y.

EXAMPLE 3 Find the domain and range for the curve $y = \sqrt{3 - x}$.

SOLUTION There is a square root, so the domain requires that

$$3 - x \geq 0$$
$$3 \geq x$$

Therefore, $D = (-\infty, 3]$.
To find the range, solve for x:

$$x = 3 - y^2 \quad \text{(where } y \geq 0\text{)} \qquad \text{\footnotesize Since } y \text{ is equal to a square root, it is nonnegative.}$$

Thus, x is defined for all positive y, so $R = [0, \infty)$.

As an aid to graphing, we will use the domain and range to shade in **excluded regions.** We will then focus our attention on the unshaded portions of the plane. For Example 3, we would shade the plane as shown in Figure 2.1. For convenience, you can draw the boundaries of all excluded regions as dashed lines.

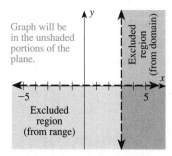

FIGURE 2.1 Excluded regions for Example 3

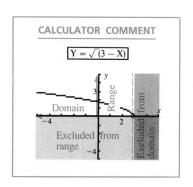

EXAMPLE 4 Find the domain and range for the curve $y^2 + x - 3 = 0$, and show the excluded regions.

SOLUTION To find the domain, solve for y:

$$y^2 = 3 - x$$
$$y = \pm\sqrt{3 - x} \qquad \text{Use the square-root property.}$$

At first glance, this example looks just like Example 3, but it is not. The domain is the same, but notice that the range is no longer restricted to positive values. Solve for x:

$$x = 3 - y^2$$

Thus, x is defined for all y, so $D = (-\infty, 3]$ and $R = (-\infty, \infty)$; see Figure 2.2.

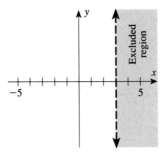

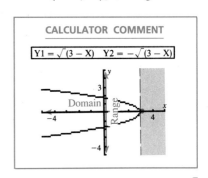

CALCULATOR COMMENT

$Y1 = \sqrt{(3 - X)} \quad Y2 = -\sqrt{(3 - X)}$

FIGURE 2.2 Excluded region for Example 4. Notice that there are no exclusions for the range.

You must be careful when finding both the domain and the range. For example, consider the following variations of curves similar to those in Examples 3 and 4:

EQUATION	DOMAIN	RANGE
$y = \sqrt{3 - x^2}$	Solve $3 - x^2 \geq 0$ to find $D = (-\sqrt{3}, \sqrt{3})$.	Solve $y^2 = 3 - x^2$, where $y \geq 0$, to find $x = \pm\sqrt{3 - y^2}$, so that $R = (0, \sqrt{3})$.
$x^2 + y^2 = 3$	Solve $y = \pm\sqrt{3 - x^2}$ to find $D = (-\sqrt{3}, \sqrt{3})$.	Solve $y^2 = 3 - x^2$ to find $x = \pm\sqrt{3 - y^2}$, so that $R = (-\sqrt{3}, \sqrt{3})$.

EXAMPLE 5 Find the domain and range for the curve $y = \sqrt{x^2 - 2x - 3}$, and show the excluded regions.

SOLUTION The domain is found by making sure the radicand is nonnegative:

$$x^2 - 2x - 3 \geq 0$$
$$(x + 1)(x - 3) \geq 0$$

$x + 1$: $-----|+++++++|++++++$
$x - 3$: $-----|------|++++++$
$\quad$ pos $\quad$ neg $\quad$ pos

$-3 \quad -2 \quad -1 \quad 0 \quad 1 \quad 2 \quad 3 \quad 4 \quad 5$

$$D = (-\infty, -1] \cup [3, \infty)$$

For the range, we see that $y \geq 0$ and $y^2 = x^2 - 2x - 3$. Solve for x to find

$$y^2 + 3 = x^2 - 2x$$
$$y^2 + 3 + 1 = x^2 - 2x + 1$$
$$y^2 + 4 = (x - 1)^2$$
$$\pm\sqrt{y^2 + 4} = x - 1$$
$$x = 1 \pm \sqrt{y^2 + 4}$$

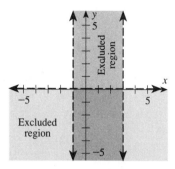

FIGURE 2.3 Excluded regions for Example 5

We see that x is defined for all values of y, so (since $y \geq 0$) the range is $R = [0, \infty)$. See Figure 2.3.

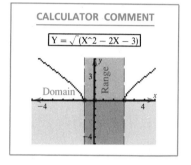

EXAMPLE 6   Find the domain and range for the curve $xy^2 - y^2 - 1 = 0$.

SOLUTION   For the domain, solve for y:

$$y^2(x - 1) - 1 = 0$$
$$y^2(x - 1) = 1$$
$$y^2 = \frac{1}{x - 1}$$
$$y = \frac{\pm 1}{\sqrt{x - 1}}$$

You can now find the domain by inspection: $D = (1, \infty)$. See Figure 2.4.
For the range, solve for x:

$$xy^2 = y^2 + 1$$
$$x = \frac{y^2 + 1}{y^2}$$

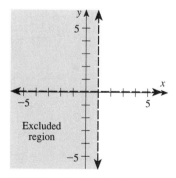

FIGURE 2.4 Excluded regions for Example 6

From this equation, we see that x is real for all y except $y = 0$ (see the dashed line in Figure 2.4). Therefore, $R = (-\infty, 0) \cup (0, \infty)$.

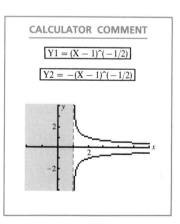

Intercepts

Several points of the graph of a function will be of particular importance to us. The first are the intercepts. These points are called intercepts because they are the places where a curve intercepts the coordinate axes.

INTERCEPTS

> If the number zero is in the domain of f, then $f(0)$ is called the **y-intercept** of the graph of f and is the point $(0, f(0))$. If a is a real number in the domain such that $f(a) = 0$, then a is called an **x-intercept** and is the point $(a, 0)$. Any number x such that $f(x) = 0$ is called a **zero of the function.**

EXAMPLE 7 Find the domain, range, and intercepts for g defined by the graph. Also find the coordinates of A and B.

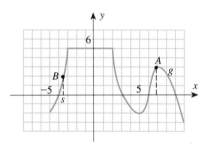

SOLUTION **Domain:** $-5 \le x \le 10$

Range: $-3 \le y \le 5$

y-intercept: $(0, 5)$; we usually say simply that the y-intercept is 5.

x-intercepts: $(-4, 0)$, $(3, 0)$, $(6, 0)$, and $(9, 0)$; the zeros of the function g are -4, 3, 6, and 9.

Point A has coordinates $(7, 3)$;
B has coordinates $(s, g(s))$.

We summarize the procedures for finding the intercepts:

To find the y-intercept: Set $x = 0$ and solve for y.

To find the x-intercept: Set $y = 0$ and solve for x.

⊘ If a graph is a function in x, then for each x value there can be associated only one y value. This means that if a curve represents a function, there can be at most one y-intercept, but if the graph is not a function, there might be more than one y-intercept. In either case, there may be several x-intercepts.

We usually begin by noting whether or not we are dealing with a function. Next, we find the y-intercept(s) and finally the x-intercepts (if any). ⊘

EXAMPLE 8 Find the intercepts and determine whether each is a function.

a. $y = \dfrac{1}{x^2 + 1}$ **b.** $y^2 = \dfrac{x^2 - 1}{x - 4}$ **c.** $|x| + |y| = 2$ **d.** $x^{2/3} + y^{2/3} = 9$

SOLUTION   **a.** This is **a function**, since for each x there is exactly one y value.

y-intercept: Let $x = 0$. $y = \dfrac{1}{0^2 + 1} = 1$ Point: **(0, 1)**

x-intercepts: Let $y = 0$. $0 = \dfrac{1}{x^2 + 1}$ Multiply both sides by $x^2 + 1$.

$0 = 1$

A false equation means that there is no point; in this example, there are **no x-intercepts.**

b. This is **not a function**, since for each x there are two possible values for y.

y-intercepts: Let $x = 0$. $y^2 = \dfrac{0^2 - 1}{0 - 4} = \dfrac{1}{4}$

$y = \pm \dfrac{1}{2}$ Points: **(0, ½), (0, −½)**

x-intercepts: Let $y = 0$. $0 = \dfrac{x^2 - 1}{x - 4}$ Multiply both sides by $x - 4$.

$0 = x^2 - 1$

$= (x - 1)(x + 1)$

$x = \pm 1$ Points: **(1, 0), (−1, 0)**

c. This is **not a function.**

y-intercepts: Let $x = 0$. $|0| + |y| = 2$

$y = \pm 2$ Points: **(0, 2), (0, −2)**

x-intercepts: Let $y = 0$. $|x| + |0| = 2$

$x = \pm 2$ Points: **(2, 0), (−2, 0)**

d. This is **not a function.**

y-intercepts: Let $x = 0$. $0^{2/3} + y^{2/3} = 9$

$(y^{2/3})^{3/2} = 9^{3/2}$

$y = \pm 27$ Points: **(0, 27), (0, -27)**

x-intercepts: Let $y = 0$. $x^{2/3} + 0^{2/3} = 9$

$x = 27$ Point: **(27, 0)**

Sometimes when you are graphing a curve, you want to find a point in a certain region or with certain properties. For example, if you want to know one point on the line $2x + 3y - 4 = 0$ where $x > 5$, then you can choose an x value satisfying $x > 5$ and find the corresponding y value. Consider the following example.

EXAMPLE 9 Find a point on each of the given curves that satisfies the given conditions.

a. $y = \dfrac{2x^2 - 3x + 5}{x^2 - x - 2}$ Find a point where $x > 5$.

You choose some value of x, say $x = 10$:

$$y = \frac{2(10)^2 - 3(10) + 5}{10^2 - 10 - 2} = \frac{175}{88} \approx 2 \qquad \text{Point:} \quad (10, \tfrac{175}{88}) \approx \textbf{(10, 2)}$$

b. Find a point (if it exists) where the curve in part **a** passes through the line $y = 2$. This means that you should substitute the value 2 for y in the given equation:

$$2 = \frac{2x^2 - 3x + 5}{x^2 - x - 2} \qquad \text{Multiply both sides by } x^2 - x - 2.$$
$$2(x^2 - x - 2) = 2x^2 - 3x + 5$$
$$2x^2 - 2x - 4 = 2x^2 - 3x + 5$$
$$x = 9$$

The point is **(9, 2)**.

c. Given:
$$y = \frac{2x^2 - 5x + 1}{x - 3}$$

find a point (if any exists) where this curve passes through the line $y = 2x + 1$. This means substitute $2x + 1$ for y in the given equation and solve for x:

$$2x + 1 = \frac{2x^2 - 5x + 1}{x - 3}$$
$$(2x + 1)(x - 3) = 2x^2 - 5x + 1$$
$$2x^2 - 5x - 3 = 2x^2 - 5x + 1$$
$$-3 = 1$$

This is a false equation, so there is **no point of intersection.** ▌

Classifications of Functions

There are several different classifications of functions, which are useful in a variety of ways.

INCREASING, DECREASING, CONSTANT FUNCTIONS; TURNING POINT

Let S be a subset of the domain of a function f. Then:

f is **increasing** on S if $f(x_1) < f(x_2)$ whenever $x_1 < x_2$ in S;
f is **decreasing** on S if $f(x_1) > f(x_2)$ whenever $x_1 < x_2$ in S;
f is **constant** on S if $f(x_1) = f(x_2)$ for every x_1 and x_2 in S.

If the line $x = a$ separates an interval over which f is increasing from an interval over which f is decreasing, then $(a, f(a))$ is a **turning point.**

These classifications are illustrated in Figure 2.5.

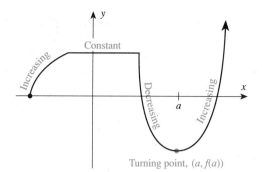

FIGURE 2.5 Classifications of functions

Another classification of functions is related to the symmetry of its graph. A function whose graph is symmetric with respect to the y-axis is called **even.** A function whose graph is symmetric with respect to the origin is called **odd.** If the function is found to be even or odd, then the symmetry of its graph helps in the graphing of the function. We now state an algebraic definition for even and odd functions.

EVEN AND ODD FUNCTIONS

A function f is called

even if $f(-x) = f(x)$ and

odd if $f(-x) = -f(x)$.

Just as not every real number is even or odd (2 is even, 3 is odd, but 2.5 is neither), not every function is even or odd.

EXAMPLE 10 Classify the given functions as even, odd, or neither.

a. $f(x) = x^2$ is **even** since

$$f(-x) = (-x)^2$$
$$= x^2$$
$$= f(x)$$

b. $g(x) = x^3$ is **odd** since

$$g(-x) = (-x)^3$$
$$= -x^3$$
$$= -[x^3]$$
$$= -g(x)$$

c. $h(x) = x^2 + 5x$ is **neither** since

$$h(-x) = (-x)^2 + 5(-x)$$
$$= x^2 - 5x$$

This is neither $h(x)$ nor $-h(x)$.

CALCULATOR COMMENT	CALCULATOR COMMENT	CALCULATOR COMMENT
Even functions are symmetric with respect to the y-axis.	Odd functions are symmetric with respect to the origin.	$Y = X^2 + 5X$

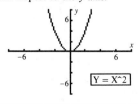

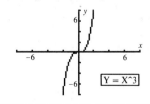

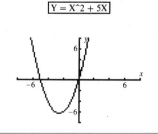

2.2 Problem Set

A *Find the domain for the functions defined by the equations in Problems 1–8 and leave your answer in interval notation.*

1. $f(x) = 3x + 1$

2. $g(x) = 3x + 1, \quad x \neq 2$

3. $h(x) = \dfrac{(3x + 1)(x + 2)}{x + 2}$

4. $F(x) = \dfrac{(2x + 1)(x - 1)}{x^2 + 1}$

5. $G(x) = \sqrt{2x + 1}$

6. $H(x) = \sqrt{1 - 3x}$

7. $f(x) = \sqrt{2 - x - x^2}$

8. $g(x) = \sqrt{2 + x - x^2}$

State whether the functions f and g are equal in Problems 9–14.

9. $f(x) = \dfrac{2x^2 + x}{x}; \quad g(x) = 2x + 1$

10. $f(x) = \dfrac{2x^2 + x}{x}; \quad g(x) = 2x + 1, x \neq 0$

11. $f(x) = \dfrac{2x^2 + x - 6}{x - 2}; \quad g(x) = 2x + 3, x \neq 2$

12. $f(x) = \dfrac{3x^2 - 7x - 6}{x - 3}; \quad g(x) = 3x + 2, x \neq 3$

13. $f(x) = \dfrac{3x^2 - 5x - 2}{x - 2}; \quad g(x) = 3x + 1$

14. $f(x) = \dfrac{(3x + 1)(x - 2)}{x - 2}, \quad x \neq 6;$

$g(x) = \dfrac{(3x + 1)(x - 6)}{x - 6}, \quad x \neq 2$

Classify the functions defined in Problems 15–20 as even, odd, or neither.

15. $f_1(x) = x^2 + 1$

16. $f_2(x) = \sqrt{x^2}$

17. $f_3(x) = \dfrac{1}{3x^3 - 4}$

18. $f_4(x) = x^3 + x$

19. $f_5(x) = |x|$

20. $f_6(x) = |x| + 3$

21. See part **a** of the figure. If point B has coordinates $(3, G(3))$, what are the coordinates of R and S?

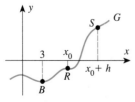

a. Graph of G

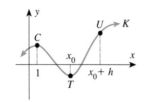

b. Graph of K

22. See part **b** of the figure in Problem 21. If point C has coordinates $(1, K(1))$, what are the coordinates of T and U?

Find the domain, range, intercepts, and turning points of the functions defined by the graphs indicated in Problems 23–28. Also tell where the function is increasing, decreasing, and constant.

23.

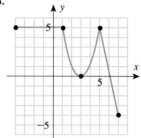

24.

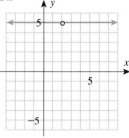

25.

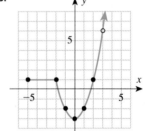

26.

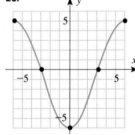

27.

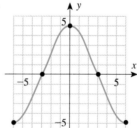

28.

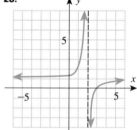

In Problems 29–50 find all intercepts.

29. $y = x + 5$

30. $y = x - 2$

31. $y = x^2 - 4$

32. $y = x^2 + 3$

33. $y = x^2$

34. $y = x^3$

35. $y = \dfrac{3}{x^2 + 1}$

36. $y = \dfrac{x^2 - 4}{x + 1}$

37. $y = \dfrac{x^2 - 9}{x - 2}$

38. $y = \dfrac{x + 1}{x^2 - 16}$

39. $y = \dfrac{x^3 - 8}{x^2 + 2x + 4}$

40. $y = \dfrac{x^3 - 1}{x^2 + x + 1}$

41. $y = \dfrac{x^3 - 1}{x - 1}$

42. $y = \dfrac{x^3 + 8}{x + 2}$

43. $y = -\dfrac{\sqrt{x + 3}}{x - 1}$

44. $y = -\dfrac{\sqrt{9 - x^2}}{x + 3}$

45. $2|x| - |y| = 5$

46. $|y| = 5 - 3|x|$

47. $xy = 1$

48. $xy + 6 = 0$

49. $x^2 + 2xy + y^2 = 4$

50. $x^{1/2} + y^{1/2} = 4$

B *Find the domain and range for the curves defined by the equations in Problems 51–68, and show the excluded regions.*

51. $y = \sqrt{x}$

52. $y = \sqrt{2x}$

53. $y = \sqrt{x - 2}$

54. $y = \sqrt{x + 2}$

55. $y = \sqrt{2x - 1}$

56. $y = \sqrt{2x + 10}$

57. $y = x\sqrt{x}$

58. $y = 2x\sqrt{x}$

59. $y = \sqrt{x^3}$

60. $y = x\sqrt{x^3}$

61. $y^2 = x^3$

62. $y^2 = 8x^3$

63. $y^2 = x^2 - 9$

64. $y^2 = x^2 - 4$

65. $y = \sqrt{x^2 - 4}$

66. $y = \sqrt{x^2 - x - 6}$

67. $y = \sqrt{x^2 + x - 12}$

68. $y = \sqrt{x^3 - 9x}$

Find a point on each of the given curves satisfying the given conditions in Problems 69–74.

69. Find the points (if any) where the curve

$$y = \frac{5x^2 - 8x}{2x + 1}$$

passes through the line $y = 3$.

70. Find the points (if any) where the curve

$$y = \frac{2x^3 + 2x}{x^2 + 1}$$

passes through the line $y = 1$.

71. Find the points (if any) where the curve

$$y = \frac{5x^2 - 8x}{2x + 1}$$

has a value $x = -4$.

72. Find the points (if any) where the curve

$$y = \frac{2x^3 + 2x}{x^2 - 2}$$

has a value $x = -1$.

73. Find the points (if any) where the curve

$$y = \frac{x^3 + 2x^2 - 2x}{x^2 - 2}$$

passes through the line $y = x + 1$.

74. Find the points (if any) where the curve

$$y = \frac{3x^3 + 4x^2 + 3}{3x^2 + 1}$$

passes through the line $y = x + 2$.

C

75. If F is a one-to-one function mapping X onto Y, and the domain of F contains exactly five elements, what can you conclude about the set Y?

76. If a function f is increasing throughout its domain, prove that f is one-to-one.

77. If a function f is decreasing throughout its domain, prove that f is one-to-one.

2.3 Transformations of Functions

Translations

In this section we look at techniques for graphing that will pay dividends in the amount of work you will need to do when graphing certain functions. For example, to graph the functions $y = x^2$ and $y - 3 = (x - 2)^2$ by plotting points, first set up tables of values and then draw the graphs, as shown on page 88.*

---

*You will not really appreciate what is being done here unless *you* carry through the arithmetic shown in the tables. We are introducing a method that makes the *arithmetic* easier, but you will not see why it is better unless you actually do the arithmetic.

Function $y = x^2$	
x	y
0	0
1	1
−1	1
2	4
−2	4
3	9
−3	9

Graph of $y = x^2$

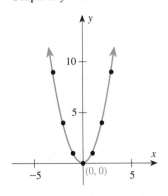

Function $y − 3 = (x − 2)^2$	
x	y
0	7
1	4
−1	12
2	3
3	4
4	7
5	12

Graph of $y − 3 = (x − 2)^2$

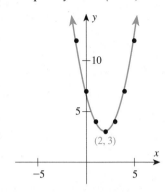

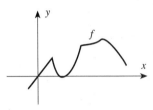

FIGURE 2.6 Graph of f

Notice that the graphs are the same, but they are in different locations. You also should have noticed (if you did the arithmetic) that the first table of values was much easier to calculate than the second. When two curves are congruent (have the same size and shape) and have the same orientation, we say that one can be found from the other by a **shift** or **translation.**

Consider the function f defined by the graph in Figure 2.6. It is possible to shift the entire curve up, down, right, or left, as shown in Figure 2.7.

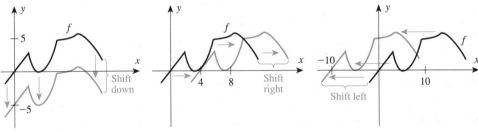

FIGURE 2.7 Shifting the graph of f

Instead of considering the curve shifting relative to fixed axes, consider the effect of shifting the axes. If the coordinate axes are shifted up k units, the origin of this new coordinate system would correspond to the point $(0, k)$ in the old coordinate system. If the axes are shifted to the right h units, the origin would correspond to the point $(h, 0)$ in the old system. A horizontal shift of h units followed by a vertical shift of k units would shift the new coordinate axes so that the origin corresponds to a point (h, k) on the old axes. Suppose a *new* coordinate system with origin at (h, k) is drawn and the new axes are labeled x' and y', as shown in Figure 2.8.

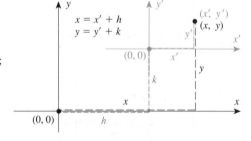

FIGURE 2.8 Shifting the axes to (h, k); comparison of coordinate axes

Every point on a given curve can now be denoted in two ways (see Figure 2.8):
1. As (x, y) measuring from the old origin
2. As (x', y') measuring from the new origin (color)

To find the relationship between (x, y) and (x', y'), consider the graph shown in Figure 2.8:

$$\begin{array}{ccc} x = x' + h & & x' = x - h \\ & \text{or} & \\ y = y' + k & & y' = y - k \end{array}$$

This says that if you are given any function

$$y - k = f(x - h)$$

the graph of this function is the same as the graph of the function

$$y' = f(x')$$

where (x', y') are measured from the new origin located at (h, k). This can greatly simplify our work since $y' = f(x')$ is usually easier to graph than $y - k = f(x - h)$.

EXAMPLE 1   Write the equations of the shifted curves in Figure 2.7.

SOLUTION   **a.** The curve has been shifted up 10 units, so $k = 10$; the equation is
$y - 10 = f(x)$; note that k is *subtracted* from y; this is sometimes written
$y = f(x) + 10$.

b. The curve has been shifted down 5 units, so $k = -5$; the equation is
$y + 5 = f(x)$; note that $y - (-5) = y + 5$.

c. The curve has been shifted to the right 4 units, so $h = 4$; the equation is
$y = f(x - 4)$.

d. The curve has been shifted to the left 10 units, so $h = -10$; the equation is
$y = f(x + 10)$.   ▍

EXAMPLE 2   Find (h, k) for each equation.

a. $y - 5 = f(x - 7)$ **b.** $y + 6 = f(x - 1)$
c. $y + 1 = f(x + 3)$ **d.** $y = f(x)$
e. $y - 6 = f(x) + 15$

SOLUTION   **a.** $(h, k) = (7, 5)$
b. Notice that $y + 6$ can be written as $y - (-6)$; $(h, k) = (1, -6)$.
c. $(h, k) = (-3, -1)$
d. This indicates no shift; $(h, k) = (0, 0)$.
e. Write the equation as $y - 21 = f(x)$; thus $(h, k) = (0, 21)$.   ▍

TRANSLATION OF AXES | The graph of the equation $y - k = f(x - h)$ is the same as the graph of $y = f(x)$ on a system of coordinate axes that has been **translated** h units horizontally and k units vertically.

EXAMPLE 3 Given f defined by $y = f(x)$ as shown by the graph in Figure 2.9, graph the following functions.

a. $y = f(x - 3)$

b. $y + 2 = f(x)$

c. $y - 4 = f(x + 5)$

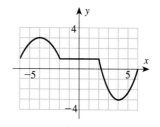

FIGURE 2.9 Graph of f

SOLUTION **a.** Since $(h, k) = (3, 0)$, the shift is 3 units to the right.

b. Since $(h, k) = (0, -2)$, the shift is 2 units down.

c. Since $(h, k) = (-5, 4)$, the shift is 5 units to the left and 4 units up.

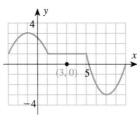 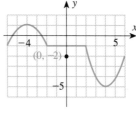 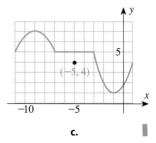

a.   b.   c.

Translations are useful because they sometimes allow us to take a computationally difficult equation and rewrite it in terms of a simpler equation. For example, consider the three curves shown in Figure 2.10.

FIGURE 2.10 Standard parabola, absolute value, and square-root curves

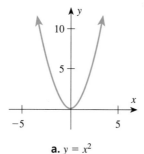 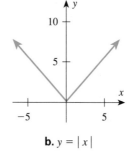 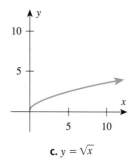

a. $y = x^2$ **b.** $y = |x|$ **c.** $y = \sqrt{x}$

We can now draw the graphs of some rather complicated equations by using these curves and the idea of translation.

EXAMPLE 4 Graph $y - \frac{1}{2} = (x - \frac{3}{2})^2$.

SOLUTION First, plot $(\frac{3}{2}, \frac{1}{2})$. Next, let $y' = y - \frac{1}{2}$ and $x' = x - \frac{3}{2}$. *Imagine* the new origin at $(h, k) = (\frac{3}{2}, \frac{1}{2})$ and plot values from the new origin using the equation $y' = x'^2$.

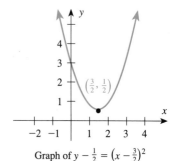

Graph of $y - \frac{1}{2} = (x - \frac{3}{2})^2$

As you can see, the graph of $y - k = f(x - h)$ is done in two steps:

TRANSLATION PROCEDURE

1. Plot (h, k).
2. Graph the simpler curve $y' = f(x')$ by using (h, k) as the new origin.

EXAMPLE 5 Graph $f(x) = |x - 3| + 2$.

SOLUTION This can be rewritten as $y - 2 = |x - 3|$, which is the graph of $y = |x|$ (see Figure 2.10b) on the system of coordinate axes that is translated to $(h, k) = (3, 2)$.

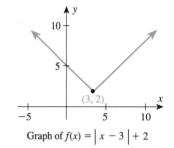

Graph of $f(x) = |x - 3| + 2$

EXAMPLE 6 Graph $y = \sqrt{x - 2} - 3$.

SOLUTION Rewrite as $y + 3 = \sqrt{x - 2}$. The graph is the same as the graph of $y = \sqrt{x}$ (see Figure 2.10c) translated to coordinate axes with origin at the point $(h, k) = (2, -3)$.

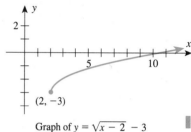

Graph of $y = \sqrt{x - 2} - 3$

Reflections

In Chapter 1 we introduced the notion of symmetry. In terms of functions and functional notation, we restate the notions of symmetry with respect to the coordinate axis as **reflections.**

REFLECTIONS

The graph of $y = -f(x)$ is a **reflection through the x-axis.** It is found by replacing each point (x, y) on the graph of $y = f(x)$ with $(x, -y)$.

The graph of $y = f(-x)$ is a **reflection through the y-axis.** It is found by replacing each point (x, y) on the graph of $y = f(x)$ with $(-x, y)$.

EXAMPLE 7 Graph **a.** $y = -|x|$, and **b.** $y = |-x|$.

SOLUTION **a.** Recall the graph of $y = |x|$ from Figure 2.10b. Recall from Chapter 1 that, if points (x, y) are replaced by the corresponding points $(x, -y)$, the result is a curve that is symmetric with respect to the x-axis, as shown at the right.

b. $y = |-x|$ is found by replacing each point on the graph of $y = |x|$ by the corresponding point $(-x, y)$; the result is a curve that is symmetric to $y = |x|$ with respect to the y-axis.

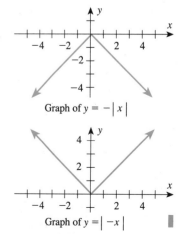

Graph of $y = -|x|$

Graph of $y = |-x|$

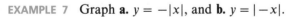

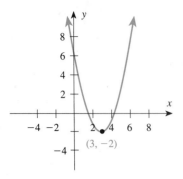

An application of this reflection property helps us to graph curves such as $y = |(x - 3)^2 - 2|$. If $y = f(x)$ where $f(x) = (x - 3)^2 - 2$, the graph we seek is of the form $y = |f(x)|$. We begin by graphing $f(x) = (x - 3)^2 - 2$, which is the standard parabola shown in Figure 2.10a translated to coordinate axes with origin at the point $(3, -2)$. Notice that part of the graph is above the x-axis and part is below the x-axis. Since the absolute value of a positive value leaves the positive value unchanged, the absolute value on $y = |f(x)|$ will leave the graph unchanged above the x-axis. However, the portion below the x-axis will be replaced by its reflection above the x-axis because of the definition of absolute value. The desired graph is shown in Figure 2.11.

FIGURE 2.11 Graph of $y = |(x - 3)^2 - 2|$

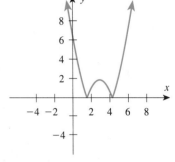

Contrast the graph shown in Figure 2.11 with the graph of the function $y = (|x| - 3)^2 - 2$. For this curve we also begin with the standard parabola shown in Figure 2.10a and translate to coordinate axes with origin at the point $(3, -2)$. In this case we note that for each value of x, there is a corresponding value $-x$ that also satisfies the equation. This means that we first sketch only values for which $x \geq 0$ and then use symmetry for the values for which $x < 0$. Thus, we see that this curve is simply a reflection about the y-axis, as shown in Figure 2.12.

FIGURE 2.12 Graph of $y = (|x| - 3)^2 - 2$

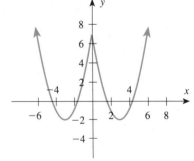

We summarize our findings involving the graph of an absolute value of a function.

ABSOLUTE VALUE REFLECTIONS

The graph of $y = |f(x)|$ is found by graphing $y = f(x)$ and then replacing all points of the graph that are below the x-axis by their reflection through the x-axis.

The graph of $y = f(|x|)$ is an even function, so the graph is found by graphing $y = f(x)$ for all $x \geq 0$ in the domain and then, *in addition to those points,* drawing all points reflected through the y-axis.

Dilations and Compressions

A curve can be compressed or dilated in either the *x*-direction, the *y*-direction, or both, as shown in Figure 2.13.

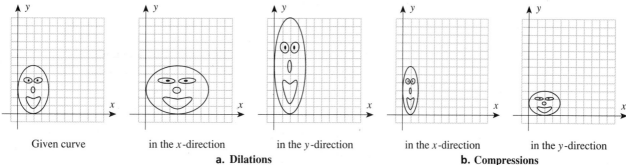

| Given curve | in the *x*-direction | in the *y*-direction | in the *x*-direction | in the *y*-direction |

a. Dilations   **b. Compressions**

FIGURE 2.13 Dilations and compressions

Dilations and Compressions in the *y*-Direction

We are interested in modifications of a known function f. Consider the graph of $y = f(x)$ as shown in Figure 2.14a. The graph of $y = 2f(x)$ has the same shape except that each *y* value is double the corresponding *y* value of f. On the other hand, the graph of $y = \frac{1}{2}f(x)$ has the same shape except that each *y* value is one-half the corresponding *y* value of f. These graphs are shown in Figures 2.14b and c.

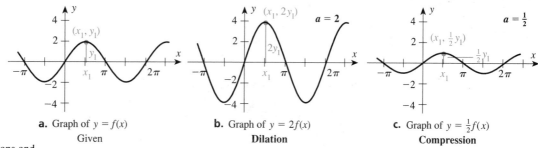

a. Graph of $y = f(x)$
Given

b. Graph of $y = 2f(x)$
Dilation

c. Graph of $y = \frac{1}{2}f(x)$
Compression

FIGURE 2.14 Dilations and compressions in the *y*-direction for a given function

Dilations and Compressions in the *x*-Direction

In order to describe a dilation and compression in the *x*-direction, we will consider the function $y = f(x)$ and examine the effect of $y = f(bx)$. If we take a particular value of *y*, then it follows that the value *bx* is plotted to the same *y* value as the value *x* in the original curve. This means that to graph $y = f(bx)$, replace each point (x, y) on the graph of $y = f(x)$ with the point $\left(\dfrac{1}{b}x, y\right)$. See Figure 2.15.

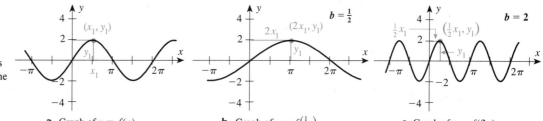

FIGURE 2.15 Dilations and compressions in the *x*-direction for a given function

a. Graph of $y = f(x)$
Given

b. Graph of $y = f\left(\frac{1}{2}x\right)$
Dilation

c. Graph of $y = f(2x)$
Compression

DILATIONS AND COMPRESSIONS

To sketch the graph of $y = af(x)$, replace each point (x, y) with (x, ay).

If $a > 1$, then we call the transformation a **y-dilation**.
If $0 < a < 1$, then we call the transformation a **y-compression**.

To sketch the graph of $y = f(bx)$, replace each point (x, y) with $\left(\dfrac{1}{b}x, y\right)$.

If $0 < b < 1$, then we call the transformation an **x-dilation**.
If $b > 1$, then we call the transformation an **x-compression**.

EXAMPLE 8   Graph: **a.** $y = 5\sqrt{x}$ **b.** $y = \sqrt{5x}$ **c.** $y = |\tfrac{1}{5}x|$ **d.** $y = \tfrac{1}{10}x^2$

SOLUTION   **a.** $y = 5\sqrt{x}$ is a y-dilation where
$f(x) = \sqrt{x}$ and $a = 5$.

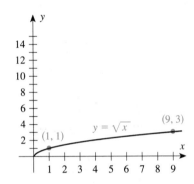

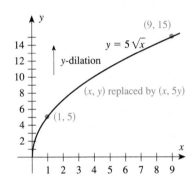

b. $y = \sqrt{5x}$ is an x-compression where
$$f(x) = \sqrt{x} \text{ and } b = 5,$$
$$\text{so } \frac{1}{b} = \frac{1}{5}.$$

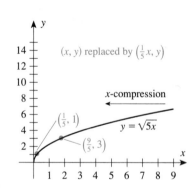

c. $y = |\tfrac{1}{5}x|$ is an x-dilation where
$f(x) = |x|$ and $b = \tfrac{1}{5}$, so $\dfrac{1}{b} = 5$.

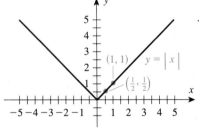

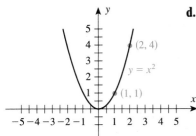

d. $y = \frac{1}{10} x^2$ is a y-compression where $f(x) = x^2$ and $a = \frac{1}{10}$.

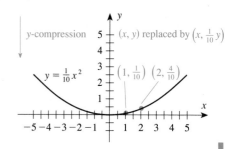

2.3 Problem Set

A Find (h, k) for each of the functions given in Problems 1–6.

1. a. $y - 3 = f(x - 6)$ **b.** $y - 5 = f(x + 3)$

2. a. $y + 1 = f(x - 6)$ **b.** $y = f(x - 4)$

3. a. $y - \sqrt{2} = f(x)$ **b.** $y = f(x) + 6$

4. a. $y = f(x - 3) - 4$ **b.** $y = f(x + 1) + 2$

5. a. $y = 4f(x)$ **b.** $y = 2f(x + \sqrt{2})$

6. a. $6y = f(x + 3)$ **b.** $y = 2f\left(x + \frac{\pi}{2}\right) - \frac{\pi}{3}$

Write the equation of the shifted curves in Problems 7–14. Notice that the curves that are shifted are those in Figure 2.10.

7. **8.**

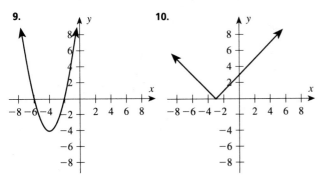

9. **10.**

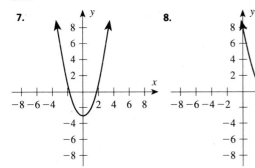

11. **12.**

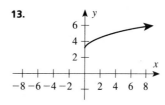

13. **14.**

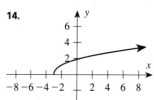

Let f, g, and h be the functions whose graphs are shown in Figure 2.16. Graph the functions indicated by the equations in Problems 15–32.

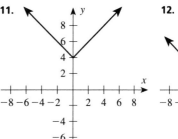

a. Graph of f

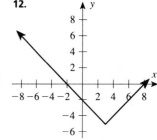

b. Graph of g

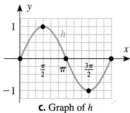

c. Graph of h

FIGURE 2.16

15. $y + 3 = f(x)$

16. $y + 2 = g(x)$

17. $y + \frac{1}{2} = h(x)$

18. $y - 5 = g(x)$

19. $y - 1 = h(x)$

20. $y + \pi = f(x)$

21. $y = h\left(x - \dfrac{\pi}{2}\right)$

22. $y = g(x + 3)$

23. $y = f(x - \sqrt{5})$

24. $y + 4 = f(x - 3)$

25. $y + 1 = g(x - 4)$

26. $y - 2 = h(x + \pi)$

27. $y - 3 = g(x + 5)$

28. $y - 2 = h(x + 3)$

29. $y + 2 = f(x - 4)$

30. $y - \dfrac{21}{5} = g\left(x - \dfrac{15}{2}\right)$

31. $y - \sqrt{3} = f(x + \sqrt{2})$

32. $y + \pi = h\left(x - \dfrac{3\pi}{2}\right)$

Let $y = c(x)$ be the function whose graph is given in Figure 2.17. Graph the curves indicated by the equations in Problems 33–40.

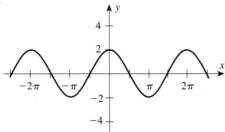

FIGURE 2.17 Graph of c

33. $y = 2c(x)$

34. $y = \frac{1}{2}c(x)$

35. $y = -c(x)$

36. $y = -2c(x)$

37. $y = c(2x)$

38. $y = c(\frac{1}{2}x)$

39. $y - 2 = c(x - 3)$

40. $y + 4 = 3c(x - 2)$

B Graph the curves defined by the equations given in Problems 41–67.

41. $y - 3 = (x - 2)^2$

42. $y = (x + 3)^2$

43. $y = x^2 - 1$

44. $y - 1 = |x - 7|$

45. $y = \sqrt{x} + 4$

46. $y = \sqrt{x + 4}$

47. $y = |x + \pi|$

48. $y = |x| + 6$

49. $y + \sqrt{3} = |x - \sqrt{2}|$

50. $y - \dfrac{\pi}{4} = (x + \pi)^2$

51. $y + 2 = (x + \sqrt{3})^2$

52. $y - \sqrt{2} = (x + \sqrt{5})^2$

53. $y - \sqrt{2} = \sqrt{x + 5}$

54. $y + \pi = \sqrt{x - 3}$

55. $y = -3|x + 5|$

56. $y = 4\sqrt{x - 2}$

57. $y = -2(x + 4)^2$

58. $y - 3 = -\frac{1}{2}(x + 2)^2$

59. $y = |(x - 4)^2 - 9|$

60. $y = |(x + 2)^2 - 4|$

61. $y = (|x| - 2)^2 - 4$

62. $y = (|x| - 4)^2 - 9$

C

63. $y + 3 = (x + 8)^2$, such that $-14 \leq x \leq -8$

64. $y - 2 = (x + 3)^2$, such that $-7 \leq x \leq -2$

65. $y + 12 = \left(x + \dfrac{25}{2}\right)^2$, such that $y > -10$

66. $y + 3 = (x + 3)^2$, such that $y < 6$

67. $y + 12 = (x - 8)^2$, such that $y < 4$

2.4 Operations and Composition of Functions

In algebra, you spent a great deal of time learning the algebra of real numbers. We can also consider an algebra of functions.

Function Addition and Subtraction

Let $f(x) = \frac{1}{2}x^2$ and $g(x) = x + 2$. These functions can be graphed by plotting points, as shown in Figure 2.18.

Next, consider the functions defined by $A(x) = f(x) + g(x)$ and $D(x) = f(x) - g(x)$. The function A is denoted by $f + g$ and D is denoted by $f - g$.

$$(f + g)(x) = f(x) + g(x)$$
$$= \tfrac{1}{2}x^2 + (x + 2)$$
$$= \tfrac{1}{2}x^2 + x + 2$$

$$(f - g)(x) = f(x) - g(x)$$
$$= \tfrac{1}{2}x^2 - (x + 2)$$
$$= \tfrac{1}{2}x^2 - x - 2$$

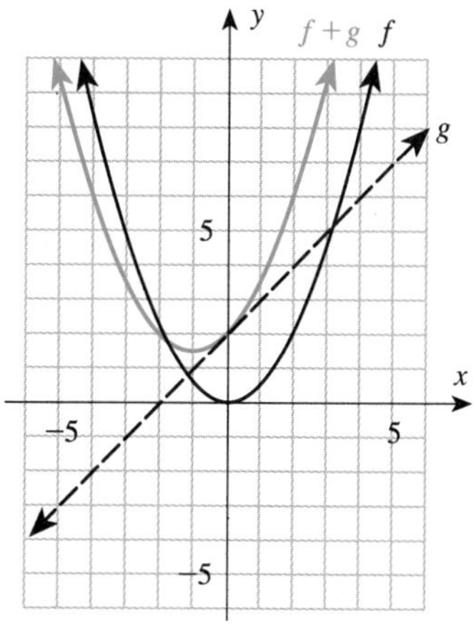

FIGURE 2.18 Graphs of $f(x)$, $g(x)$, and $(f + g)(x)$

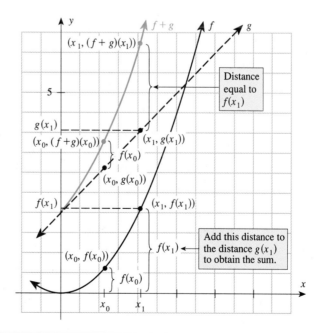

x	$f(x)$	$g(x)$	$(f + g)(x)$
-4	8	-2	6
-3	$\frac{9}{2}$	-1	$\frac{7}{2}$
-2	2	0	2
-1	$\frac{1}{2}$	1	$\frac{3}{2}$
0	0	2	2
1	$\frac{1}{2}$	3	$\frac{7}{2}$
2	2	4	6
3	$\frac{9}{2}$	5	$\frac{19}{2}$
4	8	6	14

The graph of $f + g$ can be generated by plotting points as shown in Figure 2.18 (color curve). Notice from the table that the values for $(f + g)(x)$ can be obtained by adding $f(x) + g(x)$ rather than by calculating $(f + g)(x) = \frac{1}{2}x^2 + x + 2$. This gives an idea about how $(f + g)(x)$ can be graphed.

Consider a closeup portion of Figure 2.18. It is not necessary that you actually calculate the functional values. That is, choose any x_0, as shown in Figure 2.19. Then the functional values $f(x_0)$ can be represented as distances. By choosing different x values, you can generate the entire graph of $(f + g)(x)$. This method of graphing is called *graphing by adding ordinates*.

FIGURE 2.19 Detail of Figure 2.18 showing the procedure for graphing by adding ordinates

EXAMPLE 1 Let $f(x) = \frac{1}{5}x^2$ and $g(x) = \frac{3}{2}x + 2$. The graphs are shown in Figure 2.20. Graph

$$(f + g)(x) = \frac{1}{5}x^2 + \frac{3}{2}x + 2$$

by adding ordinates.

SOLUTION The result is shown by a colored curve in Figure 2.20.

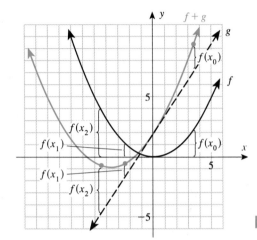

FIGURE 2.20 Graphs of f, g, and $f + g$

EXAMPLE 2 If f and g are defined by $f(x) = x^2$ and $g(x) = x + 3$, find $f + g$ and $f - g$, and evaluate these functions for $x = -1$ and $x = 5$.

SOLUTION

$$(f + g)(x) = f(x) + g(x)$$
$$= x^2 + x + 3$$

$$(f + g)(-1) = (-1)^2 + (-1) + 3$$
$$= 3$$

$$(f + g)(5) = 5^2 + 5 + 3$$
$$= 33$$

$$(f - g)(x) = f(x) - g(x)$$
$$= x^2 - (x + 3)$$
$$= x^2 - x - 3$$

$$(f - g)(-1) = (-1)^2 - (-1) - 3$$
$$= -1$$

$$(f - g)(5) = 5^2 - 5 - 3$$
$$= 17$$

The domain of both f and g is the set of real numbers, so the domains of $f + g$ and $f - g$ are also the set of real numbers. These are the usual domains for functions we consider in precalculus and calculus. However, sometimes we must work with restricted domains. The functions F and G in Example 3 are the same as f and g of Example 2 except that they have restricted domains.

EXAMPLE 3 If F and G are defined by $F(x) = x^2$ on $[-2, 3]$ and $G(x) = x + 3$ on $[-3, 4]$, find $F + G$ and $F - G$, and specify their domains.

SOLUTION
$$(F + G)(x) = x^2 + x + 3 \quad \text{on } [-2, 3]$$
$$(F - G)(x) = x^2 - x - 3 \quad \text{on } [-2, 3]$$

Note that the domain of $F + G$ and $F - G$ is the intersection of the domains of F and G, as shown in Figure 2.21.

FIGURE 2.21 Graphs of F, G, F + G, and F − G

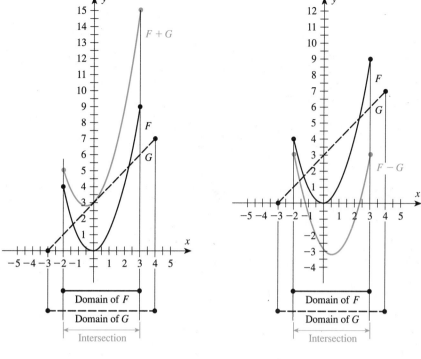

Domain of F
Domain of G
Intersection

Domain of F
Domain of G
Intersection

Function Multiplication and Division

Let $f(x) = 2x + 1$ and $g(x) = x^2$. The functions defined by $P(x) = f(x) \cdot g(x)$ and $Q(x) = f(x)/g(x)$ are denoted by $(fg)(x)$ and $(f/g)(x)$, respectively:

$$(fg)(x) = f(x) \cdot g(x) \qquad (f/g)(x) = \frac{f(x)}{g(x)}$$
$$= (2x + 1)(x^2) \qquad$$
$$= 2x^3 + x^2 \qquad = \frac{2x + 1}{x^2} \quad (x \neq 0)$$

The domain of fg is the intersection of the domains of f and g, and the domain of f/g is the intersection of the domains of f and g for which values causing $g(x) = 0$ are excluded.

EXAMPLE 4   If f and g are defined by $f(x) = x^2$ and $g(x) = x + 3$, find fg and f/g, and evaluate these functions for $x = -1$ and $x = 5$.

SOLUTION

$$(fg)(x) = f(x) \cdot g(x) \qquad (f/g)(x) = \frac{f(x)}{g(x)}, \quad \text{provided } g(x) \neq 0$$
$$= x^2(x + 3) \qquad = \frac{x^2}{x + 3} \qquad \text{It is } understood$$
$$= x^3 + 3x^2 \qquad \qquad \qquad \text{that } x \neq -3.$$

$$(fg)(-1) = (-1)^3 + 3(-1)^2 = \mathbf{2} \qquad (f/g)(-1) = \frac{(-1)^2}{-1 + 3} = \frac{\mathbf{1}}{\mathbf{2}}$$

$$(fg)(5) = 5^3 + 3(5)^2 = \mathbf{200} \qquad (f/g)(5) = \frac{5^2}{5 + 3} = \frac{25}{8} = \mathbf{3.125}$$

SPREADSHEET APPLICATION

Many of us now have access to a home computer as well as software for word processing, data bases, and spreadsheets. Spreadsheets provide a powerful, easy tool for processing data. Your first inclination when reading this will be to skip over this and say to yourself, "I don't know anything about spreadsheets, so I will not read this. Besides, my instructor is not assigning this anyway." On the contrary, regardless of whether this is assigned, chances are that sooner or later you will be using a spreadsheet.

The most commonly used spreadsheets today are versions of Excel or Lotus 1–2–3.

One of the most important features of a spreadsheet is its ability to evaluate functions (or formulas, as they are called on a spreadsheet). A spreadsheet is made up of **cells.** Each cell is a location that is referenced as points are referenced in a coordinate system. That is, cells are referenced by a row number (1, 2, 3, ...) and a column letter (A, B, C, ...). Here is a picture of a blank spreadsheet:

	A	B	C	D	E	F	G
1							
2							
3							
4							
5							

Instead of designating variables as letters (such as $x, y, z, \ldots$) as we do in algebra, a spreadsheet designates variables as cells (such as B2, A5, Z146, ...). A cell can contain a letter, word, sentence, number, or formula. If you type something into a cell, the spreadsheet program will recognize it as text if it begins with a letter, and as a number if it begins with a numeral. It also recognizes the usual mathematical symbols of $+$, $-$, $*$ for $\times$, $/$ for $\div$, and $\hat{}$ for raising to a power; parentheses are used in the usual fashion as grouping symbols. In order to enter a formula, you must begin with $+$ or @. Compare the algebraic and spreadsheet evaluations for a function shown in the table at the right.

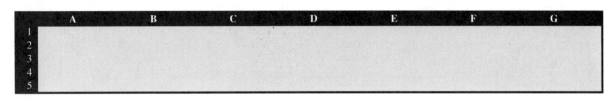

ALGEBRA		SPREADSHEET	
Function	*Evaluate*	*Formula*	*Evaluate*
$\dfrac{1}{x^2 + 2}$	For $x = 3$: $\dfrac{1}{3^2 + 2} = \dfrac{1}{11}$	$1/(A1\hat{}2 + 2)$ is written in some cell. The decimal evaluation is shown.	3 is placed in cell A1.

The advantage of a spreadsheet is that it is easy to do many evaluations. For example, let f and g be defined by $f(x) = x^2$ and $g(x) = x + 3$. Look at the spreadsheet evaluation for f, g, $f + g$, $f - g$, fg, and f/g for many different values of x:

	A	B	C	D	E	F	G
	Value of x	f	g	f + g	f − g	fg	f / g
1	−2	+A2^2	+A2+3	+B2+C2	+B2−C2	+B2*C2	+B2 / C2
2	0	+A3^2	+A3+3	+B3+C3	+B3−C3	+B3*C3	+B3 / C3
3	5	+A4^2	+A4+3	+B4+C4	+B4−C4	+B4*C4	+B4 / C4
4	−3	+A5^2	+A5+3	+B5+C5	+B5−C5	+B5*C5	+B5 / C5

On a spreadsheet it is not necessary to type in all the formulas you see in the above display. There is a very powerful command called **replicate.** We could type in cells B2 to G2 and then replicate these cells in rows 3, 4, and 5. Here is what the spreadsheet would look like:

	A	B	C	D	E	F	G
	Value of x	f	g	f + g	f − g	fg	f / g
1	−2	4	1	5	3	4	4
2	0	0	3	3	−3	0	0
3	5	25	8	33	17	200	3.125
4	−3	9	0	9	9	0	ERR

We now summarize the notation for the operations with functions.

OPERATIONS WITH FUNCTIONS

Let f and g be functions with domains D_f and D_g, respectively. Then $f + g, f - g, fg,$ and f/g are defined for the domain $D_f \cap D_g$:

$$(f + g)(x) = f(x) + g(x)$$
$$(f - g)(x) = f(x) - g(x)$$
$$(fg)(x) = f(x)g(x)$$
$$(f/g)(x) = \frac{f(x)}{g(x)} \quad \text{provided } g(x) \neq 0$$

Remember that the operation symbols on the left of the equations, namely $f + g$, $f - g$, fg, and f/g, are operations on *functions* that are being defined, whereas the operation on the right is being defined for operations on *numbers*.

Composition of Functions

There is another important operation on functions, called **composition,** which is not an elementary operation. As an example of composition, suppose a farmer sells eggs to Safeway. If the farmer's price for a dozen eggs is x dollars, then there is a function f that can be used to describe $f(x)$, the total cost of those eggs to Safeway. Note that x and $f(x)$ are not the same because Safeway must pay for ordering, shipping, and distributing the eggs. Moreover, in order to determine the price to the consumer, Safeway must add an appropriate markup. Suppose this markup function is called g. Then the price of the eggs to the consumer is

$$g[f(x)] \quad \text{and not} \quad g(x)$$

since the markup must be on Safeway's *total* cost of the eggs and not just on the price the farmer charges for the eggs. This process of evaluating a function of a function illustrates the idea of composition of functions.

Consider two functions f and g such that f is a function from X to Y and g is a function from Y to Z, as illustrated by Figure 2.22.

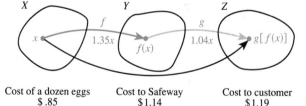

FIGURE 2.22 Composition of f and g

The image of x in the set Z is the number $g[f(x)]$ and defines a function from X to Z called the **composition of functions f and g.**

COMPOSITE FUNCTION

Let X, Y, and Z be sets of real numbers. Let f be a function from X to Y and g be a function from Y to Z. Then the **composite function $g \circ f$** is the function from X to Z defined by

$$(g \circ f)(x) = g[f(x)]$$

We can visualize composition in terms of function machines:

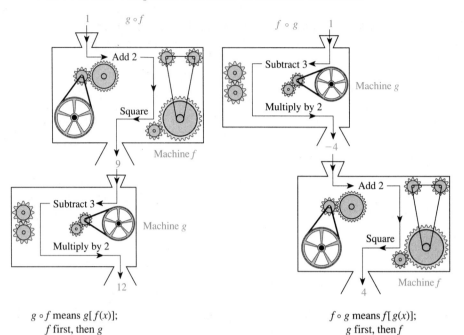

$g \circ f$ means $g[f(x)]$;
f first, then g

$f \circ g$ means $f[g(x)]$;
g first, then f

Or we can visualize composition in terms of spreadsheet cells:

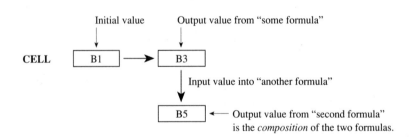

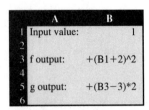

	A	B
1	Input value:	1
2		
3	f output:	+(B1+2)^2
4		
5	g output:	+(B3−3)*2
6		

	A	B
1	Input value:	1
2		
3	f output:	9
4		
5	g output:	12
6		

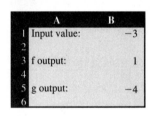

	A	B
1	Input value:	6
2		
3	f output:	64
4		
5	g output:	122
6		

	A	B
1	Input value:	−3
2		
3	f output:	1
4		
5	g output:	−4
6		

EXAMPLE 5 If $f(x) = x^2$ and $g(x) = x + 4$, find:

a. $(g \circ f)(x)$ **b.** $(f \circ g)(x)$ **c.** $(g \circ f)(-1)$ **d.** $(f \circ g)(5)$

SOLUTION

a. $(g \circ f)(x) = g[f(x)]$
$= g[x^2]$
$= x^2 + 4$

b. $(f \circ g)(x) = f[g(x)]$
$= f[x + 4]$
$= (x + 4)^2$
$= x^2 + 8x + 16$

c. $(g \circ f)(-1) = (-1)^2 + 4$
$= 5$

d. $(f \circ g)(5) = 5^2 + 8(5) + 16$
$= 81$

Notice from Examples 1a and 1b that $g \circ f \neq f \circ g$.

Some attention also must be paid to the domain of $g \circ f$. Let X be the domain of a function f, and let Y be the range of f. The situation can be viewed as shown in Figure 2.23a.

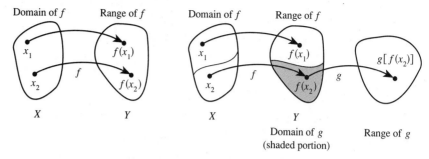

FIGURE 2.23

a. Function f **b.** Function g

If part of f maps into the shaded portion of Y and part of f maps into the portion that is not shaded, as indicated in Figure 2.23, then the domain of $g \circ f$ is just that part of X that maps into the shaded portion of Y, as shown in Figure 2.24.

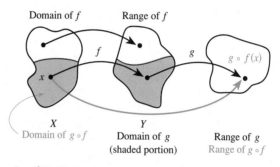

FIGURE 2.24 Composition of two functions, $g \circ f$; notice that the domain of $g \circ f$ is the subset of X for which $g \circ f$ is defined (shaded portion of Y).

EXAMPLE 6

$$f = \{(0, 0), (-1, 1), (-2, 4), (-3, 9), (5, 25)\}$$
$$g = \{(0, -5), (-1, 0), (2, -3), (4, -2), (5, -1)\}$$

a. Find $g \circ f$.   **b.** Find $f \circ g$.

SOLUTION   **a.** $D_f = \{0, -1, -2, -3, 5\}$, $D_g = \{0, -1, 2, 4, 5\}$

Think of f and g as function machines:

$$\begin{array}{cc} f & g \end{array}$$
$$0 \to \;\; 0 \cdots \to 0 \to -5$$
$$-1 \to \;\; 1 \cdots \to \text{Not defined, so go back to domain of } f \text{ and exclude } -1$$
$$-2 \to \;\; 4 \cdots \to 4 \to -2$$
$$-3 \to \;\; 9 \cdots \to \text{Not defined; exclude } -3 \text{ from the domain of } g \circ f$$
$$5 \to 25 \cdots \to \text{Not defined; exclude 5 from the domain of } g \circ f$$

$$g \circ f = \{(0, -5), (-2, -2)\}$$

The domain of the composite function $g \circ f$ is $\{0, -2\}$.

Notice that $-1, -3$, and 5 are excluded from the domain of $g \circ f$ even though they are in the domain of f.

b.

$$g \qquad\qquad f$$

$$0 \rightarrow -5 \cdots \rightarrow \text{Not defined}$$
$$-1 \rightarrow \quad 0 \cdots \rightarrow \quad 0 \rightarrow 0$$
$$2 \rightarrow -3 \cdots \rightarrow -3 \rightarrow 9$$
$$4 \rightarrow -2 \cdots \rightarrow -2 \rightarrow 4$$
$$5 \rightarrow -1 \cdots \rightarrow -1 \rightarrow 1$$

$$f \circ g = \{(-1, 0), (2, 9), (4, 4), (5, 1)\}$$

The domain of $f \circ g$ is $\{-1, 2, 4, 5\}$.

EXAMPLE 7 Express each of the following functions as the composition of two functions u and g so that $f(x) = g[u(x)]$.

a. $f(x) = (3x + 1)^3$ **b.** $f(x) = (x^2 + 5x + 1)^5$ **c.** $f(x) = \sqrt{5x^2 - x}$

SOLUTION The function $u(x)$ is the "inner function." Consider the following list:

GIVEN FUNCTION, $f(x) = g[u(x)]$	INNER FUNCTION $u(x)$	OUTER FUNCTION $g(x)$
a. $f(x) = (3x + 1)^3$	$u(x) = 3x + 1$	$g(u) = u^3$
b. $f(x) = (x^2 + 5x + 1)^5$	$u(x) = x^2 + 5x + 1$	$g(u) = u^5$
c. $f(x) = \sqrt{5x^2 - x}$	$u(x) = 5x^2 - x$	$g(u) = \sqrt{u}$

> This will be important to your future work in calculus; consider Example 7 and this paragraph carefully. ⊘

The procedure illustrated in Example 7 is useful in calculus. There are often other ways to express a composite function, but the most common procedure is to choose the function u to be the "inside" part of the given function f. Notice in parts **a** and **b** that the "inside" part is the portion inside the parentheses. In part **d** the $u(x)$ is the part "inside" the radical.

EXAMPLE 8 Environmental studies are often concerned with the relationship between the population of an urban area and the level of air pollution. Suppose it is estimated that when p hundred thousand people live in a certain city, the average daily level of carbon monoxide in the air is $L(p) = .07\sqrt{p^2 + 3}$ parts per million (ppm). Further assume that in t years, there will be $p(t) = 1 + .02t^3$ hundred thousand people in the city. Based on these assumptions, what level of air pollution should be expected in four years?

SOLUTION The level of pollution is $L(p) = .07\sqrt{p^2 + 3}$, where $p(t) = 1 + .02t^3$. Thus, the pollution level at time t is given by the composite function

$$(L \circ p)(t) = L[p(t)] = L(1 + .02t^3) = .07\sqrt{(1 + .02t^3)^2 + 3}$$

In particular, when $t = 4$, we have

$$(L \circ p)(4) = .07\sqrt{[1 + .02(4)^3]^2 + 3} \approx \textbf{.2 ppm}$$

2.4 Problem Set

A *In Problems 1–5 find the indicated values where*
$f(x) = 2x - 3$ *and* $g(x) = x^2 + 1$.

1. $(f + g)(5)$ **2.** $(f - g)(3)$
3. $(fg)(2)$ **4.** $(f/g)(4)$
5. $(f \circ g)(2)$

In Problems 6–10 find the indicated values where
$f(x) = \dfrac{x - 2}{x + 1}$ *and* $g(x) = x^2 - x - 2$.

6. $(f + g)(2)$ **7.** $(f - g)(5)$
8. $(fg)(102)$ **9.** $(f/g)(99)$
10. $(f \circ g)(1)$

In Problems 11–15 find the indicated values where
$f(x) = \dfrac{2x^2 - 5x + 2}{x - 2} - 3$ *and* $g(x) = x^2 - x - 2$.

11. $(f + g)(-1)$ **12.** $(f - g)(2)$
13. $(fg)(9)$ **14.** $(f/g)(99)$
15. $(f \circ g)(0)$

In Problems 16–20 find the indicated values where

$$f = \{(0, 1), (1, 4), (2, 7), (3, 10)\}$$

and

$$g = \{(0, 3), (1, -1), (2, 1), (3, 3)\}$$

16. $(f + g)(1)$ **17.** $(f - g)(3)$
18. $(fg)(2)$ **19.** $(f/g)(0)$
20. $(f \circ g)(2)$

In Problems 21–24 find the sum and difference of the given functions.

21. $f(x) = 2x - 3$ and $g(x) = x^2 + 1$

22. $f(x) = \dfrac{x - 2}{x + 1}$ and $g(x) = x^2 - x - 2$

23. $f(x) = \dfrac{2x^2 - x - 3}{x - 2}$ and $g(x) = x^2 - x - 2$

24. $f(x) = 4x + 2$ and $g(x) = x^3 + 3$

In Problems 25–28 find the product and quotient of the given functions.

25. $f(x) = 2x - 3$ and $g(x) = x^2 + 1$

26. $f(x) = \dfrac{x - 2}{x + 1}$ and $g(x) = x^2 - x - 2$

27. $f(x) = \dfrac{2x^2 - x - 3}{x - 2}$ and $g(x) = x^2 - x - 2$

28. $f(x) = 4x + 2$ and $g(x) = x^3 + 3$

In Problems 29–32 find $f \circ g$ and $g \circ f$ for the given functions.

29. $f(x) = 2x - 3$ and $g(x) = x^2 + 1$

30. $f(x) = \dfrac{x - 2}{x + 1}$ and $g(x) = x^2 - x - 2$

31. $f(x) = \dfrac{2x^2 - 3x - 2}{x - 2}$ and $g(x) = x^2 - x - 2$

32. $f(x) = 4x + 2$ and $g(x) = x^3 + 3$

B *In Problems 33–38 let f and g be functions defined by the graph. Graph $f + g$ by adding ordinates.*

33.

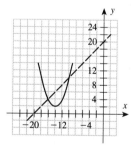

34.

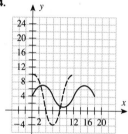

35.

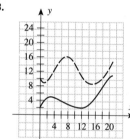

36.

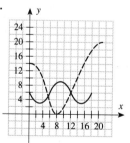

37.

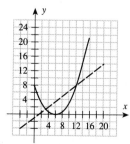

38.

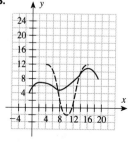

WHAT IS WRONG, *if anything, with each of the statements in Problems 39–42? Explain your reasoning.*

39. If $f(x) = \sqrt[3]{x}$ and $u(x) = 2x^2 + 1$, then
$$f[u(x)] = 2\sqrt[3]{x^2} + 1.$$

40. If $f(x) = \dfrac{1}{2x^2 + 1}$ and $u(x) = \sqrt{x - 1}$, then
$$f[u(x)] = \dfrac{1}{2x - 1}.$$

41. If $f(x) = 3x^2 + 5x + 1$ and $u(x) = x^3$, then
$$f[u(x)] = (3x^2 + 5x + 1)^3.$$

42. If $f(x) = \dfrac{x - 1}{x + 2}$ and $g(x) = \dfrac{2x + 1}{x - 3}$, then $f \circ g$ is the function defined by
$$\left(\dfrac{x - 1}{x + 2}\right)\left(\dfrac{2x + 1}{x - 3}\right)$$

43. If $f(x) = x^2$, $g(x) = 2x - 1$, and $h(x) = 3x + 2$, find $f \circ g$ and $g \circ h$.

44. If $f(x) = x^2$, $g(x) = 2x - 1$, and $h(x) = 3x + 2$, find $(f \circ g) \circ h$.

45. If $f(x) = x^2$, $g(x) = 2x - 1$, and $h(x) = 3x + 2$, find $f \circ (g \circ h)$.

46. If $f(x) = x^2$, $g(x) = 3x - 2$, and $h(x) = x^2 + 1$, find $f \circ h$ and $g \circ h$.

47. If $f(x) = x^2$, $g(x) = 3x - 2$, and $h(x) = x^2 + 1$, find $(f \circ g) \circ h$.

48. If $f(x) = x^2$, $g(x) = 3x - 2$, and $h(x) = x^2 + 1$, find $f \circ (g \circ h)$.

49. If $f(x) = \sqrt{x}$, $g(x) = x^2 - 2$, and $h(x) = x + 2$, all with domain $(0, \infty)$, find $f \circ g$ and $g \circ h$.

50. If $f(x) = \sqrt{x}$, $g(x) = x^2 - 2$, and $h(x) = x + 2$, all with domain $(0, \infty)$, find $(f \circ g) \circ h$.

51. If $f(x) = \sqrt{x}$, $g(x) = x^2 - 2$, and $h(x) = x + 2$, all with domain $(0, \infty)$, find $f \circ (g \circ h)$.

52. If $f(x) = x$, $g(x) = x$, and $h(x) = x$, find $f \circ g$ and $g \circ h$.

53. If $f(x) = x$, $g(x) = x$, and $h(x) = x$, find $(f \circ g) \circ h$.

54. If $f(x) = x$, $g(x) = x$, and $h(x) = x$, find $f \circ (g \circ h)$.

In Problems 55–60, express f as a composition of two functions u and g so that $f(x) = g[u(x)]$.

55. $f(x) = (2x^2 - 1)^4$

56. $f(x) = (3x^2 + 4x - 5)^3$

57. $f(x) = \sqrt{5x - 1}$

58. $f(x) = \sqrt[4]{x^3 - x + 1}$

59. $f(x) = (x^2 - 1)^3 + \sqrt{x^2 - 1} + 5$

60. $f(x) = |x + 1|^2 + 6$

61. In Section 1.3 we considered the volume V of a cone as a function of its height h:
$$V(h) = \dfrac{\pi h^3}{12}$$
Suppose the height is expressed as a function of time t by $h(t) = 2t$.
a. Find the volume for $t = 2$.
b. Express the volume as a function of time by finding $V \circ h$.
c. If the domain of V is $(0, 6]$, find the domain of h; that is, what are the permissible values for t?

62. The surface area S of a spherical balloon with radius r given by $S(r) = 4\pi r^2$. Suppose the radius is expressed as a function of time t by $r(t) = 3t$.
a. Find the surface area for $t = 2$.
b. Express the surface area as a function of time by finding $S \circ r$.
c. If the domain of S is $(0, 8)$, find the domain of r; that is, what are the permissible values for t?

63. If $f(x) = x^2$, then
$$f\left(\dfrac{1}{x}\right) = \left(\dfrac{1}{x}\right)^2 = \dfrac{1}{x^2} = \dfrac{1}{f(x)}$$
Give an example of a function for which
$$f\left(\dfrac{1}{x}\right) \neq \dfrac{1}{f(x)}$$

64. If $f(x) = x$, then $f(x^2) = [f(x)]^2$. Give an example of a function for which
$$f(x^2) \neq [f(x)]^2$$

C

65. Let $f(x) = 1 + \dfrac{1}{x}$. Find:
a. $(f \circ f)(x)$
b. $(f \circ f \circ f)(x)$
c. $(f \circ f \circ f \circ f)(x)$
d. Can you predict the output for further iterations of compositions?

66. Let $f(x) = \sqrt{x}$. Choose *any* positive x. Find a numerical value for $(f \circ f)(x)$, $(f \circ f \circ f)(x)$, and $(f \circ f \circ f \circ f)(x)$. If this procedure is repeated a large number of times, predict the outcome for any x.

67. Repeat Problem 66 for $f(x) = 2\sqrt{x}$.

68. Repeat Problem 66 for $f(x) = 3\sqrt{x}$.

69. Repeat Problem 66 for $f(x) = k\sqrt{x}$.

2.5 Inverse Functions

The Idea of an Inverse

In mathematics, the ideas of "opposite operations" and "inverse properties" are very important. The basic notion of an opposite operation or an inverse property is to "undo" a previously performed operation. For example, pick a number and call it x; then:

$$x \qquad \text{Think: I pick 5.}$$

$$\text{Add 7:} \quad x + 7 \qquad \text{Think: Now I have } 5 + 7 = 12.$$

The opposite operation returns you to x:

$$\text{Subtract 7:} \quad x + 7 - 7 = x \qquad \text{Think: } 12 - 7 = 5, \text{ my original number.}$$

We now want to apply this idea to functions. Pick a number and call it x:

Think: I'll pick 3 this time.

Now evaluate some given function f for the number you picked; suppose we let $f(x) = 2x + 9$:

$$f(x) = 2x + 9 \qquad \text{Think: } f(3) = 2(3) + 9 = 15$$

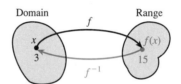

Relationship between f and f^{-1} for a one-to-one function

The *inverse function*, call it f^{-1} if it exists, is a function so that $f^{-1}(15)$ is 3, the original number. In symbols, $f(3) = 15$, so

$$f^{-1}[f(x)] = x \qquad \text{Think: } f^{-1}[f(3)] \text{ must equal 3, my original number.}$$

Of course, for f^{-1} to be an inverse function, it must "undo" the effect of f for *each and every member* of the domain. This may be impossible if f is a function such that two x values give the same y value. For example, if $g(x) = x^2$, then $g(2) = 4$ and $g(-2) = 4$. So we need a function g^{-1} so that

$$g^{-1}(4) = 2 \quad \text{and } also \quad g^{-1}(4) = -2$$

This would require the inverse to yield *two* values—namely, 2 and -2; however, this *violates* the definition of a function. So it is necessary to limit the given function so that it is *one-to-one*. Recall the *horizontal line test* from Section 2.1 to determine whether a function is one-to-one.

Inverse Functions

Each element in the domain of f^{-1} is the range element in the corresponding ordered pair for f, and each element in the domain of f is the range element in the corresponding pair for f^{-1}. Thus, for every x in the domain of f,

$$(f^{-1} \circ f)(x) = x$$

and for every x in the domain of f^{-1},

$$(f \circ f^{-1})(x) = x$$

In fact, the following relationship between functions, their inverses, and composition is used as a definition of inverse function.

INVERSE FUNCTIONS

If f is a one-to-one function with domain D and range R, and g is a function with domain R and range D such that

$$(g \circ f)(x) = x \quad \text{for every } x \text{ in } D$$
$$(f \circ g)(x) = x \quad \text{for every } x \text{ in } R$$

then f and g are called **inverses.** The function g is the **inverse function** of f and is denoted by f^{-1}.

$\otimes$ f^{-1} means inverse function—it does not mean $\dfrac{1}{f}$. $\otimes$

EXAMPLE 1 Show that f and g defined by

$$f(x) = 5x + 4 \qquad \text{and} \qquad g(x) = \frac{x - 4}{5}$$

are inverse functions.

SOLUTION You must show that f and g are inverse functions in two parts:

$$(g \circ f)(x) = g(5x + 4) \qquad \text{and} \qquad (f \circ g)(x) = f\left(\frac{x - 4}{5}\right)$$

$$= \frac{(5x + 4) - 4}{5} \qquad\qquad\qquad = 5\left(\frac{x - 4}{5}\right) + 4$$

$$= \frac{5x}{5} \qquad\qquad\qquad\qquad\qquad = (x - 4) + 4$$

$$= x \qquad\qquad\qquad\qquad\qquad\quad = x$$

Thus $(g \circ f)(x) = (f \circ g)(x) = x$, so f and g are inverse functions.

Once you are certain that a function g is the inverse of a function f, you can denote it by f^{-1} so that:

NOTATION FOR INVERSE FUNCTIONS

$(f^{-1} \circ f)(x) = x$ for x in the domain of f
$(f \circ f^{-1})(x) = x$ for x in the domain of f^{-1} (See Figure 2.25.)

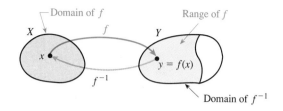

FIGURE 2.25 Inverse functions

EXAMPLE 2 Show that $f(x) = x^3 + 3$ and $g(x) = (x - 3)^{1/3}$ are inverses.

SOLUTION

$$(f \circ g)(x) = [(x - 3)^{1/3}]^3 + 3 \quad \text{and} \quad (g \circ f)(x) = [(x^3 + 3) - 3]^{1/3}$$
$$= x - 3 + 3 \qquad\qquad\qquad = [x^3]^{1/3}$$
$$= x \qquad\qquad\qquad\qquad\quad = x$$

Thus, f and g are inverse functions, and $g = f^{-1}$.

This definition can be used to show that two functions are inverses, but it *cannot* be used to *find* the inverse of a given function. To find the inverse it is helpful to visualize a function as a set of ordered pairs. Suppose we pick a number, say 3, and evaluate a function f at 3 to find $f(3) = 15$. Then $(3, 15)$ is an element of f. Now the inverse function f^{-1} requires that 15 be changed back into 3; that is, $f^{-1}(15) = 3$ so that $(15, 3)$ is an element of f^{-1}. Thus,

If function f has element (x, y),   then inverse function f^{-1} must have element (y, x).

PROCEDURE FOR FINDING INVERSE FUNCTION USING ORDERED PAIRS

If the function f is the set of ordered pairs of the form (a, b), and if f is one-to-one, then f^{-1} is the **inverse function** of f and is the set of ordered pairs of the form (b, a). This means that if $y = f(x)$, then the inverse is found by replacing x's by y's and y's by x's.

EXAMPLE 3 Let $f = \{(0, 3), (1, 5), (3, 9), (5, 13)\}$. Find the inverse of f.

SOLUTION The inverse simply reverses the ordered pairs:

$$f^{-1} = \{(3, 0), (5, 1), (9, 3), (13, 5)\}$$

It should be evident from the definition and Example 3 that the domain and range of f^{-1} are simply the range and domain, respectively, of f. Given $y = f(x)$, which is a one-to-one function, the inverse f^{-1} is defined by $x = f(y)$. However, in order to rewrite $x = f(y)$ in the form $y = f^{-1}(x)$, you must solve for y. For example, given $y = 2x - 1$, its inverse is $x = 2y - 1$ (interchanging x and y) or $y = \frac{1}{2}x + \frac{1}{2}$ (solving for y). That is, $x = f(y)$ is equivalent to $y = f^{-1}(x)$. This is used to find the inverse of given functions, as illustrated in the following example.

EXAMPLE 4 Find f^{-1} if $f(x) = 2 - \frac{1}{3}x$.

SOLUTION Let $y = 2 - \frac{1}{3}x$.

Inverse:

$$x = 2 - \frac{1}{3}y \qquad \text{Interchange } x \text{ and } y.$$
$$\frac{1}{3}y = 2 - x \qquad \text{Then solve for } y.$$
$$y = 6 - 3x$$

or

$$f^{-1}(x) = 6 - 3x$$

Let us summarize the steps illustrated by Example 4.

PROCEDURE FOR FINDING AN EQUATION FOR AN INVERSE FUNCTION

1. Let $y = f(x)$ be a given *one-to-one* function.
2. Replace all x's by y's and all y's by x's (that is, interchange x and y in the given equation).
3. Solve for y. The resulting function defined by the equation $y = f^{-1}(x)$ is the inverse of f.

The domain of f and the range of f^{-1} must be equal as well as the domain of f^{-1} and the range of f.

EXAMPLE 5 Find g^{-1} if $g(x) = x^3 + 3$.

SOLUTION Let $y = x^3 + 3$.

$$x = y^3 + 3 \qquad \text{Interchange } x \text{ and } y.$$
$$y^3 = x - 3 \qquad \text{Solve for } y.$$
$$y = (x - 3)^{1/3}$$
$$g^{-1}(x) = (x - 3)^{1/3}$$

EXAMPLE 6 Find t^{-1} if t is defined by $t(x) = x^2$.

SOLUTION $t(x) = x^2$ is not a one-to-one function, so it has **no inverse function.**

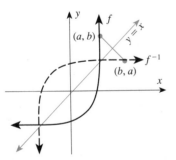

FIGURE 2.26

The Graphs of f and f^{-1}

The graphs of f and f^{-1} have an interesting relationship. Notice that the points (a, b) and (b, a) are symmetric with respect to the line $y = x$. (See Figure 2.26.) That is, the points are on opposite sides of and equidistant from the line. The points are called *reflections in the line*. Since for every ordered pair (a, b) in f, the ordered pair (b, a) is in f^{-1}, the graphs of $y = f(x)$ and $y = f^{-1}(x)$ are reflections in the line $y = x$.

EXAMPLE 7 Show that $y = 2 - \frac{1}{3}x$ and $y = 6 - 3x$ (inverses found in Example 4) are symmetric with respect to the line $y = x$ by graphing each.

SOLUTION The graphs are shown in the figure.

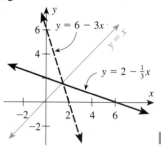

EXAMPLE 8 Given the graph of $y = f(x)$ shown here, sketch the graph of its inverse.

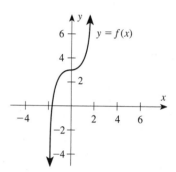

SOLUTION First, sketch the line $y = x$. Then sketch the reflection of $y = f(x)$ in the line to obtain the graph of $y = f^{-1}(x)$, as shown in the figure.

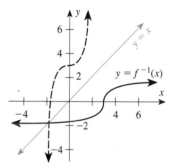

EXAMPLE 9 Consider the function f defined by the graph in Figure 2.27.
a. Find $f(5)$.   **b.** Find $f^{-1}(6)$.

SOLUTION **a.** Use the graph as shown in Figure 2.27:

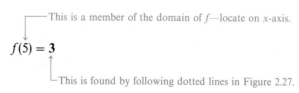

This is a member of the domain of f—locate on x-axis.

$f(5) = 3$

This is found by following dotted lines in Figure 2.27.

This is a member of the domain of f^{-1}—locate on y-axis.

b. $f^{-1}(6) = 9$

This is found by following dashed lines in Figure 2.27.

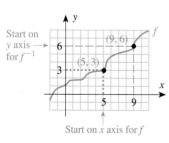

FIGURE 2.27 Graph of f

EXAMPLE 10 Let f be defined by $f(x) = x^2 - 1$ on the interval $[0, \infty)$. Find f^{-1}.

SOLUTION Do you see why some restriction on the domain is necessary? The graph of f is shown in the margin. The interval given in this example forces f to be increasing and therefore one-to-one. Without this restriction you have

$$f(3) = 9 - 1 \quad \text{and} \quad f(-3) = 9 - 1$$

which means that both $(3, 8)$ and $(-3, 8)$ belong to f—which shows that even though f is a function, it is not one-to-one. If x is a member of the domain $[0, \infty)$, however, then f is one-to-one and the inverse function is defined. Now let $y = f(x)$ so that

	DOMAIN	RANGE
$y = x^2 - 1$	$x \geq 0$	$y \geq -1$

Interchange x and y for the inverse:

$x = y^2 - 1 \qquad y \geq 0 \qquad\qquad x \geq -1$

This is now called the range because it shows the restriction on y.

This is now called the domain because it shows the restriction on x.

Solve for y:

$$y^2 = x + 1$$
$$y = \pm\sqrt{x + 1} \qquad \text{Reject } y = -\sqrt{x+1} \text{ since } y \text{ is nonnegative.}$$
$$y = \sqrt{x + 1} \qquad x \geq -1$$

Thus

$$f^{-1}(x) = \sqrt{x + 1} \quad \text{on } [-1, \infty)$$

2.5 Problem Set

A *Determine which pairs of functions defined by the equations in Problems 1–8 are inverses.*

1. $f(x) = 5x; \quad g(x) = \frac{1}{5}x$

2. $f(x) = -3x; \quad g(x) = \frac{1}{3}x$

3. $f(x) = 5x + 3; \quad g(x) = \dfrac{x - 3}{5}$

4. $f(x) = \frac{2}{3}x + 2; \quad g(x) = \frac{3}{2}x + 3$

5. $f(x) = \frac{4}{5}x + 4; \quad g(x) = \frac{5}{4}x + 3$

6. $f(x) = \dfrac{1}{x}, x \neq 0; \quad g(x) = \dfrac{1}{x}, x \neq 0$

7. $f(x) = x^2, x < 0; \quad g(x) = \sqrt{x}, x > 0$

8. $f(x) = x^2, x \geq 0; \quad g(x) = \sqrt{x}, x \geq 0$

Find the inverse function, if it exists, of each function given in Problems 9–24.

9. $f = \{(4, 5), (6, 3), (7, 1), (2, 4)\}$

10. $f = \{(1, 4), (6, 1), (4, 5), (3, 4)\}$

11. $f(x) = x + 3$ **12.** $f(x) = 2x + 3$

13. $g(x) = 5x$ **14.** $g(x) = \frac{1}{5}x$

15. $h(x) = x^2 - 5$ **16.** $h(x) = \sqrt{x} + 5$

17. $f(x) = x$

18. $f(x) = \dfrac{1}{x}$

19. $f(x) = 6$

20. $g(x) = -2$

21. $f(x) = \dfrac{1}{x-2}$

22. $f(x) = \dfrac{2x+1}{x}$

23. $f(x) = \dfrac{2x-6}{3x+3}$

24. $f(x) = \dfrac{3x+1}{2x-3}$

B If f is defined by the graph in Figure 2.28, find the values requested in Problems 25–31.

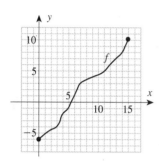

FIGURE 2.28 Graph of f

25. a. $f(3)$
 b. $f(4)$
 c. $f(7)$
 d. $f(11)$
 e. $f(14)$

26. a. $f(0)$
 b. $f^{-1}(4)$
 c. $f^{-1}(5)$
 d. $f(15)$
 e. $f^{-1}(9)$

27. a. $f(1)$
 b. $f^{-1}(-1)$
 c. $f^{-1}(-4)$
 d. $f(2)$
 e. $f^{-1}(2)$

28. a. $f^{-1}(-5)$
 b. $f(5)$
 c. $f^{-1}(3)$
 d. $f^{-1}(8)$
 e. $f(13)$

29. a. $f^{-1}(4)$
 b. $f(8)$
 c. $f^{-1}(-3)$
 d. $f(9)$
 e. $f^{-1}(7)$

30. a. What are the domain and range of the function f defined by the graph in Figure 2.28?
 b. What are the domain and range of the function f^{-1}?

31. Graph f^{-1} for the function f defined by the graph in Figure 2.28.

Determine which pairs of functions defined by the equations in Problems 32–37 are inverses.

32. $f(x) = 2x^2 + 1,\ x \ge 0;\quad g(x) = \frac{1}{2}\sqrt{2x-2},\ x \ge 1$

33. $f(x) = 2x^2 + 1,\ x \le 0;\quad g(x) = -\frac{1}{2}\sqrt{2x-2},\ x \ge 1$

34. $f(x) = (x+1)^2,\ x \ge 1;\quad g(x) = -1 - \sqrt{x},\ x \ge 0$

35. $f(x) = (x+1)^2,\ x \ge -1;\quad g(x) = -1 + \sqrt{x},\ x \ge 0$

36. $f(x) = |x+1|,\ x \ge -1;\quad g(x) = |x-1|,\ x \ge 0$

37. $f(x) = |x+1|,\ x \ge -1;\quad g(x) = x - 1,\ x \ge 0$

For each of Problems 38–51, graph the defined function and its inverse on the same coordinate axes.

38. $f(x) = x + 2;\quad f^{-1}(x) = x - 2$

39. $g(x) = x - 4;\quad g^{-1}(x) = x + 4$

40. $h(x) = 5x - 2;\quad h^{-1}(x) = \dfrac{x+2}{5}$

41. $j(x) = 7x + 3;\quad j^{-1}(x) = \dfrac{x-3}{7}$

42. $f(x) = x^2,\ \text{domain } x \ge 0$
  $f^{-1}(x) = \sqrt{x},\ \text{domain } x \ge 0$

43. $g(x) = x^2,\ \text{domain } x \le 0$
  $g^{-1}(x) = -\sqrt{x},\ \text{domain } x \ge 0$

44. $f(x) = x^2 - 2,\ \text{domain } x \ge 0$
  $f^{-1}(x) = \sqrt{x+2},\ \text{domain } x \ge -2$

45. $g(x) = |x-1|,\ \text{domain } x \ge 1$
  $g^{-1}(x) = |x+1|,\ \text{domain } x \ge 0$

46. $f(x) = \frac{1}{4}x - 2$

47. $g(x) = \frac{1}{3}x + 1$

48. $h(x) = \frac{1}{3}x - \frac{5}{3}$

49. $k(x) = -\frac{1}{4}x + \frac{3}{4}$

50. $m(x) = 4x - 2$

51. $n(x) = 3x + 6$

C Find the inverse of each function given in Problems 52–61.

52. $f(x) = x^2$ on $[0, \infty)$

53. $f(x) = x^2$ on $(-\infty, 0]$

54. $f(x) = x^2 + 1$ on $(-\infty, 0]$

55. $f(x) = x^2 + 1$ on $[0, \infty)$

56. $f(x) = 2x^2$ on $[2, 10]$

57. $f(x) = 2x^2$ on $[-10, -1]$

58. $f(x) = \dfrac{1}{3x+1}$ on $(-\frac{1}{3}, \infty)$

59. $f(x) = \dfrac{1}{2x-1}$ on $(\frac{1}{2}, \infty)$

60. $f(x) = x$ on $[0, \infty)$

61. $f(x) = x + 1$ on $[-1, \infty)$

Chapter 2 Summary

The material of this chapter is reviewed in the following list of objectives. After each objective there are some practice questions. For a sample test, select the first question of each set and check your answers with the answer section. For a sample test without answers, use the second question of each set. Additional practice is given by the other questions in each set. If you are having trouble with a particular type of problem, look back to that section for extra help.

2.1 Introduction to Functions

OBJECTIVE 1 *Be able to classify examples as one-to-one, functions, or not functions. Let X and Y be sets of real numbers such that $x \in X$ and $y \in Y$. (The symbol $\in$ means "is a member of.")*

1. $y = 3x + 3$
2. $y = 2x^2 + 3$
3. $y \leq 2x + 3$
4. The set of ordered pairs defined by the graph in Figure 2.29.

FIGURE 2.29
Graph for Problem 4

OBJECTIVE 2 *Distinguish between the f and $f(x)$ notation. Use functional notation. Let $f(x) = 3x - 1$ and $g(x) = 5 - x^2$. Find the requested values.*

5. a. $f(4)$ b. $g(4)$ 6. a. $f(-6)$ b. $g(-6)$
7. a. $f(w)$ b. $g(w + h)$ 8. a. $3f(t)$ b. $g(\sqrt{t})$

OBJECTIVE 3 *Find $\dfrac{f(x + h) - f(x)}{h}$ for a given function f.*

9. $f(x) = 5 - x^2$
10. $f(x) = 5x + 3$
11. $f(x) = 5$
12. $f(x) = 3x^2 - 1$

2.2 Graph of a Function

OBJECTIVE 4 *Find the domain and range for a curve whose equation is given.*

13. $x^2 - y + 5 = 0$
14. $y = \sqrt{10 - 2x}$
15. $y = \sqrt{\dfrac{x + 5}{1 - x}}$
16. $16x^2 - x^2y^2 + y^2 - 1 = 0$

OBJECTIVE 5 *Be able to find the x- and y-intercepts for a curve whose equation is given.*

17. $y = (x - 2)(3x + 5)(x + 2)$ 18. $y = \dfrac{5x^2 - 16x + 3}{x - 3}$
19. $2x^2 + xy - 4y^2 = 1$ 20. $\dfrac{(x - 1)^2}{4} + \dfrac{(y + 1)^2}{9} = 1$

OBJECTIVE 6 *Determine when functions are equal.*

21. $f(x) = \dfrac{x^2 + x}{x + 1}$;   22. $f(x) = \dfrac{(x + 3)(x - 2)}{x - 2}$, $x \neq 2$;
    $g(x) = x$ $g(x) = \dfrac{(x + 3)(x - 2)}{x + 3}$, $x \neq -3$

23. $f(x) = \sqrt{x^2}$;   24. $f(x) = \dfrac{(2x + 1)(3x - 2)}{3x - 2}$;
    $g(x) = x$ $g(x) = 2x + 1$, $x \neq 0$

OBJECTIVE 7 *Classify functions as even, odd, or neither.*

25. $f(x) = 3x^2 + 2x - 5$ 26. $f(x) = 2x^3 + 5x$
27. $f(x) = \dfrac{1}{(x + 4)^2}$ 28. $f(x) = 7$

2.3 Transformations of Functions

OBJECTIVE 8 *Find the shift (h, k) when given an equation $y - k = f(x - h)$.*

29. $y - 6 = f(x + \pi)$
30. $y - 1 = 3(x + 3)^2$
31. $y = 5(x - 4)^2$
32. $y = \frac{2}{3}|x - \sqrt{2}| + 5$

OBJECTIVE 9 *Given $y = f(x)$ and (h, k), write $y - k = f(x - h)$.*

33. $f(x) = 9x^2$; $(h, k) = (-\sqrt{2}, 3)$
34. $f(x) = 2|x|$; $(h, k) = (4, -3)$
35. $f(x) = -2x^2$; $(h, k) = (\pi, 0)$
36. $f(x) = 3x^2 + 2x + 5$; $(h, k) = (1, 2)$

OBJECTIVE 10 *Substitute $x' = x - h$ and $y' = y - k$ into an equation $y - k = f(x - h)$. Substitute $x' = x + 3$ and $y' = y - 1$ into the following equations.*

37. $y - 1 = 3(x + 3)^2$ **38.** $y - 1 = 2|x + 3|$

39. $y = 5(x + 3)^2 + 1$

40. $y - 1 = (x + 3)^2 + (x + 3) - 5$

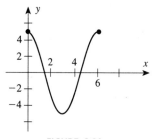

FIGURE 2.30

Graph for Problems 41–48

OBJECTIVE 11 *Given a function defined by $y = f(x)$ and shown by the graph in Figure 2.30, draw the graph of $y - k = f(x - h)$.*

41. $y - 2 = f(x - 3)$

42. $y + 1 = f(x - 3)$

43. $y = f(x - \pi)$

44. $y + \sqrt{2} = f(x - \sqrt{3})$

OBJECTIVE 12 *Given a function defined by $y = f(x)$ and shown in Figure 2.30, draw the graph's reflections, contractions, and dilations.*

45. $y = -f(x)$ **46.** $y = f(-x)$

47. $y = \frac{1}{2}f(x)$ **48.** $y = f(2x)$

2.4   Operations and Composition of Functions

OBJECTIVE 13 *Find the sum, difference, product, and quotient functions. Let $f(x) = x^2$ and $g(x) = 5x - 2$.*

49. $(f + g)(x)$

50. $(f - g)(x)$

51. $(fg)(x)$

52. $(f/g)(2)$

55.

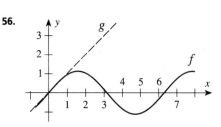

OBJECTIVE 14 *In Problems 53–56 find $f + g$ by adding the ordinates.*

53. If $f(x) = \frac{1}{2}x^2$ and $g(x) = x + 1$

54. If $f(x) = \frac{1}{2}x^2$ and $g(x) = -\frac{1}{4}x^2$

56.

2.5   Inverse Functions

OBJECTIVE 15 *Find the composition of functions. Let $f(x) = 3x - 1$, $g(x) = 5 - x^2$, and $h(x) = 6$. Find the requested values.*

57. $(f \circ g)(x)$ **58.** $(g \circ f)(x)$

59. $(f \circ h)(x)$ **60.** $(h \circ f)(x)$

63. $f(x) = 3x - 2$; $g(x) = \dfrac{x - 2}{3}$

64. $f(x) = 2x - 3$; $g(x) = \dfrac{x + 2}{3}$

OBJECTIVE 16 *Given two functions, decide whether they are inverses. Decide whether f and g are inverses.*

61. $f(x) = 3x + 6$; $g(x) = x - 2$

62. $f(x) = 2x + 1$; $g(x) = x - 2$

OBJECTIVE 17 *Given a one-to-one function, find its inverse.*

65. $f(x) = 3x - 1$ **66.** $g(x) = 5 - x^2$ on $[0, \infty)$

67. $f(x) = \frac{1}{2}x + 5$ **68.** $g(x) = \sqrt{2x}$ on $[0, \infty)$

Karl Gauss (1777–1885)

Mathematics is the queen of the sciences and arithmetic the queen of mathematics. She often condescends to render service to astronomy and other natural sciences, but in all relations she is entitled to the first rank.

KARL GAUSS

Gauss is sometimes described as the last mathematician to know everything in his subject. Such a generalization is bound to be inexact, but it does emphasize the breadth of interests Gauss displayed.

CARL B. BOYER in
A History of Mathematics,
pp. 561–562

Karl Gauss is considered to be one of the three greatest mathematicians of all time, along with Archimedes and Newton. Gauss had a great admiration for Archimedes, but could not understand how Archimedes failed to invent the positional numeration system. Howard Eves quotes Gauss as saying, "To what heights would science now be raised if Archimedes had made that discovery!" Because of the greatness of Gauss, many stories and anecdotes about him have survived. In all the history of mathematics there is nothing approaching the precocity of Gauss as a child. He showed his caliber before he was 3 years old! In later life, Gauss joked that he knew how to reckon before he could talk. It seems that when he was only 3 he corrected his father's computations on a payroll report.

Gauss graduated from college at the age of 15 and proved what was to become the Fundamental Theorem of Algebra for his doctoral thesis at the age of 22. He published only a small portion of the ideas that seemed to storm his mind because he believed that each published result had to be complete, concise, polished, and convincing. His motto was "Few, but ripe."

In this chapter you begin your study of one of the most important tools in mathematics—namely, the ability to draw a graph corresponding to a given equation. There is a quotation of Gauss in which he described the joy of learning. This quotation is from a letter dated September 2, 1808, to his friend Wolfgang Bolyai:

"*It is not knowledge, but the act of learning, not possession, but the act of getting there, which grants the greatest enjoyment. When I have clarified and exhausted a subject, then I turn away from it, in order to go into darkness again; the never-satisfied man is so strange—if he has completed a structure, then it is not in order to dwell in it peacefully, but in order to begin another. I imagine the world conqueror must feel thus, who, after one kingdom is scarcely conquered, stretches out his arms for another.*"

POLYNOMIAL FUNCTIONS

▌Contents

▌Preview

Equation solving is one of the most important topics of elementary algebra. In high school you learned about first- and second-degree equations (we reviewed these in Chapter 1). In this chapter we consider the solution of polynomial equations of degree higher than two. There are 21 objectives in this chapter, which are listed on pages 172–175.

▌Perspective

In Chapter 2 we considered general properties of functions, and in this chapter we consider graphing polynomial functions. However, you will have to wait until you take a course in calculus to be able to find the turning points and maximum and minimum values for a graph. In order to find these values in calculus you will need to analyze equations such as

$$y = x^3 - 3x^2 + 1$$

as shown on the page below from a leading calculus book. Notice that the author assumes that you know how to graph this curve (no discussion). You will learn how to do this in Section 3.5.

Solution Since f is continuous on $[-\frac{1}{2}, 4]$, we can use the procedure outlined in (3.8):

$$f(x) = x^3 - 3x^2 + 1$$
$$f'(x) = 3x^2 - 6x = 3x(x - 2)$$

Since $f'(x)$ exists for all x, the only critical numbers of f occur when $f'(x) = 0$, that is, $x = 0$ or $x = 2$. Notice that each of these critical numbers lies in the interval $[-\frac{1}{2}, 4]$. The values of f at these critical numbers are

$$f(0) = 1 \qquad f(2) = -3$$

The values of f at the endpoints of the interval are

$$f\left(-\frac{1}{2}\right) = \frac{1}{8} \qquad f(4) = 17$$

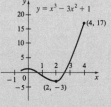

Figure 3.12

Comparing these four numbers, we see that the absolute maximum value is $f(4) = 17$ and the absolute minimum value is $f(2) = -3$.

Note that in this example the absolute maximum occurs at an endpoint, whereas the absolute minimum occurs at a critical number. The graph of f is sketched in Figure 3.12.

From James Stewart, *Calculus*, 2nd ed. (Pacific Grove, Ca.: Brooks/Cole), p. 184.

3.1 Polynomial Functions and Synthetic Division

Polynomial Functions

We defined a polynomial, along with some of the associated terminology, in Chapter 1. In this section we define a polynomial function. It is assumed that you know how to add, subtract, and multiply polynomials (these topics were reviewed in Chapter 1). In this section we review the process of dividing polynomials with a process called *synthetic division*. In the remainder of this chapter we will consider first-degree, second-degree, third-degree, and then higher-degree polynomial functions. We begin with a definition.

POLYNOMIAL FUNCTION

A function P is a **polynomial function** in x if

$$P(x) = a_n x^n + a_{n-1} x^{n-1} + a_{n-2} x^{n-2} + \cdots + a_1 x + a_0$$

where n is an integer greater than or equal to zero and the coefficients $a_0, a_1, a_2, \ldots, a_{n-1}, a_n$ are real numbers.

If $n = 0$ (degree 0), then $P(x)$ is a *constant function*; if $n = 1$ (degree 1), then $P(x)$ is a *linear function*; if $n = 2$, then $P(x)$ is a *quadratic function*; if $n = 3$, then $P(x)$ is a *cubic function*; and if $n = 4$, then $P(x)$ is called a *quartic function*. If all the coefficients of a polynomial are 0, it is called the **zero polynomial** and is denoted by 0. The zero polynomial is not assigned a degree. The domain for all polynomial functions is the set of all real numbers.

Synthetic Division

Consider positive integers P and D. If the result of P divided by D is an integer Q, then D is a factor of P. For example, if $P = 15$ and $D = 3$, then $P/D = 5$, and 5 is called the *quotient*. If D is not a factor of P, we will obtain a quotient Q and a remainder R that is less than D so that

$$\frac{P}{D} = Q + \frac{R}{D}$$

For example:

If $\dfrac{15}{3} = 5$, then $15 = 5 \cdot 3$.

If $\dfrac{17}{3} = 5 + \dfrac{2}{3}$, then $17 = 5 \cdot 3 + 2$.

Notice how division may be checked by multiplying:

If $\dfrac{P}{D} = Q$, then $P = QD$.

If $\dfrac{P}{D} = Q + \dfrac{R}{D}$, then $P = QD + R \quad (R < D)$.

Division of polynomials is similar and leads to a result called the **Division Algorithm.**

DIVISION ALGORITHM

If $P(x)$ and $D(x)$ are polynomials $[D(x) \neq 0]$, then there exist unique polynomials $Q(x)$ and $R(x)$ such that

$$\frac{P(x)}{D(x)} = Q(x) + \frac{R(x)}{D(x)}$$

where $Q(x)$ is a unique polynomial and $R(x)$ is a polynomial such that the degree of $R(x)$ is less than the degree of $D(x)$. The polynomial $Q(x)$ is called the **quotient** and $R(x)$ the **remainder.**

1. If $R(x) = 0$, then $D(x)$ is a factor of $P(x)$.
2. If the degree of $D(x)$ is greater than the degree of $P(x)$, then $Q(x) = 0$.
3. If both sides are multiplied by $D(x)$, the Division Algorithm is then stated in product form:

$$P(x) = Q(x)D(x) + R(x)$$

The question to be considered is *how to find* $Q(x)$ and $R(x)$. The first method is by long division. Let $P(x) = 3x^4 + 7x^3 + 2x^2 + 3x + 5$ and $D(x) = x + 1$. Then:

Multiply $3x^3(x + 1) = 3x^4 + 3x^3$ and write the answer so that similar terms are aligned.

$$x + 1 \overline{)3x^4 + 7x^3 + 2x^2 + 3x + 5}$$

$$\updownarrow$$

$$3x^4 + 3x^3$$

$3x^3$ was picked so that the terms shown by the arrow are identical.

$$\begin{array}{r} 3x^3 + 4x^2 \qquad\qquad \\ x + 1 \overline{)3x^4 + 7x^3 + 2x^2 + 3x + 5} \\ 3x^4 + 3x^3 \qquad\qquad\qquad\qquad \\ \hline 4x^3 + 2x^2 + 3x + 5 \\ 4x^3 + 4x^2 \qquad\qquad \end{array}$$

Multiply $4x^2(x + 1)$ and align similar terms.

Next subtract (or add the opposite).

This must be zero since the terms were identical.

$4x^2$ was picked so that these terms are identical.

Now subtract and repeat the procedure:

$$
\begin{array}{r}
3x^3 + 4x^2 - 2x + 5 \\
x + 1\overline{)3x^4 + 7x^3 + 2x^2 + 3x + 5} \\
\underline{3x^4 + 3x^3} \\
4x^3 + 2x^2 + 3x + 5 \\
\underline{4x^3 + 4x^2} \\
\end{array}
$$

Do not forget to subtract: $2x^2 - 4x^2 = -2x^2$

$$
\begin{array}{r}
-2x^2 + 3x + 5 \\
\underline{-2x^2 - 2x} \\
5x + 5
\end{array}
$$

Subtract

$3x - (-2x) = 5x$

$$
\begin{array}{r}
5x + 5 \\
\underline{} \\
0 \leftarrow 5 - 5 = 0
\end{array}
$$

The remainder is zero, so $x + 1$ is a factor of $3x^4 + 7x^3 + 2x^2 + 3x + 5$. You can check this by verifying that $P(x) = Q(x)D(x)$:

$$
\begin{aligned}
Q(x)D(x) &= (3x^3 + 4x^2 - 2x + 5)(x + 1) \\
&= 3x^4 + 4x^3 - 2x^2 + 5x + 3x^3 + 4x^2 - 2x + 5 \\
&= 3x^4 + 7x^3 + 2x^2 + 3x + 5
\end{aligned}
$$

Since this is $P(x)$, the result checks.

EXAMPLE 1 Let $P(x) = 4x^4 - 6x^2 - 10x + 3$ and $D(x) = x^2 + x - 2$. Find $P(x)/D(x)$.

SOLUTION

$$
\begin{array}{r}
4x^2 - 4x + 6 \\
x^2 + x - 2\overline{)4x^4 + 0x^3 - 6x^2 - 10x + 3} \\
\underline{4x^4 + 4x^3 - 8x^2} \\
-4x^3 + 2x^2 - 10x + 3 \\
\underline{-4x^3 - 4x^2 + 8x} \\
6x^2 - 18x + 3 \\
\underline{6x^2 + 6x - 12} \\
-24x + 15
\end{array}
$$

Notice that there is no x^3 term. Leave a space for this "missing" term, or write $0x^3$.

The remainder is $-24x + 15$. Thus

$$
\frac{P(x)}{D(x)} = 4x^2 - 4x + 6 + \frac{-24x + 15}{x^2 + x - 2}
$$

The process of long division is indeed *long* because of the duplication of symbols when carrying out this process. Consider the following example for dividing by a polynomial of the form $x - r$:

$$
\begin{array}{r}
2x^3 - 4x^2 - x + 3 \\
x - 1)\overline{2x^4 - 6x^3 + 3x^2 + 4x - 3} \\
\underline{2x^4 - 2x^3} \\
-4x^3 + 3x^2 + 4x + 3 \\
\underline{-4x^3 + 4x^2} \\
-x^2 + 4x - 3 \\
\underline{-x^2 + x} \\
3x - 3 \\
\underline{3x - 3} \\
0
\end{array}
$$

This is first degree, so the quotient is 1 degree less than the given polynomial.

Notice that each term in color is a repetition of the term directly above, so we could eliminate writing it down.

Also, the position of the term indicates the degree, so it is not even necessary to write down the variable:

The degree of the first term is 1 less than the degree of the given polynomial.

$$
\begin{array}{r}
2 - 4 - 1 + 3 \\
-1)\overline{2 - 6 + 3 + 4 - 3} \\
\underline{-2} \\
-4 \\
\underline{+4} \\
-1 \\
\underline{+1} \\
3 \\
\underline{-3} \\
0
\end{array}
$$

Interpret this answer by recognizing that it is of degree 1 less than the degree of the dividend. This means the answer is $2x^3 - 4x^2 - x + 3$. That is, these numbers are the coefficients of the quotient polynomial.

$\leftarrow$ This is the remainder.

There is no reason to spread out the array; it can be compressed into a more efficient form:

$$
\begin{array}{r}
②\quad -4 \quad -1 \quad +3 \\
-1)\,2 \quad -6 \quad +3 \quad +4 \quad -3 \\
\underline{-2 \quad +4 \quad +1 \quad -3} \\
-4 \quad -1 \quad +3 \quad 0
\end{array}
$$

$\leftarrow$ Delete these terms.

This line is no longer needed.

This is the remainder.

The top line is the same as the bottom line if you bring down the leading coefficient:

$$
\begin{array}{r}
-1|\quad 2 \quad -6 \quad +3 \quad +4 \quad -3 \\
\underline{-2 \quad +4 \quad +1 \quad -3} \\
2 \quad -4 \quad -1 \quad +3 \quad 0
\end{array}
$$

These are the coefficients of the quotient (which begins with a degree 1 less than that of the given polynomial).

This is the remainder.

The process is now fairly compact, but the most common error in this procedure is in the subtraction. *Remember*: It is usually easier to add the opposite than to subtract, so change the sign of the divisor (-1 to 1 in this example) and add:

$$
\begin{array}{r|rrrrr}
1] & 2 & -6 & 3 & 4 & -3 \\
& & 2 & -4 & -1 & 3 \\
\hline
& 2 & -4 & -1 & 3 & 0
\end{array}
$$

Change the sign of this number, and **add.**

The condensed form of the division of a polynomial by $x - r$ is called **synthetic division. Notice that synthetic division is used only when the divisor is of the form $x - r$ (r positive or negative). If the divisor is not linear, then long division must be used.**

EXAMPLE 2   Divide $x^4 + 3x^3 - 12x^2 + 5x - 2$ by $x - 2$.

SOLUTION

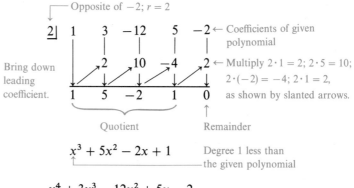

Opposite of -2; $r = 2$

$\leftarrow$ Coefficients of given polynomial

$\leftarrow$ Multiply $2 \cdot 1 = 2$; $2 \cdot 5 = 10$; $2 \cdot (-2) = -4$; $2 \cdot 1 = 2$, as shown by slanted arrows.

Quotient   Remainder

$x^3 + 5x^2 - 2x + 1$ Degree 1 less than the given polynomial

$$\frac{x^4 + 3x^3 - 12x^2 + 5x - 2}{x - 2} = \mathbf{x^3 + 5x^2 - 2x + 1}$$

EXAMPLE 3   Divide $x^4 - 2x^3 - 10x^2 + 26x + 3$ by $x + 3$.

SOLUTION

Opposite of $+3 \longrightarrow$

$$
\begin{array}{r|rrrrr}
-3] & 1 & -2 & -10 & 26 & 3 \\
& & -3 & 15 & -15 & -33 \\
\hline
& 1 & -5 & 5 & 11 & -30
\end{array}
$$

$\leftarrow$ Coefficients

Remainder

The answer is $x^3 - 5x^2 + 5x + 11 + \dfrac{-30}{x + 3}$

EXAMPLE 4   Divide $x^5 - 3x^2 + 1$ by $x - 2$.

SOLUTION

$$
\begin{array}{r|rrrrrr}
2] & 1 & 0 & 0 & -3 & 0 & 1 \\
& & 2 & 4 & 8 & 10 & 20 \\
\hline
& 1 & 2 & 4 & 5 & 10 & 21
\end{array}
$$

$\leftarrow$ Be sure to include zero coefficients for missing terms.

$\leftarrow (R)$

The answer is $x^4 + 2x^3 + 4x^2 + 5x + 10 + \dfrac{21}{x - 2}$

3.1 Problem Set

A *Fill in the blanks for the synthetic divisions in Problems 1–10.*

1. $\dfrac{3x^3 - 9x^2 + 11x - 10}{x - 2}$

$$\begin{array}{r|rrrr} 2] & 3 & \underline{\textbf{b}} & 11 & -10 \\ & & 6 & \underline{\textbf{c}} & 10 \\ \hline & \underline{\textbf{a}} & -3 & 5 & 0 \end{array}$$

The answer is **d.**

2. $\dfrac{4x^3 - 6x^2 - 8x - 5}{x - 3}$

$$\begin{array}{r|rrrr} \underline{\textbf{a}}] & 4 & -6 & \underline{\textbf{c}} & \underline{\textbf{e}} \\ & & \underline{\textbf{b}} & \underline{\textbf{d}} & \underline{\textbf{f}} \\ \hline & 4 & 6 & 10 & 25 \end{array}$$

The answer is **g.**

3. $\dfrac{2x^3 - 4x^2 - 10x + 12}{x + 2}$

$$\begin{array}{r|rrrr} \underline{\textbf{a}}] & 2 & -4 & \underline{\textbf{c}} & \underline{\textbf{e}} \\ & & \underline{\textbf{b}} & \underline{\textbf{d}} & \underline{\textbf{f}} \\ \hline & 2 & -8 & 6 & 0 \end{array}$$

The answer is **g.**

4. $\dfrac{5x^3 + 12x^2 + 2x - 4}{x + 3}$

$$\begin{array}{r|rrrr} \underline{\textbf{a}}] & 5 & 12 & \underline{\textbf{d}} & \underline{\textbf{f}} \\ & & -15 & 9 & -33 \\ \hline & \underline{\textbf{b}} & \underline{\textbf{c}} & \underline{\textbf{e}} & \underline{\textbf{g}} \end{array}$$

The answer is **h.**

5. $\dfrac{x^3 + 7x^2 + 8x - 20}{x + 4}$

$$\begin{array}{r|rrrr} \underline{\textbf{a}}] & 1 & 7 & \underline{\textbf{d}} & \underline{\textbf{f}} \\ & & -4 & -12 & 16 \\ \hline & \underline{\textbf{b}} & \underline{\textbf{c}} & \underline{\textbf{e}} & \underline{\textbf{g}} \end{array}$$

The answer is **h.**

6. $\dfrac{x^4 - 3x^2 + 2x - 7}{x - 1}$

$$\begin{array}{r|rrrrr} \underline{\textbf{a}}] & \underline{\textbf{b}} & \underline{\textbf{c}} & \underline{\textbf{d}} & \underline{\textbf{e}} & \underline{\textbf{g}} \\ & & 1 & 1 & -2 & 0 \\ \hline & 1 & 1 & -2 & \underline{\textbf{f}} & \underline{\textbf{h}} \end{array}$$

The answer is **i.**

7. $\dfrac{x^4 + 2x^3 - 5x + 2}{x - 1}$

$$\begin{array}{r|rrrrr} \underline{\textbf{a}}] & \underline{\textbf{b}} & \underline{\textbf{c}} & \underline{\textbf{d}} & \underline{\textbf{e}} & \underline{\textbf{g}} \\ & & 1 & 3 & 3 & -2 \\ \hline & 1 & 3 & 3 & \underline{\textbf{f}} & \underline{\textbf{h}} \end{array}$$

The answer is **i.**

8. $\dfrac{x^5 - 1}{x - 1}$

$$\begin{array}{r|rrrrrr} \underline{\textbf{a}}] & \underline{\textbf{b}} & \underline{\textbf{c}} & \underline{\textbf{d}} & \underline{\textbf{e}} & \underline{\textbf{f}} & \underline{\textbf{h}} \\ & & 1 & 1 & 1 & 1 & 1 \\ \hline & 1 & 1 & 1 & 1 & \underline{\textbf{g}} & \underline{\textbf{i}} \end{array}$$

The answer is **j.**

9. $\dfrac{x^5 - 32}{x + 2}$

$$\begin{array}{r|rrrrrr} \underline{\textbf{a}}] & \underline{\textbf{b}} & \underline{\textbf{c}} & \underline{\textbf{d}} & \underline{\textbf{e}} & \underline{\textbf{g}} & \underline{\textbf{j}} \\ & & -2 & 4 & -8 & \underline{\textbf{h}} & \underline{\textbf{k}} \\ \hline & 1 & -2 & 4 & \underline{\textbf{f}} & \underline{\textbf{i}} & \underline{\textbf{l}} \end{array}$$

The answer is **m.**

10. $\dfrac{x^4 + 3x^3 - 2x^2}{x + 3}$

$$\begin{array}{r|rrrrr} \underline{\textbf{a}}] & \underline{\textbf{b}} & \underline{\textbf{c}} & \underline{\textbf{e}} & \underline{\textbf{h}} & \underline{\textbf{k}} \\ & & -3 & \underline{\textbf{f}} & \underline{\textbf{i}} & \underline{\textbf{l}} \\ \hline & 1 & \underline{\textbf{d}} & \underline{\textbf{g}} & \underline{\textbf{j}} & \underline{\textbf{m}} \end{array}$$

The answer is **n.**

Use synthetic division to find the quotients in Problems 11–30.

11. $\dfrac{3x^3 - 2x^2 + 4x - 75}{x - 3}$

12. $\dfrac{3x^3 + 2x^2 - 4x + 8}{x + 2}$

13. $\dfrac{x^4 - 6x^3 + x^2 - 8}{x + 1}$

14. $\dfrac{x^4 - 3x^3 + x + 6}{x - 2}$

15. $\dfrac{2x^4 - 15x^2 + 8x - 3}{x + 3}$

16. $\dfrac{2x^4 - 15x - 2}{x - 2}$

17. $\dfrac{4x^5 - 3x^4 - 5x^3 + 4}{x - 1}$

18. $\dfrac{4x^3 - 3x^2 - 5x + 2}{x + 1}$

19. $\dfrac{3x^3 - 2x^2 + 4x - 24}{x - 2}$

20. $\dfrac{3x^3 + 2x^2 + 4x + 24}{x + 2}$

21. $\dfrac{x^3 + 5x^2 - 2x - 24}{x + 4}$

22. $\dfrac{x^3 + 3x^2 - 6x - 8}{x - 2}$

23. $\dfrac{x^3 - 4x^2 - 17x + 60}{x - 5}$

24. $\dfrac{x^5 - 3x^4 + x - 3}{x - 3}$

25. $\dfrac{4x^4 + 4x^3 - 15x^2 + 7}{x - 1}$

26. $\dfrac{5x^3 - 21x^2 - 13x - 35}{x - 5}$

27. $\dfrac{x^4 - x^3 + x^2 - x - 4}{x + 1}$

28. $\dfrac{2x^4 + 6x^3 - 4x^2 - 11x + 3}{x + 3}$

29. $\dfrac{3x^4 + 10x^3 - 8x^2 - 5x - 20}{x + 4}$

30. $\dfrac{x^4 - 9x^3 + 20x^2 - 15x + 18}{x - 6}$

B *In Problems 31–36 use long division to find Q(x) and R(x) if P(x) is divided by D(x).*

31. $P(x) = 4x^4 - 14x^3 + 10x^2 - 6x + 2; \quad D(x) = 2x - 1$

32. $P(x) = x^5 - x^3 + x^2 + x - 1; \quad D(x) = x^2 + x$

33. $P(x) = x^5 - x^3 + x^2 + 1; \quad D(x) = x^2 + x$

34. $P(x) = x^2 + 2x + 1; \quad D(x) = x^3$

35. $P(x) = 6x^4 - 11x^3 + 6x^2 - 2x + 5;$
$D(x) = 2x^2 + x + 1$

36. $P(x) = 2x^4 + 5x^3 - 16x^2 - 45x - 18$;
$D(x) = x^2 + x - 2$

In Problems 37–56 use synthetic division to find Q(x).

37. $\dfrac{3x^3 - 7x^2 - 5x + 2}{x - 3}$

38. $\dfrac{2x^3 - 3x^2 + 4x - 10}{x - 2}$

39. $\dfrac{x^4 - 3x^3 - 4x^2 + 2x - 5}{x - 4}$

40. $\dfrac{x^4 - 5x^3 + 2x^2 - x + 3}{x - 2}$

41. $\dfrac{5x^4 + 10x^3 - 20x^2 - 12x - 2}{x + 3}$

42. $\dfrac{2x^4 + 5x^3 + 2x^2 + 5x + 2}{x + 2}$

43. $\dfrac{x^5 - 3x^4 + 2x^2 - 5}{x + 2}$

44. $\dfrac{x^4 - 20x^2 - 10x - 50}{x - 5}$

45. $\dfrac{4x^3 - x^2 + 2x - 1}{x + 1}$

46. $\dfrac{6x^4 - x^2 + 1}{x - 3}$

47. $\dfrac{5x^5 - 2x + 1}{x + 2}$

48. $\dfrac{5x^4 + 7x^3 - 27x^2 + 14x - 12}{x - 1}$

49. $\dfrac{x^5 + 3x^3 - 3x^4 - 16x^2 + 21x - 6}{x - 3}$

50. $\dfrac{5x^4 - x + x^5 - 1}{x + 5}$

51. $\dfrac{x^4 - 12x^2 + 4x + 15}{(x + 1)(x - 3)}$

52. $\dfrac{x^4 - x^3 - 12x^2 + 28x - 16}{(x - 2)(x + 4)}$

53. $\dfrac{2x^3 - 3x^2 - 11x + 6}{(x - 3)(x + 2)}$

54. $\dfrac{2x^3 - 3x^2 - 2x + 3}{(2x - 3)(x + 1)}$

55. $\dfrac{2x^4 + 5x^3 - 25x^2 - 40x + 48}{(x - 3)(x + 4)}$

56. $\dfrac{2x^4 + 5x^3 - 16x^2 - 49x - 30}{(x + 1)(x + 3)}$

C

57. Devise a procedure for using synthetic division on

$$\frac{x^4 + 7x^3 + 5x^2 - 23x + 10}{x^2 + 4x - 5}$$

58. Devise a procedure for using synthetic division on

$$\frac{2x^4 - 7x^3 - 4x^2 + 27x - 18}{x^2 - x - 6}$$

59. Find K so that $x^4 + Kx^3 + 7x^2 - 2x + 8$ has no remainder when divided by $x + 2$.

60. Find h and k so that $x^4 + hx^3 - kx + 15$ has no remainder when divided by $x - 1$ and $x + 3$.

3.2 Linear Functions

Before graphing polynomial functions, we need a property that will allow us to connect known points on the curve. The ordered pairs that satisfy the equation defining a polynomial function lie on the curve representing that polynomial function, and these ordered pairs thus indicate points connected by a smooth curve. This property, called **continuity**, is studied extensively in calculus, but for our purposes we will simply state the following useful theorem (its proof requires calculus).

INTERMEDIATE-VALUE THEOREM FOR POLYNOMIAL FUNCTIONS

If f is a polynomial function such that $a \leq x \leq b$ and $f(a) \neq f(b)$, then $f(x)$ takes on every value between $f(a)$ and $f(b)$ over the interval $a \leq x \leq b$.

EXAMPLE 1 If f is a polynomial function such that $a \le x \le b$ and $f(a) \ne f(b)$, and if $f(a) > 0$, $f(b) < 0$, then there is some number c on (a, b) so that $f(c) = 0$.

SOLUTION Draw a graph of a polynomial function with $f(a) > 0$ and $f(b) < 0$. The Intermediate-Value Theorem says that $f(x)$ takes on every value between $f(a)$ and $f(b)$, so in particular must take on the value 0 since 0 is between a positive number and a negative number.

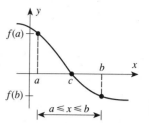

The first type of polynomial function we will consider is one with which you have had some experience in beginning algebra; it is the case in which $n = 1$.

LINEAR FUNCTION

A function f is a **linear function** if

$$f(x) = mx + b$$

where m and b are real numbers.

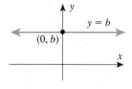

FIGURE 3.1 A horizontal line

Notice that if $m = 0$, then $f(x) = b$, which we called a constant function in Section 3.1. If the domain of a constant function is the set of real numbers, then the graph of $f(x) = b$ is a **horizontal line**, as shown in Figure 3.1.

Let $P_1(x_1, y_1)$ and $P_2(x_2, y_2)$ be any points on a line, and suppose that $x_1 = x_2$. Then the line is parallel to the y-axis and is called a **vertical line** (Figure 3.2). Notice that vertical lines are not functions.

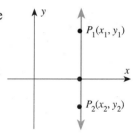

FIGURE 3.2 A vertical line is not a function.

The steepness of a line is specified by using the idea of slope. If $x_1 \ne x_2$, then the slope is defined as follows.

SLOPE OF A LINE

Let $P_1(x_1, y_1)$ and $P_2(x_2, y_2)$ be distinct points on a line such that $x_1 \ne x_2$. Then

$$\textbf{Slope} = \frac{\text{Vertical change}}{\text{Horizontal change}} = \frac{y_2 - y_1}{x_2 - x_1}$$

The numerator $y_2 - y_1$ is often called the **rise** and the denominator $x_2 - x_1$ the **run** from P_1 to P_2. If you use functional notation for the points $P_1(x_1, f(x_1))$ and $P_2(x_2, f(x_2))$, then the slope is found by

$$\text{Slope} = \frac{f(x_2) - f(x_1)}{x_2 - x_1}$$

EXAMPLE 2 Sketch the line passing through the points whose coordinates are given. Then find the slope of each line.

a. $(2, -3)$ and $(-1, 2)$ **b.** $(-4, -1)$ and $(1, 3)$
c. $(-3, 4)$ and $(5, 4)$ **d.** $(-3, 2)$ and $(-3, 4)$
e. $(3, f(3))$ and $(3 + h, f(3 + h))$ **f.** $(a, f(a))$ and $(a + h, f(a + h))$

SOLUTION **a.** $m = \dfrac{2 - (-3)}{-1 - 2} = \dfrac{5}{-3} = -\dfrac{5}{3}$ **b.** $m = \dfrac{3 - (-1)}{1 - (-4)} = \dfrac{4}{5}$

negative slope of $-\frac{5}{3}$ **positive slope of** $\frac{4}{5}$

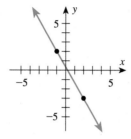 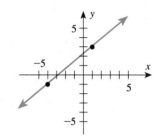

c. $m = \dfrac{4 - 4}{5 + 3} = 0$ **d.** $m = \dfrac{4 - 2}{-3 + 3}$ is undefined

0 slope; horizontal line **undefined slope; vertical line**

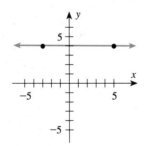 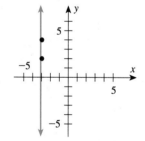

e. $m = \dfrac{f(3 + h) - f(3)}{3 + h - 3} = \dfrac{f(3 + h) - f(3)}{h}$ **f.** $m = \dfrac{f(a + h) - f(a)}{h}$

arbitrary slope of $\dfrac{f(3 + h) - f(3)}{h}$ **arbitrary slope of** $\dfrac{f(a + h) - f(a)}{h}$

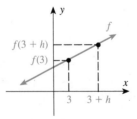 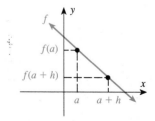

Two properties relating lines and slopes can be proved geometrically.

Let L_1 and L_2 be two nonvertical lines with slopes m_1 and m_2, respectively. Then:

1. L_1 and L_2 are parallel if and only if $m_1 = m_2$.
2. L_1 and L_2 are perpendicular if and only if $m_1 m_2 = -1$.

EXAMPLE 3   Show that the points $Q(-3, 7)$, $U(8, 2)$, $A(4, -3)$, and $D(-7, 2)$ are the corners of a parallelogram $QUAD$.

SOLUTION

$$m_{QU} = \frac{2-7}{8+3} = \frac{-5}{11}$$

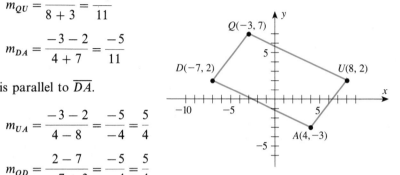

$$m_{DA} = \frac{-3-2}{4+7} = \frac{-5}{11}$$

Thus $\overline{QU}$ is parallel to $\overline{DA}$.

$$m_{UA} = \frac{-3-2}{4-8} = \frac{-5}{-4} = \frac{5}{4}$$

$$m_{QD} = \frac{2-7}{-7+3} = \frac{-5}{-4} = \frac{5}{4}$$

Thus $\overline{UA}$ is parallel to $\overline{QD}$. Since opposite sides of quadrilateral $QUAD$ are parallel, it is a parallelogram. Notice that even though it is helpful to draw the graph, you cannot use the figure to prove your argument.   ▮

EXAMPLE 4   Are the diagonals of $QUAD$ in Example 3 perpendicular?

SOLUTION

$$m_{QA} = \frac{-3-7}{4+3} = \frac{-10}{7}$$

$$m_{DU} = \frac{2-2}{8+7} = 0$$

Since $m_{QA} m_{DU} \neq -1$, the segments $\overline{QA}$ and $\overline{DU}$ are **not perpendicular.**   ▮

Consider the linear function $f(x) = mx + b$. Then

$$\text{Slope} = \frac{\text{Rise}}{\text{Run}} = \frac{f(x_2) - f(x_1)}{x_2 - x_1}$$

$$= \frac{(mx_2 + b) - (mx_1 + b)}{x_2 - x_1}$$

$$= \frac{m(x_2 - x_1)}{x_2 - x_1} = m$$

Thus the slope of the graph of a linear function is m.

The y-intercept for the linear function is found when $x = 0$:

$$f(0) = m(0) + b = b$$

Thus the y-intercept of a linear function is the point $(0, b)$. Since we know that the first component of a coordinate on the y-axis (a y-intercept) must be zero, we sometimes simplify the notation by simply saying the y-intercept of the line is b.

SLOPE–INTERCEPT FORM OF THE EQUATION OF A LINE

The graph of the equation $y = mx + b$ is a line having slope m and y-intercept b.

This form of the equation of a line can be used for graphing certain lines for which it is not convenient to plot points. This procedure is summarized in Figure 3.3.

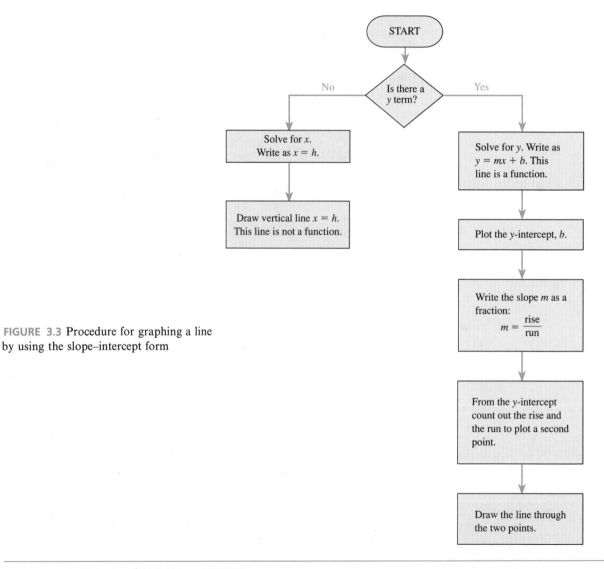

FIGURE 3.3 Procedure for graphing a line by using the slope–intercept form

EXAMPLE 5 Graph $y = \frac{1}{2}x + 3$.

SOLUTION By inspection, the y-intercept is 3 and the slope is $\frac{1}{2}$; the line is graphed by first plotting the y-intercept $(0, 3)$, then finding a second point by counting out the slope: up 1 and over 2.

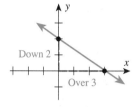

EXAMPLE 6 Graph $2x + 3y - 6 = 0$.

SOLUTION Solve for y:

$$3y = -2x + 6$$

$$y = -\frac{2}{3}x + 2$$

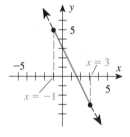

The y-intercept is 2 and the slope is $-\frac{2}{3}$; the line is graphed as shown.

EXAMPLE 7 Graph $2x + y - 3 = 0$ for $-1 \le x \le 3$.

SOLUTION Solve for y:

$$y = -2x + 3$$

The y-intercept is 3 and the slope is -2; this line is shown as a black dashed line. Because of the restriction on the domain you draw that part of the line with x values between -1 and 3 (inclusive), as shown by the colored line segment.

A variation of graphing linear functions is seen when we graph absolute value functions.

EXAMPLE 8 Graph $y = |x|$.

SOLUTION First apply the definition of absolute value:

$$y = x \qquad \text{if } x \ge 0$$

$$y = -x \quad \text{if } x < 0$$

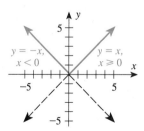

Now graph these two equations with restrictions as you did in Example 7. The graph of $y = |x|$ is the part shown in color.

In algebra you studied several forms of the equation of a line. The derivation of some of these is reviewed in the problems, and the forms are stated in the next box for review.

FORMS OF A LINEAR EQUATION

STANDARD FORM: the line	$Ax + By + C = 0$, where (x, y) is any point on the line and A, B, and C are constants, A and B not both zero.
SLOPE–INTERCEPT FORM:	$y = mx + b$, where m is the slope and b is the y-intercept.
POINT–SLOPE FORM:	$y - k = m(x - h)$, where m is the slope and (h, k) is a point on the line.
TWO-POINT FORM:	$y - y_1 = \left(\dfrac{y_2 - y_1}{x_2 - x_1}\right)(x - x_1)$, where (x_1, y_1) and (x_2, y_2) are points on the line.
INTERCEPT FORM:	$\dfrac{x}{a} + \dfrac{y}{b} = 1$, where $(a, 0)$ and $(0, b)$ are the x- and y-intercepts, respectively.
HORIZONTAL LINE:	passing through the point (h, k): $y = k$.
VERTICAL LINE:	passing through the point (h, k): $x = h$.

EXAMPLE 9 Find the equation of the line using the given information. Leave your answer in standard form.

a. y-intercept 4; slope 6   **b.** Slope 4; passing through $(-3, 2)$
c. Passing through $(2, 3)$ and $(5, 7)$ **d.** No slope; passing through $(7, -3)$

SOLUTION **a.** Since you are given the slope and the y-intercept, use the slope–intercept form: $y = mx + b$, where $b = 4$ and $m = 6$:

$$y = 6x + 4$$

In standard form, $6x - y + 4 = 0$.

b. Use the point–slope form, where $h = -3$, $k = 2$, and $m = 4$:

$$y - k = m(x - h)$$
$$y - 2 = 4(x + 3)$$
$$y - 2 = 4x + 12$$

In standard form, $4x - y + 14 = 0$.

c. Use the two-point form.

$$y - 3 = \left(\frac{7 - 3}{5 - 2}\right)(x - 2)$$

This is the point $(2, 3)$, but you could also use $(5, 7)$ to obtain the same result.

$$y - 3 = \frac{4}{3}(x - 2)$$

$$3y - 9 = 4x - 8$$

In standard form, $4x - 3y + 1 = 0$.

d. Do not confuse no slope (vertical line) with zero slope (horizontal line). This is a vertical line, so the equation has the form $x = h$ when it passes through (h, k). Thus

$$x = 7$$

In standard form, $x - 7 = 0$.

3.2 Problem Set

A *Sketch the line passing through the points whose coordinates are given in Problems 1–9. Also find the slope of each line.*

1. (2, 3) and (5, 6) **2.** (0, 7) and (3, 0)
3. $(-1, -2)$ and (4, 11) **4.** $(4, -2)$ and $(7, -3)$
5. $(-6, -4)$ and $(-9, 3)$ **6.** (6, 0) and $(-3, 0)$
7. (0, 0) and (0, 3) **8.** $(4, -3)$ and (4, 1)
9. $(-1, 2)$ and (3, 2)

Graph the lines whose equations are given in Problems 10–21 by finding the slope and y-intercept.

10. $y = 3x + 3$ **11.** $y = -4x - 1$
12. $y = \frac{2}{3}x + \frac{4}{3}$ **13.** $y = \frac{1}{5}x - \frac{6}{5}$
14. $y = 40x$ **15.** $y = 300x$
16. $x - 4 = 0$ **17.** $y + 2 = 0$
18. $5x - 4y - 8 = 0$ **19.** $x - 3y + 2 = 0$
20. $100x - 250y + 500 = 0$ **21.** $2x - 5y - 1{,}200 = 0$

Graph the line segments or the absolute-value functions in Problems 22–27.

22. $3x + y - 2 = 0, \quad -7 \le x \le 1$
23. $2x - 2y + 6 = 0, \quad 5 \le x \le 9$
24. $5x - 3y - 9 = 0, \quad -3 \le x \le 1$
25. $y = 2|x|$
26. $y = -3|x|$
27. $y = |4x|$

Use slopes to decide whether the coordinates given in Problems 28–30 are vertices of a right triangle.

28. $T(-6, -4), \quad R(6, 12), \quad I(-4, 7)$
29. $A(1, -1), \quad N(4, 1), \quad G(0, 7)$
30. $L(-4, 6), \quad E(-10, 2), \quad S(-3, -1)$

Use slopes to decide whether the coordinates given in Problems 31–34 are vertices of a parallelogram.

31. $R(3, 0), \quad E(6, 7), \quad C(2, 9), \quad T(-3, 3)$
32. $A(-1, 5), \quad N(2, 3), \quad G(-3, -4), \quad E(-6, -2)$
33. $P(1, 10), \quad A(-4, 5), \quad R(-3, -2), \quad L(2, 3)$

34. $E(-3, 6), \quad L(9, 11), \quad G(14, 1), \quad R(2, -4)$
35. Are the diagonals of the quadrilateral in Problem 33 perpendicular?
36. Are the diagonals of the quadrilateral in Problem 34 perpendicular?

B *Find the equation of the line satisfying the given conditions in Problems 37–52. Give your answer in standard form.*

37. *y*-intercept 6; slope 5
38. *y*-intercept -3; slope -2
39. *y*-intercept 0; slope 0
40. *y*-intercept 5; slope 0
41. Slope 3; passing through (2, 3)
42. Slope -1; passing through $(-4, 5)$
43. Slope $\frac{1}{2}$; passing through (3, 3)
44. Slope $\frac{2}{5}$; passing through $(5, -2)$
45. Passing through $(-4, -1)$ and (4, 3)
46. Passing through $(4, -2)$ and (4, 5)
47. Passing through (5, 6) and $(1, -2)$
48. Passing through (5, 6) and (7, 6)
49. Passing through (2, 4) parallel to $2x + 3y - 6 = 0$
50. Passing through $(-1, -2)$ parallel to $x - 2y + 4 = 0$
51. Passing through $(-1, -2)$ perpendicular to $x - 2y + 4 = 0$
52. Passing through (2, 4) perpendicular to $2x + 3y - 6 = 0$
53. Consider Figure 3.4a.
 a. What are the coordinates of *A* and *B*?
 b. What is the slope of the line passing through *A* and *B*?

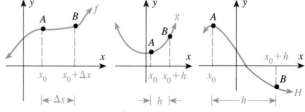

a. Graph of f **b.** Graph of g **c.** Graph of H

FIGURE 3.4

54. Consider Figure 3.4b.
  a. What are the coordinates of A and B?
  b. What is the slope of the line passing through A and B?
55. Consider Figure 3.4c.
  a. What are the coordinates of A and B?
  b. What is the slope of the line passing through A and B?

Problems 56–62 provide some real-world examples of line graphs. One way to find the equation of the line is to write two data points from the given information, and then use those two points to write the equation. Use the given information to write a standard-form equation of the line described by the problem.

56. The demand for a certain product is related to the price of the item. Suppose a new line of stationery is tested at two stores. It is found that 25 boxes are sold within a month if they are priced at $1 and 15 boxes priced at $2 are sold in the same time. Let x be the price and y be the number of boxes sold.

57. An important factor that is related to the demand for a product is the supply. The amount of the stationery in Problem 56 that can be supplied is also related to the price. At $1 each, 10 boxes can be supplied; at $2 each, 20 boxes can be supplied. Let x be the price and y be the number of boxes sold.

58. The population of Florida was roughly 9.7 million in 1980, and 12.8 million in 1990. Let x be the year (let 1980 be the base year; that is, $x = 0$ represents 1980 and $x = 10$ represents 1990) and y be the population. Use this equa tion to predict the population in 2001.

59. The population of Texas was roughly 14.2 million in 1980, and 16.8 million in 1990. Let x be the year (let 1980 be the base year; that is, $x = 0$ represents 1980 and

$x = 10$ represents 1990) and y be the population. Use this equation to predict the population in 2001.

60. It costs $90 to rent a car if you drive 100 miles and $140 if you drive 200 miles. Let x be the number of miles driven and y the total cost of the rental. Use this equation to find how much it would cost if you drove 394 miles.

61. It costs $60 to rent a car if you drive 50 miles and $60 if you drive 260 miles. Let x be the number of miles and y be the total cost of the rental. Use this equation to find how much it would cost if you drove 394 miles.

62. Suppose it costs $100 for maintenance and repairs to drive a three-year-old car 1,000 miles and $650 for maintenance and repairs to drive it 6,500 miles. Let x be the number of miles and y be the cost for repairs and maintenance.

63. Begin with the slope–intercept form and derive the point–slope form:
$$y - k = m(x - h)$$

64. Begin with the point–slope form and derive the two-point form:
$$y - y_1 = \left(\frac{y_2 - y_1}{x_2 - x_1}\right)(x - x_1)$$

65. Begin with the two-point form and derive the intercept form:
$$\frac{x}{a} + \frac{y}{b} = 1$$

66. Prove that if $m > 0$, then the linear function is an increas ing function throughout its domain.

67. Prove that if $m < 0$, then the linear function is a decreasing function throughout its domain.

3.3 Quadratic Functions

The second type of polynomial function to be considered in this chapter is the quadratic function.

QUADRATIC FUNCTION

> A function f is a **quadratic function** if
> $$f(x) = ax^2 + bx + c$$
> where a, b, and c are real numbers and $a \neq 0$.

If $b = c = 0$, however, the quadratic function has the form
$$y = ax^2$$
and has a graph called a **standard-position parabola.** You translated this function

in Section 2.3 (see Figure 2.10, page 90). We begin by sketching two additional standard-position parabolas in order to make some generalizations.

EXAMPLE 1 Sketch the graph of $y = 3x^2$.

SOLUTION Find some ordered pairs satisfying the equation:

x	0	1	2	3	...
y	0	3	12	27	...

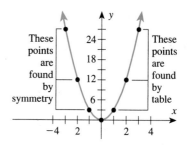

FIGURE 3.5 Graph of $y = 3x^2$

Also, if $y = f(x)$ then

$$f(-x) = 3(-x)^2 = 3x^2$$

so the graph is symmetric with respect to the y-axis. Plot the points represented by the ordered pairs, use symmetry, and draw a smooth graph as shown in Figure 3.5.

EXAMPLE 2 Sketch the graph of $y = -\frac{1}{2}x^2$.

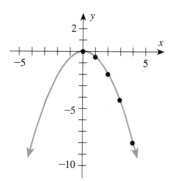

FIGURE 3.6 Graph of $y = -\frac{1}{2}x^2$

x	0	1	2	3	4
y	0	$-\frac{1}{2}$	-2	$-\frac{9}{2}$	-8

Plot these points and use symmetry as shown in Figure 3.6.

We will now make some general observations based on the special case $y = ax^2$:

1. The graph has the shape of a curve that is called a **parabola.**

2. The point $(0, 0)$ is the lowest point if the parabola opens up ($a > 0$); $(0, 0)$ is the highest point if the parabola opens down ($a < 0$). This highest or lowest point is called the **vertex.**

3. The parabola is **symmetric** with respect to the vertical line passing through the vertex.

4. Relative to a fixed scale, the magnitude of a determines the "wideness" of the parabola: Small values of $|a|$ yield "wide" parabolas; large values of $|a|$ yield "narrow" parabolas.

For graphs of parabolas of the form

$$y - k = a(x - h)^2$$

you simply translate the axes to the point (h, k) and then graph the parabola $y' = ax'^2$, as shown in Example 3. The point (h, k) is the vertex. Using functional notation, this can be written

$$f(x) = a(x - h)^2 + k$$

EXAMPLE 3 Sketch the graph of $y + 5 = 3(x + 2)^2$.

SOLUTION By inspection, $(h, k) = (-2, -5)$ and the standard-position parabola is $y' = 3x'^2$. The table of values (or points to plot) is the same as shown for Example 1, and is repeated in the margin. The only difference here is that you count out these points from $(-2, -5)$ instead of from the origin, as shown in Figure 3.7.

Table from Example 1

x	y
0	0
1	3
2	12
3	27

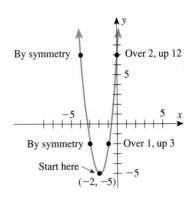

FIGURE 3.7 Graph of $y + 5 = 3(x + 2)^2$ ▌

Now you are ready to consider the general quadratic function

$$y = ax^2 + bx + c \quad (a \neq 0)$$

Consider Example 3 above: $y + 5 = 3(x + 2)^2$. This was graphed by translating the axes to $(-2, -5)$ and then considering $y' = 3x'^2$. Suppose you rewrite the given equation as

$$y + 5 = 3(x + 2)^2$$
$$y + 5 = 3(x^2 + 4x + 4)$$
$$y + 5 = 3x^2 + 12x + 12$$
$$y = 3x^2 + 12x + 7$$

The last form is the general quadratic form, where $a = 3$, $b = 12$, and $c = 7$. Suppose you are given this form and told to graph the curve. Then, to reverse the process, you must **complete the square** (review Section 1.6).

To complete the square, follow these steps:

STEP 1 Subtract c (the constant term) from both sides:

GENERAL FORM   EXAMPLE

$$y = ax^2 + bx + c \qquad\qquad y = 3x^2 + 12x + 7$$
$$y - c = ax^2 + bx \qquad\qquad y - 7 = 3x^2 + 12x$$

STEP 2 Factor the a term from the expression on the right (remember $a \neq 0$):

$$y - c = a\left(x^2 + \frac{b}{a}x\right) \qquad\qquad y - 7 = 3(x^2 + 4x)$$

STEP 3 To complete the square on the number inside parentheses:

$$x^2 + \frac{b}{a}x + ? = (x + ?)^2 \qquad\qquad x^2 + 4x + ? = (x + ?)^2$$

you square one-half the coefficient of the x term:

$$\left(\frac{b}{2a}\right)^2 = \frac{b^2}{4a^2} \qquad\qquad \left(\frac{4}{2}\right)^2 = 4$$

Then add a times this number to both sides of the original equation:

$$y - c + \frac{b^2}{4a} = a\left(x^2 + \frac{b}{a}x + \frac{b^2}{4a^2}\right) \qquad y - 7 + 12 = 3(x^2 + 4x + 4)$$

$\frac{1}{2}\cdot\frac{b}{a}$ squared $\qquad\qquad\qquad \frac{1}{2}\cdot 4$ squared

distributive property $\qquad\qquad\qquad$ distributive property

$$a\cdot\frac{b^2}{4a^2} = \frac{b^2}{4a} \qquad\qquad\qquad 3\cdot 4 = 12$$

Add $\dfrac{b^2}{4a}$ to both sides. $\qquad\qquad$ Add 12 to both sides.

STEP 4 Factor the right-hand side as a perfect square and simplify the left-hand side:

$$y + \frac{-4ac}{4a} + \frac{b^2}{4a} = a\left(x + \frac{b}{2a}\right)^2 \qquad y - 7 + 12 = 3(x + 2)^2$$

common denominator if fractions are involved

$$\left(y + \frac{b^2 - 4ac}{4a}\right) = a\left(x + \frac{b}{2a}\right)^2 \qquad y + 5 = 3(x + 2)^2$$

This is of the form

$$y - k = a(x - h)^2$$

and is called the **general form of a parabola.** This equation can be graphed by doing a translation, as shown in Example 4.

EXAMPLE 4 Sketch the graph of $y = x^2 + 6x + 10$.

SOLUTION Complete the square:

$$y - 10 = x^2 + 6x$$

$$y - 10 + 9 = x^2 + 6x + 9$$

Since $\frac{1}{2}\cdot 6 = 3$ and $3^2 = 9$, you add 9 to both sides.

$$y - 1 = (x + 3)^2$$

The vertex is at $(-3, 1)$, the parabola opens up, and the graph is shown in Figure 3.8.

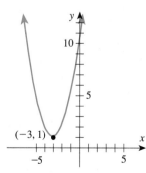

FIGURE 3.8

Graph of $y = x^2 + 6x + 10$

Notice from Example 4 above that $c = 10$ (compare the given equation with the form $y = ax^2 + bx + c$) and that the y-intercept is (0, 10). In general, if $x = 0$ then

$$y = a(0)^2 + b(0) + c$$
$$= c$$

which shows that the **y-intercept for a quadratic function is (0, c).** This often serves as a check point for graphs such as the one shown in Example 4.

EXAMPLE 5 Sketch the graph of $y = 1 - 5x - 2x^2$.

SOLUTION

$$y - 1 = -2x^2 - 5x$$

$$y - 1 = -2\left(x^2 + \frac{5}{2}x\right)$$

$$y - 1 - \frac{25}{8} = -2\left(x^2 + \frac{5}{2}x + \frac{25}{16}\right)$$

$$y - \frac{33}{8} = -2\left(x + \frac{5}{4}\right)^2$$

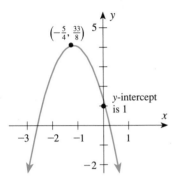

FIGURE 3.9 Graph of $y = 1 - 5x - 2x^2$

The vertex is $\left(-\frac{5}{4}, \frac{33}{8}\right)$, the parabola opens down, and the graph is shown in Figure 3.9. For fractions, you can sometimes choose a scale that is more convenient than one square per unit. Also, remember that once you have found the vertex, you simply have to graph $y' = -2x'^2$ translated to the point $\left(-\frac{5}{4}, \frac{33}{8}\right)$.

Optimization Problems

One of the most important applications of quadratic functions is finding optimum (maximum or minimum) values. It is important in business, economics, psychology, medicine, science, and social science. It is an important part of a calculus course, and for this reason we will consider some optimization problems that can be modeled by linear or quadratic equations and inequalities.

If f is a quadratic function defined by

$$y - k = a(x - h)^2$$

then the **maximum or minimum value is at the vertex.** That is, if $x = h$ then the maximum value of $f(x) = a(x - h)^2 + k$ is $y = k$. You can see this is true because $(x - h)^2$ is necessarily nonnegative for all x and zero if and only if $x = h$. For Example 5, the maximum value of y is $\frac{33}{8}$, which occurs for $x = -\frac{5}{4}$.

EXAMPLE 6 A small manufacturer of CB radios determines that the price of each item is related to the number of items produced. Suppose that x items are produced per day where the maximum number that can be produced is 10 items, and that the profit, in dollars, is determined to be

$$P(x) = -30(x - 6)^2 + 480$$

How many radios should be produced in order to maximize the profit?

SOLUTION The profit function has a graph that opens down. The maximum profit is found at the vertex of this parabola, as shown in Figure 3.10.

The vertex of the function

$$P(x) - 480 = -30(x - 6)^2$$

is (6, 480), so the **maximum profit of $480 is obtained when 6 radios are manufactured.** Notice also from the graph that if there were a strike and no radios could be produced, the daily profit would be −600 (this is a $600 per day *loss*).

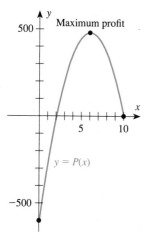

FIGURE 3.10 Graph of
$P(x) = -30(x - 6)^2 + 480$

CALCULATOR COMMENT

When using a calculator to graph polynomial functions, you will need to input the function in the form $\boxed{Y=}$. For Example 3, you would write the given form $y + 5 = 3(x + 2)^2$ as

$$y = 3(x + 2)^2 - 5.$$

Another problem you will have in graphing is in choosing an appropriate domain and range. For Example 6, when I input the functions as $Y1 = -30(x - 6)^2 + 480$ on my TI-81, I obtain the following graph with the standard plot:
The domain and range are:

Xmin = −10
Xmax = 10
Ymin = −10
Ymax = 10

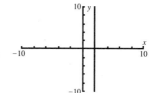

From the example we can see that proper domain is [0, 10]. To find the appropriate range, we can use the $\boxed{\text{TRACE}}$ to see $\boxed{X = .10526316 \quad Y = -562.4377}$; as we trace the curve and watch the y values, we see that they increase to a maximum of $\boxed{X = 6 \quad Y = 480}$; as we continue with the trace, the y values decrease until we obtain $\boxed{X = 10 \quad Y = 2.68E\text{-}8}$. (*Note* The number 2.68E-8 is in scientific notation and means $2.68 \times 10^{-8} = .00000\ 00268 \approx 0$.) An appropriate graph is shown:

Xmin = 0
Xmax = 10
Ymin = −600
Ymax = 500

3.3 Problem Set

A *Sketch the graph of each equation given in Problems 1–18.*

1. $y = -x^2$
2. $y = 2x^2$
3. $y = 5x^2$
4. $y = -\frac{1}{3}x^2$
5. $y = -\frac{1}{10}x^2$
6. $y = -2\pi x^2$
7. $y = (x - 3)^2$
8. $y = (x - 1)^2$
9. $y = (x + 2)^2$
10. $y = -2(x - 1)^2$
11. $y = \frac{1}{4}(x - 1)^2$
12. $y = \frac{1}{2}(x + 1)^2$
13. $y = \frac{1}{3}(x + 2)^2$
14. $y - 2 = (x - 1)^2$
15. $y - 2 = 3(x + 2)^2$
16. $y - 2 = -\frac{3}{5}(x - 1)^2$
17. $y + 3 = \frac{2}{3}(x + 2)^2$
18. $y - 1 = \frac{1}{3}(x - 4)^2$

WHAT IS WRONG, *if anything, with each of the statements in Problems 19–24? Explain your reasoning.*

19. $f(x) = 3x^2 + 5x - 3x^2 + 2$ is a quadratic function.

20. The parabola $x - 3 = 4(y + 2)^2$ opens up.

21. To complete the square on $y = 5x^2 + 10x$, you must add 5^2 to both sides.

22. To complete the square on $y = 3(x^2 + 8x)$ you must add 16 to both sides.

23. The maximum value of $y - 3 = 4(x - 5)^2$ is $y = 3$.

24. The minimum value of $y + 400 = -6(x + 4)^2$ is $y = -400$.

B *Sketch the graph of each equation given in Problems 25–40.*

25. $y = x^2 + 4x + 4$

26. $y = x^2 + 6x + 9$

27. $y = x^2 + 2x - 3$

28. $y = 2x^2 - 4x + 5$

29. $y = 2x^2 - 4x + 4$

30. $y = 2x^2 + 8x + 5$

31. $y = 3x^2 - 12x + 10$

32. $y = 3x^2 - 30x + 76$

33. $y = \frac{1}{2}x^2 + 2x - 1$

34. $y = \frac{1}{2}x^2 + 4x + 10$

35. $y = \frac{1}{2}x^2 - x + \frac{5}{2}$

36. $y = \frac{1}{2}x^2 - 3x + \frac{3}{2}$

37. $x^2 - 6x - 2y - 1 = 0$

38. $x^2 + 2x + 2y - 3 = 0$

39. $x^2 - 6x - 3y - 3 = 0$

40. $2x^2 - 4x - 3y + 11 = 0$

Find the maximum value of y for the functions defined by the equations in Problems 41–46.

41. $y = -4x^2 - 8x - 1$

42. $y = -3x^2 + 6x - 5$

43. $10x^2 - 160x + y + 655 = 0$

44. $6x^2 + 84x + y + 302 = 0$

45. $9x^2 + 6x + 81y - 53 = 0$

46. $100x^2 - 120x + 25y + 41 = 0$

47. A manufacturer produces high-quality boats at a profit, in dollars, that is determined to be

$$P(x) = -10(x - 375)^2 + 1,156,250$$

a. How many boats should be produced in order to maximize the profit?

b. What is the profit (or loss) if no boats are produced?

c. What is the maximum profit?

48. A profit function, P, is

$$P(x) = -10(x - 75)^2 + 3,750$$

Find the maximum profit.

49. The profit function for a ratchet flange is

$$P(x) = -2(x - 25)^2 + 650$$

What is the maximum profit?

50. An arch has the equation

$$y - 18 = -\frac{2}{81}x^2$$

a. What is the maximum height?

b. What is the width of the arch?

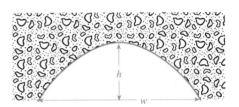

h and w are measured in feet

51. What is the height of the arch in Problem 50 at the following places?

a. Nine feet from the center

b. Eighteen feet from the center

52. One side of a storage yard is against a building. The other three sides of the rectangular yard are to be fenced with 36 ft of fencing. How long should the sides be to produce the greatest area with the given length of fence? What is the area obtained?

53. The sum of the length and the width of a rectangular area is 50 ft. Find the greatest area possible, and find the dimensions of the figure.

54. PHYSICS The highest bridge in the world is the bridge over the Royal Gorge of the Arkansas River in Colorado. It is 1,053 ft above the water. If a rock is projected vertically upward from this bridge with an initial velocity of 64 fps, the height h of the object above the river at time t is described by the function

$$h(t) = -16t^2 + 64t + 1,053$$

What is the maximum height possible for a rock projected vertically upward from the bridge with an initial velocity of 64 fps? After how many seconds does it reach that height?

55. PHYSICS In 1974, Evel Knievel attempted a skycycle ride across the Snake River. Suppose the path of the skycycle is given by the equation

$$d(x) = -.0005x^2 + 2.39x + 600$$

where $d(x)$ is the height in feet above the canyon floor for a horizontal distance of x units from the launching ramp. What was Knievel's maximum height?

C *Sketch the graph of each equation given in Problems 56–61.*

56. $y = x^2 - 5x + 2$
57. $2x^2 - x - y + 3 = 0$
58. $2x^2 - 8x + 3y + 20 = 0$
59. $4x^2 - 20x - 16y + 33 = 0$
60. $4x^2 + 24x - 27y - 17 = 0$
61. $25x^2 - 30x - 5y + 2 = 0$

62. BUSINESS A small manufacturer of CB radios determines that the price of each item is related to the number of items produced. If x items are produced per day, and the maximum number that can be produced is 10 items, then the price should be

$$400 - 25x \text{ dollars}$$

It is also determined that the overhead (the cost of producing x items) is

$$5x^2 + 40x + 600 \text{ dollars}$$

The daily profit is then found by subtracting the overhead from the revenue:

$$\begin{aligned}
\text{Profit} &= \text{Revenue} - \text{Cost} \\
&= (\text{Number of items})(\text{Price per item}) - \text{Cost} \\
&= x(400 - 25x) - (5x^2 + 40x + 600) \\
&= 400x - 25x^2 - 5x^2 - 40x - 600 \\
&= -30x^2 + 360x - 600
\end{aligned}$$

A negative profit is called a loss.
a. What is the domain for the profit function?
b. For what values of the domain does the manufacturer show a positive profit?
c. Does the manufacturer ever show a loss? For what values?
d. If there were a strike and production were brought to a halt, what would be the value of x? What would be the profit (or loss) for this situation?
e. How many items should the manufacturer produce per day, and what should be the expected daily profit in order to maximize the profit?

3.4 Cubic and Quartic Functions

Before graphing general polynomial functions in the next section, it is worthwhile to look at some general characteristics of polynomial functions. The behavior of polynomial functions depends on whether the degree is even or odd. For this reason we will look at cubic (odd example) and quartic (even example) functions separately.

Graphs of Monomials

EXAMPLE 1 Graph $y = x$, $y = x^3$, and $y = x^5$ on the same coordinate axes.

SOLUTION The graphs are shown in Figure 3.11.

FIGURE 3.11 Graphs of $y = x$, $y = x^3$, and $y = x^5$

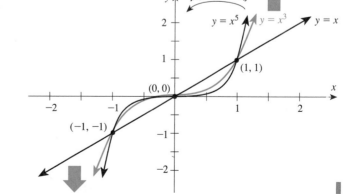

Notice that all of the curves in Example 1 pass through $(0, 0)$, $(1, 1)$, and $(-1, -1)$. This is true of $y = x^n$, for any odd integer n. Notice also that as x grows without bound in the positive direction, the values of y must be positive, and as x grows without bound in the negative direction, the values of y must be negative. We have shown this by using large arrows in Figure 3.11. In calculus, the notion of a limit is defined, and even though we are not prepared to define that concept in this course, we can anticipate the notation used in calculus. We write

$x \to \infty$ to indicate that x is growing without bound in the positive direction; for example $x = 10, 100, 1{,}000, 10{,}000, 100{,}000, \ldots$.

$x \to -\infty$ to indicate that x is growing without bound in the negative direction; for example $x = -10, -100, -1{,}000, \ldots$.

Now consider even powers of x.

EXAMPLE 2   Graph $y = x^2$, $y = x^4$, and $y = x^6$ on the same coordinate axes.

SOLUTION   The graphs are shown in Figure 3.12. As $x \to \infty$ and $x \to -\infty$, we show the direction of the graphs with arrows.

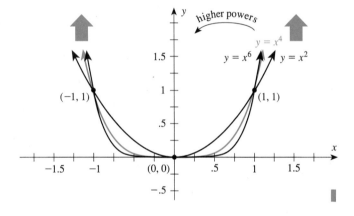

FIGURE 3.12 Graphs of $y = x^2$, $y = x^4$, and $y = x^6$

By understanding the nature of the graphs of $y = x^n$, we can also understand the nature of the graphs of

$$y - k = (x - h)^n \qquad \textbf{translations}$$

$$y = -x^n \qquad \textbf{reflections}$$

$$y = ax^n \qquad \textbf{dilations and contractions}$$

Some of these graphs are shown in Figure 3.13.

Translations

Graphs of $y = x^3$ and $y - 2 = (x + 3)^3$

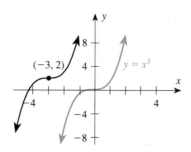

Graphs of $y = x^4$ and $y - 2 = (x + 3)^4$

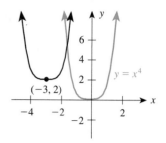

Reflections

Graphs of $y = x^3$ and $y = -x^3$

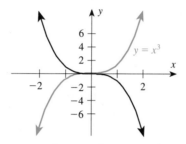

Graphs of $y = x^4$ and $y = -x^4$

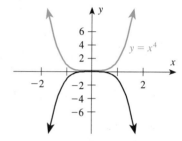

Dilations and Contractions, $y = ax^n$
For $a > 0$

Graphs of $y = x^3$, $y = 10x^3$, and $y = .1x^3$
n odd (Quadrants I and III)

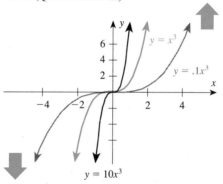

Graphs of $y = x^4$, $y = 10x^4$, and $y = .1x^4$
n even (Quadrants I and II)

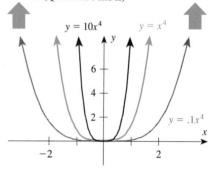

For $a < 0$

n odd (Quadrants II and III)
Graphs of $y = -x^3$, $y = -10x^3$, and $y = -.1x^3$

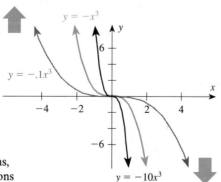

n even (Quadrants III and IV)
Graphs of $y = -x^4$, $y = -10x^4$, and $y = -.1x^4$

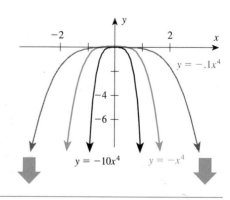

FIGURE 3.13 Graphs of translations,
reflections, dilations, and contractions

Graphs of Binomials

There are several cubic and quartic polynomial functions that are binomials. These are

$$y = ax^3 + bx^2, \quad y = ax^3 + cx, \quad \text{and}$$
$$y = ax^4 + bx^3, \quad y = ax^4 + bx^2, \quad y = ax^4 + bx$$

We begin by noting that as $x \to \infty$ or as $x \to -\infty$, the graph of y will behave like its leading term. (Remember that the domain for polynomial functions is the set of all real numbers.) The first term of a polynomial function will "overpower" all other terms for very large positive values or very large negative values. This means that the "interesting" parts of these curves will be for x values that are near the origin. The higher powers (cubes or fourth powers in these examples) will give large positive or large negative values, so it is often necessary to modify the scale on the y-axis.

EXAMPLE 3 Graph $y = x^3$, $y = x^3 + x^2$, $y = x^3 + 2x^2$, $y = x^3 + 3x^2$, and $y = x^3 + 4x^2$ on the same coordinate axes.

SOLUTION We factor each of these expressions to find the x-intercepts.

FUNCTIONS	FACTORED FORM	X-INTERCEPTS
$y = x^3$	$y = x^3$	0
$y = x^3 + x^2$	$y = x^2(x + 1)$	$0, -1$
$y = x^3 + 2x^2$	$y = x^2(x + 2)$	$0, -2$
$y = x^3 + 3x^2$	$y = x^2(x + 3)$	$0, -3$
$y = x^3 + 4x^2$	$y = x^2(x + 4)$	$0, -4$

GENERALIZATION:

$y = x^3 + nx^2$	$y = x^2(x + n)$	$0, -n$ (n positive)

> For large x values, $y = ax^3 + bx^2$ and $y = ax^3 + cx$ look like $y = ax^3$. ⊗

The graphs are shown in Figure 3.14.

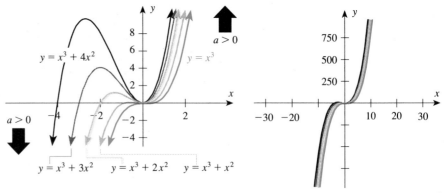

FIGURE 3.14 Graphs of typical cubic binomials

Graphs showing detail near the origin   Graphs away from the origin

Notice that each of the graphs (except for $y = x^3$) in Figure 3.14 shows two turning points. A **turning point** is a point on the graph that separates an interval over which a function f is increasing from an interval over which f is decreasing. A linear function has no turning points and a quadratic function has one turning point. In calculus it is shown that the graph of a polynomial function of degree n can have *at most* $n - 1$ turning points. The next example graphs some quartic functions which, by definition, have $n = 4$, so the maximum number of turning points is 3. However, since we are limiting our discussion to binomials, you will see only one turning point. In the next section we will be able to investigate the turning points more completely.

EXAMPLE 4 Graph each part on separate coordinate axes.

a. $y = x^4$, $y = x^4 + x$, $y = x^4 + 2x$, $y = x^4 + 3x$, and $y = x^4 + 4x$
b. $y = x^4$, $y = x^4 + x^2$, $y = x^4 + 2x^2$, $y = x^4 + 3x^2$, and $y = x^4 + 4x^2$
c. $y = x^4$, $y = x^4 + x^3$, $y = x^4 + 2x^3$, $y = x^4 + 3x^3$, and $y = x^4 + 4x^3$

SOLUTION Each of these parts can be factored to find the x-intercepts; we leave the details for you. The importance of this example is general rather than specific. By this we mean that you should look at the parts of this example for similarities and differences. Notice that the graphs away from the origin show that the leading term "overpowers" the other terms.

For large x values, $y = ax^4 + bx^3$, $y = ax^4 + bx^2$, and $y = ax^4 + bx$ look like $y = ax^4$.

Graphs showing detail near the origin

Graphs away from the origin

a.

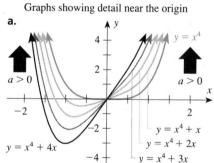

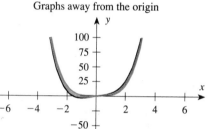

b.

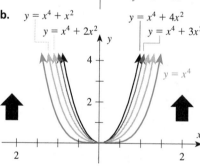

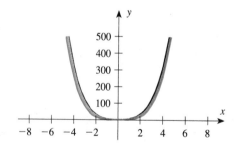

c.

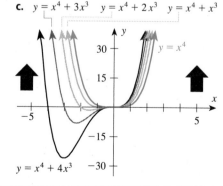

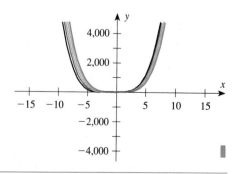

General Properties of Polynomial Functions

There are many general properties associated with the graphs of polynomial functions. We have assumed some of these properties in the graphs of linear, quadratic, cubic, and quartic functions graphed thus far. We will summarize and state these properties here, even though they require calculus for a complete discussion and justification.

PROPERTIES OF POLYNOMIAL GRAPHS

1. There are no breaks or jumps in the graph of a polynomial function. This means that the graph of a polynomial function can be sketched over any part of its domain without lifting the pencil. A function with this property is called a **continuous function.**

2. The graph of a polynomial function has no corners or abrupt changes of direction. A function with this property is called a **a smooth function**.

3. The graph of a polynomial function behaves as its leading term $(a_n x^n)$ for the extreme values of x. This means that as $x \rightarrow \infty$ or as $x \rightarrow -\infty$ the polynomial function either approaches ∞ or $-\infty$. The pattern for n odd (cubic) and n even (quartic) is:

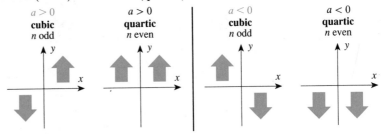

4. The graph of a polynomial function of degree n has *at most* $n - 1$ turning points. This means

   linear functions: 0 turning points
   quadratic functions: 1 turning point
   cubic function: at most 2 turning points
   quartic function: at most 3 turning points

EXAMPLE 5 Tell why the given graphs of polynomial functions are not complete, and then correctly complete the graph.

a. $y = x^3 - 8x^2 - x$ **b.** $y = -x^4 + 8x^3 + 4x^2 + 2$

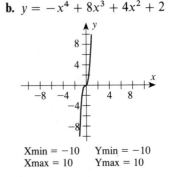

Xmin = -10 Ymin = -10 Xmin = -10 Ymin = -10
Xmax = 10 Ymax = 10 Xmax = 10 Ymax = 10

SOLUTION **a.** This graph is not correct because it is not continuous and we know that polynomial functions must be continuous. The difficulty is with the range of values shown. The graph "disappears" at about $x = 2$ and "reappears" at about $x = 8$. To find "how far down" the graph goes, we could evaluate the function at $x = 5$:

$$y = 5^3 - 8(5)^2 - 5 = -80$$

We can plot some additional values to draw the graph shown in Figure 3.15.

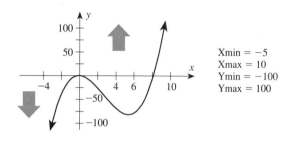

Xmin = −5
Xmax = 10
Ymin = −100
Ymax = 100

FIGURE 3.15 Graph of $y = x^3 - 8x^2 - x$

b. The graph shown is not correct. Since the power of the leading term is even and its coefficient is negative, the graph should exhibit the pattern shown at the left.

Find some additional values: If $y = f(x)$, then

$$f(2) = -2^4 + 8(2)^3 + 4(2)^2 + 2 = 66$$
$$f(4) = -4^4 + 8(4)^3 + 4(4)^2 + 2 = 322$$
$$f(6) = -6^4 + 8(6)^3 + 4(6)^2 + 2 = 578$$
$$f(8) = -8^4 + 8(8)^3 + 4(8)^2 + 2 = 258$$
$$f(10) = -10^4 + 8(10)^3 + 4(10)^2 + 2 = -1,598$$

We plot these points to draw the graph as shown in Figure 3.16.

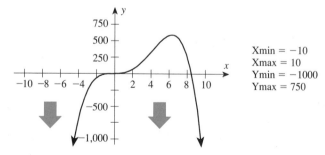

Xmin = −10
Xmax = 10
Ymin = −1000
Ymax = 750

FIGURE 3.16 Graph of $y = -x^4 + 8x^3 + 4x^2 + 2$

Evaluating functions to plot points as shown in Example 5 is very tedious. In the next section we will see how to use synthetic division to make this process much more efficient. In this section we will continue to focus on the general properties of graphing polynomial functions rather than the plotting of points.

Use your knowledge of the properties of functions and the results of Example 5 to supply the graphs in Examples 6–8.

EXAMPLE 6 *Translated polynomial functions*
  a. Graph $y = x^3 - 8x^2 - x + 25$.
  b. Graph $y = (x + 2)^3 - 8(x + 2)^2 - (x + 2)$.

SOLUTION **a.** The usual translation on a polynomial function is the addition or subtraction of a constant—in this case, $+25$. We can write this as

$$y - 25 = x^3 - 8x^2 - x$$

to see that the graph has been vertically translated. However, the usual manner of graphing a curve such as the one defined by the equation in this example is to note that if $x = 0$, then $y = 25$. That is, this is the same as Example 5a except that it has a y-intercept of 25. The graph is shown in Figure 3.17a.

b. We see that this is the same as the graph of Example 5a with $(h, k) = (-2, 0)$. That is, the graph has been translated 2 units to the left. The graph is shown in Figure 3.17b.

FIGURE 3.17 Graphs showing vertical and horizontal translations

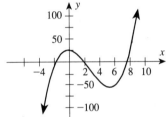

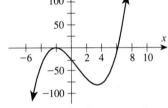

a. Graph of $y = x^3 - 8x^2 - x + 25$ **b.** Graph of $y = (x + 2)^3 - 8(x + 2)^2 - (x + 2)$

EXAMPLE 7 *Reflections, compressions, and dilations*
  a. Graph $y = x^4 - 8x^3 - 4x^2 - 2$.
  b. Graph $y = 2x^3 - 16x^2 - 2x$.

SOLUTION **a.** The graph of $y = x^4 - 8x^3 - 4x^2 - 2$ is a reflection of the graph in Example 5b. It is shown in Figure 3.18a.

b. The graph of $y = 2x^3 - 16x^2 - 2x$ is a dilation in the y-direction since we can write $y = 2(x^3 - 8x^2 - x)$ so that if $y = f(x)$ for the function in Example 5a, we see that for this example we want to graph $y = 2f(x)$. Each y value in the graph of Example 5a is doubled. The result is shown in Figure 3.18b.

FIGURE 3.18 Graphs showing a reflection and a dilation

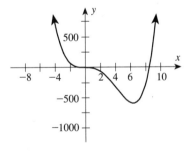

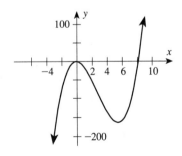

a. Graph of $y = x^4 - 8x^3 - 4x^2 - 2$ **b.** Graph of $y = 2x^3 - 16x^2 - 2x$

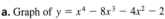

EXAMPLE 8 *Absolute value of a known function*
  a. Graph $y = |x^3 - 8x^2 - x|$.   **b.** Graph $y = |x^3| - 8|x|^2 - |x|$.

SOLUTION **a.** The graph of $y = |x^3 - 8x^2 - x|$ is the same as the graph of the function shown in Example 5a with each value below the x-axis reflected so that the y-component is positive. The result is shown in Figure 3.19a.
  b. The graph of $y = |x^3| - 8|x|^2 - |x|$ is drawn so that the part to the right of the y-axis (the part where $x > 0$) is the same as the graph in Example 5a. The part to the left of the y-axis is the same as the part to the right so it is drawn symmetric to the y-axis, as shown in Figure 3.19b. We might point out that, because of the properties of absolute value, this graph is the same as the graph of $y = |x|^3 - 8x^2 - |x|$.

FIGURE 3.19 Graph showing the absolute value of a known function

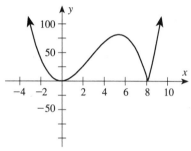
a. Graph of $y = |x^3 - 8x^2 - x|$

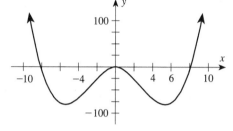
b. Graph of $y = |x|^3 - 8|x|^2 - |x|$

3.4 Problem Set

A *Each graph in Problems 1–12 is typical of a linear, quadratic, cubic, or quartic function. Classify each graph.*

1.
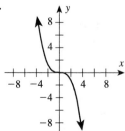

2.
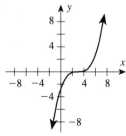

5.
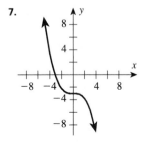

6.
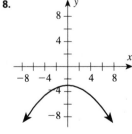

3.
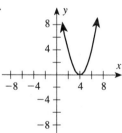

4.
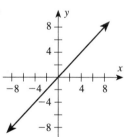

7.
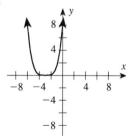

8.
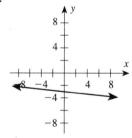

9.

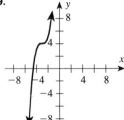

10.

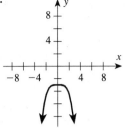

E.

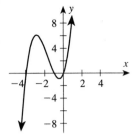

F.

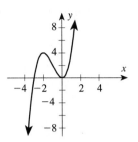

11.

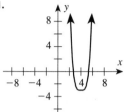

12.

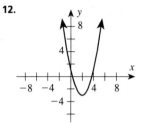

G.

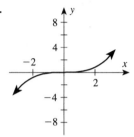

H.

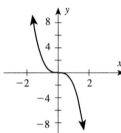

Use the properties of polynomial functions to match the equations in Problems 13–32 with the graphs labeled A–U.

13. $y = x^3$

14. $y = 10x^3$

15. $u = .1x^3$

16. $y = x^4$

17. $y = 10x^4$

18. $y = .1x^4$

19. $y = -2x^3$

20. $y = -2x^3 - 4$

21. $y = -2x^3 + 4$

22. $y = -2(x + 1)^3 + 4$

23. $y = -2(x - 1)^3 + 4$

24. $y = -\frac{1}{2}x^4$

25. $y = -\frac{1}{2}x^4 - 4$

26. $y = -\frac{1}{2}x^4 + 4$

27. $y = -\frac{1}{2}(x - 1)^4 + 4$

28. $y = -\frac{1}{2}(x + 1)^4 + 4$

29. $y = x^3 + 3x^2$

30. $y = x^4 + 2x$

31. $y = x^4 + 8x^3 - 10x^2 + 2x$

32. $y = x^3 + 5x^2 + 4x$

I.

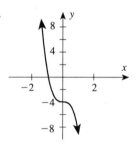

J.

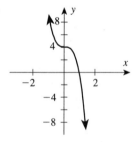

A.

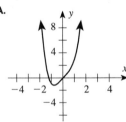

B.

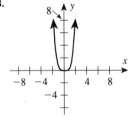

K.

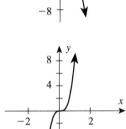

L.

C.

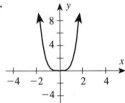

D.

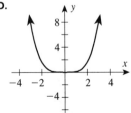

M.

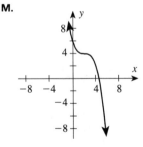

N.

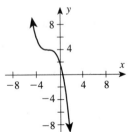

P.

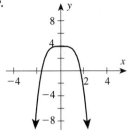

Q.

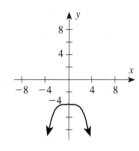

$$h(x) = x^3 - 5x^2 + 6x$$

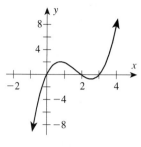

R.

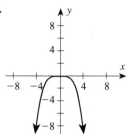

S.

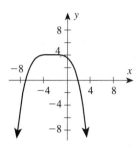

33. $y = -f(x)$

34. $y = -g(x)$

35. $y = -h(x)$

36. $y = f(x) + 4$

37. $y = g(x) - 4$

38. $y = h(x) + 2$

39. $y = f(x - 2)$

40. $y = g(x + 2)$

41. $y = h(x + 1)$

42. $y = f(x - 3) + 2$

43. $y = g(x + 2) - 4$

44. $y = 2f(x)$

45. $y = -2g(x)$

46. $y = 3h(x)$

47. $y = |f(x)|$

48. $y = |g(x)|$

49. $y = |h(x)|$

50. $y = f(|x|)$

51. $y = g(|x|)$

52. $y = h(|x|)$

T.

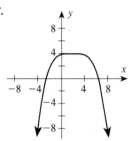

U.

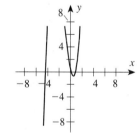

WHAT IS WRONG, *if anything, with each of the graphs shown for the equations given in Problems 53–64? Explain your reasoning and draw the correct graph.*

53. $\qquad y = x^3 + 5x^2$

54. $\qquad y = x^3 + 8x^2 - 10x$

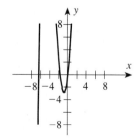

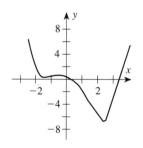

B *Graph the functions requested in Problems 33–52, using the functions f, g, and h defined by the following equations and graphs.*

$$f(x) = x^4 + 8x^3 + 16x^2$$

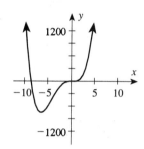

55. $\quad y = x^3 + 8x^2 + 10x$

56. $\quad y = x^4 - 8x^2 - 10x$

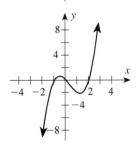

$$g(x) = x^3 - x^2 - 2x$$

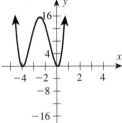

57. $y = x^3 - 4x^2 + x + 15$

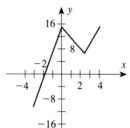

58. $y = x^3 + 6x^2$

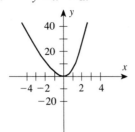

61. $y = x^4 + 5x^3$

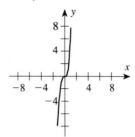

62. $y = x^4 - 8x^3$

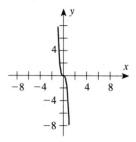

59. $y = -x^3 + 12x^2$

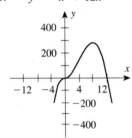

60. $y = x^4 - x^3 + x^2$

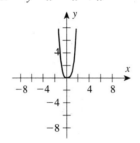

63. $y = x^4 - 8x^3 - 4x^2 + x$

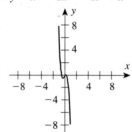

64. $y = -x^4 + 10x^3 + 4x$

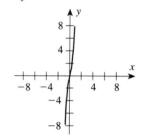

3.5 Graphing Polynomial Functions

We have learned a great deal about the graphs of functions, and also many of the properties of linear, quadratic, cubic, and quartic functions. We can now discuss the graphing of polynomial functions in general. Calculators have made it much easier to graph functions, but as we saw in the previous section, some additional considerations are often necessary. We saw that it is sometimes necessary to evaluate the function to be graphed to find some specific points. For example, suppose we wish to graph the cubic equation

$$f(x) = 2x^3 - 3x^2 - 12x + 17$$

Since there is a one-to-one correspondence between points on the graph and ordered pairs satisfying this equation, you can begin by plotting points:

$$f(0) = 2 \cdot 0^3 - 3 \cdot 0^2 - 12 \cdot 0 + 17$$
$$= 17$$

(0, 17) is on the graph.

$$f(1) = 2 \cdot 1^3 - 3 \cdot 1^2 - 12 \cdot 1 + 17$$
$$= 2 - 3 - 12 + 17$$
$$= 4$$

(1, 4) is on the graph.

$$f(-1) = 2(-1)^3 - 3(-1)^2 - 12(-1) + 17$$
$$= -2 - 3 + 12 + 17$$
$$= 24$$

(−1, 24) is on the graph.

But you can see that this procedure could be very tedious by the time you find enough points to determine the shape of the curve. Instead, consider the **Remainder Theorem.**

REMAINDER THEOREM

When a polynomial $f(x)$ is divided by $x - r$, the remainder is equal to $f(r)$.

To verify this theorem, recall the Division Algorithm:

$$\frac{P(x)}{D(x)} = Q(x) + \frac{R(x)}{D(x)} \qquad \text{or} \qquad P(x) = Q(x)D(x) + R(x)$$

In this context, $P(x) = f(x)$, $D(x) = x - r$, and $R(x)$ is a constant since the degree of $R(x)$ must be less than the degree of $D(x)$, which is 1. The Division Algorithm can be restated as

$$f(x) = Q(x)(x - r) + R$$

where R represents the remainder. Now

$$f(r) = Q(r)(r - r) + R$$
$$= R$$

since $Q(r)(r - r) = Q(r) \cdot 0 = 0$. Points on the curve can therefore be found by synthetic division.

EXAMPLE 1 Use synthetic division to find several points of the curve defined by the function.

$$f(x) = 2x^3 - 3x^2 - 12x + 17$$

SOLUTION Since the same polynomial is to be evaluated repeatedly, it is not necessary to recopy it each time. The work can be arranged as shown below:

Try to do the work mentally:
$1 \cdot 2 + (-3) = -1;$
$1 \cdot (-1) + (-12) = -13;$
and so on.

	2	−3	−12	17	POINT
1	2	−1	−13	4	(1, 4)
−1	2	−5	−7	24	(−1, 24)
2	2	1	−10	−3	(2, −3)
−2	2	−7	2	13	(−2, 13)
0				17	(0, 17) ← Why is $f(0)$ always equal to the constant term?
3	2	3	−3	8	(3, 8)
−3	2	−9	15	−28	(−3, −28)
4	2	5	8	49	(4, 49)
−4	2	−11	32	−111	(−4, −111)

CALCULATOR COMMENT

If you have a graphing calculator you can check the results of Example 1 by inputting

$Y1 = 2X^3 - 3X^2 - 12X + 17$

and then use the TRACE to find the desired points.

After you have found enough points, which is a much less tedious procedure when you use synthetic division, you can connect these points to draw a smooth curve.

EXAMPLE 2 Use the information in Example 1 to sketch the graph of the polynomial function

$$f(x) = 2x^3 - 3x^2 - 12x + 17$$

SOLUTION Plot the points from Example 1, paying attention to the scales on the axes so that they accommodate most of the values obtained. Connect the points to draw a smooth curve, as shown in Figure 3.20.

FIGURE 3.20 Graph of $f(x) = 2x^3 - 3x^2 - 12x + 17$

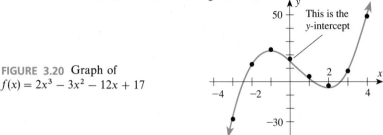

EXAMPLE 3 Sketch the graph of $f(x) = 3x^4 - 8x^3 - 30x^2 + 72x + 47$.

SOLUTION *You* select the integers that are convenient.

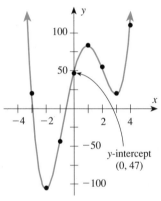

FIGURE 3.21 Graph of $f(x) = 3x^4 - 8x^3 - 30x^2 + 72x + 47$

	3	−8	−30	72	47	POINT
−4	3	−20	50	−128	559	(−4, 559)
−3	3	−17	21	9	20	(−3, 20)
−2	3	−14	−2	76	−105	(−2, −105)
−1	3	−11	−19	91	−44	(−1, −44)
0					47	(0, 47)
1	3	−5	−35	37	84	(1, 84)
2	3	−2	−34	4	55	(2, 55)
3	3	1	−27	−9	20	(3, 20)
4	3	4	−14	16	111	(4, 111)

Plot the points and draw a smooth curve as in Figure 3.21.

EXAMPLE 4 Sketch the graph of $y = (x + 1)(x - 2)(3x - 2)$.

SOLUTION Begin by locating the *critical values* for the polynomial function. Recall that these are the x values for which $y = 0$ (the x-intercepts):

$$(x + 1)(x - 2)(3x - 2) = 0 \qquad y = 0$$

$$x = -1, 2, \frac{2}{3}$$

These critical values divide the x-axis into four regions:

$$x < -1 \quad -1 < x < \tfrac{2}{3} \quad \tfrac{2}{3} < x < 2 \quad x > 2$$

Since $y = 0$ for $x = -1$, $x = \frac{2}{3}$, and $x = 2$, it follows that y must be either positive or negative in each of the four regions. That is, you need to determine whether y is positive or negative in *each* of the four regions. To do this, select an x value in *each* region and substitute that value into the function, as shown below:

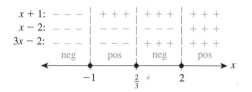

Wherever you have labeled the number line as *neg*, the graph is below the axis; and wherever you have labelled it *pos*, the graph is above the axis.

Plot some points (to the nearest tenth) using synthetic division to draw the graph, as shown in Figure 3.22. Synthetic division requires the coefficients in polynomial (not factored) form. Thus,

$$y = (x + 1)(x - 2)(3x - 2)$$
$$= 3x^3 - 5x^2 - 4x + 4$$

	3	−5	−4	4	POINT
−2	3	−11	18	−32	$(-2, -32)$ ← Need a point to the left of the critical value −1.
−.3	3	−5.9	−2.23	4.669	$(-.3, 4.7)$ ← Need a point between −1 and 0.
0				4	$(0, 4)$
1	3	−2	−6	−2	$(1, -2)$
1.5	3	−.5	−4.75	−3.125	$(1.5, -3.1)$ ← Need a point between 1 and 2.
3	3	4	8	28	$(3, 28)$ ← Need a point to the right of 2.

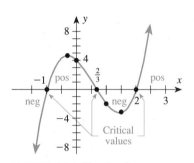

FIGURE 3.22 Graph of $y = (x + 1)(x - 2)(3x - 2)$ or $y = 3x^3 - 5x^2 - 4x + 4$

3.5 Problem Set

A *Use synthetic division to find the values specified for the functions given in Problems 1–12.*

1. $f(x) = 5x^3 - 7x^2 + 3x - 4$
   a. $f(1)$ **b.** $f(-1)$ **c.** $f(0)$ **d.** $f(6)$ **e.** $f(-4)$

2. $g(x) = 4x^3 + 10x^2 - 120x - 350$
   a. $g(1)$ **b.** $g(-1)$ **c.** $g(0)$ **d.** $g(5)$ **e.** $g(-5)$

3. $P(x) = x^4 - 10x^3 + 20x^2 - 23x - 812$
   a. $P(1)$ **b.** $P(-1)$ **c.** $P(0)$ **d.** $P(3)$ **e.** $P(7)$

4. $f(t) = 3t^4 + 5t^3 - 8t^2 - 3t$
   a. $f(0)$ **b.** $f(1)$ **c.** $f(2)$ **d.** $f(-2)$ **e.** $f(-4)$

5. $g(t) = 4t^4 - 3t^3 + 5t - 10$
   a. $g(0)$ **b.** $g(-1)$ **c.** $g(-2)$ **d.** $g(5)$ **e.** $g(-3)$

6. $P(t) = t^4 - t^3 - 39t - 70$
   a. $P(0)$ **b.** $P(1)$ **c.** $P(-1)$ **d.** $P(-5)$ **e.** $P(7)$

7. $f(h) = 8h^4 - 6h^3 + 5h^2 + 4h - 3$
   a. $f(0)$ **b.** $f(1)$ **c.** $f(\frac{1}{2})$ **d.** $f(-\frac{1}{2})$ **e.** $f(-3)$

8. $g(h) = 16h^4 + 64h^3 + 19h^2 - 81h + 18$
   a. $g(0)$ **b.** $g(-2)$ **c.** $g(-3)$ **d.** $g(\frac{1}{4})$ **e.** $g(\frac{3}{4})$

9. $P(x) = 4x^4 - 8x^3 - 43x^2 + 29x + 60$
   a. $P(-1)$ **b.** $P(1)$ **c.** $P(4)$ **d.** $P(\frac{3}{2})$ **e.** $P(-\frac{5}{2})$

10. $f(x) = (x - 2)(x + 3)(2x - 5)$
   a. $f(2)$ **b.** $f(-3)$ **c.** $f(\frac{5}{2})$ **d.** $f(1)$ **e.** $f(-2)$

11. $g(x) = (x - 1)(x - 4)(2x + 1)$
   a. $g(1)$ **b.** $g(4)$ **c.** $g(-\frac{1}{2})$ **d.** $g(-2)$ **e.** $g(-1)$

12. $h(x) = (x + 1)(x + 2)(3x - 1)$
   a. $h(-1)$ **b.** $h(1)$ **c.** $h(-2)$ **d.** $h(2)$ **e.** $h(\frac{1}{3})$

Sketch the graph of each polynomial function near the origin in Problems 13–18.

13. $f(x) = x^3 - 3x^2 + 10$
14. $f(x) = x^3 + 3x^2 + 11$
15. $f(x) = 2x^3 - 3x^2 - 12x + 3$
16. $f(x) = 2x^3 + 3x^2 - 12x + 48$
17. $f(x) = x^3 - 2x^2 + x - 5$
18. $f(x) = x^3 + 4x^2 - 3x + 2$

B

WHAT IS WRONG, *if anything, with each of the graphs shown for the equations given in Problems 19–24? Explain your reasoning and draw the correct graph.*

19. $y = x^5 + 2x^4 - 6x^3 + 2x - 3$

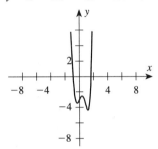

20. $y = -x^5 + 5x^4 - 6x^3 + 2x + 20$

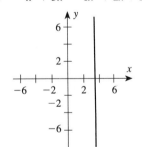

21. $y = -x^5 + 3x^4 + x + 4$ **22.** $y = x^5 - 4x^3 + x - 4$

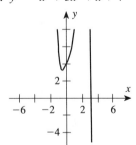 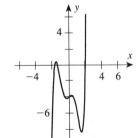

23. $y = (x + 1)(x + 4)(x - 2)(x - 3)(x - 4)$

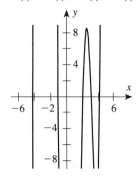

24. $y = (x + 2)(x + 1)(x - 1)(x - 3)(x - 5)$

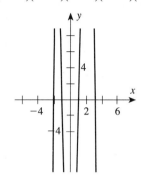

Sketch the graph of each polynomial function near the origin in Problems 25–60.

25. $f(x) = x^3 - 6x^2 + 9x - 9$
26. $f(x) = 2x^3 - 3x^2 - 36x + 78$
27. $f(x) = 4x^4 - 8x^3 - 43x^2 + 29x + 60$
28. $f(x) = 16x^4 + 64x^3 + 19x^2 - 81x + 18$
29. $f(x) = x^4 - 7x^2 - 2x + 2$
30. $f(x) = x^4 - 14x^3 + 58x^2 - 46x - 9$
31. $f(x) = x^6 - 4x^4 - 4x^2 + 4$
32. $f(x) = x^5 + 2x^4 - 5x^3 - 10x^2 + 4x + 8$
33. $y = 3x^4 - x^3 - 14x^2 + 4x + 8$
34. $y = 5x^4 + 3x^3 - 22x^2 - 12x + 8$
35. $y = x^4 - x^3 - 3x^2 + 2x + 4$
36. $y = x^4 - 2x^2 - 4x + 3$
37. $y = (x - 1)(x + 1)(x + 3)$
38. $y = (x - 1)(x - 4)(x + 3)$
39. $y = (x + 1)(x + 3)(2x - 5)$
40. $y = (x - 1)(x - 4)(2x + 1)$
41. $y = (x + 1)(x + 2)(3x - 1)$
42. $y = x(x - 3)(x + 3)$
43. $y = 3x^2(x - 3)(x + 1)$
44. $y = 5x^2(x - 4)(x + 2)$

45. $y = x^2(x^2 - 1)$

46. $y = x^2(x^2 - 4)$

47. $f(x) = 3x^4 - 7x^3 + 5x^2 + x - 10$

48. $f(x) = 8x^4 + 12x^3 - 3x^2 + 4x + 20$

49. $f(x) = x^5 - 3x^4 + 2x^3 - 7x + 15$

50. $f(x) = x^5 - 5x^4 + 3x^3 + x^2$

51. a. $y = x^5$ **b.** $y = (x - 1)^5 + 10$

52. a. $y = x^6$ **b.** $y = (x + 2)^6 - 10$

53. a. $y = x^7$ **b.** $y = -.01x^7$

54. a. $y = x^8$ **b.** $y = -\frac{1}{8}x^8$

55. $y = |x^3 + 3x^2 - x - 3|$

56. $y = |x|^3 + 3|x|^2 - |x| - 3$

57. $y = |-x|^3 + 2|x|^2 + |x| - 2$

58. $y = |-x^3 + 2x + x - 2|$

59. $y = -|x|^3 + 2|x|^2 + |x| - 2$

60. $y = x^3 + 2x^2 + |x| - 2$

3.6   Real Roots of Polynomial Equations

If

$$P(x) = a_n x^n + a_{n-1} x^{n-1} + \cdots + a_2 x^2 + a_1 x + a_0 \quad (a_n \neq 0)$$

then the *roots* or *solutions* of $P(x) = 0$ are values of x that satisfy this equation. Such an equation is called a **polynomial equation.** Recall from Chapter 2 that a is called a *zero* of a function P if $P(a) = 0$. Thus we speak of the **roots** of a polynomial equation and the **zeros** of a polynomial function. If a zero is a real number, then it is an **x-intercept** of the graph of the polynomial function.

Consider the polynomial equation

$$4x^4 - 8x^3 + 43x^2 + 29x + 60 = 0$$

To solve this equation, find the values of x that make it true. As a first step, write the equation in factored form if possible. From the Division Algorithm, we know that if $R = 0$ and

$$P(x) = Q(x)(x - r) + R$$

then $P(x) = Q(x)(x - r)$. This says that $x - r$ is a factor of $P(x)$. But since you are looking for values of x such that $P(x) = 0$,

$$0 = Q(x)(x - r)$$

Notice that this equation is satisfied by $x = r$ and leads to a result called the **Factor Theorem.** The polynomial equation $Q(x) = 0$ is called a **depressed equation** of the polynomial equation $P(x) = 0$.

FACTOR THEOREM

> If r is a root of the polynomial equation $P(x) = 0$, then $x - r$ is a factor of $P(x)$. Moreover, if $x - r$ is a factor of $P(x)$, then r is a root of the polynomial equation $P(x) = 0$.

The Factor Theorem is a generalization of the method of solving quadratic equations by factoring. *Remember:* If $P \cdot Q = 0$, then $P = 0$ or $Q = 0$ (perhaps both). Suppose you wish to solve

$$x^2 - x - 2 = 0$$

You can do so by factoring:

$$(x - 2)(x + 1) = 0$$

Then each factor is set equal to zero to find

$$x = 2, -1$$

Suppose further that the same factor appears more than once in the factoring process, as in

$$(x - 3)(x - 3) = 0$$

Then there is a single root (which occurs twice) to give

$$x = 3$$

In this chapter it will be useful to attach some terminology to a repeated root. If a factor $x - r$ occurs exactly k times in the factorization of $P(x)$, then r is called a **zero of multiplicity k.** If a factor $x - r$ occurs exactly k times in the factorization of $P(x)$ in the equation $P(x) = 0$, then r is called a **root of multiplicity k.** In the preceding illustration, 3 is a root of multiplicity 2. If

$$f(x) = (x - 1)(x - 3)(x - 4)(x - 1)(x - 3)(x - 1)$$

the **zeros** are 1 (multiplicity 3), 3 (multiplicity 2), and 4. Notice that this is a function, not an equation, so we use the word *zero*. On the other hand,

$$(x - 1)(x - 3)(x - 4)(x - 1)(x - 3)(x - 1) = 0$$

has **roots** 1 (multiplicity 3), 3 (multiplicity 2), and 4. This is an equation, so we use the word *root* (or *solution*).

The relationship between roots of the polynomial equation $f(x) = 0$ and factors of $f(x)$, as described by the Factor Theorem, leads to a statement concerning the number of roots to expect for a polynomial equation of degree n. Suppose $P(x) = 0$ is a polynomial equation in which $P(x)$ is a third-degree polynomial. It is impossible for $P(x) = 0$ to have more than three roots. If it had more, say four roots (r_1, r_2, r_3, r_4), then the Factor Theorem provides

$$P(x) = a_4(x - r_1)(x - r_2)(x - r_3)(x - r_4)$$

so that $P(x)$ is a *fourth-degree* polynomial. Thus a third-degree polynomial equation cannot have more than three roots.

The related question—Does it necessarily have three zeros—is answered by the Fundamental Theorem of Algebra, which is considered in the next section.

ROOT LIMITATION THEOREM	A polynomial function f of degree n has, at most, n distinct zeros.

As you saw in the previous section, there is a close relationship between the zeros of a polynomial function and its graph. Suppose, by synthetic division, you find for two values a and b that $P(a)$ and $P(b)$ are opposite in sign. Then from the Intermediate-Value Theorem for polynomial functions, there is a value r such that $a < r < b$ and $P(r) = 0$. We will use this important result to approximate real roots.

LOCATION THEOREM

If f is a polynomial function such that $f(a)$ and $f(b)$ are opposite in sign, then there is at least one real zero on the interval between a and b.

Next you need some reasonable method for finding the zeros, since you cannot simply find them by trial and error, even with the Location Theorem and synthetic division. The best that mathematicians can offer is a theorem that provides a list of *possible* rational roots of the polynomial equation $f(x) = 0$. Not every number on the list will be a root, but every rational root of the polynomial equation will appear someplace on the list. Given this *finite* list (which admittedly might be large), you can check values from this list using synthetic division and the Location Theorem.

RATIONAL ROOT THEOREM

If $P(x) = a_n x^n + a_{n-1} x^{n-1} + \cdots + a_1 x + a_0$ has integer coefficients and p/q (where p/q is reduced) is a rational zero, then p is a factor of a_0 and q is a positive factor of a_n.

To use this theorem, make a list of *all* factors of a_0 and divide these integers by the factors of a_n. Notice also that if $a_n = 1$, then all rational zeros are integers that divide a_0. The procedure for finding all possible rational roots is not very difficult if you work systematically.

EXAMPLE 1 List all possible rational roots of $x^3 - x^2 - 4x + 4 = 0$.

SOLUTION p: $a_0 = 4$, with factors $1, -1, 2, -2, 4, -4$ Shorten these lists by using the
q: $a_n = 1$, with factor 1 $\pm$ sign; p: $\pm 1, \pm 2, \pm 4$

Form all possible fractions:

$$\frac{p}{q}: \quad \frac{1}{1}, \frac{-1}{1}, \frac{2}{1}, \frac{-2}{1}, \frac{4}{1}, \frac{-4}{1}$$

Simplifying and not bothering to rewrite those that are repeated, we find that the possible rational roots are $\pm 1, \pm 2, \pm 4$. ∎

CALCULATOR COMMENT

By graphing

$$\boxed{Y = X^3 - X^2 - 4X + 4}$$

along with the Rational Root Theorem, you can find most roots.

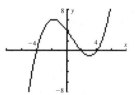

Using the $\boxed{\text{TRACE}}$, we find:

$\boxed{X = -2 \quad Y = 0}$

$\boxed{X = .94736842 \quad Y = .16328911}$

$\boxed{X = 2 \quad Y = 0}$

Since this is third degree, there are at most three rational roots. Use the list of possible roots in Example 1 as well as the $\boxed{\text{TRACE}}$ to conclude that the roots of the equation given in Example 1 are -2, 1, 2 (with the help of the Rational Root Theorem).

EXAMPLE 2 List all possible rational zeros of $P(x) = 4x^4 - 8x^3 - 43x^2 + 29x + 60$.

SOLUTION $p\ (a_0 = 60): \quad \pm 1, \pm 2, \pm 3, \pm 4, \pm 5, \pm 6, \pm 10, \pm 12, \pm 15, \pm 20, \pm 30, \pm 60$

$q\ (a_n = 4): \quad 1, 2, 4$

$$\frac{p}{q}: \pm 1, \pm \frac{1}{2}, \pm \frac{1}{4}, \pm 2, \pm 3, \pm \frac{3}{2}, \pm \frac{3}{4}, \pm 4, \pm 5,$$

$$\pm \frac{5}{2}, \pm \frac{5}{4}, \pm 6, \pm 10, \pm 12, \pm 15, \pm \frac{15}{2}, \pm \frac{15}{4},$$

$$\pm 20, \pm 30, \pm 60$$

Note: If a factor is already listed, it is not repeated. For example, $\pm \frac{2}{4}$ is not listed separately from $\pm \frac{1}{2}$.

CALCULATOR COMMENT

Can you determine which of the possible rational roots from the list in Example 2 are actually the roots? To graph P in Example 2 it is necessary to adjust the standard scale. For example, let

| Xmin = −3 | Ymax = 5 |
| Ymin = −125 | Ymax = 125 |

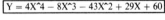

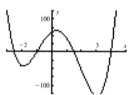

Using the TRACE we find:

| X = 1.4631579   Y = 3.6494843 | X = 3.9894737   Y = −3.393971 |
| X = −.9789474   Y = 1.5809403 | X = −2.494737   Y = −.8164336 |

From the list of possible rational roots we find the roots: $\frac{3}{2}$, 4, −1, and $-\frac{5}{2}$.

You can see from the examples that the list of possible rational zeros may be quite large, but it is a *finite* list, so with enough time and effort the entire list could be checked. Usually this is not necessary if you first pick the values that are easiest to check. (That is, do not start with the fractions or large numbers.) There is also another theorem, called the **Upper and Lower Bound Theorem,** that helps to rule out many of the values listed with the Rational Root Theorem. If f is a polynomial function, then a real number b is called an **upper bound** for the polynomial equation $f(x) = 0$ if there is no root, or solution, larger than b. A real number a is a **lower bound** if there is no solution less than a.

UPPER AND LOWER BOUND THEOREM

If $a > 0$ and, in the synthetic division of $P(x)$ by $x - a$, all the numbers in the last row have the *same sign,* then a is an *upper bound* for the roots of $P(x) = 0$.

If $b < 0$ and, in the synthetic division of $P(x)$ by $x - b$, the numbers in the last row *alternate in sign,* then b is a *lower bound* for the roots of $P(x) = 0$.

EXAMPLE 3   Solve $2x^4 - 5x^3 - 8x^2 + 25x - 10 = 0$.

SOLUTION

$$p\,(a_0 = -10): \quad \pm 1, \ \pm 2, \ \pm 5, \ \pm 10$$

$$q\,(a_n = 2): \quad 1, \ 2$$

$$\frac{p}{q}: \quad \pm 1, \ \pm \frac{1}{2}, \ \pm 2, \ \pm 5, \ \pm \frac{5}{2}, \ \pm 10$$

	2	-5	-8	25	-10
1	2	-3	-11	14	4
-1	2	-7	-1	26	-36
2	2	-1	-10	5	0

Begin with the values that are easiest to check. In Examples 3–5 we will use this shaded portion to tell you that the value checked is not a root.

$x - 2$ is a factor.

Since $x - 2$ is a factor, you can write the polynomial equation as

$$(x - 2)(2x^3 - x^2 - 10x + 5) = 0$$

Now focus your attention on the depressed equation

$$2x^3 - x^2 - 10x + 5 = 0$$

by using these coefficients in the synthetic division process. It is not necessary to recopy these coefficients in your work.

	2	-1	-10	5
-2	2	-5	0	5
5	2	9	35	180
-5	2	-11	45	-220
$\frac{1}{2}$	2	0	-10	0

← All sums are positive, so 5 is an upper bound. No larger values need to be checked.

← Sums have alternating signs, so -5 is a lower bound. No smaller values need be checked.

← $x - \frac{1}{2}$ is a factor. The resulting depressed equation is now quadratic, so you can stop the synthetic division.

$$(x - 2)(x - \tfrac{1}{2})(2x^2 - 10) = 0$$

$$2(x - 2)(x - \tfrac{1}{2})(x^2 - 5) = 0$$

Note: The resulting quadratic may not have rational roots. In fact, you may need the quadratic formula for its solution

The roots are **$2, \ \frac{1}{2}, \ \sqrt{5}, \ -\sqrt{5}$.**

CALCULATOR COMMENT

If you have a graphing calculator, you can use it to approximate the roots, but the calculator may not give you the exact roots. For Example 3, the graph of

$\boxed{Y1 = 2X^4 - 5X^3 - 8X^2 + 25X - 10}$

looks like this:

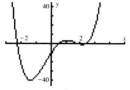

You can verify that the graph crosses the x-axis (remember that the root of the equation is a zero of the function) between -3 and -2. You can also see that the graph crosses at $\frac{1}{2}$. However, it is more difficult to see the other positive roots, so we zoom in on the graph near the point 2:

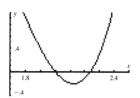

Notice that the zeros appear to be at 2 and at about 2.24. Since we know $\sqrt{5}$ (the answer found in Example 3) is approximately equal to 2.236, we see that the graph serves as a check on the work of this section, but the graph itself does not tell us that two of the roots in Example 3 are $\sqrt{5}$ and $-\sqrt{5}$.

EXAMPLE 4   Solve $4x^4 - 8x^3 - 43x^2 + 29x + 60 = 0$.

SOLUTION   The list of possible rational roots is shown in Example 2.

	4	−8	−43	29	60
1	4	−4	−47	−18	42
−1	4	−12	−31	60	0
2	4	−4	−39	−18	
−2	4	−20	9	42	
−3	4	−24	41	−63	
$-\frac{5}{2}$	4	−22	24	0	

← $x + 1$ is a factor, so −1 is a root. Do not recopy, but use these *coefficients* as you continue synthetic division.

← −3 is a lower bound. Since the Location Theorem says there is a root between −2 and −3, look on the list of possible rational roots for the next number to try.

$x + \frac{5}{2}$ is a factor. The resulting depressed equation is quadratic, so now you can write the given polynomial equation in factored form and complete the factoring process directly.

$$4x^4 - 8x^3 - 43x^2 + 29x + 60 = 0$$

$$(x + 1)(x + \tfrac{5}{2})(4x^2 - 22x + 24) = 0$$ Do you see where these factors come from in the synthetic division process?

$$(x + 1)(x + \tfrac{5}{2})(2)(2x^2 - 11x + 12) = 0$$

$$(x + 1)(x + \tfrac{5}{2})(2)(2x - 3)(x - 4) = 0$$ Now use the Factor Theorem.

The roots are **−1, $-\frac{5}{2}$, $\frac{3}{2}$ and 4.**

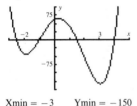

CALCULATOR COMMENT

The graph for the function

$$\boxed{Y = 4X^4 - 8X^3 - 43X^2 + 29X + 60}$$

verifies the roots found in Example 4:

Xmin = −3   Ymin = −150
Xmax = 5   Ymax = 100

EXAMPLE 5   Solve $8x^5 - 44x^4 + 86x^3 - 73x^2 + 28x - 4 = 0$.

SOLUTION

$$p\ (a_0 = -4):\quad \pm 1,\ \pm 2,\ \pm 4$$

$$q\ (a_n = 8):\quad 1,\ 2,\ 4,\ 8$$

$$\frac{p}{q}:\quad \pm 1,\ \pm \frac{1}{2},\ \pm \frac{1}{4},\ \pm \frac{1}{8},\ \pm 2,\ \pm 4$$

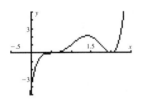

Xmin = −1   Ymin = −4
Xmax = 3   Ymax = 5

⊗ Do not forget possible multiple roots. Just because $x − 2$ was a factor once does not mean it cannot be again. ⊗

	8	−44	86	−73	28	−4
1	8	−36	50	−23	5	1
−1	8	−52	138	−211	239	−243
2	8	−28	30	−13	2	0
4	8	4	46	171	686	
2	8	−12	6	−1	0	
2	8	4	14	27		
$\frac{1}{2}$	8	−8	2	0		

← Lower bound

← $x − 2$ is a factor

← Upper bound

← $x − 2$ is a factor.

← Upper bound; this is a new upper bound because we are working with the depressed equation.

← $x − \frac{1}{2}$ is a factor.

$$(x - 2)(x - 2)(x - \tfrac{1}{2})(8x^2 - 8x + 2) = 0$$

$$2(x - 2)(x - 2)(x - \tfrac{1}{2})(4x^2 - 4x + 1) = 0 \qquad \text{Common factor}$$

$$2(x - 2)(x - 2)(x - \tfrac{1}{2})(2x - 1)(2x - 1) = 0$$

The roots are 2 and $\tfrac{1}{2}$; 2 is a root of multiplicity 2 and $\tfrac{1}{2}$ is a root of multiplicity 3. ■

In Examples 3 and 4 the degree was 4 and in both cases you found four roots. The Root Limitation Theorem tells you that there cannot be more roots. Even when you find multiple roots as illustrated by Example 5, you know you have found all the real roots. Suppose, however, that you attempt to solve a polynomial equation that has fewer *real* roots than the degree. In such a situation you will not know whether you cannot find the real roots because they *are not* rational but are still real (that is, they are *x*-intercepts) or because they are simply not real numbers. The following theorem will help you to answer this dilemma by telling you when you have found all the positive or negative real roots. When applying this theorem, remember that a root of multiplicity m is counted as m roots.

<table>
<tr><td>DESCARTES' RULE OF SIGNS</td><td>Let $P(x)$ define a polynomial function with real coefficients written in descending powers of x. Count the number of sign changes in the signs of the coefficients.

1. The number of positive real zeros is equal to the number of sign changes or is equal to that number decreased by an even integer.

2. The number of negative real zeros is equal to the number of sign changes in $P(-x)$ or is equal to that number decreased by an even integer.</td></tr>
</table>

EXAMPLE 6

$$f(x) = 2x^4 - 5x^3 - 8x^2 + 25x - 10 \qquad \text{From Example 3}$$

There are three sign changes, so there are three or one positive real zeros. Next calculate $f(-x)$:

$$f(-x) = 2(-x)^4 - 5(-x)^3 - 8(-x)^2 + 25(-x) - 10$$
$$= 2x^4 + 5x^3 - 8x^2 - 25x - 10$$

There is one sign change on the coefficients of $f(-x)$, so there is one negative real zero. Compare this result with the answer you found in Example 3:

$$2, \quad \tfrac{1}{2}, \quad \sqrt{5}, \qquad -\sqrt{5}$$

3 positive zeros   1 negative zero

■

EXAMPLE 7

$$f(x) = 8x^5 - 44x^4 + 86x^3 - 73x^2 + 28x - 4 \qquad \text{From Example 5}$$

There are five sign changes here, so there are five, three, or one positive real roots.

$$f(-x) = -8x^5 - 44x^4 - 86x^3 - 73x^2 - 28x - 4$$

Here there are no sign changes, so there are no negative real roots. Compare this result with the answer you found in Example 5: 2, $\frac{1}{2}$. There seems to be a discreplicity, but remember that 2 has multiplicity 2 and $\frac{1}{2}$ has multiplicity 3, *so think*:

$$2, \quad 2, \quad \underbrace{\frac{1}{2}, \quad \frac{1}{2}, \quad \frac{1}{2}}$$

5 positive real roots and no negative real roots

The following example ties together some of the ideas of this section and the preceding section.

EXAMPLE 8   Solve $x^4 - 3x^2 - 6x - 2 = 0$.

SOLUTION $p\ (a_0 = -2)$: $\pm 1, \pm 2$ $q\ (a_n = 1)$:   1 $\dfrac{p}{q}$: $\pm 1, \pm 2$

	1	0	−3	−6	−2	
1	1	1	−2	−8	−10	
2	1	2	1	−4	−10	
−1	1	−1	−2	−4	2	
−2	1	−2	1	−8	14	← Lower bound

There are no rational roots (we have tried all the numbers on our list). Next verify the type of roots by using Descartes' Rule of Signs:

$$f(x) = \underbrace{x^4 - 3x^2}_{1} - 6x - 2 \qquad\qquad f(-x) = \underbrace{x^4 - 3x^2}_{1} \underbrace{+\, 6x}_{2} \underbrace{-\, 2}_{3}$$

One positive root   Three or one negative roots

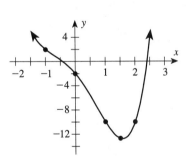

FIGURE 3.23 Graph of $y = x^4 - 3x^2 - 6x - 2$

Use synthetic division to find some additional points: $(0, -2)$ for the y-intercept; $(1.5, -12.6875)$; and $(3, 34)$ for the upper bound. Plot the known points as shown in Figure 3.23.

By looking at the graph, you would expect roots between -1 and 0 as well as between 2 and 3. Moreover, because of the upper and lower bounds, as well as Descartes' Rule of Signs, you would expect these to be the only real roots. Since these roots are not rational, you will not be able to find them exactly. You can, however, approximate them to any desired degree of accuracy by using synthetic division. For the root between -1 and 0:

A calculator would be very helpful.

	1	0	−3	−6	−2
−.5	1	−.5	−2.75	−4.625	.3125
−.4	1	−.4	−2.84	−4.864	−.0544
−.41	1	−.41	−2.8319	−4.8389	−.0160
−.42	1	−.42	−2.8236	−4.8141	.0219

Root between −.5 and −.4, since one is negative and the other is positive

Root between −.41 and −.42

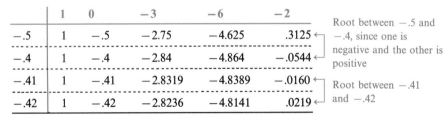

Continue in this fashion to approximate the root to any degree of accuracy desired. (This would be a good problem for a computer.) Repeat the procedure

for the root between 2 and 3 (it is about 2.41). Thus the real roots (to the nearest tenth) are **−.4 and 2.4**.

CALCULATOR COMMENT

The successive approximations shown in Example 8 can easily be done using the zoom feature on a graphing calculator.

Zoom 1 detail of Figure 3.23:   Zoom 2 detail of Figure 3.23:   Zoom 3 detail of Figure 3.23:

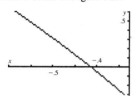   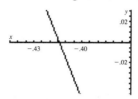   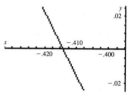

Notice that we could obtain even greater accuracy if desired. The $\boxed{\text{TRACE}}$ will show the coordinates on the curve near the x-intercept.

3.6 Problem Set

A *In Problems 1–12 find the zeros of the polynomial and state the multiplicity of each zero.*

1. $f(x) = (x - 2)(x + 3)^2$ **2.** $f(x) = (x + 1)^2(x - 3)^2$

3. $f(x) = x^3(2x - 3)^2$ **4.** $f(x) = x^2(3x + 1)^3$

5. $f(x) = (x^2 - 1)^2(x + 2)$ **6.** $f(x) = (x - 5)(x^2 - 4)^3$

7. $f(x) = (x^2 + 2x - 15)^2$ **8.** $f(x) = (6x^2 + 7x - 3)^2$

9. $f(x) = x^4 - 8x^3 + 16x^2$ **10.** $f(x) = x^4 + 6x^3 + 9x^2$

11. $f(x) = (x^3 - 9x)^2$ **12.** $f(x) = (x^3 - 25x)^2$

In Problems 13–24 use Descartes' Rule of Signs to state the number of possible positive and negative real roots.

13. $x^4 - 3x^3 + 7x^2 - 19x + 15 = 0$

14. $3x^3 - 7x^2 + 5x + 7 = 0$

15. $2x^5 + 6x^4 - 3x + 12 = 0$

16. $x^3 + 2x^2 - 5x - 6 = 0$

17. $x^3 + 3x^2 - 4x - 12 = 0$

18. $2x^3 + x^2 - 13x + 6 = 0$

19. $2x^3 - 3x^2 - 32x - 15 = 0$

20. $x^4 - 12x^3 + 54x^2 - 108x + 81 = 0$

21. $x^4 + 3x^3 - 20x^2 - 3x + 18 = 0$

22. $x^4 - 13x^2 + 36 = 0$

23. $2x^2 + 6x - 3x^3 - 4 = 0$

24. $5x^3 - 2x^4 + x^2 - 7 = 0$

Find the possible rational roots for the polynomial equations in Problems 25–36.

25. $x^4 - 3x^3 + 7x^2 - 19x + 15 = 0$

26. $3x^3 - 7x^2 + 5x + 7 = 0$

27. $2x^5 + 6x^4 - 3x + 12 = 0$

28. $x^3 + 2x^2 - 5x - 6 = 0$

29. $x^3 + 3x^2 - 4x - 12 = 0$

30. $2x^3 + x^2 - 13x + 6 = 0$

31. $2x^3 - 3x^2 - 32x - 15 = 0$

32. $x^4 - 12x^3 + 54x^2 - 108x + 81 = 0$

33. $x^4 + 3x^3 - 20x^2 - 3x + 18 = 0$

34. $x^4 - 13x^2 + 36 = 0$

35. $5x^2 + 6x^3 - 2x - 1 = 0$

36. $10x^2 - 8x^3 + 17x - 10 = 0$

B *Solve the polynomial equations in Problems 37–56.*

37. $x^3 - x^2 - 4x + 4 = 0$

38. $2x^3 - x^2 - 18x + 9 = 0$

39. $x^3 - 2x^2 - 9x + 18 = 0$

40. $x^3 + 2x^2 - 5x - 6 = 0$

41. $x^3 + 3x^2 - 4x - 12 = 0$

42. $2x^3 + x^2 - 13x + 6 = 0$

43. $2x^3 - 3x^2 - 32x - 15 = 0$

44. $x^4 - 12x^3 + 54x^2 - 108x + 81 = 0$

45. $x^4 + 3x^3 - 19x^2 - 3x + 18 = 0$

46. $x^4 - 13x^2 + 36 = 0$

47. $x^3 + 15x^2 + 71x + 105 = 0$

48. $x^3 - 15x^2 + 74x - 120 = 0$

49. $8x^3 - 12x^2 - 66x + 35 = 0$

50. $12x^3 + 16x^2 - 7x - 6 = 0$

51. $x^5 + 8x^4 + 10x^3 - 60x^2 - 171x - 108 = 0$

52. $x^5 + 6x^4 + x^3 - 48x^2 - 92x - 48 = 0$

53. $x^7 + 2x^6 - 4x^5 - 2x^4 + 3x^3 = 0$

54. $x^6 - 3x^4 + 3x^2 - 1 = 0$

55. $x^6 - 12x^4 + 48x^2 - 64 = 0$

56. $x^7 + 3x^6 - 4x^5 - 16x^4 - 13x^3 - 3x^2 = 0$

57. Does there exist a real number that exceeds its cube by 1?

58. CONSUMER The dimensions of a rectangular box are consecutive integers, and its volume is 720 cm^3. What are the dimensions of the box?

59. ENGINEERING A 2-cm-thick slice is cut from a cube, leaving a volume of 384 cm^3. What is the length of a side of the original cube?

60. ENGINEERING A rectangular sheet of tin with dimensions 3×5 m has equal squares cut from its four corners. The resulting sheet is folded up on the sides to form a topless box. Find all possible dimensions of the cutout square to the nearest .1 m such that the box has a volume of 1 m^3.

C HISTORICAL QUESTIONS *Italian mathematicians discovered the algebraic solution of cubic and quartic equations in the sixteenth century. At that time they would challenge one another to solve certain equations. Problems 61–64 were such challenge problems. Find the real roots in each problem to the nearest tenth.*

61. In 1515, Scipione del Ferro solved the cubic equation $x^3 + mx + n = 0$ and revealed his secret to his pupil Antonio Fior. At about the same time, Tartaglia solved the equation $x^3 + px^2 = n$. Fior thought Tartaglia was bluffing and challenged him to a public contest of solving cubic equations. According to the historian Howard Eves, Tartaglia triumphed completely in this contest. Solve the cubic $x^3 + px^2 = n$, where $p = 5$ and $n = 21$.

62. Girolamo Cardano stole the solution of the cubic equation from Tartaglia and published in his *Ars Magna*. Tartaglia protested, but Cardano's pupil,

Ludovico Ferrari (who solved the biquadratic equation), claimed that both Cardano and Tartaglia stole it from del Ferro. According to the historian Howard Eves, there was a dispute from which Tartaglia was lucky to have escaped alive. One of the problems in *Ars Magna* was $x^3 - 63x = 162$. Solve this cubic.

63. Cardano solved the quartic $13x^2 = x^4 + 2x^3 + 2x + 1$. Find the real roots for this equation.

64. In 1540, Cardano was given the problem "Divide 10 into three parts such that they shall be in continued proportion and that the product of the first two shall be 6." Let x, y, and z be the three parts. Then $x + y + z = 10$. Also,

$$\frac{x}{y} = \frac{y}{z} \quad \text{and} \quad xy = 6$$

Eliminating x and z, you obtain $y^4 + 6y^2 + 36 = 60y$. Find the real roots for this equation.

65. GRAPHING CALCULATOR PROBLEM Choose a polynomial such as

$$P(x) = x^4 - 3x^2 - 6x - 2$$

a. Find an upper bound for the zeros of P.

b. What happens when you use a value larger than the upper bound as a synthetic divisor? Why does this happen?

c. Does your conclusion support the Upper and Lower Bound Theorem?

d. What happens when you use a value smaller than a known lower bound for a synthetic divisor? Why does this happen?

e. Do you still agree with the Upper and Lower Bound Theorem?

66. GRAPHING CALCULATOR PROBLEM Graph

$$P(x) = 8x^5 - 44x^4 + 86x^3 - 73x^2 + 28x - 4$$

$$Q(x) = 2(x - 2)(x - \tfrac{1}{2}) = 2x^2 - 5x + 2$$

Zoom in to each of the x-intercepts. Do the graphs appear to cross the x-axis at the same points? What causes this behavior? Discuss the similarities and the differences between the zeros of P and Q.

3.7 The Fundamental Theorem of Algebra

In 1799, a 22-year-old graduate student named Karl Gauss proved in his doctoral thesis that every polynomial equation has at least one solution in the complex numbers. This, of course, is an assumption that you have made throughout your study of algebra—from the time you solved first-degree equations in beginning algebra until now. It is an idea so basic to algebra that it is called the Fundamental Theorem of Algebra. This theorem strengthens the Root Limitation Theorem of the previous section. However, in order to prove this theorem it is necessary to allow the domain to be the set of complex numbers. Remember that the set of complex numbers has the set of real numbers as a subset so that when we speak of complex coefficients of a polynomial equation, we are including all those polynomial equations previously considered in this chapter.

FUNDAMENTAL THEOREM OF ALGEBRA

> If $P(x)$ is a polynomial of degree $n \geq 1$ with complex coefficients, then $P(x) = 0$ has at least one complex root.

If an equation has one solution, a depressed equation of 1 degree less may be obtained. That new equation, according to the Fundamental Theorem, has a root. This root may now be used to obtain an equation of lower degree. The result of this process suggests the following theorem.

NUMBER OF ROOTS THEOREM

> If $P(x)$ is a polynomial of degree $n \geq 1$ with complex coefficients, then $P(x) = 0$ has exactly n roots, where roots are counted according to their multiplicity.

Of course, the roots need not be distinct or real. Consider the following example.

EXAMPLE 1 Show that $x^6 - 2x^3 + x^2 - 2x + 2 = 0$ has at least two nonreal complex roots.

SOLUTION Check Descartes' Rule of Signs:

$$P(x) = x^6 - 2x^3 + x^2 - 2x + 2$$

$$\underset{1 \quad 2 \quad 3 \quad 4}{}$$

4, 2, or 0 positive real roots

$$P(-x) = x^6 + 2x^3 + x^2 + 2x + 2$$

0 negative real roots

The polynomial equation has six roots, and at most four of these are real (positive). **Thus, at least two roots are complex and nonreal.**

If synthetic division is used on the equation in Example 1, two positive real roots are quickly found:

	1	0	0	-2	1	-2	2
1	1	1	1	-1	0	-2	0
1	1	2	3	2	2	0	

All positive, upper bound

Notice that $x = 1$ is a root of *multiplicity* 2. When counting the number of roots using Descartes' Rule, multiple roots are *not* counted as a single root. A root of multiplicity 2 counts as two roots and roots of multiplicity n count as n roots.

Note further that the depressed equation produced by the synthetic division has all positive coefficients; thus, all the positive roots have been found. Hence, the polynomial equation $x^6 - 2x^3 + x^2 - 2x + 2 = 0$ has one real root (with multiplicity 2) and four nonreal complex roots.

We can now distinguish between roots and x-intercepts of polynomial equations. There are n roots of an nth-degree polynomial equation $P(x) = 0$. The **real roots** correspond to the x-intercepts of the graph of $y = P(x)$, whereas the **imaginary roots** do not.

EXAMPLE 2 Solve $x^4 - 2x^3 + x^2 - 8x - 12 = 0$, and draw the graph of

$$y = x^4 - 2x^3 + x^2 - 8x - 12$$

SOLUTION Check Descartes' Rule of Signs:

$$P(x) = \underbrace{x^4 - 2x^3 + x^2}_{1 \quad 2 \quad 3} - 8x - 12 \qquad\qquad P(-x) = x^4 + 2x^3 + x^2 + \underbrace{8x - 12}_{1}$$

3 or 1 positive real roots   1 negative real root

$p: \quad \pm 1, \pm 2, \pm 3, \pm 4, \pm 6, \pm 12 \qquad q: \ 1 \qquad \dfrac{p}{q}: \ \pm 1, \pm 2, \pm 3, \pm 4, \pm 6, \pm 12$

	1	-2	1	-8	-12
1	1	-1	0	-8	-20
-1	1	-3	4	-12	0
-1	1	-4	8	-20	
3	1	0	4	0	

Use this depressed equation.

Since -1 could be a multiple root, try it again.

The depressed equation is

$$x^2 + 4 = 0$$
$$x = \pm 2i$$

The roots are $-1, 3, 2i, -2i$; this is consistent with the results from Descartes' Rule of Signs (one positive and one negative real root). It is also consistent with the Number of Roots Theorem (degree 4, four roots). Now you would also expect the graph to have two x-intercepts corresponding to the real roots. The graph is shown in Figure 3.24.

	1	-2	1	-8	-12
0					-12
1	1	-1	0	-8	-20
-1	1	-3	4	-12	0
2	1	0	1	-6	-24
-2	1	-4	9	-26	40
3	1	1	4	4	0
4	1	2	9	28	100

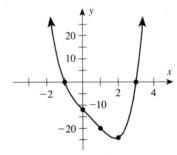

FIGURE 3.24

Graph of $y = x^4 - 2x^3 + x^2 - 8x - 12$

Another result that is sometimes helpful when finding roots is the **Conjugate Pair Theorem.**

CONJUGATE PAIR THEOREM

1. If $P(x) = 0$ is a polynomial equation with real coefficients, then when $a + bi$ is a root, $a - bi$ is also a root (a and b are real numbers).

2. If $P(x) = 0$ is a polynomial equation with rational coefficients, then when $m + \sqrt{n}$ is a root, $m - \sqrt{n}$ is also a root (m and n are rational numbers and $\sqrt{n}$ is irrational).

EXAMPLE 3   Solve $x^4 - 4x - 1 = 4x^3$ given that $2 + \sqrt{5}$ is a root.

SOLUTION $x^4 - 4x^3 - 4x - 1 = 0$

$$
\begin{array}{r|ccccc}
2+\sqrt{5} & 1 & -4 & 0 & -4 & -1 \\
 & & 2+\sqrt{5} & 1 & 2+\sqrt{5} & 1 \\
\hline
 & 1 & -2+\sqrt{5} & 1 & -2+\sqrt{5} & 0
\end{array}
$$

And $2 - \sqrt{5}$ must be a root also:

$$
\begin{array}{r|cccc}
2-\sqrt{5} & 1 & -2+\sqrt{5} & 1 & -2+\sqrt{5} \\
 & & 2-\sqrt{5} & 0 & 2-\sqrt{5} \\
\hline
 & 1 & 0 & 1 & 0
\end{array}
$$

The depressed equation is

$$x^2 + 1 = 0$$

$$x = \pm i$$

The roots are $2 \pm \sqrt{5}, \pm i.$

ALTERNATE SOLUTION   The synthetic division solution shown above is rather cumbersome, so we offer an alternate procedure. Since $2 + \sqrt{5}$ is a root, then $2 - \sqrt{5}$ must also be a root, which (by the Factor Theorem) says that the original polynomial has factors $x - (2 + \sqrt{5})$ and $x - (2 - \sqrt{5})$. Multiply these factors:

$$[x - (2 + \sqrt{5})][x - (2 - \sqrt{5})] = x^2 - (2 - \sqrt{5})x - (2 + \sqrt{5})x + (2 - \sqrt{5})(2 + \sqrt{5})$$
$$= x^2 - 4x - 1$$

Now divide (using long division) this factor into the original polynomial:

$$
\begin{array}{r}
x^2 + 1 \\
x^2 - 4x - 1 \overline{\smash{)}\, x^4 - 4x^3 + 0x^2 - 4x - 1} \\
\underline{x^4 - 4x^3 - x^2} \\
x^2 \\
\underline{x^2 - 4x - 1} \\
0
\end{array}
$$

Thus,

$$x^4 - 4x^3 - 4x - 1 = \underbrace{(x^2 - 4x - 1)}\underbrace{(x^2 + 1)}$$

Now set each factor equal to zero and solve.

$$x^2 - 4x - 1 = 0 \qquad x^2 + 1 = 0$$

$$x = 2 \pm \sqrt{5} \qquad x = \pm i$$

The roots are $2 \pm \sqrt{5},\ \pm i.$

EXAMPLE 4 Solve $x^4 - 3x^2 - 6x - 2 = 0$ given that $-1 + i$ is a root.

SOLUTION You can divide by $-1 + i$ synthetically, but instead we will multiply together the known factors:

$$[x - (-1 - i)][x - (-1 + i)] = x^2 - (-1 + i)x - (-1 - i)x$$
$$+ (-1 - i)(-1 + i)$$
$$= x^2 + x - ix + x + ix + (1 - i^2)$$
$$= x^2 + 2x + 2$$

Find the other factor(s) by long division:

$$
\require{enclose}
\begin{array}{r}
x^2 - 2x - 1 \\
x^2 + 2x + 2 \enclose{longdiv}{x^4 + 0x^3 - 3x^2 - 6x - 2} \\
\underline{x^4 + 2x^3 + 2x^2} \\
-2x^3 - 5x^2 \\
\underline{-2x^3 - 4x^2 - 4x} \\
-x^2 - 2x \\
\underline{-x^2 - 2x - 2} \\
0
\end{array}
$$

Thus, $x^4 - 3x^2 - 6x - 2 = (x^2 + 2x + 2)(x^2 - 2x - 1).$

$$x^2 + 2x + 2 = 0 \qquad\qquad x^2 - 2x - 1 = 0$$

$$x = \frac{-2 \pm \sqrt{4 - 8}}{2} \qquad\qquad x = \frac{2 \pm \sqrt{4 + 4}}{2}$$

$$x = -1 \pm i \qquad\qquad x = 1 \pm \sqrt{2}$$

The roots are $-1 \pm i,\ 1 \pm \sqrt{2}.$
You might wish to try using synthetic division to check this answer.

Graphical Analysis

We can systematically examine the relationship between the Fundamental Theorem of Algebra, the Conjugate Pair Theorem, and the graph of a function to help us understand the plausibility of these results, even though their proofs are beyond the scope of this course. We will consider the case of a general polynomial function where ax^n is the leading term and $a > 0$. In Problem Set 3.7 you are asked to carry out a similar analysis for $a < 0$. Remember that a real root to the polynomial equation $y = f(x)$ is an x-intercept.

Odd Powers   The graphing pattern is

$y = ax$, $a > 0$ **$n = 1$** (one complex root; remember that the set of real numbers is a subset of the set of complex numbers)

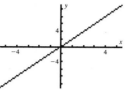 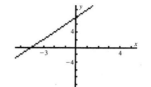 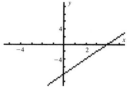

One real root   Slide up (any distance), one real root   Slide down, one real root

$y = ax^3$, $a > 0$ **$n = 3$** (three complex roots)

one turning point

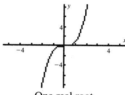 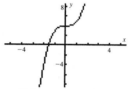 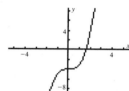

One real root   Slide up, one real root   Slide down, one real root

Since there are three roots, in each of these cases there must be two pure imaginary roots.

two turning points

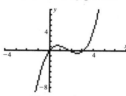 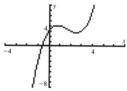 

Three real roots   Slide up, one real root   Slide down, one real root

$y = ax^5$, $a > 0$ **$n = 5$** (five complex roots)

four turning points

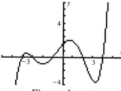 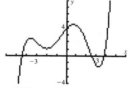 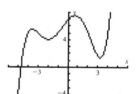

Five real roots   Slide up, three real roots   Slide up, one real root

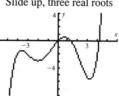 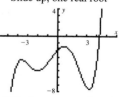

Slide down, three real roots   Slide down, one real root

The case of fewer turning points for n = 5 is left for the reader.

Even Powers   The graphing pattern is ⬆⬆

$y = ax^2, a > 0$ **$n = 2$** (one complex root; remember that the set of real numbers is a subset of the set of complex numbers)

one turning point

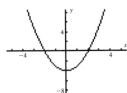
Two real roots

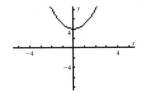
Slide up, zero real roots

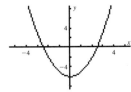
Slide down, two real roots

$y = ax^4, a > 0$ **$n = 4$**

three turning points

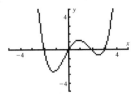
Four real roots

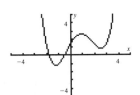
Slide up, two real roots

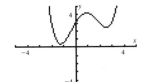
Slide up, zero real roots

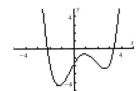
Slide down, two real roots

The case of fewer turning points for n = 4 is left for the reader.

When you imagine sliding the graph up and down you see that you gain or lose roots in pairs. This follows because polynomial graphs are continuous and "if it goes up, it must come down." You might be wondering what would happen if you slide a curve up or down so that a turning point rests on the x-axis. Three examples are shown below:

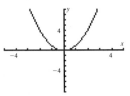
$n = 2$, two roots
One real root
Zero imaginary roots

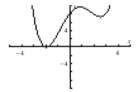
$n = 4$, four roots
One real root
Two imaginary roots

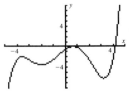
$n = 5$, five roots
Two real roots
Two imaginary roots

This seems to contradict the Number of Roots Theorem, until we remember the possibility of roots of multiplicity greater than 1. We should have written:

$n = 2$, two roots	$n = 4$, four roots	$n = 5$, five roots
one real root	one real root	two real roots
multiplicity 2	**multiplicity 2**	**one with multiplicity 2**
zero imaginary roots	two imaginary roots	two imaginary roots

You can probably see this most easily for a quadratic equation. The vertex (high or low point) will be on the x-axis when the quadratic part is a perfect square—for example,

$$y = x^2, \quad y = (x - 1)^2, \quad y = (x - 2)^2, \ldots$$

That is, if we write $y = (x - 1)^2 = (x - 1)(x - 1)$, we see that there are two roots, $r_1 = 1$ and $r_2 = 1$, which is the definition of a root of multiplicity 2. There is a multitude of possibilities for roots of multiplicity for higher powers, so we cannot show all possibilities here. You are asked to show some of these in Problem Set 3.7.

3.7 Problem Set

A *In Problems 1–6 let $f(x) = x^4 - 6x^3 + 15x^2 - 2x - 10$ to find the requested value.*

1. $f(i)$ **2.** $f(-i)$ **3.** $f(\sqrt{2})$

4. $f(-\sqrt{3})$ **5.** $f(2 + i)$ **6.** $f(1 - \sqrt{3})$

In Problems 7–12 let $f(x) = x^4 - 10x^3 + 36x^2 - 58x + 35$ to find the requested value.

7. $f(i)$ **8.** $f(-i)$ **9.** $f(2 - i)$

10. $f(2 + i)$ **11.** $f(3 + \sqrt{2})$ **12.** $f(3 - \sqrt{2})$

In Problems 13–18 let $f(x) = x^4 - 6x^3 + 18x^2 - 30x + 25$ to find the requested value.

13. $f(2)$ **14.** $f(3)$ **15.** $f(2 + i)$

16. $f(2 - i)$ **17.** $f(1 + 2i)$ **18.** $f(1 - 2i)$

In Problems 19–24 let $f(x) = 2x^4 - x^3 - 13x^2 + 5x + 15$ to find the requested value.

19. $f(2)$ **20.** $f(-2)$ **21.** $f(\sqrt{5})$

22. $f(-\sqrt{5})$ **23.** $f(i)$ **24.** $f(-i)$

25. Is $1 + \sqrt{2}$ a root of $x^3 - 2x^2 + 1 = 0$? If it is, name another root.

26. Is $1 + i$ a root of $x^3 - 4x^2 + 6x - 4 = 0$? If it is, name another root.

27. Is $1 - 2i$ a root of $x^3 - x^2 + 3x + 5 = 0$? If it is, name another root.

28. Is $1 - 2i$ a root of $x^4 - 2x^3 + 4x^2 + 2x - 5 = 0$? If it is, name another root.

29. Is $1 + 2i$ a root of $x^4 - 7x^3 + 14x^2 + 2x - 20 = 0$? If it is, name another root.

30. Is $2 + \sqrt{5}$ a root of $x^4 - 4x^3 - 5x^2 + 16x + 4 = 0$? If it is, name another root.

B *Solve the polynomial equations $P(x) = 0$ and graph the curve $y = P(x)$ in Problems 31–49.*

31. $P(x) = x^3 - 8$ **32.** $P(x) = x^3 - 64$

33. $P(x) = x^3 - 125$ **34.** $P(x) = x^4 - 64$

35. $P(x) = x^4 - 81$ **36.** $P(x) = x^4 - 625$

37. $P(x) = x^4 + 9x^2 + 20$ **38.** $P(x) = x^4 + 10x^2 + 9$

39. $P(x) = x^4 + 13x^2 + 36$

40. $P(x) = (x^2 - 4x - 1)(x^2 - 6x + 7)$

41. $P(x) = (x^2 - 6x + 10)(x^2 - 8x + 17)$

42. $P(x) = (x^2 + 2x + 5)(x^2 - 3x + 5)$

43. $P(x) = (x^2 - 4x - 1)(x^2 - 3x + 5)$

44. $P(x) = 2x^4 - x^3 + 2x - 1$

45. $P(x) = 2x^3 - 3x^2 + 4x + 3$

46. $P(x) = x^4 - 7x^3 + 14x^2 + 2x - 20$

47. $P(x) = x^4 - 2x^3 + 4x^2 + 2x - 5$

48. $P(x) = 4x^4 - 10x^3 + 10x^2 - 5x + 1$

49. $P(x) = 27x^4 - 180x^3 + 213x^2 + 62x - 10$

50. The function

$$f(x) = (x - 1)^2(x + 2)^3$$

has zeros at $x = 1$ and $x = -2$. Thus, the two real roots for the polynomial equation, $f(x) = 0$, are $x = 1$ (root of multiplicity 2) and $x = -2$ (root of multiplicity 3). Draw the graph of a fifth-degree equation that crosses the x-axis in three places.

51. Draw the graph of a fifth-degree equation that crosses the x-axis at two places.

52. Draw the graph of a fifth-degree equation that crosses the x-axis at one place.

53. Draw the graph of a fifth-degree equation that crosses the x-axis at four places.

54. Draw the graph of a sixth-degree equation with four real roots and no imaginary roots.

55. Draw the graph of a sixth-degree equation with three real roots and no imaginary roots.

C *Assuming the given value is a root, solve the equations in Problems 56–65.*

56. $x^3 - 2x^2 + 4x - 8 = 0$; $2i$

57. $x^4 + 13x^2 + 36 = 0$; $-3i$

58. $x^4 - 6x^2 + 25 = 0$; $2 + i$

59. $x^4 - 4x^3 + 3x^2 + 8x - 10 = 0$; $2 + i$

60. $2x^4 - 5x^3 + 9x^2 - 15x + 9 = 0$; $i\sqrt{3}$

61. $2x^4 - x^3 - 13x^2 + 5x + 15 = 0$; $-\sqrt{5}$

62. $3x^5 + 10x^4 - 8x^3 + 12x^2 - 11x + 2 = 0$; $-2 + \sqrt{5}$

63. $2x^5 + 9x^4 - 3x^2 - 8x - 42 = 0$; $i\sqrt{2}$

64. $x^5 - 11x^4 + 24x^3 + 16x^2 - 17x + 3 = 0$; $2 + \sqrt{3}$

65. $x^6 + x^5 - 3x^4 - 4x^3 + 4x + 4 = 0$; $-\sqrt{2}$ is a multiple root.

66. Carry out a graphical analysis for odd powers where the leading coefficient is negative.

67. Carry out a graphical analysis for even powers where the leading coefficient is negative.

3.8   Chapter 3 Summary

The material of this chapter is reviewed in the following list of objectives. After each objective there are some practice questions. For a sample test, select the first question of each set and check your answers with the answer section. For a sample test without answers, use the second question of each set. Additional practice is given by the other questions in each set. If you are having trouble with a particular type of problem, look back to that section for extra help.

3.1   Polynomial Functions and Synthetic Division

OBJECTIVE 1   *Use synthetic division.*

1. $\dfrac{x^4 + 2x^2 - x - 26}{x + 2}$

2. $\dfrac{6x^4 - x^3 + 4x^2 + 3x - 2}{x + \frac{1}{2}}$

3. $\dfrac{3x^3 + 2x^2 - 12x - 8}{x - 2}$

4. $\dfrac{4x^5 - 3x^3 + 2x - 5}{x + 1}$

3.2   Linear Functions

OBJECTIVE 2   *Define a linear function. Graph linear functions.*

5. $y = 5x - 3$

6. $5x - 2y + 8 = 0$

7. $2x + 3y + 6 = 0$

8. $y + 3 = \frac{3}{5}(x - 4)$

15. $y = \frac{3}{5}x - 3$ for $-5 \le x \le 5$

16. $y = 5$ for $-4 < x < 4$

OBJECTIVE 5   *Determine when given lines or line segments are parallel or perpendicular.*

17. Use slopes to decide whether the triangle with vertices $A(12, -7)$, $B(4, 3)$, $C(-3, -2)$ is a right triangle.

18. What is the slope of a line parallel to $3x - 4y + 7 = 0$?

19. What is the slope of a line perpendicular to $6x - 4y + 3 = 0$?

20. Are the diagonals of the quadrilateral $ABCD$ with vertices $A(3, 4)$, $B(-2, 5)$, $C(-4, -2)$, $D(3, -5)$ perpendicular?

OBJECTIVE 3   *Find the slope of a line.*

9. The line passing through $(2, 2)$ and $(6, -3)$

10. The line passing through $(a, f(a))$ and $(b, f(b))$

11. The line $7x - 5y + 3 = 0$

12. A horizontal line

OBJECTIVE 4   *Graph line segments.*

13. $4x - 3y + 9 = 0$ for $-3 < x \le 4$

14. $2x = y - 5$ for $-4 \le x < 2$

OBJECTIVE 6 *Know and use the various forms of a linear equation.*

21. State the standard-form equation and define the variables.
22. State the point–slope form and define the variables.
23. State the slope–intercept form and define the variables.
24. State the equation of a vertical line and define the variables.

OBJECTIVE 7 *Given the graph, or information about the graph, of a line, write the standard-form equation of the line.*

25. y-intercept -9; slope $\frac{2}{3}$
26. Slope -3; passing through $(3, -2)$
27. Passing through $(-3, -1)$ and $(5, -6)$
28. Passing through $(4, 5)$ and perpendicular to the line $2x - 3y + 8 = 0$.

3.3 Quadratic Functions

OBJECTIVE 8 *Define and graph quadratic functions of the form $y - k = a(x - h)^2$.*

29. $y - 1 = \frac{3}{4}(x + 3)^2$
30. $y = 10x^2$
31. $y + 1 = -2(x + 2)^2$
32. $y - 1 = 3(x + 3)^2$

OBJECTIVE 9 *Graph quadratic functions by completing the square.*

33. $x^2 + 2x - y - 1 = 0$
34. $y = x^2 - 6x + 10$

35. $x^2 + 4x - 2y - 2 = 0$
36. $x^2 - 4x + 5y + 9 = 0$

OBJECTIVE 10 *Find the maximum or minimum value for a quadratic function. Be sure to state whether your answer is a maximum or a minimum value.*

37. $2x^2 + 24x + y = 178$
38. $x^2 - 10x - y - 825 = 0$
39. $2x^2 + 20x + y + 190 = 0$
40. $x^2 - 8x - 3y + 3{,}616 = 0$

3.4 Cubic and Quartic Functions

OBJECTIVE 11 *Know the properties of polynomial graphs.* Tell what is wrong with each of the given graphs, and then draw the graph correctly.

41. $y = x^3 - 6x^2$

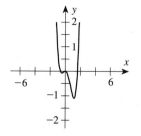

43. $y = .0016x^4 - .024x^3$

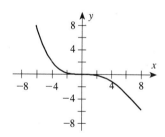

42. $y = x^4 - x^3 - x^2$

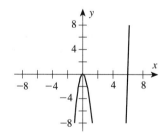

44. $y = 20x^2 - x^3$

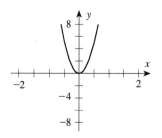

OBJECTIVE 12 *Use the properties of polynomial graphs to sketch polynomial functions. The graph of $f(x) = 2x^3 - 4x^2$ is given. Use this information to draw the given graphs.*

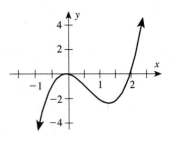

45. $y = -f(x)$

46. $y = 2f(x)$

47. $y = f(2x)$

48. $y = f(|x|)$

  Graphing Polynomial Functions

OBJECTIVE 13 *Use synthetic division and the Remainder Theorem to find points on the graph of a polynomial function.* Let $P(x) = 3x^3 + 4x^2 - 35x - 12$. Find the specified values.

49. $P(2)$ **50.** $P(-2)$

51. $P(3)$ **52.** $P(-3)$

OBJECTIVE 14 *Sketch the graph of a polynomial function with degree greater than 2.*

53. $P(x) = 3x^3 + 4x^2 - 35x - 12$

54. $Q(x) = 3x^3 + 2x^2 - 12x - 8$

55. $f(x) = 3x^4 - 8x^3 - 48x^2 + 492$

56. $g(x) = 6x^4 - x^3 + 3x^2 + 3x - 2$

3.6 **Real Roots of Polynomial Equations**

OBJECTIVE 15 *Find the zeros of polynomial functions by using the Factor Theorem, and state the multiplicity of each zero.*

57. $y = (x - 1)^3(x + 2)^3$

58. $y = (x^3 - 16x)^2$

59. $y = (x^2 - 1)^2(x + 1)^3$

60. $y = (x^2 - 4)^3$

OBJECTIVE 16 *List the possible rational roots of a polynomial equation.*

61. $3x^3 + 4x^2 - 35x - 12 = 0$

62. $3x^3 + 2x^2 - 12x - 8 = 0$

63. $6x^4 - 13x^3 + 3x^2 + 3x - 2 = 0$

64. $x^4 + 2x^2 - x - 26 = 0$

OBJECTIVE 17 *Use Descartes' Rule of Signs to determine the number of positive and negative real roots.*

65. $3x^3 + 4x^2 - 35x - 12 = 0$

66. $3x^3 + 2x^2 - 12x - 8 = 0$

67. $6x^4 - 13x^3 + 3x^2 + 9x - 5 = 0$

68. $x^4 + 11x^3 - 25x^2 + 11x - 26 = 0$

OBJECTIVE 18 *Solve polynomial equations over the set of real numbers.*

69. $3x^3 + 4x^2 - 35x - 12 = 0$

70. $3x^3 + 2x^2 - 12x - 8 = 0$

71. $6x^4 - 13x^3 + 3x^2 + 9x - 5 = 0$

72. $x^4 + 11x^3 - 25x^2 + 11x - 26 = 0$

3.7 The Fundamental Theorem of Algebra

OBJECTIVE 19 *Evaluate polynomial functions over the set of complex numbers. Let* $f(x) = x^5 + x^4 - 2x^3 - 2x^2 - 3x - 3$. *Find the requested values.*

73. $f(i)$ **74.** $f(-i)$ **75.** $f(\sqrt{3})$ **76.** $f(-\sqrt{3})$

OBJECTIVE 20 *Solve polynomial equations over the set of complex numbers and graph the corresponding polynomial functions.*

77. Solve $x^4 - 3x^3 - 9x^2 + 25x - 6 = 0$.

78. Graph $f(x) = x^4 - 3x^3 - 9x^2 + 25x - 6$.

79. Solve $x^4 - 14x^2 + x^3 - 14x = 0$.

80. Graph $g(x) = x^4 - 14x^2 + x^3 - 14x$.

OBJECTIVE 21 *Use the Conjugate Pair Theorem to solve polynomial equations. Solve the equations, assuming the given value is a root.*

81. $x^4 - 2x^3 + 5x^2 - 6x + 6 = 0$; $1 + i$

82. $x^4 + 2x^2 - 63 = 0$; $-3i$

83. $x^4 - 2x^3 - x^2 + 10x - 20 = 0$; $1 + \sqrt{3}i$

84. $x^4 - 6x^3 - 8x^2 + 62x + 15 = 0$; $2 + \sqrt{5}$

**Amalie (Emmy) Noether
(1882–1935)**

Women who have left behind a name in mathematics are very few in number—three or four, perhaps five. Does this say, as a common prejudice would tend to persuade us, that mathematics, so very abstract, is not congenial to the feminine disposition? ... The growth of female education, the overthrow of prejudices, the profound changes in the kind of life and in the role assigned to woman during the last few years will doubtless bring about a revision of her position in science. Then we shall see in what measure she can, as the equal of man, emerge from the role of the excellent pupil or the perfect collaborator, and join with those scientists whose work has opened new paths and bears the mark of genius.

MARIE-LOUISE DUBREIL-JACOTIN

Great Currents of Mathematical Thought

The greatest woman mathematician of all time was Emmy Noether, who is famous for her work in physics and algebra. She obtained her degrees at a time when it was unusual for a woman to attend college. In fact, a leading historian of the day wrote that talk of "surrendering our universities to the invasion of women ... is a shameful display of moral weakness." Nevertheless, Noether did receive her degrees and significantly influenced the development of abstract algebra. In 1935, Albert Einstein wrote the following tribute:*

"In the judgement of the most competent living mathematicians, Fraulein Noether was the most significant creative mathematical genius thus far produced since the higher education of women began. In the realm of algebra in which the most gifted of mathematicians have been busy for centuries, she discovered methods which have proved of enormous importance in the development of the present day younger generation of mathematicians."

In the nineteenth century, women were not allowed to enroll in many universities. At one school, which will remain nameless, a talented woman had persuaded the administration to let her audit a mathematics course. However, when a professor walked in and found four men and one woman in the classroom, he let her know in no uncertain terms that despite the administration's weakness and willingness to allow her in the class, he was not changing his attitude. Forthwith he began to lecture and lecture he did—with a vengeance. Week by week the lectures took their toll and by midterm only the woman and one struggling male student remained. The professor did not reduce the pace, and within another week the woman student found that she was the only person in the classroom when the professor walked in to begin a lecture. Glancing around, the professor announced, "Since there are no students here, I will not give a lecture," and walked out of the room. This story is somewhat depressing in its implication that unknown genius has been wasted or suppressed by societies that have not recognized mathematics as an appropriate activity for women.†

---

*As quoted in Howard Eves, *In Mathematical Circles Revisited* (Boston: Prindle, Weber, & Schmidt), p. 125.

†From Marie-Louise Dubreil-Jacotin, "Women Mathematicians," in Douglas M. Campbell and John C. Higgins, eds., *Mathematics: People, Problems, Results*, vol. I (Belmont, Ca.: Wadsworth), p. 166.

RATIONAL AND
RADICAL FUNCTIONS

▌Preview

The ability to sketch a curve quickly is a very important skill in mathematics. In the last chapter, we looked at the graphs of lines, parabolas, and other polynomial functions. In this chapter, we will tie some of the techniques you have already learned into a general plan for graphing, which you will be able to use not only when you study calculus, but whenever you need to sketch the graph of a given equation quickly and efficiently. There are 8 objectives in this chapter, which are listed on pages 214–215.

▌Perspective

Graphing of functions is an essential concept for the study of calculus. You can open a calculus book to almost any page at random and you will find some function whose graph is shown. For example, when finding the area bounded by two functions in calculus, it is desirable to first graph the functions in order to see which one is above and which is below and where the boundaries begin and end. The example shown below finds the area of the region bounded by the two curves

$$y = x^4 \quad \text{and} \quad y = 2x - x^2$$

We will discuss techniques for curve sketching in this chapter.

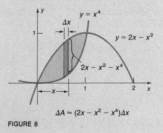

EXAMPLE 5 Find the area of the region between the curves $y = x^4$ and $y = 2x - x^2$.

Solution We start by finding where the two curves intersect and sketching the required region. This means that we need to solve $2x - x^2 = x^4$, a fourth-degree equation, which would usually be difficult to solve. However, in this case, $x = 0$ and $x = 1$ are rather obvious solutions. Our sketch of the region, together with the appropriate approximation and the corresponding integral, is shown in Figure 8.

$$\Delta A \approx (2x - x^2 - x^4)\Delta x$$

FIGURE 8

From Edwin J. Purcell and Dale Varberg, *Calculus with Analytic Geometry*, 5th ed. © 1987, p. 268. Reprinted by permission of Prentice-Hall, Englewood Cliffs, N.J.

4.1 Rational Functions

To evaluate and graph polynomial functions, we used synthetic division and the Division Algorithm. That is, we considered $P(x)/D(x)$ for polynomials $P(x)$ and $D(x)$, where $D(x) \neq 0$. Now consider this quotient from another viewpoint. An expression of the form $P(x)/D(x)$ is called a *rational function*.

RATIONAL FUNCTION

> A **rational function** f is the quotient of polynomial functions $P(x)$ and $D(x)$; that is,
>
> $$f(x) = \frac{P(x)}{D(x)} \quad \text{where } D(x) \neq 0$$

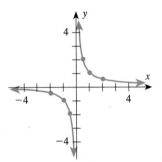

Figure 4.1 Graph of $y = \dfrac{1}{x}$

In this section, we will focus on some of the general properties of rational functions, and in the next section will discuss the graph of rational functions in more detail. We begin with the simplest rational function,

$$y = \frac{1}{x}$$

which is graphed by plotting points and is shown in Figure 4.1.

Each of the graphs in Figure 4.2 shows a variation of the graph of

$$y = \frac{1}{x}$$

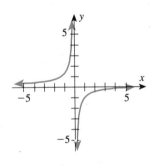

a. $y = -\dfrac{1}{x}$
The graph has been reflected about the x-axis.

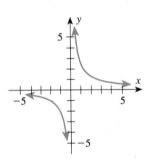

b. $y = \dfrac{2}{x}$
Each point is twice as far from the x-axis.

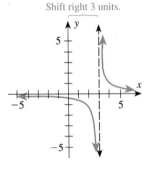

Shift right 3 units.

c. $y = \dfrac{1}{x-3}$
This curve has been shifted 3 units to the right.

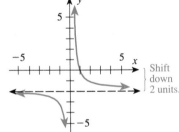

Shift down 2 units.

d. $y + 2 = \dfrac{1}{x}$
This curve has been shifted 2 units down.

FIGURE 4.2 Variations on the graph of the rational function $1/x$

EXAMPLE 1 Graph $y + 1 = \dfrac{-3}{x-2}$.

SOLUTION We can sketch this graph by comparing it to $y = 1/x$. There are four changes: It is reflected about the x-axis (negative); each point is three times as far from the x-axis; and it has been shifted 2 units to the *right* and 1 unit *down*. The graph is shown in Figure 4.3.

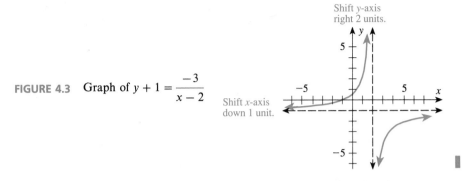

FIGURE 4.3   Graph of $y + 1 = \dfrac{-3}{x - 2}$

Shift x-axis
down 1 unit.

Shift y-axis
right 2 units.

You may have noticed that, unlike polynomial functions, rational functions cannot, in general, be drawn as a single curve. If $f(x) = P(x)/D(x)$ where there are no common factors for $P(x)$ and $D(x)$, we say that there is a **discontinuity** at a value of $x = b$ for which $D(b) = 0$. If you graph the line $x = b$, you will see that the graph of f will get closer to this line without ever touching it. We call such a line an **asymptote.**

We will investigate the nature of asymptotes in the next section. In this section we will limit our discussion to values that cause division by zero in the reduced form for a rational function. If c is a value that causes a rational function in reduced form to have a division by zero, then $x = c$ is, of course, a value that is excluded from the domain. However, the *line $x = c$* is a vertical line, and has the property (proved in the next section) that it is a **vertical asymptote.** We illustrate this property with the next example.

EXAMPLE 2   Sketch the graph of $f(x) = \dfrac{x}{(x - 2)(x + 3)}$.

SOLUTION   The rational expression is reduced, so we see that there are two vertical asymptotes ($x = 2$ and $x = -3$). Remember that $x = b$ is an asymptote if substitution of b for x results in an undefined expression. Begin by sketching the asymptotes. Then plot some points to obtain the graph shown in Figure 4.4.

x	y
-5	$-.3571429$
-4	$-.6666667$
-2	$.5$
-1	$.1666667$
0	0
1	$-.25$
3	$.5$
4	$.2857143$

Ⓩ If you do not draw the asymptotes, but simply plot and connect the points shown in the table, you will obtain an incorrect graph. Ⓩ

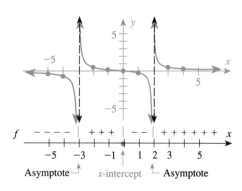

FIGURE 4.4   Graph of $f(x) = \dfrac{x}{(x - 2)(x + 3)}$

If you compare Example 2 with Example 3, you should see the effect of changing the degree of the numerator.

EXAMPLE 3 Graph $f(x) = \dfrac{x^2}{(x-2)(x+3)}$.

SOLUTION The graph is shown in Figure 4.5.

x	y
-5	1.785714
-4	2.666667
-2	-1
-1	$-.1666667$
0	0
1	$-.25$
3	1.5
4	1.142857

FIGURE 4.5 Graph of $f(x) = \dfrac{x^2}{(x-2)(x+3)}$

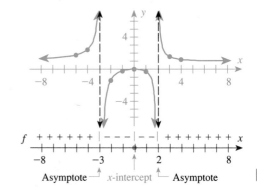

Deleted Points

A rational function such as the one defined in Example 3 has an asymptote at $x = -3$. We noted that a value that causes division by zero will define a vertical asymptote *provided the rational expression is reduced.* However, if the rational expression is not reduced, then the value that causes division by zero is simply excluded from the domain, and is therefore called a **deleted point.** This is illustrated in the following example.

EXAMPLE 4 Graph $f(x) = \dfrac{2x^2 + 5x - 3}{x + 3}$.

SOLUTION Notice that there is a value of x that would cause division by zero ($x = -3$). This value is excluded from the domain, and is a deleted point.

$$f(x) = \frac{2x^2 + 5x - 3}{x + 3}$$
$$= \frac{(2x - 1)(x + 3)}{x + 3}$$
$$= 2x - 1 \quad (x \neq -3)$$

The graph is the same as for the linear function $y = 2x - 1$, with the point at $x = -3$ deleted from the domain, as shown in Figure 4.6.

FIGURE 4.6 Graph of $f(x) = \dfrac{2x^2 + 5x - 3}{x + 3}$

EXAMPLE 5 Graph $f(x) = \dfrac{x^3 + 4x^2 + 7x + 6}{x + 2}$.

SOLUTION Simplify, if possible:

$$f(x) = \frac{x^3 + 4x^2 + 7x + 6}{x + 2}$$
$$= x^2 + 2x + 3 \quad (x \neq -2)$$

This simplification can be done by long division:

$$\begin{array}{r} x^2 + 2x + 3 \\ x + 2\overline{)x^3 + 4x^2 + 7x + 6} \\ \underline{x^3 + 2x^2} \\ 2x^2 \\ \underline{2x^2 + 4x} \\ 3x \\ \underline{3x + 6} \\ 0 \end{array}$$

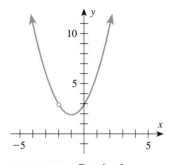

The graph of this curve is the same as for the quadratic function $y = x^2 + 2x + 3$, with the point at $x = -2$ deleted from the domain. To sketch this parabola, complete the square:

$$y = x^2 + 2x + 3$$
$$y - 3 = x^2 + 2x$$
$$y - 3 + 1 = x^2 + 2x + 1$$
$$y - 2 = (x + 1)^2$$

FIGURE 4.7 Graph of $f(x) = \dfrac{x^3 + 4x^2 + 7x + 6}{x + 2}$

The vertex is at $(-1, 2)$, and the parabola opens upward. It is drawn with the point at $x = -2$ deleted, as in Figure 4.7.

4.1 Problem Set

A *Graph the function defined by each equation given in Problems 1–30.*

1. $y = \dfrac{3}{x}$

2. $y = -\dfrac{3}{x}$

3. $y = \dfrac{-2}{x}$

4. $y = \dfrac{1}{x} + 2$

5. $y = \dfrac{1}{x} + 1$

6. $y = \dfrac{1}{x} - 3$

7. $y = -\dfrac{1}{x} + 2$

8. $y = -\dfrac{1}{x} + 1$

9. $y = -\dfrac{1}{x} - 3$

10. $y = \dfrac{1}{x + 2}$

11. $y = \dfrac{1}{x - 3}$

12. $y = \dfrac{1}{x - 4}$

13. $y = \dfrac{-1}{x + 3}$

14. $y = \dfrac{-1}{x - 2}$

15. $y = \dfrac{-2}{x - 1}$

16. $y - 3 = \dfrac{-1}{x + 1}$

17. $y + 1 = \dfrac{-1}{x + 2}$

18. $y = \dfrac{4}{x + 2} - 3$

19. $y = \dfrac{(x + 1)(x - 2)(x + 2)}{(x + 1)(x - 2)}$

20. $y = \dfrac{(x - 3)(x + 1)(x + 5)}{(x + 1)(x + 5)}$

21. $y = \dfrac{(x + 2)(x - 1)(3x + 2)}{x^2 + x - 2}$

22. $y = \dfrac{(x - 3)(2x + 3)(x + 2)}{x^2 - x - 6}$

23. $y = \dfrac{(x + 1)(x + 2)(x - 1)}{x + 2}$

24. $y = \dfrac{(x - 1)(x - 2)(x + 2)}{x - 1}$

25. $y = \dfrac{(x - 1)(x + 2)(x - 3)(x + 4)}{(x + 2)(x - 1)}$

26. $y = \dfrac{(x + 3)(x - 3)(x + 1)(2x - 3)}{(x - 3)(2x - 3)}$

27. $y = \dfrac{x^2 - x - 12}{x + 3}$

28. $y = \dfrac{x^2 + x - 2}{x - 1}$

29. $y = \dfrac{x^2 - x - 6}{x + 2}$

30. $y = \dfrac{2x^2 - 13x + 15}{x - 5}$

B *Graph the curves defined by the equations given in Problems 31–56.*

31. $y = \dfrac{4}{x^2}$

32. $y = \dfrac{-2}{x^2}$

33. $y = \dfrac{-3}{x^2}$

34. $y = \dfrac{2}{(x - 1)^2}$

35. $y = \dfrac{-1}{(x - 1)^2} + 3$

36. $y = \dfrac{1}{(x + 1)^2} - 2$

37. $y = \dfrac{2x^2 + 2}{x^2}$

38. $y = \dfrac{2x^2 - 1}{x^2}$

39. $y = \dfrac{x^2}{x - 4}$

40. $y = \dfrac{4x}{x^2 - 2}$

41. $y = \dfrac{-x^2}{x - 1}$

42. $y = \dfrac{x^2}{x - 1}$

43. $y = \dfrac{x^2}{x - 2}$

44. $y = \dfrac{x^2}{2 - x}$

45. $y = \dfrac{6x^2 - 5x - 4}{2x + 1}$

46. $y = \dfrac{x^3 + 6x^2 + 10x + 4}{x + 2}$

47. $y = \dfrac{x^3 + 12x^2 + 40x + 40}{x + 2}$

48. $y = \dfrac{15x^2 + 13x - 6}{3x - 1}$

49. $y = \dfrac{x^3 + 9x^2 + 15x - 9}{x + 3}$

50. $y = \dfrac{x^3 + 5x^2 + 6x}{x + 3}$

51. $y = \dfrac{2x^3 - 3x^2 - 32x - 15}{x^2 - 2x - 15}$

52. $y = \dfrac{3x^3 + 5x^2 - 26x + 8}{x^2 + 2x - 8}$

53. $y = \dfrac{2x + 1}{3x - 2}$

54. $y = \dfrac{4x + 3}{3x - 1}$

55. $y = \dfrac{2x^2 + 3x - 1}{x - 1}$

56. $y = \dfrac{3x^2 - 2x + 1}{x - 2}$

4.2 Asymptotes

As an additional aid to sketching certain functions, consider the notion of an asymptote, first introduced in the previous section. An **asymptote** is a line having the property that the distance from a point P on the curve to the line approaches zero as the distance from P to the origin increases without bound, and *P is on a suitable part of the curve.* This last phrase (in italics) is best illustrated by considering Figure 4.8, where L is an asymptote for the function f. Consider P and d, the distance from P to the line L, as shown in Figure 4.8a.

Now the distance from P to the origin can increase in two ways, depending on whether P moves along the curve in direction 1 or direction 2. In direction 1 the distance d increases without bound, but in direction 2 the distance d approaches zero. Thus, if you consider the portion of the curve in the shaded region of Figure 4.8b, you see that the conditions of the definition of an asymptote apply.

There are three types of asymptotes that occur frequently enough when sketching curves to merit consideration. These are **vertical, horizontal,** and **slant asymptotes.** An example of each is shown in Figure 4.9. Notice that a curve may pass through a horizontal or slant asymptote.

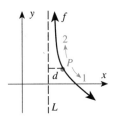

a.

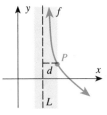

b.

FIGURE 4.8 An asymptote to a curve

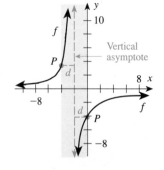

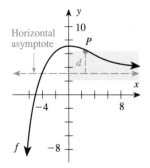

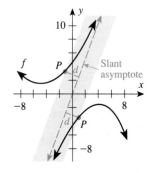

FIGURE 4.9 Asymptotes

Vertical Asymptotes

The easiest asymptotes to find are the vertical asymptotes of a function. If $f(x) = P(x)/D(x)$, where $P(x)$ and $D(x)$ have no common factors, then $|f(x)|$ must get large as x gets close to any value for which $D(x) = 0$ and $P(x) \neq 0$. This means that **if r is a root of $D(x) = 0$, then $x = r$ is the equation of a vertical asymptote.**

We found vertical asymptotes in the previous section. We review this procedure in Example 1.

EXAMPLE 1 Let $f(x) = \dfrac{x^2 + 2x - 8}{x^2 - 4}$. Find any vertical asymptotes.

SOLUTION Be careful with functions defined by equations like this one. If you do not notice the common factor, you might be led to the incorrect conclusion that $x = 2$ and $x = -2$ are the equations of the vertical asymptotes. Instead, factor *both* numerator and denominator as shown:

$$f(x) = \frac{x^2 + 2x - 8}{x^2 - 4}$$

$$= \frac{(x - 2)(x + 4)}{(x - 2)(x + 2)}$$

$$= \frac{x + 4}{x + 2} \quad (x \neq 2)$$

The vertical asymptote is $x = -2$ and the deleted point is at $x = 2$. ▌

Horizontal Asymptotes

If the equation can be solved easily for x so that

$$x = \frac{p(y)}{d(y)}$$

where $p(y)$ and $d(y)$ have no common factors, and if r is a root of $d(y) = 0$, then the line with the equation $y = r$ is a horizontal asymptote. Unfortunately, it is not always easy to solve for x, so we need to set some preliminary groundwork to find horizontal asymptotes.

To discuss horizontal asymptotes, we need to investigate the behavior of the curve as we permit x to become large (either positively or negatively). We symbolized this by writing $|x| \to \infty$. If we want to look at x becoming large positively, we write $x \to \infty$, and for x becoming large negatively we write $x \to -\infty$. There are three possibilities for what happens to y:

1. y approaches zero (written $y \to 0$), in which case there is a horizontal asymptote $y = 0$.

2. y approaches a nonzero number b (written $y \to b$), in which case there is a horizontal asymptote $y = b$.

3. y becomes numerically large (written $y \to \infty$ or $y \to -\infty$), in which case there is no horizontal asymptote.

EXAMPLE 2 Let $y = 1/x$. Find the horizontal asymptote.

SOLUTION We need to find the value of $1/x$ as x increases without bound. This is written as

$$\lim_{|x| \to \infty} \frac{1}{x}$$

Consider the following table of values for $1/x$ as x increases without bound:

x	$1/x$	x	$1/x$
1	1	-1	-1
2	.5	-2	$-.5$
10	.1	-10	$-.1$
100	.01	-100	$-.01$
1,000	.001	$-1,000$	$-.001$
10,000	.0001	$-10,000$	$-.0001$
$x \to \infty$	$1/x \to 0$	$x \to -\infty$	$1/x \to 0$

We say that $1/x \to 0$ as $|x|$ increases without bound and symbolize this by

$$\frac{1}{x} \to 0 \quad \text{as } |x| \to \infty$$

Thus, the horizontal asymptote is $y = 0$. This example illustrates the first possibility for horizontal asymptotes. ▌

The table and results of Example 2 lead to a useful theorem about limits that can help us find horizontal asymptotes.

LIMIT THEOREM Let x be any real number, k any constant, and n a natural number. Then as $|x| \to \infty$:

$$1.\ \frac{1}{x} \to 0 \qquad 2.\ \frac{1}{x^n} \to 0 \qquad 3.\ \frac{k}{x^n} \to 0$$

EXAMPLE 3 Graph $f(x) = \dfrac{x}{2x + 1}$ by finding the horizontal and vertical asymptotes.

SOLUTION First, find the vertical asymptotes by setting the denominator equal to zero and solving for x. You can often do this in your head; for this problem, we see that $2x + 1 = 0$ when $x = -\frac{1}{2}$.

The vertical asymptote is $x = -\frac{1}{2}$.

For the horizontal asymptote you could solve for x and look for the values of y for which the expression is undefined. That is, after several steps,

$$x = \frac{-y}{2y - 1} \quad \text{which is undefined when } y = \tfrac{1}{2}.$$

Solving for x is sometimes not possible (and almost always unpleasant), so we consider an alternative method for finding horizontal asymptotes. Consider

$$\lim_{|x| \to \infty} \frac{x}{2x + 1}$$

We could set up a table of values, but instead, multiply the rational expression by 1, written as $\frac{1/x}{1/x}$, and use the Limit Theorem:

$$\frac{x}{2x + 1} \cdot \frac{1/x}{1/x} = \frac{1}{2 + \frac{1}{x}}$$

Since $1/x \to 0$ as $x \to \infty$, you can see that

$$\frac{1}{2 + \frac{1}{x}} \to \frac{1}{2 + 0} \quad \text{as } |x| \to \infty$$

Thus,

$$\frac{x}{2x + 1} \to \frac{1}{2} \quad \text{as } |x| \to \infty$$

The horizontal asymptote is $y = \frac{1}{2}$.

The graph is shown in Figure 4.10.

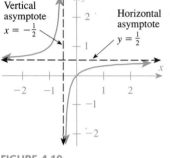

FIGURE 4.10

Graph of $f(x) = \dfrac{x}{2x + 1}$

Vertical asymptote $x = -\frac{1}{2}$

Horizontal asymptote $y = \frac{1}{2}$

EXAMPLE 4 Graph $y = \dfrac{3x^2 - 7x + 2}{7x^2 + 2x + 5}$ by finding the horizontal and vertical asymptotes as well as the x- and y-intercepts.

SOLUTION The vertical asymptotes are values for which $7x^2 + 2x + 5 = 0$. The discriminant for the quadratic formula is negative ($B^2 - 4AC = -136$) so there are no real roots. This means there are **no vertical asymptotes.**

For the horizontal asymptote we need to consider what happens as $|x| \to \infty$. The highest power of x is x^2, so multiply the numerator and denominator by $1/x^2$:

$$\frac{3x^2 - 7x + 2}{7x^2 + 2x + 5} = \frac{3x^2 - 7x + 2}{7x^2 + 2x + 5} \cdot \frac{1/x^2}{1/x^2} = \frac{3 - \frac{7}{x} + \frac{2}{x^2}}{7 + \frac{2}{x} + \frac{5}{x^2}}$$

Since k/x and k/x^2 both approach zero as x increases without bound,

$$\frac{3 - \frac{7}{x} + \frac{2}{x^2}}{7 + \frac{2}{x} + \frac{5}{x^2}} \to \frac{3 - 0 + 0}{7 + 0 + 0} = \frac{3}{7} \quad \text{as } |x| \to \infty$$

The horizontal asymptote is $y = \frac{3}{7}$.

The domain is the set of all real numbers. It is too difficult to find the range. The y-intercept is

$$y = \frac{3(0)^2 - 7(0) + 2}{7(0)^2 + 2(0) + 5} = \frac{2}{5}$$

The x-intercepts are

$$\frac{3x^2 - 7x + 2}{7x^2 + 2x + 5} = 0$$

$$3x^2 - 7x + 2 = 0$$

$$(3x - 1)(x - 2) = 0$$

$$x = \tfrac{1}{3}, \, 2$$

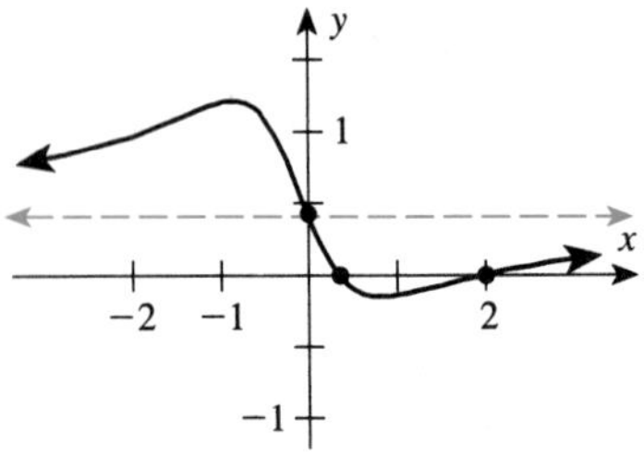

FIGURE 4.11

Graph of $y = \dfrac{3x^2 - 7x + 2}{7x^2 + 2x + 5}$

Plot the points $(0, \tfrac{2}{5})$, $(2, 0)$, and $(\tfrac{1}{3}, 0)$. By plotting a few additional points, we sketch the graph as shown in Figure 4.11.

Slant Asymptotes

The last type of asymptote we will consider is a slant asymptote. Suppose the degree of $P(x)$ is one more than the degree of $D(x)$. Then

$$f(x) = \frac{P(x)}{D(x)} = mx + b + \frac{R(x)}{D(x)}$$

where the degree of $R(x)$ is less than the degree of $D(x)$. Then

$$\lim_{|x| \to \infty} \frac{R(x)}{D(x)} = 0$$

which means that for large values of $|x|$, $f(x)$ is near the line given by $y = mx + b$. This says that the line with the equation $y = mx + b$ is a slant asymptote for the curve given by $y = f(x)$.

We can now summarize the finding of asymptotes, as given in the box.

PROCEDURE FOR FINDING ASYMPTOTES

Let

$$f(x) = \frac{P(x)}{D(x)}$$

where $P(x)$ and $D(x)$ are polynomial functions, and $P(x)$ and $D(x)$ have no common factors. Let $P(x)$ have degree M with leading coefficient p, and let $D(x)$ have degree N with leading coefficient d. Then:

VERTICAL ASYMPTOTE: $x = r$ where $D(r) = 0$

HORIZONTAL ASYMPTOTE:

$$y = 0 \text{ if } M < N \qquad y = \frac{p}{d} \text{ if } M = N \qquad \text{None if } M > N$$

SLANT ASYMPTOTE:

$$y = mx + b \text{ if } M = N + 1 \quad \text{and} \quad f(x) = mx + b + \frac{R(x)}{D(x)} \text{ (by long division)}$$

Curves may intersect horizontal and slant asymptotes, but may never intersect vertical asymptotes.

EXAMPLE 5 Graph $y = \dfrac{3x^3 - 2x^2 + 3x - 2}{x^2 + 3}$ by finding the horizontal, vertical, and slant asymptotes. You will also need to find the x- and y-intercepts. Also determine whether the curve passes through either its horizontal or slant asymptotes.

SOLUTION *Vertical asymptotes:* There are **no vertical asymptotes** since $x^2 + 3 \neq 0$ over the set of real numbers.

Horizontal asymptotes: For the horizontal asymptote, we see the highest power of x is x^3, so we multiply the numerator and denominator by $1/x^3$:

$$\frac{3x^3 - 2x^2 + 3x - 2}{x^2 + 3} \cdot \frac{\dfrac{1}{x^3}}{\dfrac{1}{x^3}} = \frac{3 - \dfrac{2}{x} + \dfrac{3}{x^2} - \dfrac{2}{x^3}}{\dfrac{1}{x} + \dfrac{1}{x^3}}$$

As $|x| \to \infty$ the denominator approaches zero, so y gets numerically large. Thus, y does not approach a constant, and there is **no horizontal asymptote.**

Slant asymptotes: For the slant asymptote, divide

$$
\begin{array}{r}
3x - 2 \\
x^2 + 3 \overline{)3x^3 - 2x^2 + 3x - 2} \\
\underline{3x^3 + 9x } \\
-2x^2 - 6x \\
\underline{-2x^2 - 6} \\
-6x + 4
\end{array}
$$

Thus,

$$y = 3x - 2 + \frac{-6x + 4}{x^2 + 3}$$

The slant asymptote is $y = 3x - 2.$

In order to see whether the curve intersects the slant asymptote, we solve the system of equations

$$
\begin{cases}
y = \dfrac{3x^3 - 2x^2 + 3x - 2}{x^2 + 3} \\[2mm]
y = 3x - 2
\end{cases}
$$

We substitute the second equation into the first equation and solve for x:

$$\frac{3x^3 - 2x^2 + 3x - 2}{x^2 + 3} = 3x - 2$$

$$3x^3 - 2x^2 + 3x - 2 = (x^2 + 3)(3x - 2)$$

$$3x^3 - 2x^2 + 3x - 2 = 3x^3 - 2x^2 + 9x - 6$$

$$-6x = -4$$

$$x = \tfrac{2}{3}$$

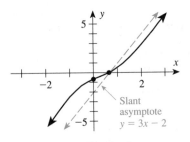

When $x = \frac{2}{3}$, then $y = 3(\frac{2}{3}) - 2 = 0$. This says that the curve **intersects its slant asymptote at $(\frac{2}{3}, 0)$.** Since the second component is 0, we see that the intersection of the curve and the asymptote is at the x-intercept.

For the y-intercept we find

$$y = \frac{3(0)^3 - 2(0)^2 + 3(0) - 2}{(0)^2 + 3} = -\frac{2}{3}; \quad \text{the point is } (0, -\frac{2}{3})$$

The graph is shown in Figure 4.12.

FIGURE 4.12 Graph of
$$y = \frac{3x^3 - 2x^2 + 3x - 2}{x^2 + 3}$$

4.2 Problem Set

A *Find the limits if they exist, in Problems 1–20.*

1. $\lim_{x \to \infty} x^2$

2. $\lim_{x \to \infty} (x^2 - 4)$

3. $\lim_{x \to \infty} \dfrac{x}{2x + 1}$

4. $\lim_{x \to \infty} \dfrac{1}{x^2 + 1}$

5. $\lim_{x \to \infty} \dfrac{x^2 + 3x - 10}{x^2 - 2}$

6. $\lim_{x \to \infty} \dfrac{x^2 - 8x + 15}{x - 3}$

7. $\lim_{x \to \infty} \dfrac{x^2 + 3x - 10}{2x^2 + 5}$

8. $\lim_{x \to \infty} \dfrac{x^2 - 1}{x - 2}$

9. $\lim_{x \to \infty} \dfrac{2x^2 - 5x - 12}{x^2 - 4}$

10. $\lim_{x \to \infty} \dfrac{x^2 - 8}{x^3 + 2x + 4}$

11. $\lim_{x \to \infty} \dfrac{x^2 + 2x + 4}{x^2 - 8}$

12. $\lim_{x \to \infty} \dfrac{x^3 + 2}{4x^3 + 5}$

13. $\lim_{x \to \infty} \dfrac{6 - x}{2x - 15}$

14. $\lim_{|x| \to \infty} \dfrac{2x^2 - 5x - 3}{x^2 - 9}$

15. $\lim_{x \to \infty} \dfrac{5x + 10,000}{x - 1}$

16. $\lim_{x \to -\infty} \dfrac{4x + 10^5}{x + 1}$

17. $\lim_{x \to -\infty} \dfrac{4x^4 - 3x^3 + 2x + 1}{3x^4 - 9}$

18. $\lim_{x \to \infty} \dfrac{x^4 + 1}{x^2 - 1}$

19. $\lim_{x \to \infty} \dfrac{3x^3 - 2x^2 + 1}{5x^3 + 3x - 100}$

20. $\lim_{x \to \infty} \dfrac{x^2 + x + 1}{x^3 - 1}$

Find the horizontal, vertical, and slant asymptotes, if any exist, for the functions given in Problems 21–40.

21. $y = \dfrac{1}{x}$

22. $y = \dfrac{1}{x} + 2$

23. $y = -\dfrac{1}{x} + 1$

24. $y = \dfrac{4}{x^2}$

25. $y = \dfrac{2x^2 + 2}{x^2}$

26. $y = \dfrac{1}{x - 4}$

27. $y = \dfrac{-1}{x + 3}$

28. $y = \dfrac{4x}{x^2 - 2}$

29. $y = \dfrac{x^2}{x - 4}$

30. $y = \dfrac{x^3}{(x - 1)^2}$

31. $y = \dfrac{-x^2}{x - 1}$

32. $y = \dfrac{x^2 + x - 2}{x - 1}$

33. $y = \dfrac{x^3 - 2x^2 + x - 2}{(x - 2)(x^2 + 1)}$

34. $y = \dfrac{(x - 3)(x^2 + 1)}{x^3 + 3x^2 + x + 3}$

35. $y = \dfrac{(15x^2 + 13x - 6)(x - 1)}{3x^2 - 4x + 1}$

36. $y = \dfrac{2x^3 - 3x^2 - 32x - 15}{x^2 - 2x - 15}$

37. $y = \dfrac{3x^3 + 5x^2 - 26x + 8}{x^2 + 2x - 8}$

38. $y = \dfrac{x^2}{x^3 - x^2 - 20x}$

39. $y = \dfrac{x^2 + 3x - 2}{x^2 + 2x - 8}$

40. $y = \dfrac{x^2}{20x - x^2 - x^3}$

B *In Problems 41–62, graph each curve by finding intercepts and asymptotes.*

41. $y = \dfrac{x + 3}{x - 2}$

42. $y = \dfrac{x - 1}{x + 1}$

43. $y = \dfrac{3x + 5}{3x - 2}$

44. $y = \dfrac{3x + 5}{2x - 3}$

45. $y = \dfrac{x^2 + x - 6}{x^2 + 2x - 8}$

46. $y = \dfrac{x^2 - x - 2}{x^2 - 2x - 3}$

47. $y = \dfrac{6x^2 - 6x - 12}{3x^2 + 4x + 5}$

48. $y = \dfrac{4x^2 + 8x - 12}{2x^2 + 3x + 5}$

49. $y = \dfrac{x}{x^2 + x - 6}$

50. $y = \dfrac{-x}{x^2 + x - 6}$

51. $y = \dfrac{x^2}{x^3 - x^2 - 20x}$

52. $y = \dfrac{x^2}{20x - x^2 - x^3}$

53. $y = \dfrac{x^2 + x - 6}{x + 3}$

54. $y = \dfrac{x^2 - x - 12}{x - 4}$

55. $y = \dfrac{x^3}{x^2 + 4}$

56. $y = \dfrac{x^2}{x^2 + 1}$

57. $y = \dfrac{x^3}{x^2 + 9}$

58. $y = \dfrac{x^3 + x^2 + 2x + 2}{x^2 + 1}$

59. $y = \dfrac{x^3 + x^2 + 2x + 2}{x^2 + 9}$

60. $y = \dfrac{x^2}{x^2 - 1}$

61. $y = \dfrac{x^3}{x^2 - 1}$

62. $y = \dfrac{x^4}{x^2 - 1}$

63. Can you see and describe a pattern in Problems 60–62?

4.3 Radical Functions

We will define a **radical function** as a function in which a variable is enclosed under a radical or includes a variable raised to a fractional exponent. As before, we begin by looking at some common properties of radical functions, then modifications such as translations, reflections, dilations, and contractions. Finally, we will graph some general radical functions using general graphical techniques.

Graph of Functions of the Form $y = x^n$

We begin with an example that examines the behavior of the function $y = x^n$, with special focus on the case where n is a rational number. Remember that if n is a rational number in reduced form p/q, then the expression x^n represents a radical expression. Furthermore, x^n is undefined when x is negative and q is even, so for convenience we will restrict x in our example to positive x values.

EXAMPLE 1 Sketch the graphs for $x \geq 0$.

  a. $y = \sqrt{x} = x^{1/2} = x^{.5}$ **b.** $y = x^{.1}$ **c.** $y = x^{3/2} = x^{1.5}$

SOLUTION We begin by plotting points and also include the graphs of $y = x$ and $y = x^2$ for reference. Notice that all of these functions pass through the points $(0, 0)$ and $(1, 1)$. The graphs are shown in Figure 4.13.

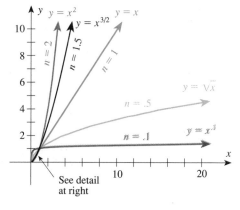

FIGURE 4.13 Graphs of functions of the form $y = x^n$

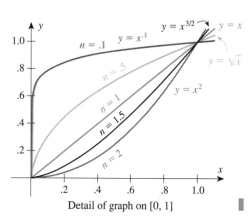

Detail of graph on [0, 1]

Notice from Figure 4.13 that the shape of the graphs of the form $y = x^n$ ($n \neq 1$) is such that they will either hold water or will not. For example, for $n = 2$ we have the shape which holds water. Such a shape, which is formally defined in calculus, is called **concave up.** The shape of the graph of $y = x^n$ for $x > 0$ is *concave up* for $n > 1$.

On the other hand, for $n = .5$, we see the shape which will not hold water. This shape, which is also defined in calculus, is called **concave down.** The shape of the graph of $y = x^n$ for $x > 0$ is *concave down* for $0 < n < 1$.

EXAMPLE 2 Draw the graphs of

a. $y = \sqrt{x}$ **b.** $y = \sqrt[3]{x}$

and label the intercepts as well as the concavity of each.

SOLUTION **a.** Since this is a square root ($y = x^{p/q}$ where p/q is reduced and q is even), we see that the domain is $[0, \infty)$. The graph is shown in Figure 4.14a.

b. For $y = \sqrt[3]{x}$ ($y = x^{p/q}$ where p/q is reduced and q is odd), we see that the domain is $(-\infty, \infty)$. First graph $y = x^{1/3}$ for $x \geq 0$, and then complete the graph for $x < 0$ by reflecting the graph through the origin. The graph is shown in Figure 4.14b.

FIGURE 4.14 Graphs showing concavity

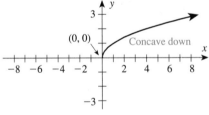

a. Graph of $y = \sqrt{x}$ **b.** Graph of $y = \sqrt[3]{x}$

EXAMPLE 3 Use Figure 4.14b to complete the following graphs.

a. $y = -\sqrt[3]{x}$ **b.** $y = \sqrt[3]{x} + 4$ **c.** $y = \sqrt[3]{x - 2} + 4$

SOLUTION **a.** This is a reflection of the graph shown in Figure 4.14b.

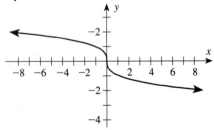

b. This is a translation up 4 units.

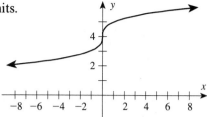

c. This is a translation to coordinate axes with origin at (2, 4).

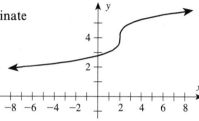

EXAMPLE 4 Graph $y = x^2 + \sqrt{x}$ on $[0, 1]$.

SOLUTION We can graph this curve by adding the ordinates. First, graph $y_1 = x^2$ on $[0, 1]$; next, graph $y_2 = \sqrt{x}$ on $[0, 1]$. Finally, add the ordinates for several x values, and then draw a smooth curve as shown in Figure 4.15.

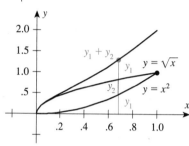

FIGURE 4.15 Graph of $y = x^2 + \sqrt{x}$

Graphs of the form $y = \sqrt{f(x)}$ where f is a polynomial or a rational function

When graphing a function of the form $y = \sqrt{f(x)}$, we need to pay particular attention to the domain, so that $f(x) \geq 0$.

EXAMPLE 5 Graph $y = \sqrt{(x - 2)(x - 4)(x - 6)}$.

SOLUTION This function is defined for $(x - 2)(x - 4)(x - 6) \geq 0$.

```
x − 2:   − − −  |  + + +  |  + + +  |  + + +
x − 4:   − − −  |  − − −  |  + + +  |  + + +
x − 6:   − − −  |  − − −  |  − − −  |  + + +
          neg   |   pos   |   neg   |   pos
```
```
          1   2   3   4   5   6   7
```

The domain is $[2, 4] \cup [6, \infty)$. The graph (with the excluded regions) is shown in Figure 4.16. Notice that the graph of $y = \sqrt{f(x)}$ exists only where f is positive.

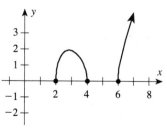

FIGURE 4.16 Graph of $y = \sqrt{(x - 2)(x - 4)(x - 6)}$

CALCULATOR COMMENT

If you have a graphing calculator, you can graph

$\boxed{\text{Y1} = (X - 2)(X - 4)(X - 6)}$ to find

the positive and negative parts of the domain, and from that can extract the positive regions for this problem. Here is the graph:

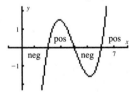

You can also graph the curve directly by setting

$\boxed{\text{Y1} = \sqrt{((X - 2)(X - 4)(X - 6))}}$

If $f(x)$ in the equation $y = \sqrt{f(x)}$ is quadratic, there is an interesting relationship between the graph of f and the graph of $y = \sqrt{f(x)}$. This relationship is explored in the next two examples.

EXAMPLE 6 Graph $y = \sqrt{x^2 + x - 12}$.

SOLUTION We begin by graphing

$$f(x) = x^2 + x - 12$$

as shown in black in Figure 4.17. We can see the regions on the x-axis where f is positive and where it is negative. The graph of $y = \sqrt{f(x)}$ exists only where f is positive. The x-intercepts of f are also the x-intercepts for $y = \sqrt{f(x)}$. For values in the domain, the value of y is the square root of the corresponding value for f. The completed graph is shown in color in Figure 4.18.

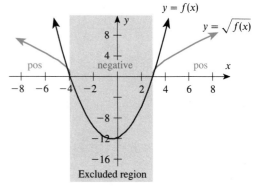

FIGURE 4.17

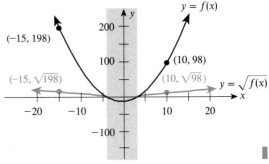

FIGURE 4.18
Graph of $y = \sqrt{x^2 + x - 12}$

EXAMPLE 7 Compare each pair of graphs and notice the relationship between the graphs of $y = f(x)$ and $y = \sqrt{f(x)}$, where f is a quadratic function.

a. $y = x^2 - 4$ and $y = \sqrt{x^2 - 4}$

b. $y = x^2$ and $y = \sqrt{x^2}$

c. $y = x^2 + 4$ and $y = \sqrt{x^2 + 4}$

d. $y = .1x^2 + 4$ and $y = \sqrt{.1x^2 + 4}$

e. $y = 10x^2 + 4$ and $y = \sqrt{10x^2 + 4}$

SOLUTION **a.** The vertex of the parabola is $(0, -4)$ and the x-intercepts are $x = 2$, $x = -2$; the domain for $y = \sqrt{x^2 - 4}$ is where $x^2 - 4 \geq 0$—namely, $(-\infty, -2] \cup [2, \infty)$.

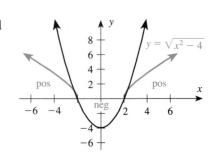

b. The vertex of the parabola is (0, 0), with y-intercept $y = 0$. As you compare this with part **a**, think of the parabola as sliding up; as that happens the radical function becomes more linear and approaches this limiting position

$$y = \sqrt{x^2} = |x|$$

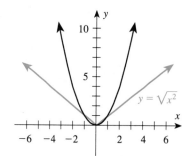

c. The vertex of the parabola is (0, 4). The parabola has now slid up to the point where there are no points of intersection with the x-axis (i.e., no real roots). Notice that the radical function is now curving toward the parabola. The concavity has changed when we compare it with part **a**. The domain for the radical function is the set of all real numbers.

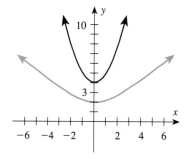

d. Compare this with the graph in part **c**. Because the coefficient of x^2 in $y = .1x^2 + 4$ is .1, we see that the parabola has been flattened out. The effect this has on the radical function is to flatten it out, too, but not so much as to change the concavity from part **c**.

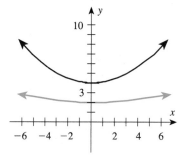

e. Compare this with the graph in parts **c** and **d**. Because the coefficient of x^2 in $y = 10x^2 + 4$ is 10, the parabola is narrower than in part **c**. The radical function "follows" the quadratic function, as shown in the graph.

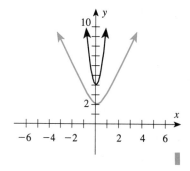

Graphs of General Radical Functions

When considering the graph of a function containing a radical, the procedure requires more analysis. In addition to comparing it to known functions, you will need to pay attention to domain, range (if possible), the intercepts, and asymptotes. We conclude this section with two examples.

EXAMPLE 8 Graph $y = \frac{2}{3}\sqrt{x^2 - 9}$.

a. by comparing to $y = \sqrt{f(x)}$ for f a quadratic function.
b. by finding the intercepts and asymptotes.

SOLUTION **a.** The function $f(x) = x^2 - 9$ is a parabola with vertex at $(0, -9)$, so the graph of $y = \frac{2}{3}(x^2 - 9)$ has vertex at $(0, -6)$ since $\frac{2}{3}(-9) = -6$. The domain for the radical function is that part for which $y = \frac{2}{3}(x^2 - 9)$ is positive. The graph is shown in Figure 4.19a.

b. The intercepts are found:

If $x = 0$, then $y = \frac{2}{3}\sqrt{0^2 - 9}$ is not a real number, so there are **no y-intercepts.**

If $y = 0$, then $\frac{2}{3}\sqrt{x^2 - 9} = 0$, which has a solution for $x = \pm 3$. The **x-intercepts are (3, 0) and (−3, 0).**

Next, find the asymptotes:

1. *Vertical:* There are no values of x that cause division by zero. There are **no vertical asymptotes.**

2. *Horizontal:* As $|x| \to \infty$, y increases without bound, so there are **no horizontal asymptotes.**

3. *Slant:* As $|x| \to \infty$, $\sqrt{x^2 - 9} \approx \sqrt{x^2} = |x|$. Thus, $y \to \pm\frac{2}{3}x$ as $x \to \infty$. The slant asymptotes are $y = \frac{2}{3}x$ and $y = -\frac{2}{3}x$, or (in standard form) **$2x - 3y = 0$ and $2x + 3y = 0$.**

We now sketch the curve using asymptotes and intercepts, as shown in Figure 4.19b.

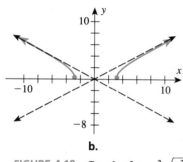

a.

b.

FIGURE 4.19 Graph of $y = \frac{2}{3}\sqrt{x^2 - 9}$

EXAMPLE 9 Graph $y = \dfrac{x}{\sqrt{x^2 - 4}}$.

SOLUTION We do not recognize this curve as a common type, so we will find the domain, intercepts, and asymptotes.

Domain: The domain is the set of all real numbers for which $x^2 - 4 > 0$ (note that we excluded division by 0). Solving this inequality, we find the domain

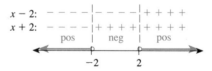

The domain is $(-\infty, -2) \cup (2, \infty)$.

y-intercept is found when $x = 0$: $y = \dfrac{0}{\sqrt{0^2 - 4}}$ does not exist in the set of real numbers, so there is **no y-intercept.**

x-intercepts are found when $y = 0$: $\dfrac{x}{\sqrt{x^2 - 4}} = 0$ when $x = 0$. However, $x = 0$ is not in the domain, so there are **no x-intercepts.**

Vertical asymptotes: Determine whether the denominator can be zero:

$$x^2 - 4 = 0$$

$$(x - 2)(x + 2) = 0$$

$$x = 2, -2$$

The vertical asymptotes are $x = 2$ and $x = -2$.

> If the denominator had been $\sqrt{x^2 + 4}$, there would have been no vertical asymptotes because $x^2 + 4 \neq 0$ for all real values of x. ⊗

Horizontal asymptote: Find $\lim\limits_{x \to \infty} \dfrac{x}{\sqrt{x^2 - 4}}$:

$$\frac{x}{\sqrt{x^2 - 4}} = \frac{x}{\sqrt{x^2\left(1 - \dfrac{4}{x^2}\right)}} = \frac{x}{|x|\sqrt{1 - \dfrac{4}{x^2}}}$$

As $x \to \infty$, $|x| = x$, and

$$\frac{x}{|x|\sqrt{1 - \dfrac{4}{x^2}}} = \frac{1}{\sqrt{1 - \dfrac{4}{x^2}}} \to 1$$

Thus, $y \to 1$ as $x \to \infty$, and the line $y = 1$ is a horizontal asymptote. As $x \to -\infty$, $|x| = -x$, and

$$\frac{x}{|x|\sqrt{1 - \dfrac{4}{x^2}}} = \frac{-1}{\sqrt{1 - \dfrac{4}{x^2}}} \to -1$$

> If the denominator had been $\sqrt{4 - x^2}$, there would have been no horizontal asymptote because the domain is $(-2, 2)$. It would be impossible for $|x| \to \infty$ and still remain in the domain $(-2, 2)$. ⊗

Thus, $y \to -1$ as $x \to -\infty$, and the line $y = -1$ is also a horizontal asymptote.

We can now sketch the function as shown in Figure 4.20.

FIGURE 4.20   Graph of $y = \dfrac{x}{\sqrt{x^2 - 4}}$

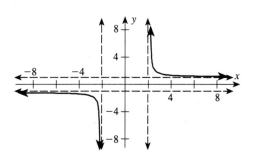

A *Graph the function defined by each equation given in Problems 1–24.*

1. $y = \sqrt[4]{x}$ on $[0, 25]$
2. $y = \sqrt[5]{x}$ on $[0, 32]$
3. $y = x^{3/5}$ on $[0, 32]$
4. $y = x^{3/4}$ on $[0, 25]$
5. $y = \sqrt[4]{x}$ on $[0, 1]$
6. $y = \sqrt[5]{x}$ on $[0, 1]$
7. $y = x^{3/5}$ on $[0, 1]$
8. $y = x^{3/4}$ on $[0, 1]$
9. $y = -2\sqrt[4]{x}$ on $[0, 1]$
10. $y = 5\sqrt[5]{x}$ on $[0, 1]$
11. $y = -\frac{1}{2}x^{3/5}$ on $[0, 1]$
12. $y = -3x^{3/4}$ on $[0, 1]$
13. $y = x^{2/3}$
14. $y = x^{2/5}$
15. $y = -2x^{1/4}$
16. $y = -5\sqrt{x}$
17. $y = x^{3/5} - 2$
18. $y = x^{3/4} - 2$
19. $y = (x - 2)^{3/5}$
20. $y = (x - 2)^{3/4}$
21. $y = (x + 1)^{2/5} + 2$
22. $y = \sqrt[4]{x} - 2 - 1$
23. $y = \sqrt[3]{x + 3} - 1$
24. $y = \sqrt{x^2 - 2}$

B *Graph the given curve by adding ordinates in Problems 25–30.*

25. $y = x + \sqrt{x}$
26. $y = x^2 + \sqrt[3]{x}$
27. $y = x^3 + \sqrt{x}$
28. $y = x + x^{2/3}$
29. $y = x^{-1} + \sqrt{x}$
30. $y = \frac{1}{2}x + \sqrt{x}$

Graph the function defined by each equation given in Problems 31–62.

31. $y = \sqrt{x^2 - 1}$
32. $y = \sqrt{x^2 - 9}$
33. $y = \sqrt{x^2 + 1}$
34. $y = \sqrt{x^2 + 9}$
35. $y = \sqrt{(x - 2)(x - 3)}$
36. $y = \sqrt{(x - 1)(x - 4)}$
37. $y = \sqrt{x^2 - 6x + 8}$
38. $y = \sqrt{x^2 - 4x + 4}$
39. $y = \sqrt{10(x^2 - 1)}$
40. $y = \sqrt{.1(x^2 - 9)}$
41. $y = \sqrt{4(x^2 + 1)}$
42. $y = \sqrt{.5(x^2 + 9)}$
43. $y = \sqrt{(x - 1)(x - 3)(x - 5)}$

44. $y = \sqrt{(x - 3)(x - 6)(x - 9)}$
45. $y = \sqrt{x(x - 2)(x - 4)}$
46. $y = \sqrt{x(x - 3)(x - 5)}$
47. $y = \dfrac{x}{\sqrt{x^2 - 9}}$
48. $y = \dfrac{x}{\sqrt{x^2 - 16}}$
49. $y = \dfrac{x}{\sqrt{x^2 + 4}}$
50. $y = \dfrac{x}{\sqrt{x^2 + 9}}$
51. $y = -\frac{2}{3}\sqrt{x^2 - 9}$
52. $y = -\frac{3}{5}\sqrt{x^2 - 4}$
53. $y = \frac{3}{5}\sqrt{x^2 - 4}$
54. $y = \frac{1}{2}\sqrt{x^2 - 5}$
55. $y = \dfrac{x}{\sqrt{1 - x^2}}$
56. $y = \dfrac{x}{\sqrt{16 - x^2}}$
57. $y = \sqrt{\dfrac{x^2 - 1}{x - 2}}$
58. $y = \sqrt{\dfrac{x + 3}{x^2 - 4}}$
59. $y = \sqrt{\dfrac{3}{x^2 - 1}}$
60. $y = \sqrt{\dfrac{1}{1 - x^2}}$
61. $y = -\sqrt{\dfrac{x^2 + 1}{x}}$
62. $y = -\sqrt{\dfrac{x^2 + 4}{x}}$

C *Use concavity to given the general shape of the curve $y = x^{p/q}$, where p/q is a reduced fraction satisfying the conditions in Problems 63–68. If q is even then the domain is the set of nonnegative numbers and if q is odd then the domain is the set of all real numbers.*

63. $p > q$, q even
64. $p > q$, q odd and p odd
65. $p > q$, q odd and p even
66. $p < q$, q even
67. $p < q$, q odd and p odd
68. $p < q$, q odd and p even

4.4 Real Roots of Rational and Radical Equations

Rational Equations

A **rational equation** is an equation with at least one variable in a denominator. The multiplication property for equality (from beginning algebra) specifies that both sides of an equation may be multiplied by any *nonzero* numbers. This means that you cannot multiply both sides of an equation by an expression containing a variable *unless* you exclude values that cause that expression to be zero.

**PROCEDURE FOR SOLVING
A RATIONAL EQUATION**

To Solve a Rational Equation:

STEP 1   Exclude values that cause division by 0.

STEP 2   Multiply both sides by the least common divisor (LCD) and simplify.

STEP 3   Solve the resulting equation.

STEP 4   Check each member of the solution set to make sure it is not one of the excluded values; that is, the values in the solution set must not cause division by 0. Any value that is excluded in this manner is called an **extraneous root.**

EXAMPLE 1 $\dfrac{1}{x} - \dfrac{1}{4} = \dfrac{x-2}{4x}$ $(x \neq 0)$

Note the restriction. It is implied by the equation, but must be stated as part of the solution.

SOLUTION

$$(4x)\frac{1}{x} - (4x)\frac{1}{4} = (4x)\frac{x-2}{4x}$$ Multiply by the LCD.

$$4 - x = x - 2$$ Simplify and solve.

$$6 = 2x$$

$$3 = x$$

By noting the restriction $x \neq 0$, or checking, $x = 3$ is verified as a solution.
The solution is $\{3\}$.

CALCULATOR COMMENT

In order to find an approximate solution to the equation in Example 1, you can graph

$$\boxed{Y1 = X^{-1} - 4^{-1} - (X - 2)/(4X)}$$

The resulting graph is:

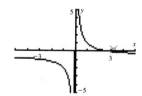

Xmin = −5
Xmax = 5
Ymin = −5
Ymax = 5

You can use the $\boxed{\text{TRACE}}$ feature to find $\boxed{X = 3 \quad Y = -3E - 14}$. Remember that $y = -3E - 14$ is scientific notation for $-3 \times 10^{-14} \approx 0$. This means the solution is $\{3\}$.

An alternate method is to treat both the left and right sides as separate functions and look for their intersection. Use this method with

$$\boxed{Y1 = X^{-1} - 4^{-1}; \quad Y2 = (X - 2)/(4X)}$$

to find:

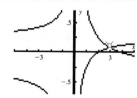

EXAMPLE 2

$$\frac{2x}{x-3} - 3 = \frac{2x - 12}{3 - x} \quad (x \neq 3)$$

SOLUTION

$$(x-3)\frac{2x}{x-3} - 3(x-3) = (x-3)\frac{2x-12}{3-x}$$

$$2x - 3(x-3) = (-1)(2x - 12)$$

$$2x - 3x + 9 = -2x + 12$$

$$-x + 9 = -2x + 12$$

$$x = 3$$

But $x \neq 3$, so the solution set is empty.

The solution set is $\emptyset$.

EXAMPLE 3 Solve $(x + 3)^{-1} + (x + 1)^{-1} = 1$.

SOLUTION Begin by noting that $x \neq -3, -1$ (these values cause division by 0). Multiply both sides by $(x + 3)(x + 1)$:

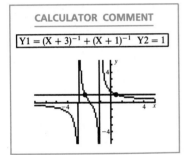
$$(x+3)(x+1)\frac{1}{x+3} + (x+3)(x+1)\frac{1}{x+1} = (x+3)(x+1)\cdot 1$$

$$(x + 1) + (x + 3) = x^2 + 4x + 3$$

$$0 = x^2 + 2x - 1$$

$$x = \frac{-2 \pm \sqrt{8}}{2}$$

$$= -1 \pm \sqrt{2}$$

The original restrictions were $x \neq -3$, $x \neq -1$, so both irrational results are solutions. The solution is $\{-1 + \sqrt{2}, -1 - \sqrt{2}\}$, or approximately $\{.4142, -2.412\}$, which is verified by the graph.

Radical Equations

A **radical equation** is an equation that contains a radical or fractional exponent with a variable as a radicand. Solving equations with rational expressions requires the following Property of Powers:

PROPERTY OF POWERS

If P and Q are algebraic expressions in a variable x, and n is any positive integer, then the solution set of

$$P = Q$$

is a subset of the solution set of

$$P^n = Q^n$$

The equation $P^n = Q^n$ is called a **derived equation** of $P = Q$.

This property implies that not all the solutions of the derived equations may be solutions of the original equation. Whenever both sides of an equation are raised to a power, the solution must be checked in the original equation. Solutions that do not check are called **extraneous solutions.** To use the property, first isolate a radical expression on one side of the equation. This ensures that raising each side to a power will eliminate that radical. The following examples illustrate the procedure.

EXAMPLE 4

$$\sqrt{3-x}+1 = x$$

SOLUTION

$$\sqrt{3-x} = x - 1$$ First, isolate the radical.

$$(\sqrt{3-x})^2 = (x-1)^2$$ Square both sides.

$$3 - x = x^2 - 2x + 1$$

$$0 = x^2 - x - 2$$

$$0 = (x+1)(x-2)$$ Simplify and solve.

$$x = -1 \quad \text{or} \quad x = 2$$

Check $x = -1$:   Check $x = 2$:

$$\sqrt{3-(-1)}+1 \overset{?}{=} -1 \qquad \sqrt{3-2}+1 \overset{?}{=} 2$$ Check the roots.

$$\sqrt{4}+1 \overset{?}{=} -1 \qquad \sqrt{1}+1 \overset{?}{=} 2$$

$$3 \neq -1 \qquad\qquad 2 = 2$$

$x = -1$ is extraneous $x = 2$ is a solution   State the solution set.

The solution is {2}.

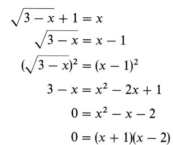

CALCULATOR COMMENT

$Y1 = \sqrt{(3-X)}+1 \quad Y2 = X$

TRACE shows $X = 2 \quad Y = 2$

EXAMPLE 5

$$2\sqrt{x+6}+3\sqrt{x+1} = 0$$

$$2\sqrt{x+6} = -3\sqrt{x+1}$$

$$4(x+6) = 9(x+1)$$ Square both sides.

$$5x = 15$$

$$x = 3$$

Check: $2\sqrt{9}+3\sqrt{4} \overset{?}{=} 0$

$$6 + 6 \overset{?}{=} 0$$

$$12 = 0$$

$x = 3$ does not check and is extraneous

The solution set is $\varnothing$. State the solution set.

EXAMPLE 6

$$\sqrt[3]{x^2 + 2} = 3$$

$$x^2 + 2 = 27 \quad \text{Cube both sides.}$$

$$x^2 = 25$$

$$x = 5 \quad \text{or} \quad x = -5$$

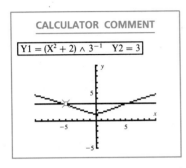
Check: $\sqrt[3]{5^2 + 2} \overset{?}{=} 3 \qquad \sqrt[3]{(-5)^2 + 2} = 3$

$$\sqrt[3]{27} \overset{?}{=} 3 \qquad\qquad \sqrt[3]{27} = 3$$

$$3 = 3 \qquad\qquad\qquad 3 = 3$$

$x = 5$ is a solution $\qquad x = -5$ is a solution

The solution is $\{5, -5\}$.

EXAMPLE 7 Solve $\sqrt{2x + 5} + \sqrt{x + 2} = 1$.

SOLUTION The procedure is to isolate one radical, square, and simplify. Be sure to check for extraneous solutions.

$$\sqrt{2x + 5} + \sqrt{x + 2} = 1$$

$$\sqrt{2x + 5} = 1 - \sqrt{x + 2}$$

$$2x + 5 = 1 - 2\sqrt{x + 2} + x + 2$$

$$x + 2 = -2\sqrt{x + 2}$$

$$x^2 + 4x + 4 = 4x + 8$$

$$x^2 = 4$$

$$x = \pm 2$$

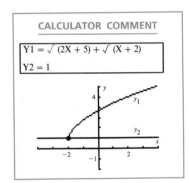
⊗ Squaring may not eliminate both radicals. When this happens, isolate the remaining radical and square again. ⊗

Check:

$$x = 2: \quad \sqrt{4 + 5} + \sqrt{2 + 2} = 3 + 2 = 5 \neq 1; \quad \text{does not check}$$

$$x = -2: \quad \sqrt{-4 + 5} + \sqrt{-2 + 2} = 1 + 0 = 1; \quad \text{checks}$$

The solution is $\{-2\}$.

4.4 Problem Set

A *Solve each equation in Problems 1–30 over the set of real numbers.*

1. $\dfrac{x^2}{12} - \dfrac{x}{3} = \dfrac{3}{4}$

2. $.09x^2 - .21x + .1 = 0$

3. $\dfrac{3}{4y} + \dfrac{7}{16} = \dfrac{4}{3y}$

4. $\dfrac{7}{4y} + \dfrac{5}{18} = \dfrac{1}{36}$

5. $\dfrac{4}{y} - \dfrac{3(2 - y)}{y - 1} = 3$

6. $\dfrac{6}{y} - \dfrac{2(y - 5)}{y + 1} = 3$

7. $\dfrac{z - 1}{z} = \dfrac{z - 1}{6}$

8. $\dfrac{3z - 2}{z - 1} = \dfrac{2z + 1}{z + 1}$

9. $\dfrac{1}{x + 2} - \dfrac{1}{2 - x} = \dfrac{3x + 8}{x^2 - 4}$

10. $\dfrac{16}{x + 5} + \dfrac{4}{5 - x} = \dfrac{5 - 3x}{x^2 - 25}$

11. $\dfrac{x+2}{3x-1} - \dfrac{1}{x} = \dfrac{x+1}{3x^2-x}$

12. $\dfrac{x}{2x-1} - \dfrac{1}{x} = \dfrac{x-1}{2x^2-x}$

13. $\dfrac{x+1}{x-1} - \dfrac{x-1}{x+1} = \dfrac{5}{6}$

14. $\dfrac{5-y}{y+1} + \dfrac{y-3}{y-1} = \dfrac{8}{15}$

15. $\dfrac{z-1}{z-3} + \dfrac{z+2}{z+3} = \dfrac{3}{4}$

16. $\dfrac{x-2}{x-3} - \dfrac{x+2}{x+3} = \dfrac{5}{8}$

17. $2\sqrt{x} = x+1$

18. $\sqrt{y} = y-6$

19. $\sqrt{z+2} = 3$

20. $\sqrt{2z+3} = 3$

21. $\sqrt[3]{4x+4} = 2$

22. $\sqrt[3]{3y+1} = 1$

23. $x - \sqrt{x-2} = 0$

24. $x + \sqrt{x+8} = -2$

25. $x = 6 - 3\sqrt{x-2}$

26. $x - \sqrt{4x-11} - 4 = 0$

27. $\sqrt{x-3} = \sqrt{4x-5}$

28. $x\sqrt{6} = \sqrt{x+2}$

29. $\sqrt{y^2+4y-5} = \sqrt{2-2y}$

30. $\sqrt{y+1} = \sqrt{y^2+3y+2}$

B *Solve each equation in Problems 31–68 over the set of real numbers.*

31. $\dfrac{3}{2x+1} + \dfrac{2x+1}{1-2x} = 1 - \dfrac{8x^2}{4x^2-1}$

32. $\dfrac{3x-5}{5x-5} + \dfrac{5x-1}{7x-7} - \dfrac{x-4}{1-x} = 2$

33. $\dfrac{x-2}{x+3} - \dfrac{1}{x-2} = \dfrac{x-4}{x^2+x-6}$

34. $\dfrac{2y+1}{y+3} + \dfrac{3-y}{2-y} = \dfrac{9-3y-y^2}{y^2+y-6}$

35. $\dfrac{2x+1}{x+2} - \dfrac{x+2}{x+1} = -1$

36. $\dfrac{x-2}{x+2} + \dfrac{1}{x-1} = \dfrac{4x-1}{x^2+x-2}$

37. $\dfrac{x-1}{x-2} + \dfrac{x+4}{2x+1} = \dfrac{1}{2x^2-3x-2}$

38. $\dfrac{x-1}{2x-1} + \dfrac{4-x}{x+1} = \dfrac{3x}{2x^2+x-1}$

39. $\dfrac{3}{x+2} + \dfrac{x-1}{x+5} = \dfrac{5x+20}{6x+24}$

40. $\dfrac{x-3}{x-2} + \dfrac{x-1}{x} = \dfrac{22x-110}{3x^2-15x}$

41. $2 - \sqrt{3x+1} = \sqrt{x-1}$

42. $\sqrt{x+3} + \sqrt{x} = \sqrt{3}$

43. $\sqrt{4x+1} - \sqrt{2x+1} = 2$

44. $\sqrt{3x+1} - \sqrt{2x-1} = 1$

45. $1 + \sqrt{x+2} = \sqrt{x}$

46. $\sqrt{3x+1} - \sqrt{2x-2} = 2$

47. $\sqrt{u+1} - \sqrt[4]{u+1} = 0$

48. $\sqrt{v-2} = \sqrt[4]{v^2-6v+1}$

49. $\sqrt{2x+3} + 3 = 3\sqrt{x+1}$

50. $\sqrt{2y-1} + \sqrt{y+4} = 6$

51. $x^6 + 7x^3 - 8 = 0$

52. $36x^4 + 4 = 25x^2$

53. $4x^{-4} - 35x^{-2} - 9 = 0$

54. $x^{-4} - 5x^{-2} - 6 = 0$

55. $(x^2+4x)^2 + 7(x^2+4x) + 12 = 0$

56. $(x^2-3x)^2 - 2(x^2-3x) - 8 = 0$

57. $\sqrt{x-1} + 2 = 3\sqrt[4]{x-1}$

58. $2\sqrt{x^2+1} = \sqrt[4]{x^2+1} + 6$

59. $x^{1/2} + 6x^{-1/2} - 5 = 0$

60. $6x^{1/2} - 6x^{-1/2} + 5 = 0$

C

61. $\dfrac{1}{\sqrt{3}} = \dfrac{\sqrt{2w+4}}{\sqrt{w}} - \dfrac{\sqrt{3w+4}}{\sqrt{3w}}$

62. $\dfrac{\sqrt{w}-3}{\sqrt{w}} - \dfrac{5-\sqrt{w}}{4} = 0$

63. $\sqrt{2x-1} = \sqrt{7x+2} - \sqrt{x+3}$

64. $\sqrt{3(y+2)} + \sqrt{y+4} = \sqrt{7y+1}$

65. $x^2 - 2x - 8 = 3\sqrt{x^2-2x+2}$

66. $5\sqrt{x^2-2x-1} = 2x^2 - 4x - 2$

67. $\sqrt{\dfrac{x}{1-x}} + \sqrt{\dfrac{1-x}{x}} = \dfrac{10}{3}$

68. $\dfrac{28x}{x^2+3} + \dfrac{2x^2+6}{x} = 15$

4.5 Curve Sketching

Up to now, we have focused our discussion on sketching functions. However, since graphing is such an important skill to acquire in mathematics, and since not everything that you must graph is a function, we need to graph some curves that may not be functions. In Chapter 1 we considered the equation of a circle centered at (h, k) with radius r, namely

$$(x - h)^2 + (y - k)^2 = r^2$$

The graph of this curve is important, even though it is not a function.

We have discussed plotting points, intercepts, symmetry, domain, range, excluded points and regions, and asymptotes. To this list of considerations, we add a question: "Is this a function?" If it is a function, then *each x* value has exactly one *y* value. On the other hand, if it is not a function, then there exists *at least one x* value that produces more than one *y* value. (Remember the vertical line test.)

This section consists of several examples of graphing curves—some are functions and some are not functions. But first, consider the following ideas, which you should keep in mind as you work through the examples.

GUIDELINES FOR SKETCHING CURVES

1. In most advanced work, we are interested in sketching a graph in the quickest, most efficient way. Usually, we need to know only a curve's general shape and location.

2. Sometimes the effort in finding some information about a curve is greater than the effort necessary to plot some points. You should exercise common sense in knowing when to abandon the effort of finding out a particular bit of information about a curve.

3. Being able to classify a curve by inspection of the equation is extremely important. You can now do this for polynomial and rational functions, but later in this book we will consider general characteristics of the graphs of polynomial, rational, trigonometric, exponential, and logarithmic functions, as well as various types of polar forms. The more you know about the general characteristics of a curve, the less work you will need to do in sketching that curve.

4. In later courses, you will add even more types of curves to your general knowledge. The most useful information about curve sketching will be discussed in a calculus course.

5. When in doubt about what a curve looks like in a certain region, you can substitute values and plot some points. In addition to knowing the intercepts, additional points that are often useful are the endpoints of the domain or range and places where a curve passes through an asymptote. Curves may intersect horizontal and slant asymptotes, but may never intersect vertical asymptotes.

In the examples that follow, we will discuss the steps that you would do as well as show you the graph, but limitations of space will not permit all the algebraic details to be shown. You should read these examples with a paper and pencil at hand so you can fill in these details.

EXAMPLE 1 Discuss and sketch $y = \dfrac{x^2 - 1}{x^2 - 4}$.

SOLUTION This curve is a function, so it will pass the vertical line test.

DOMAIN: Values that cause division by zero are excluded, so we need to determine when $x^2 - 4 = 0$. We find that $x^2 - 4 = 0$ when $x = 2$, -2. Thus, $D = (-\infty, -2) \cup (-2, 2) \cup (2, \infty)$.

RANGE: Solve for x:

$$x^2 = \frac{4y - 1}{y - 1}$$

$$x = \pm \sqrt{\frac{4y - 1}{y - 1}}$$

We need to find out when

$$\frac{4y - 1}{y - 1} \geq 0$$

We find critical values $y = \frac{1}{4}$, 1.

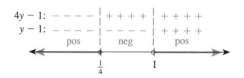

Thus, $R = (-\infty, \frac{1}{4}] \cup (1, \infty)$.

Graphing is done in stages, which is difficult to show in a textbook. I have attempted to show you these steps alongside the examples of this section. Note also that we show the boundaries of all excluded regions as dashed lines for convenience.

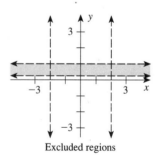

Excluded regions

SYMMETRY: It is symmetric with respect to the y-axis only.

ASYMPTOTES: $x = 2$, $x = -2$, and $y = 1$

PLOTTING POINTS:

x-intercepts: Solve $\dfrac{x^2 - 1}{x^2 - 4} = 0$ to obtain $x = 1, -1$.

Plot the points $(1, 0)$ and $(-1, 0)$.

y-intercepts: Substitute $x = 0$ to obtain $y = \frac{1}{4}$.
Plot the point $(0, \frac{1}{4})$.

Asymptote intercepts: $x = 2$, $x = -2$ are excluded values.
Substitute $y = 1$:

$$1 = \frac{x^2 - 1}{x^2 - 4}$$

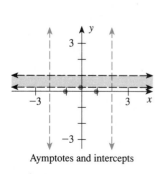

Aymptotes and intercepts

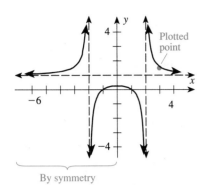

Plotted point

By symmetry

FIGURE 4.21 Graph of $y = \dfrac{x^2 - 1}{x^2 - 4}$

Solving this equation, you get $-4 = -1$, which tells you that the graph does not cross this asymptote.

Additional points: Now look at the information you have and plot some additional points using symmetry. The graph is shown in Figure 4.21. Plot as many points as you feel you need to be confident of your graph, but you do not need to plot many. Notice that this curve is a function, so each x value is associated with exactly one y value. (It will pass the vertical line test.)

To complete the graph shown in Figure 4.21, only one additional point was calculated: If $x = 3$, then calculate $y = \frac{8}{5}$.

EXAMPLE 2 Discuss and sketch $x^2 + x^2y^2 - 4y^2 - 1 = 0$.

SOLUTION This curve is not a function.

DOMAIN: Solve for y:

$$y^2 = \frac{1 - x^2}{x^2 - 4} \quad (x \neq 2, -2)$$

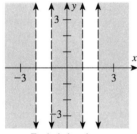

Excluded regions

Also, $\dfrac{1 - x^2}{x^2 - 4} \geq 0$ since y^2 is nonnegative.

Solving this inequality, you will find the domain to be $D = (-2, -1] \cup [1, 2)$.

RANGE: Solve for x:

$$x^2 = \frac{4y^2 + 1}{1 + y^2}$$

Also, $\dfrac{4y^2 + 1}{1 + y^2} \geq 0$ since x^2 is nonnegative.

This inequality is true for all values of y, so $R = (-\infty, \infty)$.

SYMMETRY: The graph is symmetric with respect to the x-axis, y-axis, and origin.

ASYMPTOTES: $x = 2$, $x = -2$; notice that as $x \to \infty$, we have $y^2 \to -1$, which is impossible (a square number approaching a negative number), so there are no horizontal asymptotes; no slant asymptotes.

Aymptotes and intercepts

PLOTTING POINTS:

x-intercepts: Solve $0 = \dfrac{1 - x^2}{x^2 - 4}$ to obtain $x = 1, -1$.

Plot the points $(1, 0)$ and $(-1, 0)$.

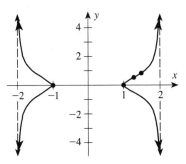

FIGURE 4.22 Graph of
$x^2 + x^2y^2 - 4y^2 - 1 = 0$

y-intercepts: Solve $0 = \dfrac{4y^2 + 1}{1 + y^2}$ (no solution).

The curve does not cross the *y*-axis.

Asymptote intercepts: $x = 2$, $x = -2$ are excluded values, so the curve does not cross its asymptotes.

Additional points: If $x = 1.5$, $y \approx .85$; plot $(1.5, .85)$ as well as three other symmetric points.
If $x = 1.3$, $y \approx .55$; plot $(1.3, .55)$ as well as three other symmetric points.

The completed graph is shown in Figure 4.22.

EXAMPLE 3   Discuss and sketch $2|x| - |3y| = 9$.

SOLUTION   This curve is not a function.

DOMAIN: Solve for *y*: $|3y| = 2|x| - 9$

This requires that $2|x| - 9 \geq 0$. Solving,

$$|x| \geq \tfrac{9}{2}$$

$$x \geq \tfrac{9}{2} \quad \text{or} \quad x \leq -\tfrac{9}{2}$$

$$D = (-\infty, -\tfrac{9}{2}] \cup [\tfrac{9}{2}, \infty)$$

RANGE: Solve for *x*: $|x| = \dfrac{9 + |3y|}{2}$

This requires that $\dfrac{9 + |3y|}{2} \geq 0$. Solving,

$$|3y| \geq -9$$

This is true for all values of *y*, so $R = (-\infty, \infty)$.

SYMMETRY: The graph is symmetric with respect to the *x*-axis, *y*-axis, and origin.

ASYMPTOTES: None

PLOTTING POINTS:

x-intercepts: If $y = 0$, then $2|x| = 9$, so $x = \pm\tfrac{9}{2}$.

y-intercepts: If $x = 0$, then $|3y| = -9$ (no values).

Additional points: Use your knowledge of lines:
If $x \geq 0$ and $y \geq 0$, sketch $2x - 3y = 9$ within domain.
If $x \geq 0$ and $y < 0$, sketch $2x + 3y = 9$ within domain.
If $x < 0$ and $y \geq 0$, sketch $-2x - 3y = 9$ within domain.
If $x < 0$ and $y < 0$, sketch $-2x + 3y = 9$ within domain.

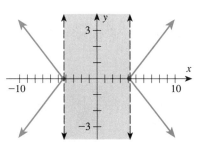

FIGURE 4.23   Graph of
$2|x| - |3y| = 9$

The graph is shown in Figure 4.23.

EXAMPLE 4 Discuss and sketch $\sqrt{x} + \sqrt{y} = 2$.

SOLUTION This curve is a function, so it will pass the vertical line test.

DOMAIN: Solve for y:

$$\sqrt{y} = 2 - \sqrt{x}$$

$x \geq 0$ and $2 - \sqrt{x} \geq 0$ or $x \leq 4$, so $D = [0, 4]$.

RANGE: Same analysis as for the domain; $R = [0, 4]$.

SYMMETRY: The curve is not symmetric with respect to the x-axis, the y-axis, or the origin.

ASYMPTOTES: None

PLOTTING POINTS:

   x-intercepts: If $y = 0$, then $x = 4$; plot $(4, 0)$.

   y-intercepts: If $x = 0$, then $y = 4$; plot $(0, 4)$.

   Additional points: If $x = 1$, then $y = 1$.
                      If $x = 2$, then $y \approx .34$.
                      If $x = 3$, then $y \approx .07$.

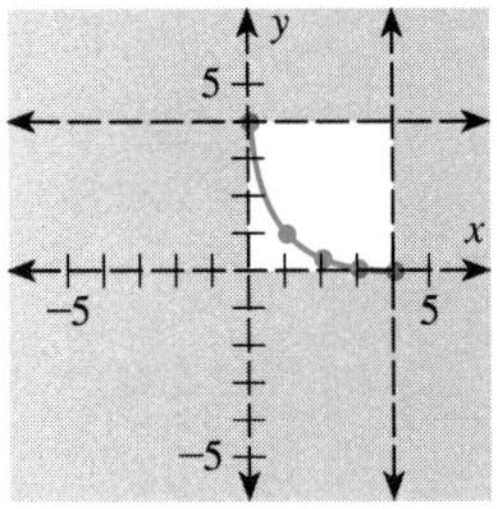

FIGURE 4.24
Graph of $\sqrt{x} + \sqrt{y} = 2$

The graph is shown in Figure 4.24.

EXAMPLE 5 Discuss and graph $y = \dfrac{2x^2 - 3x + 5}{x^2 - x - 2}$.

SOLUTION This curve is a function, so it will pass the vertical line test.

DOMAIN: $x^2 - x - 2 \neq 0$. Solving the equation $x^2 - x - 2 = 0$, you obtain $x = 2$, $x = -1$. Thus, $D = (-\infty, -1) \cup (-1, 2) \cup (2, \infty)$.

RANGE: Solve for x (you may wish to skip this step because of the amount of work involved). Multiply both sides by $x^2 - x - 2$ and simplify to obtain

$$(2 - y)x^2 + (y - 3)x + (2y + 5) = 0$$

Use the quadratic formula to solve for x (if $y \neq 2$):

$$x = \frac{-y + 3 \pm \sqrt{(y - 3)^2 - 4(2 - y)(2y + 5)}}{2(2 - y)}$$

From this we see that $(y - 3)^2 - 4(2 - y)(2y + 5) \geq 0$. Solving *this* inequality, you first find the critical values:

$$(y - 3)^2 - 4(2 - y)(2y + 5) = 0$$

$$9y^2 - 2y - 31 = 0 \qquad \text{Expand and simplify.}$$

$$y = \frac{1 \pm \sqrt{280}}{9} \qquad \text{By the quadratic formula}$$

$$\approx -1.75, 1.97$$

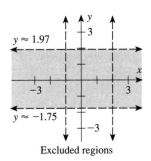

$y \approx 1.97$

$y \approx -1.75$

Excluded regions

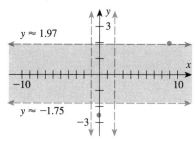

$y \approx 1.97$

$y \approx -1.75$

Asymptotes and intercepts

Finally, conclude that the range is

$$R = \left(-\infty, \frac{1 - \sqrt{280}}{9}\right] \cup \left[\frac{1 + \sqrt{280}}{9}, \infty\right)$$

SYMMETRY: Not symmetric with respect to the x-axis, the y-axis, or the origin.

ASYMPTOTES: $x = 2$, $x = -1$, and $y = 2$

PLOTTING POINTS:

x-intercepts: If $y = 0$, then $0 = \dfrac{2x^2 - 3x + 5}{x^2 - x - 2}$, which has no real roots; there are no x-intercepts.

y-intercepts: If $x = 0$, then $y = -\frac{5}{2}$. Plot the y-intercept, $(0, -\frac{5}{2})$.

Asymptote intercepts: $x = 2$ and $x = -1$ are excluded values. If $y = 2$ (notice that we considered the possibility $y \neq 2$ above), then

$$2 = \frac{2x^2 - 3x + 5}{x^2 - x - 2}$$

$$2x^2 - 2x - 4 = 2x^2 - 3x + 5 \quad (x \neq 2, -1)$$

$$-2x - 4 = -3x + 5$$

$$x = 9$$

Note: The asymptote is the line $y = 2$, but the excluded region is bounded by the line $y = (1 + \sqrt{280})/9 \approx 1.97$. You cannot see both of these lines (nor should you try to draw them both in your work).

The curve passes through $(9, 2)$. That is, $(9, 2)$ is on the curve and is also on the horizontal asymptote.

Additional points: If $x = -4$, then $y \approx 2.72$.
If $x = -2$, then $y \approx 4.75$.
If $x = .5$, then $y \approx -1.8$.
If $x = 1$, then $y = -2$.
If $x = 1.5$, then $y = -4$.
If $x = 3$, then $y \approx 3.5$.
If $x = 4$, then $y \approx 2.5$.
If $x = 12$, then $y \approx 1.98$.

The graph is shown in Figure 4.25.

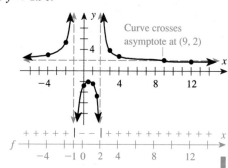

Curve crosses asymptote at $(9, 2)$

FIGURE 4.25 Graph of $y = \dfrac{2x^2 - 3x + 5}{x^2 - x - 2}$

4.5 Problem Set

A *Determine whether each is a function; then find the domain, range, intercepts, symmetry, and asymptotes for the graphs whose equations are given in Problems 1–30.*

1. $xy = 2$

2. $xy = 6$

3. $y = \dfrac{x+1}{x}$

4. $y = \dfrac{x+1}{x+2}$

5. $y = \sqrt{4-x}$

6. $y = \sqrt{x-4}$

7. $y = -\sqrt{x-6}$

8. $y = -\sqrt{6-x}$

9. $y = \dfrac{1}{x^2-4}$

10. $y = \dfrac{1}{x^2+1}$

11. $y = \dfrac{2x^2+9x+10}{x+2}$

12. $y = \dfrac{3x^2+5x-2}{x+2}$

13. $y = \dfrac{2x^3-3x^2-2x}{2x+1}$

14. $y = \dfrac{x^3+6x^2+15x+14}{x+2}$

B

15. $y = \sqrt{x}-x$

16. $y = x-\sqrt{x}$

17. $y = \sqrt{x^2+2x-3}$

18. $y = \sqrt{x^2+x-6}$

19. $y = \dfrac{-x}{\sqrt{4-x^2}}$

20. $y = \dfrac{9}{\sqrt{x^2-9}}$

21. $|x|+|y| = 5$

22. $|2x|-3|y| = 9$

23. $9x^2+16y^2 = 144$

24. $9x^2-16y^2 = 144$

25. $y = \dfrac{x^2+1}{x}$

26. $y = \dfrac{5x^2+2}{x}$

27. $y^2 = \dfrac{x^2-1}{x-4}$

28. $y^2 = \dfrac{x^2-4}{x-2}$

29. $y^2 = \dfrac{x^2-9}{x^2-2x-15}$

30. $y^2 = \dfrac{x^2-4}{x^2-3x+2}$

Sketch the graphs of the curves whose equations are given in Problems 31–60. Notice that these are the same equations as those in Problems 1–30.

31. $xy = 2$

32. $xy = 6$

33. $y = \dfrac{x+1}{x}$

34. $y = \dfrac{x+1}{x+2}$

35. $y = \sqrt{4-x}$

36. $y = \sqrt{x-4}$

37. $y = -\sqrt{x-6}$

38. $y = -\sqrt{6-x}$

39. $y = \dfrac{1}{x^2-4}$

40. $y = \dfrac{1}{x^2+1}$

41. $y = \dfrac{2x^2+9x+10}{x+2}$

42. $y = \dfrac{3x^2+5x-2}{x+2}$

43. $y = \dfrac{2x^3-3x^2-2x}{2x+1}$

44. $y = \dfrac{x^3+6x^2+15x+14}{x+2}$

45. $y = \sqrt{x}-x$

46. $y = x-\sqrt{x}$

47. $y = \sqrt{x^2+2x-3}$

48. $y = \sqrt{x^2+x-6}$

49. $y = \dfrac{-x}{\sqrt{4-x^2}}$

50. $y = \dfrac{9}{\sqrt{x^2-9}}$

51. $|x|+|y| = 5$

52. $|2x|-3|y| = 9$

53. $9x^2+16y^2 = 144$

54. $9x^2-16y^2 = 144$

55. $y = \dfrac{x^2+1}{x}$

56. $y = \dfrac{5x^2+2}{x}$

57. $y^2 = \dfrac{x^2-1}{x-4}$

58. $y^2 = \dfrac{x^2-4}{x-2}$

59. $y^2 = \dfrac{x^2-9}{x^2-2x-15}$

60. $y^2 = \dfrac{x^2-4}{x^2-3x+2}$

C *Discuss and sketch the curves whose equations are given in Problems 61–66.*

61. $x^{2/3}+y^{2/3} = 4$

62. $x^{2/3}+y^{2/3} = 1$

63. $x^2y^2-x^2-9y^2+4 = 0$ **64.** $y^2x^2-x^2-3y^2+x = 0$

65. $x^4+y^2-4x^2 = 0$

66. $x^4+4x^2+y^2 = 0$

4.6 Partial Fractions

In algebra, rational expressions are added by finding common denominators; for example,

$$\frac{5}{x-2} + \frac{3}{x+1} = \frac{5(x+1) + 3(x-2)}{(x-2)(x+1)} = \frac{8x-1}{(x-2)(x+1)}$$

In calculus, however, it is sometimes necessary to break apart the expression

$$\frac{8x-1}{(x-2)(x+1)}$$

into two fractions with denominators that are linear. The technique for doing this is called the **method of partial fractions.**

The rational expression

$$f(x) = \frac{P(x)}{D(x)}$$

can be **decomposed** into partial fractions if there are no common factors and if the degree of $P(x)$ is less than the degree of $D(x)$. If the degree of $P(x)$ is greater than or equal to the degree of $D(x)$, then use either long division or synthetic division to obtain a polynomial plus a proper fraction. For example,

$$\frac{x^4 + 2x^3 - 4x^2 + x - 3}{x^2 - x - 2} = x^2 + 3x + 1 + \frac{8x-1}{x^2 - x - 2}$$

This was found by long division.

Proper fraction:
This is the part
that is decomposed
into partial fractions.

Now look at the proper fraction. There is a theorem that says this proper fraction can be written as a sum,

$$F_1 + F_2 + \cdots + F_j$$

where *each* F_i is of the form

$$\frac{A}{(x-r)^n} \quad \text{or} \quad \frac{Ax+B}{(x^2 + sx + t)^n}$$

We begin by focusing on the first form.

Let $f(x) = P(x)/D(x)$, where $P(x)$ and $D(x)$ have no common factors and the degree of $P(x)$ is less than the degree of $D(x)$. Also suppose that $D(x) = (x - r)^n$. Then $f(x)$ can be decomposed into partial fractions:

$$\frac{A_1}{x - r} + \frac{A_2}{(x - r)^2} + \cdots + \frac{A_n}{(x - r)^n}$$

EXAMPLE 1 Decompose $\dfrac{8x - 1}{x^2 - x - 2}$ into partial fractions.

SOLUTION

$$\frac{8x - 1}{x^2 - x - 2} = \frac{8x - 1}{(x - 2)(x + 1)}$$

First, factor the denominator, if possible, and make sure there are no common factors.

$$= F_1 + F_2$$

Break up the fraction into parts, each with a linear factor.

$$F_1 + F_2 = \frac{A}{x - 2} + \frac{B}{x + 1}$$

The task is to find A and B.

$$= \frac{A(x + 1) + B(x - 2)}{(x - 2)(x + 1)}$$

Obtain a common denominator on the right.

Now, multiply both sides of this equation by the least common denominator, which is $(x - 2)(x + 1)$ for this example:

$$8x - 1 = A(x + 1) + B(x - 2)$$

Substitute, one at a time, the values that cause each of the factors in the least common denominator to be zero.

Let $x = -1$:

$$8x - 1 = A(x + 1) + B(x - 2)$$
$$8(-1) - 1 = A(-1 + 1) + B(-1 - 2)$$
$$-9 = 0 + B(-3)$$
$$-9 = -3B$$
$$\mathbf{3 = B}$$

Let $x = 2$:

$$8x - 1 = A(x + 1) + B(x - 2)$$
$$8(2) - 1 = A(2 + 1) + B(2 - 2)$$
$$15 = 3A$$
$$\mathbf{5 = A}$$

If $A = 5$ and $B = 3$, then

$$\frac{8x - 1}{(x - 2)(x + 1)} = \frac{5}{x - 2} + \frac{3}{x + 1}$$

Example 2 illustrates the process if there is a repeated linear factor.

EXAMPLE 2 Decompose $\dfrac{x^2 - 6x + 3}{(x - 2)^3}$ by using the method of partial fractions.

SOLUTION

$$\frac{x^2 - 6x + 3}{(x - 2)^3} = \frac{A}{x - 2} + \frac{B}{(x - 2)^2} + \frac{C}{(x - 2)^3}$$

Multiply both sides by $(x - 2)^3$:

$$x^2 - 6x + 3 = A(x - 2)^2 + B(x - 2) + C$$

Let $x = 2$:
$$(2)^2 - 6(2) + 3 = A(2 - 2)^2 + B(2 - 2) + C$$

$$4 - 12 + 3 = 0 + 0 + C$$

$$-5 = C$$

Notice that with repeated factors we cannot find all the numerators as we did in Example 1. Now substitute $C = -5$ into the original equation and simplify by combining terms on the right side:

$$
\begin{aligned}
x^2 - 6x + 3 &= A(x - 2)^2 + B(x - 2) - 5 \\
&= A(x^2 - 4x + 4) + B(x - 2) - 5 \\
&= Ax^2 - 4Ax + Bx + 4A - 2B - 5 \\
&= Ax^2 + (-4A + B)x + (4A - 2B - 5)
\end{aligned}
$$

If the polynomials on the left and right sides of the equality are equal, then the coefficients of the like terms must be equal. That is,

$$x^2 - 6x + 3 = Ax^2 + (-4A + B)x + (4A - 2B - 5)$$

$$A = 1$$
$$-4A + B = -6$$
$$4A - 2B - 5 = 3$$

If $A = 1$, then

$$-4A + B = -6$$
$$-4(1) + B = -6$$
$$B = -2$$

Check: If $A = 1$ and $B = -2$, then

$$
\begin{aligned}
4A - 2B - 5 &= 4(1) - 2(-2) - 5 \\
&= 4 + 4 - 5 \\
&= 3
\end{aligned}
$$

Thus,

$$\frac{x^2 - 6x + 3}{(x - 2)^3} = \frac{1}{x - 2} + \frac{-2}{(x - 2)^2} + \frac{-5}{(x - 2)^3}$$

We will now consider quadratic factors.

PARTIAL FRACTION DECOMPOSITION—QUADRATIC FACTORS

Let $f(x) = P(x)/D(x)$, where $P(x)$ and $D(x)$ have no common factors and the degree of $P(x)$ is less than the degree of $D(x)$. If $D(x) = (x^2 + sx + t)^m$, then $f(x)$ can be decomposed into partial fractions:

$$\frac{A_1x + B_1}{x^2 + sx + t} + \frac{A_2x + B_2}{(x^2 + sx + t)^2} + \cdots + \frac{A_mx + B_m}{(x^2 + sx + t)^m}$$

EXAMPLE 3   Decompose $f(x) = \dfrac{2x^3 + 3x^2 + 3x + 2}{(x^2 + 1)^2}$ Compare the denominator with the expression in the box above: $s = 0$, $t = 1$, and $m = 2$.

SOLUTION

$$\frac{2x^3 + 3x^2 + 3x + 2}{(x^2 + 1)^2} = \frac{Ax + B}{x^2 + 1} + \frac{Cx + D}{(x^2 + 1)^2}$$

Multiply by $(x^2 + 1)^2$:

$$2x^3 + 3x^2 + 3x + 2 = (Ax + B)(x^2 + 1) + Cx + D$$

This time, $x^2 + 1 \neq 0$ in the set of real numbers, so multiply out the right side:

$$2x^3 + 3x^2 + 3x + 2 = Ax^3 + Bx^2 + Ax + B + Cx + D$$
$$= Ax^3 + Bx^2 + (A + C)x + (B + D)$$

Equate the coefficients of the similar terms on the left and right:

$$A = 2$$
$$B = 3$$
$$A + C = 3 \quad \text{If } A = 2, \text{ then } 2 + C = 3 \text{ and } C = 1.$$
$$B + D = 2 \quad \text{If } B = 3, \text{ then } 3 + D = 2 \text{ and } D = -1.$$

Thus,

$$\frac{2x^3 + 3x^2 + 3x + 2}{(x^2 + 1)^2} = \frac{2x + 3}{x^2 + 1} + \frac{x - 1}{(x^2 + 1)^2}$$

4.6 Problem Set

A   List the factors of each denominator in the decomposition of the rational expressions in Problems 1–24. For Example 1 these factors are $(x - 2)$ and $(x + 1)$; for Example 2 they are $(x - 2)$, $(x - 2)^2$, and $(x - 2)^3$.

1. $\dfrac{2x + 10}{x^2 + 7x + 12}$

2. $\dfrac{2x - 14}{x^2 + x - 6}$

3. $\dfrac{7x - 7}{2x^2 - 5x - 3}$

4. $\dfrac{4(x - 1)}{x^2 - 4}$

5. $\dfrac{34 - 5x}{48 - 14x + x^2}$

6. $\dfrac{x - 7}{20 - 9x + x^2}$

7. $\dfrac{5x^2 - 5x - 4}{x^3 - x}$

8. $\dfrac{4x^2 - 7x - 3}{x^3 - x}$

9. $\dfrac{2x^2 - 18x - 12}{x^3 - 4x}$

10. $\dfrac{2x - 1}{(x - 2)^2}$

11. $\dfrac{4x - 22}{(x - 5)^2}$

12. $\dfrac{x^2 + 5x + 1}{x(x + 1)^2}$

13. $\dfrac{5x^2 - 2x + 2}{x(x - 1)^2}$

14. $\dfrac{2x^2 + 7x + 2}{(x + 1)^3}$

15. $\dfrac{7x - 3x^2}{(x - 2)^3}$

16. $\dfrac{x}{x^2 + 4x - 5}$

17. $\dfrac{x}{x^2 - 2x - 3}$

18. $\dfrac{7x - 1}{x^2 - x - 2}$

19. $\dfrac{10x^2 - 11x - 6}{x^3 - x^2 - 2x}$

20. $\dfrac{-17x - 6}{x^3 + x^2 - 6x}$

21. $\dfrac{12 + 9x - 6x^2}{x^3 - 5x^2 + 4x}$

22. $\dfrac{x^3}{(x + 1)^2(x - 2)}$

23. $\dfrac{2x^3 - 3x^2 + 6x - 1}{1 - x^4}$

24. $\dfrac{2x^3 - 7x^2 + 8x - 7}{x^2 - 4x + 4}$

33. $\dfrac{7x + 2}{(x + 2)(x - 4)}$

34. $\dfrac{2x + 10}{x^2 + 7x + 12}$

35. $\dfrac{2x - 14}{x^2 + x - 6}$

36. $\dfrac{7x - 7}{2x^2 - 5x - 3}$

37. $\dfrac{4(x - 1)}{x^2 - 4}$

38. $\dfrac{34 - 5x}{48 - 14x + x^2}$

39. $\dfrac{x - 7}{20 - 9x + x^2}$

40. $\dfrac{5x^2 - 5x - 4}{x^3 - x}$

41. $\dfrac{4x^2 - 7x - 3}{x^3 - x}$

42. $\dfrac{2x^2 - 18x - 12}{x^3 - 4x}$

43. $\dfrac{2x - 1}{(x - 2)^2}$

44. $\dfrac{4x - 22}{(x - 5)^2}$

45. $\dfrac{x^2 + 5x + 1}{x(x + 1)^2}$

46. $\dfrac{5x^2 - 2x + 2}{x(x - 1)^2}$

47. $\dfrac{2x^2 + 8x + 3}{(x + 1)^3}$

48. $\dfrac{7x - 3x^2}{(x - 2)^3}$

49. $\dfrac{x}{x^2 + 4x - 5}$

50. $\dfrac{x}{x^2 - 2x - 3}$

51. $\dfrac{7x - 1}{x^2 - x - 2}$

52. $\dfrac{10x^2 - 11x - 6}{x^3 - x^2 - 2x}$

53. $\dfrac{-17x - 6}{x^3 + x^2 - 6x}$

54. $\dfrac{12 + 9x - 6x^2}{x^3 - 5x^2 + 4x}$

B Decompose each fraction in Problems 25–60 by using the method of partial fractions.

25. $\dfrac{x^2 + 2x + 5}{x^3}$

26. $\dfrac{3x^2 - 2x + 1}{x^3}$

27. $\dfrac{2x^2 - 5x + 4}{x^3}$

28. $\dfrac{1}{(x + 2)(x + 3)}$

29. $\dfrac{1}{(x + 4)(x + 5)}$

30. $\dfrac{-4}{(x + 5)(x + 1)}$

31. $\dfrac{7x - 10}{(x - 2)(x - 1)}$

32. $\dfrac{11x - 1}{(x - 1)(x + 1)}$

C

55. $\dfrac{5x^2 - 6x + 7}{(x - 1)(x^2 + 1)}$

56. $\dfrac{x^2}{(x + 1)(x^2 + 1)}$

57. $\dfrac{x^3}{(x - 1)^2}$

58. $\dfrac{x^3}{(x + 1)^2}$

59. $\dfrac{2x^3 - 3x^2 + 6x - 1}{1 - x^4}$

60. $\dfrac{2x^3 - 7x^2 + 8x - 7}{x^2 - 4x + 4}$

4.7 Chapter 4 Summary

The material of this chapter is reviewed in the following list of objectives. After each objective there are some practice questions. For a sample test, select the first question of each set and check your answers with the answer section. For a sample test without answers, use the second question of each set. Additional practice is given by the other questions in each set. If you are having trouble with a particular type of problem, look back to that section for extra help.

Rational Functions

OBJECTIVE 1 *Graph rational functions.*

1. $f(x) = \dfrac{1}{x-2} + 2$

2. $f(x) = \dfrac{3x^2 + 2x - 5}{x - 1}$

3. $f(x) = \dfrac{2x^2 - 3x - 1}{x^2 - x - 2}$

4. $f(x) = \dfrac{x^3 - x - 6}{x - 2}$

4.2 **Asymptotes**

OBJECTIVE 2 *Find limits as $|x| \to \infty$.*

5. $\displaystyle \lim_{|x| \to \infty} \dfrac{3x^5 - 2x^3 + 1}{4x^5 - 1}$

6. $\displaystyle \lim_{|x| \to \infty} \dfrac{3x^5 - 2x^3 + 1}{5x^4 - 1}$

7. $\displaystyle \lim_{|x| \to \infty} \dfrac{5x^3 - 2x^2 + 1}{4x^5 - 1}$

8. $\displaystyle \lim_{|x| \to \infty} \dfrac{(x-1)(3x-2)(5x-3)}{x^3}$

OBJECTIVE 3 *Find vertical, horizontal, and slant asymptotes.*

9. $y = \dfrac{6x^2 - 11x}{2x - 1}$

10. $y = \dfrac{2x^2 - 5x - 3}{3x^2 - 7x - 6}$

11. $y = \dfrac{3x}{\sqrt{9 - x^2}}$

12. $x^2 - x^2 y^2 + 9y^2 - 4 = 0$

4.3 **Radical Functions**

OBJECTIVE 4 *Graph radical functions.*

13. $y = \sqrt{x^2 - 4x + 3}$

14. $y = \sqrt[3]{x - 2} - 3$

15. $y = \dfrac{x}{\sqrt{x^2 + 25}}$

16. $y = \dfrac{x}{\sqrt{x^2 - 25}}$

4.4 **Real Roots of Rational and Radical Equations**

OBJECTIVE 5 *Solve rational equations for real numbers.* State the solution set and give the excluded values and extraneous roots, if any.

17. $\dfrac{x+1}{x+3} = \dfrac{2x-1}{2x+1}$

18. $\dfrac{2x-1}{x+3} = \dfrac{x+1}{2x+1}$

19. $\dfrac{5}{x} - \dfrac{x-5}{x-3} = \dfrac{x-6}{2x-12}$

20. $\dfrac{8}{x+6} - \dfrac{1}{x-4} = \dfrac{1}{3}$

OBJECTIVE 6 *Solve radical equations for real numbers.* State the solution set and give the extraneous roots, if any.

21. $x - \sqrt{3 - x} = 3$

22. $x - \sqrt{2x - 3} = 1$

23. $\sqrt{x + 5} = \sqrt{x} + 1$

24. $\sqrt{x + 2} = \sqrt{x + 3} - 1$

4.5 **Curve Sketching**

OBJECTIVE 7 *Discuss and sketch curves, given an equation.*

25. $x^2 - x^2 y^2 - y^2 + 25 = 0$ **26.** $|y| - |x| = 5$

27. $x^{2/3} + y^{2/3} = 1$

28. $y = \sqrt{x(x^2 - 9)}$

4.6 Partial Fractions

OBJECTIVE 8 *Decompose rational expressions by using the method of partial fractions.*

29. $\dfrac{5x^2 - 19x + 17}{(x - 1)(x - 2)^2}$

30. $\dfrac{7x - 7}{2x^2 - 9x + 4}$

31. $\dfrac{2x^2 + 13x - 9}{x^2 + 2x - 15}$

32. $\dfrac{2x^3 + 2x + 3}{(x^2 + 1)^2}$

John Napier (1550–1617)

John Napier was the Isaac Asimov of his day, having envisioned the tank, the machine gun, and the submarine. He also predicted that the end of the world would occur between 1688 and 1700. He is best known today as the inventor of logarithms. Napier, however, believed that his reputation would rest ultimately on his predictions about the end of the world. He considered logarithms merely an interesting recreational diversion.

The word *logarithm* means "ratio number" and was adopted by Napier after he had first chosen the term *artificial number*. As you will see in this chapter, today we define a logarithm as an exponent, but historically logarithms were discovered before exponents were in use. Napier's original tables were very cumbersome. They did not give the logarithm of x but $10^7 \ln(10^7 x^{-1})$. Common logarithms, or logs to the base 10, were introduced by a professor at Oxford named Henry Briggs (1516–1631). Briggs saw Napier's book, immediately recognized the importance of logarithms, and began working on them himself. Within months he joined Napier in Scotland and convinced him that tables for log x would be more convenient.

"Today, when logarithms appear so easy, the student may have trouble understanding how they could ever have been difficult. But in Napier's time the notion of a function was unknown, and even the simple exponential notation $a^n = a \cdot a \cdot a \cdots a$, introduced later by René Descartes, did not exist. Napier invented logarithms to ease computation, and by so doing became known as the founder of applied mathematics. Although the computational efficiency of calculators is now greater than that of logarithms, Napier's place in mathematics remains secure, for from his work came the logarithm functions and their inverses, the exponentials. These functions occupy a central role in the mathematical sciences."*

Until the advent of the low-cost calculator, logarithms were used extensively in complicated mathematical calculation. The mathematician F. Cajori said that the powers of modern calculations are due to three inventions: "arabic notation, decimal fractions, and logarithms." Today we would have to add another to those three inventions: the hand-held calculator. This, however, does not minimize the importance of the logarithm. It simply changes the emphasis from calculation to application. This chapter reflects the recent change in emphasis brought about by the calculator.

---

*From R. A. Bonic, G. Hajian, E. DuCasse, and M. M. Lipschutz, *Freshman Calculus*, 2nd ed. (Lexington, Mass.: D.C. Heath).

5

EXPONENTIAL AND LOGARITHMIC FUNCTIONS

Preview

The important ideas of this chapter are solving exponential and logarithmic equations. In order to do this, two new and important functions are defined and discussed. The 13 objectives of this chapter are listed on pages 245–246. Following these objectives is a Cumulative Review for Chapters 2–5. It is a natural breaking point for the functions and graphing in the first part of the book. Following the Cumulative Review you will find the first Extended Application in the book, which concerns population growth. This is introduced with a newspaper article, and a mathematical discussion follows, in which the techniques of this part of the book are used to answer questions related to the article.

Perspective

Transcendental functions form a major category of functions discussed and developed in calculus. The logarithmic and exponential functions introduced in this chapter fall into this category, and are used in calculus in many applications including growth, decay, population growth, evaluation of integrals, and solving equations called *differential equations*. The following excerpt is from the introduction of transcendental functions in a calculus book.

6.0 Introduction

A function *f* is *algebraic* if its formula involves only the operations of addition, subtraction, multiplication, division, absolute value, and exponentiation to rational powers. Otherwise the function is called *transcendental*. The six trigonometric functions, which we have been working with since Chapter 2, are familiar transcendental functions. This chapter develops the calculus of additional transcendental functions.

The first five sections are devoted to logarithmic and exponential functions. Following a review of the basic properties of logarithms, Section 1 defines the natural logarithm function *ln* as an antiderivative. Properties of integration are then used to show that this function has all the usual logarithmic properties, and to derive its calculus. The next section defines and studies the natural exponential function *exp* as the inverse function of *ln*. Section 3 uses *ln* and *exp* to analyze several important types of growth and decay. The following section discusses further exponential and logarithmic functions, which arise in several areas outside mathematics.

Section 5 introduces the inverse trigonometric functions and their calculus. These functions are at the heart of one of the techniques of integration presented in the next chapter. Section 6 is devoted to hyperbolic functions, a class of exponential functions with properties similar to those of the trigonometric functions. Section 7 discusses an important application of the hyperbolic functions to physics and engineering, as well as some applications of trigonometric functions to motion problems.

Section 8 presents a technique named for the author of the first calculus textbook, written in 1696. That technique, l'Hôpital's rule, can be used to evaluate a number of limits that, up to now, can only be classified as *indeterminate*, not capable of being determined by earlier methods. Many of those limits involve transcendental functions. The final section gives an optional discussion of some

---

From James Hurley, *Calculus* (Belmont, Ca.: Wadsworth), p. 335.

5.1 Exponential Functions

The linear, quadratic, polynomial, and rational functions considered in the first part of this book are all called **algebraic functions**. An algebraic function is a function that can be expressed in terms of algebraic operations alone. If a function is not algebraic, it is called a **transcendental function**. In this chaper, two examples of transcendental functions, *exponential* and *logarithmic* functions, are considered. In the next chapter, other types of transcendental functions will be considered.

Definition and Graphs of Exponential Functions

In order to define an exponential function, we need to recall the definition of an exponent. In Chapter 1 we defined an expression with a rational exponent. We need to give meaning to exponents that are irrational. For example, the expression

$$2^{\sqrt{3}}$$

has not yet been defined. We might turn to a calculator with an exponent key to find

$$2^{\sqrt{3}} \approx 3.321997085$$

but what does this mean? Where does it come from if it is not defined? Since we can approximate $\sqrt{3}$ to any degree of accuracy, we can use the definition of a rational exponent to express $2^{\sqrt{3}}$ to any degree of accuracy. Consider the following sequence:

$$\sqrt{3} \approx 1.7, \quad \text{so} \quad 2^{\sqrt{3}} \approx 2^{1.7} \quad \text{and } \textit{by definition} \quad 2^{1.7} = 2^{17/10}$$

The number $2^{17/10}$ is defined because 17/10 is a rational number. Thus, we can use a calculator to obtain $2^{17/10} \approx 3.249009585$.

If we could analyze what is happening inside your calculator, we would see that when we input $2^{\sqrt{3}}$, what is really being evaluated (to the best accuracy available with your calculator) is an expression such as

$$2^{\sqrt{3}} \approx 2^{1.732050808} \approx 3.321997085$$

However, this discussion still does not give a definition for an expression with an irrational exponent. To do this we need calculus and the limiting process, but the following theorem will give us a foundation for considering exponential functions.

SQUEEZE THEOREM FOR EXPONENTS

Suppose b is a real number greater than 1. Then for any real number x there is a unique real number b^x. Moreover, if h and k are any two rational numbers such that $h < x < k$, then

$$b^h < b^x < b^k$$

The Squeeze Theorem will give meaning to expressions such as $2^{\sqrt{3}}$. Consider the graph of the function $f(x) = 2^x$ by plotting the points shown in the table as in Figure 5.1.

x	$y = f(x) = 2^x$
-3	$2^{-3} = \frac{1}{8}$
-2	$2^{-2} = \frac{1}{4}$
-1	$2^{-1} = \frac{1}{2}$
0	$2^0 = 1$
1	$2^1 = 2$
2	$2^2 = 4$
3	$2^3 = 8$

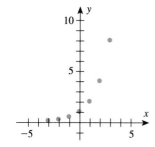

FIGURE 5.1 Selected points that satisfy $f(x) = 2^x$

If these points are connected with a smooth curve, as shown in Figure 5.2, you can see that $2^{\sqrt{3}}$ is defined and is between 2^1 and 2^2.

The number $2^{\sqrt{3}}$ can be approximated to any desired degree of accuracy:

FIGURE 5.2 Graph of $f(x) = 2^x$

$$1 < \sqrt{3} < 2 \qquad\qquad 2^1 < 2^{\sqrt{3}} < 2^2$$

$$1.7 < \sqrt{3} < 1.8 \qquad\qquad 2^{1.7} < 2^{\sqrt{3}} < 2^{1.8}$$

Since $\quad 1.73 < \sqrt{3} < 1.74 \quad$ we have $\quad 2^{1.73} < 2^{\sqrt{3}} < 2^{1.74}$

$$1.732 < \sqrt{3} < 1.733 \qquad\qquad 2^{1.732} < 2^{\sqrt{3}} < 2^{1.733}$$

$$\vdots \qquad\qquad\qquad\qquad \vdots$$

Base with irrational exponent is squeezed between the same base with rational exponents.

EXPONENTIAL FUNCTION

The function f is an **exponential function** if

$$f(x) = b^x$$

where b is a positive constant other than 1 and x is any real number. The number x is called the **exponent** and b is called the **base**.

Because it is beyond the scope of this book to prove that the usual laws of exponents hold for all real exponents in exponential functions, we will accept them as axioms. Because of the restrictions on b, however, the hypotheses for these laws of exponents with any real exponents must be changed so that they apply only when the bases are positive numbers not equal to 1. Suppose we sketch several exponential functions and observe their behavior.

EXAMPLE 1 Sketch $f(x) = (\frac{1}{2})^x$.

SOLUTION Notice that

$$y = \left(\frac{1}{2}\right)^x$$
$$= (2^{-1})^x$$

The points are plotted and connected by a smooth curve in Figure 5.3. Note that f is a decreasing function and there is a horizontal asymptote at $y = 0$.

x	$f(x) = (\frac{1}{2})^x$
-3	$(2^{-1})^{-3} = 8$
-2	$(2^{-1})^{-2} = 4$
-1	$(2^{-1})^{-1} = 2$
0	$(2^{-1})^0 = 1$
1	$(\frac{1}{2})^1 = \frac{1}{2}$
2	$(\frac{1}{2})^2 = \frac{1}{4}$
3	$(\frac{1}{2})^3 = \frac{1}{8}$

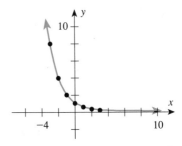

FIGURE 5.3 Graph of $f(x) = (\frac{1}{2})^x$

EXAMPLE 2 Sketch $f(x) = 10^x$.

SOLUTION The function f is an increasing function, with a horizontal asymptote at $y = 0$. The graph, along with a table of values, is shown in Figure 5.4. Notice that it is often necessary to alter the scale for exponential functions.

x	$f(x) = 10^x$
-3	$10^{-3} = \frac{1}{1,000}$
-2	$10^{-2} = \frac{1}{100}$
-1	$10^{-1} = \frac{1}{10}$
0	$10^0 = 1$
1	$10^1 = 10$
2	$10^2 = 100$
3	$10^3 = 1,000$

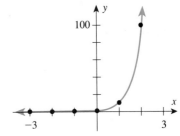

FIGURE 5.4 Graph of $f(x) = 10^x$

By looking at Figures 5.2, 5.3 and 5.4, we can make some observations regarding the graph of $f(x) = b^x$:

1. It passes through the point $(0, 1)$.
2. $f(x) > 0$ for all x.
3. If $b > 1$, f is an increasing function.
4. If $b < 1$, f is a decreasing function.

EXAMPLE 3 Sketch $f(x) = 10^{x-3} + 50$.

SOLUTION This is the graph shown in Figure 5.4 translated to $(h, k) = (3, 50)$, as shown in Figure 5.5. Notice that the curve passes through the point $(3, 51)$.

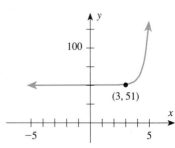

FIGURE 5.5 Graph of $f(x) = 10^{x-3} + 50$

EXAMPLE 4 Sketch $f(x) = 2^{-x^2}$.

SOLUTION Notice that $2^{-x^2} = \dfrac{1}{2^{x^2}} \to 0$ as $|x| \to \infty$, so $y = 0$ is the equation of a horizontal asymptote. Also,

$$f(-x) = 2^{-(-x)^2} = 2^{-x^2} = f(x)$$

so the curve is symmetric with respect to the y-axis. The graph is shown in Figure 5.6.

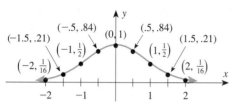

FIGURE 5.6 Graph of $f(x) = 2^{-x^2}$

x	$f(x) = 2^{-x^2}$
-3	$2^{-9} = \frac{1}{512}$
-2	$2^{-4} = \frac{1}{16}$
-1	$2^{-1} = \frac{1}{2}$
0	$2^{0} = 1$
1	$2^{-1} = \frac{1}{2}$
2	$2^{-4} = \frac{1}{16}$
3	$2^{-9} = \frac{1}{512}$

A calculator could be used to estimate additional points.

x	$f(x) = 2^{-x^2}$
-1.5	.21
$-.5$	.84
$.5$	.84
1.5	.21

Compound Interest

Many natural phenomena follow patterns of exponential growth or decay. Human population growth exhibits exponential growth, for example, and is considered at length in a special Extended Application following this chapter. Compo interest provides another application of exponential growth that is very import in business. If a sum of money, called the **principal** and denoted by P, is invested at an annual interest rate of r for t years, then the amount of money present is denoted by A and is found by

$$A = P + I$$

where I denotes the interest. **Interest** is an amount of money paid for the use of another's money. **Simple interest** is found by multiplication:

$$I = Prt$$

For example, $1000 invested for 3 years at 15% simple interest would generate an interest of $\$1,000(.15)(3) = \$450$, so the amount present after 3 years is $1,450.

Most businesses, however, pay interest on the interest, as well as the principal, after a certain period of time. When this is done, it is called **compound interest**. For example, $1,000 invested at 15% annual interest compounded annually for 3 years can be found as follows:

First year:
$$A = P + I$$
$$= P + Pr \qquad \text{\small $I = Prt$ and $t = 1$}$$
$$= P(1 + r)$$
$$= \$1,000(1 + .15)$$
$$= \$1,150$$

Second year: The amount from the first year becomes the principal for the second year:

$$A = \$1{,}150(1 + .15)$$
$$= \$1{,}322.50$$

$\downarrow$

Third year: $A = \$1{,}322.50(1 + .15)$
$$= \$1{,}520.88 \qquad \text{Rounded to the nearest cent}$$

Notice that the amount with simple interest is \$1,450, whereas with compound interest it is \$1,520.88. The calculation for compound interest, shown above, can become very tedious (especially for large t), so it is desirable to derive a general formula:

First year: $A = P + I$
$$= P + Pr \qquad I = Prt \text{ and } t = 1$$
$$= \underbrace{P(1 + r)}$$

Second year: $A = P(1 + r) + I$
$$= P(1 + r) + P(1 + r)r$$
$$= P(1 + r)(1 + r) \qquad \text{Common factor } P(1 + r)$$
$$= \underbrace{P(1 + r)^2}$$

Third year: $A = P(1 + r)^2 + P(1 + r)^2 r$
$$= P(1 + r)^2(1 + r) \qquad \text{Common factor } P(1 + r)^2$$
$$= P(1 + r)^3$$
$$\vdots$$

This pattern leads to the compound interest formula. The proof requires mathematical induction, which is discussed in Chapter 10.

COMPOUND INTEREST

If a principal (or present value) of P dollars is invested at an annual interest rate of r for t years compounded n times per year, then the amount present (or future value), A, is given by the formula

$$A = P(1 + i)^N \quad \text{where} \quad i = \frac{r}{n} \quad \text{and} \quad N = nt$$

EXAMPLE 5   If \$12,000 is invested for 5 years at 18% compounded annually, what is the amount present at the end of 5 years?

SOLUTION $P = \$12{,}000; r = .18; t = 5$ and $n = 1$. Then $i = \dfrac{.18}{1} = .18$ and $N = nt = 5(1) = 5$.

$$A = \$12{,}000(1 + .18)^5$$
$$\approx \$12{,}000(2.2877578) \qquad \text{Use a calculator.}$$
$$\approx \mathbf{\$27{,}453.09}$$

EXAMPLE 6 If the interest in Example 5 is compounded monthly, find the amount present.

SOLUTION $P = \$12,000$ and $n = 12$.

$$i = \frac{r}{n} = \frac{.18}{12}$$ *r* is the rate per period; in this case the period is monthly, so divide by 12.

$$= .015$$

$$N = tn = 5(12)$$ *t* is the number of periods.

$$= 60$$

$$A = \$12,000(1 + .015)^{60}$$

$$\approx \$12,000(2.4432198)$$ Use a calculator.

$$\approx \mathbf{\$29,318.64}$$ ▮

If \$12,000 is deposited at 18% and is compounded annually for 5 years, the amount present is \$27,453.09 (Example 5), and if it is compounded monthly the amount present is \$29,318.64 (Example 6). A reasonable extension is to ask what happens if the interest is compounded even more frequently than monthly. Can we compound daily, hourly, every minute, or every split second? We certainly can; in fact, money can be compounded **continuously,** which means that every instant the newly accumulated interest is used as part of the principal for the next instant. In order to understand these concepts, consider the following contrived example. Suppose \$1 is invested at 100% interest for 1 year compounded at different intervals. The compound interest formula for this example is

$$A = \left(1 + \frac{1}{n}\right)^n$$

where *n* is the number of times for compounding in 1 year. The calculations of this formula for different values of *n* are shown in Table 5.1.

Table 5.1 Effect of compounding on \$1 investment

NUMBER OF PERIODS	FORMULA	AMOUNT
Annually, $n = 1$	$\left(1 + \dfrac{1}{1}\right)^1$	\$2.00
Semiannually, $n = 2$	$\left(1 + \dfrac{1}{2}\right)^2$	2.25
Quarterly, $n = 4$	$\left(1 + \dfrac{1}{4}\right)^4$	2.44
Monthly, $n = 12$	$\left(1 + \dfrac{1}{12}\right)^{12}$	2.61
Daily, $n = 360$	$\left(1 + \dfrac{1}{360}\right)^{360}$	2.715
Hourly, $n = 8,640$	$\left(1 + \dfrac{1}{8,640}\right)^{8,640}$	2.7181

If you continue these calculations for even larger n, you will obtain the following results:

$$n = 10,000 \qquad \text{the formula yields} \qquad 2.718145927$$
$$n = 100,000 \qquad\qquad\qquad\qquad 2.718268237$$
$$n = 1,000,000 \qquad\qquad\qquad\qquad 2.718280469$$
$$n = 10,000,000 \qquad\qquad\qquad\qquad 2.718281693$$
$$n = 100,000,000 \qquad\qquad\qquad\qquad 2.718281815$$

The calculator can no longer distinguish the values of $(1 + 1/n)^n$ for larger n. These values are approaching a particular number. This number, it turns out, is an irrational number so it does not have a convenient decimal representation. (That is, its decimal representation does not terminate and does not repeat.) Mathematicians, therefore, have agreed to denote this number by the symbol e, which is defined as a limit.

THE NUMBER e

$$\left(1 + \frac{1}{n}\right)^n \to e \quad \text{as } n \to \infty$$

EXAMPLE 7   Find e, e^2, and e^{-3}.

SOLUTION   On a scientific calculator, locate a key labeled e^x. First enter the value for x, then press $\boxed{e^x}$. On a graphing calculator enter $\boxed{e^x}$ and then enter the value of x:

$$e \approx 2.718 \boxed{1}\,\boxed{e^x} \quad \text{or} \quad \boxed{e^x}\,\boxed{1} \qquad\qquad \text{DISPLAY:} \quad \boxed{2.718281828}$$

$$e^2 \approx 7.389 \qquad\qquad\qquad\qquad\qquad\quad \text{DISPLAY:} \quad \boxed{7.389056099}$$

$$e^{-3} \approx .050 \qquad\qquad\qquad\qquad\qquad\quad \text{DISPLAY:} \quad \boxed{.0497870684} \quad \blacksquare$$

For interest compounded continuously, the following formula is used:

CONTINUOUS INTEREST

$$A = Pe^{rt}$$

EXAMPLE 8   If the interest in Example 5 is compounded continuously, find the amount present.

SOLUTION   Since $P = \$12,000$, $r = .18$, and $t = 5$,

$$A = \$12,000e^{.18(5)}$$
$$\approx \$12,000(2.4596031)$$
$$\approx \mathbf{\$29,515.24} \qquad\qquad\qquad\qquad \blacksquare$$

You should memorize at least the first six digits of e.

APPROXIMATION OF e

$$e \approx 2.71828$$

A *Sketch the graph of each function given in Problems 1–30.*

1. $y = 3^x$
2. $y = 4^x$
3. $y = 5^x$
4. $y = (\frac{1}{3})^x$
5. $y = 4^{-x}$
6. $y = 5^{-x}$
7. $y = 2^{x-2}$
8. $y = 2^{x-1}$
9. $f(x) = e^{-x}$
10. $f(x) = e^x$
11. $y = e^{x+1}$
12. $y = e^x + 1$
13. $y - 2 = e^{x+2}$
14. $y + 1 = e^{x+3}$
15. $y = 2^{x+3}$
16. $y - 2 = 2^x$
17. $y - 3 = 2^x$
18. $y + 5 = 2^x$
19. $y - 5 = 2^{x+4}$
20. $y - 10 = 2^{x+3}$
21. $y = 2^{x-3} - 10$
22. $y = 2^{|x|}$
23. $y = 3^{|x|}$
24. $y = 2^{-|x|}$
25. $y = 3^{x^2}$
26. $y = 2^{x^2}$
27. $y = 10^{x^2}$
28. $y = 3^{-x^2}$
29. $y = 5^{-x^2}$
30. $y = 10^{-x^2}$

Use a calculator to evaluate the expressions given in Problems 31–42.

31. e^3
32. e^{-2}
33. $e^{.12}$
34. $e^{.08}$
35. $e^{.15(5)}$
36. $856e^{.05}$
37. $(1 + \frac{.08}{12})^{24}$
38. $(1 + \frac{.05}{6})^{72}$
39. $(1 + \frac{.12}{365})^{365}$
40. $(1 + \frac{.18}{360})^{720}$
41. $(1 + \frac{1}{1,000})^{1,000}$
42. $(1 + \frac{1}{100,000})^{100,000}$

B

43. Use graphical methods to estimate the value of $2^{\sqrt{2}}$.

44. Use graphical methods to estimate the value of $3^{\sqrt{2}}$.

45. Use graphical methods to estimate the value of $10^{\sqrt{2}}$.

46. Graph $y = 10^x$, $-1 \le x \le 1$ and approximate x if:
    a. $10^x = 5$ b. $10^x = .5$ c. $10^x = 2$
    d. $10^x = 8.4$ e. $10^x = -1$

47. In the definition of the exponential function $f(x) = b^x$, we require that b is a positive constant.
    a. What happens if $b = 1$? Draw the graph of $f(x) = b^x$, where $b = 1$. Is this an algebraic or a transcendental function?
    b. What happens if $b = 0$? Draw a graph of $f(x) = b^x$, where $b = 0$. Is this an algebraic or a transcendental function?

48. In the definition of the exponential function $f(x) = b^x$, we require that b is a positive constant. What happens if $b < 0$, say $b = -2$? For what values of x is f defined? Describe the graph of $f(x)$ in this case.

49. If $1,000 is invested at 7% compounded annually, how much money will there be in 25 years?

50. If $1,000 is invested at 12% compounded semiannually, how much money will there be in 10 years?

51. If $1,000 is invested at 16% interest compounded continuously, how much money will there be in 25 years?

52. If $1,000 is invested at 14% interest compounded continuously, how much money will there be in 10 years?

53. If $8,500 is invested at 18% interest compounded monthly, how much money will there be in 3 years?

54. If $3,600 is invested at 15% interest compounded daily, how much money will there be in 7 years? (Use a 365-day year; this is called *exact* interest.)

55. If $10,000 is invested at 14% interest compounded daily, how much money will there be in 6 months? (Use a 360-day year; this is called *ordinary* interest.)

BUSINESS   *If P dollars are borrowed for N months at an annual interest rate of r then the monthly payment is found by the formula*

$$M = \frac{Pi}{1 - (1 + i)^{-N}} \quad \text{where } i = \frac{r}{12}$$

To use this formula to find M with a calculator with algebraic logic, press:

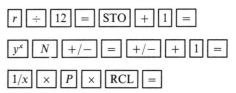

Use this information in Problems 56–59.

56. What is the monthly car payment for a new car costing $12,487 with a down payment of $2,487? The car is financed for 4 years at 12%.

57. A home loan is made for $110,000 at 12% interest for 30 years. What is the monthly payment?

58. A purchase of $2,430 is financed at 23% for 3 years. What is the monthly payment?

59. A home is financed at $14\frac{1}{2}$% for 30 years. If the amount financed is $125,000, what is the monthly payment?

60. Graph $y = \left(1 + \frac{1}{x}\right)^x$ for $x > 0$.

61. Graph: a. $c(x) = \dfrac{e^x + e^{-x}}{2}$

    b. $s(x) = \dfrac{e^x - e^{-x}}{2}$

c

62. ENGINEERING The functions c and s of Problem 61 are called the *hyperbolic cosine* and *hyperbolic sine* functions. These are defined by

$$\cosh x = \frac{e^x + e^{-x}}{2} \quad \text{and} \quad \sinh x = \frac{e^x - e^{-x}}{2}$$

a. Show that $\cosh^2 x - \sinh^2 x = 1$.
b. Show that the hyperbolic cosine is an even function and that the hyperbolic sine is an odd function.
c. Show that $\sinh 2x = 2 \sinh x \cosh x$.
d. Graph $y = \cosh x + \sinh x$.
e. Define the *hyperbolic tangent* function as

$$\tanh x = \frac{\sinh x}{\cosh x}$$

Graph $\tanh x$.

63. PHYSICS Radioactive argon-39 has a half-life of 4 min. This means that the time required for half of the argon-39 to decompose is 4 min. If we start with 100 milligrams (mg) of argon-39, the amount (A) left after t min is given by

$$A = 100\left(\frac{1}{2}\right)^{t/4}$$

Graph this function for $t \geq 0$.

64. EARTH SCIENCE Carbon-14, used for archaeological dating, has a half-life of 5,700 years. This means that the time required for half of the carbon-14 to decompose is 5,700 years. If we start with 100 mg of carbon-14, the amount (A) left after t years is given by

$$A = 100\left(\frac{1}{2}\right)^{t/5,700}$$

Graph this function for $t \geq 0$.

65. SOCIAL SCIENCE In 1990 the world population was about 5.3 billion. If we assume a growth rate of 2%, the formula expressing the world population for t years after 1990 is given by

$$P = 5.3(1 + .02)^t$$
$$= 5.3(1.02)^t$$

where P is the population in billions. Graph this function for 1990–2010.

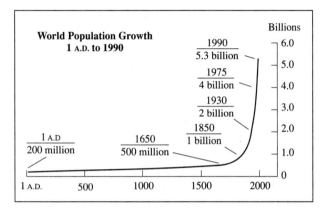

66. Use graphical methods to determine which is larger:
a. $(\sqrt{3})^\pi$ or $\pi^{\sqrt{3}}$ **b.** $(\sqrt{5})^\pi$ or $\pi^{\sqrt{5}}$ **c.** $(\sqrt{6})^\pi$ or $\pi^{\sqrt{6}}$
d. Consider $(\sqrt{N})^\pi = \pi^{\sqrt{N}}$, where N is a positive real number. From parts **a–c,** notice that $(\sqrt{N})^\pi$ is larger for some values of N and $\pi^{\sqrt{N}}$ is larger for others. For $N = \pi^2$,

$$(\sqrt{N})^\pi = \pi^{\sqrt{N}}$$

is obviously true. Using a graphic method, find another value (approximately) for which the given statement is true.

5.2 Introduction to Logarithms

Consider an exponential function with base b ($b > 1$) so that

$$A = b^x$$

An equation for which the exponent is a variable, it is called an **exponential equation.** Suppose we wish to solve this equation for x. The algebraic techniques we have been using do not help us solve this equation. (Dividing both sides by b or by x does not help, for example.) We begin by writing this exponential equation in words, as if we were reading it out loud to someone else:

$$x \text{ is the exponent of } b \text{ that yields } A$$

This can be rewritten $x = \text{exponent of } b \text{ to get } A$

It appears that the equation is now solved for x, but this is simply a notational change. The expression "exponent of b to get A" is called, for historical reasons, "the log of A to the base b." That is,

$$x = \log A \text{ to the base } b$$

But this phrase is shortened to the notation

$$x = \log_b A$$

The term *log* is an abbreviation for **logarithm,** but even the introduction of this notation does not give us the numerical answer we are seeking. It does solve for x algebraically, however, which is a first step in solving exponential equations. It is important to recognize this as a notational change only:

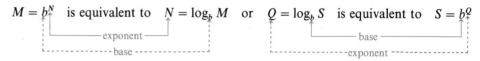

EXAMPLE 1   Change from exponential form to logarithmic form.

a. Remember: The log form solves for the exponent.

$$5^2 = 25 \Leftrightarrow \log_5 25 = 2 \qquad \text{Use the symbol} \Leftrightarrow \text{to mean "is equivalent to."}$$

Remember: This is the base.

b. $3^2 = 9 \Leftrightarrow \log_3 9 = 2$

base   exponent

c. $\frac{1}{8} = 2^{-3} \Leftrightarrow \log_2 \frac{1}{8} = -3$

d. $\sqrt{16} = 4 \Leftrightarrow \log_{16} 4 = \frac{1}{2}$ Remember: $\sqrt{16} = 16^{1/2}$.

EXAMPLE 2   Change from logarithmic form to exponential form.
a. $\log_{10} 100 = 2 \Leftrightarrow 10^2 = 100$

base   exponent

b. $\log_{10} \frac{1}{1,000} = -3 \Leftrightarrow 10^{-3} = \frac{1}{1,000}$
c. $\log_3 1 = 0 \Leftrightarrow 3^0 = 1$

To **evaluate a logarithm** means to find a numerical value for the given logarithm. The first ones we evaluate require that we know under what conditions $x = y$ when

$$b^x = b^y$$

If $b = 1$, you cannot conclude that $x = y$ since

$$1^5 = 1^4$$

but $5 \neq 4$. If $b \neq 1$, however, it can be proved true for all positive real numbers b and is called the **exponential property of equality.**

For positive real b ($b \neq 1$):

$$\text{If } b^x = b^y, \text{ then } x = y.$$

We now use this property to evaluate certain logarithms, as shown in Example 3.

EXAMPLE 3   Evaluate the given logarithms.

a. $\log_2 64$. Since it is usually necessary to supply a variable to convert to exponential form, we will use N in these examples.

$$\text{Let } N = \log_2 64 \quad \text{or} \quad 2^N = 64$$
$$2^N = 2^6$$
$$N = 6$$

Thus **$\log_2 64 = 6$.**

b. $\log_3 \frac{1}{9}$

$$\text{Let} \quad N = \log_3 \tfrac{1}{9}$$
$$3^N = \tfrac{1}{9}$$
$$3^N = 3^{-2}$$
$$N = -2$$

Thus **$\log_3 \frac{1}{9} = -2$.**

c. $\log_9 27$

$$\text{Let} \quad N = \log_9 27$$
$$9^N = 27$$
$$3^{2N} = 3^3$$
$$2N = 3$$
$$N = \tfrac{3}{2}$$

Thus **$\log_9 27 = \frac{3}{2}$.**

d. $\log_{10} 1 = 0$ Can you do this mentally?

e. $\log_{10} 10 = 1$

f. $\log_{10} 100 = 2$

g. $\log_{10} .1 = -1$

Consider Example 3d–g; suppose you want to find $\log_{10} 5.03$:

$$\log_{10} 5.03 = x \Leftrightarrow 10^x = 5.03$$

Since 5.03 is between 1 and 10 and

$$10^0 = 1$$
$$10^x = 5.03 \qquad \text{You want to find this } x.$$
$$10^1 = 10$$

the number x should be between 0 and 1 by the Squeeze Theorem for Exponents. Although you *still* do not have the value of x, all is not lost because tables showing approximations for these exponents have been prepared. Base 10 is fairly common, and if a logarithm is to the base 10 it is called a **common logarithm** and written without the subscript 10. Logarithms to the base e are called **natural logarithms** and are denoted by ln x. The expression ln x is often pronounced "ell en" or "lon x."

LOGARITHM NOTATIONS

> Common logarithm: log x means $\log_{10} x$
> Natural logarithm: ln x means $\log_e x$

Calculators have, to a large extent, eliminated the need for extensive log tables. You should find two logarithm keys on your calculator. One is labeled $\boxed{\text{LOG}}$ for common logarithm and the other is labeled $\boxed{\text{LN}}$ for natural logarithm.

EXAMPLE 4   Evaluate log 7.680 correct to four significant digits.*

SOLUTION   By calculator,   DISPLAY: $\boxed{.88536122}$
To four significant digits, $x = \mathbf{.8854}$.

EXAMPLE 5   Evaluate ln 3.49.

SOLUTION   By calculator,   DISPLAY: $\boxed{1.249901736}$
Calculator answers are more accurate than table answers. However, it is important to realize that any answer (whether from the table or a calculator) is only as accurate as the input number (3.49 in this problem). So the answer is $x = \mathbf{1.25}$ to three significant digits.

EXAMPLE 6   Find x, where log .00728 $= x$.

SOLUTION   DISPLAY: $\boxed{-2.137868621}$ To three significant digits, $x = \mathbf{-2.14}$.

When evaluating logarithms for answers (as in Examples 4–6), we round our answers to an appropriate number of significant digits. However, if you are using logarithms for calculations, keep all of the accuracy your calculator offers for all calculations and round only one time at the end of the problem when you are stating your answer.

Examples 4–6 are fairly straightforward evaluations since they are common or natural logarithms and since your calculator has both $\boxed{\text{LOG}}$ and $\boxed{\text{LN}}$ keys. However, suppose we wish to evaluate a logarithm to some base *other than* base 10 or base e. The first method uses the definition of logarithm, and the second method uses what we will call the **Change of Base Theorem**. Before we state this theorem, we consider its plausibility with the following example.

---

*Significant digits are discussed in Appendix A.

EXAMPLE 7 Evaluate the given expressions.

a. $\log_2 8,$ $\dfrac{\log 8}{\log 2},$ and $\dfrac{\ln 8}{\ln 2}$ **b.** $\log_3 9,$ $\dfrac{\log 9}{\log 3},$ and $\dfrac{\ln 9}{\ln 3}$

c. $\log_5 625,$ $\dfrac{\log 625}{\log 5},$ and $\dfrac{\ln 625}{\ln 5}$

SOLUTION   **a.** $\log_2 8 = x$ means (from the definition of logarithm),

$$2^x = 8$$
$$2^x = 2^3$$
$$x = 3$$

Thus, $\log_2 8 = 3.$

By calculator, $\dfrac{\log 8}{\log 2} \approx \dfrac{.903089987}{.3010299957} \approx 3.$

Also, $\dfrac{\ln 8}{\ln 2} \approx \dfrac{2.079441542}{.6931471806} \approx 3.$

b. $\log_3 9 = x$ means $3^x = 3^2$ so that $x = \log_3 9 = 2.$

By calculator, $\dfrac{\log 9}{\log 3} \approx \dfrac{.9542425094}{.4771212547} \approx 2.$

Also, $\dfrac{\ln 9}{\ln 3} \approx \dfrac{2.197224577}{1.098612289} \approx 2.$

c. $\log_5 625 = x$ means $5^x = 5^4$ so that $\log_5 625 = 4.$

By calculator, $\dfrac{\log 625}{\log 5} \approx \dfrac{2.795880017}{.6989700043} \approx 4$

Also, $\dfrac{\ln 625}{\ln 5} \approx \dfrac{6.43775165}{1.609437912} \approx 4.$

You no doubt noticed that the answers within each part of Example 7 were identical. This result is summarized with the following theorem.

CHANGE OF BASE THEOREM

$$\log_a x = \frac{\log_b x}{\log_b a}$$

EXAMPLE 8 Change $\log_7 3$ to logarithms with base 10 and evaluate to four significant digits.

SOLUTION $\log_7 3 = \dfrac{\log 3}{\log 7} \approx \underbrace{\dfrac{.4771212547}{.84509804}}_{\text{All this should be done by calculator.}} \approx .5645750341$

Answer: **.5646**

EXAMPLE 9 Change $\log_3 3.84$ to logarithms with base e and evaluate.

SOLUTION
$$\log_3 3.84 = \underbrace{\frac{\ln 3.84}{\ln 3} \approx \frac{1.345472367}{1.098612289} \approx 1.224701726}_{\text{By calculator}}$$

Answer: **1.22**

5.2 Problem Set

A *Write the equations in Problems 1–12 in logarithmic form.*

1. $64 = 2^6$ **2.** $100 = 10^2$ **3.** $1{,}000 = 10^3$

4. $64 = 8^2$ **5.** $125 = 5^3$ **6.** $a = b^c$

7. $m = n^p$ **8.** $1 = e^0$ **9.** $9 = \left(\frac{1}{3}\right)^{-2}$

10. $8 = \left(\frac{1}{2}\right)^{-3}$ **11.** $\frac{1}{3} = 9^{-1/2}$ **12.** $\frac{1}{2} = 4^{-1/2}$

Write the equations in Problems 13–28 in exponential form.

13. $\log_{10} 10{,}000 = 4$ **14.** $\log .01 = -2$

15. $\log 1 = 0$ **16.** $\log x = 2$

17. $\log_e e^2 = 2$ **18.** $\ln e^3 = 3$

19. $\ln x = 5$ **20.** $\ln x = .03$

21. $\log_2 \frac{1}{8} = -3$ **22.** $\log_2 32 = 5$

23. $\log_4 2 = \frac{1}{2}$ **24.** $\log_{1/2} 16 = -4$

25. $\log_m n = p$ **26.** $\log_a b = c$

27. $\log_x 8 = 3$ **28.** $\log_x 52 = 2$

Use the definition of logarithm or a calculator to evaluate the expressions in Problems 29–60.

29. $\log_b b^2$ **30.** $\log_t t^3$ **31.** $\log_e e^4$

32. $\log_\pi \sqrt{\pi}$ **33.** $\log_\pi \left(\frac{1}{\pi}\right)$ **34.** $\log_2 8$

35. $\log_3 9$ **36.** $\log_{19} 1$ **37.** $\log 4.27$

38. $\log 1.08$ **39.** $\log 8.43$ **40.** $\log 9{,}760$

41. $\log 71{,}605$ **42.** $\log .042$ **43.** $\log 1.321$

44. $\log 2.0532$ **45.** $\ln 2.27$ **46.** $\ln 16.77$

47. $\ln 2.431$ **48.** $\ln .125$ **49.** $\ln 13$

50. $\ln .15$ **51.** $\ln 7.3$ **52.** $\ln 10.57$

53. $\log_5 304$ **54.** $\log_2 1583$ **55.** $\log_6 .10$

56. $\log_4 3.05$ **57.** $\log_\pi 100.0$ **58.** $\log_{\sqrt{2}} 8.5$

59. $\log_\pi 25$ **60.** $\log_{1.08} 164$

B

WHAT IS WRONG, *if anything, with each of the statements in Problems 61–68? Explain your reasoning.*

61. $\log_b N$ is negative when N is negative.

62. $\log N$ is negative when $N > 1$.

63. In $\log_b N$, the exponent is N.

64. $\log 500$ is the exponent on 10 that gives 500.

65. A common logarithm is a logarithm in which the base is 2.

66. A natural logarithm is a logarithm in which the base is 10.

67. $\dfrac{\log A}{\log B} = \dfrac{\ln A}{\ln B}$

68. To evaluate $\log_5 N$, divide $\log 5$ by $\log N$.

69. BUSINESS An advertising agency conducted a survey and found that the number of units sold, N, is related to the amount a spent on advertising (in dollars) by the following formula:

$$N = 1{,}500 + 300 \ln a \quad (a \geq 1)$$

a. How many units are sold after spending $1,000?
b. How many units are sold after spending $50,000?
c. If the company wants to sell 5000 units, how much money will it have to spend?

70. CHEMISTRY The pH of a substance measures its acidity or alkalinity. It is found by the formula

$$pH = -\log[H^+]$$

where $[H^+]$ is the concentration of hydrogen ions in an aqueous solution given in moles per liter.

a. What is the pH (to the nearest tenth) of a lemon for which $[H^+] = 2.86 \times 10^{-4}$?
b. What is the pH (to the nearest tenth) of rainwater for which $[H^+] = 6.31 \times 10^{-7}$?
c. If a shampoo advertises that it has a pH of 7, what is $[H^+]$?

71. EARTH SCIENCE   The Richter scale for measuring earthquakes was developed by Gutenberg and Richter. It relates the energy E (in ergs) to the magnitude of the earthquake, M, by the formula

$$M = \frac{\log E - 11.8}{1.5}$$

a. A small earthquake is one that releases 15^{15} ergs of energy. What is the magnitude of such an earthquake on the Richter scale?

b. A large earthquake is one that releases 10^{25} ergs of energy. What is the magnitude of such an earthquake on the Richter scale?

72. PHYSICS   The intensity of sound is measured in decibels D and is given by

$$D = \log\left(\frac{I}{I_0}\right)^{10}$$

where I is the power of the sound in watts per cubic centimeter (W/cm^3) and $I_0 = 10^{-16}$ W/cm^3 (the power of sound just below the threshold of hearing). Find the intensity in decibels of the given sound:

a. A whisper, 10^{-13} W/cm^3

b. Normal conversation, $3.16 \cdot 10^{-10}$ W/cm^3

c. The world's loudest shout by Skipper Kenny Leader, 10^{-5} W/cm^3

d. A rock concert, $5.23 \cdot 10^{-6}$ W/cm^3

5.3   Logarithmic Functions and Equations

Definition and Graphs of Logarithmic Functions

The notion of a logarithm, which was introduced in the previous section, can also be considered as a special type of function.

LOGARITHMIC FUNCTION

The function f defined by

$$f(x) = \log_b x$$

where $b > 0$, $b \neq 1$, is called the **logarithmic function with base b.**

By relating logarithmic and exponential functions, you will notice that you already know a great deal about the logarithmic function. For example, the graph of $y = \log_2 x$ is the same as the graph of $2^y = x$, as shown in Figure 5.7.

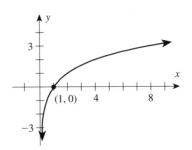

y	x
-3	$2^{-3} = \frac{1}{8}$
-2	$2^{-2} = \frac{1}{4}$
-1	$2^{-1} = \frac{1}{2}$
0	$2^0 = 1$
1	$2^1 = 2$
2	$2^2 = 4$
3	$2^3 = 8$

FIGURE 5.7   Graph of $y = \log_2 x$

The shape of this graph is concave down and is characteristic of all graphs of $f(x) = \log_b x$ for which $b > 1$. On the other hand, if $0 < b < 1$, then the shape of a logarithmic graph is concave up and is shown in Example 1b. Other characteristics of logarithmic functions are:

The **domain** of $f(x) = \log_b x$ is all positive real numbers.

There is a **vertical asymptote**, $x = 0$:

$$\log_b x \to \infty \quad \text{as } x \to 0 \quad (\text{if } 0 < b < 1)$$
$$\log_b x \to -\infty \quad \text{as } x \to 0 \quad (\text{if } b > 1)$$

$f(x) = \log_b x$ is a one-to-one function.

EXAMPLE 1 Graph: **a.** $y = \log_2(x + 2) - 3$ **b.** $y = \log_{.5} x$

SOLUTION **a.** This graph is the same as the one shown in Figure 5.7, except it is translated to coordinate axes with origin at the point $(-2, -3)$. The graph is shown in Figure 5.8a.

b. The graph of $y = \log_{.5} x$ is the same as that of $.5^y = x$, and is shown in Figure 5.8b. Notice that the shape of this graph is concave up, because the base .5 is between 0 and 1.

FIGURE 5.8 Logarithmic graphs

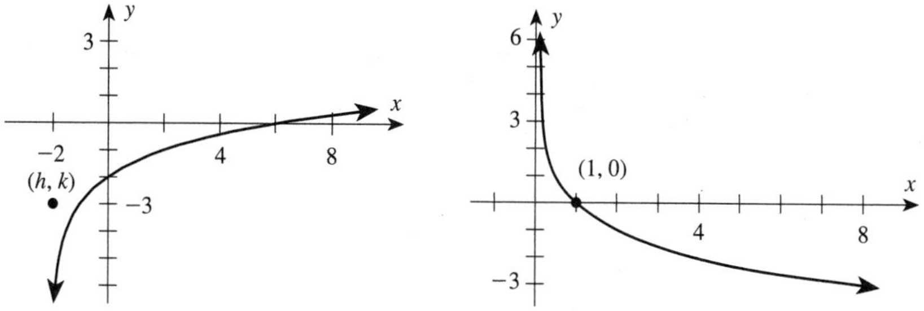

a. Graph of $y = \log_2(x + 2) - 3$

b. Graph of $y = \log_{.5} x$

Logarithmic and Exponential Functions as Inverses

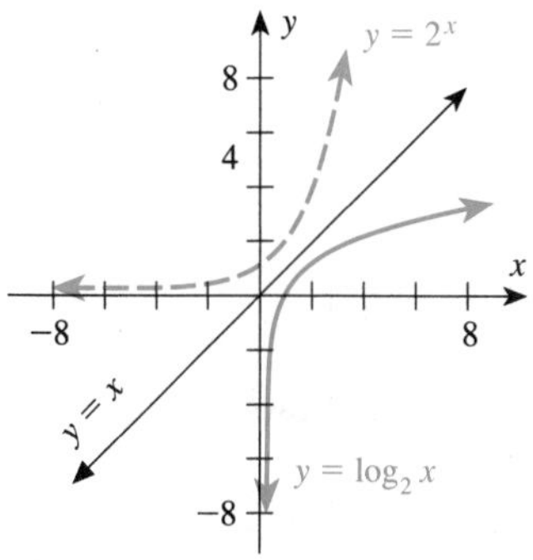

Compare the graph in Figure 5.2 with the graph of $y = \log_2 x$, as shown along with $y = x$ in Figure 5.9.

The functions $f(x) = 2^x$ and $g(x) = \log_2 x$ appear to be inverse functions. To prove this, we will need to consider two properties of logarithms that follow from the definition of logarithm:

1. $\log_b b^x = x$ 2. $b^{\log_b x} = x$

FIGURE 5.9 Graphs of $y = 2^x$, $y = \log_2 x$, and $y = x$

Both of these properties follow directly from the definition of logarithms:

$$b^M = N \Leftrightarrow \log_b N = M$$

$$\overbrace{\text{exponent}}$$

Definition: $b^M = N \Leftrightarrow \log_b N = M$

Property 1: $b^x = b^x \Leftrightarrow \log_b b^x = x$

exponent

Property 1 is an exact statement of the definition, only with $M = x$ and $N = b^x$. Since $b^x = b^x$ is true, then $\log_b b^x = x$ is true by the definition of logarithm.

$$\overbrace{}^{\text{exponent}}$$

Definition: $b^M = N \Leftrightarrow \log_b N = M$

Property 2: $b^{\log_b x} = x \Leftrightarrow \log_b x = \log_b x$

exponent

Property 2 is an exact statement of the definition. Let $M = \log_b x$ and $N = x$. Since $\log_b x = \log_b x$, then $b^{\log_b x} = x$ is true by the definition of logarithm.

To show that $f(x) = \log_b x$ and $g(x) = b^x$ are inverse functions, show that $(f \circ g)(x) = (g \circ f)(x) = x$:

$$\begin{aligned}(f \circ g)(x) &= f[g(x)] \\ &= f(b^x) \\ &= \log_b b^x \\ &= x \quad \text{By Property 1}\end{aligned}$$

and

$$\begin{aligned}(g \circ f)(x) &= g[f(x)] \\ &= g(\log_b x) \\ &= b^{\log_b x} \\ &= x \quad \text{By Property 2}\end{aligned}$$

This proves the following theorem.

EXPONENTIAL AND LOGARITHMIC FUNCTIONS ARE INVERSES

The exponential and logarithmic functions with base b are inverse functions of one another.

This relationship of inverse functions is needed to find e on several brands of calculators. If a calculator has

$$\boxed{\ln x} \quad \text{and} \quad \boxed{\text{INV}}$$

keys, but no e^x key, how can you find e or e^x? Since

$$y = \ln x \quad \text{and} \quad y = e^x$$

are inverse functions, to find e (or e^1) you can press

$$\boxed{1}\,\boxed{\text{INV}}\,\boxed{\ln x} \quad \text{DISPLAY:} \quad \boxed{2.718281828}$$

This gives the inverse of the ln x function; that is, it is e^x where the x value is input just prior to pressing these buttons.

If you need $e^{5.2}$ on such a calculator, press

$$\boxed{5.2}\,\boxed{\text{INV}}\,\boxed{\ln x} \quad \text{DISPLAY:} \quad \boxed{181.2722419}$$

Properties 1 and 2 also lead to another theorem that allows us to solve logarithmic equations.

LOG OF BOTH SIDES THEOREM

If A, B, and b are positive real numbers with $b \neq 1$, then

$$\log_b A = \log_b B \text{ is equivalent to } A = B$$

PROOF If $A = B$, then

$$\begin{aligned}\log_b A &= \log_b A \\ &= \log_b B \quad \text{By substitution}\end{aligned}$$

If $\log_b A = \log_b B$, then

$$b^{\log_b B} = A \qquad \text{By definition of logarithm}$$
$$b^{\log_b B} = B \qquad \text{By Property 2}$$
$$B = A \qquad \text{By substitution} \qquad \square\square\square$$

We now use this theorem and the definition of logarithm to solve logarithmic equations.

Solving Logarithmic Equations

EXAMPLE 2 Solve $\log_2 \sqrt{2} = x$ for x.

SOLUTION For this logarithmic equation, the exponent is the unknown. Apply the definition of logarithm:

$$2^x = \sqrt{2}$$
$$2^x = 2^{1/2}$$
$$x = \tfrac{1}{2} \qquad \text{Exponential property of equality}$$

EXAMPLE 3 Solve $\log_x 25 = 2$.

SOLUTION For this logarithmic equation, the base is the unknown. Apply the definition of logarithm:

$$x^2 = 25$$
$$x = \pm 5$$

Be sure the values you obtain are permissible values for the definition of a logarithm. In this case, $x = -5$ is not a permissible value since a logarithm with a negative base is not defined. Therefore, the solution is **$x = 5$.**

EXAMPLE 4 Solve $\ln x = 5$.

SOLUTION The power itself is the unknown in this logarithmic equation. Apply the definition of logarithm (to the base e):

$$e^5 = x \qquad \text{This is an exact solution.}$$

An approximate solution (by calculator) is

$$\text{DISPLAY:} \quad \boxed{148.4131591}$$

EXAMPLE 5 Solve $\log_5 x = \log_5 72$.

SOLUTION Use the Log of Both Sides Theorem: **$x = 72$.**

Basically, all logarithmic equations fall into one of the four types illustrated by Examples 2–5. For problems like those in Examples 2–4, apply the definition of logarithm and for those like Example 5, apply the Log of Both Sides Theorem.

More difficult examples require algebraic simplification using three laws of logarithms.

FIRST LAW OF LOGARITHMS	$\log_b AB = \log_b A + \log_b B$	The log of the product of two numbers is the sum of the logs of those numbers.
SECOND LAW OF LOGARITHMS	$\log_b \dfrac{A}{B} = \log_b A - \log_b B$	The log of the quotient of two numbers is the log of the numerator minus the log of the denominator.
THIRD LAW OF LOGARITHMS	$\log_b A^p = p \log_b A$	The log of the pth power of a number is p times the log of that number.

The proofs of these laws of logarithms are easy if you remember that logarithmic equations are equivalent to exponential equations and that the properties of exponents can be applied. The first law of logarithms is a restatement of the first law of exponents:

$$b^x b^y = b^{x+y}$$

Let $A = b^x$ and $B = b^y$. Then $\log_b A = x$ and $\log_b B = y$. The first law concerns the product of A and B, so

$$AB = b^x b^y$$
$$= b^{x+y}$$

Therefore, by changing to logarithmic form,

$$\log_b AB = x + y$$
$$= \log_b A + \log_b B \qquad \text{By substitution}$$

The second law concerns the quotient A/B $(B \neq 0)$, so

$$\frac{A}{B} = \frac{b^x}{b^y}$$

$$\frac{A}{B} = b^{x-y} \qquad \text{Fifth law of exponents}$$

Thus

$$\log_b \frac{A}{B} = x - y$$

$$= \log_b A - \log_b B \qquad \text{By substitution}$$

The proof of the third law of logarithms follows from the second law of exponents and is left as an exercise.

EXAMPLE 6

$$\log_8 3 + \tfrac{1}{2}\log_8 25 = \log_8 x$$

$\log_8 3 + \log_8 25^{1/2} = \log_8 x$ Third law of logarithms

$\log_8 3 + \log_8 5 = \log_8 x$ $25^{1/2} = \sqrt{25}$ $= 5$

$\log_8(3 \cdot 5) = \log_8 x$ First law of logarithms

$15 = x$ Log of Both Sides Theorem

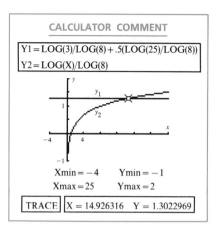
The solution is **15**.

EXAMPLE 7

$$5\log_x 2 - \tfrac{1}{2}\log_x 8 = 2 - \tfrac{1}{2}\log_x 2$$

$$\log_x 2^5 - \log_x \sqrt{8} = 2 - \log_x \sqrt{2}$$

$$\log_x 32 - \log_x 2\sqrt{2} + \log_x \sqrt{2} = 2$$

$$\log_x\left(\frac{32}{2\sqrt{2}} \cdot \sqrt{2}\right) = 2$$

$$\log_x 16 = 2$$

$$x^2 = 16 \qquad \text{Definition of logarithm}$$

$$x = \pm 4$$

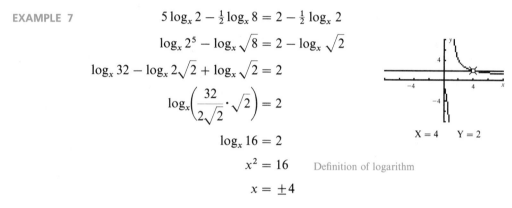

By the definition of a logarithm, x must be positive, so the solution is **4**.

EXAMPLE 8

$$\ln x - \tfrac{1}{2}\ln 2 = \tfrac{1}{2}\ln(x + 4)$$

$$\ln x - \ln \sqrt{2} = \ln \sqrt{x + 4}$$

$$\ln\left(\frac{x}{\sqrt{2}}\right) = \ln \sqrt{x + 4}$$

$$\frac{x}{\sqrt{2}} = \sqrt{x + 4}$$

$$\frac{x^2}{2} = x + 4$$

$$x^2 - 2x - 8 = 0$$

$$(x - 4)(x + 2) = 0$$

$$x = 4, \; -2$$

Notice that $\ln(-2)$ is not defined, so $x = -2$ is an extraneous root. Therefore the solution is **4**.

EXAMPLE 9 What is $\log_2 p + \log_2 q$, where p and q are roots of $2x^2 - mx + 1 = 0$?

SOLUTION If you begin by solving the quadratic equation you will have some difficulties with the constant m. However, consider the general quadratic equation

$$ax^2 + bx + c = 0 \qquad a \neq 0$$

Suppose we know that r_1 and r_2 are roots of this general quadratic equation. Then $x = r_1$ and $x = r_2$ or $x - r_1 = 0$ and $x - r_2 = 0$. This means

$$(x - r_1)(x - r_2) = 0$$
$$x^2 - r_1 x - r_2 x + r_1 r_2 = 0$$
$$x^2 - (r_1 + r_2)x + r_1 r_2 = 0$$

Compare this with the general quadratic after dividing both sides by a ($a \neq 0$) to find

$$\frac{b}{a} = -(r_1 + r_2) \quad \text{and} \quad \frac{c}{a} = r_1 r_2$$

Thus, for this example $pq = \frac{1}{2}$. If we let $a = \log_2 p + \log_2 q$, then $a = \log_2 pq$ so that $2^a = pq$. This means

$$2^a = \tfrac{1}{2} = 2^{-1}$$
$$a = -1$$

Since a is the desired value, we have $\log_2 p + \log_2 q = -1$. ∎

5.3 Problem Set

A *Solve the equations in Problems 1–16.*

1. a. $\log_5 25 = x$ **b.** $\log_2 128 = x$
2. a. $\log_3 81 = x$ **b.** $\log_4 64 = x$
3. a. $\log \frac{1}{10} = x$ **b.** $\log 10,000 = x$
4. a. $\log 1,000 = x$ **b.** $\log \frac{1}{1,000} = x$
5. a. $\log_x 28 = 2$ **b.** $\log_x 81 = 4$
6. a. $\log_x 50 = 2$ **b.** $\log_x 84 = 2$
7. a. $\log_x e = 2$ **b.** $\log_x e = 1$
8. a. $\ln x = 2$ **b.** $\ln x = 3$
9. a. $\ln x = 4$ **b.** $\ln x = \ln 14$
10. a. $\ln 9.3 = \ln x$ **b.** $\ln 109 = \ln x$
11. a. $\log_3 x^2 = \log_3 125$ **b.** $\ln x^2 = \ln 12$
12. a. $\log x^2 = \log 70$ **b.** $\log_2 8\sqrt{2} = x$
13. a. $\log_3 27\sqrt{3} = x$ **b.** $\log_x 1 = 0$

14. a. $\log_x 10 = 0$ **b.** $\log_e x = 3$
15. a. $\log_2 x = 5$ **b.** $\log_{10} x = 5$
16. a. $\ln x^2 = 4$ **b.** $\ln(x + 1) = 0$

WHAT IS WRONG, *if anything, with each of the statements in Problems 17–22? Explain your reasoning.*

17. $1.5^6 = x$ is equivalent to $\log_{1.5} x = 6$.
18. $1.004^{72} = y$ is equivalent to $\log_{1.004} 72 = y$.
19. The graph of $f(x) = \log_2 x$ is concave up.
20. The graph of $f(x) = \log_{.2} x$ is concave up.
21. $\log_b AB = (\log_b A)(\log_b B)$
22. $\dfrac{\log_b A}{\log_b B} = \log_b \dfrac{A}{B}$

By classifying the concavity of each graph of a logarithmic function in Problems 23–26, give the possible values for the base, b.

23.

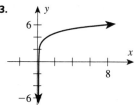

24.

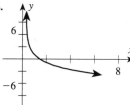

25.

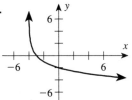

26.

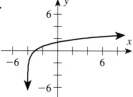

B *Graph the functions whose equations are given in Problems 27–44.*

27. $y = \log_3 x$ **28.** $y = \log_{1/3} x$

29. $y = \log_\pi x$ **30.** $y = \log_{e/4} x$

31. $y = \log(x - 4)$ **32.** $y = \log(x - 2)$

33. $y = \log x + 3$ **34.** $y = \log x + 4$

35. $y - 1 = \log(x + 2)$ **36.** $y + 2 = \log(x - 1)$

37. $y = \log_{1/2} x + 3$ **38.** $y = \log_{1/2}(x + 3)$

39. $y = \log |x|$ **40.** $y = |\log x|$

41. $y = -3 \log |x|$ **42.** $y = \log |x| + 1$

43. $y - 1 = \log x^2$ **44.** $y - 1 = (\log x^2)^2$

Solve the logarithmic equations in Problems 45–61.

45. $\log 2 = \frac{1}{4} \log 16 - x$

46. $\frac{1}{2} \log x - \log 100 = 2$

47. $\ln 2 + \frac{1}{2} \ln 3 + 4 \ln 5 = \ln x$

48 $\ln x = \ln 5 - 3 \ln 2 + \frac{1}{4} \ln 3$

49. $1 = 2 \log x - \log 1{,}000$

50. $\log_8 5 + \frac{1}{2} \log_8 9 = \log_8 x$

51. $\log_7 x - \frac{1}{2} \log_7 4 = \frac{1}{2} \log_7(2x - 3)$

52. $\ln 10 - \frac{1}{2} \ln 25 = \ln x$

53. $\frac{1}{2} \ln x = 3 \ln 5 - \ln x$

54. $\ln x - \frac{1}{2} \ln 3 = \frac{1}{2} \ln(x + 6)$

55. $2 \ln x - \frac{1}{2} \ln 9 = \ln 3(x - 2)$

56. $\log 2 - \frac{1}{2} \log 9 + 3 \log 3 = x$

57. $3 \log 3 - \frac{1}{2} \log 3 = \log \sqrt{x}$

58. $3 \ln \dfrac{e}{\sqrt[3]{5}} = 3 - \ln x$

59. $5 \ln \dfrac{e}{\sqrt[5]{5}} = 3 - \ln x$

60. $2 \ln \dfrac{e}{\sqrt{7}} = 2 - \ln x$

61. $\log_x(x + 6) = 2$

62. CHEMISTRY The pH (hydrogen potential) of a solution is given by

$$pH = \log \frac{1}{[H^+]}$$

where $[H^+]$ is the concentration of hydrogen ions in a water solution in moles per liter. Find the pH (to the nearest tenth) for the substances with the $[H^+]$ given.
a. Lemon juice, $5.01 \cdot 10^{-3}$
b. Milk, $3.98 \cdot 10^{-7}$
c. Vinegar, $7.94 \cdot 10^{-4}$
d. Rainwater, $5.01 \cdot 10^{-7}$
e. Seawater, $4.35 \cdot 10^{-9}$

63. PSYCHOLOGY A learning curve describes the rate at which a person learns certain tasks. If a person sets a goal of typing N words per minute (wpm), the length of time t, in days, to achieve this goal is given by

$$t = -62.5 \ln\left(1 - \frac{N}{80}\right)$$

a. How long would it take to learn to type 30 wpm?
b. If we accept this formula, is it possible to learn to type 80 wpm?
c. Solve for N.

64. PSYCHOLOGY In Problem 63 an equation for learning was given. Psychologists are also concerned with forgetting. In a certain experiment, students were asked to remember a set of nonsense syllables, such as "hom." The students then had to recall the syllables after t seconds. The equation that was found to describe forgetting was

$$R = 80 - 27 \ln t \quad (t \geq 1)$$

where R is the percentage of students who remember the syllables after t sec.
a. What percentage of the students remembered the syllables after 3 sec?
b. In how many seconds would only 10% of the students remember?
c. Solve for t.

65. The equation for the Richter scale relating energy E (in ergs) to the magnitude of the earthquake, M, is given by the formula

$$M = \frac{\log E - 11.8}{1.5}$$

a. Solve for E.
b. What was the energy of the 1988 Armenian earthquake, which measured 7.5 on the Richter scale?

66. Prove the third law of logarithms, $\log_b A^p = p \log_b A$.

67. Prove the Change of Base Theorem:

$$\log_b x = \frac{\log_a x}{\log_a b}$$

C

68. If p and q are roots of $5x^2 - mx + 1 = 0$, what is $\log_5 p + \log_5 q$?

69. If p and q are roots of $3x^2 - nx + 27 = 0$, what is $\log_3 p + \log_3 q$?

70. __WHAT IS WRONG?__ Consider the following argument:

$4 > 3$	Obviously true
$4 \log_{10} \frac{1}{3} > 3 \log_{10} \frac{1}{3}$	Multiply both sides by $\log_{10} \frac{1}{3}$.
$\log_{10}(\frac{1}{3})^4 > \log_{10}(\frac{1}{3})^3$	Property of exponents
$(\frac{1}{3})^4 > (\frac{1}{3})^3$	Theorem of logarithms
$\frac{1}{81} > \frac{1}{27}$	Obviously false

Can you find the error?

5.4 Exponential Equations

In this section, we return to the question that prompted our discussion of logarithms in the first place: How do you solve an exponential equation? The most straightforward method for solving exponential equations applies the exponential property of equality stated in Section 5.2. Recall that if equal bases are raised to some power and the results are equal, then the exponents must also be equal. This is a consequence of the exponential function being one-to-one.

EXAMPLE 1 Solve $7^x = 343$ for x.

SOLUTION Write both sides with the same base if possible:

$$7^x = 7^3$$

$$x = 3 \qquad \text{Exponential property of equality}$$

EXAMPLE 2 Solve $25^x = 125$.

SOLUTION
$$(5^2)^x = 5^3$$
$$5^{2x} = 5^3$$
$$2x = 3 \qquad \text{Exponential property of equality}$$
$$x = \tfrac{3}{2}$$

EXAMPLE 3 Solve $36^x = \frac{1}{6}$.

SOLUTION
$$(6^2)^x = 6^{-1}$$
$$6^{2x} = 6^{-1}$$
$$2x = -1$$
$$x = -\tfrac{1}{2}$$

Not all exponential equations are as easy to solve as those in Examples 1–3. If the bases cannot be forced to be the same, then the exponential equations will fall into one of three types:

Base:	10 (common log)	e (natural log)	b (arbitrary base)
Example:	$10^x = 5$	$e^{.06t} = 3.456$	$8^x = 156.8$

We will work two examples of each of these types. First, consider base 10.

EXAMPLE 4 Solve $10^x = 5$.

SOLUTION Use the definition of logarithm.

$$x = \log 5 \qquad \text{Exact answer}$$

$$x \approx .699 \qquad \text{Approximate answer, found by using tables or a calculator}$$

EXAMPLE 5 Solve $10^{5x+3} = 195$.

SOLUTION Rewrite in logarithmic form: $\log 195 = 5x + 3$. Solve for x:

$$\log 195 - 3 = 5x$$

$$x = \tfrac{1}{5}(\log 195 - 3) \qquad \text{Exact answer}$$

By calculator: DISPLAY: $\boxed{-.1419930777}$

The next two examples involve base e.

EXAMPLE 6 Solve $e^{.06t} = 3.456$ for t.

SOLUTION Write in logarithmic form:

$$e^{.06t} = 3.456$$

$$.06t = \ln 3.456$$

$$t = \frac{\ln 3.456}{.06}$$

By calculator: DISPLAY: $\boxed{20.66853085}$

EXAMPLE 7 Solve $\tfrac{1}{2} = e^{-.000425t}$

SOLUTION In logarithmic form:

$$\ln .5 = -.000425t$$

$$t = \frac{\ln .5}{-.000425}$$

By calculator: DISPLAY: $\boxed{1630.934542}$

We now consider exponential equations with an arbitrary base.

EXAMPLE 8 Solve $8^x = 156.8$.

SOLUTION Write in logarithmic form:

$$8^x = 156.8$$

$$x = \log_8 \textbf{156.8} \qquad \text{Exact answer}$$

By calculator: DISPLAY: $\boxed{2.43092725}$

Do not forget change of base: $\log_8 156.8 = \dfrac{\log 156.8}{\log 8}$

Note: Many people will solve this by "taking the log of both sides":

$$8^x = 156.8$$

$$\log 8^x = \log 156.8$$

$$x \log 8 = \log 156.8$$

$$x = \frac{\log 156.8}{\log 8}$$

You should notice that this result is the same, but the solution involves several extra steps. Before the ready availability of calculators there was a reason for avoiding a representation such as $\log_8 156.8$, but now it does not need to be avoided. When you *see* expressions such as $\log_8 156.8$, you *know* how to calculate it: evaluate $\boxed{\log 156.8}$ $\boxed{\div}$ $\boxed{\log 8}$.

EXAMPLE 9 Solve $6^{3x+2} = 200$.

SOLUTION Method 1: $\log_6 200 = 3x + 2$

$$\log_6 200 - 2 = 3x$$

$$x = \frac{\log_6 200 - 2}{3}$$

$$= \frac{\dfrac{\ln 200}{\ln 6} - 2}{3}$$

By calculator: DISPLAY: $\boxed{.3190157417}$

Method 2: $\log 6^{3x+2} = \log 200$ Log of Both Sides Theorem

$$(3x + 2)\log 6 = \log 200$$

$$3x \log 6 + 2 \log 6 = \log 200$$

$$(3 \log 6)x = \log 200 - 2 \log 6$$

$$x = \frac{\log 200 - 2 \log 6}{3 \log 6}$$

In Section 5.1 we considered continuous compounding of interest. This is an example of continuous *growth*. A similar application involves continuous *decay*.

DECAY FORMULA $$A = A_0 e^{-kt}$$

where an initial quantity A_0 decays to an amount A after a time t. The positive constant k depends on the substance.

EXAMPLE 10 If 100.0 mg of neptunium-239 (^{239}Np) decays to 73.36 mg after 24 hr, find the value of k (to four significant digits) in the decay formula for t expressed as days.

SOLUTION Since $A = 73.36$, $A_0 = 100$, and $t = 1$, we have

$$73.36 = 100e^{-k(1)}$$

$$.7336 = e^{-k}$$

$$k = -\ln .7336$$

$$\approx .30979136$$

$$\boldsymbol{k = .3098} \qquad \text{Rounded to four significant digits}$$

Radioactive decay is usually specified in terms of its **half-life,** which means the amount of time necessary for one-half of its substance to disintegrate into another substance. The decay formula for half-life is

HALF-LIFE DECAY $$\tfrac{1}{2} = e^{-kt}$$

EXAMPLE 11 What is the half-life of ^{239}Np? Use the value of k found in Example 10 ($k = .3098$).

SOLUTION $$\tfrac{1}{2} = e^{-.3098t}$$

$$\ln .5 = -.3098t$$

$$t = \frac{\ln .5}{-.3098}$$

$$\approx 2.2374021$$

The half-life is about **2.24 days.**

A very important application of both growth and decay is the prediction of the size of various populations. Population growth is considered at length in a special Extended Application following the Cumulative Review at the end of this chapter.

5.4 Problem Set

A *Solve the exponential equations in Problems 1–22.*

1. $2^x = 128$

2. $3^x = 243$

3. $8^x = 32$

4. $9^x = 27$

5. $125^x = 25$

6. $216^x = 36$

7. $128^x = 8$

8. $2{,}187^x = 81$

9. $4^x = \frac{1}{16}$

10. $27^x = \frac{1}{81}$

11. $(\frac{1}{2})^x = 8$

12. $(\frac{1}{2})^x = \frac{1}{8}$

13. $(\frac{2}{3})^x = \frac{9}{4}$

14. $(\frac{3}{4})^x = \frac{16}{9}$

15. $8^{3x} = 2$

16. $64^{2x} = 2$

17. $2^{3x+1} = \frac{1}{2}$

18. $3^{4x-3} = \frac{1}{9}$

19. $27^{2x+1} = 3$

20. $8^{5x+2} = 16$

21. $125^{2x+1} = 25$

22. $8^{2x+3} = 4$

B *Solve the exponential equations in Problems 23–46. Show the approximation you obtain with your calculator without rounding.*

23. $10^x = 42$ 24. $10^x = 126$ 25. $10^x = .00325$

26. $10^x = .0234$ 27. $10^{x+3} = 214$ 28. $10^{x-5} = .036$

29. $10^{x-1} = .613$ 30. $10^{4x+1} = 719$ 31. $10^{5-3x} = .041$

32. $10^{2x-1} = 515$ 33. $e^{2x} = 10$ 34. $e^{5x} = \frac{1}{4}$

35. $e^{4x} = \frac{1}{10}$ 36. $e^{2x+1} = 5.474$ 37. $e^{1-2x} = 3$

38. $e^{1-5x} = 15$ 39. $8^x = 300$ 40. $2^x = 1,000$

41. $5^x = 10$ 42. $9^x = .045$ 43. $2^{-x} = 5$

44. $4^x = .82$ 45. $5^{-x} = 8$ 46. $7^{-x} = 125$

47. If $1,000 is invested at 12% compounded semiannually, how long will it take (to the nearest year) for the money to double?

48. If $1,000 is invested at 16% interest compounded annually, how long will it take (to the nearest year) for the money to quadruple?

49. If $1,000 is invested at 12% interest compounded quarterly, how long will it take (to the nearest quarter) for the money to reach $2,500?

50. If $1,000 is invested at 15% interest compounded continuously, how long will it take (to the nearest year) for the money to triple?

51. If the half-life of cesium-137 is 30 years, find the constant k for which $A = A_0e^{-kt}$ where t is expressed in years.

52. Find the half-life of strontium-90 if $k = .0246$, where $A = A_0e^{-kt}$ and t is expressed in years.

53. Find the half-life of krypton if $k = .0641$, where $A = A_0e^{-kt}$ and t is expressed in years.

54. The formula used for carbon-14 dating in archaeology is

$$A = A_0\left(\frac{1}{2}\right)^{t/5,700} \quad \text{or} \quad P = \left(\frac{1}{2}\right)^{t/5,700}$$

where P is the percentage of carbon-14 present after t years. Solve for t.

55. Some bone artifacts were found at the Lindenmeier site in northeastern Colorado and tested for their carbon-14 content. If 25% of the original carbon-14 was still present, what is the probable age of the artifacts? Use the formula given in Problem 54.

56. An artifact was discovered at the Debert site in Nova Scotia. Tests showed that 28% of the original carbon-14 was still present. What is the probable age of the artifact? Use the formula given in Problem 54.

57. An artifact was found and tested for its carbon-14 content. If 12% of the original carbon-14 was still present, what is its probable age? Use the formula given in Problem 54.

58. An artifact was found and tested for its carbon-14 content. If 85% of the original carbon-14 was still present, what is its probable age? Use the formula given in Problem 54.

59. PHYSICS The atmospheric pressure P in pounds per square inch (psi) is given by

$$P = 14.7e^{-.21a}$$

where a is the altitude above sea level (in miles). If a city has an atmospheric pressure of 13.23 psi, what is its altitude?

60. SPACE SCIENCE A satellite has an inital radioisotope power supply of 50 watts (W). The power output in watts is given by the equation

$$P = 50e^{-t/250}$$

where t is the time in days. Solve for t.

C

61. PHYSICS Newton's law of cooling states that an object at temperature B surrounded by air temperature A will cool to a temperature T after t min according to the equation

$$T = A + (B - A)e^{-kt}$$

where k is a constant depending on the item being cooled.

a. If you draw a tub of 120°F water for a bath and let it stand in a 75°F room, what is the temperature of the water after 30 min if $k = .01$?

b. What is k for an apple pie taken from a 375°F oven and cooled to 75°F after it is left in a 72°F room for 1 hr?

c. Solve the given equation for t.

62. In calculus it is shown that

$$e^x = 1 + x + \frac{x^2}{2} + \frac{x^3}{2\cdot 3} + \frac{x^4}{2\cdot 3\cdot 4} + \cdots$$

a. What are the next two terms?

b. What is the rth term?

c. Calculate e correct to the nearest thousandth using this equation.

d. Calculate $\sqrt{e} = e^{.5}$ correct to the nearest thousandth using this equation.

63. Solve $10^{5x+1} = e^{2-3x}$.

64. Solve $5^{2+x} = 6^{3x+2}$.

65. Solve $e^x + e^{-x} = 10$. (*Hint:* Multiply both sides by e^x and use the quadratic formula.)

66. Solve $e^x - e^{-x} = 100$.

67. If p and q are roots of $3x^2 + mx + 4 = 0$, what is $\log_2 p + \log_2 q$?

Chapter 5 Summary

The material of this chapter is reviewed in the following list of objectives. After each objective there are some practice questions. For a sample test, select the first question of each set and check your answers with the answer section. For a sample test without answers, use the second question of each set. Additional practice is given by the other questions in each set. If you are having trouble with a particular type of problem, look back to that section for extra help.

5.1 **Exponential Functions** ─────────────────────

OBJECTIVE 1 *Sketch exponential functions.*

1. $y = (\frac{1}{2})^x$ **2.** $y = -2^x$ **3.** $y = 2^{-x}$ **4.** $y = e^{-x/2}$

5.2 **Introduction to Logarithms** ─────────────────

OBJECTIVE 2 *Write an exponential equation in logarithmic form.*

5. $10^{.5} = \sqrt{10}$ **6.** $e^0 = 1$

7. $9^3 = 729$ **8.** $(\sqrt{2})^3 = 2\sqrt{2}$

OBJECTIVE 3 *Write a logarithmic equation in exponential form.*

9. $\log 1 = 0$ **10.** $\ln \frac{1}{e} = -1$

11. $\log_2 64 = 6$ **12.** $\log_\pi \pi = 1$

OBJECTIVE 4 *Evaluate common logarithms. Show calculator answer.*

13. $\log 3$ **14.** $\log 451$

15. $\log .0021$ **16.** $\log 3^4$

OBJECTIVE 5 *Evaluate natural logarithms. Show calculator answer.*

17. $\ln 3$ **18.** $\ln 451$

19. $\ln .013$ **20.** $\ln 3^4$

OBJECTIVE 6 *Evaluate logarithms to an arbitrary base. Show calculator answer.*

21. $\log_4 15$ **22.** $\log_{1/4} 65.3$

23. $\log_{1.06} 14{,}650$ **24.** $\log_2 5^9$

5.3 **Logarithmic Functions and Equations** ──────────

OBJECTIVE 7 *Graph logarithmic functions.*

25. $y = \log x$ **26.** $y = \ln x$

27. $y = \log_3 x$ **28.** $y + 3 = \log(x - 2)$

OBJECTIVE 8 *State and prove the laws of exponents.* Fill in the blanks.

29. $\log_b AB = $ _____

30. $\log_b A - \log_b B = $ _____

31. $\log_b A^p = $ _____

32. If $A = b^x$, then $x = $ _____

OBJECTIVE 9 *Solve logarithmic equations.* Solve for x.

33. $\log_5 25 = x$

34. $\log_x(x + 6) = 2$

35. $3 \log 3 - \frac{1}{2} \log 3 = \log \sqrt{x}$

36. $2 \ln \frac{e}{\sqrt{7}} = 2 - \ln x$

5.4 **Exponential Equations** ─────────────────────

OBJECTIVE 10 *Solve exponential equations with base 10.*

Give calculator answer.

37. $10^{x+2} = 125$ **38.** $10^{2x-3} = .5$

39. $10^{-x^2} = .45$ **40.** $10^{4-3x} = 15$

OBJECTIVE 11 *Solve exponential equations with base e.* Give calculator answer.

41. $e^{3x} = 50$ **42.** $e^{x-5} = 0.49$ **43.** $e^{1-2x} = 690$

44. $e^{x^2} = 9$

OBJECTIVE 12 *Solve exponential equations with arbitrary bases.* Give calculator answer.

45. $5^x = 125$ **46.** $5^{2x+1} = .2$ **47.** $3^x = 7$

48. $7^{1-3x} = .048$

OBJECTIVE 13 *Solve applied problems involving exponential equations.*

49. The half-life of arsenic-76 is 26.5 hr. Find the constant k for which $A = A_0 e^{-kt}$, where t is expressed in hours.

50. If $1,500 is placed in a $2\frac{1}{2}$-year time certificate paying 8.5% compounded monthly, what is the amount in the account when the certificate matures in $2\frac{1}{2}$ years?

51. If a person's present salary is $20,000 per year, use the formula $A = P(1 + r)^n$ to determine the salary necessary to equal this salary 15 years from now if you assume an inflation rate of 13.4%.

52. Solve the formula given in Problem 51 for n.

53. Repeat Problem 51 for the 1990 rate of 4%.

Cumulative Review I

Suggestions for study of Chapters 2–5:

MAKE A LIST OF IMPORTANT IDEAS FROM CHAPTERS 2–5. Study this list. Use the objectives at the end of each chapter to help you make up this list.

WORK SOME PRACTICE PROBLEMS. A good source of problems is the set of chapter objectives at the end of each chapter. You should try to work at least one problem from each objective:

Chapter 2, pp. 114–115; work 17 problems

Chapter 3, pp. 172–175; work 21 problems

Chapter 4, pp. 214–215; work 8 problems

Chapter 5, pp. 245–246; work 13 problems

Check the answers for the practice problems you worked. The first and third problems for each objective have their answers listed in the back of the book.

ADDITIONAL PROBLEMS. Work additional odd-numbered problems (answers in the back of the book) from the problem sets as needed. Focus on the problems you missed in the chapter summaries.

WORK THE PROBLEMS IN THE FOLLOWING CUMULATIVE REVIEW. These problems should be done after you have studied the material. They should take you about 1 hour and will serve as a sample test for Chapters 2–5. Assume that all variables are restricted so that each expression is defined. All the answers for these questions are provided in the back of the book for self-checking.

Practice Test for Chapters 2–5

1. If $f(x) = 3x^2 + 2$ and $g(x) = 2x - 1$, find:
  a. $f(-2)$ **b.** $g(t)$ **c.** $f(t + h)$
  d. $f[g(x)]$ **e.** $(g \circ f)(x)$ **f.** $\dfrac{f(x + h) - f(x)}{h}$

2. Find the inverse of the given functions, if possible.
  a. $f(x) = 3x^2 + 2$ **b.** $g(x) = 2x - 1$

3. Graph the given functions.
  a. $y + 2 = -\frac{1}{2}(x + 3)$ **b.** $y + 2 = -\frac{1}{2}(x + 3)^2$

4. Let $F(x) = 4x^3 - 20x^2 + 3x + 27$. Find:
  a. $F(-1)$ **b.** $F(0)$ **c.** $F(3)$ **d.** $F(\frac{3}{2})$
  e. the zeros for F

5. Graph $F(x) = 4x^3 - 20x^2 + 3x + 27$. You might want to look at your work for Problem 4.

6. Solve $4x^4 - 20x^3 - 19x^2 + 26x - 6 = 0$, given that $3 + \sqrt{7}$ is a root.

7. Let $G(x) = \dfrac{2x^2 - 3x + 1}{x^2 - x - 2}$.

  a. What is the domain?
  b. Is the curve defined by $G(x)$ symmetric with respect to the x-axis, y-axis, or origin?

  c. Are there any horizontal, vertical, or slant asymptotes?
  d. What are the x-intercepts and y-intercepts?
  e. Graph G.

8. Graph the following rational functions.
  a. $y = \dfrac{-10}{x}$ **b.** $y = \dfrac{-3x}{x + 4}$

  c. $y = \dfrac{x^2}{x(x - 3)}$

9. Solve the given equations.
  a. $8^x = 32$ **b.** $\log_3 x = 4.12$
  c. $\log_8 x = \log_8(4 - 3x)$ **d.** $2 \log 3 = \frac{1}{2} \log x$

10. If the original quantity of a radioactive substance is A_0, then the amount present, A, after t years is given by

$$A = A_0 2^{-kt}$$

  a. Carbon-14 has a half-life of 5,700 years. Find k.
  b. If an ancient scroll is found and it is estimated that 72.4% of the original carbon-14 is still present, how old is the scroll? (Use the k you found in part **a.**)
  c. Solve the equation for t.

Population Growth

World Population 4 Billion Tonight

BY EDWARD K. DELONG

WASHINGTON (UPI) By midnight tonight, the Earth's population will reach the 4 billion mark, twice the number of people living on the planet just 46 years ago, the Population Reference Bureau said Saturday.

The bureau expressed no joy at the new milestone.

It said global birth rates are too high, placing serious pressures on all aspects of future life and causing "major concern" in the world scientific community, and more than one-third of the present population has yet to reach child-bearing age.

The PRB found cause for optimism, however, in that some governments are stressing birth control to blunt the impact of "explosive growth" and the population growth rate dropped slightly in the past year.

"In 1976, each new dawn brings a formidable increase of approximately 195,000 newborn infants to share the resources of our finite world," it said.

One expert warned that a lack of jobs, rather than too little food, may be the "ultimate threat" facing society as the planet becomes more and more crowded.

It took between two and three million years for the human race to hit the one billion mark in 1850, the PRB said. By 1930, 80 years later, the population stood at 2 billion. A mere 31 years after that, in 1961, it was 3 billion. The growth from 3 to the present 4 billion took just 15 years.

The world could find it has 5 billion people by 1989—just 13 years from now—if population growth continues at the present rate of 1.8 per cent a year, said Dr. Leon F. Bouvier, vice president of the private, nonprofit PRB.

Bouvier said the newly calculated growth rate is a little lower than the 1.9 per cent estimated last year. Thanks to that slowdown, the passing of the 4 billion milestone came a year later than some demographers had predicted.

"I really think the rate of growth is going to start declining ever so slightly because of declining fertility," Bouvier said. "I think there is some evidence of progress—ever so slow, much too slow."

The new PRB figures show there were 3,982,815,000 people on Earth on Jan. 1. By March 1 the number had grown to 3,994,812,000, the organization said, and by April 1 the total will by 4,000,824,000.

The bureau said its calculations are based on estimates of 328,000 live births per day minus 133,000 deaths.

A growing number of governments are taking steps to slow growth rates, the PRB said.

Singapore appears likely to meet the goal of the two-child family "well before the target date of 1980," it said, and several states in India, which yearly adds the equivalent of the population of Australia, are considering financial incentives to birth control and mandatory sterilization after the birth of two children.

Dr. Paul Ehrlich of Stanford University, one of several population experts contacted by PRB, said he was sad to realize at the age of 44 he had lived through a doubling of Earth's population. He expressed fear the next 44 years could see population growth halted "by a horrifying increase in death rates."

"At this point, hunger does not seem the greatest issue presented by the ever growing number of people," said Dr. Louis M. Hellman, chief of population staff at the Health, Eduation and Welfare Department.

"Rather, the threat appears to lie in the increasing numbers who can find no work. As these masses of unemployed migrate toward the cities, they create a growing impetus toward political unrest and instability."

SOURCE: Courtesy of *The Press Democrat*, Santa Rosa, California, March 28, 1976. Reprinted by permission.

Human populations grow according to the exponential equation

$$P = P_0 e^{rt}$$

where P_0 is the size of the initial population, r is the growth rate, t is the length of time, and P is the size of the population after time t.

EXAMPLE 1 On March 28, 1976, the world population reached 4 billion. If the annual growth rate is 2%, what was the expected population on March 28, 1990?

SOLUTION The initial population P_0 is 4 (billion), $r = .02$, and $t = 14$. Then

$$P = 4e^{.02(14)} \approx 5.29$$

The expected population was **5.3 billion**.

EXAMPLE 2 When is the world population expected to reach 6 billion, given a growth rate of 2%?

SOLUTION
$$r = .02, \quad P_0 = 4, \quad P = 6$$
$$6 = 4e^{.02t}$$
$$e^{.02t} = 1.5$$
$$.02t = \ln 1.5$$
$$t = 50 \ln 1.5 \qquad \text{Divide both sides by .02} \ (\tfrac{1}{.02} = 50).$$
$$\approx 20.2733$$

This is 20 years, 100 days after March 28, 1976, or **July 6, 1996.**

If you can find the present world population (consult a recent almanac or call your library), you can compare the *predicted* population from Example 1 (using 1976 figures) with the actual figures. Suppose they differ; what conclusion can you make?

If you know the actual population of a city, state, country, or even the world for two dates, you can use the population formula to find the **growth rate** over that period. If factors influencing population growth do not change significantly, this growth rate can be used to predict future population growth. The question asked at the end of the preceding paragraph would be answered by saying that the *actual* growth rate was not the 2% assumed in Example 1.

EXAMPLE 3 The population actually reached 5 billion on July 7, 1986. Find the growth rate since it was reported in the news article.

SOLUTION $P_0 = 4, \quad P = 5, \quad t \approx 10.2767$

To find t, note that it is 10 years from March 28, 1976 to March 28, 1986; from March 28 to July 7 is 101 days; $101/365 \approx .2767123$. Now solve $P = P_0 e^{rt}$ for r:

$$\frac{P}{P_0} = e^{rt}$$

$$rt = \ln \frac{P}{P_0}$$

$$r = \frac{1}{t} \ln \frac{P}{P_0}$$

$$= \frac{1}{10.2767123} \ln \tfrac{5}{4}$$

$$\approx .0217135$$

The growth rate is about **2.2%**.

EXAMPLE 4 The 1990 population of Texas was 16,825,000 and the 1980 population was 14,226,000. What was the yearly growth rate of the Texas population for this period?

SOLUTION Since $P_0 = 14.226$, $P = 16.825$, and $t = 10$, we have

$$r = \frac{1}{t} \ln \frac{P}{P_0} \qquad \text{From Example 3}$$

$$= \frac{1}{10} \ln \left(\frac{16.825}{14.226} \right) \qquad \text{DISPLAY:} \quad \boxed{.0167794599}$$

The yearly growth rate is about **1.7%**.

EXAMPLE 5 Using the growth rate found in Example 4, predict the population of Texas in 2000.

SOLUTION Find P when $P_0 = 16.825$, $r = .0167794599$, and $t = 10$:

$$P = P_0 e^{rt}$$

$$= 16.825 e^{.0167794599(10)}$$

$$\approx 19.89882082$$

The predicted population of Texas in 2000 is about **19,900,000**.

Extended Application Problems—Population Growth

1. The world population was 1 billion in 1850 and it took 80 years to reach the 2 billion mark. What was the annual growth rate for this period from 1850 to 1930?

2. The world population was 2 billion in 1930 and it took 31 years to reach 3 billion. What was the annual growth rate for this period from 1930 to 1961?

3. The world population was 3 billion in 1961 and it took 15 years to reach 4 billion. What was the annual growth rate for this period from 1961 to 1976?

4. The world population was 4 billion in 1976 and it took 10 years to reach 5 billion. What is the growth rate for this period from 1976 to 1986?

5. On July 7, 1986, the world population reached 5 billion. If the annual growth rate is 2.2%, when would you expect the population to reach 6 billion?

6. On July 7, 1986, the world population reached 5 billion. If the annual growth rate is 2.2%, when would you expect the population to reach 7 billion?

7. On July 7, 1986, the world population reached 5 billion. If the annual growth rate is 2.2%, when would you expect the population to reach 8 billion?

8. a. If the population of California grew from 23,668,000 in 1980 to 29,279,000 in 1990, what was the yearly growth rate for this period?

   b. If you use the growth rate found in part **a**, what is the expected population in 2000?

9. a. If the population of Iowa declined from 2,914,000 in 1980 to 2,767,000 in 1990, what was the growth rate for this period? (*Note:* A negative growth rate means the population declined.)

   b. If you use the growth rate found in part **a**, what is the expected population in 2000?

10. a. If the population of Washington, D.C. declined from 638,000 in 1980 to 575,000 in 1990, what was the growth rate for this period? (*Note:* A negative growth rate means the population declined.)

   b. If you use the growth rate found in part **a**, what is the expected population in 2000?

11. RESEARCH PROBLEM Use the information from the news article and Problems 1–7 to graph the world population from 1800 to 2020.

12. RESEARCH PROBLEM From your local Chamber of Commerce, obtain the population figures for your city for 1970, 1980, and 1990.

a. Find the rate of growth for each period.

b. Forecast the population of your city for the year 2000. Which of the rates you obtained in part **a** is the most accurate for this purpose?

13. List some factors, such as new zoning laws, that could change the growth rate of your city.

14. List some factors, such as a change in the tax laws of your state, that could change the growth rate of your state.

15. List some factors, such as war, that could change the growth rate of a country or the world.

Hipparchus (c. 180–125 B.C.)

Trigonometry contains the science of continually undulating magnitude: meaning magnitude which becomes alternately greater and less, without any termination to succession of increase or decrease .... All trigonometric functions are not undulating; but it may be stated that in common algebra nothing but infinite series undulate; in trigonometry nothing but infinite series do not undulate.

AUGUSTUS DE MORGAN
Trigonometry and Double Algebra

▌Sin²φ is odious to me, even though Laplace made use of it; should it be feared that sin ϕ^2 might become ambiguous, which would perhaps never occur, or at most very rarely when speaking of $\sin(\phi^2)$, well then, let us write $(\sin \phi)^2$ but not $\sin^2\phi$, which by analogy should signify $\sin(\sin \phi)$.

KARL GAUSS
Gauss–Shumacher Briefweches, vol. 3, p. 292

We might add that the notation we use today does not conform to Gauss' suggestion, because $\sin^2\phi$ is used to mean $(\sin \phi)^2$.

The origins of trigonometry are obscure, but we do know that it began more than 2,000 years ago with the Mesopotamian, Babylonian, and Egyptian civilizations. Much of the knowledge from these civilizations was passed on to the Greeks, who formally developed many of the ideas of this chapter. The word *trigonometry* comes from the words *trigonon* (triangle) and *metron* (measurement), and the ancients used trigonometry in a very practical way to measure triangles, as we will in Chapter 8.

During the second half of the second century B.C., the astronomer Hipparchus of Nicaea compiled the first trigonometric tables in 12 books. It was he who used the 360° of the Babylonians and thus introduced trigonometry with the degree angle measure that we still use today. Hipparchus' work formed the foundation for Claudius Ptolemy's *Mathematical Syntaxis*, the most significant early work in trigonometry.

The Greek discoveries were later lost in Europe. Fortunately, they had been translated into Arabic, and the Arabs took them as far as India, which accounts for the interesting origin of the word *sine*. The Hindu mathematician Aryabhata (c. A.D. 476–550) called it *jyā-adhā* (chord half) and abbreviated it *jyā*, which the Arabs wrote as *jiba* but shortened to *jb*. Later writers erroneously interpreted *jb* to stand for *jaib*, which means cove or bay. When Europeans rediscovered trigonometry through the Arab tradition, they translated the texts into Latin. The Latin equivalent for *jaib* is *sinus*, from which our present word *sine* is derived. Several of our mathematical terms, including the word *algebra*, are based on misunderstandings of Arabic words—but in this case the Arabs themselves were confused. The words *tangent* and *secant* come from the relationship of the ratios to the lengths of the tangent and secant lines drawn on a circle. Finally, Edmund Gunter (1581–1626) first used the prefix *co* to invent the words *cosine*, *cotangent*, and *cosecant*. In 1620, Gunter published a seven-place table of the common logarithms of the sines and tangents of angles for intervals of one minute of an arc. Gunter originally entered the ministry, but later decided on astronomy as a career. He was such a poor preacher that historian Howard Eves stated that Gunter left the ministry in 1619 to the "benefit of both occupations."

6

TRIGONOMETRIC FUNCTIONS

Contents

Preview

This chapter introduces you to six very important functions in mathematics, called the trigonometric functions. After considering the definition, we consider their evaluation as well as their graphs. Finally, we look at the inverse trigonometric functions. You will need to be very familiar with these functions in order to continue with your studies in mathematics. The 25 objectives of this chapter are listed on pages 293–295.

Perspective

Trigonometric functions are tools in calculus just as are linear or quadratic functions. For example, trigonometric substitutions form a common method of integration, which is an operation considered in integral calculus. In the page below, taken from a calculus book, notice that the example uses the definitions of both the secant and the tangent functions as we introduce them in Section 6.2. You may also notice, toward the bottom of this extract, that the range of the inverse function is used, so you would need the material we introduce in Section 6.5. (Notice also the use of ln, which we introduced in Chapter 5.)

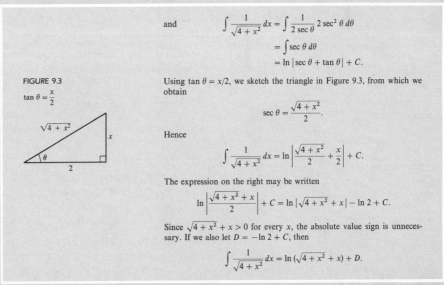

and
$$\int \frac{1}{\sqrt{4+x^2}}\,dx = \int \frac{1}{2\sec\theta}\,2\sec^2\theta\,d\theta$$
$$= \int \sec\theta\,d\theta$$
$$= \ln|\sec\theta + \tan\theta| + C.$$

FIGURE 9.3

$$\tan\theta = \frac{x}{2}$$

Using $\tan\theta = x/2$, we sketch the triangle in Figure 9.3, from which we obtain

$$\sec\theta = \frac{\sqrt{4+x^2}}{2}.$$

Hence

$$\int \frac{1}{\sqrt{4+x^2}}\,dx = \ln\left|\frac{\sqrt{4+x^2}}{2} + \frac{x}{2}\right| + C.$$

The expression on the right may be written

$$\ln\left|\frac{\sqrt{4+x^2}+x}{2}\right| + C = \ln\left|\sqrt{4+x^2}+x\right| - \ln 2 + C.$$

Since $\sqrt{4+x^2}+x > 0$ for every x, the absolute value sign is unnecessary. If we also let $D = -\ln 2 + C$, then

$$\int \frac{1}{\sqrt{4+x^2}}\,dx = \ln\left(\sqrt{4+x^2}+x\right) + D.$$

From Earl Swokowski, *Calculus with Analytic Geometry*, 5th ed. (Boston: Prindle, Weber, & Schmidt), p. 470.

6.1   Angles and the Unit Circle

The circle is of primary importance in the study of trigonometry. Although you are no doubt familiar with a circle, it is worthwhile to review circles briefly at this time. A **circle** is the set of points a given distance, called the **radius**, from a given point called the **center**. Since a circle is not the graph of a function, we will delay discussion of the general equation and properties of a circle until Chapter 11. We do, however, have need for the equation of a circle with radius r and center at the origin. The equation of this circle is derived by using the distance formula (from Section 1.3). Let (x, y) be any point on a circle of radius r. Then

$$(x_2 - x_1)^2 + (y_2 - y_1)^2 = d^2 \qquad \text{This is the distance formula with both sides squared.}$$

$$(x - 0)^2 + (y - 0)^2 = r^2 \qquad \text{The center is } (0, 0), \text{ so let } (x_1, y_1) = (0, 0);$$
$$\text{let } (x_2, y_2) = (x, y) \text{ and } d = r.$$

$$x^2 + y^2 = r^2$$

If $r = 1$, this is called the *unit circle*.

UNIT CIRCLE

> The **unit circle** is the circle with radius 1 and center at the origin. The equation of the unit circle is
>
> $$x^2 + y^2 = 1$$

In mathematics it is often useful to consider functions of angles; however, the definition of an angle depends on the context in which it is being used. In geometry, an angle is usually defined as the union of two rays with a common endpoint. In advanced mathematics courses, a more general definition is used.

ANGLE

> An **angle** is formed by rotating a ray about its endpoint (called the **vertex**) from some initial position (called the **initial side**) to some terminal position (called the **terminal side**). The measure of an angle is the amount of rotation. An angle is also formed if a line segment is rotated about one of its endpoints.

Commonly used Greek letters

SYMBOL	NAME
α	alpha
β	beta
γ	gamma
δ	delta
θ	theta
λ	lambda
ϕ or φ	phi
ω	omega

π (pi) is a lowercase Greek letter that will not be used to represent an angle. It denotes an irrational number approximately equal to 3.141592654.

If the rotation of the ray is in a counterclockwise direction, the measure of the angle is called **positive.** If the rotation is in a clockwise direction, the measure is called **negative.** The notation $\angle ABC$ means the measure of an angle with vertex B and points A and C (different from B) on the sides; $\angle B$ denotes the measure of an angle with vertex at B, and a curved arrow is used to denote the direction and amount of rotation, as shown in Figure 6.1. If no arrow is shown, the measure of the angle is considered to be the smallest positive rotation. Lowercase Greek letters are also used to denote the angles as well as the measure of angles. For example, θ may represent the angle or the measure of the angle called θ; you will know which is meant by the context in which it is used. Some examples are shown in Figure 6.1.

A Cartesian coordinate system may be superimposed on an angle so that the vertex is at the origin and the initial side is along the positive x-axis. In this case

FIGURE 6.1 Examples of angles

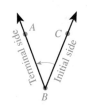
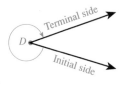

∠*ABC* is a positive angle   ∠*D* is a negative angle   α is a positive angle
β is a negative angle

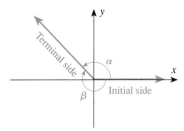

FIGURE 6.2 Standard-position angles α and β; α is a positive angle and β is a negative angle; α and β are coterminal angles.

the angle is in **standard position.** Angles in standard position having the same terminal sides are **coterminal angles.** Given any angle α, there is an unlimited number of coterminal angles (some positive and some negative). In Figure 6.2, β is coterminal with α. Can you find other angles coterminal with α?

Several units of measurement are used for measuring angles. Let α be an angle in standard position with a point *P* not the vertex but on the terminal side. As this side is rotated through one revolution, the trace of the point *P* forms a circle. The measure of the angle is one revolution, but since much of our work will be with amounts less than one revolution, we need to define measures of smaller angles. Historically the most common scheme divides one revolution into 360 equal parts with each part called a **degree.** Sometimes even finer divisions are necessary, so a degree is divided into 60 equal parts, each called a **minute** ($1° = 60'$). Furthermore, a minute is divided into 60 equal parts, each called a **second** ($1' = 60''$). For most applications, we will write decimal parts of degrees instead of minutes and seconds. That is, $32.5°$ is preferred over $32°30'$.

In calculus and scientific work, another measure for angles is defined. This method uses real numbers to measure angles. Draw a circle with any nonzero radius *r*. Next measure out an arc with length *r*. Figure 6.3a shows the case in which $r = 1$ and Figure 6.3b shows $r = 2$. Regardless of your choice for *r*, the angle determined by this arc of length *r* is the same. (It is labeled θ in Figure 6.3.) This angle is used as a basic unit of measurement and is called a **radian.**

FIGURE 6.3 Radian measure

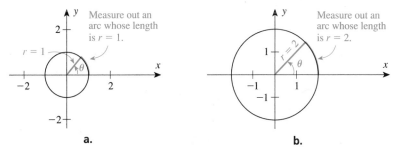

a.   b.

Notice that the circumference *C* generates an angle of one revolution. Since $C = 2\pi r$, and since the basic unit of measurement on the circle is *r*,

$$\text{One revolution} = \frac{C}{r}$$

$$= \frac{2\pi r}{r}$$

$$= 2\pi$$

Thus $\frac{1}{2}$ revolution is $\frac{1}{2}(2\pi) = \pi$ radians; $\frac{1}{4}$ revolution is $\frac{1}{4}(2\pi) = \frac{\pi}{2}$ radians.

Notice that when measuring angles in radians, you are using *real numbers*. Because radian measure is used so frequently, we agree that **radian measure is understood when no units of measure for an angle are indicated**. Figure 6.4 shows a protractor for measuring angles using radian measure.

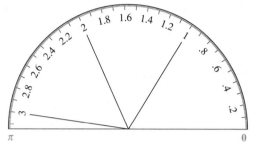

FIGURE 6.4 Protractor for radian measure

Example 1 asks you to draw several angles using radian measures. Do not try to change these angles to degree measure. Work them directly as shown in the examples. You will have to work at thinking in terms of radian measure. You should memorize the approximate size of an angle of measure 1 radian in much the same way you have memorized the approximate size of an angle of measure 45°.

EXAMPLE 1   Let $r = 1$ and draw the angles with the given measures.

a. $\theta = 2$ (2 radians is understood)   **b.** $\theta = 3$ (3 radians is understood)

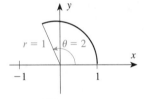   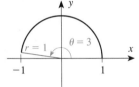

c. $\theta = \sqrt{19}$
$\sqrt{19} \approx 4.36$; since a straight angle is $\pi \approx 3.14$, use the radian protractor for $\sqrt{19} - \pi \approx 1.22$.

d. $\theta = \sqrt{50}$
$\sqrt{50} \approx 7.07$; since one revolution is $2\pi \approx 6.28$, use the radian protractor for $\sqrt{50} - 2\pi \approx .79$.

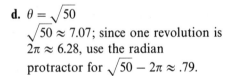

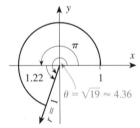   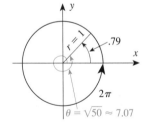

e. $\theta = .5$

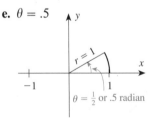

f. $\theta = -\dfrac{\pi}{4}$

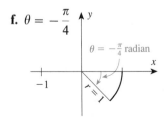

Next consider the relationship between degree and radian measure:

One revolution is measured by 360° or by 2π radians.

Then

$$\text{Number of revolutions} = \frac{\text{Angle in degrees}}{360}$$

$$\text{Number of revolutions} = \frac{\text{Angle in radians}}{2\pi}$$

Therefore:

RELATIONSHIP BETWEEN DEGREE AND RADIAN MEASURE

$$\frac{\text{Angle in degrees}}{360} = \frac{\text{Angle in radians}}{2\pi}$$

EXAMPLE 2   **a.** Change 45° to radians.

$$\frac{45}{360} = \frac{\theta}{2\pi}$$

$$\frac{90\pi}{360} = \theta$$

$$\frac{\pi}{4} = \theta$$

An alternative method is to remember that π radians is 180°. That is, since 45° is $\frac{1}{4}$ of 180°, you know that the radian measure is $\frac{\pi}{4}$. As a decimal, θ can be approximated by calculator:

$$\theta \approx .78539816$$

b. Change 2.30 to degrees.

$$\theta = \frac{180}{\pi}(2.30)$$

$$\approx (57.296)(2.30)$$

$$= 131.7808$$

To the nearest hundredth, the angle is 131.78°. If you have a calculator, you can obtain a much more accurate answer by using a better approximation for $\frac{180}{\pi}$. For example, if 2.30 is exact, then by calculator:

Algebraic logic: $\boxed{180}\ \boxed{\div}\ \boxed{\pi}\ \boxed{\times}\ \boxed{2.3}\ \boxed{=}$

RPN logic: $\boxed{180}\ \boxed{\text{ENTER}}\ \boxed{\pi}\ \boxed{\div}\ \boxed{2.3}\ \boxed{\times}$

The result is approximately **131.7802929°.**

c. Change 1° to radians.

$$\frac{1}{360} = \frac{\theta}{2\pi}$$

$$\frac{\pi}{180} = \theta$$

Even though this solution is in the desired form, you might be interested in performing the division on a calculator:

1° ≈ .0174532925 radian

d. Change 1 to degrees.

$$\frac{\theta}{360} = \frac{1}{2\pi}$$

$$\theta = \frac{180}{\pi}$$

By calculator:

θ ≈ 57.29577951° or 57°17′45″

Decimal degrees is the preferred form.

For the more common measures of angles, it is a good idea to memorize the equivalent degree and radian measures. If you keep in mind that 180° in radian measure is π, the rest of the values will be easy to remember.

COMMONLY USED DEGREE AND RADIAN MEASURES

DEGREES	RADIANS
0°	0
30°	$\frac{\pi}{6}$
45°	$\frac{\pi}{4}$
60°	$\frac{\pi}{3}$
90°	$\frac{\pi}{2}$
180°	π
270°	$\frac{3\pi}{2}$
360°	2π

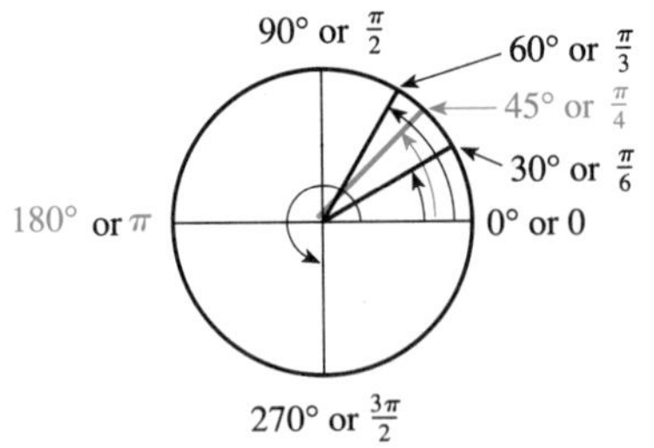

We now relate the radian measure of an angle to a circle to find the arc length. An **arc** is part of a circle; thus **arc length** is the distance around part of a circle. The arc length corresponding to one revolution is the **circumference** of the circle. Let s be the length of an arc and let θ be the angle measured in radians. Then

$$\text{Angle in revolutions} = \frac{s}{2\pi r}$$

since one revolution has an arc length (circumference of the circle) of $2\pi r$. Also,

$$\text{Angle in radians} = (\text{Angle in revolutions})(2\pi)$$

Substituting,

$$\theta = \frac{s}{2\pi r}(2\pi)$$

$$= \frac{s}{r}$$

From this result, we derive the following formula.

ARC LENGTH FORMULA

The **arc length** cut by a central angle θ (measured in radians) from a circle of radius r is denoted by s and is found by

$$s = r\theta$$

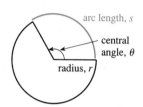

arc length, s

central angle, θ

radius, r

EXAMPLE 3   The length of the arc subtended (cut off) by a central angle of 36° in a circle with a radius of 20 centimeters (cm) is found as follows. First change 36° to radians so that you can use the formula given above.

s is the arc length

$$\frac{36}{360} = \frac{\theta}{2\pi}$$

Solving for θ yields

$$\frac{\pi}{5} = \theta$$

Thus

$$s = 20\left(\frac{\pi}{5}\right)$$

$$= 4\pi$$

The length of the arc is **4π cm,** or about 12.6 cm.

If the terminal side of an angle coincides with a coordinate axis, the angle is called a **quadrantal angle.** If the angle θ is not a quadrantal angle then we refer to its *reference angle*, which is denoted by θ' throughout this book.

REFERENCE ANGLE

Given an angle θ in standard position, the **reference angle** θ' is defined as the acute angle the terminal side makes with the x-axis.

The procedure for finding the reference angle depends on the quadrant of θ. One example for each quadrant is shown in Figure 6.5.

FIGURE 6.5 Reference angles

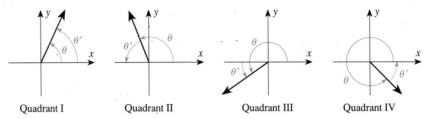

Quadrant I   Quadrant II   Quadrant III   Quadrant IV

EXAMPLE 4 Find the reference angle and draw both the given angle and the reference angle.

a. 210°   **b.** 150°

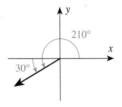   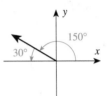

$210° - 180° = 30°$
Reference angle is **30°**.

$180° - 150° = 30°$
Reference angle is **30°**.

c. $-\dfrac{5\pi}{3}$ **d.** 2.5

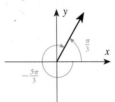   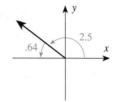

$2\pi - \dfrac{5\pi}{3} = \dfrac{\pi}{3}$
Reference angle is $\frac{\pi}{3}$.

$\pi - 2.5 \approx .64$
Reference angle is approximately **.64**.

e. 812°. If the angle is more than one revolution, first find a nonnegative coterminal angle that is less than one revolution. 812° is coterminal with 92°.

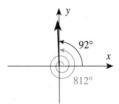   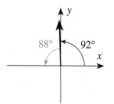

It is not correct to say $812° = 92°$.
812° and 92° are coterminal.

$180° - 92° = 88°$
Reference angle is **88°**.

f. The angle whose measure is 30; 30 is coterminal with 4.867: $(30 - 8\pi \approx 4.867)$.

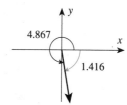

$2\pi - 4.867 \approx 1.416$
Reference angle is
approximately **1.42.**

The purpose of this example is to
remind you that $30 \neq 30°$.

Keep in mind the differences between coterminal angles and reference angles. Note also that reference angles are always between 0 and 90° or 0 and $\frac{\pi}{2}$ (about 1.57).

6.1 Problem Set

A *From memory, give the radian measure for each of the angles whose degree measure is stated, and the degree measure for each of the angles whose radian measure is stated, in Problems 1–4.*

1. a. 30° **b.** 90° **c.** 270° **d.** 45°
2. a. 360° **b.** 60° **c.** 180° **d.** 0°
3. a. π **b.** $\frac{\pi}{4}$ **c.** $\frac{\pi}{3}$ **d.** 2π
4. a. $\frac{\pi}{2}$ **b.** 0 **c.** $\frac{3\pi}{2}$ **d.** $\frac{\pi}{6}$

Use the radian protractor in Figure 6.4 to help you sketch each of the angles in Problems 5–10.

5. a. $\frac{\pi}{2}$ **b.** $\frac{\pi}{6}$ **c.** -2.5 **d.** $\sqrt{17}$
6. a. $\frac{2\pi}{3}$ **b.** $\frac{3\pi}{4}$ **c.** -1 **d.** $\sqrt{95}$
7. a. $-\frac{\pi}{4}$ **b.** $\frac{7\pi}{6}$ **c.** -2.76 **d.** $\sqrt{115}$
8. a. $-\frac{3\pi}{2}$ **b.** $\frac{13\pi}{3}$ **c.** -1.2365 **d.** $\sqrt{23}$
9. a. $\frac{\pi}{15}$ **b.** $-\frac{9\pi}{7}$ **c.** -4 **d.** $\sqrt{10}$
10. a. $-\frac{5\pi}{6}$ **b.** $\frac{5\pi}{4}$ **c.** -3 **d.** $\sqrt{19}$

Find the exact value of a positive angle less than one revolution that is coterminal with each of the angles in Problems 11–16.

11. a. 400° **b.** 540° **c.** 750° **d.** 1,050°
12. a. $-30°$ **b.** $-200°$ **c.** $-55°$ **d.** $-320°$
13. a. $-120°$ **b.** 500° **c.** $-180°$ **d.** 1,000°
14. a. 3π **b.** $\frac{13\pi}{6}$ **c.** $-\pi$ **d.** 7
15. a. $-\frac{\pi}{4}$ **b.** $\frac{17\pi}{4}$ **c.** $\frac{11\pi}{3}$ **d.** -2
16. a. $-\frac{\pi}{6}$ **b.** $-\frac{5\pi}{4}$ **c.** $\frac{15\pi}{6}$ **d.** 8

Find a positive angle less than one revolution correct to four decimal places so that it is coterminal with each of the angles in Problems 17–19.

17. a. 9 **b.** -5 **c.** $\sqrt{50}$ **d.** -6
18. a. 6.2832 **b.** -3.1416 **c.** 30 **d.** $3\sqrt{5}$

19. a. 6.8068 **b.** $-.7854$ **c.** 150 **d.** 9.4247

Find the reference angles for the angles given in Problems 20–27. Use the unit of measurement (degree or radians) given in the problem.

20. a. 150° **b.** 210° **c.** 240° **d.** 330°
21. a. 60° **b.** 120° **c.** 300° **d.** 135°
22. a. $\frac{5\pi}{3}$ **b.** $\frac{7\pi}{6}$ **c.** $\frac{4\pi}{3}$ **d.** $\frac{5\pi}{4}$
23. a. $\frac{11\pi}{12}$ **b.** $\frac{2\pi}{3}$ **c.** $\frac{11\pi}{6}$ **d.** $\frac{\pi}{4}$
24. a. $-30°$ **b.** $-200°$ **c.** $-55°$ **d.** $-320°$
25. a. $-\frac{\pi}{4}$ **b.** $-\pi$ **c.** $-\frac{13\pi}{6}$ **d.** $-\frac{5\pi}{3}$
26. a. 7 **b.** 9 **c.** -5 **d.** -6
27. a. $\sqrt{50}$ **b.** $3\sqrt{5}$ **c.** 6.8068 **d.** $-.7854$

B *Change the angles in Problems 28–39 to decimal degrees correct to the nearest hundredth.*

28. $\frac{2\pi}{9}$ **29.** $\frac{\pi}{10}$ **30.** $\frac{\pi}{30}$ **31.** $\frac{5\pi}{3}$
32. $-\frac{11\pi}{12}$ **33.** $\frac{3\pi}{18}$ **34.** 2 **35.** -3
36. $-.25$ **37.** -2.5 **38.** .4 **39.** .51

Change the angles in Problems 40–45 to radians using exact values.

40. 40° **41.** 20° **42.** $-64°$ **43.** $-220°$
44. 254° **45.** 85°

Change the angles in Problems 46–51 to radians correct to the nearest hundredth.

46. 112° **47.** 314° **48.** $-62.8°$ **49.** 350°
50. $-480°$ **51.** 985°

In Problems 52–59, find the intercepted arc to the nearest hundredth if the central angle and radius are as given.

52. Angle 1, radius 1 m **53.** Angle 2.34, radius 6 cm
54. Angle 3.14, radius 10 m **55.** Angle $\frac{\pi}{3}$, radius 4 m

56. Angle $\frac{3\pi}{2}$, radius 15 cm   **57.** Angle 40°, radius 7 ft

58. Angle 72°, radius 10 ft   **59.** Angle 112°, radius 7.2 cm

60. How far does the tip of an hour hand on a clock move in 3 hr if the hour hand is 2.00 cm long?

61. A 50-cm pendulum on a clock swings through an angle of 100°. How far does the tip travel in one arc?

C

62. SURVEYING In about 230 B.C., a mathematician named Eratosthenes estimated the radius of the earth using the following information: Syene and Alexandria in Egypt are on the same line of longitude. They are also 800 km apart. At noon on the longest day of the year, when the sun was directly overhead in Syene, Eratosthenes measured the sun to be 7.2° from the vertical in Alexandria. Because of the distance of the earth from the sun, he assumed that the rays were parallel. Thus he concluded that the arc from Syene to Alexandria is subtended by a central angle of 7.2° measured at the center of the earth. Using this information, find the approximate radius of the earth.

63. GEOGRAPHY Omaha, Nebraska, is located at approximately 97° west longitude, 41° north latitude; Wichita, Kansas, is located at approximately 97° west longitude, 37° north latitude. Notice that these two cities have about the same longitude. If we know that the radius of the earth is about 6,370 kilometers (km), what is the distance between these cities to the nearest 10 km?

64. GEOGRAPHY Entebbe, Uganda, is located at approximately 33° east longitude, and Stanley Falls in Zaire is located at 25° east longitude. Both these cities lie approximately on the equator. If we know that the radius of the earth is about 6,370 km, what is the distance between the cities to the nearest 10 km?

65. ASTRONOMY Suppose it is known that the moon subtends an angle of 45.75′ at the center of the earth (see Figure 6.6). It is also known that the center of the moon is 384,417 km from the surface of the earth. What is the diameter of the moon to the nearest 10 km?

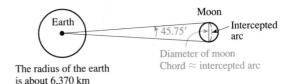

The radius of the earth is about 6,370 km

For small central angles with large radii, the intercepted arc is approximately equal to its chord.

FIGURE 6.6

66. One side of a triangle is 20 cm longer than another, and the angle between them is 60°. If two circles are drawn with these sides as diameters, one of the points of intersection of the two circles is the common vertex. How far from the third side is the other point of intersection?

6.2 Trigonometric Functions

To introduce you to the trigonometric functions, we will consider a relationship between angles and circles. Draw a unit circle with an angle θ in standard position, as in Figure 6.7.

FIGURE 6.7 The unit circle $x^2 + y^2 = 1$ with an angle θ in standard position

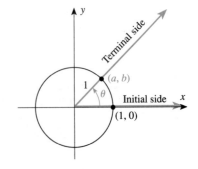

The initial side of θ intersects the unit circle at $(1, 0)$, and the terminal side intersects the unit circle at (a, b). Define functions of θ as follows:

$$c(\theta) = a \quad \text{and} \quad s(\theta) = b$$

EXAMPLE 1 Find: **a.** $s(90°)$ **b.** $c(90°)$ **c.** $s(-270°)$ **d.** $s(-\frac{\pi}{2})$ **e.** $c(-3.1416)$

SOLUTION **a.** $s(90°)$. This is the second component of the ordered pair (a, b), where (a, b) is the point of intersection of the terminal side of a $90°$ standard-position angle and the unit circle. By inspection, it is 1. Thus, $s(90°) = $ **1.**

b. $c(90°) = $ **0.**

c. $s(-270°) = $ **1.** This is the same as part **a** because the angles $90°$ and $-270°$ are coterminal.

d. $s(-\frac{\pi}{2}) = $ **−1.**

e. $c(-3.1416) \approx c(-\pi) = $ **−1.** ∎

The function $c(\theta)$ is called the **cosine function,** and the function $s(\theta)$ is called the **sine function.** These functions, along with four others, make up the **trigonometric functions** or **trigonometric ratios,** which are further examples of *transcendental functions.* One way to define the trigonometric functions is as follows.

UNIT CIRCLE DEFINITION OF THE TRIGONOMETRIC FUNCTIONS

Let θ be an angle in standard position with the pont (a, b) the intersection of the terminal side of θ and the unit circle. Then the six trigonometric functions, with their standard abbreviations, are defined as follows:

cosine: $\cos \theta = a$	**secant:** $\sec \theta = \dfrac{1}{a}$ $(a \neq 0)$
sine: $\sin \theta = b$	**cosecant:** $\csc \theta = \dfrac{1}{b}$ $(b \neq 0)$
tangent: $\tan \theta = \dfrac{b}{a}$ $(a \neq 0)$	**cotangent:** $\cot \theta = \dfrac{a}{b}$ $(b \neq 0)$

Notice the condition on the tangent, secant, cosecant, and cotangent functions. These exclude division by 0; for example, $a \neq 0$ means that θ cannot be $90°$, $270°$, or any angle coterminal to these angles. We summarize this condition by saying that *the tangent and secant are not defined for $90°$ or $270°$.* If $b \neq 0$, then $\theta \neq 0°$, $180°$, or any angle coterminal to these angles; thus, *cosecant and contangent are not defined for $0°$ or $180°$.*

The angle θ in the above definition is called the **argument** of the function. The argument, of course, does not need to be the same as the variable. For example, in $\cos(2\theta + 1)$, the argument is $2\theta + 1$ and the variable is θ.

In many applications, you will know the angle measure and will want to find one or more of its trigonometric functions. To do this you will carry out a process called **evaluation of the trigonometric functions**. In order to help you see the relationship between the angle and the function, consider Figure 6.8 and Example 2.

EXAMPLE 2 Evaluate the trigonometric functions of $\theta = 110°$ by drawing the unit circle and approximating the pont (a, b).

SOLUTION Draw a unit circle as shown in Figure 6.8.

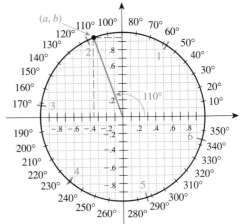

FIGURE 6.8 Approximate values of circular functions

Next draw the terminal side of the angle 110°. Estimate $a \approx -.35$ and $b \approx .95$. Thus

$$\cos 110° \approx -.35 \qquad\qquad \sec 110° \approx \frac{1}{-.35} \approx -2.9$$

$$\sin 110° \approx .95 \qquad\qquad \csc 110° \approx \frac{1}{.95} \approx 1.1$$

$$\tan 110° \approx \frac{.95}{-.35} \approx -2.7 \qquad \cot 110° \approx \frac{-.35}{.95} \approx -.37$$

Notice from Example 2 that some of the trigonometric functions of 110° are positive and others are negative. We can make certain predictions about the signs of these functions, as summarized in Table 6.1.

Table 6.1 Signs of trigonometric functions

	QUADRANT I *a* pos *b* pos	QUADRANT II *a* neg *b* pos	QUADRANT III *a* neg *b* neg	QUADRANT IV *a* pos *b* neg
$\cos \theta = a$	pos	neg	neg	pos
$\sin \theta = b$	pos	pos	neg	neg
$\tan \theta = b/a$	pos	neg	pos	neg
Summary	**All pos**	**Sine pos**	**Tangent pos**	**Cosine pos**

PROOF In Quadrant I, a and b are both positive, so all six trigonometric functions must be positive. In Quadrant II, a is negative and b is positive, so from the definition of trigonometric functions, all are negative except the sine and

Table 6.1 can be summarized by remembering the following:

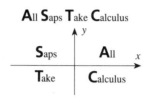

Easy-to-remember form:

All **S**aps **T**ake **C**alculus

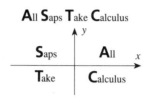

cosecant. In Quadrant III, a and b are both negative, so all the functions are negative except tangent and cotangent because those are ratios of two negatives, which are positive. In Quadrant IV, a is positive and b is negative, so all are negative except cosine and secant. ⬜⬜⬜

You may have noticed certain relationships among these functions. The first is that since the tangent is the ratio b to a, it can be found by dividing the sine by the cosine. That is,

$$\tan \theta = \frac{\sin \theta}{\cos \theta}$$

whenever $\cos \theta \neq 0$. Note also that the secant, cosecant, and cotangent functions are reciprocals of the cosine, sine, and tangent functions. This means that if values of θ that cause division by zero are excluded, then

$$\sec \theta = \frac{1}{\cos \theta} \qquad \csc \theta = \frac{1}{\sin \theta} \qquad \cot \theta = \frac{1}{\tan \theta}$$

These are called the **reciprocal** relationships, or **identities.** The term *identity* is defined and discussed in the next chapter, so for now we will just remember which pairs of functions are called reciprocals. This can be confusing because the functions are also paired according to their names: sine and cosine, secant and cosecant, and tangent and cotangent. These pairs are called **cofunctions**. Study Table 6.2 until this terminology is clear to you.

Table 6.2   Reciprocal and cofunction relationships

RECIPROCALS	COFUNCTIONS
cosine and secant	cosine and sine
sine and cosecant	cosecant and secant
tangent and cotangent	cotangent and tangent

Since the method shown in Example 2 for evaluating the trigonometric functions is not very practical, many additional procedures for finding them have been discovered. The most common method today, however, is to use a calculator. The procedure is straightforward, but you must note several details:

1. Note the unit of measure used: degree or radian. Calculators have a variety of ways of changing from radian to degree format, so consult your owner's manual to find out how your particular calculator does it. Most, however, simply have a switch (similar to an on/off switch) that sets the calculator in either degree or radian mode. From now on, we will assume that you are working in the appropriate radian/degree mode. *Remember:*   If later in the course you suddenly start obtaining strange answers and have no idea what you are doing wrong, double-check to make sure you are using the proper mode.

2. With most scientific calculators: ⬚Angle⬚ ⬚Trig function⬚
   With most graphing calculators: ⬚Trig function⬚ ⬚Angle⬚

3. You must remember which functions are reciprocals since a normal scientific calculator does not have sec θ, csc θ, and cot θ keys. It does, however, have a reciprocal key, which is labeled

$$\boxed{1/x} \quad \text{or} \quad \boxed{x^{-1}}$$

EXAMPLE 3 Find the trigonometric functions of 110° by calculator.

SOLUTION
$$\cos 110° \approx -.34202014$$
$$\sin 110° \approx .93969262$$
$$\tan 110° \approx -2.74747742$$
$$\sec 110° \approx -2.9238044$$
$$\csc 110° \approx 1.06417777$$
$$\cot 110° \approx -.36397023$$

EXAMPLE 4 Find csc(π/12) by calculator.

SOLUTION The answer displayed is **3.8637033.**

EXAMPLE 5 Find tan 70°23′40″.

SOLUTION Switch key to degrees; this measure must first be changed to a decimal degree measure:

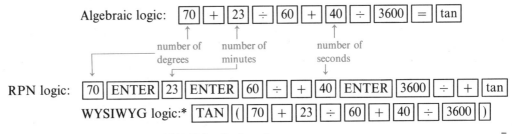

Algebraic logic: $\boxed{70}\ \boxed{+}\ \boxed{23}\ \boxed{\div}\ \boxed{60}\ \boxed{+}\ \boxed{40}\ \boxed{\div}\ \boxed{3600}\ \boxed{=}\ \boxed{\tan}$

number of degrees   number of minutes   number of seconds

RPN logic: $\boxed{70}\ \boxed{\text{ENTER}}\ \boxed{23}\ \boxed{\text{ENTER}}\ \boxed{60}\ \boxed{\div}\ \boxed{+}\ \boxed{40}\ \boxed{\text{ENTER}}\ \boxed{3600}\ \boxed{\div}\ \boxed{+}\ \boxed{\tan}$

WYSIWYG logic:* $\boxed{\text{TAN}}\ \boxed{(}\ \boxed{70}\ \boxed{+}\ \boxed{23}\ \boxed{\div}\ \boxed{60}\ \boxed{+}\ \boxed{40}\ \boxed{\div}\ \boxed{3600}\ \boxed{)}$

The answer **2.807464818** is displayed.

Suppose you want to find the trigonometric functions of an angle whose terminal side passes through some known point, say (3, 4). To apply the definition, you need to find the point (a, b), as shown in Figure 6.9.

Let (3, 4) be denoted by P and (a, b) by A. Let B be the point $(a, 0)$ and Q be the point (3, 0): ($OA = 1$, $OP = \sqrt{3^2 + 4^2} = 5$). Now consider $\triangle AOB$ and $\triangle POQ$. Recall from geometry that two triangles are **similar** if two angles of one are congruent to two angles of the other. For these triangles, $\angle OBA$ is congruent to $\angle OQP$ because they are both right angles; also $\angle O$ is congruent to $\angle O$ because equal angles are congruent. Thus these triangles are similar, which is denoted by

$$\triangle AOB \sim \triangle POQ$$

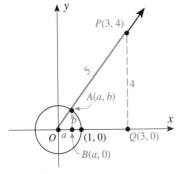

FIGURE 6.9 Finding (a, b) when a point on the terminal side is known

The important property of similar triangles is that corresponding parts of similar

---

*This is the logic used on most graphing calculators.

triangles are proportional. Thus

$$a = \frac{a}{1} = \frac{3}{5} \quad \text{and} \quad b = \frac{b}{1} = \frac{4}{5}$$

Thus $\cos \theta = \frac{3}{5}$, $\sin \theta = \frac{4}{5}$, and $\tan \theta = \frac{4/5}{3/5} = \frac{4}{3}$. The reciprocals are $\sec \theta = \frac{5}{3}$, $\csc \theta = \frac{5}{4}$, and $\cot \theta = \frac{3}{4}$.

If you carry out these steps for a point $P(x, y)$ instead of $(3, 4)$, as shown in Figure 6.10, it is still true that

$$\triangle AOB \sim \triangle POQ$$

Let r be the distance from O to P. That is, let

$$r = \sqrt{x^2 + y^2}$$

Then

$$a = \frac{a}{1} = \frac{x}{r} \qquad \frac{1}{a} = \frac{r}{x}$$

$$b = \frac{b}{1} = \frac{y}{r} \qquad \frac{1}{b} = \frac{r}{y}$$

$$\frac{b}{a} = \frac{y/r}{x/r} = \frac{y}{x} \qquad \frac{a}{b} = \frac{x}{y}$$

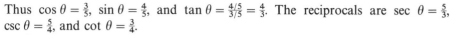

FIGURE 6.10

These ratios lead to an alternative definition of the trigonometric functions that allows you to choose *any* point (x, y). In practice, it is this definition that is most frequently used.

RATIO DEFINITION OF THE TRIGONOMETRIC FUNCTIONS

Let θ be an angle in standard position with any point $P(x, y)$ on the terminal side a distance r from the origin ($r \neq 0$). Then the six trigonometric functions are defined as follows:

$$\cos \theta = \frac{x}{r} \qquad\qquad \sec \theta = \frac{r}{x} \quad (x \neq 0)$$

$$\sin \theta = \frac{y}{r} \qquad\qquad \csc \theta = \frac{r}{y} \quad (y \neq 0)$$

$$\tan \theta = \frac{y}{x} \quad (x \neq 0) \qquad \cot \theta = \frac{x}{y} \quad (y \neq 0)$$

EXAMPLE 6 Find the values of the six trigonometric functions for an angle θ in standard position with the terminal side passing through $(-5, -2)$.

SOLUTION $x = -5$, $y = -2$, and $r = \sqrt{25 + 4} = \sqrt{29}$. Thus

$$\cos \theta = \frac{-5}{\sqrt{29}} = \frac{-5}{29}\sqrt{29} \qquad \sec \theta = \frac{-\sqrt{29}}{5}$$

$$\sin \theta = \frac{-2}{\sqrt{29}} = \frac{-2}{29}\sqrt{29} \qquad \csc \theta = \frac{-\sqrt{29}}{2}$$

$$\tan \theta = \frac{-2/\sqrt{29}}{-5/\sqrt{29}} = \frac{2}{5} \qquad \cot \theta = \frac{5}{2}$$

Both the unit circle and ratio definitions of the trigonometric functions were given using angle domains. We can extend the definition to include real number domains since radian measure is in terms of real numbers. That is, for

$$\cos \frac{\pi}{2} \quad \text{or} \quad \cos 2$$

it does not matter whether $\frac{\pi}{2}$ and 2 are considered as radian measures of angles or simply as real numbers—the functional values are the same.

TRIGONOMETRIC FUNCTIONS OF REAL NUMBERS	For any real number x, let $x = \theta$, where θ is a standard-position angle measured in radians. Then

$$\cos x = \cos \theta \qquad \sin x = \sin \theta \qquad \tan x = \tan \theta$$

$$\sec x = \sec \theta \qquad \csc x = \csc \theta \qquad \cot x = \cot \theta$$

6.2 Problem Set

A *From the unit circle definition of the trigonometric functions, estimate to one decimal place the numbers in Problems 1–8.*

1. $\cos 50°$ **2.** $\sin 20°$ **3.** $\sin 320°$

4. $\tan 80°$ **5.** $\tan(-20°)$ **6.** $\cos(-340°)$

7. $\sec 70°$ **8.** $\csc 190°$

Use a calculator to evaluate the functions given in Problems 9–27.

9. a. $\cos 50°$ **b.** $\sin 20°$

10. a. $\sec 70°$ **b.** $\csc 150°$

11. a. $\cot 250°$ **b.** $\sec 135°$

12. a. $\cos(-34°)$ **b.** $\sin(-95°)$

13. a. $-\cot(-18°)$ **b.** $-\sec(-213°)$

14. a. $\tan 56.2°$ **b.** $\cot 78.4°$

15. a. $\sin 1$ **b.** $\sin 1°$

16. a. $\tan 15$ **b.** $\tan 15°$

17. a. $\cot 2.5$ **b.** $-\sec 1.5$

18. a. $\cos(-.48)$ **b.** $\sec(-21.3°)$

19. a. $\tan 129°9'12''$ **b.** $\cos 240°8''$

20. a. $\dfrac{3}{5 \sin 2}$ **b.** $\frac{3}{5} \sin 2$

21. a. $\frac{3}{5} \sin \frac{1}{2}$ **b.** $\dfrac{3 \sin 2}{5}$

22. a. $\dfrac{1}{2 \sec 3}$ **b.** $\frac{1}{2} \sec 3$

23. a. $2 \sec \frac{1}{3}$ **b.** $\dfrac{2}{\sec \frac{1}{3}}$

24. a. $(-3) \cos 2$ **b.** $3 \cos(-2)$

25. a. $3 \cos(-\frac{1}{2})$ **b.** $\dfrac{-3}{\cos \frac{1}{2}}$

26. a. $(-3) \cot 2$ **b.** $3 \cot(-2)$

27. a. $3 \cot(-\frac{1}{2})$ **b.** $\dfrac{-3}{\cot \frac{1}{2}}$

Tell whether each of the functions in Problems 28–39 is positive or negative. You should be able to do this without tables or a calculator.

28. sine, Quadrant I **29.** cosine, Quadrant I

30. tangent, Quadrant II **31.** secant, Quadrant II

32. cosecant, Quadrant III **33.** cotangent, Quadrant IV

34. $\sin 1$ **35.** $\cos 2$ **36.** $\tan 3$

37. $\sec 4$ **38.** $\sin(-1)$ **39.** $\cos(-2)$

Tell in which quadrant(s) a standard-position angle θ could lie if the conditions in Problems 40–45 are true.

40. $\sin \theta > 0$ **41.** $\cos \theta < 0$ **42.** $\tan \theta < 0$

43. $\sin \theta < 0$ and $\tan \theta > 0$

44. $\sin \theta > 0$ and $\tan \theta < 0$

45. $\cos \theta < 0$ and $\sin \theta < 0$

B *Find the values of the six trigonometric functions for an angle θ in standard position with terminal side passing through the points given in Problems 46–54. Draw a picture showing θ and the reference angle θ'.*

46. $(3, 4)$
47. $(-3, 4)$
48. $(-3, -4)$
49. $(5, 12)$
50. $(-5, -12)$
51. $(5, -12)$
52. $(2, -5)$
53. $(-6, 1)$
54. $(-4, -5)$

In the next section you will need to simplify some radical expressions. Recall that $\sqrt{x^2} = |x|$. This means that $\sqrt{x^2} = x$ if x is positive and $\sqrt{x^2} = -x$ if x is negative. Simplify the expressions in Problems 55–64.

55. $\sqrt{2x^2}$
56. $\sqrt{9x^2}$
57. $\sqrt{2x^2}$ if x is negative
58. $\sqrt{2x^2}$ if x is positive
59. $\sqrt{9x^2}$ if x is positive
60. $\sqrt{9x^2}$ if x is negative
61. $\dfrac{x}{\sqrt{4x^2}}$ if x is positive
62. $\dfrac{x}{\sqrt{4x^2}}$ if x is negative
63. $\dfrac{\sqrt{16x^2}}{x}$ if x is negative
64. $\dfrac{\sqrt{16x^2}}{x}$ if x is positive

C

65. **a.** Let $P(x, y)$ be any point in the plane. Show that $P(r \cos \theta, r \sin \theta)$ is a representation for P, where θ is the standard-position angle formed by drawing ray $\overrightarrow{OP}$.
    b. Let $A(\cos \alpha, \sin \alpha)$ and $B(\cos \beta, \sin \beta)$ be any two points on a unit circle. Use the distance formula to show that

$$|AB| = \sqrt{2 - 2(\cos \alpha \cos \beta + \sin \alpha \sin \beta)}$$

66. You will learn in calculus that

$$\sin x = x - \frac{x^3}{3!} + \frac{x^5}{5!} - \frac{x^7}{7!} + \cdots$$

where $n! = n(n - 1)(n - 2) \cdots 3 \cdot 2 \cdot 1$. Find sin 1 correct to four decimal places by using this equation. (Remember that the 1 in sin 1 refers to radian measure.)

67. You will learn in calculus that

$$\cos x = 1 - \frac{x^2}{2!} + \frac{x^4}{4!} - \frac{x^6}{6!} + \cdots$$

Use this equation to find cos 1 correct to four decimal places. $(n! = n(n - 1)(n - 2) \cdots 3 \cdot 2 \cdot 1)$

6.3 Values of the Trigonometric Functions

If an angle has a terminal side that coincides with one of the coordinate axes, it is easy to evaluate the trigonometric functions by using the unit circle definition.

EXAMPLE 1 Evelute the trigonometric functions for $-\frac{5\pi}{2}$.

SOLUTION Since $-\frac{5\pi}{2}$ has a terminal side coinciding with the negative y-axis, the intersection of this terminal side and the unit circle is $(0, -1)$. This means that $a = 0$ and $b = -1$. Hence

$$\cos\left(\frac{-5\pi}{2}\right) = a = \mathbf{0} \qquad \sin\left(\frac{-5\pi}{2}\right) = b = \mathbf{-1}$$

$$\csc\left(\frac{-5\pi}{2}\right) = \frac{1}{b} = \mathbf{-1} \qquad \cot\left(\frac{-5\pi}{2}\right) = \frac{a}{b} = \mathbf{0}$$

$$\tan\left(\frac{-5\pi}{2}\right) = \frac{b}{a} \quad \text{and} \sec\left(\frac{-5\pi}{2}\right) = \frac{1}{a} \quad \text{are } \mathbf{undefined} \text{ since } a = 0. \qquad \blacksquare$$

There are times when you will not be able to rely on calculator approximations for the trigonometric functions but instead will need to find **exact values.**

EXAMPLE 2 Evaluate the trigonometric functions for $\frac{\pi}{4}$.

SOLUTION
$$\cos\frac{\pi}{4} = \frac{x}{r} \qquad \text{From the ratio definition (this is true for any angle)}$$

$$= \frac{x}{\sqrt{x^2 + y^2}} \qquad r = \sqrt{x^2 + y^2} \text{ for any angle}$$

If $\theta = \frac{\pi}{4}$, then $x = y$ since $\frac{\pi}{4}$ bisects Quadrant I. By substitution,

$$\cos\frac{\pi}{4} = \frac{x}{\sqrt{x^2 + x^2}}$$

$$= \frac{x}{\sqrt{2x^2}}$$

$$= \frac{x}{x\sqrt{2}} \qquad \sqrt{x^2} = |x| = x \text{ since } x \text{ is positive in Quadrant I}$$

$$= \frac{1}{\sqrt{2}}$$

$$= \frac{1}{2}\sqrt{2} \qquad \frac{1}{\sqrt{2}} = \frac{1}{\sqrt{2}} \cdot \frac{\sqrt{2}}{\sqrt{2}} = \frac{\sqrt{2}}{2} = \frac{1}{2}\sqrt{2}$$

Similarly, $\sin\frac{\pi}{4} = \frac{\sqrt{2}}{2}$, $\tan\frac{\pi}{4} = 1$, $\sec\frac{\pi}{4} = \sqrt{2}$, $\csc\frac{\pi}{4} = \sqrt{2}$, and $\cot\frac{\pi}{4} = 1$.

EXAMPLE 3 Find the exact values for the trigonometric functions of $30°$.

SOLUTION Consider not only the standard-position angle $30°$ but also the standard-position angle $-30°$. Choose $P_1(x, y)$ and $P_2(x, -y)$, respectively, on the terminal sides. (See Figure 6.11.)

Angles OQP_1 and OQP_2 are right angles, so $\angle OP_1Q = 60°$ and $\angle OP_2Q = 60°$. Thus $\triangle OP_1P_2$ is an equiangular triangle. (All angles measure $60°$.) From geometry we know that an equiangular triangle has sides the same length. Thus $2y = r$. Notice the following relationship between x and y:

$$r^2 = x^2 + y^2$$
$$(2y)^2 = x^2 + y^2 \qquad 2y = r$$
$$3y^2 = x^2$$
$$\sqrt{3}\,|y| = |x|$$

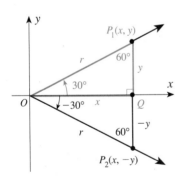

FIGURE 6.11

For $30°$, x and y are both positive, so $x = \sqrt{3}y$.

$$\cos 30° = \frac{x}{r} = \frac{\sqrt{3}y}{2y} = \frac{\sqrt{3}}{2} \qquad\qquad \sec 30° = \frac{2}{\sqrt{3}} = \frac{2}{3}\sqrt{3}$$

$$\sin 30° = \frac{y}{r} = \frac{y}{2y} = \frac{1}{2} \qquad\qquad \csc 30° = 2$$

$$\tan 30° = \frac{y}{x} = \frac{y}{\sqrt{3}y} = \frac{1}{\sqrt{3}} = \frac{\sqrt{3}}{3} \qquad\qquad \cot 30° = \sqrt{3}$$

The derivation in Example 3 leads to the following result from plane geometry.

30°–60°–90° TRIANGLE THEOREM

In a 30°–60°–90° triangle, the leg opposite the 30° angle equals one-half the hypotenuse and the leg opposite the 60° angle equals one-half the hypotenuse times the square root of 3.

EXAMPLE 4 Evaluate the trigonometric functions for 60°.

SOLUTION Using the 30°–60°–90° triangle theorem, the hypotenuse r is twice the length of the shorter leg, as shown in Figure 6.12. Thus $r = 2x$ and $y = \sqrt{3}x$. Then

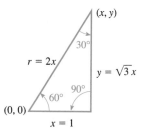

$$\cos 60° = \frac{x}{2x} = \frac{1}{2} \qquad \sec 60° = 2$$

$$\sin 60° = \frac{\sqrt{3}x}{2x} = \frac{1}{2}\sqrt{3} \qquad \csc 60° = \frac{2}{\sqrt{3}} = \frac{2}{3}\sqrt{3}$$

$$\tan 60° = \frac{\frac{1}{2}\sqrt{3}}{\frac{1}{2}} = \sqrt{3} \qquad \cot 60° = \frac{1}{\sqrt{3}} = \frac{1}{3}\sqrt{3}$$

FIGURE 6.12 30°–60°–90° triangle

In a manner similar to Examples 1–4, a **table of exact values** is constructed (see Table 6.3). Since this table of values is used extensively, you should memorize at least the values for $\cos\theta$, $\sin\theta$, and $\tan\theta$ as you did multiplication tables in elementary school.

Table 6.3 Exact values

FUNCTION	$0 = 0°$	$\frac{\pi}{6} = 30°$	$\frac{\pi}{4} = 45°$	$\frac{\pi}{3} = 60°$	$\frac{\pi}{2} = 90°$	$\pi = 180°$	$\frac{3\pi}{2} = 270°$
			ANGLE θ				
$\cos\theta$	1	$\frac{\sqrt{3}}{2}$	$\frac{\sqrt{2}}{2}$	$\frac{1}{2}$	0	-1	0
$\sin\theta$	0	$\frac{1}{2}$	$\frac{\sqrt{2}}{2}$	$\frac{\sqrt{3}}{2}$	1	0	-1
$\tan\theta$	0	$\frac{\sqrt{3}}{3}$	1	$\sqrt{3}$	undef.	0	undef.
$\sec\theta$	1	$\frac{2}{\sqrt{3}} = \frac{2}{3}\sqrt{3}$	$\frac{2}{\sqrt{2}} = \sqrt{2}$	$\frac{2}{1} = 2$	undef.	$\frac{1}{-1} = -1$	undef.
$\csc\theta$	undef.	$\frac{2}{1} = 2$	$\frac{2}{\sqrt{2}} = \sqrt{2}$	$\frac{2}{\sqrt{3}} = \frac{2}{3}\sqrt{3}$	1	undef.	-1
$\cot\theta$	undef.	$\frac{3}{\sqrt{3}} = \sqrt{3}$	1	$\frac{1}{\sqrt{3}} = \frac{\sqrt{3}}{3}$	0	undef.	0

These are the reciprocals (which is why the exact values are given in reciprocal form as well as in simplified form). In a problem you would use the rationalized form. The reciprocal form makes it easy to remember them.

The values for sec θ, csc θ, and cot θ do not need to be memorized as separate entries because they are simply the reciprocals of cos θ, sin θ, and tan θ.

You can find exact values of the trigonometric functions that are multiples of those in Table 6.3 by using the idea of a reference angle and the **reduction principle**:

REDUCTION PRINCIPLE

If t represents any of the six trigonometric functions, then

$$t(\theta) = \pm t(\theta')$$

where θ' is the reference angle of θ and the sign plus or minus depends on the quadrant of the terminal side of the angle θ.

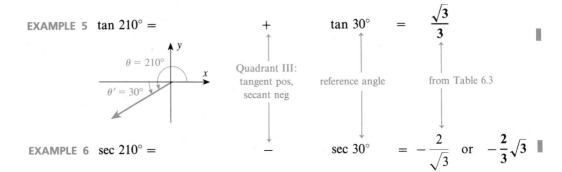

EXAMPLE 5 $\tan 210° = \qquad + \qquad \tan 30° = \dfrac{\sqrt{3}}{3}$

Quadrant III:
tangent pos,
secant neg

reference angle

from Table 6.3

EXAMPLE 6 $\sec 210° = \qquad - \qquad \sec 30° = -\dfrac{2}{\sqrt{3}} \quad \text{or} \quad -\dfrac{2}{3}\sqrt{3}$

EXAMPLE 7 $\csc \dfrac{3\pi}{2} = -\mathbf{1}$

If it is a quadrantal angle, then it comes directly from the memorized table.

EXAMPLE 8 $\cos 405° = \cos 45° = \tfrac{1}{2}\sqrt{2}$

Quadrant I:
cosine positive

Sketch the angle if necessary to find the quadrant and the reference angle.

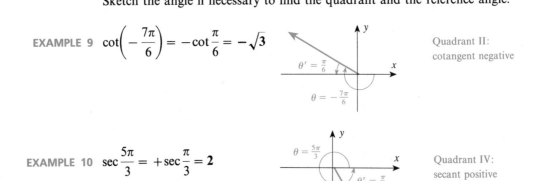

EXAMPLE 9 $\cot\left(-\dfrac{7\pi}{6}\right) = -\cot\dfrac{\pi}{6} = -\sqrt{3}$

Quadrant II:
cotangent negative

EXAMPLE 10 $\sec \dfrac{5\pi}{3} = +\sec\dfrac{\pi}{3} = \mathbf{2}$

Quadrant IV:
secant positive

A *In Problems* 1–10 *give the exact values in simplified form.*

1. a. $\tan \frac{\pi}{4}$ **b.** $\cos 0$
    c. $\sin 60°$ **d.** $\cos 30°$

2. a. $\cos 270°$ **b.** $\tan \frac{\pi}{6}$
    c. $\tan 180°$ **d.** $\sin 45°$

3. a. $\sin \pi$ **b.** $\sin \frac{\pi}{2}$
    c. $\tan 0$ **d.** $\cos \frac{\pi}{4}$

4. a. $\sec \frac{\pi}{6}$ **b.** $\csc 0$
    c. $\sec \frac{\pi}{4}$ **d.** $\sec 0°$

5. a. $\csc \frac{\pi}{4}$ **b.** $\cot \pi$
    c. $\sec \frac{\pi}{3}$ **d.** $\cot \frac{\pi}{6}$

6. a. $\tan 90°$ **b.** $\tan 60°$
    c. $\cos \frac{\pi}{3}$ **d.** $\sec \frac{\pi}{3}$

7. a. $\cot 45°$ **b.** $\cos \pi$
    c. $\sin \frac{3\pi}{2}$ **d.** $\sin 0°$

8. a. $\sec \pi$ **b.** $\tan 270°$
    c. $\sin \frac{\pi}{6}$ **d.** $\csc \frac{3\pi}{2}$

9. a. $\cos(-300°)$ **b.** $\sin 390°$
    c. $\sin \frac{17\pi}{4}$ **d.** $\cos(-6\pi)$

10. a. $\cos \frac{9\pi}{2}$ **b.** $\sin(-765°)$
    c. $\tan(-765°)$ **d.** $\cos 495°$

Use a calculator to evaluate the functions in Problems 11–26. *Round your answers to four decimal places.*

11. a. $\sin 34.4°$ **b.** $\cos 54.2°$

12. a. $\tan 70.2°$ **b.** $\cot 46.7°$

13. a. $\cos 50°$ **b.** $\cot 80°$

14. a. $\sin 70°$ **b.** $\tan 20°$

15. a. $\tan(-20°)$ **b.** $\sin 190°$

16. a. $\cos(-340°)$ **b.** $\cot(-213°)$

17. a. $\sin 132.8°$ **b.** $\tan(-25.6°)$

18. a. $\cot(-125.6°)$ **b.** $\cos 163.4°$

19. a. $\sin 1.20$ **b.** $\cos .65$

20. a. $\tan .51$ **b.** $\cot 1.85$

21. a. $\tan 1$ **b.** $\cot 1.5$

22. a. $\sin .8$ **b.** $\cos .5$

23. a. $\tan 2.5$ **b.** $\sin 3$

24. a. $\cos 4.5$ **b.** $\cot 6$

25. a. $\cos(-.45)$ **b.** $\tan(-2.8)$

26. a. $\sin(-3.9)$ **b.** $\cot 10$

B

27. Verify the entries in Table 6.3 for the angle $\frac{\pi}{3}$.

28. Find $\cos \frac{3\pi}{4}$ by using the procedure illustrated in Example 2.

29. Find $\cos \frac{5\pi}{4}$ by using the procedure illustrated in Example 2.

30. Find $\cos 135°$ by choosing an arbitrary point (x, y) on the terminal side of $135°$ and applying the ratio definition of the trigonometric functions.

31. Find $\sin(-\frac{\pi}{4})$ by choosing an arbitrary point (x, y) on the terminal side of $-\frac{\pi}{4}$ and applying the ratio definition of the trigonometric functions.

32. Find $\sin 210°$ by choosing an arbitrary point (x, y) on the terminal side of $210°$ and applying the ratio definition of the trigonometric functions.

33. Find $\cos 210°$ by choosing an arbitrary point (x, y) on the terminal side of $210°$ and applying the ratio definition of the trigonometric functions.

Substitute the exact values for the trigonometric functions in the expressions in Problems 34–59 *and simplify. When a trigonometric function is raised to a power, such as* $(\sin x)^2$, *it is written as* $\sin^2 x$.

34. $\sin 30° + \cos 0°$ **35.** $\sin \frac{\pi}{2} + 3 \cos \frac{\pi}{2}$

36. $2 \cos \frac{\pi}{2}$ **37.** $\cos \frac{2\pi}{2}$

38. $\sin \frac{2\pi}{4}$ **39.** $2 \sin \frac{\pi}{4}$

40. $\sin^2 60°$ **41.** $\cos^2 \frac{\pi}{4}$

42. $\sin^2 \frac{\pi}{6} + \cos^2 \frac{\pi}{2}$ **43.** $\sin^2 \frac{\pi}{2} + \cos^2 \frac{\pi}{2}$

44. $\sin^2 \frac{\pi}{3} + \cos^2 \frac{\pi}{3}$ **45.** $\sin^2 \frac{\pi}{6} + \cos^2 \frac{\pi}{3}$

46. $\sin \frac{\pi}{6} \csc \frac{\pi}{6}$ **47.** $\csc \frac{\pi}{2} \sin \frac{\pi}{2}$

48. $\cos(\frac{\pi}{4} - \frac{\pi}{2})$ **49.** $\cos \frac{\pi}{4} - \cos \frac{\pi}{2}$

50. $\tan(2 \cdot 30°)$ **51.** $2 \tan 30°$

52. $\csc(\frac{1}{2} \cdot 60°)$ **53.** $\dfrac{\csc 60°}{2}$

54. $\cos(\frac{1}{2} \cdot 60°)$ **55.** $\sqrt{\dfrac{1 + \cos 60°}{2}}$

56. $\tan(2 \cdot 60°)$ **57.** $\dfrac{2 \tan 60°}{1 - \tan^2 60°}$

58. $\cos(\frac{\pi}{2} - \frac{\pi}{6})$ **59.** $\cos \frac{\pi}{2} \cos \frac{\pi}{6} + \sin \frac{\pi}{2} \sin \frac{\pi}{6}$

C

60. What is the smaller angle between the hands of a clock at 12:25 P.M.?

61. a. If θ is in Quadrant I, then $\theta + \pi$ is in Quadrant III with a reference angle θ. Use this fact and the reduction principle to show that $\sin(\theta + \pi) = -\sin \theta$ if θ is in Quadrant I.

b. Show that $\sin(\theta + \pi) = -\sin \theta$ if θ is in Quadrant II.
c. Show that $\sin(\theta + \pi) = -\sin \theta$ if θ is in Quadrant III.
d. Show that $\sin(\theta + \pi) = -\sin \theta$ if θ is in Quadrant IV.
e. By considering parts **a–d**, show that $\sin(\theta + \pi) = -\sin \theta$ for any angle θ.

62. Show that $\cos(\theta + \pi) = -\cos \theta$ for any angle θ. (*Hint:* See Problem 61.)

63. COMPUTER If you have access to a computer, write a program that will output a table of trigonometric values for the sine, cosine, and tangent for every degree from $0°$ to $45°$.

6.4 Graphs of the Trigonometric Functions

As with the polynomial and rational functions, we are interested in the graphs of the trigonometric functions. We will first determine the general shape of the trigonometric functions by plotting points and then generalize so we can graph the functions without too many calculations concerning points.

Graph of Standard Sine Function

To graph $y = \sin x$, begin by plotting familiar values for the sine:

$x = $ REAL NUMBER	0	$\frac{\pi}{2}$	π	$\frac{3\pi}{2}$	2π
$y = \sin x$	0	1	0	-1	1

We are using exact values here, but you could also use a calculator to generate other values. The hardest part of graphing a sine function is deciding on the scales to use for the x- and y-axes. If you are graphing with a calculator or computer, you will need to specify a "frame" of values. If you are graphing with a pencil and paper, you may find it convenient to choose 12 intervals on the x-axis for π units, and 10 intervals on the y-axis for 1 unit. You can then plot additional values by using the reduction principle. Continue to include x in Quadrants II, III, and IV and plot the points (x, y) as shown in Figure 6.13. The smooth curve that connects these points is called the **sine curve**.

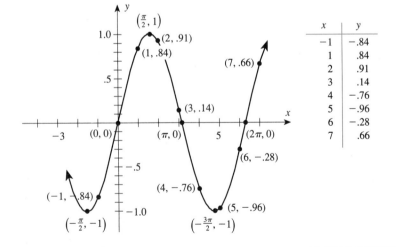

FIGURE 6.13 Graph of $y = \sin x$ by plotting points

x	y
-1	$-.84$
1	$.84$
2	$.91$
3	$.14$
4	$-.76$
5	$-.96$
6	$-.28$
7	$.66$

The domain for x is all real numbers, so what about values other than $0 \leq x \leq 2\pi$? Using the reduction principle, we known that

$$\sin x = \sin(x + 2\pi) = \sin(x - 2\pi) = \sin(x + 4\pi) = \sin(x + 6\pi) = \cdots$$

More generally,

$$\sin(\theta + 2n\pi) = \sin \theta$$

for any integer n. In other words, the values of the sine function repeat themselves after 2π. We describe this by saying that sine is **periodic** with period 2π. The sine curve is shown in Figure 6.14. Notice that, even though the domain of the sine function is all real numbers, the range is restricted to values between -1 and 1 (inclusive).

Graph of $y = \sin x$

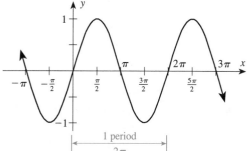

Framing a Sine Curve

Notice that, for the period labeled in Figure 6.14, the sine curve starts at $(0, 0)$, goes *up* to $(\frac{\pi}{2}, 1)$ and then *down* to $(\frac{3\pi}{2}, -1)$ passing through $(\pi, 0)$, and then goes back up to $(2\pi, 0)$, which completes one period. This graph shows that the range of the sine function is $-1 \leq y \leq 1$. We can summarize a technique for sketching the sine curve (see Figure 6.15) called **framing the curve.**

FRAMING A SINE CURVE

The standard sine function

$$y = \sin x$$

has domain $-\infty < x < \infty$ and range $-1 \leq y \leq 1$, and is periodic with period 2π. One period of this curve can be sketched by framing, as follows:

1. *Start* at the origin $(0, 0)$.
2. *Height* of this frame is two units: one unit up and one unit down from the starting point ($-1 \leq y \leq 1$).
3. *Length* of this frame is 2π units (about 6.28) from the starting point (the period is 2π).
4. The curve is now framed. Plot five critical points within the frame:
   a. Endpoints (along axis)
   b. Midpoint (along axis)
   c. Quarterpoints (up first, then down)
5. Draw the curve through the critical points, remembering the shape of the sine curve.

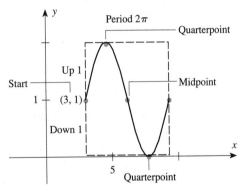

Step 3: Length of frame is 2π.

Step 4c: Quarterpoint of interval is midpoint of this interval.

Step 4: The curve has been "framed"; now plot five critical points.

Step 4b: Plot midpoint from Step 1.

Step 2: Height of frame

Frame is up one unit.

Step 1:

$(0, 0)$

$(2\pi, 0)$

Step 4a

Frame is down one unit.

Step 5: Draw the curve passing through the endpoint, midpoint, and quarterpoints.

Step 4c: Quarterpoint

FIGURE 6.15 Procedure for framing the sine curve

You can draw a standard sine curve using any point as a starting point; the procedure is always the same.

EXAMPLE 1 Draw one period of a standard sine curve using (3, 1) as the starting point for building a frame.

SOLUTION Plot the point (3, 1) as the starting point. Draw the standard frame as shown at the right.

Period 2π

Quarterpoint

Start

Up 1

(3, 1)

Midpoint

Down 1

5

Quarterpoint

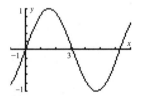

Framing a Cosine Curve

We can graph the cosine curve by plotting points, as we did with the sine curve. The details of plotting these points are left as an exercise. The cosine curve $y = \cos x$ is "framed" as described in Figure 6.16.

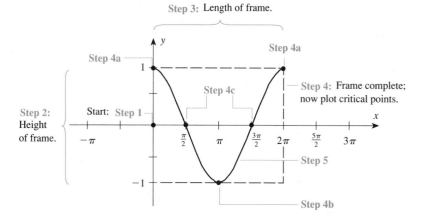

FIGURE 6.16 Procedure for framing the cosine curve

Notice that the only difference in the steps for graphing the cosine and the sine curves is in Step 4; the procedures for building the frame are identical.

FRAMING A COSINE CURVE

The standard cosine function

$$y = \cos x$$

has domain $-\infty < x < \infty$ and range $-1 \le y \le 1$, and is periodic with period 2π. One period of this curve can be sketched by framing, as follows:

1. *Start* at the origin $(0, 0)$.
2. *Height* of the frame is two units: one unit up and one unit down from the starting point ($-1 \le y \le 1$).
3. *Length* of the frame is 2π units (about 6.28) from the starting point (the period is 2π).
4. The curve is now framed. Plot five critical points within the frame:
   a. Endpoints (at the top corners of the frame)
   b. Midpoint (at the bottom of the frame)
   c. Quarterpoints (along the axis)
5. Draw the curve through the critical points, remembering the shape of the cosine curve.

Since values for x greater than 2π or less than 0 are coterminal with those already considered, we see that **the period of the cosine function is 2π.** The cosine curve is shown in Figure 6.17. Notice that the domain and range of the cosine function are the same as they were for the sine function; D: all reals; R: $-1 \le \cos x \le 1$.

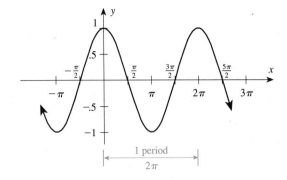

FIGURE 6.17   Graph of $y = \cos x$;
domain: all real x; range: $-1 \le y \le 1$

Framing a Tangent Curve

By setting up a table of values and plotting points (the details are left as an exercise), we notice that $y = \tan x$ does not exist at $\frac{\pi}{2}, \frac{3\pi}{2}$, or $\frac{\pi}{2} + n\pi$ for any integer n. The lines $x = \frac{\pi}{2}, x = \frac{3\pi}{2}, \ldots, x = \frac{\pi}{2} + n\pi$ for which the tangent is not defined are vertical asymptotes. The procedure is summarized in Figure 6.18.

FRAMING A TANGENT CURVE

The standard tangent function
$$y = \tan x$$
has as domain and range all real numbers except $x \ne \frac{\pi}{2} + n\pi$, and is periodic with period π. One period of this curve can be sketched by framing, as follows:

1. *Start* at the origin $(0, 0)$; for the tangent curve this is the *center* of the frame.

2. *Height* of the frame is two units: one unit up and one unit down from the starting point.

3. *Length* of this frame is π units (about 3.14) and is drawn so that it is $\frac{\pi}{2}$ (about 1.57) units on each side of the starting point (the period is π).

4. The curve is now framed. Draw the asymptotes and plot three critical points within the frame:
   a. Extend the vertical sides of the frame; these are the asymptotes.
   b. Midpoint (this was the starting point)
   c. Quarterpoints (down first; then up on the frame)

5. Draw the curve through the critical points, using the asymptotes as guides and remembering the shape of the tangent curve.

The tangent curve is indicated in Figure 6.19. Even though the curve repeats for values of x greater than 2π or less than 0, it also repeats after it has passed through an interval with length π. For this reason, $\tan(\theta + n\pi) = \tan\theta$ for any integer n, and we see that **the tangent has a period of π.** The domain of the tangent function is restricted so that multiples of π added to $\frac{\pi}{2}$ are excluded, because the tangent is not defined for these values. The range, on the other hand, is unrestricted; it is the set of all real numbers.

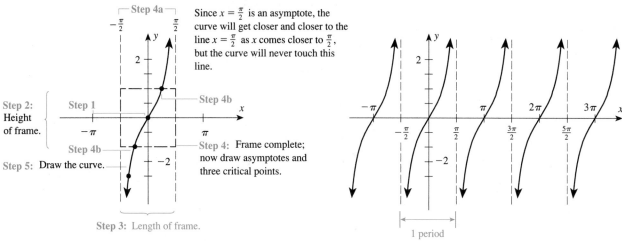

Since $x = \frac{\pi}{2}$ is an asymptote, the curve will get closer and closer to the line $x = \frac{\pi}{2}$ as x comes closer to $\frac{\pi}{2}$, but the curve will never touch this line.

Step 4a

Step 4b

Step 2: Height of frame.

Step 1

Step 4b

Step 4: Frame complete; now draw asymptotes and three critical points.

Step 5: Draw the curve.

Step 3: Length of frame.

FIGURE 6.18   Procedure for framing a tangent curve

FIGURE 6.19   Graph of $y = \tan x$;
domain: all real x, $x \neq \frac{n\pi}{2}$ (n any integer); range: all real y

1 period

Graphs of Reciprocal Functions

The graphs of the other three trigonometric functions can be done in the same fashion. Instead, however, we will make use of the reciprocal relationships and graph them as shown in Example 2.

EXAMPLE 2   Sketch $y = \sec x$ by first sketching the reciprocal $y = \cos x$.

SOLUTION   Begin by sketching the reciprocal, $y = \cos x$ (black curve in Figure 6.20). Wherever $\cos x = 0$, $\sec x$ is undefined; draw asymptotes at these places. Now plot points by finding the reciprocals of the ordinates of points previously plotted. When $y = \cos x = \frac{1}{2}$, for example, the reciprocal is

$$y = \sec x = \frac{1}{\cos x} = \frac{1}{\frac{1}{2}} = 2$$

The completed graph is shown in Figure 6.20.

FIGURE 6.20   Graph of $y = \sec x$

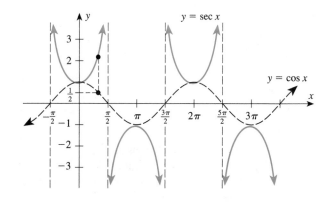

$y = \sec x$

$y = \cos x$

The graphs of the other reciprocal trigonometric functions are shown in Figure 6.21.

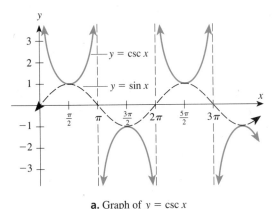

a. Graph of $y = \csc x$

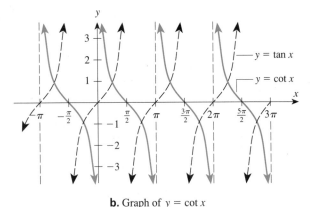

b. Graph of $y = \cot x$

FIGURE 6.21

Translations and Phase Shift

In Section 2.3 we saw that $y - k = f(x - h)$ can be sketched by translating the origin of the coordinate axes to a point (h, k) and then graphing the related function $y = f(x)$ on this new coordinate system. Thus if $f(x) = \sin x$, then $f(x - h) = \sin(x - h)$ is a sine curve shifted h units to the right. This shifting is called a **phase shift** but, in this book, we will treat it as a **translation.**

EXAMPLE 3   Graph one period of $y = \sin(x + \frac{\pi}{2})$.

SOLUTION   STEP 1   Frame the curve as in Figure 6.22.
    a. Plot $(h, k) = (-\frac{\pi}{2}, 0)$.
    b. The period of the sine curve is 2π, and it has a high point up one unit and a low point down one unit.

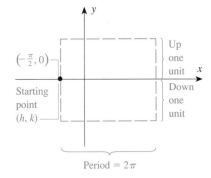

FIGURE 6.22   Framing the curve: This step is the same regardless of whether you are graphing a sine or a cosine.

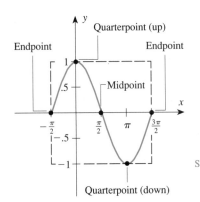

FIGURE 6.23   Graph of one period of $y = \sin(x + \frac{\pi}{2})$

STEP 2   Plot the five critical points (two endpoints, the midpoint, and two quarterpoints). For the sine curve, plot the endpoint (h, k) and use the frame to plot the other endpoint and the midpoint. For the quarterpoints, remember that the sine curve is "up–down"; use the frame to plot the quarterpoints as shown in Figure 6.23.

STEP 3   Remembering the shape of the sine curve, sketch one period of $y = \sin(x + \frac{\pi}{2})$ using the frame and the five critical points. If you want to show more than one period, just repeat the same pattern.

Notice from Figure 6.23 that the graph of $y = \sin(x + \frac{\pi}{2})$ is the same as the graph of $y = \cos x$. Thus

$$\sin\left(x + \frac{\pi}{2}\right) = \cos x$$

EXAMPLE 4 Graph one period of $y - 2 = \cos(x - \frac{\pi}{6})$.

SOLUTION STEP 1 Frame the curve as shown in Figure 6.24. Notice that $(h, k) = (\frac{\pi}{6}, 2)$.

STEP 2 Plot the five critical points. For the cosine curve the left and right endpoints are at the upper corners of the frame; the midpoint is the bottom of the frame; the quarterpoints are on a line through the middle of the frame.

STEP 3 Draw one period of the curve, as shown in Figure 6.24.

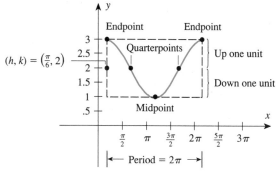

FIGURE 6.24 Graph of one period of $y - 2 = \cos(x - \frac{\pi}{6})$

EXAMPLE 5 Sketch one period of $y + 3 = \tan(x + \frac{\pi}{3})$.

SOLUTION STEP 1 Frame the curve as shown in Figure 6.25. Notice that $(h, k) = (-\frac{\pi}{3}, -3)$, and remember that the period of the tangent is π.

STEP 2 For the tangent curve (h, k) is the midpoint of the frame. The endpoints, each of which is a distance of one-half the period from the midpoint, determine the location of the asymptotes. The top and bottom of the frame are one unit from (h, k). Locate the quarterpoints at the top and bottom of the frame, as shown in Figure 6.25.

STEP 3 Sketch one period of the curve as shown in Figure 6.25. Remember that the tangent curve is not contained entirely within the frame.

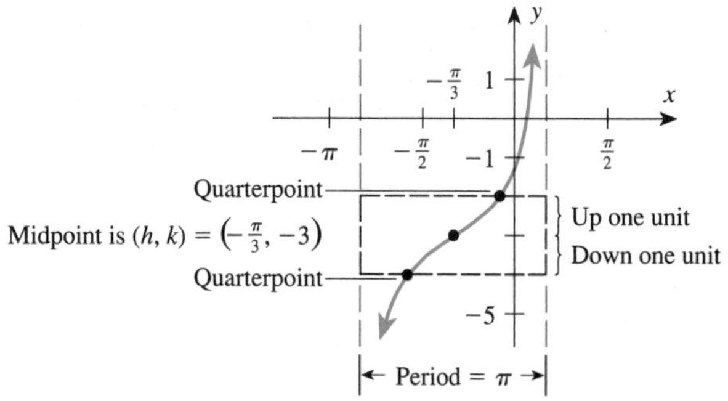

FIGURE 6.25 Graph of one period of $y + 3 = \tan(x + \frac{\pi}{3})$

Quarterpoint

Midpoint is $(h, k) = \left(-\frac{\pi}{3}, -3\right)$

Quarterpoint

Up one unit
Down one unit

$\leftarrow$ Period $= \pi \rightarrow$

Changes in Amplitude and Period

We will now discuss two additional changes for the function defined by $y = f(x)$. The first, $y = af(x)$, changes the scale on the y-axis; the second, $y = f(bx)$, changes the scale on the x-axis.

For a function $y = af(x)$, it is clear that the y value is a times the corresponding value of $f(x)$, which means that $f(x)$ is stretched or shrunk in the y-direction by the factor of a. For example, if $y = f(x) = \cos x$, then $y = 3f(x) = 3\cos x$ is the graph of $\cos x$ that has been stretched so that the high point is at 3 units and the low point is at negative 3 units. In general, given

$$y = af(x)$$

where f represents a trigonometric function, $2|a|$ gives the height of the frame for f. To graph $y = 3\cos x$, frame the cosine curve using 3 units rather than 1 (see Figure 6.26). For the sine and cosine curves, $|a|$ is the **amplitude** of the function. When $a = 1$, the amplitude is 1, so $y = \cos x$ and $y = \sin x$ are said to have amplitude 1.

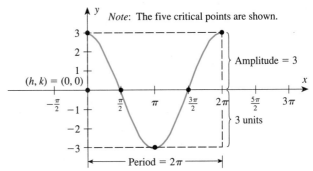

FIGURE 6.26 Graph of one period of $y = 3\cos x$

For a function $y = f(bx)$, $b > 0$, b affects the scale on the x-axis. Recall that $y = \sin x$ has a period of 2π ($f(x) = \sin x$, so $b = 1$). A function $y = \sin 2x$ ($f(x) = \sin x$ and $f(2x) = \sin 2x$) must complete one period as $2x$ varies from zero to 2π. This means that one period is completed as x varies from zero to π. (Remember that for each value of x the result is doubled *before* we find the sine of that number.) In general, the period of $y = \sin bx$ is $\frac{2\pi}{b}$ and the period of $y = \cos bx$ is $\frac{2\pi}{b}$. Since the period of $y = \tan x$ is π, however, $y = \tan bx$ has a period of $\frac{\pi}{b}$. Therefore, when framing the curve, use $\frac{2\pi}{b}$ for the sine and cosine and $\frac{\pi}{b}$ for the tangent.

EXAMPLE 6 Graph one period of $y = \sin 2x$.

SOLUTION The period is $\frac{2\pi}{2} = \pi$; thus the endpoints of the frame are $(0, 0)$ and $(\pi, 0)$, as shown in Figure 6.27.

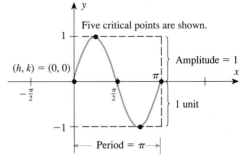

FIGURE 6.27 Graph of one period of $y = \sin 2x$ ▮

Graphs of General Trigonometric Curves

Summarizing all the preceding results, we have the *general* cosine, sine, and tangent curves:

GENERAL COSINE, SINE, AND
TANGENT CURVES

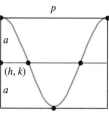

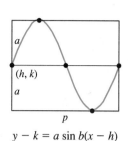

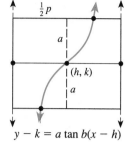

$$y - k = a \cos b(x - h) \qquad y - k = a \sin b(x - h) \qquad y - k = a \tan b(x - h)$$

1. Algebraically, put the equation into one of the general forms shown above.

2. Identify (by inspection) the following values: (h, k), a, and b. Then calculate p. Remember that $p = \frac{2\pi}{b}$ for cosine and sine, and $p = \frac{\pi}{b}$ for tangent.

3. Draw the frame:
   a. Plot (h, k); this is the starting point.
   b. Draw a; the *height of the frame* is $2|a|$—up a units from (h, k) and down a units from (h, k).*
   c. Draw p; the *length of the frame* is p.

4. Locate the critical values using the frame as a guide, and then sketch the appropriate graphs. You do not need to know the coordinates of these critical values.

EXAMPLE 7 Graph $y + 1 = 2 \sin \frac{2}{3}(x - \frac{\pi}{2})$.

SOLUTION Notice that $(h, k) = (\frac{\pi}{2}, -1)$ and that the amplitude is 2; the period is $2\pi/(\frac{2}{3}) = 3\pi$. Now plot (h, k) and frame the curve. Then plot the five critical points (two endpoints, the midpoint, and two quarterpoints). Finally, after sketching one period, draw the other periods as in Figure 6.28.

FIGURE 6.28 Graph of $y + 1 = 2 \sin \frac{2}{3}(x - \frac{\pi}{2})$; one period inside the frame is drawn first, and then the curve is extended outside the frame

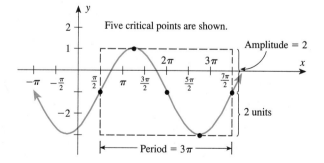

*In this section we are assuming $a > 0$. The graphs of these functions where $a < 0$ are considered in Section 7.4.

EXAMPLE 8   Graph $y = 3\cos(2x + \frac{\pi}{2}) - 2$.

SOLUTION   Rewrite in standard form to obtain $y + 2 = 3\cos 2(x + \frac{\pi}{4})$. Notice that $(h, k) = (-\frac{\pi}{4}, -2)$; the amplitude is 3 and the period is $\frac{2\pi}{2} = \pi$. Plot (h, k) and frame the curve, as shown in Figure 6.29.

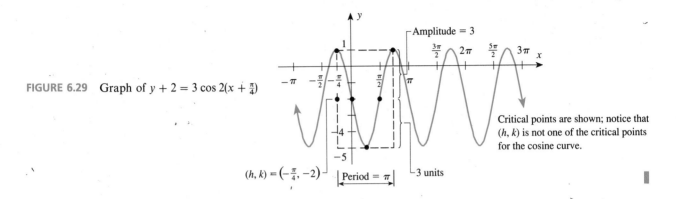

FIGURE 6.29   Graph of $y + 2 = 3\cos 2(x + \frac{\pi}{4})$

EXAMPLE 9   Graph $y - 2 = 3\tan \frac{1}{2}(x - \frac{\pi}{3})$.

SOLUTION   Notice that $(h, k) = (\frac{\pi}{3}, 2)$, $a = 3$, and the period is $\pi/(\frac{1}{2}) = 2\pi$. Plot (h, k) and frame the curve, as shown in Figure 6.30.

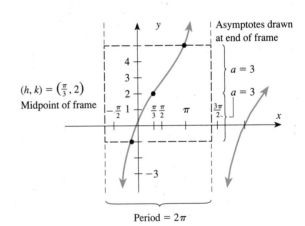

FIGURE 6.30   Graph of $y - 2 = 3\tan \frac{1}{2}(x - \frac{\pi}{3})$

A

1. Complete the following table of values for $y = \cos x$:

$x =$ angle	$\frac{2\pi}{3}$	$\frac{3\pi}{4}$	$\frac{5\pi}{6}$	$\frac{7\pi}{6}$	$\frac{5\pi}{4}$	$\frac{4\pi}{3}$	$\frac{7\pi}{4}$	$\frac{11\pi}{6}$
Quadrant: sign of cos x								
$y = \cos x$								
y (approximate)								

Use this table, along with other values if necessary, to plot $y = \cos x$.

2. Complete a table of values like the one in Problem 1 for $y = \tan x$. Use this table, along with other values if necessary, to plot $y = \tan x$.

3. We have emphasized the fact that the sine function can be considered as a function of a real number, x. Instead of using units of π, graph the sine function by finding additional values for the table shown here:

$x =$ real number	0	1	2	3	4	5	6
$y = \sin x$	0	.84	.91	.14	−.76	−.96	−.28

Plot the ordered pairs, $(0, 0)$, $(1, .84)$, $(2, .91)$, ..., and complete the graph of $y = \sin x$.

4. We have emphasized the fact that the cosine function can be considered as a function of a real number, x. Instead of using units of π, graph the cosine function by finding additional values for the table shown here:

$x =$ real number	0	1	2	3	4	5	6
$y = \cos x$	1	.54	−.42	−.99	−.65	.28	.96

Plot the ordered pairs, $(0, 1)$, $(1, .54)$, $(2, −.42)$, ..., and complete the graph of $y = \cos x$.

Graph one period of each function given in Problems 5–18.

5. $y = \sin(x + \pi)$

6. $y = \cos(x + \frac{\pi}{2})$

7. $y = \tan(x + \frac{\pi}{3})$

8. $y = 3 \sin x$

9. $y = 2 \cos x$

10. $y = \frac{1}{2} \sin x$

11. $y = \sin 3x$

12. $y = \cos 2x$

13. $y = \cos \frac{1}{2}x$

14. $y = \tan(x - \frac{3\pi}{2})$

15. $y = \tan(x + \frac{\pi}{6})$

16. $y = \frac{1}{3} \sin x$

17. $y = \frac{1}{3} \tan x$

18. $y = 4 \tan x$

B *Graph one period of each function given in Problems 19–30.*

19. $y - 2 = \sin(x - \frac{\pi}{2})$

20. $y + 1 = \cos(x + \frac{\pi}{3})$

21. $y - 3 = \tan(x + \frac{\pi}{6})$

22. $y - \frac{1}{2} = \frac{1}{2} \cos x$

23. $y - 1 = 2 \sin x$

24. $y + 2 = 3 \cos x$

25. $y - 1 = 2 \cos(x - \frac{\pi}{4})$

26. $y - 1 = \cos 2(x - \frac{\pi}{4})$

27. $y + 2 = 3 \sin(x + \frac{\pi}{6})$

28. $y + 2 = \sin 3(x + \frac{\pi}{6})$

29. $y = 1 + \tan 2(x - \frac{\pi}{4})$

30. $y + 2 = \tan(x - \frac{\pi}{4})$

Graph the curves given in Problems 31–48.

31. $y = \sin(4x + \pi)$

32. $y = \sin(3x + \pi)$

33. $y = \tan(2x - \frac{\pi}{2})$

34. $y = \tan(\frac{x}{2} + \frac{\pi}{3})$

35. $y = \frac{1}{2} \cos(x + \frac{\pi}{6})$

36. $y = \cos(\frac{1}{2}x + \frac{\pi}{12})$

37. $y = 3 \cos(3x + 2\pi) - 2$

38. $y = 4 \sin(\frac{1}{2}x + 2)$

39. $y = \sqrt{2} \cos(x - \sqrt{2}) - 1$

40. $y = \sqrt{3} \sin(\frac{1}{3}x - \sqrt{\frac{1}{3}})$

41. $y = 2 \sin 2\pi x$

42. $y = 3 \cos 3\pi x$

43. $y = 4 \tan \frac{\pi x}{5}$

44. $y + 2 = \frac{1}{2} \cos(\pi x + 2\pi)$

45. $y = |\sin x|$

46. $y = |\cos x|$

47. $y = \sin|x|$

48. $y = \cos|x|$

So far we have limited ourselves to $a > 0$. If $a < 0$, the curve is reflected through the x-axis. Graph the curves in Problems 49–54.

49. $y = -\sin x$

50. $y = -\cos x$

51. $y = -\tan x$

52. $y = -3 \sin x$

53. $y = -2 \cos x$

54. $y = -\sin 3x$

Use your knowledge of the nature of functions, as well as the information of this section, to sketch the graphs of the functions given in Problems 55–60.

55. $y = 2 \sec x$

56. $y = \csc 2x$

57. $y = 2 \cos x - 1$

58. $y = \sin x + \cos x$

59. $y = \sin 2x + \cos x$

60. $y = 2 \cos x + \sin 2x$

Graph the functions given in each of Problems 61–64 on the same coordinate axes. If you are using a graphing calculator, you will need to pay particular attention to the domain and range.

61. a. $y = x$
    b. $y = -x$
    c. $y = \dfrac{x}{\sin x}$

62. a. $y = x$
    b. $y = -x$
    c. $y = x \sin x$

63. a. $y = x^2$
  b. $y = -x^2$
  c. $y = x^2 \sin x$

64. a. $y = \sqrt{x}$
  b. $y = -\sqrt{x}$
  c. $y = \sqrt{x} \sin x$

C

65. ELECTRICAL ENGINEERING The current I (in amperes) in a certain circuit is given by

$$I = 60 \cos(120\pi t - \pi)$$

where t is time in seconds. Graph this equation for $0 \le t \le \frac{1}{30}$.

66. ENGINEERING Suppose a point P on a waterwheel with a 30-ft radius is d units from the water, as shown in Figure 6.31. If it turns at 6 revolutions per minute, then

$$d = 29 + 30 \cos(\tfrac{\pi}{5}t - \pi)$$

Graph this equation for $0 \le t \le 20$.

67. SPACE SCIENCE The distance (in kilometers) that a certain satellite is north or south of the equator is given by

$$y = 3{,}000 \cos(\tfrac{\pi}{60}t + \tfrac{\pi}{5})$$

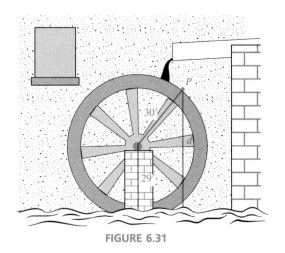

FIGURE 6.31

where t is the number of minutes that have elapsed since liftoff.
a. Graph the equation for $0 \le t \le 120$.
b. What is the farthest distance that the satellite ever reaches north of the equator?
c. How long does it take to complete one orbit?

6.5 Inverse Trigonometric Functions

Inverse Sine Function

In Section 2.5 the notion of inverse functions was introduced. In this section, that idea is applied to the trigonometric functions. Recall that a function f must be one-to-one in order to have an inverse function. This means that the trigonometric functions do not have inverse functions. We can, however, restrict the domains of the trigonometric functions so that they become one-to-one. We will illustrate with the sine function. Figure 6.32 shows the graph of $y = \sin x$.

Notice that if $y = \sin x$, then each x, say $\frac{5\pi}{6}$, is associated with exactly one y value, $\frac{1}{2}$ in this case. But it is *not* one-to-one because for a given y value, say $\frac{1}{2}$, there are infinitely many x values, as shown in Figure 6.32.

FIGURE 6.32 Graph of $y = \sin x$

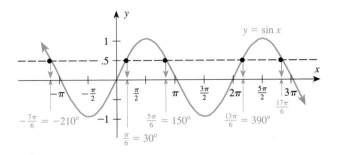

Define a new function related to $y = \sin x$ but having the property that it is one-to-one. We do this simply by restricting its domain to the first and fourth quadrants. That is, define

The capital S in $y = \operatorname{Sin} x$ is important; $\operatorname{Sin} x \neq \sin x$ because of their different domains. This function is shown in color in Figure 6.33a; notice that this function is one-to-one. ⊗

$y = \operatorname{Sin} x$ so that x is on the interval $\left[-\dfrac{\pi}{2}, \dfrac{\pi}{2} \right]$

This means that $\operatorname{Sin} x = \sin x$ for $-\pi/2 \leq x \leq \pi/2$ and is not defined elsewhere.

FIGURE 6.33 Comparison graphs of $y = \sin x$, $y = \operatorname{Sin} x$, and $y = \operatorname{Sin}^{-1}x$

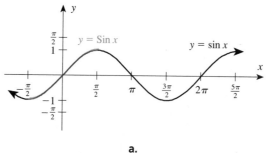

a.

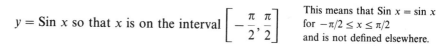

b.

Now we can define the inverse by using the notation introduced in Section 2.5. *Remember*: The inverse of $y = \operatorname{Sin} x$ is found by interchanging the x and y components to $x = \operatorname{Sin} y$.

INVERSE SINE

> The **inverse sine function,** denoted by $\operatorname{Sin}^{-1}x$ or by Arcsin x and called "arcsine of x," is defined by
>
> $$y = \operatorname{Sin}^{-1}x \quad \text{if and only if} \quad x = \sin y$$
>
> where $-1 \leq x \leq 1$ and $-\frac{\pi}{2} \leq y \leq \frac{\pi}{2}$. See Figure 6.33b.

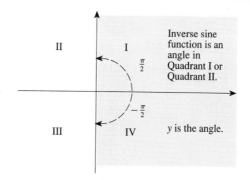

EXAMPLE 1 Find $\operatorname{Sin}^{-1}(\frac{1}{2}\sqrt{3})$.

SOLUTION Let $\theta = \operatorname{Sin}^{-1}(\frac{1}{2}\sqrt{3})$. (*Remember*: An inverse sine is an angle; so we denote it by θ.) Find the angle or real number θ with sine equal to $\frac{1}{2}\sqrt{3}$ so that $-\frac{\pi}{2} \leq \theta \leq \frac{\pi}{2}$. From the memorized table of exact values you know that $\sin(\frac{\pi}{3}) = \frac{1}{2}\sqrt{3}$. And since $\frac{\pi}{3}$ is between $-\frac{\pi}{2}$ and $\frac{\pi}{2}$, you have $\operatorname{Sin}^{-1}(\frac{1}{2}\sqrt{3}) = \frac{\pi}{3}$. ∎

EXAMPLE 2 Find $\operatorname{Sin}^{-1}(-\frac{1}{2}\sqrt{3})$.

SOLUTION You will find it easier to work with reference angles when finding inverse trigonometric functions. That is, because the table of exact values was memorized for the first quadrant, work with the reference angle. Let

$$\theta' = \mathrm{Sin}^{-1}(\tfrac{1}{2}\sqrt{3})$$

reference angle   absolute value of the given number

$$\theta' = \frac{\pi}{3} \qquad \text{From Example 1}$$

Now place θ in the appropriate quadrant. The sine is negative in both the third and fourth quadrants, but you choose the fourth quadrant because of the restrictions on $y = \mathrm{Sin}\ x$. Thus θ is the fourth-quadrant angle with its reference angle $\theta' = \frac{\pi}{3}$. Therefore $\mathrm{Sin}^{-1}(-\frac{1}{2}\sqrt{3}) = -\frac{\pi}{3}$.

Inverse Trigonometric Functions

The other trigonometric functions are handled similarly:

GIVEN FUNCTION	INVERSE	OTHER NOTATIONS FOR INVERSE	
$y = \mathrm{Cos}\ x$	$x = \mathrm{Cos}\ y$	$y = \mathrm{Cos}^{-1}x$	$y = \mathrm{Arccos}\ x$
$y = \mathrm{Sin}\ x$	$x = \mathrm{Sin}\ y$	$y = \mathrm{Sin}^{-1}x$	$y = \mathrm{Arcsin}\ x$
$y = \mathrm{Tan}\ x$	$x = \mathrm{Tan}\ y$	$y = \mathrm{Tan}^{-1}x$	$y = \mathrm{Arctan}\ x$
$y = \mathrm{Sec}\ x$	$x = \mathrm{Sec}\ y$	$y = \mathrm{Sec}^{-1}x$	$y = \mathrm{Arcsec}\ x$
$y = \mathrm{Csc}\ x$	$x = \mathrm{Csc}\ y$	$y = \mathrm{Csc}^{-1}x$	$y = \mathrm{Arccsc}\ x$
$y = \mathrm{Cot}\ x$	$x = \mathrm{Cot}\ y$	$y = \mathrm{Cot}^{-1}x$	$y = \mathrm{Arccot}\ x$

These are the same.

FIGURE 6.34 Graphs of inverse trigonometric functions

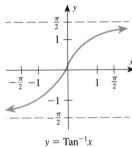

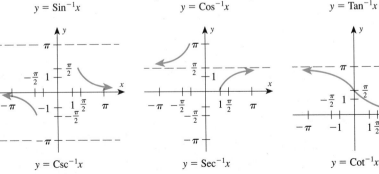

$y = \mathrm{Sin}^{-1}x$

$y = \mathrm{Cos}^{-1}x$

$y = \mathrm{Tan}^{-1}x$

$y = \mathrm{Csc}^{-1}x$

$y = \mathrm{Sec}^{-1}x$

$y = \mathrm{Cot}^{-1}x$

Consider the graphs in Figure 6.34. We have to restrict each trigonometric function so it is one-to-one, but we also want to include all possible values in the range of the original function. For the sine curve, x was restricted so that $-\frac{\pi}{2} \le x \le \frac{\pi}{2}$. Then the inverse is the function

$$y = \text{Sin}^{-1}x \quad \text{where } -\frac{\pi}{2} \le y \le \frac{\pi}{2}$$

The same restrictions (leaving out the values $-\frac{\pi}{2}$ and $\frac{\pi}{2}$) apply for the tangent and arctangent curves. For the cosine function, however, notice that by restricting x to the same interval you obtain only positive values for $y = \cos x$. Thus, to include the entire range of the cosine curve, x can be restricted so that $0 \le x \le \pi$. Then the inverse is the function

$$y = \text{Cos}^{-1}x \quad \text{where } 0 \le y \le \pi$$

The cotangent function is restricted in almost the same way, and the results are summarized in the following definitions.

The following box lists the domain, range, and quadrants for the range of the six inverse trigonometric functions.

INVERSE TRIGONOMETRIC FUNCTIONS

INVERSE FUNCTION	DOMAIN	RANGE	QUADRANTS OF THE RANGE
$y = \text{Arcsin } x$ or $y = \text{Sin}^{-1}x$	$-1 \le x \le 1$	$-\frac{\pi}{2} \le y \le \frac{\pi}{2}$	I and IV
$y = \text{Arccos } x$ or $y = \text{Cos}^{-1}x$	$-1 \le x \le 1$	$0 \le y \le \pi$	I and II
$y = \text{Arctan } x$ or $y = \text{Tan}^{-1}x$	All reals	$-\frac{\pi}{2} < y < \frac{\pi}{2}$	I and IV
$y = \text{Arccot } x$ or $y = \text{Cot}^{-1}x$	All reals	$0 < y < \pi$	I and II
$y = \text{Arcsec } x$ or $y = \text{Sec}^{-1}x$	$x \ge 1$ or $x \le -1$	$0 \le y \le \pi,$ $y \ne \frac{\pi}{2}$	I and II
$y = \text{Arccsc } x$ or $y = \text{Csc}^{-1}x$	$x \ge 1$ or $x \le -1$	$-\frac{\pi}{2} \le y \le \frac{\pi}{2},$ $y \ne 0$	I and IV

EXAMPLE 3 Find Arctan 1.

SOLUTION Let Arctan $1 = \theta$. You are looking for an angle θ with tangent equal to 1. Since this is an exact value, you know that $\theta = \frac{\pi}{4}$ **or 45°.** ▌

EXAMPLE 4 Find Arccot$(-\sqrt{3})$.

SOLUTION Let Arccot$(-\sqrt{3}) = \theta$. Find θ so that $\cot \theta = -\sqrt{3}$; the reference angle is 30°, and the cotangent is negative in Quadrants II and IV. Since Arccot x is defined in Quadrant II but not in Quadrant IV, Arccot$(-\sqrt{3}) = \frac{5\pi}{6}$ **or 150°.** ▌

EXAMPLE 5 Find Arcsin$(-.4695)$.

SOLUTION You might use

$$\boxed{\text{ARC}}\,\boxed{\text{TAN}}, \quad \boxed{\text{INV}}\,\boxed{\text{TAN}} \quad \text{or} \quad \boxed{2^{nd}}\,\boxed{\text{TAN}^{-1}}$$

The display, $\boxed{-28.00184535}$, is the decimal representation of the angle in degrees (if the calculator is set to degrees) or $\boxed{-.488724398}$ in radians (if it is set to radians). ▌

EXAMPLE 6   Find Arcsec(−2.747).

SOLUTION   Let Arcsec(−2.747) = θ. Since $0 < \text{Sec}^{-1}x < \pi$, we must make sure that θ is in Quadrant II. Because calculators have no secant function, note that if sec $y = x$, then cos $y = \frac{1}{x}$. Thus,

$$y = \text{Sec}^{-1}x \quad \text{and} \quad y = \text{Cos}^{-1}\left(\frac{1}{x}\right) \quad \text{so} \quad \text{Sec}^{-1}x = \text{Cos}^{-1}\left(\frac{1}{x}\right)$$

This tells us that to find the inverse secant on a calculator we must *first take the reciprocal* of the given value and then complete the problem.

   Procedure: Find the reciprocal of −2.747 and then find the inverse cosine of that reciprocal.   DISPLAY: $\boxed{1.943391209}$. Notice that an angle of **1.9** radians is in Quadrant II. ▌

The procedure for finding Arcsec x or Arccsc x for any x in the domain is the same as shown in Example 6. It is also the same for Arccot x where Arccot x is positive. That is, we have the following properties:

$$\text{Sec}^{-1}x = \text{Cos}^{-1}\left(\frac{1}{x}\right) \quad \text{for } x \geq 1 \text{ or } x \leq -1$$

$$\text{Csc}^{-1}x = \text{Sin}^{-1}\left(\frac{1}{x}\right) \quad \text{for } x \geq 1 \text{ or } x \leq -1$$

$$\text{Cot}^{-1}x = \text{Tan}^{-1}\left(\frac{1}{x}\right) \quad \text{for } x > 0$$

If $x < 0$, then $\text{Tan}^{-1}x$ is in Quadrant IV, whereas $\text{Cot}^{-1}x$ is in Quadrant II. This means

$$\text{Cot}^{-1}x = \text{Tan}^{-1}\left(\frac{1}{x}\right) + \pi \quad \text{for } x < 0$$

We can now summarize the calculator steps for inverse trigonometric functions.

INVERSE FUNCTION	CALCULATOR Enter the value of x, then press:
$y = \text{Arccos } x$	inv cos
$y = \text{Arcsin } x$	inv sin
$y = \text{Arctan } x$	inv tan
$y = \text{Arcsec } x$	1/x inv cos
$y = \text{Arccsc } x$	1/x inv sin
$y = \text{Arccot } x$	If $x > 0$: 1/x inv tan
	If $x < 0$: 1/x inv tan + π = *

*Use 180 instead of π if working in degrees.

EXAMPLE 7   Find Arccot(−2.747).

SOLUTION   First find the reciprocal of −2.747; *also note that −2.747 is negative*. Next find the inverse tangent of that reciprocal. Finally, since −2.747 is negative, we add

π (or 180 if we are working in degrees).

DISPLAY: | 2.79247095 | Note that this answer is in Quadrant II.   ❚

Functions of Inverse Functions

A final word of caution is in order regarding the inverse trigonometric functions, especially when you are using a calculator. Recall from Section 2.5 that

$$(f^{-1} \circ f)(x) = x \text{ and } (f \circ f^{-1})(x) = x$$

In the context of this section, this means

$$\text{Cos}^{-1}(\text{Cos } x) = x \quad \text{and} \quad \text{Cos}(\text{Cos}^{-1}x) = x$$

$$\text{Sin}^{-1}(\text{Sin } x) = x \quad \text{and} \quad \text{Sin}(\text{Sin}^{-1}x) = x$$

$$\text{Tan}^{-1}(\text{Tan } x) = x \quad \text{and} \quad \text{Tan}(\text{Tan}^{-1}x) = x$$

$$\text{Cot}^{-1}(\text{Cot } x) = x \quad \text{and} \quad \text{Cot}(\text{Cot}^{-1}x) = x$$

However, you should not forget the appropriate restrictions that are also part of the definition. Consider Examples 8 to 13.

EXAMPLE 8 $\text{Cos}^{-1}(\cos 2.2) = \mathbf{2.2}$

Calculator check: DISPLAY: | 2.2 |

(Some calculators may show a display such as 2.1999997; this should be considered a proper check.)   ❚

EXAMPLE 9 $\text{Sin}^{-1}(\sin 2.2) \approx .9415926536 \approx \mathbf{.9}$ (Not 2.2)
The reason for this answer is that Sin x is defined in Quadrants I and IV, whereas 2.2 is not an angle in these quadrants.
    Since sin 2.2 ≈ sin .9 (by the reduction principle) and since .9 is an angle in Quadrants I or IV,
$$\text{Sin}^{-1}(\sin 2.2) \approx \text{Sin}^{-1}(\sin .9)$$
$$= .9$$ ❚

EXAMPLE 10   Arccos(cos 4) ≈ **2.3**
Since the angle 4 radians is in Quadrant III, cos 4 is negative. The Arccosine of a negative angle will be a Quadrant II angle. This is the Quadrant II angle having the same reference angle as 4. The reference angle is $4 - \pi \approx .86$. In Quadrant II, $\pi - $ reference angle ≈ 2.28. Therefore

Arccos(cos 4) ≈ 2.28

By calculator: DISPLAY: | 2.283185307 | ❚

EXAMPLE 11 $\sin(\text{Sin}^{-1} .463) = \mathbf{.463}$ ❚

EXAMPLE 12 $\sin(\text{Sin}^{-1} 2.463)$ is **not defined** since 2.463 is not between -1 and $+1$.   ❚

EXAMPLE 13 $\tan(\text{Tan}^{-1} 2.463) = \mathbf{2.463}$ ❚

Problem Set

___**A**___ *In Problems 1–7, obtain the given angle (in radians) from memory.*

1. a. Arcsin 0
   c. Arccot $\sqrt{3}$
   b. $\text{Tan}^{-1}(\frac{\sqrt{3}}{3})$
   d. Arccos 1

2. a. $\text{Cos}^{-1}(\frac{\sqrt{3}}{2})$
   c. $\text{Tan}^{-1}1$
   b. Arcsin $\frac{1}{2}$
   d. $\text{Sin}^{-1}1$

3. a. Arctan $\sqrt{3}$
   c. $\text{Arcsin}(\frac{1}{2}\sqrt{2})$
   b. $\text{Cos}^{-1}(\frac{\sqrt{2}}{2})$
   d. Arccot 1

4. a. Arcsin(-1)
   c. $\text{Arcsin}(-\frac{\sqrt{3}}{2})$
   b. $\text{Cot}^{-1}(-1)$
   d. $\text{Cos}^{-1}(-1)$

5. a. $\text{Cot}^{-1}(-\sqrt{3})$
   c. $\text{Sin}^{-1}(-\frac{1}{2}\sqrt{2})$
   b. Arctan(-1)
   d. $\text{Cos}^{-1}(-\frac{1}{2})$

6. a. $\text{Arccos}(-\frac{\sqrt{2}}{2})$
   c. $\text{Sin}^{-1}(-\frac{1}{2})$
   b. $\text{Cot}^{-1}(-\frac{\sqrt{3}}{3})$
   d. $\text{Arctan}(-\frac{\sqrt{3}}{3})$

7. a. $\text{Tan}^{-1}0$
   c. Arccos $\frac{1}{2}$
   b. $\text{Arccot}(\frac{\sqrt{3}}{3})$
   d. $\text{Sin}^{-1}(\frac{\sqrt{3}}{2})$

Use a calculator to find the values (in radians correct to the nearest hundredth) for the functions in Problems 8–19.

8. Arcsin .20846
9. $\text{Cos}^{-1}.83646$
10. Arctan 1.1156
11. $\text{Cot}^{-1}(-.08097)$
12. $\text{Tan}^{-1}(-3.7712)$
13. Arccos$(-.94604)$
14. $\text{Sin}^{-1}.75$
15. Arccos .25
16. Arctan 2
17. $\text{Tan}^{-1}1.489$
18. $\text{Cot}^{-1}3.451$
19. $\text{Sec}^{-1}4.315$

Use a calculator to find the values (to the nearest degree) given in Problems 20–33.

20. $\text{Sin}^{-1}.3584$
21. $\text{Cos}^{-1}.3584$
22. Arccos .9455
23. $\text{Sin}^{-1}(-.4695)$
24. $\text{Tan}^{-1}2.050$
25. Arctan 1.036
26. $\text{Tan}^{-1}(-3.732)$
27. $\text{Cot}^{-1}.0875$
28. $\text{Csc}^{-1}2.816$
29. Arccot(-1)
30. Arccsc 3.945
31. Arccot(-2)
32. Arcsec(-6)
33. Arctan(-3)

___**WHAT IS WRONG**___, *if anything, with each of the statements in Problems 34–43? Explain your reasoning.*

34. In $y = \cos x$, the angle is y.

35. In $y = \cos^{-1}x$, the angle is y.

36. $\text{Tan}^{-1}(-2.5)$ is in Quadrant II.

37. $\text{Cot}^{-1}(-2.5)$ is in Quadrant IV.

38. The domain in $y = \sin x$ is $[-1, 1]$.

39. The domain in $y = \text{Sin}^{-1}x$ is $[-1, 1]$.

40. $\text{Tan}^{-1}x = \dfrac{1}{\tan x}$
   41. $\text{Cot}^{-1}x = \text{Tan}^{-1}\left(\dfrac{1}{x}\right)$

42. $\text{Sec}^{-1}x = \text{Cos}^{-1}\left(\dfrac{1}{x}\right)$
   43. $\text{Csc}^{-1}x = \text{Cos}^{-1}\left(\dfrac{1}{x}\right)$

___**B**___ *Simplify the expressions in Problems 44–59.*

44. cot(Arccot 1)
45. $\text{Arccos}[\cos(\frac{\pi}{6})]$
46. sin(Arcsin $\frac{1}{3}$)
47. $\text{Tan}^{-1}[\tan(\frac{\pi}{15})]$
48. cos(Arccos $\frac{2}{3}$)
49. $\text{Arcsin}[\sin(\frac{2\pi}{15})]$
50. Arccot(cot 35°)
51. tan(Arctan .4163)
52. Arcsin(sin 4)
53. Arccos(cos 5)
54. sin(Arcsin .7568)
55. cos(Arccos .2836)
56. tan(Arctan .2910)
57. Arctan(tan 2.5)
58. $\sin(\text{Tan}^{-1}\frac{2}{3})$
59. $\cos(\text{Sin}^{-1}.4)$

In Problems 60–62 graph the given pair of curves on the same axes.

60. $y = \text{Sin } x$;
   $y = \text{Sin}^{-1}x$
61. $y = \text{Cos } x$;
   $y = \text{Cos}^{-1}x$
62. $y = \text{Tan } x$;
   $y = \text{Tan}^{-1}x$

___**C**___ *Graph the curves given in Problems 63–68.*

63. $y + 2 = \text{Arctan } x$
   64. $y - 1 = \text{Arcsin } x$
65. $y = 2 \text{ Cos}^{-1}x$
   66. $y = 3 \text{ Sin}^{-1}x$
67. $y = \text{Arcsin}(x - 2)$
   68. $y = \text{Arcsin}(x + 1)$

6.6 Chapter 6 Summary

The material of this chapter is reviewed in the following list of objectives. After each objective there are some practice questions. For a sample test, select the first question of each set and check your answers with the answer section. For a sample test without answers, use the second question of each set. Additional practice is given by the other questions in each set. If you are having trouble with a particular type of problem, look back to that section for extra help.

6.1 Angles and the Unit Circle

OBJECTIVE 1 *Know the definition and equation of a unit circle. Know the definition and notation of an angle, including positive and negative angles. Know the Greek letters. Know what it means for an angle to be in standard position.*

1. Name the Greek letter used for each angle:
   a. λ **b.** θ **c.** ϕ **d.** α **e.** β

Fill in the blanks.

2. A unit circle is —————————.
3. The equation of a unit circle is —————————.
4. An angle is in standard position if —————————.

OBJECTIVE 2 *Find angles coterminal with a given angle.* Find the positive angle coterminal with the given angle and less than one revolution.

5. $-215°$ 6. $\frac{11\pi}{3}$ 7. $-\frac{5\pi}{6}$ 8. $1,000°$

OBJECTIVE 3 *Be familiar with the degree measure of an angle and be able to approximate the angle associated with a given degree measure without using any measuring devices.* Draw the indicated angles.

9. $180°$ 10. $120°$ 11. $-30°$ 12. $135°$

OBJECTIVE 4 *Be familiar with the radian measure of an angle and be able to approximate the angle associated with a given radian measure without using any measuring devices.* Draw the indicated angles.

13. $\frac{\pi}{3}$ 14. $\frac{\pi}{4}$ 15. $\frac{5\pi}{6}$ 16. 2

OBJECTIVE 5 *Change from radian to degree measure; know the commonly used degree and radian measure equivalences.*

17. $\frac{3\pi}{2}$ 18. 2
19. $-\frac{7\pi}{4}$ 20. $\frac{5\pi}{6}$

OBJECTIVE 6 *Change from degree measure to radian measure; know the commonly used radian and degree measure equivalences.* Use exact values when possible.

21. $300°$ 22. $-45°$
23. $54°$ 24. $-210°$

OBJECTIVE 7 *Know the arc length formula and be able to apply it.*

25. The arc length formula is —————————, where s is the arc length, r is the —————————, and θ is —————————.

26. If the radius is 1 and the angle is 1, then what is the arc length?

27. If the minute hand on a clock is 15 cm long, how far does the tip move in 10 minutes?

28. A curve on a highway is laid out as the arc of a circle of radius 500 m. If the curve subtends a central angle of $18°$, what is the distance around this section of road? Give the exact answer and an answer rounded off to the nearest meter.

OBJECTIVE 8 *Find the reference angle θ for a given angle θ.*

29. $300°$ 30. $\frac{11\pi}{3}$
31. -4 32. $-215°$

6.2 Trigonometric Functions

OBJECTIVE 9 *Know the unit circle definition of the trigonometric functions. Fill in the blanks.*

33. Let θ be —————————. Then the trigonometric functions are defined as follows for a point (a, b) on a unit circle:

34. ————————— —————————
35. ————————— —————————
36. ————————— —————————

OBJECTIVE 10 *Know the signs of the six trigonometric functions in each of the four quadrants. Name the function(s) that is (are) positive in the given quadrant.*

37. I   38. II
39. III   40. IV

OBJECTIVE 11   *Know that* $\tan \theta = \dfrac{\sin \theta}{\cos \theta}$; *know the reciprocal relationships.*

41. If $\sin \theta = \frac{4}{5}$ and $\cos \theta = \frac{3}{5}$, then what is $\tan \theta$?

42. What is the reciprocal function of tangent?

43. What is the reciprocal function of cosine?

44. What is the reciprocal function of sine?

OBJECTIVE 12   *Be able to evaluate the trigonometric functions.*

45. $\sec 23.4°$ **46.** $\cot 2.5$ **47.** $\csc 43.28°$ **48.** $\sin 7$

OBJECTIVE 13   *Know the ratio definition of the trigonometric functions.*

49. If _____ , then

50. _____   _____

51. _____   _____

52. _____   _____

OBJECTIVE 14   *Use the definition of the trigonometric functions to approximate their values for a given angle or for an angle passing through a given point. Assume that the terminal side passes through the given point.*

53. $(5, -12)$ **54.** $(3, -4)$ **55.** $(-5, 2)$ **56.** $(4, 5)$

6.3   Values of the Trigonometric Functions

OBJECTIVE 15   *Know and be able to derive the table of exact values. Complete the table.*

	FUNCTION	0	$\dfrac{\pi}{6}$	$\dfrac{\pi}{4}$	$\dfrac{\pi}{3}$	$\dfrac{\pi}{2}$	π	$\dfrac{3\pi}{2}$
57.	$\cos \theta$	a.	b.	c.	d.	e.	f.	g.
58.	$\csc \theta$	a.	b.	c.	d.	e.	f.	g.
59.	$\tan \theta$	a.	b.	c.	d.	e.	f.	g.
60.	$\sin \theta$	a.	b.	c.	d.	e.	f.	g.

OBJECTIVE 16   *Use the reduction principle, along with the table of exact values, to evaluate certain trigonometric functions.*

61. $\cos\left(-\frac{5\pi}{3}\right)$ **62.** $\sin\left(\frac{11\pi}{6}\right)$ **63.** $\tan 135°$

64. $\cos(-210°)$

OBJECTIVE 17   *Use a calculator to approximate values of the trigonometric functions.*

65. $\csc 43.28°$ **66.** $\sin 9$ **67.** $\sec 23.4°$

68. $\cot 2.5$

6.4   Graphs of the Trigonometric Functions

OBJECTIVE 18   *Graph the trigonometric functions, or variations, by plotting points.*

69. $y = 2 \cot x$

70. $y = 2 \sec x$

71. $y = \frac{1}{2} \csc x$

72. $y = 2 \cos x + \sin 2x$

OBJECTIVE 19   *Sketch $y = \cos x$, $y = \sin x$, and $y = \tan x$ from memory; know their periods and amplitudes.*

73. $y = \sin x$ **74.** $y = \cos x$ **75.** $y = \tan x$

76. Fill in the blanks:

FUNCTION	PERIOD	AMPLITUDE
$\sin x$	a. _____	b. _____
$\cos x$	c. _____	d. _____
$\tan x$	e. _____	f. _____

OBJECTIVE 20   *Graph the general cosine, sine, and tangent curves.*

77. $y = 2 \cos \frac{2}{3}x$ **78.** $y = \cos(x + \frac{\pi}{4})$

79. $y - 2 = \sin(x - \frac{\pi}{6})$ **80.** $y = \tan(x - \frac{\pi}{3}) - 2$

Inverse Trigonometric Functions

OBJECTIVE 21 *Know the definition of the inverse cosine, sine, tangent, and cotangent functions, especially the range of each. Fill in the blanks.*

INVERSE FUNCTION	DOMAIN	RANGE
81. $y = \text{Arctan } x$	All reals	_____
82. $y = \text{Cos}^{-1}x$	$-1 \leq x \leq 1$	_____
83. $y = \text{Cot}^{-1}x$	All reals	_____
84. $y = \text{Arcsin } x$	$-1 \leq x \leq 1$	_____

OBJECTIVE 22 *Evaluate inverse cosine, sine, tangent, and cotangent functions using exact values (in radians).*

85. $\text{Arcsin } \frac{1}{2}$

86. $\text{Cot}^{-1}(\frac{1}{3}\sqrt{3})$

87. $\text{Arccos}(\frac{\sqrt{3}}{2})$

88. $\text{Tan}^{-1}(-\sqrt{3})$

OBJECTIVE 23 *Evaluate inverse cosine, sine, tangent, and cotangent functions using a calculator. Answer in radians.*

89. $\text{Arcsin } .3140$

90. $\text{Arccos}(-.6494)$

91. $\text{Arctan } 3.271$

92. $\text{Arccot } 2$

OBJECTIVE 24 *Graph the inverse cosine, sine, and tangent functions.*

93. $y = \text{Sin}^{-1}x$

94. $y = \text{Arccos } x$

95. $y = \text{Arctan } x$

96. $y - 1 = \text{Sin}^{-1}x$

OBJECTIVE 25 *Simplify a function and its inverse function.*

97. $\text{Arccos}(\cos 1)$

98. $\text{Arcsin}(\sin 1)$

99. $\text{Arcsin}(\cos 2)$

100. $\text{Arcsin}(\sin 2)$

Nicholas Copernicus (1473–1543)

There is perhaps nothing which so occupies, as it were, the middle position of mathematics, as trigonometry.

J. F. HERBART
Idee eines ABC der Anschauung

Trigonometry was invented by Ptolemy, known as Claudius Ptolemy, who worked with Hipparchus and Menelaus. Their goal was to build a quantitative astronomy that could be used to predict the paths and positions of the heavenly bodies and to aid in telling time, calendar reckoning, navigating, and studying geography. In his book *Syntaxis Mathematica*, Ptolemy derived many of the trigonometric identities of this chapter. Although his purposes were more related to solving triangles (Chapter 8), today a major focus of trigonometry is on the relationships of the functions themselves, and it is important to be able to change the form of a trigonometric expression in many different ways.

Nicholas Copernicus is probably best known as the astronomer who revolutionized the world with his heliocentric theory of the universe, but in his book *De revolutionibus orbium coelestium* he also developed a substantial amount of trigonometry. This book was published in the year of his death; as a matter of fact, the first copy off the press was rushed to him as he lay on his deathbed. It was on Copernicus' work that his student Rheti-cus based his ideas, which soon brought trigonometry into full use. In a two-volume work, *Opus palatinum de triangulis*, Rheticus used and calculated elaborate tables for all six trigonometric functions.

The transition of trigonometry from the Renaissance to the modern world is due to a Frenchman, François Viète (1540–1603). He was a lawyer, not a mathematician, and served as a member of the king's council under Henry III and Henry IV. Viète spent his leisure time working on mathematics. He was the first to link trigonometry to the solution of algebraic problems. In 1593, a Belgian ambassador to the court of Henry IV boasted that France had no mathematician capable of solving the equation

$$x^{45} - 45x^{43} + 945x^{41} - \cdots$$
$$- 3{,}795x^3 + 45x = K$$

Viète used trigonometry to find that this equation results when expressing $K = \sin 45\theta$ in terms of $x = 2 \sin \theta$. He was then able to find the positive roots of this equation. It was during this period that the word *trigonometry* was first used.

7

TRIGONOMETRIC EQUATIONS AND IDENTITIES

Contents

Preview

Equation solving is an important skill to be learned in mathematics. In this chapter we consider solving equations in which the unknown is related to the angle in a trigonometric function. A **solution** of a trigonometric equation has the same meaning as the solution of any equation, namely value(s) that make a given equation true. Remember, **to solve an equation** means to find all replacements for the variable that makes the equation true.

This chapter also introduces you to *eight fundamental identities* that can be used to change the form of a trigonometric expression or equation. These identities will then be used to derive a variety of other useful identities in the next chapter. The 17 objectives of this chapter are listed on pages 335–337.

Perspective

Trigonometric identities are frequently used in a variety of calculus applications. In the example shown on this page, the application is finding the length of part of a curve. Notice that first the identity $\cos^2 t + \sin^2 t = 1$ is used (without a remark or warning), and then some half-angle identities are used. We introduce the fundamental identities in Section 7.2 and the half-angle identities in Section 7.5.

Solution First we notice that

$$\frac{dx}{dt} = r - r\cos t \quad \text{and} \quad \frac{dy}{dt} = r\sin t$$

Therefore by (4) we have

$$\mathscr{L} = \int_0^{2\pi} \sqrt{(r - r\cos t)^2 + (r\sin t)^2}\, dt$$

$$= \int_0^{2\pi} \sqrt{r^2 - 2r^2\cos t + r^2\cos^2 t + r^2\sin^2 t}\, dt$$

$$= \int_0^{2\pi} \sqrt{2r^2 - 2r^2\cos t}\, dt$$

$$= r\int_0^{2\pi} \sqrt{2(1 - \cos t)}\, dt$$

By the half-angle formula for $\sin t/2$,

$$\frac{1 - \cos t}{2} = \sin^2\frac{t}{2}, \quad \text{so that} \quad \sqrt{2(1 - \cos t)} = \sqrt{4\sin^2\frac{t}{2}}$$

Since $\sin t/2 \geq 0$ for $0 \leq t \leq 2\pi$, we conclude that

From Fig. 8–29, "Solution" from *Calculus and Analytic Geometry*, Third Edition, by Robert Ellis and Denny Gulick, copyright © 1986 by Harcourt Brace Jovanovich, Inc., reprinted by permission of the publisher.

7.1 Trigonometric Equations

Section 6.5 introduced notation for inverse functions. In this section we will use a similar notation to solve equations.

If $\cos x = \frac{1}{2}$, then write $x = \cos^{-1}(\frac{1}{2})$ to mean that x is *any* angle or real number whose cosine is $\frac{1}{2}$. Note the use of the small letter on $\cos^{-1}(\frac{1}{2})$ rather than the capital C we used for the inverse cosine function. The procedure for finding $\cos^{-1}(\frac{1}{2})$ relies on knowing $\text{Cos}^{-1}(\frac{1}{2})$. The steps for solving $\cos x = y$ are summarized:

1. Find $\text{Cos}^{-1}|y|$; this will give you the reference angle. It can be found by using the table of exact values or a calculator.

2. Find the principal values; use the sign of y to determine the proper quadrant placement. Use reference angles for finding the values of x less than one revolution that satisfy the equation.

   For cosine:
   - y positive: Quadrants I and IV
   - y negative: Quadrants II and III

   For sine:
   - y positive: Quadrants I and II
   - y negative: Quadrants III and IV

   For tangent:
   - y positive: Quadrants I and III
   - y negative: Quadrants II and IV

   Use:

Sine pos	All pos
Tangent pos	Cosine pos

3. For the entire solution, use the period of the function:

   For cosine and sine: Add multiples of 2π or $360°$.
   For tangent: Add multiples of π or $180°$.

EXAMPLE 1 Solve $\cos \theta = \frac{1}{2}$.

SOLUTION STEP 1 $\theta' = \text{Cos}^{-1}|\frac{1}{2}| = 60°$ or $\frac{\pi}{3}$.

STEP 2 $\frac{1}{2}$ is positive; cosine is positive in Quadrants I and IV; find the angles less than one revolution whose reference angle is $60°$ or $\frac{\pi}{3}$.

	DEGREES	RADIANS
Quadrant I:	$60°$	$\dfrac{\pi}{3}$
Quadrant IV:	$300°$	$\dfrac{5\pi}{3}$

STEP 3 Add multiples (let k be any integer):

$$\theta = \begin{cases} 60° + 360°k \\ 300° + 360°k \end{cases} \quad \text{or} \quad \theta = \begin{cases} \dfrac{\pi}{3} + 2k\pi \\ \dfrac{5\pi}{3} + 2k\pi \end{cases}$$

The solution set is infinite. To check, select *any* integral value of k, say $k = 5$. From the solution, $300° + 360°(5) = 2,100°$. If this is a solution, it must satisfy the given equation: $\cos 2,100° = \frac{1}{2}$, which checks.

You can use a calculator with graphics capability to approximate solutions to equations. For example, in the text we are looking for the solution of the equation $\cos x = \frac{1}{2}$. On a graphing calculator you can first graph $y = \cos x$ and then graph the function $y = \frac{1}{2}$. This is shown at the right. The solution of the equation $\cos x = \frac{1}{2}$ is found using the trace function. Watch the cursor move along the line $y = .5$; the displays when it is closest to the points of intersection are

X = −64.42105	X = 421.57895
X = 63.473684	X = 660.31579
X = 302.21053	

On the other hand, if you graph $y = \cos x$ and then trace along the cosine curve you obtain the following (x, y) values:

$$(-64.42105°, .43175435)$$

$$(63.473684°, .44660881)$$

$$(302.21053°, .53303173)$$

$$(421.57895°, .47594739)$$

$$(660.31579°, .50476554)$$

Since $y = .5$, you can see from the above components that there is a great deal of approximation. Do not be misled about a calculator's accuracy because it shows you 10 or 12 decimal places. You could redraw these curves using the ZOOM key to obtain better accuracy. However, a picture such as the one shown here can help to give you a good intuitive idea about what you are looking for when solving an equation.

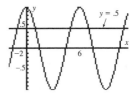

Calculator graph of the function $y = \cos x$ and $y = \frac{1}{2}$

For the graph shown above the following choices were made for the domain and range values (we are working in degree mode for this problem):

Xmin: −90	Ymin: −1.1
Xmax: 720	Ymax: 1.1
Xscl: 30	Yscl: .1

EXAMPLE 2 Solve $\cos \theta = -\frac{1}{2}$.

SOLUTION $\theta' = \text{Cos}^{-1}|-\frac{1}{2}| = 60°$ or $\frac{\pi}{3}$; this time the reference angle is in Quadrants II and III (since cosine is negative):

$$\theta = \begin{cases} 120° + 360°k \\ 240° + 360°k \end{cases} \quad \text{or} \quad \theta = \begin{cases} \dfrac{2\pi}{3} + 2k\pi \\ \dfrac{4\pi}{3} + 2k\pi \end{cases}$$

Graph in radian mode

In trigonometry it is customary to delete the step in which multiples of 2π (for cosine and sine) or π (for tangent) are added to the principal values. In the examples that follow, you need find only the positive solutions less than one revolution. Give your solutions in radians unless specifically asked for in degrees. The reason for this is that radians occur in calculus more often than degrees.

EXAMPLE 3 Solve $\tan \theta = -.66956$ for $0 \le \theta < 2\pi$.

SOLUTION By calculator: .66956 inv tan or tan⁻¹ .66956

gives the reference angle $.59000301$; when placed in the proper quadrants (II and IV), the solution is

2.5515896, 5.6931823

$$\underset{\pi - \theta'}{\uparrow} \qquad \underset{2\pi - \theta}{\uparrow}$$

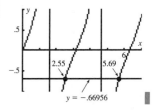

EXAMPLE 4   Solve $\sin \theta = -.4446$ for $0° \le \theta < 360°$ (in degrees).

SOLUTION   By calculator: $\boxed{.4446}\ \boxed{\text{inv}}\ \boxed{\sin}$ or $\boxed{\sin^{-1}}\ \boxed{.4446}$

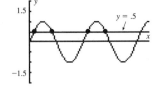

gives the reference angle 26.397749720; when placed in the proper quadrants (III and IV), the solution is

$$\underbrace{206.3977497°}, \ \underbrace{333.6022503°}$$

$$\uparrow \qquad\qquad\qquad \uparrow$$

Quadrant III: $180° + \theta'$ $\qquad$ Quadrant IV: $360° - \theta'$

If the angle is a multiple of θ, the following inequalities hold true:

$$0° \le \theta < 360° \qquad\qquad 0 \le \theta < 2\pi$$
$$0° \le 2\theta < 720° \qquad\qquad 0 \le 2\theta < 4\pi$$
$$0° \le 3\theta < 1{,}080° \qquad\quad 0 \le 3\theta < 6\pi$$
$$0° \le 4\theta < 1{,}440° \qquad\quad 0 \le 4\theta < 8\pi$$
$$\vdots \qquad\qquad\qquad\qquad \vdots$$

EXAMPLE 5   Solve $\sin 2\theta = \frac{1}{2}$ for $0 \le \theta < 2\pi$.

SOLUTION   Since $0 \le \theta < 2\pi$, solve for 2θ such that $0 \le 2\theta < 4\pi$ and $\theta' = \frac{\pi}{6}$.

$$2\theta = \frac{\pi}{6}, \ \frac{5\pi}{6}, \ \frac{13\pi}{6}, \ \frac{17\pi}{6}$$

$$\left.\begin{array}{c} \dfrac{5\pi}{6} + 2\pi \\[4pt] \dfrac{\pi}{6} + 2\pi \end{array}\right\} \begin{array}{l} \text{Add } 2\pi \text{ to each of} \\ \text{the principal values.} \end{array}$$

Quadrant I   Quadrant II (sine is positive in Quadrants I and II)

Mentally solve each equation for θ:

$$2\theta = \frac{\pi}{6} \qquad 2\theta = \frac{5\pi}{6} \qquad 2\theta = \frac{13\pi}{6} \qquad 2\theta = \frac{17\pi}{6}$$

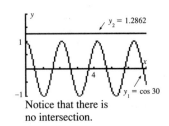

As you solve these equations, notice that in each case θ is between zero and 2π. The solution is

$$\frac{\pi}{12}, \ \frac{5\pi}{12}, \ \frac{13\pi}{12}, \ \frac{17\pi}{12}$$

EXAMPLE 6   Solve $\cos 3\theta = 1.2862$.

SOLUTION   The solution is **empty** because $-1 \le \cos 3\theta \le 1$, which is why $\cos 3\theta \ne 1.2862$.

Notice that there is no intersection.

EXAMPLE 7 Solve $\cos 3\theta = -.68222$ for $0 \leq \theta < 2\pi$ and round your answer to 2 decimal places.

SOLUTION $\theta' = \text{Cos}^{-1}|-.68222| \approx .820001652$ This is the reference angle for 3θ.

Since cosine is negative in quadrants II and III, we have

<table>
<tr><td></td><td>3θ is in Quad II (note negative value):</td><td>3θ is in Quad III:</td></tr>
<tr><td></td><td>$3\theta = 2.321591002$</td><td>$3\theta = 3.961594305$</td></tr>
<tr><td></td><td>$\theta = .7738636675$</td><td>$\theta = 1.32053145$</td></tr>
<tr><td>Add 2π</td><td>$3\theta = 8.60477631$</td><td>$3\theta = 10.24477961$</td></tr>
<tr><td></td><td>$\theta = 2.86825877$</td><td>$\theta = 3.414926537$</td></tr>
<tr><td>Add 4π</td><td>$3\theta = 14.88796162$</td><td>$3\theta = 16.52796492$</td></tr>
<tr><td></td><td>$\theta = 4.962653872$</td><td>$\theta = 5.50932164$</td></tr>
</table>

The rounded solution is

$$.77, \quad 1.32, \quad 2.87, \quad 3.41, \quad 4.96, \quad 5.51$$

$y = -.68222$

Algebraic Solution for the Angle

When solving trigonometric equations, you need to distinguish the unknown from the angle, and the angle from the function. Consider

The unknown is x.

$$\cos(2x + 1) = 0$$

The function is cosine.   The angle, or argument, is $2x + 1$.

The steps in solving a trigonometric equation are now given.

PROCEDURE FOR SOLVING TRIGONOMETRIC EQUATIONS

1. Solve for a single trignometric function. You may use identities, factoring, or the quadratic formula.
2. Solve for the argument (angle). You will use the definition of the inverse trigonometric functions for this step.
3. Solve for the unknown.

EXAMPLE 8 Solve $2 \cos \theta \sin \theta = \sin \theta$ for $0 \leq \theta < 2\pi$.

SOLUTION This problem is solved by factoring:

$$2 \cos \theta \sin \theta - \sin \theta = 0$$

$$\sin \theta(2 \cos \theta - 1) = 0$$

$$\sin \theta = 0 \qquad 2 \cos \theta - 1 = 0$$

$$\theta = 0, \pi \qquad \cos \theta = \frac{1}{2}$$

$$\theta = \frac{\pi}{3}, \frac{5\pi}{3}$$

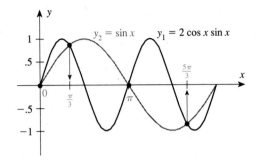

CALCULATOR COMMENT

If you have a graphing calculator, you can rewrite the given equation so that it is equal to zero. Then, you can graph a single function and look for the places where it crosses the x-axis.

Xmin = 0
Xmax = 7
Xscl = 1
Ymin = -3
Ymax = 3
Yscl = 1

$Y = 2 \cos X \sin X - \sin X$

The ⬜TRACE feature will then approximate the roots. We find (in addition to the obvious solution of $x = 0$):

| X = 1.0315789, X = 3.1684211, X = 5.2315789 |

Solution: **0, π, $\frac{\pi}{3}$, $\frac{5\pi}{3}$**

EXAMPLE 9 Solve $2 \sin^2\theta = 1 + 2 \sin \theta$ for $0 \le \theta < 2\pi$.

SOLUTION This problem is solved by the quadratic formula: If $ax^2 + bx + c = 0, a \ne 0$, then

$$x = \frac{-b \pm \sqrt{b^2 - 4ac}}{2a}$$

Since $2 \sin^2\theta - 2 \sin \theta - 1 = 0$, let $x = \sin \theta$ in the quadratic formula:

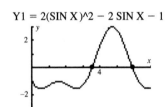

$$\sin \theta = \frac{2 \pm \sqrt{4 - 4(2)(-1)}}{2(2)}$$

$$= \frac{1 \pm \sqrt{3}}{2} \approx 1.366, \ -.366025$$

Reject since $-1 \le \sin \theta \le 1$.

Solve $\sin \theta \approx -.366025$ by using a calculator to find a reference angle of .3747. Since the sine is negative in Quadrants III and IV, the solutions are $\pi + .3747 \approx 3.5163$ and $2\pi - .3747 \approx 5.9085$. To four decimal places: **3.5163, 5.9085.**

EXAMPLE 10 Solve the equation $\sin x - 1 = \sqrt{1 - \sin^2 x}$ on $[0, 2\pi)$.

SOLUTION In order to avoid radicals, we square both sides of the given equation:

$$(\sin x - 1)^2 = 1 - \sin^2 x$$

$$\sin^2 x - 2 \sin x + 1 = 1 - \sin^2 x$$

$$2 \sin^2 x - 2 \sin x = 0$$

$$\sin x(\sin x - 1) = 0$$

$$\sin x = 0 \quad \text{or} \quad \sin x - 1 = 0$$

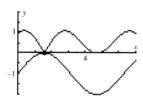

From these equations we see $x = 0$, π, $\frac{\pi}{2}$. Since we squared both sides we may have introduced extraneous roots. We must check each of these values in the original equation.

Check $x = 0$: $\sin 0 - 1 = \sqrt{1 - \sin^2 0}$; this is false.

Check $x = \pi$: $\sin \pi - 1 = \sqrt{1 - \sin^2 \pi}$; this is false.

Check $x = \frac{\pi}{2}$: $\sin \frac{\pi}{2} - 1 = \sqrt{1 - \sin^2 \frac{\pi}{2}}$; this is true.

The root is $\frac{\pi}{2}$.

EXAMPLE 11 Find the zeros of the function $f(x) = \cos x + x^2$.

SOLUTION Since $\cos x$ is positive in the first quadrant ($0 < x < \frac{\pi}{2}$) and x^2 is positive in the first quadrant, we see $f > 0$. For $x > \frac{\pi}{2}$ we see $x^2 > 1$ and since $-1 \le \cos x \le 1$ we see $f > 0$ for all $x \ge 0$. Since f is an even function, the graph is symmetric with respect to the y-axis so we see that $f > 0$ for all x. Thus, there are **no zeros**.

CALCULATOR COMMENT

We can use a calculator to estimate the roots of equations that would otherwise be difficult to solve. For example, solve $\cos 2x + x^2 - 2 = 0$. The graph is shown.

Using the $\boxed{\text{TRACE}}$ we approximate the positive root.

| X = 1.8315789   Y = .48764106 |

| X = 1.7052632   Y = −.0561323 |

Xmin = −6   Ymin = −6
Xmax = 6   Ymax = 6
Xscl = 2   Yscl = 2

The Location Theorem tells us there is a root between these x values. Use the $\boxed{\text{ZOOM}}$ to enlarge the graph in a BOX.

We can obtain any desired degree of accuracy.

| X = 1.7203048   Y = .0038221 |

| X = 1.710558   Y = −.0351782 |

And again,

| X = 1.7195866   Y = 9.2949E − 4 |

| X = 1.7191762   Y = −7.221E − 4 |

Since the graph is symmetric with respect to the y-axis, there is a negative root to match the positive root we have just found. Thus, we estimate the roots to be ± 1.719.

EXAMPLE 12 Solve $\text{Sin}^{-1}\frac{x}{2} + \text{Cos}^{-1}\frac{x}{3} = \frac{2\pi}{3}$ where $x > 0$ using a graphing calculator.

SOLUTION Graph

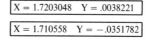

$$Y1 = \text{Sin}^{-1}(X/2) + \text{Cos}^{-1}(X/3) \quad \text{and} \quad Y2 = (2\pi/3)$$

Use the $\boxed{\text{TRACE}}$ on the line to estimate the real roots:

| X = 1.8631579   Y = 2.0943951 |

By using the $\boxed{\text{ZOOM}}$, you can obtain any reasonable degree of accuracy. In particular, to two decimal places we find

| X = 1.8581717   Y = 2.0943951 |

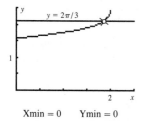

Xmin = 0   Ymin = 0
Xmax = 3   Ymax = 3
Xscl = .5   Yscl = .5

so $x \approx 1.86$.

A *Solve each of the equations in Problems 1–12. Use exact values.*

1. $\sin x = \frac{1}{2}$ $(0 \le x < \frac{\pi}{2})$
2. $\sin x = \frac{1}{2}$ $(0° \le x < 90°)$
3. $\sin x = -\frac{1}{2}$ $(-\frac{\pi}{2} \le x \le \frac{\pi}{2})$
4. $\sin x = -\frac{1}{2}$ $(0 \le x \le \pi)$
5. $\sin x = \frac{1}{2}$
6. $\sin x = -\frac{1}{2}$
7. $\cos x = \frac{1}{2}$ $(0 \le x < \frac{\pi}{2})$
8. $\cos x = \frac{1}{2}$ $(0° \le x < 90°)$
9. $\cos x = -\frac{1}{2}$ $(0° \le x \le 180°)$
10. $\cos x = -\frac{1}{2}$ $(0 \le x \le \pi)$
11. $\cos x = \frac{1}{2}$
12. $\cos x = -\frac{1}{2}$

Solve each of the equations in Problems 13–27 for exact values such that $0 \le x < 2\pi$.

13. $\cos 2x = \frac{1}{2}$
14. $\cos 3x = \frac{1}{2}$
15. $\cos 2x = -\frac{1}{2}$
16. $\sin 2x = \frac{\sqrt{2}}{2}$
17. $\sin 2x = -\frac{\sqrt{3}}{2}$
18. $\sin 3x = \frac{\sqrt{2}}{2}$
19. $\tan 3x = 1$
20. $\tan 3x = -1$
21. $\sec 2x = -\frac{2\sqrt{3}}{3}$
22. $(\sin x)(\cos x) = 0$
23. $(\sec x)(\tan x) = 0$
24. $(\sin x)(\cot x) = 0$
25. $(\cot x)(\cos x) = 0$
26. $(\csc x - 2)(2 \cos x - 1) = 0$
27. $(\sec x - 2)(2 \sin x - 1) = 0$

B *Solve each of the equations in Problems 28–41 for $0 \le x < 2\pi$. Use exact values where possible, but state approximate answers to four decimal places.*

28. $\tan^2 x = \sqrt{3} \tan x$
29. $\tan^2 x = \tan x$
30. $\sin^2 x = \frac{1}{2}$
31. $\cos^2 x = \frac{1}{2}$
32. $3 \sin x \cos x = \sin x$
33. $2 \cos x \sin x = \sin x$
34. $\sin^2 x - \sin x - 2 = 0$
35. $\cos^2 x - 1 - \cos x = 0$
36. $4 \cot^2 x - 8 \cot x + 3 = 0$
37. $\tan^2 x - 3 \tan x + 1 = 0$
38. $\sec^2 x - \sec x - 1 = 0$
39. $\csc^2 x - \csc x - 1 = 0$
40. $\cos x + 1 = \sqrt{3}$
41. $\sin x + 1 = \sqrt{3}$

Solve each of the equations in Problems 42–53 for $0 \le x < 2\pi$ correct to two decimal places.

42. $2 \cos 2x \sin 2x = \sin 2x$
43. $\sin 2x + 2 \cos x \sin 2x = 0$
44. $\cos 3x + 2 \sin 2x \cos 3x = 0$
45. $\sin 2x + 1 = \sqrt{3}$
46. $\cos 3x - 1 = \sqrt{2}$
47. $\tan 2x + 1 = \sqrt{3}$
48. $1 - 2 \sin^2 x = \sin x$
49. $2 \cos^2 x - 1 = \cos x$
50. $2 \cos x \sin x + \cos x = 0$
51. $1 - \sin x = 1 - 2 \sin^2 x$
52. $\cos x = 2 \sin x \cos x$
53. $\sin^2 3x + \sin 3x + 1 = 1 - \sin^2 3x$

Find all solutions (in radian measure) of the equations given in Problems 54–59.

54. $\tan x = -\sqrt{3}$
55. $\cos x = -\frac{\sqrt{3}}{2}$
56. $\sin x = -\frac{\sqrt{2}}{2}$
57. $\sin x = .3907$
58. $\cos x = .2924$
59. $\tan x = 1.376$

C

60. SOUND WAVES A tuning fork vibrating at 264 Hz (frequency $f = 264$) with an amplitude of .0050 cm produces C on the musical scale and can be described by an equation of the form

$$y = .0050 \sin 528\pi x$$

Find the smallest positive value of x (correct to four decimal places) for which $y = .0020$.

61. ELECTRICAL In a certain electric circuit, the electromotive force V (in volts) and the time t (in seconds) are related by an equation of the form

$$V = \cos 2\pi t$$

Find the smallest positive value for t (correct to three decimal places) for which $V = .400$.

62. SPACE SCIENCE The orbit of a certain satellite alternates above and below the equator according to the equation

$$y = 4,000 \sin\left(\frac{\pi}{45} t + \frac{5\pi}{18}\right)$$

where t is the time (in minutes) and y is the distance (in kilometers) from the equator. Find the times at which the satellite crosses the equator during the first hour and a half (that is, for $0 \le t \le 90$).

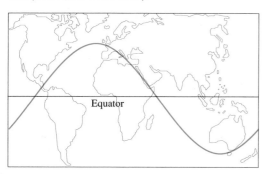

Equator

Graphing Calculator Problems

Solve the trigonometric equations (to the nearest hundredth) in Problems 63–68 for $0 \le x < 2\pi$.

63. $\cos 2x - 1 = \sin 2x$

64. $\sin 2x + x^2 = 0$

65. $x + \cos x = 0$

66. $\sin^{-1}x + 2 \cos^{-1}x = \pi$

67. $\sin^{-1}\frac{x}{3} + \cos^{-1}\frac{x}{2} = \frac{\pi}{3}$

68. $\sin^{-1}\frac{x}{2} + \tan^{-1}x = \frac{\pi}{2}$

7.2 Fundamental Identities

In the previous section we solved some trigonometric equations. In this section we focus on trigonometric identities. The procedure for proving identities is quite different from the procedure for solving equations. Compare the following two examples from algebra:

SOLVE: $2x + 3x = 5$

SOLUTION: $2x + 3x = 5$ Given

$5x = 5$ Combining similar terms

$x = 1$ Multiply both sides by $\frac{1}{5}$

PROVE: $2x + 3x = 5x$

SOLUTION: $2x + 3x = (2 + 3)x$ Distributive property

$= 5x$ Closure

$2x + 3x = 5x$ Transitive property

Notice that when *solving* the equation, $2x + 3x = 5$ was given and *used as a starting point*. On the other hand, when *proving* the identity, $2x + 3x = 5x$ *could not be used as a starting point*. Indeed $2x + 3x = 5x$ was the *last step*, not the first step. The reason for this difference in procedure is apparent if you look at the addition and multiplication principles:

Addition principle: If $a = b$, then $a + c = b + c$.

Multiplication principle: If $a = b$, then $ac = bc$.

In both cases you must *know* that $a = b$ before you can use the addition or multiplication principles. If you are asked to *prove* that $a = b$, you cannot assume $a = b$ to work the problem. You must begin with what is known to be true and *end* with the given identity.

All our work with trigonometric identities is ultimately based on eight basic identities called the **fundamental identities.** Notice that these identities are classified into three categories and numbered for later reference. Values of θ that cause division by zero are excluded.

FUNDAMENTAL IDENTITIES

RECIPROCAL IDENTITIES:

1. $\sec \theta = \dfrac{1}{\cos \theta}$

2. $\csc \theta = \dfrac{1}{\sin \theta}$

3. $\cot \theta = \dfrac{1}{\tan \theta}$

RATIO IDENTITIES:

4. $\tan \theta = \dfrac{\sin \theta}{\cos \theta}$

5. $\cot \theta = \dfrac{\cos \theta}{\sin \theta}$

PYTHAGOREAN IDENTITIES:

6. $\sin^2\theta + \cos^2\theta = 1$ 7. $1 + \tan^2\theta = \sec^2\theta$ 8. $\cot^2\theta + 1 = \csc^2\theta$

The proofs of these identities follow directly from the definitions of the trigonometric functions.

CALCULATOR COMMENT

You can verify each of these identities on a graphing calculator. For example, for Identity 6, graph $\boxed{\text{Y1} = (\text{SIN X})^2 + (\text{COS X})^2}$. The graph is a constant function:

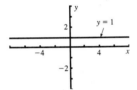

You can similarly check the other Pythagorean identities or the reciprocal identities. For the ratio identities, graph $\boxed{\text{Y1} = \text{SIN X/COS X}}$ and the result is recognized as a tangent curve:

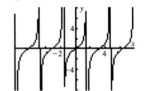

EXAMPLE 1   Write all six trigonometric functions in terms of $\sin \theta$.

SOLUTION   **a.** $\sin \theta = \boldsymbol{\sin \theta}$

b. $\cos \theta = \pm\sqrt{1 - \sin^2\boldsymbol{\theta}}$ From Identity 6

c. $\tan \theta = \dfrac{\sin \theta}{\cos \theta}$ Identity 4

$\quad\quad\quad = \dfrac{\boldsymbol{\sin \theta}}{\pm\sqrt{1 - \sin^2\boldsymbol{\theta}}}$ From part b

d. $\cot \theta = \dfrac{1}{\tan \theta}$ Identity 3

$\quad\quad\quad = \dfrac{\pm\sqrt{1 - \sin^2\boldsymbol{\theta}}}{\boldsymbol{\sin \theta}}$ From part c

e. $\csc \theta = \dfrac{1}{\boldsymbol{\sin \theta}}$ From Identity 2

f. $\sec \theta = \dfrac{1}{\cos \theta}$ From Identity 1

$\quad\quad\quad = \dfrac{1}{\pm\sqrt{1 - \sin^2\boldsymbol{\theta}}}$ From part b

---

The $\pm$ sign we have been using, as in

$$\cos\theta = \pm\sqrt{1 - \sin^2\theta}$$

means that $\cos\theta$ is positive for some values of θ and negative for other values of θ. The plus sign or the minus sign is chosen by determining the proper quadrant, as shown in Example 2.

EXAMPLE 2 Given $\sin\theta = \frac{3}{5}$ and $\tan\theta < 0$, find the other functions of θ.

SOLUTION Since the tangent is negative and the sine is positive, the quadrant is II. Thus

$$\cos\theta = -\sqrt{1 - \sin^2\theta} \qquad \text{Since the cosine is negative in Quadrant II}$$
$$= -\sqrt{1 - (\tfrac{3}{5})^2}$$
$$= -\sqrt{1 - \tfrac{9}{25}}$$
$$= -\sqrt{\tfrac{16}{25}}$$
$$= -\tfrac{4}{5}$$

Also, $$\tan\theta = \frac{\sin\theta}{\cos\theta} = \frac{3/5}{-4/5} = -\frac{3}{4}$$

Using the reciprocal identities, **cot $\theta = -\frac{4}{3}$, sec $\theta = -\frac{5}{4}$, and csc $\theta = \frac{5}{3}$.**

7.2 Problem Set

A

1. State from memory the eight fundamental identities.

In Problems 2–9 state the quadrant or quadrants in which θ may lie to make the expression true.

2. $\sin\theta = \sqrt{1 - \cos^2\theta}$

3. $\sin\theta = -\sqrt{1 - \cos^2\theta}$

4. $\sec\theta = -\sqrt{1 + \tan^2\theta}$

5. $\sec\theta = \sqrt{1 + \tan^2\theta}$

6. $\csc\theta = \sqrt{1 + \cot^2\theta}$; $\tan\theta < 0$

7. $\cos\theta = -\sqrt{1 - \sin^2\theta}$; $\sin\theta > 0$

8. $\tan\theta = \sqrt{\sec^2\theta - 1}$; $\cos\theta < 0$

9. $\csc\theta = \sqrt{1 + \cot^2\theta}$; $\cos\theta > 0$

Write each of the expressions in Problems 10–18 as a single trigonometric function of some angle by using one of the eight fundamental identities.

10. $\dfrac{\sin 50°}{\cos 50°}$

11. $\dfrac{\cos(A + B)}{\sin(A + B)}$

12. $\dfrac{1}{\sec 75°}$

13. $\dfrac{1}{\cot(\frac{\pi}{15})}$

14. $\tan 42° \cos 42°$

15. $\cot\frac{\pi}{8} \sin\frac{\pi}{8}$

16. $1 - \cos^2 18°$

17. $-\sqrt{1 - \sin^2 127°}$

18. $\sec^2(\frac{\pi}{6}) - 1$

Evaluate the expressions in Problems 19–24 by using one of the eight fundamental identities.

19. $\cos 128° \sec 128°$

20. $\sin^2\frac{\pi}{3} + \cos^2\frac{\pi}{3}$

21. $\sec^2\frac{\pi}{6} - \tan^2\frac{\pi}{6}$

22. $\cot^2 45° - \csc^2 45°$

23. $\tan^2 135° - \sec^2 135°$

24. $\csc 85° \sin 85°$

25. Prove that $\csc\theta = 1/\sin\theta$.

26. Prove that $\cot\theta = 1/\tan\theta$.

27. Prove that $\cot\theta = \dfrac{\cos\theta}{\sin\theta}$.

28. Prove that $\cot^2\theta + 1 = \csc^2\theta$.

29. Prove that $1 + \tan^2\theta = \sec^2\theta$.

B *In Problems 30–33, write all the trigonometric functions in terms of the given function.*

30. $\tan\theta$

31. $\cot\theta$

32. $\sec\theta$

33. $\csc\theta$

In Problems 34–45, find the other functions of θ using the given information.

34. $\cos \theta = \frac{3}{5}$; $\tan \theta > 0$

35. $\cos \theta = \frac{3}{5}$; $\csc \theta < 0$

36. $\cos \theta = \frac{5}{13}$; $\tan \theta < 0$

37. $\cos \theta = \frac{5}{13}$; $\tan \theta > 0$

38. $\tan \theta = \frac{5}{12}$; $\sin \theta > 0$

39. $\tan \theta = \frac{5}{12}$; $\sin \theta < 0$

40. $\sin \theta = \frac{2}{3}$; $\sec \theta > 0$

41. $\sin \theta = \frac{2}{3}$; $\sec \theta < 0$

42. $\sec \theta = \frac{\sqrt{34}}{5}$; $\tan \theta < 0$

43. $\sec \theta = \frac{\sqrt{34}}{5}$; $\tan \theta > 0$

44. $\csc \theta = -\frac{\sqrt{10}}{3}$; $\cos \theta > 0$

45. $\csc \theta = -\frac{\sqrt{10}}{3}$; $\cos \theta < 0$

Simplify the expressions in Problems 46–55 using only sines, cosines, and the fundamental identities.

46. $\dfrac{1 - \sin^2\theta}{\cos \theta}$

47. $\dfrac{1 - \cos^2\theta}{\sin \theta}$

48. $\dfrac{\sin \theta}{\cos \theta} + \dfrac{\cos \theta}{\sin \theta}$

49. $\dfrac{1}{1 + \cos \theta} + \dfrac{1}{1 - \cos \theta}$

50. $\dfrac{\dfrac{\sin \theta}{\cos \theta} + \dfrac{\cos \theta}{\sin \theta}}{\dfrac{1}{\sin \theta \cos \theta}}$

51. $\sin \theta + \dfrac{\cos^2\theta}{\sin \theta}$

52. $\dfrac{\cos \theta + \dfrac{\sin^2\theta}{\cos \theta}}{\sin \theta}$

53. $\dfrac{\sin \theta - \dfrac{\cos^2\theta}{\sin \theta}}{\cos \theta}$

54. $\dfrac{\dfrac{\cos^4\theta}{\sin^2\theta} + \cos^2\theta}{\dfrac{\cos^2\theta}{\sin^2\theta}}$

55. $\dfrac{\dfrac{\sin \theta}{\cos \theta} + \dfrac{\cos \theta}{\sin \theta}}{\dfrac{1}{\sin \theta}}$

Reduce the expressions in Problems 56–63 so that they involve only sines and cosines, and then simplify.

56. $\sin \theta + \cot \theta$

57. $\sec \theta + \tan \theta$

58. $\dfrac{\tan \theta + \cot \theta}{\sec \theta \csc \theta}$

59. $\dfrac{\sec \theta + \csc \theta}{\tan \theta \cot \theta}$

60. $\sec^2\theta + \tan^2\theta$

61. $\csc^2\theta + \cot^2\theta$

62. $(\cot \theta - \sec \theta)(\sin \theta \cos \theta)$

63. $(\tan \theta - \csc \theta)(\cos \theta \sin \theta)$

7.3   Proving Identities

In Section 7.2 we considered eight fundamental identities, which are used to simplify and change the form of a variety of trigonometric expressions. Suppose you are given a trigonometric equation such as

$$\tan \theta + \cot \theta = \sec \theta \csc \theta$$

and are asked to show that it is an identity. You must be careful not to treat this problem as though it were an algebraic equation. When asked to prove an identity, do *not* start with the given expression, since you cannot assume it is true. You should *begin* with what you know is true and *end* with the given identity. There are three ways to proceed:

1. Reduce the left-hand side to the right-hand side by using algebra and the fundamental identities.

2. Reduce the right-hand side to the left-hand side.

3. Reduce both sides independently to the same expression.

EXAMPLE 1 Prove that $\tan \theta + \cot \theta = \sec \theta \csc \theta$.

SOLUTION Begin with either the left- or right-hand side:

$$\tan \theta + \cot \theta = \frac{\sin \theta}{\cos \theta} + \frac{\cos \theta}{\sin \theta}$$

$$= \frac{\sin^2\theta + \cos^2\theta}{\cos \theta \sin \theta}$$

$$= \frac{1}{\cos \theta \sin \theta}$$

This is algebraically simplified, so return to the other side and begin anew:

$$\sec \theta \csc \theta = \frac{1}{\cos \theta} \cdot \frac{1}{\sin \theta}$$

$$= \frac{1}{\cos \theta \sin \theta}$$

This too is simplified, but notice that the simplified forms for both the left and right sides are the same. Therefore

$$\tan \theta + \cot \theta = \sec \theta \csc \theta$$

CALCULATOR COMMENT

You can use a graphing calculator to verify the given equations are identities. Let Y1 be the function on the left of the equation, and Y2 be the function on the right of the equation. Graph both Y1 and Y2, and if the graphs are identical, then you can assume that the given equation is an identity. For Example 1, we found both graphs looked the same:

$$Y1 = TAN\,X + COT\,X$$
$$Y2 = SEC\,X + CSC\,X$$

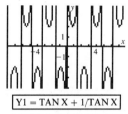

Y1 = TAN X + 1/TAN X

Y2 = 1/COS X + 1/SIN X

Usually it is easier to begin with the more complicated side and try to reduce it to the simpler side. If both sides seem equally complex, you might change all the functions to sines and cosines and then simplify.

EXAMPLE 2 Prove that $2 \csc^2\theta = \dfrac{1}{1 + \cos \theta} + \dfrac{1}{1 - \cos \theta}$.

SOLUTION Begin with the more complicated side:

$$\frac{1}{1 + \cos \theta} + \frac{1}{1 - \cos \theta} = \frac{(1 - \cos \theta) + (1 + \cos \theta)}{(1 + \cos \theta)(1 - \cos \theta)}$$

$$= \frac{2}{1 - \cos^2\theta}$$

$$= \frac{2}{\sin^2\theta}$$

$$= 2 \csc^2\theta$$

EXAMPLE 3 Prove that $\dfrac{\sec 2\lambda + \cot 2\lambda}{\sec 2\lambda} = 1 + \csc 2\lambda - \sin 2\lambda$.

SOLUTION Begin with the left-hand side. When working with a fraction consisting of a single function as a denominator, it is often helpful to separate the fraction into the sum of several fractions:

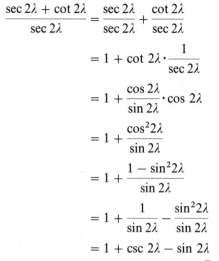

$$\frac{\sec 2\lambda + \cot 2\lambda}{\sec 2\lambda} = \frac{\sec 2\lambda}{\sec 2\lambda} + \frac{\cot 2\lambda}{\sec 2\lambda}$$

$$= 1 + \cot 2\lambda \cdot \frac{1}{\sec 2\lambda}$$

$$= 1 + \frac{\cos 2\lambda}{\sin 2\lambda} \cdot \cos 2\lambda$$

$$= 1 + \frac{\cos^2 2\lambda}{\sin 2\lambda}$$

$$= 1 + \frac{1 - \sin^2 2\lambda}{\sin 2\lambda}$$

$$= 1 + \frac{1}{\sin 2\lambda} - \frac{\sin^2 2\lambda}{\sin 2\lambda}$$

$$= 1 + \csc 2\lambda - \sin 2\lambda$$

EXAMPLE 4 Prove that $\dfrac{\cos \theta}{1 - \sin \theta} = \dfrac{1 + \sin \theta}{\cos \theta}$.

SOLUTION Sometimes, when there is a binomial in the numerator or denominator, the identity can be proved by multiplying one side by 1, where 1 is written in the form of the conjugate of the binomial. When changing one side, keep a sharp eye on the other side, since it often gives a clue about what to do. Thus in this example we can multiply the numerator and denominator of the left-hand side by $1 + \sin \theta$, as shown at the right.

$$\frac{\cos \theta}{1 - \sin \theta} = \frac{\cos \theta}{1 - \sin \theta} \cdot \frac{1 + \sin \theta}{1 + \sin \theta}$$

$$= \frac{\cos \theta(1 + \sin \theta)}{1 - \sin^2 \theta}$$

$$= \frac{\cos \theta(1 + \sin \theta)}{\cos^2 \theta}$$

$$= \frac{1 + \sin \theta}{\cos \theta}$$

EXAMPLE 5 Prove that $\dfrac{\sec^2 2\theta - \tan^2 2\theta}{\tan 2\theta + \sec 2\theta} = \dfrac{\cos 2\theta}{1 + \sin 2\theta}$.

SOLUTION Sometimes the identity can be proved by factoring:

$$\frac{\sec^2 2\theta - \tan^2 2\theta}{\tan 2\theta + \sec 2\theta} = \frac{(\sec 2\theta + \tan 2\theta)(\sec 2\theta - \tan 2\theta)}{\tan 2\theta + \sec 2\theta}$$

$$= \sec 2\theta - \tan 2\theta$$

$$= \frac{1}{\cos 2\theta} - \frac{\sin 2\theta}{\cos 2\theta}$$

$$= \frac{1 - \sin 2\theta}{\cos 2\theta}$$

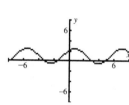

$$= \frac{1 - \sin 2\theta}{\cos 2\theta} \cdot \frac{1 + \sin 2\theta}{1 + \sin 2\theta}$$

$$= \frac{1 - \sin^2 2\theta}{\cos 2\theta(1 + \sin 2\theta)}$$

$$= \frac{\cos^2 2\theta}{\cos 2\theta(1 + \sin 2\theta)}$$

$$= \frac{\cos 2\theta}{1 + \sin 2\theta}$$

EXAMPLE 6 Prove that $\dfrac{-2 \sin \theta \cos \theta}{1 - \sin \theta - \cos \theta} = 1 + \sin \theta + \cos \theta$.

SOLUTION Sometimes, when there is a fraction on one side, the identity can be proved by multiplying the other side by 1 written so that the desired denominator is obtained. Thus, for this example,

$$1 + \sin \theta + \cos \theta = (1 + \sin \theta + \cos \theta) \cdot \frac{1 - \sin \theta - \cos \theta}{1 - \sin \theta - \cos \theta}$$

$$= \frac{(1 + \sin \theta + \cos \theta)(1 - \sin \theta - \cos \theta)}{1 - \sin \theta - \cos \theta}$$

$$= \frac{1 - \sin \theta - \cos \theta + \sin \theta - \sin^2\theta - \sin \theta \cos \theta + \cos \theta - \cos \theta \sin \theta - \cos^2\theta}{1 - \sin \theta - \cos \theta}$$

$$= \frac{1 - (\sin^2\theta + \cos^2\theta) - 2 \sin \theta \cos \theta}{1 - \sin \theta - \cos \theta}$$

$$= \frac{-2 \sin \theta \cos \theta}{1 - \sin \theta - \cos \theta}$$

In summary, there is no single method that is best for proving identities. However, the following hints should help:

PROCEDURES FOR PROVING IDENTITIES

1. If one side contains one function only, write all the trigonometric functions on the other side in terms of that function.
2. If the denominator of a fraction consists of only one function, break up the fraction.
3. Simplify by combining fractions.
4. Factoring is sometimes helpful.
5. Change all trigonometric functions to sines and cosines and simplify.
6. Multiply by the conjugate of either the numerator or the denominator.
7. If there are squares of functions, look for alternate forms of the Pythagorean identities.
8. Avoid the introduction of radicals.
9. Keep your destination in sight. Watch where you are going, and know when you are finished.

7.3 Problem Set

A *Prove that the equations in Problems 1–38 are identities.*

1. $\sin \theta = \sin^3\theta + \cos^2\theta \sin \theta$

2. $\sec \theta = \sec \theta \sin^2\theta + \cos \theta$

3. $\tan \theta = \cot \theta \tan^2\theta$

4. $\dfrac{\sin \theta \cos \theta + \sin^2\theta}{\sin \theta} = \cos \theta + \sin \theta$

5. $\tan^2\theta - \sin^2\theta = \tan^2\theta \sin^2\theta$

6. $\cot^2\theta \cos^2\theta = \cot^2\theta - \cos^2\theta$

7. $\tan A + \cot A = \sec A \csc A$

8. $\cot A = \csc A \sec A - \tan A$

9. $\sin x + \cos x = \dfrac{\sec x + \csc x}{\csc x \sec x}$

10. $\dfrac{\cos \gamma + \tan \gamma \sin \gamma}{\sec \gamma} = 1$

11. $\dfrac{1 - \sec^2 t}{\sec^2 t} = -\sin^2 t$

12. $\dfrac{1 + \cot^2 t}{\cot^2 t} = \sec^2 t$

13. $(\sec \theta - \cos \theta)^2 = \tan^2\theta - \sin^2\theta$

14. $\dfrac{\sin \theta}{\csc \theta} + \dfrac{\cos \theta}{\sec \theta} = 1$

15. $1 - \sin 2\theta = \dfrac{1 - \sin^2 2\theta}{1 + \sin 2\theta}$

16. $\dfrac{1 - \tan^2 3\theta}{1 - \tan 3\theta} = 1 + \tan 3\theta$

17. $\sin \lambda = \dfrac{\sin^2\lambda + \sin \lambda \cos \lambda + \sin \lambda}{\sin \lambda + \cos \lambda + 1}$

18. $\dfrac{1 + \cot 2\lambda \sec 2\lambda}{\tan 2\lambda + \sec 2\lambda} = \cot 2\lambda$

19. $\sin 2\alpha \cos 2\alpha(\tan 2\alpha + \cot 2\alpha) = 1$

20. $(\sin \beta - \cos \beta)^2 + (\sin \beta + \cos \beta)^2 = 2$

21. $\csc 3\beta - \cos 3\beta \cot 3\beta = \sin 3\beta$

22. $\dfrac{1 + \cot^2 A}{1 + \tan^2 A} = \cot^2 A$

23. $\dfrac{\sin^2 B - \cos^2 B}{\sin B + \cos B} = \sin B - \cos B$

24. $\dfrac{\tan^2\gamma - \cot^2\gamma}{\tan \gamma + \cot \gamma} = \tan \gamma - \cot \gamma$

25. $\tan^2 2\gamma + \sin^2 2\gamma + \cos^2 2\gamma = \sec^2 2\gamma$

26. $\cot^2 C + \cos^2 C + \sin^2 C = \csc^2 C$

27. $\dfrac{\tan \theta + \cot \theta}{\sec \theta \csc \theta} = 1$

28. $\dfrac{\tan \theta - \cot \theta}{\sec \theta \csc \theta} = \sin^2\theta - \cos^2\theta$

29. $1 + \sin^2\lambda = 2 - \cos^2\lambda$

30. $2 - \sin^2 3\lambda = 1 + \cos^2 3\lambda$

31. $\dfrac{\sin \alpha}{\tan \alpha} + \dfrac{\cos \alpha}{\cot \alpha} = \cos \alpha + \sin \alpha$

32. $\dfrac{1}{1 + \cos 2\alpha} + \dfrac{1}{1 - \cos 2\alpha} = 2 \csc^2 2\alpha$

33. $\sec \beta + \cos \beta = \dfrac{2 - \sin^2\beta}{\cos \beta}$

34. $2 \sin^2 3\beta - 1 = 1 - 2 \cos^2 3\beta$

35. $\dfrac{\tan 2\theta + \cot 2\theta}{\sec 2\theta} = \csc 2\theta$

36. $\dfrac{\tan 3\theta + \cot 3\theta}{\csc 3\theta} = \sec 3\theta$

37. $\dfrac{\sec \lambda + \tan^2\lambda}{\sec \lambda} = 1 + \sec \lambda - \cos \lambda$

38. $\dfrac{\sin 2\lambda}{\tan 2\lambda} + \dfrac{\cos 2\lambda}{\cot 2\lambda} = \cos 2\lambda + \sin 2\lambda$

B *Prove that the equations in Problems 39–60 are identities.*

39. $\dfrac{1 + \tan C}{1 - \tan C} = \dfrac{\sec^2 C + 2 \tan C}{2 - \sec^2 C}$

40. $(\cot x + \csc x)^2 = \dfrac{\sec x + 1}{\sec x - 1}$

41. $\dfrac{\sin^3 x - \cos^3 x}{\sin x - \cos x} = 1 + \sin x \cos x$

42. $\dfrac{\tan^3 t - \cot^3 t}{\tan t - \cot t} = \sec^2 t + \cot^2 t$

43. $\dfrac{1 - \cos \theta}{1 + \cos \theta} = \left(\dfrac{1 - \cos \theta}{\sin \theta}\right)^2$

44. $\dfrac{(\sec^2\gamma + \tan^2\gamma)^2}{\sec^4\gamma - \tan^4\gamma} = 1 + 2\tan^2\gamma$

45. $\dfrac{(\cos^2\gamma - \sin^2\gamma)^2}{\cos^4\gamma - \sin^4\gamma} = 2\cos^2\gamma - 1$

46. $(\sec 2\theta + \csc 2\theta)^2 = \dfrac{1 + 2\sin 2\theta \cos 2\theta}{\cos^2 2\theta \sin^2 2\theta}$

47. $\dfrac{1}{\sec\theta + \tan\theta} = \sec\theta - \tan\theta$

48. $\csc\theta + \cot\theta = \dfrac{1}{\csc\theta - \cot\theta}$

49. $\sec^2 2\lambda + \csc^2 2\lambda = \csc^2 2\lambda \sec^2 2\lambda$

50. $\dfrac{1 + \tan^3\theta}{1 + \tan\theta} = \sec^2\theta - \tan\theta$

51. $\dfrac{1 - \sec^3\theta}{1 - \sec\theta} = \tan^2\theta + \sec\theta + 2$

52. $\dfrac{\cos^2\theta - \cos\theta \csc\theta}{\cos^2\theta \csc\theta - \cos\theta \csc^2\theta} = \sin\theta$

53. $\dfrac{\tan^2\theta - 2\tan\theta}{2\tan\theta - 4} = \dfrac{1}{2}\tan\theta$

54. $\dfrac{\tan\theta}{\cot\theta} - \dfrac{\cot\theta}{\tan\theta} = \sec^2\theta - \csc^2\theta$

55. $\sqrt{(3\cos\theta - 4\sin\theta)^2 + (3\sin\theta + 4\cos\theta)^2} = 5$

56. $\dfrac{\cos\theta + \cos^2\theta}{\cos\theta + 1} = \dfrac{\cos\theta \sin\theta + \cos^2\theta}{\sin\theta + \cos\theta}$

57. $\sec^2\lambda - \csc^2\lambda = (2\sin^2\lambda - 1)(\sec^2\lambda + \csc^2\lambda)$

58. $2\csc A = 2\csc A - \cot A \cos A + \cos^2 A \csc A$

59. $\dfrac{\csc\theta + 1}{\cot^2\theta + \csc\theta + 1} = \dfrac{\sin^2\theta + \sin\theta \cos\theta}{\sin\theta + \cos\theta}$

60. $\dfrac{\cos^4\theta - \sin^4\theta}{(\cos^2\theta - \sin^2\theta)^2} = \dfrac{\cos\theta}{\cos\theta + \sin\theta} + \dfrac{\sin\theta}{\cos\theta - \sin\theta}$

C *Prove that the equations in Problems 61–72 are identities.*

61. $(\cos\alpha - \cos\beta)^2 + (\sin\alpha - \sin\beta)^2$
$= 2 - 2(\cos\alpha \cos\beta + \sin\alpha \sin\beta)$

62. $(\sec\alpha + \sec\beta)^2 - (\tan\alpha - \tan\beta)^2$
$= 2 + 2(\sec\alpha \sec\beta + \tan\alpha \tan\beta)$

63. $\tan A + \cot B = (\sin A \sin B + \cos A \cos B)\sec A \csc B$

64. $\sec A + \csc B = (\cos A \sin^2 B + \sin B \cos^2 A)\sec^2 A \csc^2 B$

65. $(\sin A \cos A \cos B + \sin B \cos B \cos A)\sec A \sec B$
$= \sin A + \sin B$

66. $(\cos A \cos B \tan A + \sin A \sin B \cot B)\csc A \sec B = 2$

67. $\sin\theta + \cos\theta + 1 = \dfrac{2\sin\theta \cos\theta}{\sin\theta + \cos\theta - 1}$

68. $\dfrac{2\tan^2\theta + 2\tan\theta \sec\theta}{\tan\theta + \sec\theta - 1} = \tan\theta + \sec\theta + 1$

69. $\dfrac{\csc\theta + 1}{\csc\theta - 1} - \dfrac{\sec\theta - \tan\theta}{\sec\theta + \tan\theta} = 4\tan\theta \sec\theta$

70. $\dfrac{\cos\theta + \sin\theta}{\cos\theta - \sin\theta} + \dfrac{\cot\theta - 1}{\cot\theta + 1} = \dfrac{-2}{\sin^2\theta - \cos^2\theta}$

71. $\dfrac{\cos\theta + 1}{\cos\theta - 1} + \dfrac{1 - \sec\theta}{1 + \sec\theta} = -2\cot^2\theta - 2\csc^2\theta$

72. $\dfrac{\sin\theta}{1 - \cos\theta} + \dfrac{\cos\theta}{1 - \sin\theta} = (1 + \sin\theta + \cos\theta)(\sec\theta \csc\theta)$

7.4 Addition Laws

When proving identities, it is sometimes necessary to simplify the functional value of the sum or difference of two angles. If α and β represent any two angles,

$$\cos(\alpha - \beta) \neq \cos\alpha - \cos\beta$$

For example, if $\alpha = 60°$ and $\beta = 30°$ then

$$\cos(60° - 30°) = \cos 30° \quad \text{and} \quad \cos 60° - \cos 30° = \dfrac{1}{2} - \dfrac{\sqrt{3}}{2}$$

$$= \dfrac{\sqrt{3}}{2} \qquad\qquad\qquad = \dfrac{1 - \sqrt{3}}{2}$$

Thus $\cos(60° - 30°) \neq \cos 60° - \cos 30°$.

In this section, we discuss twelve more identities. First we consider $\cos(\alpha - \beta)$ expanded in terms of trigonometric functions of single angles. We shall later list this as Identity 16, but it is convenient to discuss it first because it provides the cornerstone for building a great many additional identities.

DIFFERENCE OF ANGLES
IDENTITY

$$\cos(\alpha - \beta) = \cos \alpha \cos \beta + \sin \alpha \sin \beta$$

PROOF   Find the length of any chord in a unit circle with a corresponding arc intercepted by the central angle θ, where θ is in standard position. Let A be the point $(1, 0)$ and P be the point on the intersection of the terminal side of angle θ and the unit circle. This means that the coordinates of P are $(\cos \theta, \sin \theta)$. Now find the length of the chord AP (see Figure 7.1) by using the distance formula:

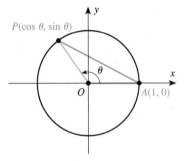

$$|AP| = \sqrt{(1 - \cos \theta)^2 + (0 - \sin \theta)^2}$$
$$= \sqrt{1 - 2 \cos \theta + \cos^2\theta + \sin^2\theta}$$
$$= \sqrt{1 - 2 \cos \theta + 1}$$
$$= \sqrt{2 - 2 \cos \theta}$$

FIGURE 7.1   Length of a chord determined by an angle θ

Next apply this result to a chord determined by any two angles α and β, as shown in Figure 7.2. Let P_α and P_β be the points on the unit circle determined by the angles α and β, respectively. By the previous result,

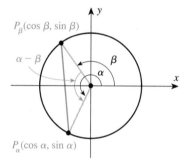

$$|P_\alpha P_\beta| = \sqrt{2 - 2 \cos(\alpha - \beta)}$$

But you could also have found this distance directly via the distance formula:

$$|P_\alpha P_\beta| = \sqrt{(\cos \beta - \cos \alpha)^2 + (\sin \beta - \sin \alpha)^2}$$
$$= \sqrt{\cos^2\beta - 2 \cos \alpha \cos \beta + \cos^2\alpha + \sin^2\beta - 2 \sin \alpha \sin \beta + \sin^2\alpha}$$
$$= \sqrt{(\cos^2\beta + \sin^2\beta) + (\cos^2\alpha + \sin^2\alpha) - 2(\cos \alpha \cos \beta + \sin \alpha \sin \beta)}$$
$$= \sqrt{2 - 2(\cos \alpha \cos \beta + \sin \alpha \sin \beta)}$$

FIGURE 7.2   Distance between P_α and P_β

Finally, equate these quantities since they both represent the distance between P_α and P_β:

$$\sqrt{2 - 2 \cos(\alpha - \beta)} = \sqrt{2 - 2(\cos \alpha \cos \beta + \sin \alpha \sin \beta)}$$
$$2 - 2 \cos(\alpha - \beta) = 2 - 2(\cos \alpha \cos \beta + \sin \alpha \sin \beta)$$
$$-2 \cos(\alpha - \beta) = -2(\cos \alpha \cos \beta + \sin \alpha \sin \beta)$$
$$\cos(\alpha - \beta) = \cos \alpha \cos \beta + \sin \alpha \sin \beta \qquad \square \square \square$$

You can use this identity to find the exact values of functions of angles that are multiples of $15°$, as shown by Example 1.

EXAMPLE 1

$$\cos 345° = \cos 15° \qquad \text{Reduction principle}$$
$$= \cos(45° - 30°) \qquad \text{Since } 45° - 30° = 15°$$
$$= \cos 45° \cos 30° + \sin 45° \sin 30°$$
$$= \frac{1}{2}\sqrt{2} \cdot \frac{1}{2}\sqrt{3} + \frac{1}{2}\sqrt{2} \cdot \frac{1}{2} \qquad \text{Using exact values}$$
$$= \frac{\sqrt{6}}{4} + \frac{\sqrt{2}}{4}$$
$$= \frac{\sqrt{6} + \sqrt{2}}{4}$$

Even though this identity is helpful for making evaluations (as in the preceding example), its real value lies in the fact that it is true for *any* choice of α and β. By making some particular choices for α and β, we find several useful special cases of this identity.

EXAMPLE 2 Prove $\cos(\frac{\pi}{2} - \theta) = \sin \theta$.

SOLUTION This proof is based on the identity

$$\cos(\alpha - \beta) = \cos \alpha \cos \beta + \sin \alpha \sin \beta$$

Let $\alpha = \frac{\pi}{2}$ and $\beta = \theta$:

$$\cos(\frac{\pi}{2} - \theta) = \cos \frac{\pi}{2} \cos \theta + \sin \frac{\pi}{2} \sin \theta$$
$$= 0 \cdot \cos \theta + 1 \cdot \sin \theta$$
$$= \sin \theta$$

Example 2 is one of three identities known as the **cofunction identities**. (Remember that Identities 1–8 are the fundamental identities discussed in Section 7.2.)

COFUNCTION IDENTITIES

For any real number (or angle) θ,

9. $\cos(\frac{\pi}{2} - \theta) = \sin \theta$ 10. $\sin(\frac{\pi}{2} - \theta) = \cos \theta$ 11. $\tan(\frac{\pi}{2} - \theta) = \cot \theta$

The proof of Identity 9 was done in Example 2, and the proof of Identity 10 depends on Identity 9, and is shown below.

PROOF OF IDENTITY 10

$$\cos \theta = \cos[\frac{\pi}{2} - (\frac{\pi}{2} - \theta)]$$
$$= \sin(\frac{\pi}{2} - \theta) \qquad \text{This is Identity 9.}$$

Therefore $\sin(\frac{\pi}{2} - \theta) = \cos \theta$.

Identities involving the tangent are usually proved after proving similar identities for cosine and sine. The fundamental identity $\tan \theta = \sin \theta / \cos \theta$ is applied first, allowing you then to use the appropriate identities for cosine and sine. This process is illustrated with the following proof.

$$\tan(\tfrac{\pi}{2} - \theta) = \frac{\sin(\tfrac{\pi}{2} - \theta)}{\cos(\tfrac{\pi}{2} - \theta)}$$

$$= \frac{\cos \theta}{\sin \theta}$$

$$= \cot \theta \qquad \qquad \square\square\square$$

The cofunction identities allow us to change a trigonometric function to the cofunction of its complement.

EXAMPLE 3 Write each function in terms of its cofunction.

a. $\sin 28°$ **b.** $\cos 43°$ **c.** $\cot 9°$ **d.** $\sin \tfrac{\pi}{6}$

SOLUTION **a.** $\sin 28° = \cos(90° - 28°)$ **b.** $\cos 43° = \sin(90° - 43°)$

$\qquad\qquad = \mathbf{\cos\ 62°}$ $\qquad\qquad\qquad = \mathbf{\sin\ 47°}$

c. $\cot 9° = \tan(90° - 9°)$ **d.** $\sin \tfrac{\pi}{6} = \cos(\tfrac{\pi}{2} - \tfrac{\pi}{6})$

$\qquad\quad = \mathbf{\tan\ 81°}$ $\qquad\qquad\quad = \mathbf{\cos\ \tfrac{2\pi}{6}}$

$\qquad\qquad\qquad\qquad\qquad\qquad\qquad = \mathbf{\cos\ \tfrac{\pi}{3}}$ ▌

Suppose the given angle in Example 3 is larger than 90°; for example,

$$\cos 125° = \sin(90° - 125°)$$

$$= \sin(-35°)$$

This result can be further simplified using the following **opposite-angle identities.**

OPPOSITE-ANGLE IDENTITIES

For any real number (or angle) θ,

12. $\cos(-\theta) = \cos \theta$ 13. $\sin(-\theta) = -\sin \theta$ 14. $\tan(-\theta) = -\tan \theta$

PROOF OF IDENTITY 12 Let $\alpha = 0$ and $\beta = \theta$ in the identity $\cos(\alpha - \beta) = \cos \alpha \cos \beta + \sin \alpha \sin \beta$. Then

$$\cos(0 - \theta) = \cos 0 \cos \theta + \sin 0 \sin \theta$$

$$= 1 \cdot \cos \theta + 0 \cdot \sin \theta$$

$$= \cos \theta$$

But, if you simplify directly,

$$\cos(0 - \theta) = \cos(-\theta)$$

Therefore $\cos(-\theta) = \cos \theta$. $\qquad\qquad\qquad \square\square\square$

The proofs of Identities 13 and 14 are left for the reader.

EXAMPLE 4 Write each as a function of a positive angle or number.

a. $\cos(-19°)$ **b.** $\sin(-19°)$ **c.** $\tan(-2)$

SOLUTION **a.** $\cos(-19°) = \mathbf{\cos\ 19°}$ **b.** $\sin(-19°) = \mathbf{-\sin\ 19°}$ **c.** $\tan(-2) = \mathbf{-\tan\ 2}$ ▌

EXAMPLE 5 Write the given functions in terms of their cofunctions.

a. cos 125°   **b.** sin 102°   **c.** cot 2.5

SOLUTION   **a.** $\cos 125° = \sin(90° - 125°)$ **b.** $\sin 102° = \cos(90° - 102°)$

$$= \sin(-35°)$$
$$= \mathbf{-\sin 35°}$$

$$= \cos(-12°)$$
$$= \mathbf{\cos 12°}$$

c. $\cot 2.5 = \tan(\frac{\pi}{2} - 2.5)$

$$\approx \tan(-.9292)$$
$$= \mathbf{-\tan .9292}$$

You may also need to use opposite-angle identities together with other identities. For example, you know from algebra that

$$a - b \quad \text{and} \quad b - a$$

are opposites. This means that $a - b = -(b - a)$. In trigonometry you often see angles such as $\frac{\pi}{2} - \theta$ and want to write $\theta - \frac{\pi}{2}$. This means that $\frac{\pi}{2} - \theta = -(\theta - \frac{\pi}{2})$. In particular,

$$\cos(\tfrac{\pi}{2} - \theta) = \cos[-(\theta - \tfrac{\pi}{2})]$$
$$= \cos(\theta - \tfrac{\pi}{2}) \qquad \text{By Identity 12}$$

EXAMPLE 6 Write the given functions using the opposite-angle identities.

a. $\sin(\frac{\pi}{2} - \theta)$ **b.** $\tan(\frac{\pi}{2} - \theta)$ **c.** $\cos(\pi - \theta)$

SOLUTION   **a.** $\sin(\frac{\pi}{2} - \theta) = \sin[-(\theta - \frac{\pi}{2})]$ **b.** $\tan(\frac{\pi}{2} - \theta) = \tan[-(\theta - \frac{\pi}{2})]$

$$= \mathbf{-\sin(\theta - \tfrac{\pi}{2})}$$

$$= \mathbf{-\tan(\theta - \tfrac{\pi}{2})}$$

c. $\cos(\pi - \theta) = \cos[-(\theta - \pi)]$

$$= \mathbf{\cos(\theta - \pi)}$$

EXAMPLE 7 Graph $y = \sin(-\theta)$.

SOLUTION   First use an opposite-angle identity, if necessary: $\sin(-\theta) = -\sin \theta$. To graph $y = -\sin \theta$, build the frame as before; the endpoints and midpoints are the same, but the quarterpoints are reversed, as shown in Figure 7.3.

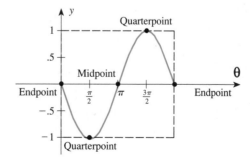

FIGURE 7.3   One period of $y = -\sin \theta$

The difference-of-angles identity proved at the beginning of this section is one of six identities known as the **addition laws**. Since subtraction can easily be written as a sum, the designation *addition laws* refers to both addition and subtraction.

ADDITION LAWS

15. $\cos(\alpha + \beta) = \cos \alpha \cos \beta - \sin \alpha \sin \beta$

16. $\cos(\alpha - \beta) = \cos \alpha \cos \beta + \sin \alpha \sin \beta$

17. $\sin(\alpha + \beta) = \sin \alpha \cos \beta + \cos \alpha \sin \beta$

18. $\sin(\alpha - \beta) = \sin \alpha \cos \beta - \cos \alpha \sin \beta$

19. $\tan(\alpha + \beta) = \dfrac{\tan \alpha + \tan \beta}{1 - \tan \alpha \tan \beta}$

20. $\tan(\alpha - \beta) = \dfrac{\tan \alpha - \tan \beta}{1 + \tan \alpha \tan \beta}$

EXAMPLE 8 Write $\cos(\frac{2\pi}{3} + \theta)$ as a function of θ only.

SOLUTION Use Identity 15:

$$\cos\left(\frac{2\pi}{3} + \theta\right) = \cos \frac{2\pi}{3} \cos \theta - \sin \frac{2\pi}{3} \sin \theta$$

$$= \left(-\frac{1}{2}\right)\cos \theta - \left(\frac{\sqrt{3}}{2}\right)\sin \theta \qquad \text{Substitute exact values where possible.}$$

$$= -\frac{1}{2}(\cos \theta + \sqrt{3}\sin \theta) \qquad \text{Simplify.}$$

EXAMPLE 9 Evaluate

$$\frac{\tan 18° - \tan 40°}{1 + \tan 18° \tan 40°}$$

using a calculator.

SOLUTION You can do a lot of arithmetic, or use Identity 20:

$$\frac{\tan 18° - \tan 40°}{1 + \tan 18° \tan 40°} = \tan(18° - 40°)$$

$$= \tan(-22°)$$

$$= -\tan 22° \qquad \text{Identity 14}$$

$$\approx -.4040 \qquad \text{By calculator}$$

We will conclude this section by proving some of the addition laws; others are left as problems.

PROOF OF IDENTITY 15 $\cos(\alpha + \beta) = \cos[\alpha - (-\beta)]$

$= \cos \alpha \cos(-\beta) + \sin \alpha \sin(-\beta) \qquad$ By Identity 16, proved earlier

$= \cos \alpha \cos \beta - \sin \alpha \sin \beta \qquad$ By Identities 12 and 13

Identity 16 is the main cosine identity proved at the beginning of this section. It is numbered and included here for the sake of completeness.

PROOF OF IDENTITY 17

$$\begin{aligned}
\sin(\alpha + \beta) &= \cos[\tfrac{\pi}{2} - (\alpha + \beta)] \\
&= \cos[(\tfrac{\pi}{2} - \alpha) - \beta] \\
&= \cos(\tfrac{\pi}{2} - \alpha)\cos\beta + \sin(\tfrac{\pi}{2} - \alpha)\sin\beta \\
&= \sin\alpha\cos\beta + \cos\alpha\sin\beta \qquad \Box\Box\Box
\end{aligned}$$

To prove Identity 18, replace β by $-\beta$ in Identity 17; the details are left as an exercise.

PROOF OF IDENTITY 19

$$\begin{aligned}
\tan(\alpha + \beta) &= \frac{\sin(\alpha + \beta)}{\cos(\alpha + \beta)} \\
&= \frac{\sin\alpha\cos\beta + \cos\alpha\sin\beta}{\cos\alpha\cos\beta - \sin\alpha\sin\beta} \\
&= \frac{\sin\alpha\cos\beta + \cos\alpha\sin\beta}{\cos\alpha\cos\beta - \sin\alpha\sin\beta} \cdot \frac{\dfrac{1}{\cos\alpha\cos\beta}}{\dfrac{1}{\cos\alpha\cos\beta}} \qquad \text{Multiply by 1.} \\
&= \frac{\dfrac{\sin\alpha\cos\beta}{\cos\alpha\cos\beta} + \dfrac{\cos\alpha\sin\beta}{\cos\alpha\cos\beta}}{\dfrac{\cos\alpha\cos\beta}{\cos\alpha\cos\beta} - \dfrac{\sin\alpha\sin\beta}{\cos\alpha\cos\beta}} \\
&= \frac{\tan\alpha + \tan\beta}{1 - \tan\alpha\tan\beta} \qquad \Box\Box\Box
\end{aligned}$$

The proof of Identity 20 is left as an exercise.

7.4 Problem Set

A Change each of the expressions in Problems 1–9 to functions of θ only.

1. $\cos(30° + \theta)$ **2.** $\sin(\theta - 45°)$ **3.** $\cos(\theta - \tfrac{\pi}{4})$

4. $\sin(\theta + \theta)$ **5.** $\tan(45° + \theta)$ **6.** $\sin(\tfrac{2\pi}{3} + \theta)$

7. $\tan(\theta + \theta)$ **8.** $\cos(\tfrac{\pi}{3} - \theta)$ **9.** $\cos(\theta + \theta)$

Write each function in Problems 10–15 in terms of its cofunction by using the cofunction identities.

10. $\cos 15°$ **11.** $\sin 38°$ **12.** $\cot 41°$

13. $\sin \tfrac{\pi}{6}$ **14.** $\cos \tfrac{5\pi}{6}$ **15.** $\cot \tfrac{2\pi}{3}$

Write the functions in Problems 16–21 as functions of positive angles.

16. $\cos(-18°)$ **17.** $\tan(-49°)$ **18.** $\sin(-41°)$

19. $\sin(-31°)$ **20.** $\cos(-39°)$ **21.** $\tan(-24°)$

Evaluate each of the expressions in Problems 22–27 to four decimal places.

22. $\sin 158° \cos 92° - \cos 158° \sin 92°$

23. $\cos 114° \cos 85° + \sin 114° \sin 85°$

24. $\cos 30° \cos 48° - \sin 30° \sin 48°$

25. $\sin 18° \cos 23° + \cos 18° \sin 23°$

26. $\dfrac{\tan 32° + \tan 18°}{1 - \tan 32° \tan 18°}$ **27.** $\dfrac{\tan 59° - \tan 25°}{1 + \tan 59° \tan 25°}$

B Using the identities of this section, find the exact values of the sine, cosine, and tangent of each of the angles given in Problems 28–33.

28. $15°$ **29.** $-15°$ **30.** $195°$

31. $75°$ **32.** $165°$ **33.** $105°$

34. Write each as a function of $\theta - \frac{\pi}{3}$.
  a. $\cos(\frac{\pi}{3} - \theta)$ **b.** $\sin(\frac{\pi}{3} - \theta)$ **c.** $\tan(\frac{\pi}{3} - \theta)$
35. Write each as a function of $(\theta - \frac{2\pi}{3})$.
  a. $\cos(\frac{2\pi}{3} - \theta)$ **b.** $\sin(\frac{2\pi}{3} - \theta)$ **c.** $\tan(\frac{2\pi}{3} - \theta)$
36. Write each as a function of $\alpha - \beta$.
  a. $\cos(\beta - \alpha)$ **b.** $\sin(\beta - \alpha)$ **c.** $\tan(\beta - \alpha)$

Use the opposite-angle identities, if necessary, to graph the functions in Problems 37–46.

37. $y = -2 \cos x$ **38.** $y = -3 \sin x$
39. $y = \tan(-x)$ **40.** $y = \cos(-2x)$
41. $y = \sin(-3x)$ **42.** $y - 1 = \sin(2 - x)$
43. $y - 2 = \cos(\pi - x)$ **44.** $y - 3 = \sin(\pi - x)$
45. $y - 1 = \tan(\frac{\pi}{6} - x)$ **46.** $y + 2 = \cos(3 - x)$

47. Decide whether each of the following functions is even, odd or neither.
  a. $s(x) = \sin x$ **b.** $c(x) = \cos x$
48. Decide whether each of the following functions is even, odd, or neither.
  a. $t(x) = \tan x$ **b.** $f(x) = \sec x$
49. Decide whether each of the following functions is even, odd, or neither:
  a. $f(x) = \dfrac{\sin x}{x}$, if $x \neq 0$ **b.** $g(x) = \dfrac{x}{\cos x}$, if $\cos x \neq 0$
50. Decide whether each of the following functions is even, odd, or neither:
  a. $f(x) = \sin x \cos x$ **b.** $g(x) = \sin x \tan x$

Prove the identities in Problems 51–71.

51. $\sin(\alpha - \beta) = \sin \alpha \cos \beta - \cos \alpha \sin \beta$

52. $\tan(\alpha - \beta) = \dfrac{\tan \alpha - \tan \beta}{1 + \tan \alpha \tan \beta}$

53. $\cot(\alpha + \beta) = \dfrac{\cot \alpha \cot \beta - 1}{\cot \beta + \cot \alpha}$

54. $\cot(\alpha - \beta) = \dfrac{\cot \alpha \cot \beta + 1}{\cot \beta - \cot \alpha}$

55. $\dfrac{\cos 5\theta}{\sin \theta} - \dfrac{\sin 5\theta}{\cos \theta} = \dfrac{\cos 6\theta}{\sin \theta \cos \theta}$

56. $\dfrac{\sin 6\theta}{\sin 3\theta} - \dfrac{\cos 6\theta}{\cos 3\theta} = \sec 3\theta$

57. $\sec(\frac{\pi}{2} - \theta) = \csc \theta$ **58.** $\cot(\frac{\pi}{2} - \theta) = \tan \theta$
59. $\csc(\frac{\pi}{2} - \theta) = \sec \theta$ **60.** $\cot(-\theta) = -\cot \theta$
61. $\sec(-\theta) = \sec \theta$ **62.** $\csc(-\theta) = -\csc \theta$
63. $\sin(\alpha + \beta) \cos \beta - \cos(\alpha + \beta) \sin \beta = \sin \alpha$
64. $\cos(\alpha - \beta) \cos \beta - \sin(\alpha - \beta) \sin \beta = \cos \alpha$

C

65. $\dfrac{\tan(\alpha + \beta) - \tan \beta}{1 + \tan(\alpha + \beta) \tan \beta} = \tan \alpha$

66. $\dfrac{\sin(\theta + h) - \sin \theta}{h} = \cos \theta \left(\dfrac{\sin h}{h} \right) - \sin \theta \left(\dfrac{1 - \cos h}{h} \right)$

67. $\dfrac{\cos(\theta + h) - \cos \theta}{h} = -\sin \theta \left(\dfrac{\sin h}{h} \right) - \cos \theta \left(\dfrac{1 - \cos h}{h} \right)$

68. $\sin(\alpha + \beta + \gamma) = \sin \alpha \cos \beta \cos \gamma + \cos \alpha \sin \beta \cos \gamma$
    $+ \cos \alpha \cos \beta \sin \gamma - \sin \alpha \sin \beta \sin \gamma$
69. $\cos(\alpha + \beta + \gamma) = \cos \alpha \cos \beta \cos \gamma - \cos \alpha \sin \beta \sin \gamma$
    $- \sin \alpha \cos \beta \sin \gamma - \sin \alpha \sin \beta \cos \gamma$
70. $\tan(\alpha + \beta + \gamma)$
    $= \dfrac{\tan \alpha + \tan \beta + \tan \gamma - \tan \alpha \tan \beta \tan \gamma}{1 - \tan \beta \tan \gamma - \tan \alpha \tan \gamma - \tan \alpha \tan \beta}$
71. $\cot(\alpha + \beta + \gamma)$
    $= \dfrac{\cot \alpha \cot \beta \cot \gamma - \cot \alpha - \cot \beta - \cot \gamma}{\cot \beta \cot \gamma + \cot \alpha \cot \gamma + \cot \alpha \cot \beta - 1}$

7.5   Double-Angle and Half-Angle Identities

Two additional, special cases of the addition laws of Section 7.4 are now considered. The first is that of the **double-angle identities.**

DOUBLE-ANGLE IDENTITIES

21. $\cos 2\theta = \cos^2\theta - \sin^2\theta$
      $= 2\cos^2\theta - 1$
      $= 1 - 2\sin^2\theta$

22. $\sin 2\theta = 2 \sin \theta \cos \theta$

23. $\tan 2\theta = \dfrac{2 \tan \theta}{1 - \tan^2\theta}$

Use the addition laws where $\alpha = \theta$ and $\beta = \theta$.

$$\cos 2\theta = \cos(\theta + \theta) = \cos\theta \cos\theta - \sin\theta \sin\theta$$
$$= \cos^2\theta - \sin^2\theta$$
$$= \cos^2\theta - (1 - \cos^2\theta)$$
$$= 2\cos^2\theta - 1$$
$$= 2(1 - \sin^2\theta) - 1$$
$$= 1 - 2\sin^2\theta$$

$$\sin 2\theta = \sin(\theta + \theta) = \sin\theta \cos\theta + \cos\theta \sin\theta$$
$$= 2\sin\theta \cos\theta$$

$$\tan 2\theta = \tan(\theta + \theta) = \frac{\tan\theta + \tan\theta}{1 - \tan\theta \tan\theta}$$
$$= \frac{2\tan\theta}{1 - \tan^2\theta}$$

EXAMPLE 1 $\cos 100x = \cos(2 \cdot 50x) = \mathbf{cos^2 50x - sin^2 50x}$

EXAMPLE 2 $\sin 120° = \sin 2(60°) = \mathbf{2\ sin\ 60°\ cos\ 60°}$

EXAMPLE 3 Write $\cos 3\theta$ in terms of $\cos\theta$.

SOLUTION
$$\cos 3\theta = \cos(2\theta + \theta) = \cos 2\theta \cos\theta - \sin 2\theta \sin\theta$$
$$= (\cos^2\theta - \sin^2\theta)\cos\theta - (2\sin\theta \cos\theta)\sin\theta$$
$$= \cos^3\theta - \sin^2\theta \cos\theta - 2\sin^2\theta \cos\theta$$
$$= \cos^3\theta - 3\sin^2\theta \cos\theta$$
$$= \cos^3\theta - 3(1 - \cos^2\theta)\cos\theta$$
$$= \cos^3\theta - 3\cos\theta + 3\cos^3\theta$$
$$= \mathbf{4\ cos^3\theta - 3\ cos\ \theta}$$

EXAMPLE 4 Evaluate $\dfrac{2\tan\frac{\pi}{16}}{1 - \tan^2\frac{\pi}{16}}$.

SOLUTION Notice this is the right-hand side of Identity 23 so it is the same as $\tan(2 \cdot \frac{\pi}{16}) = \tan\frac{\pi}{8}$. Now use a calculator to find $\tan\frac{\pi}{8} \approx \mathbf{.4142.}$

EXAMPLE 5 If $\cos\theta = \frac{3}{5}$ and θ is in Quadrant IV, find $\cos 2\theta$, $\sin 2\theta$, and $\tan 2\theta$.

SOLUTION Since $\cos 2\theta = 2\cos^2\theta - 1$,

$$\cos 2\theta = 2(\tfrac{3}{5})^2 - 1$$
$$= 2(\tfrac{9}{25}) - 1$$
$$= -\tfrac{7}{25}$$

For the other functions of 2θ, you need to know $\sin\theta$. Begin with the fundamental identity relating cosine and sine:

$$\sin^2\theta = 1 - \cos^2\theta$$

$$\sin\theta = -\sqrt{1 - \cos^2\theta} \qquad \text{Negative since the sine is negative in Quadrant IV}$$

$$= -\sqrt{1 - (\tfrac{3}{5})^2} = -\tfrac{4}{5}$$

Now, $\sin 2\theta = 2\sin\theta\cos\theta$ Finally, $\tan 2\theta = \dfrac{\sin 2\theta}{\cos 2\theta}$

$$= 2\left(-\frac{4}{5}\right)\left(\frac{3}{5}\right)$$

$$= -\frac{24}{25}$$

$$= \frac{-\frac{24}{25}}{-\frac{7}{25}}$$

$$= \frac{24}{7}$$ ∎

Identity 21 leads us to the second important special case of the addition laws, called the **half-angle identities.** We wish to solve $\cos 2\alpha = 2\cos^2\alpha - 1$ for $\cos^2\alpha$:

$$2\cos^2\alpha - 1 = \cos 2\alpha$$

$$2\cos^2\alpha = 1 + \cos 2\alpha$$

$$\cos^2\alpha = \frac{1 + \cos 2\alpha}{2}$$

Now, if $\alpha = \tfrac{1}{2}\theta$, then $2\alpha = \theta$ and

$$\cos^2\tfrac{1}{2}\theta = \frac{1 + \cos\theta}{2}$$

If $\tfrac{1}{2}\theta$ is in Quadrant I or IV, then   If $\tfrac{1}{2}\theta$ is in Quadrant II or III, then

$$\cos\tfrac{1}{2}\theta = \sqrt{\frac{1 + \cos\theta}{2}} \qquad\qquad \cos\tfrac{1}{2}\theta = -\sqrt{\frac{1 + \cos\theta}{2}}$$

These results are summarized by writing

$$\cos\tfrac{1}{2}\theta = \pm\sqrt{\frac{1 + \cos\theta}{2}}$$

⊗ **Notice $\pm$ usage and meaning.** ⊗

However, *you must be careful.* The sign $+$ or $-$ is chosen according to which quadrant $\tfrac{1}{2}\theta$ is in. The formula requires either $+$ or $-$, but not both. This use of $\pm$ is different from the use of $\pm$ in algebra. For example, when using $\pm$ in the quadratic formula, we are indicating *two* possible correct roots. In this trigonometric identity we will obtain *one* correct value depending on the quadrant of $\tfrac{1}{2}\theta$.

For the sine, solve $\cos 2\alpha = 1 - 2\sin^2\alpha$ for $\sin^2\alpha$.

$$\cos 2\alpha = 1 - 2\sin^2\alpha$$

$$2\sin^2\alpha = 1 - \cos 2\alpha$$

$$\sin^2\alpha = \frac{1 - \cos 2\alpha}{2}$$

Replace $\alpha = \frac{1}{2}\theta$, and

$$\sin^2 \tfrac{1}{2}\theta = \frac{1 - \cos\theta}{2} \quad \text{or} \quad \sin\tfrac{1}{2}\theta = \pm\sqrt{\frac{1 - \cos\theta}{2}}$$

where the sign depends on the quadrant of $\frac{1}{2}\theta$. If $\frac{1}{2}\theta$ is in Quadrant I or II, you use $+$; if it is in Quadrant III or IV, you use $-$.

Finally, to find the half-angle identity for the tangent, write

$$\tan\tfrac{1}{2}\theta = \frac{\sin\tfrac{1}{2}\theta}{\cos\tfrac{1}{2}\theta}$$

$$= \frac{\pm\sqrt{\dfrac{1 - \cos\theta}{2}}}{\pm\sqrt{\dfrac{1 + \cos\theta}{2}}}$$

$$= \pm\sqrt{\frac{1 - \cos\theta}{1 + \cos\theta}}$$

$$= \pm\sqrt{\frac{1 - \cos\theta}{1 + \cos\theta} \cdot \frac{1 - \cos\theta}{1 - \cos\theta}}$$

$$= \pm\sqrt{\frac{(1 - \cos\theta)^2}{\sin^2\theta}} \qquad \text{Remember that } 1 - \cos^2\theta = \sin^2\theta$$

$$= \frac{1 - \cos\theta}{\sin\theta} \qquad \begin{array}{l}\text{Notice that } 1 - \cos\theta \text{ is positive. Also, since } \tan\tfrac{1}{2}\theta \text{ and } \sin\theta \text{ have the} \\ \text{same sign regardless of the quadrant of } \theta, \text{ the desired result follows.}\end{array}$$

You can also show that $\tan\tfrac{1}{2}\theta = \dfrac{\sin\theta}{1 + \cos\theta}$.

HALF-ANGLE IDENTITIES

24. $\cos\tfrac{1}{2}\theta = \pm\sqrt{\dfrac{1 + \cos\theta}{2}}$ 26. $\tan\tfrac{1}{2}\theta = \dfrac{1 - \cos\theta}{\sin\theta}$

25. $\sin\tfrac{1}{2}\theta = \pm\sqrt{\dfrac{1 - \cos\theta}{2}}$ $= \dfrac{\sin\theta}{1 + \cos\theta}$

To help you remember the correct sign between the first two half-angle identities, remember "*sinus-minus*"—the sine is minus.

EXAMPLE 6   Find the exact value of $\cos\frac{9\pi}{8}$.

SOLUTION $\cos\dfrac{9\pi}{8} = \cos\left(\dfrac{1}{2} \cdot \dfrac{9\pi}{4}\right) = -\sqrt{\dfrac{1 + \cos\frac{9\pi}{4}}{2}}$ Choose a negative sign, since $\frac{9\pi}{8}$ is in Quadrant III and the cosine is negative in this quadrant.

$$= -\sqrt{\frac{1 + \cos\frac{\pi}{4}}{2}} = -\sqrt{\frac{1 + \frac{\sqrt{2}}{2}}{2}}$$

$$= -\sqrt{\frac{2 + \sqrt{2}}{4}} = -\frac{1}{2}\sqrt{2 + \sqrt{2}}$$

∎

EXAMPLE 7   If $\cot 2\theta = \frac{3}{4}$, find $\cos \theta$, $\sin \theta$, and $\tan \theta$, where 2θ is in Quadrant I.

SOLUTION   You need to find $\cos 2\theta$ so that you can use it in the half-angle identities. To do this, first find $\tan 2\theta$:

$$\tan 2\theta = \frac{1}{\cot 2\theta} = \frac{4}{3}$$

Next, find $\sec 2\theta$:

$$\sec 2\theta = \pm\sqrt{1 + \tan^2 2\theta} \qquad \text{From Identity 7}$$
$$= \sqrt{1 + \tfrac{16}{9}} \qquad \text{It is positive because } 2\theta \text{ is in Quadrant I.}$$
$$= \tfrac{5}{3}$$

Finally, $\cos 2\theta$ is the reciprocal of $\sec 2\theta$: $\cos 2\theta = \frac{3}{5}$
Next use the half-angle identities:

$$\cos \theta = \pm\sqrt{\frac{1 + \cos 2\theta}{2}} \quad \text{and} \quad \sin \theta = \pm\sqrt{\frac{1 - \cos 2\theta}{2}}$$

Do you see that θ is one-half of 2θ in these formulas?

$$\cos \theta = +\sqrt{\frac{1 + \tfrac{3}{5}}{2}} \qquad\qquad \sin \theta = +\sqrt{\frac{1 - \tfrac{3}{5}}{2}}$$

Positive value chosen because θ is in Quadrant I.

$$\cos \theta = \frac{2}{\sqrt{5}} \qquad\qquad\qquad \sin \theta = \frac{1}{\sqrt{5}}$$

Finally,

$$\tan \theta = \frac{\sin \theta}{\cos \theta} = \frac{1/\sqrt{5}}{2/\sqrt{5}} = \frac{1}{2}$$

EXAMPLE 8   Prove that $\sin \theta = \dfrac{2 \tan \frac{1}{2}\theta}{1 + \tan^2 \frac{1}{2}\theta}$.

SOLUTION   When proving identities involving functions of different angles, you should write all the trigonometric functions in the problem as functions of a single angle.

$$\frac{2 \tan \frac{1}{2}\theta}{1 + \tan^2 \frac{1}{2}\theta} = \frac{2\,\dfrac{\sin \frac{1}{2}\theta}{\cos \frac{1}{2}\theta}}{\sec^2 \frac{1}{2}\theta} = 2\,\frac{\sin \frac{1}{2}\theta}{\cos \frac{1}{2}\theta} \cdot \cos^2 \tfrac{1}{2}\theta = 2 \sin \tfrac{1}{2}\theta \cos \tfrac{1}{2}\theta = \sin \theta$$

It is sometimes convenient, or even necessary, to write a trigonometric sum as a product or a product as a sum. We conclude this section with eight additional identities.

PRODUCT IDENTITIES

27. $2 \cos \alpha \cos \beta = \cos(\alpha - \beta) + \cos(\alpha + \beta)$
28. $2 \sin \alpha \sin \beta = \cos(\alpha - \beta) - \cos(\alpha + \beta)$
29. $2 \sin \alpha \cos \beta = \sin(\alpha + \beta) + \sin(\alpha - \beta)$
30. $2 \cos \alpha \sin \beta = \sin(\alpha + \beta) - \sin(\alpha - \beta)$

The proofs of the product identities involve systems of equations, and will therefore be delayed until Chapter 9, Problem Set 9.1.

EXAMPLE 9 Write $2 \sin 3 \sin 1$ as the sum of two functions.*

SOLUTION Use Identity 28, where $\alpha = 3$ and $\beta = 1$:

$$2 \sin 3 \sin 1 = \cos(3 - 1) - \cos(3 + 1)$$
$$= \mathbf{cos\ 2 - cos\ 4}$$

∎

EXAMPLE 10 Write $\sin 40° \cos 12°$ as the sum of two functions.

SOLUTION Use Identity 29 where $\alpha = 40°$ and $\beta = 12°$:

$$2 \sin 40° \cos 12° = \sin(40° + 12°) + \sin(40° - 12°)$$
$$= \sin 52° + \sin 28°$$

But what about the coefficient 2? Since you know that the preceding is an *equation* that is true, you can divide both sides by 2 to obtain:

$$\sin 40° \cos 12° = \frac{1}{2} (\mathbf{sin\ 52° + sin\ 28°})$$

∎

By making appropriate substitutions and again using systems (as shown in Chapter 9), Identities 27–30 can be rewritten in a form known as the **sum identities**.

SUM IDENTITIES

31. $\cos x + \cos y = 2 \cos\left(\dfrac{x + y}{2}\right) \cos\left(\dfrac{x - y}{2}\right)$

32. $\cos x - \cos y = -2 \sin\left(\dfrac{x + y}{2}\right) \sin\left(\dfrac{x - y}{2}\right)$

33. $\sin x + \sin y = 2 \sin\left(\dfrac{x + y}{2}\right) \cos\left(\dfrac{x - y}{2}\right)$

34. $\sin x - \sin y = 2 \sin\left(\dfrac{x - y}{2}\right) \cos\left(\dfrac{x + y}{2}\right)$

*Remember that sum also includes difference, because $a - b = a + (-b)$.

EXAMPLE 11 Write $\sin 35° + \sin 27°$ as a product.

SOLUTION $x = 35°$, $y = 27°$, and

$$\frac{x + y}{2} = \frac{35° + 27°}{2} = 31°; \quad \frac{x - y}{2} = 4°$$

Therefore, $\sin 35° + \sin 27° = \mathbf{2 \sin 31° \cos 4°}$. ∎

You sometimes will use these product and sum identities to prove other identities.

EXAMPLE 12 Prove $\dfrac{\sin 7\gamma + \sin 5\gamma}{\cos 7\gamma - \cos 5\gamma} = -\cot \gamma$.

SOLUTION

$$\frac{\sin 7\gamma + \sin 5\gamma}{\cos 7\gamma - \cos 5\gamma} = \frac{2 \sin\left(\dfrac{7\gamma + 5\gamma}{2}\right) \cos\left(\dfrac{7\gamma - 5\gamma}{2}\right)}{-2 \sin\left(\dfrac{7\gamma + 5\gamma}{2}\right) \sin\left(\dfrac{7\gamma - 5\gamma}{2}\right)}$$

$$= \frac{2 \sin 6\gamma \cos \gamma}{-2 \sin 6\gamma \sin \gamma}$$

$$= -\frac{\cos \gamma}{\sin \gamma}$$

$$= -\cot \gamma$$ ∎

In calculus you frequently need to solve trigonometric equations that are sums or differences. The procedure is to change a sum to a product and then use the Zero Factor Theorem, as illustrated by Example 13.

EXAMPLE 13 Solve $\cos 5x - \cos 3x = 0$ $(0 \le x < 2\pi)$.

SOLUTION $\cos 5x - \cos 3x = 0$

$$-2 \sin\left(\frac{5x + 3x}{2}\right) \sin\left(\frac{5x - 3x}{2}\right) = 0 \qquad \text{Use Identity 32.}$$

$$\sin 4x \sin x = 0 \qquad \text{Divide both sides by } -2 \text{ and simplify the argument.}$$

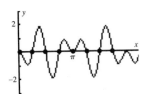

$\sin 4x = 0 \qquad\qquad\qquad \sin x = 0 \qquad \text{Use the Zero Factor Theorem.}$

$4x = 0, \pi, 2\pi, 3\pi, \qquad\qquad x = 0, \pi$
$\quad\;\; 4\pi, 5\pi, 6\pi, 7\pi$

$x = 0, \frac{\pi}{4}, \frac{\pi}{2}, \frac{3\pi}{4},$
$\quad\; \pi, \frac{5\pi}{4}, \frac{3\pi}{2}, \frac{7\pi}{4}$

The solution is: $\mathbf{0, \frac{\pi}{4}, \frac{\pi}{2}, \frac{3\pi}{4}, \pi, \frac{5\pi}{4}, \frac{3\pi}{2}, \frac{7\pi}{4}}$ ∎

Problem Set

A *Use the double-angle or half-angle identities to evaulate each of Problems 1–6 using exact values.*

1. $2\cos^2 22.5° - 1$ **2.** $\dfrac{2\tan\frac{\pi}{8}}{1 - \tan^2\frac{\pi}{8}}$ **3.** $\sqrt{\dfrac{1 - \cos 60°}{2}}$

4. $\sin 22.5°$ **5.** $\cos\frac{\pi}{8}$ **6.** $\tan 22.5°$

__WHAT IS WRONG,__ *if anything, with each of the statements in Problems 7–12? Explain your reasoning.*

7. $\cos(\theta + \phi) = \cos\theta + \cos\phi$

8. $\cot(45° + \theta) = 1 + \cot\theta$

9. $\dfrac{\sin 2\theta}{2} = \sin\theta$ **10.** $\cos\frac{1}{2}\theta = \frac{1}{2}\cos\theta$

11. $\cos\dfrac{\theta}{2} = \sqrt{\dfrac{1 - \cos\theta}{2}}$ **12.** $\sin\dfrac{1}{2}\theta = \sqrt{\dfrac{1 - \cos\theta}{2}}$

In each of Problems 13–18 find the exact values of cosine, sine, and tangent of 2θ.

13. $\sin\theta = \frac{3}{5}$; θ in Quadrant I
14. $\sin\theta = \frac{5}{13}$; θ in Quadrant II
15. $\tan\theta = -\frac{5}{12}$; θ in Quadrant IV
16. $\tan\theta = -\frac{3}{4}$; θ in Quadrant II
17. $\cos\theta = \frac{5}{9}$; θ in Quadrant I
18. $\cos\theta = -\frac{5}{13}$; θ in Quadrant III

In each of Problems 19–24, find the exact values of cosine, sine, and tangent of $\frac{1}{2}\theta$.

19. $\sin\theta = \frac{3}{5}$; θ in Quadrant I
20. $\sin\theta = \frac{5}{13}$; θ in Quadrant II
21. $\tan\theta = -\frac{5}{12}$; θ in Quadrant IV
22. $\tan\theta = -\frac{3}{4}$; θ in Quadrant II
23. $\cos\theta = \frac{5}{9}$; θ in Quadrant I
24. $\cos\theta = -\frac{5}{13}$; θ in Quadrant III

Write each of the expressions in Problems 25–30 as the sum of two functions.

25. $2\sin 35°\sin 24°$ **26.** $\sin 41°\cos 19°$
27. $\sin 225°\sin 300°$ **28.** $\cos\theta\cos 3\theta$
29. $\sin 2\theta\sin 5\theta$ **30.** $\cos 3\theta\sin 2\theta$

Write each of the expressions in Problems 31–36 as a product of functions.

31. $\sin 43° + \sin 63°$ **32.** $\cos 81° - \cos 79°$
33. $\sin 215° + \sin 300°$ **34.** $\cos 25° - \cos 100°$
35. $\cos 6x + \cos 2x$ **36.** $\sin 3x - \sin 2x$

B *Solve each of the equations in Problems 37–44 for $0 \le x < 2\pi$.*

37. $\sin x - \sin 3x = 0$ **38.** $\sin x + \sin 3x = 0$
39. $\cos 5y + \cos 3y = 0$ **40.** $\cos 3y + \cos 3y = 0$
41. $\cos 5x - \cos 3x = 0$ **42.** $\cos 3\theta + \cos\theta = 0$
43. $\sin 3z - \sin 5z = 0$ **44.** $\sin 4\theta + \sin 4\theta = 0$

In each of Problems 45–50 find $\cos\theta$, $\sin\theta$, and $\tan\theta$ when θ is in Quadrant I and $\cot 2\theta$ is given.

45. $\cot 2\theta = -\frac{3}{4}$ **46.** $\cot 2\theta = \frac{1}{\sqrt{3}}$
47. $\cot 2\theta = -\frac{4}{3}$ **48.** $\cot 2\theta = 0$
49. $\cot 2\theta = -\frac{1}{\sqrt{3}}$ **50.** $\cot 2\theta = \frac{4}{3}$

51. AVIATION An airplane flying faster than the speed of sound is said to have a speed greater than Mach 1. The Mach number is the ratio of the speed of the plane to the speed of sound and is denoted by M. When a plane flies faster than the speed of sound, a sonic boom is heard, created by sound waves that form a cone with a vertex angle θ, as shown in Figure 7.4. It can be shown that, if $M > 1$, then

$$\sin\frac{\theta}{2} = \frac{1}{M}$$

a. If $\theta = \frac{\pi}{6}$, find the Mach number to the nearest tenth.
b. Find the exact Mach number for part **a.**

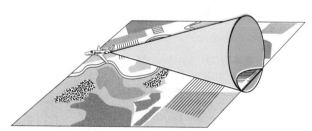

FIGURE 7.4
Pattern of sound waves creating a sonic boom

Prove each of the identities in Problems 52–60.

52. $\sin\alpha = 2\sin\frac{\alpha}{2}\cos\frac{\alpha}{2}$

53. $\cos 4\theta = \cos^2 2\theta - \sin^2 2\theta$

54. $\sin 2\theta = \dfrac{2\tan\theta}{1 + \tan^2\theta}$

55. $\tan\frac{3}{2}\beta = \dfrac{2\tan\frac{3\beta}{4}}{1 - \tan^2\frac{3\beta}{4}}$

56. $\tan \frac{1}{2}\theta = \dfrac{1 - \cos\theta}{\sin\theta}$

57. $\tan \frac{1}{2}\theta = \dfrac{\sin\theta}{1 + \cos\theta}$

58. $\dfrac{\sin 5\theta + \sin 3\theta}{\cos 5\theta + \cos 3\theta} = \tan 4\theta$

59. $\dfrac{\cos 5w + \cos w}{\cos w - \cos 5w} = \dfrac{\cot 2w}{\tan 3w}$

60. $\dfrac{\cos 3\theta - \cos\theta}{\sin\theta - \sin 3\theta} = \tan 2\theta$

C *Prove each of the identities in Problems 61–65.*

61. $\sin 3\theta = 3\sin\theta - 4\sin^3\theta$

62. $\tan \frac{B}{2} = \csc B - \cot B$

63. $\sin 4\theta = 4\sin\theta\cos\theta - 8\sin^3\theta\cos\theta$

64. $\frac{1}{2}\cot x - \frac{1}{2}\tan x = \cot 2x$

65. $\cos^4\theta = \frac{1}{8}(3 + 4\cos 2\theta + \cos 4\theta)$

7.6 De Moivre's Theorem*

Trigonometric Form of Complex Numbers

Consider a graphical representation of a complex number. To give a graphic representation of complex numbers, such as

$$2 + 3i, \quad -i, \quad -3 - 4i, \quad 3i, \quad -2 + \sqrt{2}i, \quad \tfrac{3}{2} - \tfrac{5}{2}i,$$

a two-dimensional coordinate system is used. The horizontal axis represents the **real axis** and the vertical axis is the **imaginary axis**, so that $a + bi$ is represented by the ordered pair (a, b). Remember that a and b represent real numbers, so we plot (a, b) in the usual manner, as shown in Figure 7.5. The coordinate system in Figure 7.5 is called the **complex plane** or the **Gaussian plane,** in honor of Karl Friedrich Gauss.

The **absolute value** of a complex number z is, graphically, the distance between z and the origin (just as it is for real numbers). The absolute value of a complex number is also called the **modulus.** The distance formula leads to the following definition.

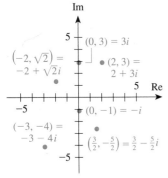

FIGURE 7.5 Complex plane

ABSOLUTE VALUE, OR MODULUS, OF A COMPLEX NUMBER

If $z = a + bi$, then the **absolute value,** or **modulus,** of z is denoted by $|z|$ and is defined by

$$|z| = \sqrt{a^2 + b^2}$$

EXAMPLE 1 Find the absolute value.

a. $3 + 4i$ Absolute value: $\ |3 + 4i| = \sqrt{3^2 + 4^2}$
$$= \sqrt{25}$$
$$= 5$$

b. $-2 + \sqrt{2}i$ Absolute value: $\ |-2 + \sqrt{2}i| = \sqrt{4 + 2}$
$$= \sqrt{6}$$

*This is an optional section that requires complex numbers from Section 1.5.

c. -3
$$\text{Absolute value:} \quad |-3 + 0i| = \sqrt{9 + 0}$$
$$= 3$$

The form $a + bi$ is called the **rectangular form,** but another useful representation uses trigonometry. Consider the graphical representation of a complex number $a + bi$, as shown in Figure 7.6. Let r be the distance from the origin to (a, b) and let θ be the angle the segment makes with the real axis. Then

$$r = \sqrt{a^2 + b^2}$$

and θ, called the **argument**, is chosen so that it is the smallest nonnegative angle the terminal side makes with the positive real axis. From the definition of the trigonometric functions,

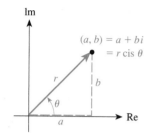

FIGURE 7.6 Trigonometric form and rectangular form of a complex number

$$\cos \theta = \frac{a}{r} \qquad \sin \theta = \frac{b}{r} \qquad \tan \theta = \frac{b}{a}$$

$$a = r \cos \theta \qquad b = r \sin \theta$$

Therefore
$$a + bi = r \cos \theta + ir \sin \theta$$
$$= r(\cos \theta + i \sin \theta)$$

Sometimes $r(\cos \theta + i \sin \theta)$ is abbreviated by

$$r(\cos \theta + i \sin \theta)$$
$$\downarrow\downarrow \qquad \downarrow\downarrow \; \downarrow$$
$$r(c \qquad i s \quad \theta) \qquad \text{or} \quad r \text{ cis } \theta$$

TRIGONOMETRIC FORM OF A COMPLEX NUMBER

The **trigonometric form** of a complex number $z = a + bi$ is
$$r(\cos \theta + i \sin \theta) = r \text{ cis } \theta$$
where $r = \sqrt{a^2 + b^2}$; $\tan \theta = b/a$ if $\quad a = r \cos \theta$; $b = r \sin \theta$. This representation is unique for $0° \leq \theta < 360°$ for all z except $0 + 0i$.

The placement of θ in the proper quadrant is an important consideration because there are two values of $0° \leq \theta < 360°$ that will satisfy the relationship

$$\tan \theta = \frac{b}{a}$$

For example, compare the following:

1. $-1 + i \qquad a = -1, b = 1 \qquad \tan \theta = \dfrac{1}{-1}$ or $\tan \theta = -1$

2. $1 - i \qquad a = 1, b = -1 \qquad \tan \theta = \dfrac{-1}{1}$ or $\tan \theta = -1$

Notice that the trigonometric equation is the same for both complex numbers, even though $-1 + i$ is in Quadrant II and $1 - i$ is in Quadrant IV. This consideration of quadrants is even more important when you are doing the problem on a

CALCULATOR COMMENT

Many calculators have a built-in conversion from rectangular to trigonometric form. Most calculators call this **rectangular–polar** conversion $(x, y) \Leftrightarrow (r, \theta)$ where $x + yi \Leftrightarrow r$ cis θ. For example, on the TI-81, these conversions are found when you press the $\boxed{\text{MATH}}$ key; they are called

R → P (rectangular to trig form)

and

P → R (trig to rectangular form)

calculator, since the proper sequence of steps for this example gives the result $-45°$, which is not true for either example since $0° \leq \theta < 360°$. The entire process can be dealt with quite simply if you let θ' be the reference angle for θ. Then find the reference angle

$$\theta' = \tan^{-1}\left|\frac{b}{a}\right|$$

After you know the reference angle and the quadrant, it is easy to find θ. For these examples,

$$\theta' = \tan^{-1}|-1|$$
$$= 45°$$

For Quadrant II, $\theta = 135°$; for Quadrant IV, $\theta = 315°$. $-1 + i$ is in Quadrant II (plot it), so $\theta = 135°$; $1 - i$ is in Quadrant IV, so $\theta = 315°$.

EXAMPLE 2 Change the complex numbers to trigonometric form.

  a. $1 - \sqrt{3}i$ $a = 1$ and $b = -\sqrt{3}$; the number is in Quadrant IV.

$$r = \sqrt{1^2 + (-\sqrt{3})^2} \qquad \theta' = \tan^{-1}\left|\frac{-\sqrt{3}}{1}\right|$$
$$= \sqrt{4} \qquad\qquad\qquad = 60°$$
$$= 2$$

Thus $1 - \sqrt{3}i = 2$ cis $300°$.   The reference angle is $60°$; in Quadrant IV: $\theta = 300°$

  b. $6i$ $a = 0$ and $b = 6$; notice that $\tan \theta$ is not defined for $\theta = 90°$. By inspection, $6i = 6$ cis $90°$.

  c. $4.310 + 5.516i$ $a = 4.310$ and $b = 5.516$; the number is in Quadrant I.

$$r = \sqrt{(4.310)^2 + (5.516)^2} \qquad \theta = \tan^{-1}\left|\frac{5.516}{4.310}\right|$$
$$\approx \sqrt{49} \qquad\qquad\qquad \approx \tan^{-1}(1.2798)$$
$$= 7 \qquad\qquad\qquad\qquad \approx 52°$$

Thus $4.310 + 5.516i \approx 7$ cis $52°$.

EXAMPLE 3 Change the complex numbers to rectangular form.

  a. 4 cis $330° = 4(\cos 330° + i \sin 330°) = 4[\frac{\sqrt{3}}{2} + i(-\frac{1}{2})] = 2\sqrt{3} - 2i$
  $r = 4$ and $\theta = 330°$

  b. $5(\cos 38° + i \sin 38°) = 5 \cos 38° + (5 \sin 38°)i \approx 3.94 + 3.08i$
  $r = 5$ and $\theta = 38°$

Multiplication and Division of Complex Numbers

The great advantage of trigonometric form over rectangular form is the ease with which you can multiply and divide complex numbers.

PRODUCTS AND QUOTIENTS OF COMPLEX NUMBERS IN TRIGONOMETRIC FORM

Let $z_1 = r_1$ cis θ_1 and $z_2 = r_2$ cis θ_2 be nonzero complex numbers. Then

$$z_1 z_2 = r_1 r_2 \text{ cis}(\theta_1 + \theta_2) \qquad \frac{z_1}{z_2} = \frac{r_1}{r_2} \text{ cis}(\theta_1 - \theta_2)$$

PROOF

$$
\begin{aligned}
z_1 z_2 &= (r_1 \text{ cis } \theta_1)(r_2 \text{ cis } \theta_2) \\
&= [r_1(\cos \theta_1 + i \sin \theta_1)][r_2(\cos \theta_2 + i \sin \theta_2)] \\
&= r_1 r_2 (\cos \theta_1 + i \sin \theta_1)(\cos \theta_2 + i \sin \theta_2) \\
&= r_1 r_2 (\cos \theta_1 \cos \theta_2 + i \cos \theta_1 \sin \theta_2 - i \sin \theta_1 \cos \theta_2 - \sin \theta_1 \sin \theta_2) \\
&= r_1 r_2 [(\cos \theta_1 \cos \theta_2 - \sin \theta_1 \sin \theta_2) + i(\cos \theta_1 \sin \theta_2 + \sin \theta_1 \cos \theta_2)] \\
&= r_1 r_2 [\cos(\theta_1 + \theta_2) + i \sin(\theta_1 + \theta_2)] \\
&= r_1 r_2 \text{ cis}(\theta_1 + \theta_2)
\end{aligned}
$$

The proof of the quotient form is similar and is left as a problem. ▯▯▯

EXAMPLE 4   Simplify:

a. 5 cis $38° \cdot 4$ cis $75° = 5 \cdot 4$ cis$(38° + 75°) =$ **20 cis 113°**

b. $\sqrt{2}$ cis $188° \cdot 2\sqrt{2}$ cis $310° = 4$ cis $498° =$ **4 cis 138°**

c. $(2 \text{ cis } 48°)^3 = (2 \text{ cis } 48°)(2 \text{ cis } 48°)^2$

$= (2 \text{ cis } 48°)(4 \text{ cis } 96°)$ Notice that this result is the same as

$= $ **8 cis 144°** $(2 \text{ cis } 48°)^3 = 2^3 \text{ cis}(3 \cdot 48°) = 8 \text{ cis } 144°$ ▮

EXAMPLE 5   Find $\dfrac{15(\cos 48° + i \sin 48°)}{5(\cos 125° + i \sin 125°)}$.

SOLUTION $\dfrac{15 \text{ cis } 48°}{5 \text{ cis } 125°} = 3 \text{ cis}(48° - 125°) = 3 \text{ cis}(-77°) =$ **3 cis 283°**

Remember that arguments should be between $0°$ and $360°$. ▮

EXAMPLE 6   Simplify $(1 - \sqrt{3}i)^5$.

SOLUTION   First change to trigonometric form:

$$a = 1;\ b = -\sqrt{3};\quad \text{Quadrant IV}$$
$$r = \sqrt{1 + 3} \qquad \theta' = \tan^{-1}\left|-\tfrac{\sqrt{3}}{1}\right|$$
$$= 2 \qquad\qquad\quad = 60°$$
$$\theta = 300° \quad \text{(Quadrant IV)}$$
$$(1 - \sqrt{3}i)^5 = (2 \text{ cis } 300°)^5 = 2^5 \text{ cis}(5 \cdot 300°)$$
$$= 32 \text{ cis } 1{,}500° = \textbf{32 cis 60°}$$

If you want the answer in rectangular form, you can now change back:

$$32 \text{ cis } 60° = 32 \cos 60° + i\,(32 \sin 60°) = \mathbf{16 + 16\sqrt{3}i}$$ ▮

Powers of Complex Numbers—De Moivre's Theorem

As you can see from Example 6, multiplication in trigonometric form extends quite nicely to any positive integral power in a result called **De Moivre's Theorem**. This theorem is proved by mathematical induction (Chapter 10).

DE MOIVRE'S THEOREM

If n is a natural number, then

$$(r \text{ cis } \theta)^n = r^n \text{ cis } n\theta$$

for a complex number $r \text{ cis } \theta = r(\cos \theta + i \sin \theta)$.

Although De Moivre's Theorem is useful for powers, as illustrated by Example 6, its real usefulness is in finding the complex roots of numbers. Recall from algebra that $\sqrt[n]{r} = r^{1/n}$ is used to denote the principal nth root of r. However, $r^{1/n}$ is only *one* of the nth roots of r. How do you find *all* nth roots of r? To find the principal root, you can use a calculator or logarithms along with the following theorem, which follows directly from De Moivre's Theorem.

nth ROOT THEOREM

If n is any positive integer, then the nth roots of $r \text{ cis } \theta$ are given by

$$\sqrt[n]{r} \text{ cis}\left(\frac{\theta + 360°k}{n}\right) \quad \text{or} \quad \sqrt[n]{r} \text{ cis}\left(\frac{\theta + 2\pi k}{n}\right)$$

for $k = 0, 1, 2, 3, \ldots, n - 1$.

The proof of this theorem is left as a problem.

EXAMPLE 7 Find the square roots of $-\frac{9}{2} + \frac{9}{2}\sqrt{3}i$.

SOLUTION First change to trigonometric form:

$$r = \sqrt{(-\tfrac{9}{2})^2 + (\tfrac{9}{2}\sqrt{3})^2} = 9$$

Thus $\quad -\frac{9}{2} + \frac{9}{2}\sqrt{3}i = 9 \text{ cis } 120°$

The square roots are:

$$9^{1/2} \text{ cis}\left(\frac{120° + 360°k}{2}\right) = 3 \text{ cis}(60° + 180°k)$$

$k = 0$: $\quad 3 \text{ cis } 60° = \frac{3}{2} + \frac{3}{2}\sqrt{3}i$

$k = 1$: $\quad 3 \text{ cis } 240° = -\frac{3}{2} - \frac{3}{2}\sqrt{3}i$

$$\theta' = \tan^{-1}\left|\frac{\frac{9}{2}\sqrt{3}}{-\frac{9}{2}}\right|$$
$$= \tan^{-1}(\sqrt{3})$$
$$= 60°$$
$$\theta = 120° \quad \text{(Quadrant II)}$$

All other integral values of k repeat one of the previously found roots. For example,

$$k = 2: \quad 3 \text{ cis } 420° = \frac{3}{2} + \frac{3}{2}\sqrt{3}i$$

Check:

$$(\tfrac{3}{2} + \tfrac{3}{2}\sqrt{3}i)^2 = \tfrac{9}{4} + \tfrac{9}{2}\sqrt{3}i + \tfrac{9}{4} \cdot 3i^2 \qquad (-\tfrac{3}{2} - \tfrac{3}{2}\sqrt{3}i)^2 = \tfrac{9}{4} + \tfrac{9}{2}\sqrt{3}i + \tfrac{9}{4} \cdot 3i^2$$
$$= -\tfrac{9}{2} + \tfrac{9}{2}\sqrt{3}i \qquad\qquad\qquad = \tfrac{9}{2} + \tfrac{9}{2}\sqrt{3}i$$

EXAMPLE 8 Find the fifth roots of 32.

SOLUTION Begin by writing 32 in trigonometric form: $32 = 32 \text{ cis } 0°$. The fifth roots are found by

$$32^{1/5} \text{ cis}\left(\frac{0° + 360°k}{5}\right) = 2 \text{ cis } 72°k$$

$k = 0$: $2 \text{ cis } 0°$ $= 2$

$k = 1$: $2 \text{ cis } 72°$ $= .6180 + 1.9021i$

$k = 2$: $2 \text{ cis } 144° = -1.6180 + 1.1756i$

$k = 3$: $2 \text{ cis } 216° = -1.6180 - 1.1756i$

$k = 4$: $2 \text{ cis } 288° = .6180 - 1.9021i$

The first root, which is located so that its argument is θ/n, is called the **principal nth root.**

All other integral values for k repeat those listed here. ▌

If all the fifth roots of 32 are represented graphically, as shown in Figure 7.7, notice that they all lie on a circle of radius 2 and are equally spaced.

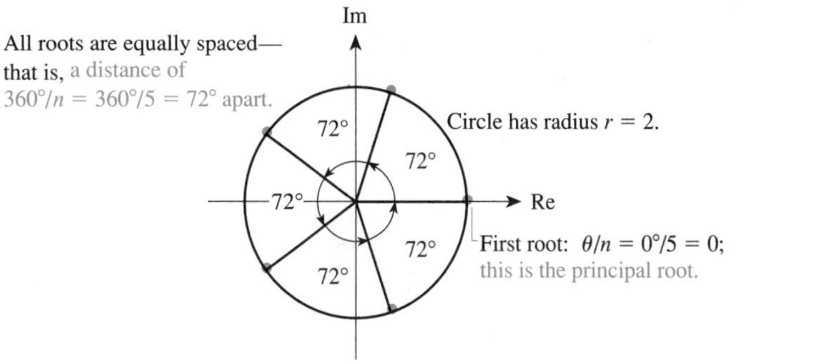

FIGURE 7.7 Graphical representation of the fifth roots of 32

All roots are equally spaced—
that is, a distance of
$360°/n = 360°/5 = 72°$ apart.

Circle has radius $r = 2$.

First root: $\theta/n = 0°/5 = 0$;
this is the principal root.

If n is a positive integer, then the nth roots of a complex number $a + bi = r \text{ cis } \theta$ are equally spaced on the circle of radius r centered at the origin.

7.6 Problem Set

A

WHAT IS WRONG, if anything, with each of the statements in Problems 1–6? Explain your reasoning.

1. cis 90° does not exist because tan 90° is undefined.

2. All negative real numbers have $\theta = 180°$.

3. If a point is on the coordinate axis, then θ is 0°, 90°, 180°, or 270°.

4. If $\theta = 90°$, then the complex number is plotted on the negative imaginary axis.

5. $2 - 3i$ is a point in Quadrant II.

6. 4 cis 250° is a point in Quadrant IV.

Plot the complex numbers given in Problems 7–9. Find the modulus of each.

7. **a.** $3 + i$ **b.** $7 - i$ **c.** $3 + 2i$ **d.** $-3 - 2i$

8. **a.** $-1 + 3i$ **b.** $2 + 4i$ **c.** $5 + 6i$ **d.** $2 - 5i$

9. **a.** $-2 + 5i$ **b.** $-5 + 4i$
   c. $4 - i$ **d.** $-1 + i$

Plot and then change to trigonometric form each of the numbers in Problems 10–19.

10. **a.** $1 + i$ **b.** $1 - i$

11. **a.** $\sqrt{3} - i$ **b.** $\sqrt{3} + i$

12. **a.** $1 - \sqrt{3}i$ **b.** $-1 - \sqrt{3}i$

13. a. 1   **b.** 2   **c.** 3   **d.** 4

14. a. i **b.** $2i$ **c.** $3i$ **d.** $4i$

15. a. $-5i$ **b.** $-6i$ **c.** $-7i$ **d.** $-8i$

16. a. -5 **b.** -6 **c.** -7 **d.** -8

17. $5.7956 - 1.5529i$ **18.** $1.5321 - 1.2856i$

19. $-.6946 + 3.9392i$

Plot and then change to rectangular form each of the numbers in Problems 20–25. Use exact values whenever possible.

20. $2(\cos 45° + i \sin 45°)$ **21.** $3(\cos 60° + i \sin 60°)$

22. $\text{cis}(\frac{5\pi}{6})$ **23.** $5 \text{ cis}(\frac{3\pi}{2})$

24. $6 \text{ cis } 247°$ **25.** $9 \text{ cis } 190°$

Perform the indicated operations in Problems 26–37.

26. $2 \text{ cis } 60° \cdot 3 \text{ cis } 150°$ **27.** $3 \text{ cis } 48° \cdot 5 \text{ cis } 92°$

28. $4(\cos 65° + i \sin 65°) \cdot 12(\cos 87° + i \sin 87°)$

29. $\dfrac{5(\cos 315° + i \sin 315°)}{2(\cos 48° + i \sin 48°)}$

30. $\dfrac{8 \text{ cis } 30°}{4 \text{ cis } 15°}$ **31.** $\dfrac{12 \text{ cis } 250°}{4 \text{ cis } 120°}$

32. $(2 \text{ cis } 50°)^3$ **33.** $(3 \text{ cis } 60°)^4$

34. $(\cos 210° + i \sin 210°)^5$ **35.** $(2 - 2i)^4$

36. $(1 + i)^6$ **37.** $(\sqrt{3} - i)^8$

Find the indicated roots of the numbers in Problems 38–55. Leave your answers in trigonometric form.

38. Square roots of $16 \text{ cis } 100°$

39. Fourth roots of $81 \text{ cis } 88°$

40. Cube roots of $64 \text{ cis } 216°$

41. Cube roots of -1

42. Cube roots of 8

43. Fourth roots of $1 + i$

44. Cube roots of $8 \text{ cis } 240°$

45. Fifth roots of $32 \text{ cis } 200°$

46. Fifth roots of $32 \text{ cis } 160°$

47. Cube roots of 27

48. Fourth roots of i

49. Fourth roots of $-1 - i$

50. Sixth roots of $64i$

51. Ninth roots of $-1 + i$

52. Tenth roots of 1

53. Sixth roots of -64

54. Ninth roots of 1

55. Tenth roots of i

Find the indicated roots of the numbers in Problems 56–63. Leave your answers in rectangular form. Show the roots graphically.

56. Cube roots of 1

57. Cube roots of -8

58. Fourth roots of 1

59. Cube roots of $4\sqrt{3} - 4i$

60. Square roots of $\dfrac{25}{2} - \dfrac{25\sqrt{3}}{2} i$

61. Fourth roots of $12.2567 + 10.2846i$

C
____

62. Find the cube roots of $(4\sqrt{2} + 4\sqrt{2}i)^2$.

63. Find the fifth roots of $(-16 + 16\sqrt{3}i)^3$.

64. Prove that

$$\frac{r_1 \text{ cis } \theta_1}{r_2 \text{ cis } \theta_2} = \frac{r_1}{r_2} \text{cis}(\theta_1 - \theta_2)$$

65. Prove that

$$\sqrt[n]{r} \text{ cis}\left(\frac{\theta + 360°k}{n}\right) = (r \text{ cis } \theta)^{1/n}$$

66. If $z_1 = a + bi$ and $z_2 = c + di$, show that

$$|z_1 + z_2| \le |z_1| + |z_2|.$$

This relationship is called the *triangle inequality*.

67. If

$$\cos \theta = 1 - \frac{\theta^2}{2!} + \frac{\theta^4}{4!} - \frac{\theta^6}{6!} + \cdots + \frac{(-1)^n \theta^{2n}}{(2n)!} + \cdots$$

$$\sin \theta = \theta - \frac{\theta^3}{3!} + \frac{\theta^5}{5!} - \frac{\theta^7}{7!} + \cdots + \frac{(-1)^n \theta^{2n+1}}{(2n+1)!} + \cdots$$

and

$$e^{i\theta} = 1 + (i\theta) + \frac{(i\theta)^2}{2!} + \frac{(i\theta)^3}{3!} + \frac{(i\theta)^4}{4!} + \cdots + \frac{(i\theta)^n}{n!} + \cdots$$

show that $e^{i\theta} = \cos \theta + i \sin \theta$. This equation is called *Euler's formula*.

68. Using Problem 67, show that $e^{i\pi} = -1$.

Chapter 7 Summary

The material of this chapter is reviewed in the following list of objectives. After each objective there are some practice questions. For a sample test, select the first question of each set and check your answers with the answer section. For a sample test without answers, use the second question of each set. Additional practice is given by the other questions in each set. If you are having trouble with a particular type of problem, look back to that section for extra help.

7.1 **Trigonometric Equations**

OBJECTIVE 1 *Solve first-degree trigonometric equations.*
Solve for $0 \le \theta < 2\pi$.

1. $3 \tan 2\theta - \sqrt{3} = 0$

2. $\sin^2\theta + 2 \cos \theta = 1 + 3 \cos \theta + \sin^2\theta$

3. $2 \sin 3\theta = \sqrt{2}$

4. $3 \sec 2\theta - 2 = 0$

OBJECTIVE 2 *Solve second-degree trigonometric equations by factoring or by using the quadratic formula.* Solve for $0 \le \theta < 2\pi$.

5. $4 \cos^2\theta = 1$　　　　**6.** $3 \cos^2\theta = 1 + \cos \theta$

7. $\frac{1}{2} \cos^2\theta = 1$　　　　**8.** $\tan^2 2\theta = 3 \tan 2\theta$

7.2 **Fundamental Identities**

OBJECTIVE 3 *State and prove the eight fundamental identities.*

9. State the reciprocal identities.

10. State the ratio identities.

11. State the Pythagorean identities.

12. Prove one of the Pythagorean identities.

OBJECTIVE 4 *Use the fundamental identities to find the other trigonometric functions if you are given the value of one function and want to find the others.*

13. Find the other trigonometric functions so that $\sin \delta = \frac{3}{5}$ when $\tan \delta < 0$.

14. Find the other trigonometric functions so that $\cos \beta = \frac{3}{5}$ and $\tan \beta < 0$.

15. Find the other trigonometric functions so that $\sin \omega = -\frac{3}{5}$ when $\tan \omega > 0$.

16. Find the other trigonometric functions so that $\cos \omega = -\frac{3}{5}$ when $\tan \omega < 0$.

OBJECTIVE 5 *Alebgraically simplify expressions involving the trigonometric functions.* Leave your answer in terms of sines and cosines.

17. $\dfrac{\sin \theta}{\cos \theta} + \dfrac{1}{\sin \theta}$　　　　**18.** $\dfrac{1}{\sin \theta + \cos \theta} + \dfrac{1}{\sin \theta - \cos \theta}$

19. $\tan^2\theta + \sec^2\theta$　　　　**20.** $\dfrac{\tan \theta + \cot \theta}{\sec \theta}$

7.3 **Proving Identities**

OBJECTIVE 6 *Prove identities using algebraic simplification and the eight fundamental identities.*

21. $\dfrac{\csc^2\alpha}{1 + \cot^2\alpha} = 1$　　　　**22.** $\dfrac{1 + \tan^2\theta}{\csc \theta} = \sec \theta \tan \theta$

23. $\dfrac{\cos \theta}{\sec \theta} - \dfrac{\sin \theta}{\cot \theta} = \dfrac{\cos \theta \cot \theta - \tan \theta}{\csc \theta}$

24. $\tan \beta + \sec \beta = \dfrac{1 + \csc \beta}{\cos \beta \csc \beta}$

OBJECTIVE 7 *Prove a given identity by using various "tricks of the trade."*

25. $\dfrac{\sin^2\theta - \cos^2\theta}{\sin \theta + \cos \theta} = \sin \theta - \cos \theta$

26. $(\tan \theta + \cot \theta)^2 = \sec^2\theta + \csc^2\theta$

27. $\cos^2x \tan x \csc x \sec x = 1$

28. $\dfrac{1}{\sin \theta + \cos \theta} + \dfrac{1}{\sin \theta - \cos \theta} = \dfrac{2 \sin \theta}{\sin^4\theta - \cos^4\theta}$

7.4 Addition Laws

7.5 Double-Angle and Half-Angle Identities

7.6 De Moivre's Theorem

OBJECTIVE 17 *State De Moivre's Theorem and the nth root theorem, and find all roots of a complex number.*

65. State the *n*th root theorem.

66. State De Moivre's Theorem.

67. Find and plot the square roots of $\frac{7}{2}\sqrt{3} - \frac{7}{2}i$. Leave your answer in rectangular form.

68. Find the cube root of *i*. Leave your answer in trigonometric form.

Benjamin Banneker (1731–1806)

I apprehend you will embrace every opportunity to eradicate that train of absurd and false ideas and opinions which so generally prevail with respect to us [black people] and that your sentiments are concurrent with mine, which are: that one universal Father hath given being to us all; and He not only made us all of one flesh, but that He hath also without partiality afforded us all with these same faculties and that, however diversified in situation or color, we are all the same family and stand in the same relation to Him.

BENJAMIN BANNEKER IN
Black Mathematicians and Their Works

Analytic trigonometry has its roots in geometry and in the practical application of surveying. Surveying is the technique of locating points on or near the surface of the earth. The Egyptians are credited with being the first to do surveying. The annual flooding of the Nile and its constant destruction of property markers led the Egyptians to the principles of surveying, as noted by the Greek historian Herodotus:

"Sesostris . . . made a division of the soil of Egypt among the inhabitants . . . . If the river carried away any portion of a man's lot . . . the king sent persons to examine, and determine by measurement the exact extent of the loss . . . . From this practice, I think, geometry first came to be known in Egypt, whence it passed into Greece."

The first recorded measurements to determine the size of the earth were also made in Egypt when Eratosthenes measured a meridian arc in 230 B.C. (See Problem 62 of Section 6.1.) A modern-day example of a surveyor's dream is the city of Washington D.C. The center of the city is the Capitol, and the city is divided into four sections: Northwest, Northeast, Southwest, and Southeast. These sections are separated by North, South, East, and Capitol streets, all of which converge at the Capitol. The initial surveying of this city was done by a group of mathematicians and surveyors who included the first distinguished black mathematician, Benjamin Banneker.

*"Benjamin Banneker, born in 1731 in Maryland, was the first black to be recognized as a mathematician and astronomer. He exhibited unusual talent in mathematics, and, with little formal education, produced an almanac and invented a clock. Banneker was born free in a troubled time of black slavery. The right to vote was his by birth. It is important to note that this right to vote was denied him as well as other free black Americans by 1803 in his native state of Maryland. Yet, history must acknowledge him as the first American black man of science."**

---

*From Virginia K. Newell, Joella H. Gipson, L. Waldo Rich, and Beauregard Stubblefield, *Black Mathematicians and Their Works* (Admore: Dorrance), p. xiv.

8

ANALYTIC
TRIGONOMETRY

Preview

In this chapter, we are concerned with another use of trigonometry—solving triangles to find unknown parts. We begin by solving right triangles using the Pythagorean theorem. We then turn to the solution of *oblique* triangles (triangles that are not right triangles). We will derive two trigonometric laws, the *Law of Cosines* and *the Law of Sines*, which allow us to solve certain triangles using a calculator or trigonometric tables and logarithms. There is a third trigonometric law that lends itself to logarithmic calculation, the *Law of Tangents*, which is useful for times when no calculator is at hand. This chapter has 6 objectives, which are listed at the end of the chapter on pages 363–364. Following these objectives is a Cumulative Review for Chapters 6–8 and an Extended Application about solar power.

Perspective

The first step in carrying out the operation of integration in calculus is frequently solving a triangle. This technique, called *trigonometric substitution*, involves expressions of the form $a^2 + u^2$, $\sqrt{a^2 - u^2}$, $\sqrt{a^2 + u^2}$, and $\sqrt{u^2 - a^2}$. In the page below, taken from a calculus book, two reference triangles are shown in Figures 7.1 and 7.2.

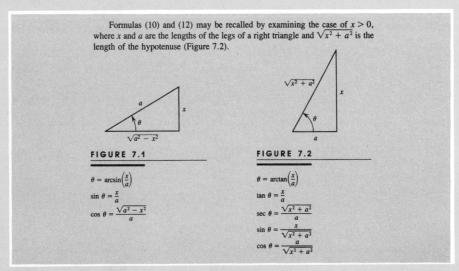

Formulas (10) and (12) may be recalled by examining the case of $x > 0$, where x and a are the lengths of the legs of a right triangle and $\sqrt{x^2 + a^2}$ is the length of the hypotenuse (Figure 7.2).

FIGURE 7.1

$\theta = \arcsin\left(\dfrac{x}{a}\right)$

$\sin \theta = \dfrac{x}{a}$

$\cos \theta = \dfrac{\sqrt{a^2 - x^2}}{a}$

FIGURE 7.2

$\theta = \arctan\left(\dfrac{x}{a}\right)$

$\tan \theta = \dfrac{x}{a}$

$\sec \theta = \dfrac{\sqrt{x^2 + a^2}}{a}$

$\sin \theta = \dfrac{x}{\sqrt{x^2 + a^2}}$

$\cos \theta = \dfrac{a}{\sqrt{x^2 + a^2}}$

From *Calculus and Analytic Geometry*, 4th Ed., by A. Shenk, p. 451. Copyright © 1984 by Scott, Foresman and Company. Reprinted by permission.

8.1 Right Triangles

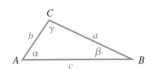

FIGURE 8.1
Correctly labeled triangle

One of the most important uses of trigonometry is in solving triangles. Recall from geometry that every **triangle** has three sides and three angles, which are called the six *parts* of the triangle. We say that a **triangle is solved** if all six parts are known. Typically three parts will be given, or known, and you will want to find the other three parts. Label a triangle as shown in Figure 8.1. The vertices are labeled A, B, and C, with the sides opposite those vertices a, b, and c, respectively. The angles are labeled α, β, and γ, respectively. In this section the examples are limited to right triangles in which γ denotes the right angle and c the **hypotenuse**.

According to the definition of the trigonometric functions, the angle under consideration must be in standard position. This requirement is sometimes inconvenient, so we use that definition to create a special case that applies to any acute angle θ of a right triangle. Notice that in Figure 8.1, θ might be α or β, but it would not be γ since γ is not a right angle. Also notice that the hypotenuse is one of the sides of both acute angles. The other side making up the angle is called the **adjacent side**. Thus side a is adjacent to β and side b is adjacent to α. The third side of the triangle (the one not making up the angle) is called the **opposite side**. Thus side a is opposite α and side b is opposite β.

RIGHT-TRIANGLE DEFINITION OF THE TRIGONOMETRIC FUNCTIONS

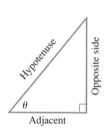

If θ is an acute angle in a right triangle, then

$$\cos \theta = \frac{\text{adjacent side}}{\text{hypotenuse}}$$

$$\sin \theta = \frac{\text{opposite side}}{\text{hypotenuse}}$$

$$\tan \theta = \frac{\text{opposite side}}{\text{adjacent side}}$$

The other trigonometric functions are the reciprocals of these relationships.

We can now use this definition to solve some given triangles.

EXAMPLE 1   Solve the triangle with $a = 50$, $\alpha = 35°$, and $\gamma = 90°$. (*Note:* $\alpha = 35°$ means the measure of angle α is 35°.)

SOLUTION

$\alpha = 35°$ Given

$\beta = 55°$ Since $\alpha + \beta = 90°$ for any right triangle with right angle at C

$\gamma = 90°$ Given

$a = 50$ Given

$$b: \quad \tan 35° = \frac{50}{b}$$

$$b = \frac{50}{\tan 35°} \approx 71.4074$$

To two significant digits, $b = 71$.

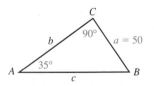

$$c: \quad \sin 35° = \frac{50}{c}$$

$$c = \frac{50}{\sin 35°} \approx 87.1723 \qquad \text{To two significant digits, } c = 87. \quad \blacksquare$$

EXAMPLE 2   Solve the triangle with $a = 32$, $b = 58$, and $\gamma = 90°$.

SOLUTION $\alpha: \quad \tan \alpha = \dfrac{32}{58}$ or $\alpha = \text{Tan}^{-1}\left(\dfrac{32}{58}\right) \approx 28.9°$

To the nearest degree,

$$\alpha = 29°$$
$$\beta = 61° \qquad \beta = 90° - \alpha$$
$$\gamma = 90° \qquad \text{Given}$$
$$a = 32 \qquad \text{Given}$$
$$b = 58 \qquad \text{Given}$$

$c:$ To find c we write

$$\sin \alpha = \frac{32}{c}$$

$$c = \frac{32}{\sin \alpha} \approx 66.24198 \qquad \text{To two significant digits, } c = 66. \quad \blacksquare$$

When solving triangles, you are dealing with measurements that are never exact. In real life, measurements are made with a certain number of digits of precision, and results are not claimed to have more digits of precision than the least accurate number in the input data. There are rules for working with significant digits (see Appendix A), but it is important for you to focus first on the trigonometry. Just remember that your answers should not be more accurate than any of the given measurements. However, be careful not to round your answers when doing your calculations; you should round only when stating your answers.

⊘ Note rounding agreement. ⊘

The solution of right triangles is necessary in a variety of situations. The first one we will consider concerns an observer looking at an object. The **angle of depression** is the acute angle measured down from the horizontal line to the line of sight, whereas the **angle of elevation** is the acute angle measured up from a horizontal line to the line of sight. These ideas are illustrated in Examples 3 and 4.

EXAMPLE 3   The angle of elevation of a tree from a point on the ground 42 m from its base is 33°. Find the height of the tree.

SOLUTION   Let $\theta = $ angle of elevation and $h = $ height of tree. Then

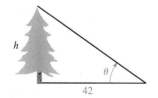

$$\tan \theta = \frac{h}{42}$$

$$h = 42 \tan 33° \approx 27.28$$

The tree is 27 m tall. $\qquad \blacksquare$

EXAMPLE 4 The distance across a canyon can be determined from an airplane. Suppose the angles of depression to the two sides of the canyon are 43° and 55°, as shown in Figure 8.2. If the altitude of the plane is 20,000 ft, how far is it across the canyon?

SOLUTION Label parts x, y, θ, and ϕ as shown in Figure 8.2.

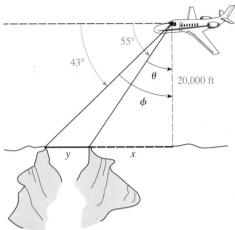

$$\theta = 90° - 55° \quad \text{and} \quad \phi = 90° - 43°$$
$$= 35° \qquad\qquad\qquad = 47°$$

First find x: $\quad \tan 35° = \dfrac{x}{20,000}$

$$x = 20,000 \tan 35° \approx 14,004$$

Next find $x + y$:

$$\tan 47° = \frac{x + y}{20,000}$$

$$x + y = 20,000 \tan 47° \approx 21,447$$

Thus $\qquad\qquad y \approx 21,447 - 14,004 = 7,443$

FIGURE 8.2 Determining the distance across a canyon from the air

To two significant digits, **the distance across the canyon is 7,400 ft.** ▮

Remember, when you are working these problems on your calculator, do not work with the rounded results, but with the entire accuracy possible with your calculator. You should round only once and that is with your final answer.

A second application of the solution of right triangles involves the **bearing** of a line, which is defined as an acute angle made with a north–south line. When giving the bearing of a line, first write N or S to determine whether to measure the angle from the north or the south side of a point on the line. Then give the measure of the angle followed by E or W, denoting on which side of the north–south line the angle is to be measured. Some examples are shown in Figure 8.3.

FIGURE 8.3 Bearing of an angle

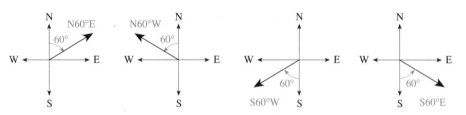

EXAMPLE 5 To find the width AB of a canyon (see Figure 8.4), a surveyor measures 100 m from A in the direction of N42.6°W to locate point C. The surveyor then determines that the bearing of CB is N73.5°E. Find the width of the canyon if point B is situated so that $\angle BAC = 90.0°$.

SOLUTION Let $\theta = \angle BCA$ in Figure 8.4.

$\angle BCE' = 16.5°$	Complementary angles
$\angle ACS' = 42.6°$	Alternate interior angles
$\angle E'CA = 47.4°$	Complementary angles

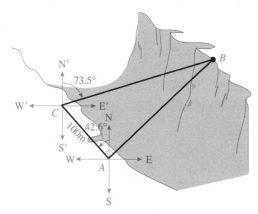

$$\theta = \angle BCA = \angle BCE' + \angle E'CA$$
$$= 16.5° + 47.4°$$
$$= 63.9°$$

Since $\tan \theta = \dfrac{|AB|}{|AC|}$,

$$|AB| = |AC| \tan \theta$$
$$= 100 \tan \theta$$
$$= 100 \tan 63.9°$$
$$\approx 204.125$$

FIGURE 8.4 Surveying problem

The canyon is 204 m across.

8.1 Problem Set

A *Solve the right triangles ($\gamma = 90°$) in Problems 1–30.*

1. $a = 80$; $\beta = 60°$
2. $a = 18$; $\beta = 30°$
3. $a = 9.0$; $\beta = 45°$
4. $b = 37$; $\alpha = 65°$
5. $b = 15$; $\alpha = 37°$
6. $b = 50$; $\alpha = 53°$
7. $a = 69$; $c = 73$
8. $b = 23$; $c = 45$
9. $b = 13$; $c = 22$
10. $a = 68$; $b = 83$
11. $a = 24$; $b = 29$
12. $a = 12$; $b = 8.0$
13. $a = 29$; $\alpha = 76°$
14. $a = 93$; $\alpha = 26°$
15. $a = 49$; $\beta = 45°$
16. $b = 13$; $\beta = 65°$
17. $b = 90$; $\beta = 13°$
18. $b = 47$; $\beta = 108°$
19. $b = 82$; $\alpha = 50°$
20. $c = 28.3$; $\alpha = 69.2°$
21. $c = 36$; $\alpha = 6°$
22. $\beta = 57.4°$; $a = 70.0$
23. $\alpha = 56.00°$; $b = 2,350$
24. $\beta = 23°$; $a = 9,000$
25. $b = 3,100$; $c = 3,500$
26. $\beta = 16.4°$; $b = 2,580$
27. $\alpha = 42°$; $b = 350$
28. $b = 3,200$; $c = 7,700$
29. $b = 4,100$; $c = 4,300$
30. $c = 75.4$; $\alpha = 62.5°$

B

31. SURVEYING The angle of elevation of a building from a point on the ground 30 m from its base is 38°. Find the height of the building.

32. SURVEYING Repeat Problem 31 for 150 ft instead of 30 m.

33. SURVEYING From a cliff 150 m above the shoreline, the angle of depression of a ship is 37°. Find the distance of the ship from a point directly below the observer.

34. SURVEYING Repeat Problem 33 for 450 ft instead of 150 m.

35. From a police helicopter flying at 1,000 ft, a stolen car is sighted at an angle of depression of 71°. Find the distance of the car from a point directly below the helicopter.

36. Repeat Problem 35 for an angle of depression of 64° instead of 71°.

37. SURVEYING To find the east–west boundary of a piece of land, a surveyor must divert his path from point C on the boundary by proceeding due south for 300 ft to point A. Point B, which is due east of point C, is now found to be in the direction of N49°E from point A. What is the distance CB?

38. SURVEYING To find the distance across a river that runs east–west, a surveyor locates points P and Q on a north–south line on opposite sides of the river. She then paces out 150 ft from Q due east to a point R. Next she determines that the bearing of RP is N58°W. How far is it across the river?

39. A 16-ft ladder on level ground is leaning against a house. If the angle of elevation of the ladder is 52°, how far above the ground is the top of the ladder?

40. How far is the base of the ladder in Problem 39 from the house?

41. If the ladder in Problem 39 is moved so that the bottom is 9 ft from the house, what will be the angle of elevation?

42. Find the height of the Barrington Space Needle if the angle of elevation at 1,000 ft from a point on the ground directly below the top is 58.15°.

43. The world's tallest chimney is the stack of the International Nickel Company. Find its height if the angle of elevation at 1,000 ft from a point on the ground directly below the top of the stack is 51.36°.

44. In the movie *Close Encounters of the Third Kind*, there was a scene in which the star, Richard Dreyfuss, was approaching Devil's Tower in Wyoming. He could have determined his distance from Devil's Tower by first

stopping at a point P and estimating the angle P, as shown in Figure 8.5. After moving 100 m towards Devil's Tower, he could have estimated the angle N, as shown in the figure. How far away from Devil's Tower is point N?

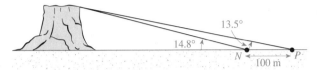

FIGURE 8.5   Procedure for determing the height of Devil's Tower

45. Determine the height of Devil's Tower in Problem 44.

46. SURVEYING To find the boundary of a piece of land, a surveyor must divert his path from a point A on the boundary for 500 ft in the direction S50°E. He then determines that the bearing of a point B located directly south of A is S40°W. Find the distance AB.

47. SURVEYING To find the distance across a river, a surveyor locates points P and Q on either side of the river. Next she measures 100 m from point Q in the direction S35°E to point R. Then she determines that point P is now in the direction of N25.0°E from point R and that $\angle PQR$ is a right angle. Find the distance across the river.

48. SURVEYING If the Empire State Building and the Sears Tower were situated 1,000 ft apart, the angle of depression from the top of the Sears Tower to the top of the Empire State Building would be 11.53°, and the angle of depression to the foot of the Empire State Building would be 55.48°. Find the heights of the buildings.

49. SURVEYING On the top of the Empire State Building is a television tower. From a point 1,000 ft from a point on the ground directly below the top of the tower, the angle of elevation to the bottom of the tower is 51.34° and to the top of the tower is 55.81°. What is the length of the television tower?

50. PHYSICS A wheel 5.00 ft in diameter rolls up a 15.0° incline. What is the height of the center of the wheel above the base of the incline after the wheel has completed one revolution?

51. PHYSICS What is the height of the center of the wheel in Problem 50 after three revolutions?

52. ASTRONOMY If the distance from the earth to the sun is 92.9 million mi and the angle formed between Venus, the earth, and the sun (as shown in Figure 8.6) is 47.0°, find the distance from the sun to Venus.

53. ASTRONOMY Use the information in Problem 52 to find the distance from the earth to Venus.

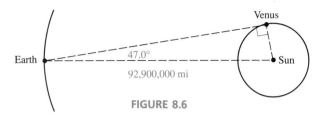

FIGURE 8.6

54. The largest ground area covered by any office building is that of the Pentagon in Arlington, Virginia. If the radius of the circumscribed circle is 783.5 ft, find the length of one side of the Pentagon.

55. Use the information in Problem 54 to find the radius of the circle inscribed in the Pentagon.

56. SURVEYING To determine the height of the building shown in Figure 8.7, we select a point P and find that the angle of elevation is 59.64°. We then move out a distance of 325.4 ft (on a level plane) to point Q and find that the angle of elevation is now 41.32°. Find the height h of the building.

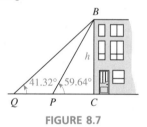

FIGURE 8.7

57. Using Figure 8.7, let the angle of elevation at P be α and at Q be β, and let the distance from P to Q be d. If h is the height of the building, show that

$$h = \frac{d \sin \alpha \sin \beta}{\sin(\alpha - \beta)}$$

58. A 6.0-ft person is casting a shadow of 4.2 ft. What time of the morning is it if the sun rose at 6:15 A.M. and is directly overhead at 12:15 P.M.?

59. How long will the shadow of the person in Problem 58 be at 8:00 A.M.?

60. SURVEYING From the top of a tower 100 ft high, the angles of depression to two landmarks on the plane upon which the tower stands are 18.5° and 28.4°.
   a. Find the distance between the landmarks when they are on the same side of the tower.
   b. Find the distance between the landmarks when they are on opposite sides of the tower.

61. Show that in every right triangle the value of c lies between $(a + b)/\sqrt{2}$ and $a + b$.

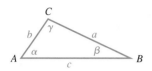

FIGURE 8.8
Correctly labeled triangle

Law of Cosines

In the previous section we solved right triangles. We now extend that study to triangles with no right angles. Such triangles are called **oblique triangles**. Consider a triangle labeled as in Figure 8.8, and notice that γ is not now restricted to 90°.

In general, you will be given three parts of a triangle and be asked to find the remaining three parts. But can you do so given *any* three parts? Consider the possibilities:

1. SSS: By SSS we mean that you are given three sides and want to find the three angles.

2. SAS: You are given two sides and an included angle.

3. AAA: You are given three angles.

4. ASA or AAS: You are given two angles and a side.

5. SSA: You are given two sides and the angle opposite one of them.

We will consider these possibilities one at a time.

SSS

To solve a triangle given SSS, it is necessary for the sum of the lengths of the two smaller sides to be greater than the length of the largest side. In this case, we use a generalization of the Pythagorean theorem called the **Law of Cosines:**

LAW OF COSINES

In triangle ABC,

$$c^2 = a^2 + b^2 - 2ab \cos \gamma$$

PROOF   Let γ be an angle in standard position with A on the positive x-axis, as shown in Figure 8.9. The coordinates of the vertices are as follows:

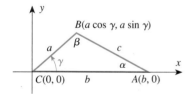

FIGURE 8.9

$C(0, 0)$ Since C is in standard position

$A(b, 0)$ Since A is on the x-axis a distance of b units from the origin

$B(a \cos \gamma, a \sin \gamma)$ Let $B = (x, y)$; then by definition of the trigonometric functions, $\cos \gamma = x/a$ and $\sin \gamma = y/a$. Thus, $x = a \cos \gamma$ and $y = a \sin \gamma$.

Use the formula for the distance between the point $A(b, 0)$ and the point $B(a \cos \gamma, a \sin \gamma)$:

$$\begin{aligned}
c^2 &= (a \cos \gamma - b)^2 + (a \sin \gamma - 0)^2 \\
&= a^2 \cos^2\gamma - 2ab \cos \gamma + b^2 + a^2 \sin^2\gamma \\
&= a^2(\cos^2\gamma + \sin^2\gamma) + b^2 - 2ab \cos \gamma \\
&= a^2 + b^2 - 2ab \cos \gamma
\end{aligned}$$

Notice that for a right triangle, $\gamma = 90°$. This means that

$$c^2 = a^2 + b^2 - 2ab \cos 90° \quad \text{or} \quad c^2 = a^2 + b^2$$

since $\cos 90° = 0$. But this last equation is simply the Pythagorean theorem.

By letting A and B, respectively, be in standard position, it can also be shown that

$$a^2 = b^2 + c^2 - 2bc \cos \alpha \quad \text{and} \quad b^2 = a^2 + c^2 - 2ac \cos \beta$$

To find the angles when you are given three sides, solve for α, β, or γ.

LAW OF COSINES, ALTERNATE FORMS

$$a^2 = b^2 + c^2 - 2bc \cos \alpha \qquad \cos \alpha = \frac{b^2 + c^2 - a^2}{2bc}$$

$$b^2 = a^2 + c^2 - 2ac \cos \beta \qquad \cos \beta = \frac{a^2 + c^2 - b^2}{2ac}$$

$$c^2 = a^2 + b^2 - 2ab \cos \gamma \qquad \cos \gamma = \frac{a^2 + b^2 - c^2}{2ab}$$

EXAMPLE 1 What is the smallest angle of a triangular patio whose sides measure 25, 18, and 21 ft?

SOLUTION If γ represents the smallest angle, then c (the side opposite γ) must be the smallest side, so $c = 18$. Then:

$$\cos \gamma = \frac{a^2 + b^2 - c^2}{2ab}$$

$$= \frac{25^2 + 21^2 - 18^2}{2(25)(21)}$$

$$\gamma = \cos^{-1} \left[\frac{25^2 + 21^2 - 18^2}{2(25)(21)} \right]$$

To two significant digits, the answer is **45°**.

SAS

The second possibility listed for solving oblique triangles is that of being given two sides and an included angle. It is necessary that the given angle be less than 180°. Again use the Law of Cosines for this possibility, as shown by Example 2.

EXAMPLE 2 Find c where $a = 52.0$, $b = 28.3$, and $\gamma = 28.5°$.

SOLUTION By the Law of Cosines:

$$c^2 = a^2 + b^2 - 2ab \cos \gamma$$

$$= (52.0)^2 + (28.3)^2 - 2(52.0)(28.3) \cos 28.5° \approx 918.355474$$

To three significant digits, $c = $ **30.3**.

AAA

The third case supposes that three angles are given. However, from what you know of similar triangles (see Figure 8.10)—that they have the same shape but

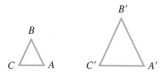

FIGURE 8.10 Similar triangles

not necessarily the same size, and thus their corresponding angles have equal measure—you can conclude that the triangle cannot be solved without knowing the length of at least one side.

We will discuss ASA and AAS in the next section. The last possibility, SSA, is discussed in Section 8.4.

8.2 Problem Set

A. *Solve △ABC for the parts requested in Problems 1–24. If the triangle cannot be solved, tell why.*

1. $a = 7.0$; $b = 8.0$; $c = 2.0$. Find α.
2. $a = 7.0$; $b = 5.0$; $c = 4.0$. Find β.
3. $a = 11$; $b = 9.0$; $c = 8.0$. Find α.
4. $a = 4.0$; $b = 5.0$; $c = 6.0$. Find α.
5. $a = 15$; $b = 8.0$; $c = 20$. Find β.
6. $a = 25$; $b = 18$; $c = 40$. Find γ.
7. $a = 18$; $b = 25$; $\gamma = 30°$. Find c.
8. $a = 18$; $c = 11$; $\beta = 63°$. Find b.
9. $a = 15$; $b = 8.0$; $\gamma = 38°$. Find c.
10. $b = 21$; $c = 35$; $\alpha = 125°$. Find a.
11. $b = 14$; $c = 12$; $\alpha = 82°$. Find a.
12. $a = 31$; $b = 24$; $\gamma = 120°$. Find c.
13. $b = 18$; $c = 15$; $\alpha = 50°$. Find a.
14. $a = 20$; $c = 45$; $\beta = 85°$. Find b.
15. $a = 241$; $b = 187$; $c = 100$. Find β.
16. $a = 38.2$; $b = 14.8$; $\gamma = 48.2°$. Find c.
17. $b = 123$; $c = 485$; $\alpha = 163.0°$. Find a.
18. $a = 48.3$; $c = 35.1$; $\beta = 215.0°$. Find b.
19. $a = 11$; $b = 9.0$; $c = 8.0$. Find the largest angle.
20. $a = 12$; $b = 6.0$; $c = 15$. Find the smallest angle.
21. $a = 123$; $b = 310$; $c = 250$. Find the smallest angle.
22. $a = 123$; $b = 310$; $c = 250$. Find the largest angle.
23. $a = 38$; $b = 41$; $c = 25$. Find the largest angle.
24. $a = 45$; $b = 92$; $c = 41$. Find the smallest angle.

Solve △ABC in Problems 25–36. If the triangle cannot be solved, tell why.

25. $b = 5.2$; $c = 3.4$; $\alpha = 54.4°$
26. $a = 81$; $b = 53$; $\gamma = 85.2°$
27. $a = 214$; $b = 320$; $\gamma = 14.8°$
28. $a = 18$; $b = 12$; $c = 23.4$
29. $a = 140$; $b = 85.0$; $c = 105$
30. $\alpha = 83°$; $\beta = 52°$; $\gamma = 45°$
31. $\alpha = 35°$; $\beta = 123°$; $\gamma = 22°$

32. $a = 4.82$; $b = 3.85$; $\gamma = 34.2°$
33. $b = 520$; $c = 235$; $\alpha = 110.5°$
34. $a = 31.2$; $c = 51.5$; $\beta = 109.5°$
35. $a = 341$; $b = 340$; $\gamma = 23.4°$
36. $b = 1,234$; $c = 3,420$; $\alpha = 24.58°$

Problems 37–42 refer to the baseball field shown in Figure 8.11. Find the requested distances to the nearest foot.

37. Right field (point A) to left field (point C)
38. Right field (point A) to center field (point B)
39. First base (point F) to center field (point B)
40. Third base (point T) to right field (point A)
41. Second base (point S) to center field (point B)
42. First base (point F) to third base (point T)

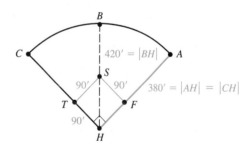

FIGURE 8.11 Baseball field

B

43. Prove that $a^2 = b^2 + c^2 - 2bc \cos \alpha$.
44. Prove that $b^2 = a^2 + c^2 - 2ac \cos \beta$.
45. Prove that $c^2 = a^2 + b^2 - 2ab \cos \gamma$.
46. Prove that $\cos \alpha = \dfrac{b^2 + c^2 - a^2}{2bc}$.
47. Prove that $\cos \beta = \dfrac{a^2 + c^2 - b^2}{2ac}$.
48. Prove that $\cos \gamma = \dfrac{a^2 + b^2 - c^2}{2ab}$.

49. AVIATION New York City is approximately 600 mi N9°E of Washington, D.C., and Buffalo, New York, is N49°W of Washington, D.C. How far is Buffalo from New York City if the distance from Buffalo to Washington, D.C., is approximately 800 mi?

50. AVIATION New Orleans, Louisiana, is approximately 3,000 mi S56°E of Denver, Colorado, and Chicago, Illinois, is N76°E of Denver. How far is Chicago from New Orleans if it is approximately 2,500 mi from Denver to Chicago?

51. AVIATION An airplane flies N51°W for a distance of 345 mi and then alters course and flies 150 mi in the direction N25°W. How far is the airplane from the starting point?

52. AVIATION An airplane flies due south for a distance of 135 mi and then alters course and flies S46°W for 82 mi before landing. How far from the starting point is the plane upon landing?

53. GEOMETRY A dime, a penny, and a quarter are placed on a table so that they just touch each other, as shown in Figure 8.12. Let D, P, and Q be the respective centers of the coins. Solve $\triangle DPQ$. (*Hint:* If you measure the coins, the diameters are 1.75 cm, 2.00 cm, and 2.50 cm.)

FIGURE 8.12

54. GEOMETRY An equilateral triangle is inscribed in a circle of radius 5.0 in., as shown in Figure 8.13. Find the perimeter of the triangle.

55. GEOMETRY An equilateral triangle is inscribed in a circle of radius 3.0 in., as shown in Figure 8.13. Find the perimeter of the triangle.

FIGURE 8.13

56. GEOMETRY A square is inscribed in a circle of radius 5.0 in. Find the area of the square.

57. GEOMETRY A square is inscribed in a circle of radius 15.0 in. Find the area of the square.

58. In $\triangle ABC$, use the Law of Cosines to prove

$$a = b \cos \gamma + c \cos \beta$$

59. In $\triangle ABC$, use the Law of Cosines to prove

$$b = a \cos \gamma + c \cos \alpha$$

60. In $\triangle ABC$, use the Law of Cosines to prove

$$c = a \cos \beta + b \cos \alpha$$

C

61. Using the Law of Cosines, show that

$$\cos \alpha + 1 = \frac{(b + c - a)(a + b + c)}{2bc}$$

62. Using the Law of Cosines, show that

$$\cos \beta + 1 = \frac{(a + c - b)(a + b + c)}{2ac}$$

63. Using the Law of Cosines, show that

$$\cos \gamma + 1 = \frac{(a + b - c)(a + b + c)}{2ab}$$

8.3 Law of Sines

In Section 8.2, we listed five possibilities for given information in solving oblique triangles: SSS, SAS, AAA, ASA or AAS, and SSA. We discussed three of those cases: SSS, SAS, and AAA. In this section we will consider the fourth case, ASA or AAS. Notice that ASA and AAS both give us the same information about a triangle, because knowing any two angles is the same as knowing all three angles (the sum of the angles of any triangle is always 180°).

ASA or AAS

Case 4 supposes that two angles and a side are given. For a triangle to be formed, the sum of the two given angles must be less than 180°, and the given side must be greater than 0. The Law of Cosines is not sufficient in this case because at least two sides are needed for the application of that law. So we state and prove a result called the **Law of Sines.**

LAW OF SINES

In any $\triangle ABC$,

$$\frac{\sin \alpha}{a} = \frac{\sin \beta}{b} = \frac{\sin \gamma}{c}$$

The equation in the Law of Sines means that you can use any of the following equations to solve a triangle:

$$\frac{\sin \alpha}{a} = \frac{\sin \beta}{b} \qquad \frac{\sin \alpha}{a} = \frac{\sin \gamma}{c} \qquad \frac{\sin \beta}{b} = \frac{\sin \gamma}{c}$$

PROOF Consider any oblique triangle, as shown in Figure 8.14.

FIGURE 8.14 Oblique triangles for Law of Sines

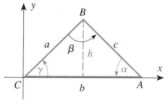
a. γ is acute

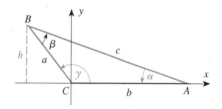
b. γ is obtuse

Let h = height of the triangle with base CA. Then $\sin \alpha = \dfrac{h}{c}$ and $\sin \gamma = \dfrac{h}{a}$.

Solving for h,

$$h = c \sin \alpha \quad \text{and} \quad h = a \sin \gamma$$

Thus, $c \sin \alpha = a \sin \gamma$. Dividing by ac,

$$\frac{\sin \alpha}{a} = \frac{\sin \gamma}{c}$$

Repeat these steps for the height of the same triangle with base AB to get

$$\frac{\sin \alpha}{a} = \frac{\sin \beta}{b}$$

EXAMPLE 1 Solve the triangle in which $a = 20$, $\alpha = 38°$, and $\beta = 121°$.

SOLUTION

$\alpha = 38°$ Given

$\beta = 121°$ Given

$\gamma = 21°$ Since $\alpha + \beta + \gamma = 180°$, then $\gamma = 180° - 38° - 121° = 21°$.

$a = 20$ Given

b: Use the Law of Sines: $\dfrac{\sin 38°}{20} = \dfrac{\sin 121°}{b}$

$$b = \frac{20 \sin 121°}{\sin 38°} \approx 27.8454097$$

Give answer to two significant digits. $b \approx 28$

c: Use the Law of Sines: $\dfrac{\sin 38°}{20} = \dfrac{\sin 21°}{c}$

$$c \approx \frac{20 \sin 21°}{\sin 38°} \approx 11.64172078$$

Give answer to two significant digits. $c \approx 12$

EXAMPLE 2 A boat traveling at a constant rate due west passes a buoy that is 1.0 kilometers from a lighthouse. The lighthouse is N30°W of the buoy. After the boat has traveled for $\frac{1}{2}$ hour, its bearing to the lighthouse is N74°E. How fast is the boat traveling?

SOLUTION The angle at the lighthouse (see Figure 8.15) is $180° - 60° - 16° = 104°$.

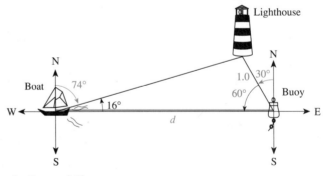

FIGURE 8.15 Finding the rate of travel

Therefore, by the Law of Sines,

$$\frac{\sin 104°}{d} = \frac{\sin 16°}{1.0}$$

$$d \approx 3.520189502$$ Do not round answers in your work; round only when stating the answer.

After 1 hour the distance is $2d$, so the rate of the boat is about **7.0 kph** (since $d = rt$).

8.3 Problem Set

A *Solve △ABC with the parts given in Problems 1–40. If the triangle cannot be solved, tell why.*

1. $a = 10; \alpha = 48°; \beta = 62°$
2. $a = 18; \alpha = 65°; \gamma = 15°$
3. $a = 30; \beta = 50°; \gamma = 100°$
4. $b = 23; \alpha = 25°; \beta = 110°$
5. $b = 40; \alpha = 50°; \gamma = 60°$
6. $b = 90; \alpha = 85°; \gamma = 25°$
7. $c = 43; \alpha = 120°; \gamma = 7°$
8. $c = 115; \beta = 81.0°; \gamma = 64.0°$
9. $a = 107; \alpha = 18.3°; \beta = 54.0°$
10. $b = 223; \beta = 85°; \gamma = 24°$
11. $a = 85; \alpha = 48.5°; \gamma = 72.4°$
12. $a = 183; \alpha = 65.0°; \gamma = 105.2°$
13. $a = 105; \alpha = 48.5°; \beta = 62.7°$
14. $b = 128; \alpha = 165°; \beta = 15°$
15. $b = 10.3; \alpha = 78°; \beta = 102°$
16. $c = 110; \beta = 105.0°; \gamma = 15.5°$
17. $c = 105; \beta = 148.0°; \gamma = 22.5°$
18. $c = 1,823; \beta = 65.25°; \gamma = 15.55°$
19. $a = 41.0; \alpha = 45.2°; \beta = 21.5°$
20. $b = 55.0; c = 92.0°; \alpha = 98.0°$
21. $b = 58.3; \alpha = 120°; \gamma = 68.0°$
22. $c = 123; \alpha = 85.2°; \beta = 38.7°$
23. $a = 26; b = 71; c = 88$
24. $a = 38; b = 82; c = 115$
25. $\alpha = 48°; \beta = 105°; \gamma = 27°$
26. $\alpha = 38°; \beta = 100°; \gamma = 42°$
27. $a = 80.6; b = 23.2; \gamma = 89.2°$
28. $b = 1,234; \alpha = 85.26°; \beta = 24.45°$
29. $c = 28.36; \beta = 42.10°; \gamma = 102.30°$
30. $a = 481; \beta = 28.6°; \gamma = 103.0°$
31. $a = 10; b = 48; c = 52$
32. $a = 58; b = 165; c = 180$
33. $a = 48.1; \alpha = 4.8°; \beta = 82.0°$
34. $c = 101; \beta = 35.0°; \gamma = 25.85°$
35. $c = 83.1; \beta = 81.5°; \gamma = 162°$
36. $a = 381; \alpha = 25°; \gamma = 90.00°$
37. $a = 8.1; \alpha = 28.5°; \gamma = 90.00°$
38. $b = 23.1; c = 16.5; \alpha = 15.25°$
39. $b = 10.9; c = 4.45; \alpha = 16.2°$
40. $a = 18.55; b = 16.54; c = 51.45$

B

41. SURVEYING In San Francisco, a certain hill makes an angle of 20° with the horizontal and has a tall building at the top. At a point 100 ft down the hill from the base of the building, the angle of elevation to the top of the building is 72°. What is the height of the building?

42. SURVEYING A certain hill makes an angle of 15° with the horizontal and has a tall building at the top. At a point 100 ft down the hill from the base of the building, the angle of elevation to the top of the building is 68°. What is the height of the building?

43. SURVEYING A hill makes an angle of 8° with the horizontal and has a tower at the top. At a point 500 ft down the hill from the base of the building, the angle of elevation to the top of the tower is 35°. What is the height of the tower?

44. NAVIGATION In the movie *Star Wars*, the hero, Luke, must hit a small target on the Death Star by flying a horizontal distance to reach the target. When the target is sighted, the onboard computer calculates the angle of depression to be 28.0°. If, after 150 km, the target has an angle of depression of 42.0°, how far is the target from Luke's spacecraft at that instant?

45. NAVIGATION The Galactic Empire's computers on the Death Star are monitoring the positions of the invading forces (see Problem 44). At a particular instant, two observation points 2,500 m apart make a fix on Luke's spacecraft, which is between the observation points and in the same vertical plane. If the angle of elevation from the first observation point is 3.00° and the angle of elevation from the second is 1.90°, find the distance from Luke's spacecraft to each of the observation points.

46. NAVIGATION Solve Problem 45 if both observation points are on the same side of the spacecraft and all the other information is unchanged.

47. Mr. T, whose eye level is 6 ft, is standing on a hill with an inclination of 5° and needs to throw a rope to the top of a nearby building. If the angle of elevation to the top of the building is 20° and the angle of depression to the base of the building is 11°, how tall is the building and how far is it from Mr. T to the top of the building? See Figure 8.16 on page 352.

48. Repeat Problem 47 with all the information unchanged except assume that the hill has an inclination of 9°.

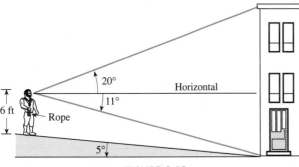

FIGURE 8.16

49. Jay Leno is standing on the deck of his Malibu home and is looking down a stairway with an inclination of 35° to a boat on the beach. If the angles of depression to the bow (point *B*) and stern (point *S*) are $\beta = 32°$ and $\theta = 28°$, how long is the boat if it is known that the distance *JF* is 50 ft, and *F*, *B*, and *S* are on level ground? See Figure 8.17.

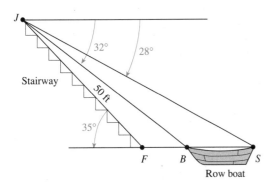

FIGURE 8.17

50. Repeat Problem 49, except assume that *JF* is 100 ft and all of the other information is unchanged.

51. Given $\triangle ABC$, show that

$$\frac{\sin \alpha}{a} = \frac{\sin \beta}{b}$$

52. Given $\triangle ABC$, show that

$$\frac{\sin \beta}{b} = \frac{\sin \gamma}{c}$$

53. Give $\triangle ABC$, show that

$$\frac{\sin \alpha}{\sin \beta} = \frac{a}{b}$$

C

54. Using Problem 53, show that

$$\frac{\sin \alpha - \sin \beta}{\sin \alpha + \sin \beta} = \frac{a - b}{a + b}$$

55. Using Problem 54 and the formula for the sum and difference of sines, show that

$$\frac{2 \cos \frac{1}{2}(\alpha + \beta) \sin \frac{1}{2}(\alpha - \beta)}{2 \sin \frac{1}{2}(\alpha + \beta) \cos \frac{1}{2}(\alpha - \beta)} = \frac{a - b}{a + b}$$

56. Using Problem 55, show that

$$\frac{\tan \frac{1}{2}(\alpha - \beta)}{\tan \frac{1}{2}(\alpha + \beta)} = \frac{a - b}{a + b}$$

This result is known as the **Law of Tangents.**

57. Another form of the Law of Tangents (see Problem 56) is

$$\frac{\tan \frac{1}{2}(\beta - \gamma)}{\tan \frac{1}{2}(\beta + \gamma)} = \frac{b - c}{b + c}$$

Derive this formula.

58. A third form of the Law of Tangents (see Problems 56 and 57) is

$$\frac{\tan \frac{1}{2}(\alpha - \gamma)}{\tan \frac{1}{2}(\alpha + \gamma)} = \frac{a - c}{a + c}$$

Derive this formula.

59. Newton's formula involves all six parts of a triangle. It is not useful in solving a triangle, but it is helpful in checking your results after you have done so. Show that

$$\frac{a + b}{c} = \frac{\cos \frac{1}{2}(\alpha - \beta)}{\sin \frac{1}{2}\gamma}$$

60. Mollweide's formula involves all six parts of a triangle. It is not useful in solving a triangle, but it is helpful in checking your results after you have done so. Show that

$$\frac{a - b}{c} = \frac{\sin \frac{1}{2}(\alpha - \beta)}{\cos \frac{1}{2}\gamma}$$

61. The letter to the editor shown at the top of page 353 appeared in the February 1977 issue of *Popular Science*. See if you can solve this puzzle. As a hint, you might guess from its placement in this book that it uses the Law of Cosines or the Law of Sines.

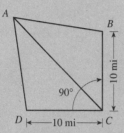
8.4 Ambiguous Case: SSA

The remaining case of solving oblique triangles as given in Section 8.2 is case 5, in which we are given two sides and an angle that is not an included angle. Call the given angle θ (which may be α, β, or γ, depending on the problem). Since we are given SSA, one of the given sides must not be one of the sides of θ; that is, it is opposite θ. Call this side OPP. The given side that is one of the sides of θ is called ADJ. (You might find it helpful to refer to Table 8.1 on page 357 as you read this section.)

a. Suppose that $\theta > 90°$. There are two possibilities:

i. OPP $\leq$ ADJ

No triangle is formed (see Figure 8.18).

ii. OPP $>$ ADJ

One triangle is formed (see Figure 8.19). Use the Law of Sines as shown in Example 1.

FIGURE 8.18 $\theta > 90°$, OPP $\leq$ ADJ

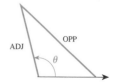

FIGURE 8.19 $\theta > 90°$, OPP $>$ ADJ

*Reprinted with permission from *Popular Science*. © 1977, Times Mirror Magazines, Inc.

EXAMPLE 1 Let $a = 3.0$, $b = 2.0$, and $\alpha = 110°$. Solve the triangle.

SOLUTION

$\alpha = 110°$ Given

$\beta = 39°$ Use the Law of Sines:

$$\frac{\sin \alpha}{a} = \frac{\sin \beta}{b}$$

$$\frac{\sin 110°}{3.0} = \frac{\sin \beta}{2.0}$$

$$\sin \beta = \frac{2.0}{3.0} \sin 110°$$

$$\beta \approx 38.78955642°$$

$\gamma = 31°$ $\gamma = 180° - 110° - \beta$

$$\approx 31.2104436°$$

$a = 3.0$ Given

$b = 2.0$ Given

$c = 1.7$ Use the Law of Sines:

$$\frac{\sin 110°}{3.0} = \frac{\sin \gamma}{c}$$

$$c = \frac{3.0 \sin \gamma}{\sin 110°}$$

$$\approx 1.654$$

Remember: Do not work with rounded results.

Notice, when finding c in Example 1, if you work with

$$\gamma \approx 31.2104436°$$

you obtain $c \approx 1.654$, or 1.7, to two significant digits. However, if you work with $\gamma \approx 31°$, you obtain $c \approx 1.644$, or 1.6, to two significant digits. The proper procedure is to round only when stating answers.

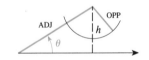

FIGURE 8.20 $\theta < 90°$, OPP $< h <$ ADJ

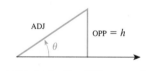

FIGURE 8.21 $\theta < 90°$, OPP $= h <$ ADJ

b. Suppose that $\theta < 90°$. There are four possibilities. Let h be the altitude of the triangle drawn from the vertex connecting the OPP and ADJ sides. Now, to find h use the right-triangle definition of sine:

$$h = (\text{ADJ}) \sin \theta$$

i. OPP $< h <$ ADJ
No triangle is formed (see Figure 8.20).

ii. OPP $= h <$ ADJ
A right triangle is formed (see Figure 8.21). Use the methods of Section 8.1 to solve the triangle.

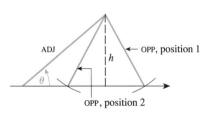

FIGURE 8.22   The ambiguous case

iii. $h <$ OPP $<$ ADJ
This situation is called the ambiguous case and is really the only special case you must watch for. All the other cases can be determined from the calculations without any special consideration. Notice from Figure 8.22 that two *different* triangles are formed with the given information. The process for finding both solutions is shown in Example 2.

EXAMPLE 2 Solve the triangle with $a = 1.50$, $b = 2.00$, and $\alpha = 40.0°$.

SOLUTION

$$\frac{\sin \alpha}{a} = \frac{\sin \beta}{b}$$

$$\frac{\sin 40.0°}{1.50} = \frac{\sin \beta}{2.00}$$

$$\sin \beta = \frac{2.00 \sin 40.0°}{1.50}$$

$$\beta \approx 58.98696953°$$

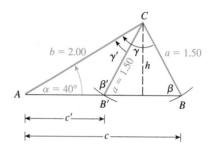

FIGURE 8.23

But from Figure 8.23, you can see that this is only the acute angle solution. For the obtuse angle solution—call it β'—find

$$\beta' = 180° - \beta$$
$$\approx 121.0°$$

Finish the problem by working two calculations:

Solution 1

$\alpha = 40.0°$	Given	
$\beta = 59.0°$	See above	
$\gamma = 81.0°$	$\gamma = 180° - \alpha - \beta$	
$a = 1.50$	Given	
$b = 2.00$	Given	

$c = 2.30$

$$\frac{\sin \alpha}{a} = \frac{\sin \gamma}{c}$$

$$c = \frac{1.50 \sin \gamma}{\sin 40.0°}$$

$$\approx 2.30493839$$

Solution 2

$\alpha = 40.0°$	Given	
$\beta' = 121.0°$	See prior calculations	
$\gamma' = 19.0°$	$\gamma' = 180° - \alpha - \beta'$	
$a = 1.50$	Given	
$b = 2.00$	Given	

$c' = .759$

$$\frac{\sin \alpha}{a} = \frac{\sin \gamma'}{c'}$$

$$c' = \frac{1.50 \sin \gamma'}{\sin 40.0°}$$

$$\approx .7592393826$$

iv. OPP ≥ ADJ

One triangle is formed, as shown in Figure 8.24.

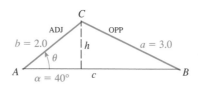

FIGURE 8.24 $\theta < 90°$, OPP ≥ ADJ

EXAMPLE 3   Solve the triangle given by $a = 3.0$, $b = 2.0$, and $\alpha = 40°$, as shown in Figure 8.24.

SOLUTION

$$\frac{\sin \alpha}{a} = \frac{\sin \beta}{b}$$

$$\frac{\sin 40°}{3.0} = \frac{\sin \beta}{2.0}$$

$$\sin \beta = \frac{2}{3} \sin 40°$$

$$\approx .4285250731$$

$$\beta \approx 25.37399394$$

$\alpha = 40°$ Given

$\beta = 25°$ See work shown above.

$\gamma = 115°$ $\gamma = 180° - \alpha - \beta$

$a = 3.0$ Given

$b = 2.0$ Given

$c = 4.2$ $\dfrac{\sin 40°}{3.0} = \dfrac{\sin \gamma}{c}$

$c = \dfrac{3.0 \sin \gamma}{\sin 40°}$

≈ 4.2427

The most important skill to be learned from this section is the ability to select the proper trigonometric law when given a particular problem. In the rest of this section, you will encounter applications of right triangles, the Law of Cosines, and the Law of Sines. A review of various types of problems may be helpful. Table 8.1 summarizes the solution of oblique triangles.

Table 8.1 Summary for solving triangles

GIVEN	CONDITIONS ON GIVEN INFORMATION	LAW TO USE FOR SOLUTION
1. SSS	**a.** The sum of the lengths of the two smaller sides is less than or equal to the length of the larger side.	No solution
	b. The sum of the lengths of the two smaller sides is greater than the length of the larger side.	**Law of Cosines**
2. SAS	**a.** The angle is greater than or equal to 180°.	No solution
	b. The angle is less than 180°.	**Law of Cosines**
3. AAA		**No solution**
4. ASA or AAS	**a.** The sum of the angles is greater than or equal to 180°.	No solution
	b. The sum of the angles is less than 180°.	**Law of Sines**
5. SSA	Let θ be the given angle with adjacent (ADJ) and opposite (OPP) sides given; the height h is found by $$h = (\text{ADJ})\sin\theta$$ **a.** $\theta > 90°$ **i.** OPP $\leq$ ADJ **ii.** OPP $>$ ADJ **b.** $\theta < 90°$ **i.** OPP $< h <$ ADJ **ii.** OPP $= h <$ ADJ **iii.** $h <$ OPP $<$ ADJ **iv.** OPP $\geq$ ADJ	No solution **Law of Sines** No solution Right-triangle solution *Ambiguous case*: Use Law of Sines to find two solutions **Law of Sines**

Remember: When given two sides and an angle that is not an included angle, check to see whether one side is between the height of the triangle and the length of the other side. If it is, there will be two solutions.

EXAMPLE 4 An airplane is 100 km N40°E of a Loran station, traveling due west at 240 kph. How long will it be (to the nearest minute) before the plane is 90 km from the Loran station? (See Figure 8.25).

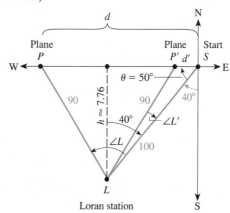

FIGURE 8.25

You are given SSA; check the other conditions for the ambiguous case.
Since the heading is 40°, $\angle PSL = 50°$; call this angle θ. Notice that 100 km is the side adjacent to θ; 90 km is the side opposite to θ; thus, we have

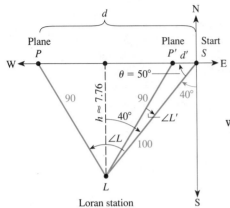

Plane P

d

N

Plane P' d' Start S

W ← → E

$\theta = 50°$

90    $h \approx 7.76$ 90   40°

40° $\angle L'$

$\angle L$ 100

L

Loran station

S

$$h = (\text{ADJ}) \sin \theta$$
$$= 100 \sin 50°$$
$$\approx 76.6$$
$$h < \text{OPP} < \text{ADJ}$$
$$76.6 < 90 < 100$$

which is the ambiguous case.

Solution 1

$\angle S$: 50°

$\angle P$: $\dfrac{\sin 50°}{90} = \dfrac{\sin P}{100}$

$$\sin P = \dfrac{100 \sin 50°}{90}$$

$$P \approx 58.33811615°$$

$\angle L$: $L = 180° - S - P$
$$\approx 71.66188385°$$

d: $\dfrac{\sin L}{d} = \dfrac{\sin 50°}{90}$

$$d = \dfrac{90 \sin L}{\sin 50°}$$
$$\approx 111.5202587$$

Solution 2

$\angle S$: 50°

$\angle P'$: $P' = 180° - P$
$$\approx 121.6618839°$$

$\angle L'$: $L' = 180 - S - P'$
$$\approx 8.3338116149°$$

d: $\dfrac{\sin L'}{d'} = \dfrac{\sin 50°}{90}$

$$d' = \dfrac{90 \sin L'}{\sin 50°}$$
$$\approx 17.0372632$$

At 240 kph, the times are

$$\dfrac{d}{240} \approx .4646677447 \quad \text{and} \quad \dfrac{d'}{240} \approx .0709885967$$

To convert these to minutes, multiply by 60 and round to the nearest minute. The times are **28 min** and **4 min**.

In elementary school you learned that the area, K, of a triangle is $K = \frac{1}{2}bh$.* Now, you sometimes need to find the area of a triangle given the measurements for the sides and angles but not the height. Then you need trigonometry to find the area.

---

*In elementary school you no doubt used A for area. In trigonometry, we use K for area because we have already used A to represent a vertex of a triangle.

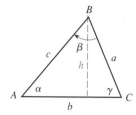
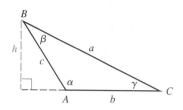

FIGURE 8.26   Oblique triangles

Using the orientation shown in Figure 8.26,

$$\sin \alpha = \frac{h}{c} \quad \text{or} \quad h = c \sin \alpha$$

Thus,

$$K = \frac{1}{2} bh = \frac{1}{2} bc \sin \alpha$$

We could use any other pair of sides to derive the following area formulas:

AREA OF A TRIANGLE (TWO SIDES AND AN INCLUDED ANGLE ARE KNOWN)

$$K = \frac{1}{2} bc \sin \alpha \qquad K = \frac{1}{2} ac \sin \beta \qquad K = \frac{1}{2} ab \sin \gamma$$

EXAMPLE 5   Find the area of $\triangle ABC$ where $\alpha = 18.4°$, $b = 154$ ft, and $c = 211$ ft.

SOLUTION $K = \frac{1}{2}(154)(211) \sin 18.4° \approx 5,128.349903$

To three significant figures, the area is **5,130 ft^2**.

The area formulas given above are useful for finding the area of a triangle when two sides and an included angle are known. Suppose, however, that you know the angles but only one side. Say you know side a. Then you can use either formula involving a and replace the other variable by using the Law of Sines:

$$\frac{\sin \alpha}{a} = \frac{\sin \beta}{b} \quad \text{so} \quad b = \frac{a \sin \beta}{\sin \alpha}$$

Therefore,

$$K = \frac{1}{2} ab \sin \gamma = \frac{1}{2} a \frac{a \sin \beta}{\sin \alpha} \sin \gamma = \frac{a^2 \sin \beta \sin \gamma}{2 \sin \alpha}$$

The other area formulas shown in the following box can be similarly derived:

AREA OF A TRIANGLE (TWO ANGLES AND AN INCLUDED SIDE ARE KNOWN)

$$K = \frac{a^2 \sin \beta \sin \gamma}{2 \sin \alpha} \qquad K = \frac{b^2 \sin \alpha \sin \gamma}{2 \sin \beta} \qquad K = \frac{c^2 \sin \alpha \sin \beta}{2 \sin \gamma}$$

EXAMPLE 6 Find the area of a triangle with angles 20°, 50°, and 110° if the side opposite the 50° angle is 24 in. long.

SOLUTION
$$K = \frac{24^2 \sin 20° \sin 110°}{2 \sin 50°} \approx 120.8303469$$

To two significant figures, the area is **120 m².**

If three sides (but none of the angles) are known, you will need another formula to find the area of a triangle. The formula is derived from the Law of Cosines; this derivation is left as an exercise but is summarized in the following box. The result is known as **Heron's (or Hero's) Formula.**

AREA OF A TRIANGLE (THREE SIDES KNOWN)

$$K = \sqrt{s(s-a)(s-b)(s-c)} \quad \text{where } s = \tfrac{1}{2}(a+b+c)$$

EXAMPLE 7 Find the area of a triangle having sides of 43 ft, 89 ft, and 120 ft.

SOLUTION Let $a = 43$, $b = 89$, and $c = 120$. Then $s = \tfrac{1}{2}(43 + 89 + 120) = 126$. Thus,

$$K = \sqrt{126(126 - 43)(126 - 89)(126 - 120)} \approx 1{,}523.704696$$

To two significant digits, the area is **1,500 ft².**

FIGURE 8.27 Sector of a circle (shaded portion)

Another type of area problem that requires trigonometry is the procedure for finding the area of a sector of a circle. A **sector of a circle** is the portion of the interior of a circle cut by a central angle, θ (Figure 8.27). Since the area of a circle is $A = \pi r^2$, where r is the radius, the area of a sector is the fraction $\theta/(2\pi)$ of the entire circle. In general,

$$\text{Area of sector} = (\text{Fractional part of circle})(\text{Area of circle})$$
$$= \frac{\theta}{2\pi} \cdot \pi r^2$$

AREA OF A SECTOR

$$\text{Area of sector} = \tfrac{1}{2}\theta r^2 \quad \text{where } \theta \text{ is measured in radians}$$

EXAMPLE 8 What is the area of the sector of a circle of radius 12 in. whose central angle is 2?

SOLUTION
$$\text{Area of sector} = \tfrac{1}{2} \cdot 2 \cdot (12 \text{ in.})^2 = \mathbf{144 \text{ in.}^2}$$

EXAMPLE 9 What is the area of the sector of a circle of radius 420 m whose central angle is 2°?

SOLUTION First change 2° to radians: $2° = 2\left(\dfrac{\pi}{180}\right) = \dfrac{\pi}{90}$ radian

$$\text{Area of sector} = \frac{1}{2} \cdot \left(\frac{\pi}{90}\right)(420 \text{ m})^2$$
$$= 980\pi \text{ m}^2$$
$$\approx \mathbf{3{,}079 \text{ m}^2}$$

A third application in this section is finding the volume of a cone when you do not know its height. From geometry, the volume is

$$V = \frac{\pi r^2 h}{3}$$

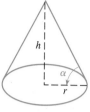

FIGURE 8.28

where r is the radius of the base and h is the height of the cone. Now suppose you know the radius of the base but not the height. If you know the angle of elevation α (see Figure 8.28), then

$$\tan \alpha = \frac{h}{r} \quad \text{or} \quad h = r \tan \alpha$$

Thus, the volume is

$$V = \frac{\pi r^3 \tan \alpha}{3} = \tfrac{1}{3}\pi r^3 \tan \alpha$$

EXAMPLE 10 If sand is dropped from the end of a conveyor belt, the sand will fall in a conical heap such that the angle of elevation α is about 33°. Find the volume of sand when the radius r is exactly 10 ft. (See Figure 8.29.)

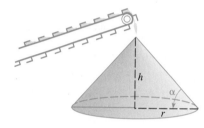

SOLUTION $V = \tfrac{1}{3}\pi(10)^3 \tan 33° \approx 680.0580413$

FIGURE 8.29

The volume (to two significant figures) is **680 ft³.**

8.4 Problem Set

A *Solve △ABC in Problem 1–12. If two triangles are possible, give both solutions. If the triangle does not have a solution, tell why.*

1. $a = 3.0;\ b = 4.0;\ \alpha = 125°$
2. $a = 3.0;\ b = 4.0;\ \alpha = 80°$
3. $a = 5.0;\ b = 7.0;\ \alpha = 75°$
4. $a = 5.0;\ b = 7.0;\ \alpha = 135°$
5. $a = 5.0;\ b = 4.0;\ \alpha = 125°$
6. $a = 5.0;\ b = 4.0;\ \alpha = 80°$
7. $a = 9.0;\ b = 7.0;\ \alpha = 75°$
8. $a = 9.0;\ b = 7.0;\ \alpha = 135°$
9. $a = 7.0;\ b = 9.0;\ \alpha = 52°$
10. $a = 12.0;\ b = 9.00;\ \alpha = 52.0°$
11. $a = 8.629973679;\ b = 11.8;\ \alpha = 47.0°$
12. $a = 10.2;\ b = 11.8;\ \alpha = 47.0°$

Solve △ABC with the parts given in Problems 13–24. If the triangle cannot be solved, tell why.

13. $a = 14.2;\ b = 16.3;\ \beta = 115.0°$
14. $a = 14.2;\ c = 28.2;\ \gamma = 135.0°$
15. $\beta = 15.0°;\ \gamma = 18.0°;\ b = 23.5$
16. $b = 45.7;\ \alpha = 82.3°;\ \beta = 61.5°$
17. $b = 82.5;\ c = 52.2;\ \gamma = 32.1°$
18. $a = 151;\ b = 234;\ c = 416$
19. $a = 68.2;\ \alpha = 145°;\ \beta = 52.4°$
20. $\alpha = 82.5°;\ \beta = 16.9°;\ \gamma = 80.6°$
21. $a = 123;\ b = 225;\ c = 351$
22. $a = 27.2;\ c = 35.7;\ \alpha = 43.7°$
23. $c = 196;\ \alpha = 54.5°;\ \gamma = 63.0°$
24. $b = 428;\ c = 395;\ \gamma = 28.4°$

Find the area of each triangle in Problems 25–36.

25. $a = 15; b = 8.0; \gamma = 38°$

26. $a = 18; c = 11; \beta = 63°$

27. $b = 14; c = 12; \alpha = 82°$

28. $b = 21; c = 35; \alpha = 125°$

29. $a = 30; \beta = 50°; \gamma = 100°$

30. $b = 23; \alpha = 25°; \beta = 110°$

31. $b = 40; \alpha = 50°; \gamma = 60°$

32. $b = 90; \beta = 85°; \gamma = 25°$

33. $a = 7.0; b = 8.0; c = 2.0$

34. $a = 10; b = 4.0; c = 8.0$

35. $a = 11; b = 9.0; c = 8.0$

36. $a = 12; b = 6.0; c = 15$

B *Find the area of each triangle in Problems 37–48. If two triangles are formed, give the area of each, and if no triangle is formed, so state.*

37. $a = 12.0; b = 9.00; \alpha = 52.0°$

38. $a = 7.0; b = 9.0; \alpha = 52°$

39. $a = 10.2; b = 11.8; \alpha = 47.0°$

40. $a = 8.629973679; b = 11.8; \alpha = 47.0°$

41. $b = 82.5; c = 52.2; \gamma = 32.1°$

42. $a = 352; b = 230; c = 418$

43. $\beta = 15.0°; \gamma = 18.0°; b = 23.5$

44. $b = 45.7; \alpha = 82.3°; \beta = 61.5°$

45. $a = 68.2; \alpha = 145°; \beta = 52.4°$

46. $a = 151; b = 234; c = 416$

47. $a = 124; b = 325; c = 351$

48. $a = 27.2; c = 35.7; \alpha = 43.7°$

49. BALLISTICS An artillery-gun observer must determine the distance to a target at point T. He knows that the target is 5.20 mi from point I on a nearby island. He also knows that he (at point H) is 4.30 mi from point I. If $\angle HIT$ is 68.4°, how far is he from the target? (See Figure 8.30.)

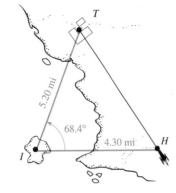

FIGURE 8.30
Determining the distance to a target

50. CONSUMER A buyer is interested in purchasing a triangular lot with vertices LOT, but unfortunately, the marker at point L has been lost. The deed indicates that TO is 453 ft and LO is 112 ft and that the angle at L is 82.6°. What is the distance from L to T?

51. NAVIGATION A UFO is sighted by people in two cities 2.300 mi apart. The UFO is between and in the same vertical plane as the two cities. The angle of elevation of the UFO from the first city is 10.48° and from the second is 40.79°. At what altitude is the UFO flying? What is the actual distance of the UFO from each city?

52. ENGINEERING At 500 ft in the direction that the Tower of Pisa is leaning, the angle of elevation is 20.24°. If the tower leans at an angle of 5.45° from the vertical, what is the length of the tower?

53. ENGINEERING What is the angle of elevation of the leaning Tower of Pisa (described in Problem 52) if you measure from a point 500 ft in the direction exactly opposite from the way it is leaning?

54. SURVEYING The world's longest deepwater jetty is at Le Havre, France. Since access to the jetty is restricted, it was necessary for me to calculate its length by noting that it forms an angle of 85.0° with the shoreline. After pacing out 1,000 ft along the line making an 85.0° angle with the jetty, I calculated the angle to the end of the jetty to be 83.6°. What is the length of the jetty?

55. If the central angle subtended by the arc of a segment of a circle is 1.78 and the area is 54.5 cm², what is the radius of the circle?

56. If the area of a sector of a circle is 162.5 cm² and the angle of the sector is .52, what is the radius of the circle?

57. A field is in the shape of a sector of a circle with a central angle of 20° and a radius of 320 m. What is the area of the field?

58. CONSUMER A level lot has the dimensions shown in Figure 8.31. What is the total cost of treating the area for poison oak if the fee is $45 per acre (1 acre = 43,560 ft²)?

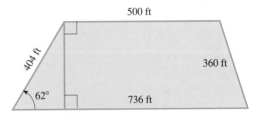

FIGURE 8.31 Area of a lot

59. If vulcanite is dropped from a conveyor belt, it will fall in a conical heap such that the angle of elevation is about 36°. Find the volume of vulcanite when the radius is 30 ft.

60. The volume of a slice cut from a cylinder is found from the formula

$$V = hr^2\left(\frac{\theta}{2} - \sin\frac{\theta}{2}\cos\frac{\theta}{2}\right)$$

for θ measured in radians such that $0 \le \theta \le \pi$. See Figure 8.32. Find the volume of a slice cut from a log with a 6.0-in. radius that is 3.0 ft long when $\theta = \pi/3$.

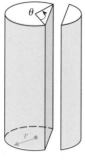

FIGURE 8.32

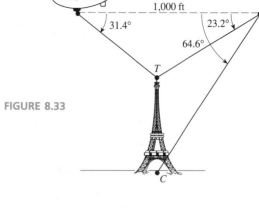

FIGURE 8.33

C

61. NAVIGATION From a blimp, the angle of depression to the top of the Eiffel Tower is 23.2° and to the bottom is 64.6°. After flying over the tower at the same height and at a distance of 1,000 ft from the first location, you determine that the angle of depression to the top of the tower is now 31.4°. What is the height of the Eiffel Tower given that these measurements are in the same vertical plane? (See Figure 8.33.)

62. HISTORICAL QUESTION In January 1978, Martin Cohen, Terry Goodman, and John Benard published an article,

"SSA and the Law of Cosines," in *The Mathematics Teacher*. In this article, they used the quadratic formula to solve the Law of Cosines for c:

$$c = b\cos\alpha \pm \sqrt{a^2 - b^2\sin^2\alpha}$$

Show this derivation.

63. Suppose $\alpha < 90°$. If $d = a^2 - b^2\sin^2\alpha$ in Problem 62, show that if $d < 0$, there is no solution and this equation corresponds to no triangle; if $d = 0$, there is one solution, which corresponds to one triangle; if $d > 0$, there are two solutions and two triangles are formed.

64. CONSTRUCTION A 50-ft culvert carries water under a road. If there is 2 ft of water in a culvert with a 3-ft radius, use Problem 60 to find the volume of water in the culvert.

65. Derive the formula for finding the area when you know three sides of a triangle.

8.5 **Chapter 8 Summary**

The material of this chapter is reviewed in the following list of objectives. After each objective there are some practice questions. For a sample test, select the first question of each set and check your answers with the answer section. For a sample test without answers, use the second question of each set. Additional practice is given by the other questions in each set. If you are having trouble with a particular type of problem, look back to that section for extra help.

8.1 **Right Triangles**

OBJECTIVE 1 *Know the right-triangle definition of the trigonometric functions and solve right triangles.*

1. State the right-triangle definition of the trigonometric functions.

2. Solve $\triangle ABC$, where $a = 7.3$, $c = 15$, and $\gamma = 90°$.

3. Solve $\triangle ABC$, where $b = 678$, $\beta = 55.0°$, and $\gamma = 90.0°$.

4. Solve $\triangle ABC$, where $a = 3.0$, $b = 4.0$, and $c = 5.0$.

8.2 Law of Cosines
8.3 Law of Sines

OBJECTIVE 2   *Know the Laws of Cosines and Sines, as well as the proof for each.*

5. State the Law of Cosines.

6. State the Law of Sines.

7. Prove the Law of Cosines.

8. Prove the Law of Sines.

8.4 Ambiguous Case: SSA

OBJECTIVE 3   *Know when to apply the Laws of Cosines and Sines, as well as recognizing the ambiguous case.* Complete the table for $\triangle ABC$ labeled in the usual fashion.

	TO FIND	KNOWN	PROCEDURE	SOLUTION
	β	a, b, α	$\dfrac{\sin \alpha}{a} = \dfrac{\sin \beta}{b}$	$\beta = \sin^{-1}\left(\dfrac{b \sin \alpha}{a}\right)$
9.	α	a, b, c		
10.	α	a, b, β		
11.	α	a, β, γ		
12.	b	a, α, β		

OBJECTIVE 4   *Solve oblique triangles.*

13. $a = 24$; $c = 61$; $\beta = 58°$

14. $a = 6.8$; $b = 12.2$; $c = 21.5$

15. $b = 34$; $c = 21$; $\gamma = 16°$

16. $b = 4.6$; $\alpha = 108°$; $\gamma = 38°$

OBJECTIVE 5   *Find the area of a given triangle, including the ambiguous case.*

17. $\alpha = 48.0°$; $b = 25.5$ ft; $c = 48.5$ ft

18. $a = 275$ ft; $b = 315$ ft; $\alpha = 50.0°$

19. $a = 14.5$ in.; $b = 17.2$ in.; $\alpha = 35.5°$

20. $a = 6.50$ m; $b = 8.30$ m; $c = 12.6$ m

OBJECTIVE 6   *Solve applied problems using either right triangles or oblique triangles.*

21. A mine shaft is dug into the side of a sloping hill. The shaft is dug horizontally for 485 ft. Next, a turn is made so that the angle of elevation of the second shaft is 58.0°, thus forming a 58.0° angle between the shafts. The shaft is then continued for 382 ft before exiting, as shown in Figure 8.34. How far is it along a straight line from the entrance to the exit, assuming that all tunnels are in a single plane? If the slope of the hill follows the line from the entrance to the exit, what is the angle of elevation from the entrance to the exit?

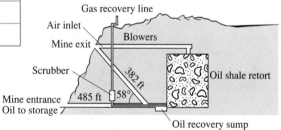

FIGURE 8.34   Determing the exit of a mine shaft

22. Ferndale is 7 mi N50°W of Fortuna. If I leave Fortuna at noon and travel due west at 2 mph, when will I be exactly 6 mi from Ferndale?

23. To measure the span of the Rainbow Bridge in Utah, a surveyor selected two points, P and Q, on either end of the bridge. From point Q, the surveyor measured 500 ft in the direction N38.4°E to point R. Point P was then determined to be in the direction S67.5°W. What is the span of the Rainbow Bridge if all the preceding measurements are in the same plane and $\angle PQR$ is a right angle?

24. When viewing Angel Falls (the world's highest waterfall) from Observation Platform A, located on the same level as the bottom of the falls, we calculate the angle of elevation to the top of the falls to be 69.30°. From Observation Platform B, which is located on the same level exactly 1,000 ft from the first observation point, we calculate the angle of elevation to the top of the falls to be 52.90°. How high are the falls?

Cumulative Review II

Suggestions for study of Chapters 6–8:

MAKE A LIST OF IMPORTANT IDEAS FROM CHAPTERS 6–8. Study this list. Use the objectives at the end of each chapter to help you make up this list.

WORK SOME PRACTICE PROBLEMS. A good source of problems is the set of chapter objectives at the end of each chapter. You should try to work at least one problem from each objective:

    Chapter 6, pp. 293–295; work 25 problems

    Chapter 7, pp. 335–337; work 17 problems

    Chapter 8, pp. 363–364; work 6 problems

Check the answers for the practice problems you worked. The first and third problems of each objective have their answers listed in the back of the book.

ADDITIONAL PROBLEMS. Work additional odd-numbered problems (answers in the back of the book) from the problem sets as needed. Focus on the problems you missed in the chapter summaries.

WORK ON THE PROBLEMS IN THE FOLLOWING CUMULATIVE REVIEW. These problems should be done after you have studied the material. They should take you about 1 hour and will serve as a sample test for Chapters 6–8. Assume that all variables are restricted so that each expression is defined. All the answers for these questions are provided in the back of the book for self-checking.

Practice Test for Chapters 6–8

1. Define the six trigonometric functions of any angle θ.

2. Evaluate the given functions over the set of real numbers (use exact values wherever possible).
  a. $\sin 300°$ **b.** $\cos \frac{5\pi}{6}$
  c. $\tan 3\pi$ **d.** $\text{Sin}^{-1}(-\frac{\sqrt{3}}{2})$
  e. $\text{Arccos}\frac{1}{2}$ **f.** $\text{Arccos}(-4.521)$
  g. $\text{Tan}^{-1}2.310$ **h.** $\tan(\text{Sec}^{-1}3)$

3. Graph each curve
  a. $y = \sqrt{2}\,\cos\frac{1}{2}x$
  b. $y + 3 = 2\sin(3x + \pi)$
  c. $y - 1 = \tan(x - \frac{\pi}{3})$

4. a. State and prove one of the reciprocal identities.
  b. State and prove one of the ratio identities.
  c. State and prove one of the Pythagorean identities.

5. Solve the given equations for $0 \le \theta < 2\pi$.
  a. $3\tan 3\theta = \sqrt{3}$ (exact values)
  b. $4\sin^2\theta + 8\sin\theta - 1 = 0$ (four decimal places)

6. Prove any *two* of the following identities:
  a. $\dfrac{\sin 3\theta}{\sin \theta} - \dfrac{\cos 3\theta}{\cos \theta} = 2$

  b. $\dfrac{\sin \theta}{\csc \theta - \cot \theta} = 1 + \cos \theta$

  c. $\cos 2\theta = 1 - 2\sin^2\theta$

  (Do not assume the identity for $\cos 2\theta$.)

7. For the given triangle, state (one form for each is sufficient):
  a. Law of Sines   **b.** Law of Cosines

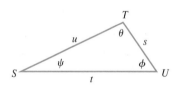

8. Solve each of the given triangles.
  a. $a = 14; b = 27; c = 19$
  b. $b = 7.2; c = 15; \alpha = 113°$
  c. $a = 35; b = 45; \beta = 35°$
  d. $c = 85.7; \alpha = 50.8°; \beta = 83.5°$

9. Find the area of each of the given triangles.
  a. $b = 16$ ft, $c = 43$ ft, $\alpha = 113°$
  b. $\alpha = 40.0°, \beta = 51.8°, c = 14.3$ in.
  c. $a = 121$ cm, $b = 46$ cm, $c = 92$ cm

10. a. The world's largest pyramid is located 63 mi S45°E of Mexico City. If I leave Mexico City in a jeep and travel due east, how far from Mexico City will I be when I am 50 mi from the pyramid?
  b. The longest bridge in the world is over the Humber Estuary in England. To measure its length I paced out 100.0 ft in a direction that formed a right angle with one end. I then determined that the angle formed when sighting the other end was 88.76°. How long is the bridge?

Solar Power

Energy from the sun provides enough power to heat water or even a home

BY NEALE LESLIE A revolutionary solar-powered water heater that aligns itself facing directly into the sun now is available on the market.

Unlike flat plate collectors which have to be installed in a southward direction, usually on roofs of the buildings being served, this device can be installed on any part of a north or south-facing roof—or any portion of the yard—as long as the sun's rays reach it.

This device, called the SAV, manufactured by SAV Solar Systems Inc. of Los Angeles, follows the sun. It is both a parabolic reflector and storage tank combined.

The tank—riding piggyback above the parabolic mirror—is a collector itself. It is coated with black chrome, one of the most superior materials known for absorbing heat. Actually, it absorbs 95 percent of all incoming radiation, and only emits 10 percent of the heat collected. The result: A trapping and holding of 85 percent of the solar radiation.

The tank is further encased in cylindrical double glazing made of clear plastic, with one inch of air space between the tank and inner glazing and another inch of air space between the inner and outer glazing.

This provides a "glass house" heating and insulating effect. At the same time the tank is tilted to the most direct angle to the sun, an angle the homeowner can adjust to conform to his latitude and the change of the season.

The tank, designed by Jon Makeever, SAV Solar Systems president, permits the sun's heat to set up "thermosyphonic" circulation patterns, greatly increasing the efficiency of the unit.

Makeever also incorporated a system of parabolic surfaces concentrating the energy of the sun directly onto the solar collector and storage tank. And working in conjunction with this is his tracking system which causes the entire mechanism, including the tank, to follow the sun across the sky.

Makeever chose Freon 12, available anywhere in refrigeration service shops, as the agent to provide the power for the tracking mechanism. The Freon 12 is energized by the sun. It is maintenance free, requires no wires, no electricity, timbers or other apparatus. It is nonexplosive, providing enough pressure to operate a couple of hydraulic pistons, but not enough to burst any of the tracker's components. The Freon 12 has to be replaced once every four years.

The unit, approximately 4 by 4 feet, provides up to 98 percent more heat per collector area than flat plate systems used in a test, Makeever said.

One unit would supply the hot water needs of a typical family, but three would be needed for space heating a house. Empty, the unit weighs 140 pounds; filled another 100, but way under the minimal weight standard for roofs.

The unit is said to be adaptable to any part of a sloping or flat roof, and capable of withstanding wind velocities up to 120 m.p.h.

Retail price of the unit is about $2,500 installed, available through plumbing, heating and air conditioning firms and solar equipment dealers.

SOURCE: Courtesy of *The Press Democrat*, Santa Rosa, California, July 29, 1979. Reprinted by permission.

The efficient use of a solar collector requires knowledge of the length of daylight and the angle of the sun throughout the year at the location of the collector. Table 1 gives the times of sunrise and sunset for various latitudes. Figure 1 shows a graph of the sunrise and sunset times for latitude 35°N. These are definitely not graphs of sine or cosine functions. If you plot the length of daylight (see Problem 7), however, you will obtain a curve that is nearly a sine curve.

If you wish to use Table 1 for your own town's latitude and it is not listed, you can use a procedure called **linear interpolation**. Your local Chamber of Commerce

Table 1 Times of sunrise and sunset for various latitudes

Date	Time of Sunrise							Time of Sunset						
	20°N. Latitude (Hawaii)	30°N. Latitude (New Orleans)	35°N. Latitude (Albuquerque)	40°N. Latitude (Philadelphia)	45°N. Latitude (Minneapolis)	60°N. Latitude (Alaska)		20°N. Latitude (Hawaii)	30°N. Latitude (New Orleans)	35°N. Latitude (Albuquerque)	40°N. Latitude (Philadelphia)	45°N. Latitude (Minneapolis)	60°N. Latitude (Alaska)	
	h m	h m	h m	h m	h m	h m		h n	h m	h m	h m	h m	h m	
Jan. 1	6 35	6 56	7 08	7 22	7 38	9 03		17 31	17 10	16 58	16 44	16 28	15 03	
Jan. 15	6 38	6 57	7 08	7 20	7 35	8 48		17 41	17 22	17 12	16 59	16 44	15 31	
Jan. 30	6 36	6 52	7 01	7 11	7 23	8 19		17 51	17 35	17 27	17 17	17 05	16 09	
Feb. 14	6 30	6 41	6 48	6 55	7 03	7 42		17 59	17 48	17 42	17 34	17 26	16 48	
Mar. 1	6 20	6 26	6 30	6 34	6 39	6 59		18 05	17 59	17 56	17 52	17 47	17 27	
Mar. 16	6 08	6 09	6 10	6 11	6 11	6 15		18 10	18 09	18 08	18 08	18 07	18 04	
Mar. 31	5 55	5 51	5 49	5 46	5 43	5 29		18 14	18 18	18 20	18 23	18 26	18 41	
Apr. 15	5 42	5 34	5 28	5 23	5 16	4 44		18 18	18 27	18 32	18 38	18 45	19 18	
Apr. 30	5 32	5 18	5 11	5 02	4 51	4 01		18 23	18 37	18 44	18 53	19 04	19 55	
May 15	5 24	5 07	4 57	4 45	4 31	3 23		18 29	18 46	18 56	19 08	19 22	20 31	
May 30	5 20	5 00	4 48	4 34	4 18	2 53		18 35	18 55	19 07	19 21	19 38	21 04	
June 14	5 20	4 58	4 45	4 30	4 13	2 37		18 40	19 02	19 15	19 30	19 48	21 24	
June 29	5 23	5 02	4 49	4 34	4 16	2 40		18 43	19 05	19 18	19 33	19 51	21 26	
July 14	5 29	5 08	4 56	4 43	4 26	3 01		18 43	19 03	19 15	19 29	19 45	21 09	
July 29	5 34	5 17	5 07	4 55	4 41	3 33		18 39	18 56	19 06	19 17	19 31	20 38	
Aug. 13	5 39	5 26	5 18	5 09	4 59	4 09		18 30	18 43	18 51	19 00	19 10	19 59	
Aug. 28	5 43	5 35	5 29	5 24	5 17	4 45		18 19	18 27	18 33	18 38	18 45	19 16	
Sept. 12	5 47	5 43	5 40	5 38	5 35	5 20		18 06	18 10	18 12	18 14	18 17	18 31	
Sept. 27	5 50	5 51	5 51	5 52	5 53	5 55		17 52	17 51	17 50	17 49	17 49	17 45	
Oct. 12	5 54	6 00	6 03	6 07	6 11	6 31		17 39	17 33	17 29	17 26	17 21	17 01	
Oct. 22	5 57	6 06	6 12	6 18	6 25	6 56		17 32	17 22	17 17	17 11	17 04	16 32	
Nov. 6	6 04	6 18	6 26	6 35	6 45	7 35		17 23	17 10	17 02	16 52	16 42	15 52	
Nov. 21	6 12	6 30	6 40	6 52	7 05	8 12		17 19	17 02	16 51	16 40	16 26	15 19	
Dec. 6	6 22	6 42	6 54	7 07	7 23	8 44		17 20	17 00	16 48	16 35	16 19	14 58	
Dec. 21	6 30	6 52	7 04	7 18	7 35	9 02		17 26	17 05	16 52	16 38	16 21	14 54	

Table courtesy of U.S. Naval Observatory. This table of sunrise and sunset may be used in any year of the 20th century with an error not exceeding two minutes and generally less than one minute. It may also be used anywhere in the vicinity of the stated latitude with an additional error of less than one minute for each 9 miles.

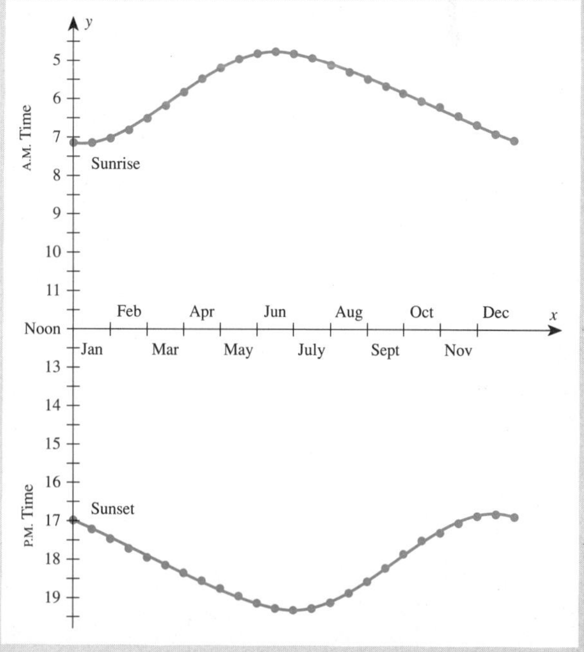

FIGURE 1 Sunrise and sunset times for latitude 35°N

will probably be able to tell you your town's latitude. The latitude for Santa Rosa, California, is about 38°N, which falls between 35° and 40° on the table (see Figure 2).

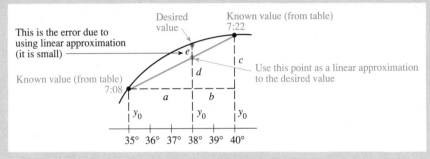

FIGURE 2 Linear interpolation

ANGLE	TIME	
35°	7:08	Known, from Table 1
38°	x	Unknown; this is the latitude of the town for which the times of sunrise and sunset are not available.
40°	7:22	Known, from Table 1

Use the known information and a proportion to find x:

$$5\left[3\begin{bmatrix} 35° & 7:08 \\ 38° & x \\ 40° & 7:22 \end{bmatrix}14\right.$$

$$\frac{3}{5} = \frac{x}{14}$$

$$x = \frac{3}{5}(14) = 8.4 \qquad \text{To the nearest minute this is :08.}$$

Since 38° is $\frac{3}{5}$ of the way between 35° and 40°, the sunrise time will be $\frac{3}{5}$ of the way between 7:08 and 7:22, namely 7:16.

One type of solar collector requires the construction of a central cylinder. The construction of a template for cutting a 45° angle in a cardboard tube to make a rotating mirror is described in the May 1978 issue of *Byte* magazine. The template is shown in Figure 3, and the details of construction are given in Problem 12.

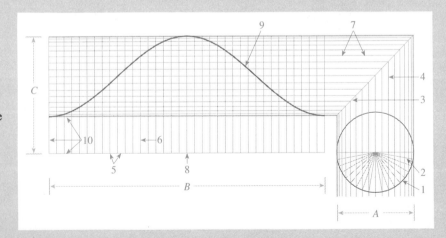

FIGURE 3 Construction of a template for cutting a 45° angle in a cardboard tube to make a rotating mirror

Extended Application Problems—Solar Power

Use linear interpolation and Table 1 to approximate the times requested in each of Problems 1–6.

1. Sunset for Santa Rosa, California (lat. 38°N) on January 1.
2. Sunrise for Tampa, Florida (lat. 28°N), on June 29.
3. Sunset for Tampa, Florida (lat. 28°N), on June 29.
4. Sunset for Winnipeg, Canada (lat. 50°N), on October 12.
5. Sunrise for Winnipeg, Canada (lat. 50°N), on October 12.
6. Sunrise for Juneau, Alaska (lat. 58°N), on March 1.
7. Plot the length of daylight in Chattanooga, Tennessee (lat. 35°N).
8. Plot the time of sunrise and sunset for Seward, Alaska (lat. 60°).
9. Plot the length of daylight for Seward, Alaska (lat. 60°N).
10. Plot the sunrise and sunset for your town's latitude by drawing a graph similar to that in Figure 1.
11. Plot the length of daylight for your town from the data in Problem 10.
12. Carry out the following steps from *Byte* magazine (May 1978, p. 129) to cut a 45° angle in a tube of diameter A, circumference B, and finished height C:*

    1. Draw a circle the same size as the outside diameter of the tube.
    2. Divide the circle into 36 equal parts of $11\frac{1}{4}$ in. each.
    3. Draw a 45° angle above the circle.
    4. Carry the points at the outside of the circle straight up to the 45° angle.
    5. Find the circumference of the circle, draw a straight line, and divide it into 36 equal parts.
    6. Carry lines from those divisions straight upward.
    7. Bring lines straight across from the intersections at the 45° angle line and intersect with the vertical lines.
    8. Starting at the centerline, mark points of intersection.
    9. Fill in between points of intersection.
    10. Cut out pattern and wrap around tube, lining up straight circumference side of pattern with (cut) end of tube square.
    11. Carefully trace curved line end of pattern onto tube.
    12. Remove pattern from tube and save for making black paper cover for tube.
    13. Cut along traced line on tube with an X-acto (or similar) knife.
    14. Tape large sheet of sandpaper to a table top or other flat surface.
    15. Sand end of tube until flat.
    16. Tube is now ready for application of Mylar mirror surface. Construct the tube.

---

*From "How to Multiply in a Wet Climate," by J. Bryant and M. Swasdee, May, 1978, p. 129. Copyright © Byte Publications, Inc. Reprinted by permission.

**Gottfried Wilhelm von Leibniz
(1646–1716)**

I have so many ideas that may perhaps be of some use in time if others more penetrating than I go deeply into them some day and join the beauty of their minds to the labor of mine.

LEIBNIZ

For what is the theory of determinants? It is an algebra upon algebra; a calculus which enables us to combine and foretell the results of algebraical operations, in the same way as special operations of arithmetic. All analysis must ultimately clothe itself under this form.

J. J. SYLVESTER
Philosophical Magazine, vol. 1 (1851)

Gottfried Wilhelm von Leibniz has been described as a true example of a "universal genius." Morris Kline described Leibniz as a philosopher, lawyer, historian, philologist, and pioneer geologist who did important work in logic, mechanics, optics, mathematics, hydrostatics, pneumatics, nautical science, and calculating machines. He was a professional lawyer who was employed as a diplomat and historian. By the time he was 20, he had mastered most of the important works on mathematics and had set the stage for modern logic. At 30, he invented calculus, and at 36, he invented a reckoning machine, a forerunner of the computer. He tried, without success, to reconcile the Catholic and Protestant faiths. Throughout his life he was searching for a *universal mathematics* in which he aimed to create "a general method in which all truths of the reason would be reduced to a kind of calculation." Because of his vast talent, his search for a universal mathematics led him to many applied problems.

In this chapter, we investigate an important tool in solving a wide variety of applied problems. That tool—*matrices*—will provide procedures for solving *systems of equations and inequalities*. Systems of equations were solved by the Babylonians as early as 1600 B.C. By 1683, the Japanese mathematician Seki Kōwa had updated an old Chinese method of solving simultaneous linear equations that involved rearrangements of rods similar to the way we simplify *determinants* in this chapter. It was Leibniz, however, who originated the notation of determinants.

Leibniz' later years were dimmed by a bitter controversy with Isaac Newton concerning whether he had discovered calculus independently of Newton. When he died, in 1716, his funeral was attended only by his secretary. Today we ascribe the independent discovery of calculus to both Leibniz and Newton.

9

SYSTEMS AND MATRICES

Contents

Preview

This chapter introduces a very important tool in mathematics—namely, the matrix. We are motivated by our desire to solve systems of equations, but we soon learn that matrices have many other applications and uses. This chapter concludes with an introduction to a very significant new branch of mathematics, linear programming. There are 12 specific objectives in this chapter, which are listed on pages 425–426.

Perspective

We live in a three-dimensional world, and surfaces in three dimensions are studied in calculus. In order to analyze a surface you need to look at what are called *tangent planes*. In the discussion taken from a leading calculus book shown below, the following determinants is used to find the tangent plane $\mathbf{T}_x \times \mathbf{T}_y$:

$$\begin{vmatrix} \mathbf{i} & \mathbf{j} & \mathbf{k} \\ 1 & 0 & \dfrac{\partial f}{\partial x} \\ 0 & 1 & \dfrac{\partial f}{\partial y} \end{vmatrix}$$

We introduce determinants in Section 9.4.

Solution   We first compute the tangent vectors:

$$\mathbf{r}_u = \frac{\partial x}{\partial u}\mathbf{i} + \frac{\partial y}{\partial u}\mathbf{j} + \frac{\partial z}{\partial u}\mathbf{k} = 2u\mathbf{i} + \mathbf{k}$$

$$\mathbf{r}_v = \frac{\partial x}{\partial v}\mathbf{i} + \frac{\partial y}{\partial v}\mathbf{j} + \frac{\partial z}{\partial v}\mathbf{k} = 2v\mathbf{j} + 2\mathbf{k}$$

Thus the normal vector is

$$\mathbf{r}_u \times \mathbf{r}_v = \begin{vmatrix} \mathbf{i} & \mathbf{j} & \mathbf{k} \\ 2u & 0 & 1 \\ 0 & 2v & 2 \end{vmatrix} = -2v\mathbf{i} - 4u\mathbf{j} + 4uv\mathbf{k}$$

Notice that the point (1, 1, 3) corresponds to the parameter values $u = 1$ and $v = 1$, so the normal vector there is

$$-2\mathbf{i} - 4\mathbf{j} + 4\mathbf{k}$$

From James Stewart, *Calculus*, 2nd Edition (Pacific Grove, Ca.: Brooks/Cole), p. 891.

9.1 Systems of Equations

Solving systems of equations is a procedure that arises throughout mathematics. We will begin our study of this topic by considering an arbitrary system of two equations with two variables:

$$\begin{cases} a_{11}x_1 + a_{12}x_2 = b_1 \\ a_{21}x_1 + a_{22}x_2 = b_2 \end{cases}$$

The notation may seem strange, but it will prove to be useful. The variables are x_1 and x_2; the constants are a_{11}, a_{12}, a_{21}, a_{22}, b_1, and b_2. By a **system** of two equations with two variables, we mean any two equations in those variables. The **simultaneous solution** of a system is the intersection of the solution sets of the individual equations. The brace is used to show that this intersection is desired. If all the equations in a system are linear, it is called a **linear system.**

You solved linear systems with two variables in elementary algebra, so we will begin by reviewing the methods used there. Then we will generalize first to nonlinear systems and finally to more complicated linear systems.

Graphing Method

Since the graph of each equation in a system of linear equations in two variables is a line, the solution set is the intersection of two lines. In two dimensions, two lines must be related to each other in one of three possible ways:

1. They intersect at a single point.

2. The graphs are parallel lines. In this case, the solution set is empty and the system is called *inconsistent*. In general, any system that has an empty solution set is referred to as an **inconsistent system.**

3. The graphs are the same line. In this case, there are infinitely many points in the solution set and any solution of one equation is also a solution of the other. Such a system is called **dependent.** This word is also used to describe nonlinear systems.

EXAMPLE 1 Relate the given specific system to the general system and solve by graphing.

$$\begin{cases} 2x - 3y = -8 \\ x + y = 6 \end{cases} \qquad \begin{cases} a_{11}x_1 + a_{12}x_2 = b_1 \\ a_{21}x_1 + a_{22}x_2 = b_2 \end{cases}$$

SOLUTION The variables are $x_1 = x$ and $x_2 = y$. The subscripts in a_{11}, a_{12}, a_{21}, and a_{22} are called **double subscripts** and indicate *position*. That is, a_{12} should not be read "a sub twelve" but rather "a sub one two." It represents the second constant in the first equation:

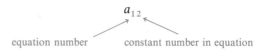

This notation may seem unnecessarily complicated when dealing with only two equations and two variables, but we want to develop a notation that can be used with many variables and many equations. For this example,

$$a_{11} = 2 \qquad a_{12} = -3 \qquad b_1 = -8$$
$$a_{21} = 1 \qquad a_{22} = 1 \qquad b_2 = 6$$

The solution is $x = 2$, $y = 4$, or **(2, 4)** as shown in Figure 9.1.

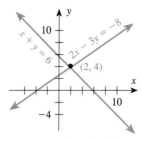

FIGURE 9.1 Graph of the system
$$\begin{cases} 2x - 3y = -8 \\ x + y = 6 \end{cases}$$

EXAMPLE 2 Relate the system

$$\begin{cases} 2x - 3y = -8 \\ 4x - 6y = 0 \end{cases}$$

to the general system and solve by graphing.

SOLUTION The variables are $x_1 = x$, $x_2 = y$, and the constants are

$$a_{11} = 2 \qquad a_{12} = -3 \qquad b_1 = -8$$
$$a_{21} = 4 \qquad a_{22} = -6 \qquad b_2 = 0$$

The graphs of the lines are parallel since there is no point of intersection (Figure 9.2). This is an **inconsistent system.**

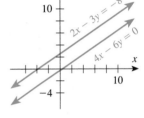

FIGURE 9.2 Graph of the system
$$\begin{cases} 2x - 3y = -8 \\ 4x - 6y = 0 \end{cases}$$

EXAMPLE 3 Relate the system

$$\begin{cases} 2x - 3y = -8 \\ y = \frac{2}{3}x + \frac{8}{3} \end{cases}$$

to the general system and solve by graphing.

SOLUTION To relate to the general system, both equations must be algebraically in the usual form; the second equation needs to be rewritten:

$$y = \tfrac{2}{3}x + \tfrac{8}{3}$$
$$3y = 2x + 8$$
$$-2x + 3y = 8$$

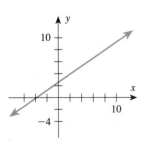

FIGURE 9.3 Graph of the system
$$\begin{cases} 2x - 3y = -8 \\ y = \frac{2}{3}x + \frac{8}{3} \end{cases}$$

This means that the variables are $x_1 = x$, $x_2 = y$, and the constants are

$$a_{11} = 2 \qquad a_{12} = -3 \qquad b_1 = -8$$
$$a_{21} = -2 \qquad a_{22} = 3 \qquad b_2 = 8$$

The equations represent the same line, as shown in Figure 9.3. This is a **dependent system.**

In your previous algebra courses it was probably sufficient to simply state that the system was dependent. However, to say that there are infinitely many solutions does not describe the situation. Look at Example 3 and the graph of the system in Figure 9.3. A solution is any point on the line—that is, any point satisfying the equation $2x - 3y = -8$. This means we can choose *any* value for y and calculate the corresponding value for x. We call the value chosen for y a **parameter,** and usually represent this parameter by t. Thus, if $y = t$, then $2x - 3t = -8$ so that $x = \frac{3}{2}t - 4$. We now can say that the solution for Example 3 is $(\frac{3}{2}t - 4, t)$. We will discuss parametric solutions to systems of equations more completely in the next section.

EXAMPLE 4 Solve the system

$$\begin{cases} x - y = 3 \\ 6x - y = x^2 + 7 \end{cases}$$

SOLUTION This is a nonlinear system. The graphs of $y = x - 3$ and $y = -x^2 + 6x - 7$ are shown in Figure 9.4. Remember that to graph this second-degree equation you need to complete the square:

$$y = -x^2 + 6x - 7$$

$$y + 7 - 9 = -(x^2 - 6x + 9)$$

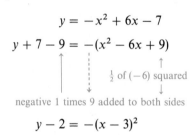

$\frac{1}{2}$ of (-6) squared

negative 1 times 9 added to both sides

$$y - 2 = -(x - 3)^2$$

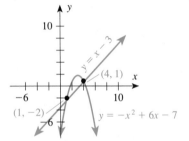

FIGURE 9.4 Graph of
$$\begin{cases} x - y = 3 \\ 6x - y = x^2 + 7 \end{cases}$$

This is a parabola that opens downward with vertex at $(3, 2)$. By inspection, the solution is **(4, 1) and (1, −2).**

To check an answer, make sure that every member of the solution set satisfies all the equations of the system:

Check $(4, 1)$: $x - y = 4 - 1 = 3$ (checks)

$6x - y = 6(4) - 1$ and $x^2 + 7 = (4)^2 + 7$
$= 23$ $= 23$
            checks

Check $(1, -2)$: $x - y = 1 - (-2) = 3$ (checks)

$6x - y = 6(1) - (-2)$ and $x^2 + 7 = (1)^2 + 7$
$= 8$ $= 8$
            checks

Both $(4, 1)$ and $(1, -2)$ check.

Graphing calculators make it very easy to find approximate solutions to systems of equations. For Example 4 we see:

$$\boxed{Y1 = X - 3 \quad Y2 = -X{\char`\^}2 + 6X - 7}$$

The $\boxed{\text{TRACE}}$ on the line shows

$$\boxed{X = 1.0315789 \quad Y = -1.968421}$$

and $\boxed{X = 3.9789474 \quad Y = .97894737}$

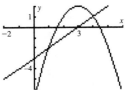

Xmin $= -2$ Ymin $= -6$

Xmax $= 6$ Ymax $= 2$

Remember, if you want to obtain more accurate approximations with a calculator, you can use its $\boxed{\text{ZOOM}}$ feature. An advantage of using a calculator is that it can be used as a tool for solving systems of equations which would be very difficult to solve otherwise. For example, consider the system

Xmin $= -10$ Ymin $= -14$ $\begin{cases} y = x^2 + 3x - 10 \\ y = 2\cos 2x + \sin^2 3x \end{cases}$

Xmax $= 6$ Ymax $= 7$

$$\boxed{X = -4.778947 \quad Y = -1.498504}$$

and

$$\boxed{X = 1.7894737 \quad Y = -1.429363}$$

Substitution Method

The graphing method can give solutions only as accurate as the graphs you can draw, and consequently it is inadequate for most applications. There is therefore a need for more efficient methods.

In general, given a system, the procedure is to write a simpler equivalent system. Two systems are said to be **equivalent** if they have the same solution set. In this chapter we will limit ourselves to finding only real roots. There are several ways to go about writing equivalent systems. The first nongraphical method we will consider comes from the substitution property of real numbers and leads to a **substitution method** for solving systems.

SUBSTITUTION METHOD FOR SOLVING SYSTEMS OF EQUATIONS

1. *Solve* one of the equations for one of the variables.
2. *Substitute* the expression that you obtain into the other equation.
3. *Solve* the resulting equation in a single variable for the value of that variable.
4. *Substitute* that value into either of the original equations to determine the value of the other variable.
5. *State* the solution.

EXAMPLE 5 Solve $\begin{cases} 2p + 3q = 5 \\ q = -2p + 7 \end{cases}$ by substitution.

SOLUTION Since $q = -2p + 7$, substitute $-2p + 7$ for q in the other equation:

$$2p + 3q = 5$$
$$2p + 3(-2p + 7) = 5$$
$$2p - 6p + 21 = 5$$
$$-4p = -16$$
$$p = 4$$

Substitute 4 for p in either of the given equations:

$$q = -2p + 7 = -2(4) + 7 = -1$$

The solution is $(p, q) = (4, -1)$.

Note: If the variables are not x and y, then you must also show the variables along with the ordered pairs. This establishes which variable is associated with which numbers. ▌

EXAMPLE 6 Solve $\begin{cases} x - y + 7 = 0 \\ y = x^2 + 4x + 3 \end{cases}$

SOLUTION Solve one of the equations for one of the variables. The second equation is solved for y, so substitute $x^2 + 4x + 3$ for y in the first equation:

$$x - y + 7 = 0 \qquad \text{If } x = -4, \text{ then}$$
$$x - (x^2 + 4x + 3) + 7 = 0 \qquad y = (-4)^2 + 4(-4) + 3$$
$$-x^2 - 3x + 4 = 0 \qquad\qquad = 3$$
$$x^2 + 3x - 4 = 0 \qquad\quad \text{If } x = 1, \text{ then}$$
$$(x + 4)(x - 1) = 0 \qquad\quad y = 1^2 + 4(1) + 3$$
$$x = -4, 1 \qquad\qquad\quad = 8$$

Solution: $(-4, 3)$ and $(1, 8)$. ▌

EXAMPLE 7 Solve $\begin{cases} x^2 + 2xy = 0 \\ x^2 - 5xy = 14 \end{cases}$

SOLUTION You want to solve the first equation for y, but that requires $x \neq 0$. If $x = 0$, then the first equation is satisfied but the second is not. Thus you can say that $x \neq 0$. Then

$$y = \frac{-x^2}{2x} = \frac{-x}{2}$$

Substitute into the second equation:

$$x^2 - 5xy = 14 \qquad \text{If } x = 2, \text{ then}$$

$$x^2 - 5x\left(\frac{-x}{2}\right) = 14 \qquad y = \frac{-2}{2} = -1$$

$$2x^2 + 5x^2 = 28 \qquad \text{If } x = -2, \text{ then}$$

$$7x^2 = 28 \qquad y = \frac{-(-2)}{2} = 1$$

$$x^2 = 4$$

$$x^2 - 4 = 0$$

$$(x - 2)(x + 2) = 0$$

$$x = 2, -2$$

Thus the solution is **(2, −1) and (−2, 1).**

EXAMPLE 8 Solve $\begin{cases} y = x^2 - x - 1 \\ 4x = 7 + 2y - y^2 \end{cases}$

SOLUTION Substitute the first equation into the second one:

$$4x = 7 + \qquad 2y \qquad - \qquad y^2$$

$$4x = 7 + 2(x^2 - x - 1) - (x^2 - x - 1)^2$$

$$4x = 7 + 2x^2 - 2x - 2 - x^4 + 2x^3 + x^2 - 2x - 1$$

$$4x = 4 + 3x^2 - 4x - x^4 + 2x^3$$

$$x^4 - 2x^3 - 3x^2 + 8x - 4 = 0$$

The possible rational roots are: $\pm 1, \pm 2, \pm 4.$

	1	−2	−3	8	−4	The depressed equation is
1	1	−1	−4	4	0	$x^2 - 4 = 0$
1	1	0	−4	0		$(x - 2)(x + 2) = 0$
						$x = 2, -2$

The roots of this fourth-degree equation are $x = 1, 2,$ and $-2.$

If $x = 1$, then $y = 1^2 - 1 - 1 = -1.$

If $x = 2$, then $y = 2^2 - 2 - 1 = 1.$

If $x = -2$, then $y = (-2)^2 - (-2) - 1 = 5.$

The solution is **(1, −1), (2, 1), and (−2, 5).**

Linear Combination Method

A third method for solving systems is called the **linear combination method.** It involves substitution and the idea that if equal quantities are added to equal quantities, the resulting equation is equivalent to the original system. In general, such addition will not simplify matters unless the numerical coefficients of one or more terms are opposites. However, you can often force them to be opposites by multiplying one or both of the given equations by nonzero constants.

LINEAR COMBINATION METHOD FOR SOLVING SYSTEMS OF EQUATIONS

1. *Multiply* one or both of the equations by a constant or constants, so that the coefficients of one of the variables become opposites.
2. *Add* corresponding members of the equations to obtain a new equation in a single variable.
3. *Solve* the derived equation for that variable.
4. *Substitute* the value of the found variable into either of the original equations and solve for the second variable.
5. *State* the solution.

EXAMPLE 9   Solve $\begin{cases} 3x + 5y = -2 \\ 2x + 3y = 0 \end{cases}$

SOLUTION   Multiply both sides of the first equation by 2 and both sides of the second equation by -3. This procedure, denoted as shown below, forces the coefficients of x to be opposites:

$$\begin{matrix} 2 \\ -3 \end{matrix} \begin{cases} 3x + 5y = -2 \\ 2x + 3y = 0 \end{cases}$$

This means you should add the equations of the system.

$$+ \begin{cases} 6x + 10y = -4 \\ -6x - 9y = 0 \end{cases}$$

$$[6x + (-6x)] + [10y + (-9y)] = -4 + 0 \leftarrow \text{This step should be done}$$
$$y = -4 \qquad \text{mentally.}$$

If $y = -4$, then $2x + 3y = 0$ means $2x + 3(-4) = 0$, or $x = 6$. The solution is **(6, -4).**

EXAMPLE 10   Solve $\begin{cases} y^2 - 5xy = 3 \\ 5xy + 4 = y^2 \end{cases}$

SOLUTION

$$-1 \begin{cases} y^2 - 5xy = 3 \\ y^2 - 5xy = 4 \end{cases}$$

$$+ \begin{cases} -y^2 + 5xy = -3 \\ y^2 - 5xy = 4 \end{cases}$$

$$0 = 1$$

This is an **inconsistent system.**

A *Solve the systems in Problems 1–9 by graphing.*

1. $\begin{cases} y = 3x - 7 \\ y = -2x + 8 \end{cases}$

2. $\begin{cases} x - y = -1 \\ 3x - y = 5 \end{cases}$

3. $\begin{cases} y = \frac{2}{3}x - 7 \\ 2x + 3y = 3 \end{cases}$

4. $\begin{cases} y = \frac{3}{5}x + 2 \\ 3x - 5y = -10 \end{cases}$

5. $\begin{cases} 2x - 3y = 9 \\ y = \frac{2}{3}x - 3 \end{cases}$

6. $\begin{cases} x + y = 0 \\ y = -x^2 - 6x - 4 \end{cases}$

7. $\begin{cases} x + y = 4 \\ y = x^2 + 6x + 14 \end{cases}$

8. $\begin{cases} y = x^2 + 8x + 11 \\ x + y = -7 \end{cases}$

9. $\begin{cases} y = x^2 - 4x \\ y = x^2 - 4x + 8 \end{cases}$

Solve the systems in Problems 10–18 by substitution. Relate each system to the general system

$$\begin{cases} a_{11}x_1 + a_{12}x_2 = b_1 \\ a_{21}x_1 + a_{22}x_2 = b_2 \end{cases}$$

10. $\begin{cases} a = \quad 3b - 7 \\ a = -2b + 8 \end{cases}$

11. $\begin{cases} s + t = 1 \\ 3s + t = -5 \end{cases}$

12. $\begin{cases} m = \frac{2}{3}n - 7 \\ 2n + 3m = 3 \end{cases}$

13. $\begin{cases} v = \frac{3}{5}u + 2 \\ 3u - 5v = 10 \end{cases}$

14. $\begin{cases} 2p - 3q = 9 \\ q = \frac{2}{3}p - 3 \end{cases}$

15. $\begin{cases} 3t_1 + 5t_2 = 1{,}541 \\ t_2 = 2t_1 + 160 \end{cases}$

16. $\begin{cases} \alpha = -7\beta - 3 \\ 2\alpha + 5\beta = 3 \end{cases}$

17. $\begin{cases} \gamma = 3\delta - 4 \\ 5\gamma - 4\delta = -9 \end{cases}$

18. $\begin{cases} \theta + 3\phi = 0 \\ \theta = 5\phi + 16 \end{cases}$

Solve the systems in Problems 19–27 by linear combinations. Relate each system to the general system

$$\begin{cases} a_{11}x_1 + a_{12}x_2 = b_1 \\ a_{21}x_1 + a_{22}x_2 = b_2 \end{cases}$$

19. $\begin{cases} c + d = 2 \\ 2c - d = 1 \end{cases}$

20. $\begin{cases} 2s_1 + s_2 = 10 \\ 5s_1 - 2s_2 = 16 \end{cases}$

21. $\begin{cases} 3q_1 - 4q_2 = 3 \\ 5q_1 + 3q_2 = 5 \end{cases}$

22. $\begin{cases} 9x + 3y = 5 \\ 3x + 2y = 2 \end{cases}$

23. $\begin{cases} 7x + \quad y = 5 \\ 14x - 2y = -2 \end{cases}$

24. $\begin{cases} 2x + 3y = 1 \\ 3x - 2y = 0 \end{cases}$

25. $\begin{cases} \alpha + \quad \beta = 12 \\ \alpha - 2\beta = -4 \end{cases}$

26. $\begin{cases} 2\gamma - 3\delta = 16 \\ 5\gamma + 2\delta = 21 \end{cases}$

27. $\begin{cases} 2\theta + 5\phi = 7 \\ 3\theta + 4\phi = 0 \end{cases}$

B *Solve the systems in Problems 28–45 for x and y by any method. Limit your answers to the set of real numbers.*

28. $\begin{cases} 5x + 4y = 5 \\ 15x - 2y = 8 \end{cases}$

29. $\begin{cases} 3x + 2y = 1 \\ 6x + 4y = 2 \end{cases}$

30. $\begin{cases} 4x - 2y = -28 \\ y = \frac{1}{2}x + 5 \end{cases}$

31. $\begin{cases} 12x - 5y = -39 \\ y = 2x + 9 \end{cases}$

32. $\begin{cases} y = 2x - 1 \\ y = -3x - 9 \end{cases}$

33. $\begin{cases} y = \quad \frac{2}{3}x - 5 \\ y = -\frac{4}{3}x + 7 \end{cases}$

34. $\begin{cases} x + y = \alpha \\ x - y = \beta \end{cases}$

35. $\begin{cases} x - \quad y = \gamma \\ x - 2y = \delta \end{cases}$

36. $\begin{cases} x + y = 2\alpha \\ x - y = 2\beta \end{cases}$

37. $\begin{cases} x + y = a \\ x - y = b \end{cases}$

38. $\begin{cases} x + 2y = a \\ x - 3y = b \end{cases}$

39. $\begin{cases} 2x - y = c \\ 3x + y = d \end{cases}$

40. $\begin{cases} x^2 - 10x = -5 - 2y \\ y = x + 1 \end{cases}$

41. $\begin{cases} y = 2x + 6 \\ 2x^2 + 8x = -2 - 3y \end{cases}$

42. $\begin{cases} 3x^2 + 4y^2 = 12 \\ x^2 + \quad y^2 = -8 \end{cases}$

43. $\begin{cases} 3x^2 + 4y^2 = 19 \\ x^2 + \quad y^2 = 5 \end{cases}$

44. $\begin{cases} x^2 + y^2 = 25 \\ y^2 = 5 - x \end{cases}$

45. $\begin{cases} x^2 - 12y - 1 = 0 \\ x^2 - 4y^2 - 9 = 0 \end{cases}$

46. Use the system

$$\begin{cases} \cos(x + y) = \cos x \cos y - \sin x \sin y \\ \cos(x - y) = \cos x \cos y + \sin x \sin y \end{cases}$$

to prove the identity

$$2 \cos x \cos y = \cos(x + y) + \cos(x - y)$$

47. Use the system

$$\begin{cases} \sin(x + y) = \sin x \cos y + \cos x \sin y \\ \sin(x - y) = \sin x \cos y - \cos x \sin y \end{cases}$$

to prove the identity

$$2 \sin x \cos y = \sin(x + y) + \sin(x - y)$$

48. Use the system in Problem 47 to prove the identity

$$2 \sin x \sin y = \cos(x - y) - \cos(x + y)$$

49. Use the system in Problem 46 to prove the identity

$$2 \cos x \sin y = \sin(x + y) - \sin(x - y)$$

50. Find two numbers whose sum is 9 and whose product is 18.

51. Find two numbers whose difference is 4 and whose product is -3

52. Find the lengths of the legs of a right triangle whose area is 60 ft^2 and whose hypotenuse is 17 ft.

53. Find the length and width of a rectangle whose area is 60 ft^2 with a diagonal of length 13 ft.

C *Solve the systems in Problems 54–63 for x and y. For Problems 54–56 give only those solutions between 0 and 6.*

54. $\begin{cases} \sin x + \cos y = 1 \\ \sin x - \cos y = 0 \end{cases}$

55. $\begin{cases} \cos x - \sin y = 1 \\ \cos x + \sin y = 0 \end{cases}$

56. $\begin{cases} x + 3y = \cos^2 60° \\ x + y = -\sin^2 60° \end{cases}$

57. $\begin{cases} x^2 - xy + y^2 = 21 \\ xy - y^2 = 15 \end{cases}$

58. $\begin{cases} x^2 - xy + y^2 = 3 \\ x^2 + y^2 = 6 \end{cases}$

59. $\begin{cases} x^2 + 2xy = 8 \\ x^2 - 4y^2 = 8 \end{cases}$

60. $\begin{cases} \dfrac{3}{x} + \dfrac{4}{y} = 3 \\ \dfrac{9}{x} - \dfrac{2}{y} = 2 \end{cases}$

61. $\begin{cases} \dfrac{2}{x - 1} - \dfrac{5}{y + 2} = 12 \\ \dfrac{4}{x - 1} - \dfrac{2}{y + 2} = -12 \end{cases}$

62. $\begin{cases} y^2 - 5xy = 1 \\ y^2 - 4xy = 2y \end{cases}$

63. $\begin{cases} x^2 - 3xy = 2 \\ x^2 - 4xy = 3x \end{cases}$

64. If the parabola $(x - h)^2 = y - k$ passes through the points $(-2, 6)$ and $(-4, 2)$, find (h, k).

65. If the parabola $-3(x - h)^2 = y - k$ passes through the points $(2, 5)$ and $(-1, -4)$, find (h, k).

66. HISTORICAL QUESTION The Babylonians knew how to solve systems of equations as early as 1600 B.C. One tablet, called the Yale Tablet, shows a system equivalent to

$$\begin{cases} xy = 600 \\ (x + y)^2 - 150(x - y) = 100 \end{cases}$$

Find a positive solution for this system correct to the nearest tenth.

67. HISTORICAL QUESTION The Louvre Tablet from the Babylonian civilization is dated at about 1500 B.C. It shows a system equivalent to

$$\begin{cases} xy = 1 \\ x + y = a \end{cases}$$

Solve this system for x and y in terms of a.

9.2   Matrix Solution of a System of Equations

Definition of a Matrix

One of the most common types of problems for which you can apply mathematics in a variety of different disciplines is in solving systems of equations. In fact, far more real-world problems require the solution to systems involving 3, 4, 5, or even 10 or 20 unknowns, than require the solution to the systems of the last section. In this section, we will introduce a way of solving large systems of equations in a general way so that we can handle the solution of a system of m equations and n unknowns.

In the last section we used subscripts to find our way around a system of equations:

$$\begin{cases} a_{11}x_1 + a_{12}x_2 = b_1 \\ a_{21}x_1 + a_{22}x_2 = b_2 \end{cases}$$

We will now separate the parts of this system of equations into rectangular

arrays of numbers. An **array** of numbers is called a **matrix**. A matrix is denoted by enclosing the array in large brackets.

Let A be the matrix (array) of coefficients: $\begin{bmatrix} a_{11} & a_{12} \\ a_{21} & a_{22} \end{bmatrix}$

Let X be the matrix of unknowns: $\begin{bmatrix} x_1 \\ x_2 \end{bmatrix}$

Let B be the matrix of constants: $\begin{bmatrix} b_1 \\ b_2 \end{bmatrix}$

We will write the system of equations as a **matrix equation** $AX = B$, but before we do this we will do some preliminary work with matrices (plural of *matrix*).

Matrices are classified by the number of rows (horizontal) and columns (vertical). The number of rows and columns of a matrix need not be the same; but if they are, the matrix is called a **square matrix**. The **order** or **dimension** of a matrix is given by an expression **$m \times n$** (pronounced "*m* by *n*"), where m is the number of rows and n is the number of columns. For example, A (shown above) is a matrix of order 2×2; matrices X and B have order 2×1.

$$C = \begin{bmatrix} 7 & 3 & 2 \\ 5 & -4 & -3 \end{bmatrix} \begin{matrix} \leftarrow \text{Row 1} \\ \leftarrow \text{Row 2} \end{matrix} \qquad D = \begin{bmatrix} 4 & -2 & 1 & 6 \\ -3 & 3 & -1 & 3 \\ 2 & 4 & -1 & 5 \end{bmatrix} \begin{matrix} \leftarrow \text{Row 1} \\ \leftarrow \text{Row 2} \\ \leftarrow \text{Row 3} \end{matrix} \qquad G = [g_{ij}]_{m,n}$$

(Columns labeled: Column 1, Column 2, Column 3 for C; Column 1, Column 2, Column 3, Column 4 for D.)

Matrix G is an arbitrary m by n matrix; it is shorthand notation for

$$\begin{bmatrix} g_{11} & g_{12} & g_{13} & \cdots & g_{1n} \\ g_{21} & g_{22} & g_{23} & \cdots & g_{2n} \\ g_{31} & g_{32} & g_{33} & \cdots & g_{3n} \\ & & \vdots & & \\ g_{m1} & g_{m2} & g_{m3} & \cdots & g_{mn} \end{bmatrix}$$

Capital letters are generally used to denote matrices and the same lowercase letters for entries of that matrix.

Augmented Matrices

Consider a system of m equations and n variables:

$$\begin{cases} a_{11}x_1 + a_{12}x_2 + \cdots + a_{1n}x_n = b_1 \\ a_{21}x_1 + a_{22}x_2 + \cdots + a_{2n}x_n = b_2 \\ \quad \vdots \qquad \vdots \qquad \vdots \qquad \vdots \\ a_{m1}x_1 + a_{m2}x_2 + \cdots + a_{mn}x_n = b_m \end{cases}$$

Now consider what is called the **augmented matrix** for this system:

$$\begin{bmatrix} a_{11} & a_{12} & \cdots & a_{1n} & b_1 \\ a_{21} & a_{22} & \cdots & a_{2n} & b_2 \\ \vdots & \vdots & & \vdots & \vdots \\ a_{m1} & a_{m2} & \cdots & a_{mn} & b_m \end{bmatrix}$$

EXAMPLE 1   Write the given systems in augmented matrix form.

a. $\begin{cases} 2x + 3y = 8 \\ 3x + 2y = 7 \end{cases}$

b. $\begin{cases} 2x + y = 3 \\ 3x - y = 2 \\ 4x + 3y = 7 \end{cases}$

c. $\begin{cases} 3x - 2y + z = -2 \\ 4x - 5y + 3z = -9 \\ 2x - y + 5z = -5 \end{cases}$

d. $\begin{cases} 5x - 3y + z = -3 \\ 2x + 5z = 14 \end{cases}$

e. $\begin{cases} x_1 - 3x_3 + x_5 = -3 \\ x_2 + x_4 = -1 \\ x_3 + x_5 = 7 \\ x_1 + x_2 - x_3 + 4x_4 = -8 \\ x_1 + x_2 + x_3 + x_4 + x_5 = 8 \end{cases}$

SOLUTION   Note that some coefficients are negative and some are zero.

a. $\begin{bmatrix} 2 & 3 & 8 \\ 3 & 2 & 7 \end{bmatrix}$

b. $\begin{bmatrix} 2 & 1 & 3 \\ 3 & -1 & 2 \\ 4 & 3 & 7 \end{bmatrix}$

c. $\begin{bmatrix} 3 & -2 & 1 & -2 \\ 4 & -5 & 3 & -9 \\ 2 & -1 & 5 & -5 \end{bmatrix}$

d. $\begin{bmatrix} 5 & -3 & 1 & -3 \\ 2 & 0 & 5 & 14 \end{bmatrix}$

e. $\begin{bmatrix} 1 & 0 & -3 & 0 & 1 & -3 \\ 0 & 1 & 0 & 1 & 0 & -1 \\ 0 & 0 & 1 & 0 & 1 & 7 \\ 1 & 1 & -1 & 4 & 0 & -8 \\ 1 & 1 & 1 & 1 & 1 & 8 \end{bmatrix}$

EXAMPLE 2   Write a system of equations (use $x_1, x_2, \ldots, x_n$ for the variables) that has the given augmented matrix.

a. $\begin{bmatrix} 2 & 1 & -1 & -3 \\ 3 & -2 & 1 & 9 \\ 1 & -4 & 3 & 17 \end{bmatrix}$

b. $\begin{bmatrix} 1 & 0 & 0 & 3 \\ 0 & 1 & 0 & -2 \\ 0 & 0 & 1 & -5 \end{bmatrix}$

c. $\begin{bmatrix} 1 & 0 & 4 \\ 0 & 1 & 3 \end{bmatrix}$

d. $\begin{bmatrix} 1 & 0 & -5 \\ 0 & 1 & 2 \\ 2 & 1 & -3 \end{bmatrix}$

e. $\begin{bmatrix} 1 & 0 & 0 & 3 \\ 0 & 1 & 0 & -3 \\ 0 & 0 & 1 & 6 \\ 0 & 0 & 0 & 1 \end{bmatrix}$

f. $\begin{bmatrix} 1 & 0 & 0 & 1 \\ 0 & 1 & 0 & 5 \\ 0 & 0 & 1 & -2 \\ 0 & 0 & 0 & 0 \end{bmatrix}$

SOLUTION   a. $\begin{cases} 2x_1 + x_2 - x_3 = -3 \\ 3x_1 - 2x_2 + x_3 = 9 \\ x_1 - 4x_2 + 3x_3 = 17 \end{cases}$

b. $\begin{cases} x_1 = 3 \\ x_2 = -2 \\ x_3 = -5 \end{cases}$

c. $\begin{cases} x_1 = 4 \\ x_2 = 3 \end{cases}$

d. $\begin{cases} x_1 = -5 \\ x_2 = 2 \\ 2x_1 + x_2 = -3 \end{cases}$

e. $\begin{cases} x_1 = 3 \\ x_2 = -3 \\ x_3 = 6 \\ 0 = 1 \end{cases}$

f. $\begin{cases} x_1 = 1 \\ x_2 = 5 \\ x_3 = -2 \\ 0 = 0 \end{cases}$

The goal of this section is to solve a system of m equations with n unknowns. We have already looked at systems of two equations with two unknowns. In high school you may have solved three equations with three unknowns. Now, however, we want to be able to solve problems with two equations and five unknowns, or three equations and two unknowns, or any mixture of linear equations or unknowns. The procedure of this section—**Gauss–Jordan elimination**—is a general method for solving all of these types of systems. We write the system in augmented matrix form (as in Example 1), then carry out a process that transforms the matrix until the solution is obvious. Look back at Example 2—the solutions to parts **b** and **c** are obvious. Part **e** shows $0 = 1$ in the last equation, so this system has no solution (0 cannot equal 1), and part **f** shows $0 = 0$ (which is true for all replacements of the variable), which means that the solution is found by looking at the other equations (namely, $x_1 = 1$, $x_2 = 5$, and $x_3 = -2$). The terms with nonzero coefficients in these examples are arranged on a diagonal, and such a system is said to be in *diagonal form*.

Elementary Row Operations and Pivoting

What process will allow us to transform a matrix into diagonal form? We begin with some steps called **elementary row operations.** Elementary row operations change the *form* of a matrix, but the new form represents an equivalent system. Matrices that represent equivalent systems are called **equivalent matrices** and we introduce these elementary row operations in order to write equivalent matrices. Let us work with a system with three equations and three unknowns (any size will work the same way).

SYSTEM FORMAT

$$\begin{cases} 3x - 2y + 4z = 11 \\ x - y - 2z = -7 \\ 2x - 3y + z = -1 \end{cases}$$

MATRIX FORMAT

$$\left[\begin{array}{ccc|c} 3 & -2 & 4 & 11 \\ 1 & -1 & -2 & -7 \\ 2 & -3 & 1 & -1 \end{array}\right]$$

Interchanging two equations is equivalent to interchanging two rows in matrix format, and, certainly, if we do this, the solution to the system will be the same:

SYSTEM FORMAT

$$\begin{cases} x - y - 2z = -7 \\ 3x - 2y - 4z = 11 \\ 2x - 3y + z = -1 \end{cases}$$

MATRIX FORMAT

$$\left[\begin{array}{ccc|c} 1 & -1 & -2 & -7 \\ 3 & -2 & 4 & 11 \\ 2 & -3 & 1 & -1 \end{array}\right]$$

The first and second equations (rows) are interchanged

Elementary Row Operation 1: Interchange any two rows.

Since multiplying or dividing both sides of any equation by any nonzero number does not change the solution, then, in matrix format, the solution will not be changed if any row is multiplied or divided by a nonzero constant. For example, multiply both sides of the first equation by -3:

SYSTEM FORMAT

$$\begin{cases} -3x + 3y + 6z = 21 \\ 3x - 2y - 4z = 11 \\ 2x - 3y + z = -1 \end{cases}$$

MATRIX FORMAT

$$\left[\begin{array}{ccc|c} -3 & 3 & 6 & 21 \\ 3 & -2 & 4 & 11 \\ 2 & -3 & 1 & -1 \end{array}\right]$$

Both sides of the first equation are multiplied by -3; the first row is multiplied by -3.

Because it is easier to program software by separating multiplication and division into two distinct row operations, we have done the same.

Elementary Row Operation 2: Multiply all the elements of a row by the same nonzero real number.

Elementary Row Operation 3: Divide all the elements of a row by the same nonzero real number.

The last property we need to carry out Gauss–Jordan elimination rests on the property we used with the linear combination method—namely, adding equations to eliminate a variable. In terms of the system format this means that one equation can be replaced by its sum with another equation in the system. For example, if we add the first equation to the second equation we have:

SYSTEM FORMAT   MATRIX FORMAT

$$\begin{cases} -3x + 3y + 6z = 21 \\ \quad\quad y + 10z = 32 \\ 2x - 3y + \quad z = -1 \end{cases} \qquad \begin{bmatrix} -3 & 3 & 6 & \vdots & 21 \\ 0 & 1 & 10 & \vdots & 32 \\ 2 & -3 & 1 & \vdots & -1 \end{bmatrix}$$

In terms of matrix format, any row can be replaced by its sum with some other row.

This process is usually used in conjunction with Elementary Row Operation 2. That is, an equation is changed by adding to it a nonzero multiple of another equation in the system. Go back to the original system:

SYSTEM FORMAT   MATRIX FORMAT

$$\begin{cases} 3x - 2y + 4z = 11 \\ x - y - 2z = -7 \\ 2x - 3y + z = -1 \end{cases} \qquad \begin{bmatrix} 3 & -2 & 4 & \vdots & 11 \\ 1 & -1 & -2 & \vdots & -7 \\ 2 & -3 & 1 & \vdots & -1 \end{bmatrix}$$

We can change this system by multiplying the second equation by -3 and adding the result to the first equation. In matrix terminology we would say multiply the second row by -3 and add it to the first row:

SYSTEM FORMAT   MATRIX FORMAT

$$\begin{cases} \quad\quad y + 10z = 32 \\ x - y - 2z = -7 \\ 2x - 3y + z = -1 \end{cases} \qquad \begin{bmatrix} 0 & 1 & 10 & \vdots & 32 \\ 1 & -1 & -2 & \vdots & -7 \\ 2 & -3 & 1 & \vdots & -1 \end{bmatrix}$$

Once again, multiply the second row, this time by -2, and add it to the third row. Wait! Why -2? Where did that come from? The idea is the same one we used in the linear combination method—we use a number that will give a zero coefficient to the x in the third equation.

SYSTEM FORMAT   MATRIX FORMAT

$$\begin{cases} \quad\quad y + 10z = 32 \\ x - y - 2z = -7 \\ \quad -y + 5z = 13 \end{cases} \qquad \begin{bmatrix} 0 & 1 & 10 & \vdots & 32 \\ 1 & -1 & -2 & \vdots & -7 \\ 0 & -1 & 5 & \vdots & 13 \end{bmatrix}$$

Note that the multiplied row is not changed; instead, the changed row is the one to which the multiplied row is added. We call the original row the **pivot row** and the changed row the **target row.**

Elementary Row Operation 4: **Multiply all the entries of a row (the pivot row) by a nonzero real number and add each resulting product to the corresponding entry of another specified row (the target row).** *Note that this operation changes only the target row.*

There you have it! You should carry out these four elementary row operations until you have a system for which the solution is obvious, as illustrated in Example 3.

EXAMPLE 3 Solve $\begin{cases} 2x - 5y = 5 \\ x - 2y = 1 \end{cases}.$

SOLUTION

SYSTEM NOTATION

$\begin{cases} 2x - 5y = 5 \\ x - 2y = 1 \end{cases}$

MATRIX NOTATION

$\begin{bmatrix} 2 & -5 & \vdots & 5 \\ 1 & -2 & \vdots & 1 \end{bmatrix}$

Elementary Row Operation 1   Interchange the first and second equations

$\begin{cases} x - 2y = 1 \\ 2x - 5y = 5 \end{cases}$

Interchange the first and second rows

$\begin{bmatrix} 1 & -2 & \vdots & 1 \\ 2 & -5 & \vdots & 5 \end{bmatrix}$ ← Pivot row ← Target row

Elementary Row Operation 4   Add -2 times the first equation to the second.

$\begin{cases} x - 2y = 1 \\ - y = 3 \end{cases}$

Add -2 times the first row to the second row.

$\begin{bmatrix} 1 & -2 & \vdots & 1 \\ 0 & -1 & \vdots & 3 \end{bmatrix}$ ← Pivot row remains unchanged ← Target row changes

Elementary Row Operation 2   Multiply both sides of the second equation by -1

$\begin{cases} x - 2y = 1 \\ y = -3 \end{cases}$

Multiply row 2 by -1

$\begin{bmatrix} 1 & -2 & \vdots & 1 \\ 0 & 1 & \vdots & -3 \end{bmatrix}$ ← New pivot row

Elementary Row Operation 4   Add 2 times the second equation to the first.

$\begin{cases} x = -5 \\ y = -3 \end{cases}$

Add 2 times the second row to the first row.

$\begin{bmatrix} 1 & 0 & \vdots & -5 \\ 0 & 1 & \vdots & -3 \end{bmatrix}$ This is called the **reduced row-echelon form.**

The solution, $(-5, -3)$, is now obvious. ▌

As you study Example 3, first look at how the elementary row operations led to a system equivalent to the first—but one for which the solution is obvious. Next, try to decide *why* a particular row operation was chosen when it was. Most students quickly learn the elementary row operations, but then use a series of (almost random) steps until the obvious solution results. This often works, but is not very efficient. The steps chosen in Example 3 illustrate a very efficient method of using the elementary row operations to determine a system whose solution is obvious. The method was discovered independently by two mathematicians, Karl

Friedrich Gauss (1777–1855) and Camille Jordan (1838–1922), so today the process is known as the Gauss–Jordan method. (Since the process was only recently attributed to Jordan, many books refer to the method simply as *Gaussian elimination*.) Before stating the process, however, we will consider a procedure called **pivoting.**

PIVOTING

1. Divide all entries in the row in which the pivot appears (called the *pivot row*) by the pivot element so that the pivot entry becomes a 1. This uses Elementary Row Operation 3.

2. Obtain zeros above and below the pivot element by using Elementary Row Operation 4.

EXAMPLE 4 Pivot the given matrix about the circled element

SOLUTION

$$
\begin{bmatrix} 15 & \text{⑤} & 35 \\ 5 & 2 & -3 \end{bmatrix}
\xrightarrow{\text{R1} \div 5}
\begin{bmatrix} 3 & 1 & 7 \\ 5 & 2 & -3 \end{bmatrix}
\xrightarrow{-2\text{R1} + \text{R2}}
\begin{bmatrix} 3 & 1 & 7 \\ -1 & 0 & -17 \end{bmatrix}
$$

This shorthand notation shows what we did: divide row 1 by 5.

Multiply row 1 (pivot row) by −2 and add it to row 2 (target row).

Consider another matrix.

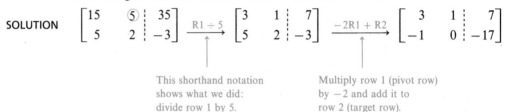

$$
\begin{bmatrix} 2 & 3 & 4 & 1 \\ -1 & \text{②} & 3 & -2 \\ 0 & 1 & -1 & 3 \end{bmatrix}
\xrightarrow{\frac{1}{2}\text{R2}}
\begin{bmatrix} 2 & 3 & 4 & 1 \\ -\frac{1}{2} & 1 & \frac{3}{2} & -1 \\ 0 & 1 & -1 & 3 \end{bmatrix}
$$

$$
\xrightarrow{-3\text{R2} + \text{R1}}
\begin{bmatrix} \frac{7}{2} & 0 & -\frac{1}{2} & 4 \\ -\frac{1}{2} & 1 & \frac{3}{2} & -1 \\ 0 & 1 & -1 & 3 \end{bmatrix}
\xrightarrow{-1\text{R2} + \text{R3}}
\begin{bmatrix} \frac{7}{2} & 0 & -\frac{1}{2} & 4 \\ -\frac{1}{2} & 1 & \frac{3}{2} & -1 \\ \frac{1}{2} & 0 & -\frac{5}{2} & 4 \end{bmatrix}
$$

The four elementary row operations are listed below for easy reference.

ELEMENTARY ROW OPERATIONS

There are **four elementary row operations** for producing equivalent matrices:

1. Interchange any two rows.

2. Multiply all the elements of a row by the same nonzero real number.

3. Divide all the elements of a row by the same nonzero real number.

4. Multiply all the entries of a row (*pivot row*) by a nonzero real number and add each resulting product to the corresponding entry of another specified row (*target row*). (Note that this operation changes only the target row.)

Gauss–Jordan Elimination

You are now ready to see the method worked out by Gauss and Jordan. It efficiently uses the elementary row operations to diagonalize the matrix. That is, the first pivot is the first entry in the first row, first column; the second is the entry in the second row, second column; and so on until the solution is obvious.

GAUSS–JORDAN ELIMINATION

1. Select the element in the first row, first column, as a pivot.
2. Pivot.
3. Select the element in the second row, second column, as a pivot.
4. Pivot.
5. Repeat the process until you arrive at the last row, or until the pivot element is a zero. If it is a zero and you can interchange that row with a row below it, so that the pivot element is no longer a zero, do so and continue. If it is a zero and you cannot interchange rows so that it is not a zero, continue with the next column.

EXAMPLE 5   Solve $\begin{cases} 3x - 2y + z = -2 \\ 4x - 5y + 3z = -9. \\ 2x - y + 5z = -5 \end{cases}$

SOLUTION   We will solve this system by choosing the steps according to the Gauss–Jordan method.

$$\begin{bmatrix} ③ & -2 & 1 & \vdots & -2 \\ 4 & -5 & 3 & \vdots & -9 \\ 2 & -1 & 5 & \vdots & -5 \end{bmatrix} \xrightarrow{R1 \div 3} \begin{bmatrix} 1 & -\frac{2}{3} & \frac{1}{3} & \vdots & -\frac{2}{3} \\ 4 & -5 & 3 & \vdots & -9 \\ 2 & -1 & 5 & \vdots & -5 \end{bmatrix} \xrightarrow{-4R1 + R2} \begin{bmatrix} 1 & -\frac{2}{3} & \frac{1}{3} & \vdots & -\frac{2}{3} \\ 0 & -\frac{7}{3} & \frac{5}{3} & \vdots & -\frac{19}{3} \\ 2 & -1 & 5 & \vdots & -5 \end{bmatrix}$$

3 is the pivot. **Pivot** Obtain a 1 in the pivot position. Pivot row is 1; target row is 2; -4 is the opposite of the corresponding element in the target row so that a zero is obtained.

$$\xrightarrow{-2R1 + R3} \begin{bmatrix} 1 & -\frac{2}{3} & \frac{1}{3} & \vdots & -\frac{2}{3} \\ 0 & ⟨-\frac{7}{3}⟩ & \frac{5}{3} & \vdots & -\frac{19}{3} \\ 0 & \frac{1}{3} & \frac{13}{3} & \vdots & -\frac{11}{3} \end{bmatrix} \xrightarrow{R2 \div -\frac{7}{3}} \begin{bmatrix} 1 & -\frac{2}{3} & \frac{1}{3} & \vdots & -\frac{2}{3} \\ 0 & 1 & -\frac{5}{7} & \vdots & \frac{19}{7} \\ 0 & \frac{1}{3} & \frac{13}{3} & \vdots & -\frac{11}{3} \end{bmatrix} \xrightarrow{\frac{2}{3}R2 + R1} \begin{bmatrix} 1 & 0 & -\frac{1}{7} & \vdots & \frac{8}{7} \\ 0 & 1 & -\frac{5}{7} & \vdots & \frac{19}{7} \\ 0 & \frac{1}{3} & \frac{13}{3} & \vdots & -\frac{11}{3} \end{bmatrix}$$

Pivot

$$\xrightarrow{-\frac{1}{3}R2 + R3}
\begin{bmatrix} 1 & 0 & -\frac{1}{7} & \vdots & \frac{8}{7} \\ 0 & 1 & -\frac{5}{7} & \vdots & \frac{19}{7} \\ 0 & 0 & \boxed{\frac{32}{7}} & \vdots & -\frac{32}{7} \end{bmatrix}
\xrightarrow[\textbf{Pivot}]{R3 \div \frac{32}{7}}
\begin{bmatrix} 1 & 0 & -\frac{1}{7} & \vdots & \frac{8}{7} \\ 0 & 1 & -\frac{5}{7} & \vdots & \frac{19}{7} \\ 0 & 0 & 1 & \vdots & -1 \end{bmatrix}$$

Select a new pivot

$$\xrightarrow{\frac{5}{7}R3 \div R2}
\begin{bmatrix} 1 & 0 & -\frac{1}{7} & \vdots & \frac{8}{7} \\ 0 & 1 & 0 & \vdots & 2 \\ 0 & 0 & 1 & \vdots & -1 \end{bmatrix}
\xrightarrow{\frac{1}{7}R3 + R1}
\begin{bmatrix} 1 & 0 & 0 & \vdots & 1 \\ 0 & 1 & 0 & \vdots & 2 \\ 0 & 0 & 1 & \vdots & -1 \end{bmatrix}$$

The solution, **(1, 2, −1)**, is found by inspection since

$$\begin{bmatrix} 1 & 0 & 0 & \vdots & 1 \\ 0 & 1 & 0 & \vdots & 2 \\ 0 & 0 & 1 & \vdots & -1 \end{bmatrix} \quad \text{means} \quad
\begin{aligned} 1x + 0y + 0z &= 1 \\ 0x + 1y + 0z &= 2 \\ 0x + 0y + 1z &= -1 \end{aligned}$$

For these variables we agree that $(1, 2, -1)$ denotes (x, y, z) unless otherwise stated.

Note that the Gauss–Jordan method usually introduces (often ugly) fractions. For this reason, many people are turning to readily available computer programs to carry out the drudgery of this method. However, if you do not have access to a computer, you can often reduce the amount of arithmetic by forcing the pivot elements to be 1 by using elementary row operations other than division. Also, you can combine the pivoting steps. In this example we could have forced the first pivot to be 1 by multiplying row 3 by −1 and adding it to the first row instead of dividing by 3. Let us look at the arithmetic in this simplified version.

$$\begin{bmatrix} 3 & -2 & 1 & \vdots & -2 \\ 4 & -5 & 3 & \vdots & -9 \\ 2 & -1 & 5 & \vdots & -5 \end{bmatrix}
\longrightarrow
\begin{bmatrix} \boxed{1} & -1 & -4 & \vdots & 3 \\ 4 & -5 & 3 & \vdots & -9 \\ 2 & -1 & 5 & \vdots & -5 \end{bmatrix}
\longrightarrow
\begin{bmatrix} 1 & -1 & -4 & \vdots & 3 \\ 0 & -1 & 19 & \vdots & -21 \\ 0 & 1 & 13 & \vdots & -11 \end{bmatrix}
\longrightarrow$$

$-1R3 + R1$ Pivot; combine two steps in one: $-4R1 + R2; \; -2R1 + R3$

$$\begin{bmatrix} 1 & -1 & -4 & \vdots & 3 \\ 0 & \boxed{1} & -19 & \vdots & 21 \\ 0 & 1 & 13 & \vdots & -11 \end{bmatrix}
\longrightarrow
\begin{bmatrix} 1 & 0 & -23 & \vdots & 24 \\ 0 & 1 & -19 & \vdots & 21 \\ 0 & 0 & \boxed{32} & \vdots & -32 \end{bmatrix}
\longrightarrow$$

New pivot is $R2 \div (-1)$ Pivot second row (combine two steps): $R2 + R1; \; -1R2 + R3$ New pivot is $R3/32$

$$\begin{bmatrix} 1 & 0 & -23 & \vdots & 24 \\ 0 & 1 & -19 & \vdots & 21 \\ 0 & 0 & 1 & \vdots & -1 \end{bmatrix}
\longrightarrow
\begin{bmatrix} 1 & 0 & 0 & \vdots & 1 \\ 0 & 1 & 0 & \vdots & 2 \\ 0 & 0 & 1 & \vdots & -1 \end{bmatrix}$$

Pivot third row: $19R3 + R2$ $23R3 + R1$

The real beauty of Gauss–Jordan elimination is that it works with all sizes of linear systems. Consider Example 6, consisting of five equations and five unknowns.

EXAMPLE 6 Solve
$$\begin{cases} x_1 - 3x_3 + x_5 = -3 \\ x_2 + x_4 = -1 \\ x_3 + x_5 = 7 \\ x_1 + x_2 - x_3 + 4x_4 = -8 \\ x_1 + x_2 + x_3 + x_4 + x_5 = 8 \end{cases}$$

SOLUTION

$$\left[\begin{array}{ccccc|c} ① & 0 & -3 & 0 & 1 & -3 \\ 0 & 1 & 0 & 1 & 0 & -1 \\ 0 & 0 & 1 & 0 & 1 & 7 \\ 1 & 1 & -1 & 4 & 0 & -8 \\ 1 & 1 & 1 & 1 & 1 & 8 \end{array}\right] \xrightarrow[\;-R1+R5\;]{-R1+R4} \left[\begin{array}{ccccc|c} 1 & 0 & -3 & 0 & 1 & -3 \\ 0 & ① & 0 & 1 & 0 & -1 \\ 0 & 0 & 1 & 0 & 1 & 7 \\ 0 & 1 & 2 & 4 & -1 & -5 \\ 0 & 1 & 4 & 1 & 0 & 11 \end{array}\right]$$

$$\xrightarrow[\;-R2+R5\;]{-R2+R4} \left[\begin{array}{ccccc|c} 1 & 0 & -3 & 0 & 1 & -3 \\ 0 & 1 & 0 & 1 & 0 & -1 \\ 0 & 0 & ① & 0 & 1 & 7 \\ 0 & 0 & 2 & 3 & -1 & -4 \\ 0 & 0 & 4 & 0 & 0 & 12 \end{array}\right] \xrightarrow[\substack{-2R3+R4\\-4R3+R5}]{3R3+R1} \left[\begin{array}{ccccc|c} 1 & 0 & 0 & 0 & 4 & 18 \\ 0 & 1 & 0 & 1 & 0 & -1 \\ 0 & 0 & 1 & 0 & 1 & 7 \\ 0 & 0 & 0 & ③ & -3 & -18 \\ 0 & 0 & 0 & 0 & -4 & -16 \end{array}\right]$$

$$\xrightarrow[R4 \div 3]{} \left[\begin{array}{ccccc|c} 1 & 0 & 0 & 0 & 4 & 18 \\ 0 & 1 & 0 & 1 & 0 & -1 \\ 0 & 0 & 1 & 0 & 1 & 7 \\ 0 & 0 & 0 & ① & -1 & -6 \\ 0 & 0 & 0 & 0 & -4 & -16 \end{array}\right] \xrightarrow[\;(-1)R4+R2\;]{} \left[\begin{array}{ccccc|c} 1 & 0 & 0 & 0 & 4 & 18 \\ 0 & 1 & 0 & 0 & 1 & 5 \\ 0 & 0 & 1 & 0 & 1 & 7 \\ 0 & 0 & 0 & 1 & -1 & -6 \\ 0 & 0 & 0 & 0 & \boxed{-4} & -16 \end{array}\right]$$

$$\xrightarrow[R5 \div (-4)]{} \left[\begin{array}{ccccc|c} 1 & 0 & 0 & 0 & 4 & 18 \\ 0 & 1 & 0 & 0 & 1 & 5 \\ 0 & 0 & 1 & 0 & 1 & 7 \\ 0 & 0 & 0 & 1 & -1 & -6 \\ 0 & 0 & 0 & 0 & 1 & 4 \end{array}\right] \xrightarrow[\substack{-R5+R3\\-R5+R2\\-4R5+R1}]{R5+R4} \left[\begin{array}{ccccc|c} 1 & 0 & 0 & 0 & 0 & 2 \\ 0 & 1 & 0 & 0 & 0 & 1 \\ 0 & 0 & 1 & 0 & 0 & 3 \\ 0 & 0 & 0 & 1 & 0 & -2 \\ 0 & 0 & 0 & 0 & 1 & 4 \end{array}\right]$$

The solution is $(x_1, x_2, x_3, x_4, x_5) = (2, 1, 3, -2, 4)$. ∎

Example 7 is an example of a system of three equations with two unknowns that has a solution. Example 8 shows what Gauss–Jordan elimination looks like when there are three equations with two unknowns and no solution. (Remember, if there is no solution, the system is inconsistent.)

EXAMPLE 7 Solve $\begin{cases} 2x + y = 3 \\ 3x - y = 2. \\ 4x + 3y = 7 \end{cases}$

SOLUTION

$$\left[\begin{array}{rr:r} 2 & 1 & 3 \\ 3 & -1 & 2 \\ 4 & 3 & 7 \end{array}\right] \xrightarrow{-1R1 + R2} \left[\begin{array}{rr:r} 2 & 1 & 3 \\ 1 & -2 & -1 \\ 4 & 3 & 7 \end{array}\right] \xrightarrow[R_1 \leftrightarrow R_2]{\substack{\text{Interchange} \\ \text{rows 1 and 2}}} \left[\begin{array}{rr:r} 1 & -2 & -1 \\ 2 & 1 & 3 \\ 4 & 3 & 7 \end{array}\right]$$

$$\xrightarrow[\substack{-4R1 + R3}]{-2R1 + R2} \left[\begin{array}{rr:r} ① & -2 & -1 \\ 0 & 5 & 5 \\ 0 & 11 & 11 \end{array}\right] \xrightarrow{R2 \div 5} \left[\begin{array}{rr:r} 1 & -2 & -1 \\ 0 & ① & 1 \\ 0 & 11 & 11 \end{array}\right] \xrightarrow[\substack{-11R2 + R3}]{2R2 + R1} \left[\begin{array}{rr:r} 1 & 0 & 1 \\ 0 & 1 & 1 \\ 0 & 0 & 0 \end{array}\right]$$

This final matrix is equivalent to the system: $\begin{cases} x = 1 \\ y = 1. \\ 0 = 0 \end{cases}$

The solution is **(1, 1)**.

$$\text{Check:} \quad \begin{aligned} 2x + y &= 2(1) + 1 = 3 \quad \checkmark \\ 3x - y &= 3(1) - 1 = 2 \quad \checkmark \\ 4x + 3y &= 4(1) + 3(1) = 7 \quad \checkmark \end{aligned}$$

EXAMPLE 8 Solve $\begin{cases} 2x + y = 3 \\ 3x - y = 2 \\ x - 2y = 4 \end{cases}$

SOLUTION

$$\left[\begin{array}{rr:r} 2 & 1 & 3 \\ 3 & -1 & 2 \\ 1 & -2 & 4 \end{array}\right] \xrightarrow{R1 \leftrightarrow R3} \left[\begin{array}{rr:r} 1 & -2 & 4 \\ 3 & -1 & 2 \\ 2 & 1 & 3 \end{array}\right] \xrightarrow[\substack{-2R1 + R3}]{-3R1 + R2} \left[\begin{array}{rr:r} 1 & -2 & 4 \\ 0 & 5 & -10 \\ 0 & 5 & -5 \end{array}\right]$$

$$\xrightarrow{R2 \div 5} \left[\begin{array}{rr:r} 1 & -2 & 4 \\ 0 & 1 & -2 \\ 0 & 5 & -5 \end{array}\right] \xrightarrow[\substack{-5R2 + R3}]{2R2 + R1} \left[\begin{array}{rr:r} 1 & 0 & 0 \\ 0 & 1 & -2 \\ 0 & 0 & 5 \end{array}\right]$$

This is equivalent to $\begin{cases} x = 0 \\ y = -2. \text{ But, since } 0 \neq 5 \text{ regardless of the values of } x \\ 0 = 5 \end{cases}$
and y, this is an **inconsistent system.**

A dependent system has infinitely many solutions, but just because a system has infinitely many solutions does not mean that it is satisfied by any set of values. For example, $x + y = 5$ has infinitely many solutions, but not just *any* replacements for x and y will satisfy that equation. In fact, we might say if we choose any value, say, t, for x, then y is fixed to be $5 - t$. In the last section we called t a *parameter*, and we could say that any ordered pair of the form $(t, 5 - t)$ will satisfy the equation $x + y = 5$. Example 9 illustrates a dependent system and the use of a parameter in specifying the solution.

EXAMPLE 9   Solve $\begin{cases} 3x - 2y + 4z = 8 \\ x - y - 2z = 5 \\ 4x - 3y + 2z = 13 \end{cases}$.

SOLUTION

$$\begin{bmatrix} 3 & -2 & 4 & \vdots & 8 \\ 1 & -1 & -2 & \vdots & 5 \\ 4 & -3 & 2 & \vdots & 13 \end{bmatrix} \xrightarrow{R1 \leftrightarrow R2} \begin{bmatrix} 1 & -1 & -2 & \vdots & 5 \\ 3 & -2 & 4 & \vdots & 8 \\ 4 & -3 & 2 & \vdots & 13 \end{bmatrix}$$

$$\xrightarrow[-4R1 + R3]{-3R1 + R2} \begin{bmatrix} 1 & -1 & -2 & \vdots & 5 \\ 0 & 1 & 10 & \vdots & -7 \\ 0 & 1 & 10 & \vdots & -7 \end{bmatrix} \xrightarrow[-R2 + R3]{R2 + R1} \begin{bmatrix} 1 & 0 & 8 & \vdots & -2 \\ 0 & 1 & 10 & \vdots & -7 \\ 0 & 0 & 0 & \vdots & 0 \end{bmatrix}$$

This is equivalent to the system $\begin{cases} x + 8z = -2 \\ y + 10z = -7 \\ 0 = 0 \end{cases}$

Since the third equation $(0 = 0)$ is true for all values of x, y, and z, we focus on the first and second equations; there are two equations with three unknowns. This is a *dependent system*. In your previous courses you probably finished by saying the system is dependent. In this course, as well as in more advanced mathematics, we will find the set of all values that make the system true. By looking at the solved system we see that z appears in both equations, so we can pick *any* value for z, say $z = 2$; then

$$x = -2 - 8z = -2 - 8(2) = -18$$
$$y = -7 - 10z = -7 - 10(2) = -27$$

Thus, a solution is $(-18, -27, 2)$. But why did we pick $z = 2$? What if we picked $z = 0$? We would find another solution, namely

$$x = -2 - 8(0) = -2$$
$$y = -7 - 10(0) = -7$$

for a solution $(-2, -7, 0)$. But we do not want only two solutions; we want *all* solutions, so we use the concept of a *parameter*, which is an arbitrary constantused to distinguish various specific cases. For this example, let $z = t$ (the variable t is the *parameter*) to find

$$x = -2 - 8t$$
$$y = -7 - 10t$$

This gives a complete **parametric solution** $(-2 - 8t, -7 - 10t, t)$.   ▮

You may also need a parameter when there are more unknowns than equations. For example, the system

$$\begin{cases} 5x - 3y + z = -3 \\ 2x + 5z = 14 \end{cases}$$

has more unknowns than equations so it is a *dependent system*. In more advanced work, the Gauss–Jordan method could be used to find a parameter and a parametric solution to this system.

EXAMPLE 10 Solve
$$\begin{cases} x + y + 2z + 3w = 5 \\ 4z + 2w = -32 \\ z + 6w = 3 \\ 2z + 12w = 6 \end{cases}$$

SOLUTION Write the augmented matrix:

$$\begin{bmatrix} ① & 1 & 2 & 3 & \vdots & 5 \\ 0 & 0 & 4 & 2 & \vdots & -32 \\ 0 & 0 & \boxed{1} & 6 & \vdots & 3 \\ 0 & 0 & 2 & 12 & \vdots & 6 \end{bmatrix}$$

The first pivot operation (first pivot is circled) is completed. When we move to the second row, second column, there is a zero in the pivot position that cannot be eliminated, so we move to the third row, third column. The second pivot is boxed. Carry out the pivot operation.

$$\begin{bmatrix} 1 & 1 & 2 & 3 & \vdots & 5 \\ 0 & 0 & 4 & 2 & \vdots & -32 \\ 0 & 0 & \boxed{1} & 6 & \vdots & 3 \\ 0 & 0 & 2 & 12 & \vdots & 6 \end{bmatrix} \longrightarrow \begin{bmatrix} 1 & 1 & 0 & -9 & \vdots & -1 \\ 0 & 0 & 0 & -22 & \vdots & -44 \\ 0 & 0 & 1 & 6 & \vdots & 3 \\ 0 & 0 & 0 & 0 & \vdots & 0 \end{bmatrix}$$

The final row has all zeros, but the zero in the fourth row, fourth column can be eliminated by interchanging rows 2 and 4.

$$\longrightarrow \begin{bmatrix} 1 & 1 & 0 & -9 & \vdots & -1 \\ 0 & 0 & 0 & 0 & \vdots & 0 \\ 0 & 0 & 1 & 6 & \vdots & 3 \\ 0 & 0 & 0 & -22 & \vdots & -44 \end{bmatrix} \longrightarrow \begin{bmatrix} 1 & 1 & 0 & -9 & \vdots & -1 \\ 0 & 0 & 0 & 0 & \vdots & 0 \\ 0 & 0 & 1 & 6 & \vdots & 3 \\ 0 & 0 & 0 & 1 & \vdots & 2 \end{bmatrix}$$

$$\longrightarrow \begin{bmatrix} 1 & 1 & 0 & 0 & \vdots & 17 \\ 0 & 0 & 0 & 0 & \vdots & 0 \\ 0 & 0 & 1 & 0 & \vdots & -9 \\ 0 & 0 & 0 & 1 & \vdots & 2 \end{bmatrix}$$

This is equivalent to the system

$$\begin{cases} x + y = 17 \\ 0 = 0 \\ z = -9 \\ w = 2 \end{cases}$$

We see that even though there are values for w and z, the equation $x + y = 17$ has infinitely many solutions. This means it is a dependent system. We state a parametric solution by selecting $y = t$ to give $x = 17 - t$ for the complete solution $(x, y, z, w) = (17 - t, t, -9, 2)$.

Our discussion concludes with two applied problems.

EXAMPLE 11 A rancher has to mix three types of feed for her cattle. The following analysis shows the amounts per bag (100 lb) of grain:

GRAIN	PROTEIN	CARBOHYDRATES	SODIUM
A	7 lb	88 lb	1 lb
B	6 lb	90 lb	1 lb
C	10 lb	70 lb	2 lb

How many bags of each type of grain should she mix to provide 71 lb of protein, 854 lb of carbohydrates, and 12 lb of sodium?

SOLUTION Let a, b, and c be the number of bags of grains A, B, and C, respectively, that are needed for the mixture. Then:

GRAIN	PROTEIN	CARBOHYDRATES	SODIUM
A	$7a$	$88a$	a
B	$6b$	$90b$	b
C	$10c$	$70c$	$2c$
Total	71	854	12

Thus

$$\begin{cases} 7a + 6b + 10c = 71 \\ 88a + 90b + 70c = 854 \\ a + b + 2c = 12 \end{cases}$$

$$\begin{bmatrix} 7 & 6 & 10 & \vdots & 71 \\ 88 & 90 & 70 & \vdots & 854 \\ 1 & 1 & 2 & \vdots & 12 \end{bmatrix} \longrightarrow \begin{bmatrix} 1 & 1 & 2 & \vdots & 12 \\ 88 & 90 & 70 & \vdots & 854 \\ 7 & 6 & 10 & \vdots & 71 \end{bmatrix}$$

$$\longrightarrow \begin{bmatrix} 1 & 1 & 2 & \vdots & 12 \\ 0 & 2 & -106 & \vdots & -202 \\ 0 & -1 & -4 & \vdots & -13 \end{bmatrix} \longrightarrow \begin{bmatrix} 1 & 1 & 2 & \vdots & 12 \\ 0 & 1 & -53 & \vdots & -101 \\ 0 & -1 & -4 & \vdots & -13 \end{bmatrix}$$

$$\longrightarrow \begin{bmatrix} 1 & 0 & 55 & \vdots & 113 \\ 0 & 1 & -53 & \vdots & -101 \\ 0 & 0 & -57 & \vdots & -114 \end{bmatrix} \longrightarrow \begin{bmatrix} 1 & 0 & 55 & \vdots & 113 \\ 0 & 1 & -53 & \vdots & -101 \\ 0 & 0 & 1 & \vdots & 2 \end{bmatrix}$$

$$\longrightarrow \begin{bmatrix} 1 & 0 & 0 & \vdots & 3 \\ 0 & 1 & 0 & \vdots & 5 \\ 0 & 0 & 1 & \vdots & 2 \end{bmatrix}$$

Mix three bags of grain A, five bags of grain B, and two bags of grain C.

EXAMPLE 12 Suppose the equation of a certain parabola has the form

$$y = ax^2 + bx + c$$

Find the equation of the parabola passing through $(-1, -6)$, $(2, 9)$, and $(-2, -3)$.

SOLUTION If a curve passes through a point, then the coordinates of that point must satisfy the equation.

$$(-1, -6): \quad -6 = a(-1)^2 + b(-1) + c$$
$$(2, 9): \quad 9 = a(2)^2 + b(2) + c$$
$$(-2, -3): \quad -3 = a(-2)^2 + b(-2) + c$$

In standard form the system is

$$\begin{cases} a - b + c = -6 \\ 4a + 2b + c = 9 \\ 4a - 2b + c = -3 \end{cases}$$

$$\begin{bmatrix} 1 & -1 & 1 & \vdots & -6 \\ 4 & 2 & 1 & \vdots & 9 \\ 4 & -2 & 1 & \vdots & -3 \end{bmatrix} \xrightarrow{-R3 + R2} \begin{bmatrix} 1 & -1 & 1 & \vdots & -6 \\ 0 & 4 & 0 & \vdots & 12 \\ 4 & -2 & 1 & \vdots & -3 \end{bmatrix} \xrightarrow[R2 \div 4]{-4R1 + R3} \begin{bmatrix} 1 & -1 & 1 & \vdots & -6 \\ 0 & 1 & 0 & \vdots & 3 \\ 0 & 2 & -3 & \vdots & 21 \end{bmatrix}$$

$$\xrightarrow{-2R2 + R3} \begin{bmatrix} 1 & -1 & 1 & \vdots & -6 \\ 0 & 1 & 0 & \vdots & 3 \\ 0 & 0 & -3 & \vdots & 15 \end{bmatrix} \xrightarrow{-\frac{1}{3}R3} \begin{bmatrix} 1 & -1 & 1 & \vdots & -6 \\ 0 & 1 & 0 & \vdots & 3 \\ 0 & 0 & 1 & \vdots & -5 \end{bmatrix}$$

$$\xrightarrow{-R3 + R1} \begin{bmatrix} 1 & -1 & 0 & \vdots & -1 \\ 0 & 1 & 0 & \vdots & 3 \\ 0 & 0 & 1 & \vdots & -5 \end{bmatrix} \xrightarrow{R2 + R1} \begin{bmatrix} 1 & 0 & 0 & \vdots & 2 \\ 0 & 1 & 0 & \vdots & 3 \\ 0 & 0 & 1 & \vdots & -5 \end{bmatrix}$$

Thus $a = 2$, $b = 3$, and $c = -5$, so the equation of the parabola is

$$y = 2x^2 + 3x - 5$$

9.2 Problem Set

A

1. Write each system in augmented matrix form.

a. $\begin{cases} 4x + 5y = -16 \\ 3x + 2y = 5 \end{cases}$

b. $\begin{cases} x + y + z = 4 \\ 3x + 2y + z = 7 \\ x - 3y + 2z = 0 \end{cases}$

c. $\begin{cases} x_1 + 3x_2 + x_3 + x_4 = 3 \\ x_1 - 2x_3 + 2x_4 = 0 \\ x_3 + 5x_4 = -14 \\ x_2 - 3x_3 - x_4 = 2 \end{cases}$

2. Write a system of equations that has the given augmented matrix.

a. $\begin{bmatrix} 2 & 1 & 4 & \vdots & 3 \\ 6 & 2 & -1 & \vdots & -4 \\ -3 & -1 & 0 & \vdots & 1 \end{bmatrix}$

b. $\begin{bmatrix} 1 & 0 & 0 & \vdots & 5 \\ 0 & 1 & 2 & \vdots & 4 \end{bmatrix}$

c. $\begin{bmatrix} 1 & 0 & 0 & \vdots & 3 \\ 0 & 1 & 0 & \vdots & 2 \\ 0 & 0 & 1 & \vdots & -8 \\ 0 & 0 & 0 & \vdots & 1 \end{bmatrix}$

Given the matrices in Problems 3–8, perform elementary row operations to obtain a 1 in the row 1, column 1 position.

3. $\begin{bmatrix} 3 & 1 & 2 & | & 1 \\ 0 & 2 & 4 & | & 5 \\ 1 & 3 & -4 & | & 9 \end{bmatrix}$ **4.** $\begin{bmatrix} -2 & 3 & 5 & | & 9 \\ 1 & 0 & 2 & | & -8 \\ 0 & 1 & 0 & | & 5 \end{bmatrix}$

5. $\begin{bmatrix} 2 & 4 & 10 & | & -12 \\ 6 & 3 & 4 & | & 6 \\ 10 & -1 & 0 & | & 1 \end{bmatrix}$

6. $\begin{bmatrix} 5 & 20 & 15 & | & 6 \\ 7 & -5 & 3 & | & 2 \\ 12 & 0 & 1 & | & 4 \end{bmatrix}$

7. $\begin{bmatrix} 5 & 6 & -3 & | & 4 \\ 4 & 1 & 9 & | & 2 \\ 7 & 6 & 1 & | & 3 \end{bmatrix}$

8. $\begin{bmatrix} 4 & 8 & 5 & | & 9 \\ 3 & 2 & 1 & | & -4 \\ -2 & 5 & 0 & | & 1 \end{bmatrix}$

Given the matrices in Problems 9–14, perform elementary row operations to obtain zeros under the 1 in the first column.

9. $\begin{bmatrix} 1 & 2 & -3 & | & 0 \\ 0 & 3 & 1 & | & 4 \\ 2 & 5 & 1 & | & 6 \end{bmatrix}$ **10.** $\begin{bmatrix} 1 & 3 & -5 & | & 6 \\ -3 & 4 & 1 & | & 2 \\ 0 & 5 & 1 & | & 3 \end{bmatrix}$

11. $\begin{bmatrix} 1 & 2 & 4 & | & 1 \\ -2 & 5 & 0 & | & 2 \\ -4 & 5 & 1 & | & 3 \end{bmatrix}$ **12.** $\begin{bmatrix} 1 & 5 & 3 & | & 2 \\ 2 & 3 & -1 & | & 4 \\ 3 & 2 & 1 & | & 0 \end{bmatrix}$

13. $\begin{bmatrix} 1 & 4 & -1 & 3 & | & 3 \\ -3 & 4 & 6 & 4 & | & 0 \\ 5 & 1 & 9 & 1 & | & -2 \\ 1 & 0 & 2 & 0 & | & 0 \end{bmatrix}$

14. $\begin{bmatrix} 1 & 8 & 5 & 2 & | & -1 \\ 2 & 0 & 7 & 5 & | & -1 \\ 6 & -2 & 7 & 7 & | & 5 \\ 3 & 1 & 0 & 0 & | & 0 \end{bmatrix}$

Given the matrices in Problems 15–20, perform elementary row operations to obtain a 1 in the second row, second column without changing the entries in the first column.

15. $\begin{bmatrix} 1 & 3 & 5 & | & 2 \\ 0 & 2 & 6 & | & -8 \\ 0 & 3 & 4 & | & 1 \end{bmatrix}$ **16.** $\begin{bmatrix} 1 & 4 & -1 & | & 6 \\ 0 & 5 & 1 & | & 3 \\ 0 & 4 & 6 & | & 5 \end{bmatrix}$

17. $\begin{bmatrix} 1 & 3 & -2 & | & 4 \\ 0 & 5 & 1 & | & 3 \\ 0 & 7 & 9 & | & 2 \end{bmatrix}$ **18.** $\begin{bmatrix} 1 & 5 & -3 & | & 5 \\ 0 & 3 & 9 & | & -15 \\ 0 & 2 & 1 & | & 5 \end{bmatrix}$

19. $\begin{bmatrix} 1 & 3 & -2 & | & 0 \\ 0 & 4 & 2 & | & 9 \\ 0 & 3 & 6 & | & 1 \end{bmatrix}$ **20.** $\begin{bmatrix} 1 & 3 & -2 & | & 0 \\ 0 & 4 & 2 & | & 9 \\ 0 & 10 & -3 & | & 4 \end{bmatrix}$

Given the matrices in Problems 21–23, perform elementary row operations to obtain a zero (or zeros) under the 1 in the second column without changing the entries in the first column.

21. $\begin{bmatrix} 1 & 5 & -3 & | & 2 \\ 0 & 1 & 4 & | & 5 \\ 0 & 3 & 4 & | & 2 \end{bmatrix}$

22. $\begin{bmatrix} 1 & 6 & -3 & 4 & | & 1 \\ 0 & 1 & 7 & 3 & | & 0 \\ 0 & 3 & 4 & 0 & | & -2 \\ 0 & -2 & 3 & 1 & | & 0 \end{bmatrix}$

23. $\begin{bmatrix} 1 & 7 & 6 & 6 & | & 2 \\ 0 & 1 & 9 & 2 & | & 1 \\ 0 & -4 & 6 & 1 & | & 1 \\ 0 & 5 & 8 & 10 & | & 3 \end{bmatrix}$

Given the matrices in Problems 24–26, perform elementary row operations to obtain a 1 in the third row, third column without changing the entries in the first two columns.

24. $\begin{bmatrix} 1 & 3 & 4 & | & 5 \\ 0 & 1 & -3 & | & 6 \\ 0 & 0 & 5 & | & 10 \end{bmatrix}$ **25.** $\begin{bmatrix} 1 & -2 & 5 & | & 6 \\ 0 & 1 & 9 & | & 1 \\ 0 & 0 & -3 & | & 5 \end{bmatrix}$

26. $\begin{bmatrix} 1 & -3 & 4 & | & -5 \\ 0 & 1 & 3 & | & 6 \\ 0 & 0 & 8 & | & 12 \end{bmatrix}$

Given the matrices in Problems 27–29, perform elementary row operations to obtain zeros above the 1 in the third column without changing the entries in the first or second columns.

27. $\begin{bmatrix} 1 & 3 & -1 & | & 5 \\ 0 & 1 & 2 & | & 6 \\ 0 & 0 & 1 & | & 4 \end{bmatrix}$ **28.** $\begin{bmatrix} 1 & 6 & -3 & | & -2 \\ 0 & 1 & 4 & | & 5 \\ 0 & 0 & 1 & | & 3 \end{bmatrix}$

29. $\begin{bmatrix} 1 & 6 & 3 & 0 & | & -2 \\ 0 & 1 & -2 & 0 & | & 1 \\ 0 & 0 & 1 & 0 & | & 0 \\ 0 & 0 & 0 & 1 & | & -6 \end{bmatrix}$

Given the matrices in Problems 30–32 perform elementary row operations to obtain a zero above the 1 in the second column and interpret the solution to the system if the variables are x, y, and z. For example,

$$\begin{bmatrix} 1 & 3 & 0 & | & 5 \\ 0 & 1 & 0 & | & 2 \\ 0 & 0 & 1 & | & 3 \end{bmatrix} \longrightarrow \begin{bmatrix} 1 & 0 & 0 & | & -1 \\ 0 & 1 & 0 & | & 2 \\ 0 & 0 & 1 & | & 3 \end{bmatrix}$$

so the solution is $(x, y, z) = (-1, 2, 3)$.

30. $\begin{bmatrix} 1 & 3 & 0 & \vdots & 4 \\ 0 & 1 & 0 & \vdots & -5 \\ 0 & 0 & 1 & \vdots & 3 \end{bmatrix}$ **31.** $\begin{bmatrix} 1 & -4 & 0 & \vdots & 3 \\ 0 & 1 & 0 & \vdots & -\frac{8}{3} \\ 0 & 0 & 1 & \vdots & \frac{3}{5} \end{bmatrix}$

32. $\begin{bmatrix} 1 & -8 & 0 & \vdots & 3 \\ 0 & 1 & 0 & \vdots & 4 \\ 0 & 0 & 1 & \vdots & -21 \end{bmatrix}$

Solve the systems given in Problems 33–46 by using the Gauss–Jordan method of solution.

33. $\begin{cases} 4x - 7y = -2 \\ -x + 2y = 1 \end{cases}$ **34.** $\begin{cases} 4x - 7y = -5 \\ 2y - x = 1 \end{cases}$

35. $\begin{cases} 8x + 6y = 17 \\ y - x = \frac{1}{2} \end{cases}$ **36.** $\begin{cases} 8x + 6y = 14 \\ 4y - 2x = 24 \end{cases}$

37. $\begin{cases} y = -\frac{2}{3}x + 3 \\ x - 6y = -3 \end{cases}$ **38.** $\begin{cases} y = \frac{1}{2}x + 1 \\ x + 3y = 18 \end{cases}$

39. $\begin{cases} x + y + z = 6 \\ 2x - y + z = 3 \\ x - 2y - 3z = -12 \end{cases}$

40. $\begin{cases} 2x - y + z = 3 \\ x - 3y + 2z = 7 \\ x - y - z = -1 \end{cases}$

41. $\begin{cases} x + y + z = 4 \\ x + 3y + 2z = 4 \\ x - 2y + z = 7 \end{cases}$ **42.** $\begin{cases} x + 2z = 13 \\ 2x + y = 8 \\ -2y + 9z = 41 \end{cases}$

43. $\begin{cases} x + 2z = 7 \\ x + y = 11 \\ -2y + 9z = -3 \end{cases}$ **44.** $\begin{cases} 4x + y + 2z = 7 \\ x + 2y = 0 \\ 3x - y - z = 7 \end{cases}$

45. $\begin{cases} 6x + y + 20z = 27 \\ x - y = 0 \\ y + z = 2 \end{cases}$ **46.** $\begin{cases} 2x - y + 4z = 13 \\ 3x + 6y = 0 \\ 2y - 3z = 3 + 3x \end{cases}$

B

Solve the systems in Problems 47–54 by using the Gauss–Jordan method of solution.

47. $\begin{cases} 2x + 2y + 3z = 1 \\ 2x - z = -11 \\ 3y + 2z = 6 \end{cases}$ **48.** $\begin{cases} x + 2y + z = 1 \\ x - 3y - 2z = 2 \\ 3x - 2y + z = 3 \end{cases}$

49. $\begin{cases} 2x - y + z = 4 \\ 3x - 2y + 2z = 3 \\ x - y + 3z = 2 \end{cases}$ **50.** $\begin{cases} 2x + 10y - 6z = 28 \\ 4x - 3y + z = -14 \\ x + 5y - 3z = 14 \end{cases}$

51. $\begin{cases} x + 5y - 3z = 2 \\ 2x + 2y + z = -1 \\ 3x - y + 5z = 3 \end{cases}$ **52.** $\begin{cases} x + y + z = 1 \\ 2x - y + z = 2 \\ 3x + 2y - 3z = 3 \end{cases}$

53. $\begin{cases} x + 2y = 5 \\ z = 3 \\ w + x + 3y = 6 \\ 2w + 4x = 2 \end{cases}$ **54.** $\begin{cases} x + y + z + w = -2 \\ 2x + y + w = 2 \\ w + 3x + z = 5 \\ 2x + y + 3z = -5 \end{cases}$

Suppose the equation of a certain parabola has the form $y = ax^2 + bx + c$. Find the equation of the parabola passing through the points given in Problems 55–57.

55. $(0, 5), (-1, 2), (3, 26)$

56. $(1, -2), (-2, -14), (3, -4)$

57. $(4, -4), (5, -5), (7, -1)$

C

58. AGRICULTURE In order to control a certain type of crop disease, it is necessary to use 23 gal of chemical A and 34 gal of chemical B. The dealer can order commercial spray I, each container of which holds 5 gal of chemical A and 2 gal of chemical B, and commercial spray II, each container of which holds 2 gal of chemical A and 7 gal of chemical B. How many containers of each type of commercial spray should be used to attain exactly the right proportion of chemicals needed?

59. BUSINESS In order to manufacture a certain alloy, it is necessary to use 33 kg (kilograms) of metal A and 56 kg of metal B. It is cheaper for the manufacturer if she buys and mixes an alloy, each bar of which contains 3 kg of metal A and 5 kg of metal B, along with another alloy, each bar of which contains 4 kg of metal A and 7 kg of metal B. How much of the two alloys should she use to produce the alloy desired?

60. BUSINESS A candy maker mixes chocolate, milk, and almonds to produce three kinds of candy—I, II, and III—with the following proportions:

    I: 7 lb chocolate, 5 gal milk, 1 oz almonds
    II: 3 lb chocolate, 2 gal milk, 2 oz almonds
    III: 4 lb cholocate, 3 gal milk, 3 oz almonds

If 67 lb of chocolate, 48 gal of milk, and 32 oz of almonds are available, how much of each kind of candy can be produced?

61. BUSINESS Using the data from Problem 60, how much of each type of candy can be produced with 62 lb of chocolate, 44 gal of milk, and 32 oz of almonds?

9.3 Inverse Matrices

The availability of computer software and calculators that can carry out matrix operations has considerably increased the importance of a matrix solution to systems of equations. If we let A be the matrix of coefficients of a system of equations, X the matrix of unknowns, and B the matrix of constants, we can then represent the system of equations by the matrix equation

$$AX = B$$

If we can define an inverse matrix, denoted by A^{-1}, we should be able to solve the *system* by multiplying both sides by A^{-1}:

$$AX = B$$
$$A^{-1}(AX) = A^{-1}B$$
$$X = A^{-1}B$$

We will now consider how to find $A^{-1}B$. In order to understand this simple process we need to understand what it means for matrices to be inverses, as well as a basic algebra for matrices. Even though these matrix operations are difficult with a pencil and paper, they are easy with the aid of computer software or a calculator that does matrix operations.

Matrix Operations

Following is a definition of matrix equality along with the fundamental matrix operations.

MATRIX OPERATIONS

EQUALITY: $M = N$ if and only if matrices M and N have the same dimension and $m_{ij} = n_{ij}$ for all i and j.

ADDITION: $M + N = P$ if and only if M and N have the same dimension and $p_{ij} = m_{ij} + n_{ij}$ for all i and j.

MULTIPLICATION OF A MATRIX AND A REAL NUMBER: $cM = Mc = [cm_{ij}]$ for any real number c.

SUBTRACTION: $M - N = P$ if and only if matrices M and N have the same dimension and $p_{ij} = m_{ij} - n_{ij}$ for all i and j.

MULTIPLICATION: $MN = P$ if and only if the number of columns of matrix M is the same as the number of rows of matrix N and

$$p_{ij} = m_{i1}n_{1j} + m_{i2}n_{2j} + m_{i3}n_{3j} + \cdots + m_{in}n_{nj}$$

The matrix P has the same number of rows as M and the same number of columns as N.

All of these definitions, except multiplication, are straighforward, so we will consider multiplication separately after Example 1. If an addition or multiplication cannot be performed because of the order of the given matrices, the matrices are said to be **not conformable**.

EXAMPLE 1 Let $A = [5 \quad 2 \quad 1]$, $B = [4 \quad 8 \quad -5]$, $C = \begin{bmatrix} 7 & 3 & 2 \\ 5 & -4 & -3 \end{bmatrix}$,

$$D = \begin{bmatrix} 4 & -2 & 1 \\ -3 & 3 & -1 \\ 2 & 4 & -1 \end{bmatrix} \qquad E = \begin{bmatrix} 3 & 4 & -1 \\ 2 & 0 & 5 \\ -4 & 2 & 3 \end{bmatrix}$$

Find **a.** $A + B$ **b.** $A + C$ **c.** $E + D$ **d.** $(-5)C$ **e.** $2A - 3B$

SOLUTION **a.** $A + B = [5 \quad 2 \quad 1] + [4 \quad 8 \quad -5]$
$= [5 + 4 \quad 2 + 8 \quad 1 + (-5)]$
$= [9 \quad 10 \quad -4]$

b. $A + C$ is not defined because A and C are not conformable.

c. $E + D = \begin{bmatrix} 3 & 4 & -1 \\ 2 & 0 & 5 \\ -4 & 2 & 3 \end{bmatrix} + \begin{bmatrix} 4 & -2 & 1 \\ -3 & 3 & -1 \\ 2 & 4 & -1 \end{bmatrix} = \begin{bmatrix} 7 & 2 & 0 \\ -1 & 3 & 4 \\ -2 & 6 & 2 \end{bmatrix}$

Add entry by entry.

d. $(-5)C = (-5)\begin{bmatrix} 7 & 3 & 2 \\ 5 & -4 & -3 \end{bmatrix} = \begin{bmatrix} -35 & -15 & -10 \\ -25 & 20 & 15 \end{bmatrix}$

Multiply each entry by (-5).

e. $2A - 3B = 2[5 \quad 2 \quad 1] + (-3)[4 \quad 8 \quad -5]$
$= [10 \quad 4 \quad 2] + [-12 \quad -24 \quad 15]$
$= [-2 \quad -20 \quad 17]$ ∎

You will need to study the definition of matrix multiplication carefully to understand the process. Consider the following specific example:

$$M = \begin{bmatrix} m_{11} & m_{12} & m_{13} & m_{14} \\ m_{21} & m_{22} & m_{23} & m_{24} \end{bmatrix} = \begin{bmatrix} 1 & 2 & 3 & 4 \\ 5 & 6 & 7 & 8 \end{bmatrix}$$

$$N = \begin{bmatrix} n_{11} & n_{12} & n_{13} \\ n_{21} & n_{22} & n_{23} \\ n_{31} & n_{32} & n_{33} \\ n_{41} & n_{42} & n_{43} \end{bmatrix} = \begin{bmatrix} -3 & 1 & -2 \\ 0 & -1 & 5 \\ -4 & 3 & -1 \\ 2 & 3 & -2 \end{bmatrix}$$

The matrix N and its entries are shown in color so you can keep track of them.

$$M N = \begin{bmatrix} m_{11} & m_{12} & m_{13} & m_{14} \\ m_{21} & m_{22} & m_{23} & m_{24} \end{bmatrix} \begin{bmatrix} n_{11} & n_{12} & n_{13} \\ n_{21} & n_{22} & n_{23} \\ n_{31} & n_{32} & n_{33} \\ n_{41} & n_{42} & n_{43} \end{bmatrix} = \begin{bmatrix} p_{11} & p_{12} & p_{13} \\ p_{21} & p_{22} & p_{23} \end{bmatrix}$$

These are the entries of the product.

These must be the same for multiplication to be conformable.

Order of M is 2×4 Order of N is 4×3

Order of P is 2×3; this is the order of the product.

ENTRY IN PRODUCT	ROW	COLUMN	NOTATION
Row 1 $MN = \begin{bmatrix} 1 & 2 & 3 & 4 \\ 5 & 6 & 7 & 8 \end{bmatrix} \begin{bmatrix} -3 & 1 & -2 \\ 0 & -1 & 5 \\ -4 & 3 & -1 \\ 2 & 3 & -2 \end{bmatrix} = \begin{bmatrix} p_{11} & p_{12} & p_{13} \\ p_{21} & p_{22} & p_{23} \end{bmatrix}$ Row 2 Column 1 Column 2 Column 3	1 1 1 2 2 2	1 2 3 1 2 3	p_{11} p_{12} p_{13} p_{21} p_{22} p_{23}

$p_{11} = m_{11}n_{11} + m_{12}n_{21} + m_{13}n_{31} + m_{14}n_{41}$
$= 1(-3) + 2(0) + 3(-4) + 4(2) = -7$

$p_{12} = m_{11}n_{12} + m_{12}n_{22} + m_{13}n_{32} + m_{14}n_{42}$
$= 1(1) + 2(-1) + 3(3) + 4(3) = 20$

$p_{13} = m_{11}n_{13} + m_{12}n_{23} + m_{13}n_{33} + m_{14}n_{43}$
$= 1(-2) + 2(5) + 3(-1) + 4(-2) = -3$

$p_{21} = m_{21}n_{11} + m_{22}n_{21} + m_{23}n_{31} + m_{24}n_{41}$
$= 5(-3) + 6(0) + 7(-4) + 8(2) = -27$

$p_{22} = m_{21}n_{12} + m_{22}n_{22} + m_{23}n_{32} + m_{24}n_{42}$
$= 5(1) + 6(-1) + 7(3) + 8(3) = 44$

$p_{23} = m_{21}n_{13} + m_{22}n_{23} + m_{23}n_{33} + m_{24}n_{43}$
$= 5(-2) + 6(5) + 7(-1) + 8(-2) = -3$

This example makes the process seem very lengthy, but it has been written out so that you could study the *process*. The way it would look in your work is shown in Example 2.

EXAMPLE 2 Find the products.

a. $\begin{bmatrix} 1 & 2 & 3 & 4 \\ 5 & 6 & 7 & 8 \end{bmatrix} \begin{bmatrix} -3 & 1 & -2 \\ 0 & -1 & 5 \\ -4 & 3 & -1 \\ 2 & 3 & -2 \end{bmatrix}$

$= \begin{bmatrix} 1(-3) + 2(0) + 3(-4) + 4(2) & 1(1) + 2(-1) + 3(3) + 4(3) & 1(-2) + 2(5) + 3(-1) + 4(-2) \\ 5(-3) + 6(0) + 7(-4) + 8(2) & 5(1) + 6(-1) + 7(3) + 8(3) & 5(-2) + 6(5) + 7(-1) + 8(-2) \end{bmatrix}$

$= \begin{bmatrix} -7 & 20 & -3 \\ -27 & 44 & -3 \end{bmatrix}$

b. $\begin{bmatrix} 3 & -1 & 4 \\ 2 & 1 & 0 \\ -1 & 3 & 2 \end{bmatrix} \begin{bmatrix} 5 & 1 & -1 \\ 2 & 3 & -2 \\ 0 & 3 & 4 \end{bmatrix}$

$= \begin{bmatrix} 3(5) + (-1)(2) + 4(0) & 3(1) + (-1)(3) + 4(3) & 3(-1) + (-1)(-2) + 4(4) \\ 2(5) + (1)(2) + 0(0) & 2(1) + (1)(3) + 0(3) & 2(-1) + (1)(-2) + 0(4) \\ (-1)(5) + 3(2) + 2(0) & (-1)(1) + 3(3) + 2(3) & (-1)(-1) + 3(-2) + 2(4) \end{bmatrix}$

$= \begin{bmatrix} 13 & 12 & 15 \\ 12 & 5 & -4 \\ 1 & 14 & 3 \end{bmatrix}$

Systems of equations can be written in the form of matrix multiplication, as shown by Example 3.

EXAMPLE 3   Let

$$A = \begin{bmatrix} 1 & 2 & 3 \\ 4 & -1 & 5 \\ 3 & 2 & -1 \end{bmatrix} \qquad X = \begin{bmatrix} x \\ y \\ z \end{bmatrix} \qquad B = \begin{bmatrix} 3 \\ 16 \\ 5 \end{bmatrix}$$

What is $AX = B$?

SOLUTION $AX = \begin{bmatrix} x + 2y + 3z \\ 4x - y + 5z \\ 3x + 2y - z \end{bmatrix}$

so $AX = B$ is a matrix equation representing the system

$$\begin{cases} x + 2y + 3z = 3 \\ 4x - y + 5z = 16 \\ 3x + 2y - z = 5 \end{cases}$$

Properties for an algebra of matrices can also be developed. The $m \times n$ **zero matrix,** denoted by 0, is the matrix with m rows and n columns in which each entry is 0. The **identity matrix,** denoted by I_n, is the square matrix with n rows and n columns consisting of a 1 in each position on the **main diagonal** (entries $m_{11}, m_{22}, m_{33}, \ldots$) and zeros elsewhere:

$$I_2 = \begin{bmatrix} 1 & 0 \\ 0 & 1 \end{bmatrix} \qquad I_3 = \begin{bmatrix} 1 & 0 & 0 \\ 0 & 1 & 0 \\ 0 & 0 & 1 \end{bmatrix} \qquad I_4 = \begin{bmatrix} 1 & 0 & 0 & 0 \\ 0 & 1 & 0 & 0 \\ 0 & 0 & 1 & 0 \\ 0 & 0 & 0 & 1 \end{bmatrix}$$

The **additive inverse** of a matrix M is denoted by $-M$ and is defined by $(-1)M$; the **multiplicative inverse** of M is denoted by M^{-1} if it exists. (*Note:* $M^{-1} \neq 1/M$.) The following list summarizes the properties of matrices. Assume that matrices M, N, and P all have order n, which forces them to be conformable for the given operations. Also, if the dimension of the identity matrix is clear from the context, we will simply write I.

Table 9.1   Properties of matrices

PROPERTY	ADDITION	MULTIPLICATION	
Commutative	$M + N = N + M$	$MN \neq NM$	Matrix multiplication is not commutative.
Associative	$(M + N) + P = M + (N + P)$	$(MN)P = M(NP)$	
Identity	$M + 0 = 0 + M = M$	$I_n M = M I_n = M$	For a square matrix M of order n
Inverse	$M + (-M) = (-M) + M = 0$	$M(M^{-1}) = (M^{-1})M$ $= I_n$	For a square matrix M for which M^{-1} exists
Distributive		$M(N + P) = MN + MP$ and $(N + P)M = NM + PM$	

Inverse Matrices

These properties are straightforward except for the inverse property. There seem to be two unanswered questions. Given a square matrix M, when does M^{-1} exist? And if it exists, how do you find it?

INVERSE OF A MATRIX

If A is a square matrix and if there exists a matrix A^{-1} such that

$$A^{-1}A = AA^{-1} = I$$

then A^{-1} is called the **inverse of A** for multiplication.

Usually, in the context of matrices, when we talk simply of the inverse of A we mean the inverse of A for multiplication.

EXAMPLE 4 Verify that the inverse of $A = \begin{bmatrix} 2 & 1 \\ 3 & 2 \end{bmatrix}$ is $B = \begin{bmatrix} 2 & -1 \\ -3 & 2 \end{bmatrix}$.

SOLUTION

$AB = \begin{bmatrix} 2 & 1 \\ 3 & 2 \end{bmatrix}\begin{bmatrix} 2 & -1 \\ -3 & 2 \end{bmatrix}$

$= \begin{bmatrix} 4-3 & -2+2 \\ 6-6 & -3+4 \end{bmatrix}$

$= \begin{bmatrix} 1 & 0 \\ 0 & 1 \end{bmatrix} = I$

$BA = \begin{bmatrix} 2 & -1 \\ -3 & 2 \end{bmatrix}\begin{bmatrix} 2 & 1 \\ 3 & 2 \end{bmatrix}$

$= \begin{bmatrix} 4-3 & 2-2 \\ -6+6 & -3+4 \end{bmatrix}$

$= \begin{bmatrix} 1 & 0 \\ 0 & 1 \end{bmatrix} = I$

Thus $B = A^{-1}$.

EXAMPLE 5 Show that A and B are inverses of each other, where

$$A = \begin{bmatrix} 0 & 1 & 2 \\ -1 & 1 & 2 \\ 1 & -2 & -5 \end{bmatrix} \quad \text{and} \quad B = \begin{bmatrix} 1 & -1 & 0 \\ 3 & 2 & 2 \\ -1 & -1 & -1 \end{bmatrix}$$

SOLUTION

$AB = \begin{bmatrix} 0 & 1 & 2 \\ -1 & 1 & 2 \\ 1 & -2 & -5 \end{bmatrix}\begin{bmatrix} 1 & -1 & 0 \\ 3 & 2 & 2 \\ -1 & -1 & -1 \end{bmatrix}$

$= \begin{bmatrix} 0+3-2 & 0+2-2 & 0+2-2 \\ -1+3-2 & 1+2-2 & 0+2-2 \\ 1-6+5 & -1-4+5 & 0-4+5 \end{bmatrix} = \begin{bmatrix} 1 & 0 & 0 \\ 0 & 1 & 0 \\ 0 & 0 & 1 \end{bmatrix} = I$

$BA = \begin{bmatrix} 1 & -1 & 0 \\ 3 & 2 & 2 \\ -1 & -1 & -1 \end{bmatrix}\begin{bmatrix} 0 & 1 & 2 \\ -1 & 1 & 2 \\ 1 & -2 & -5 \end{bmatrix}$

$= \begin{bmatrix} 0+1+0 & 1-1+0 & 2-2+0 \\ 0-2+2 & 3+2-4 & 6+4-10 \\ 0+1-1 & -1-1+2 & -2-2+5 \end{bmatrix} = \begin{bmatrix} 1 & 0 & 0 \\ 0 & 1 & 0 \\ 0 & 0 & 1 \end{bmatrix} = I$

Since $AB = I = BA$, then $B = A^{-1}$.

If a given matrix has an inverse, we say that it is **nonsingular.** The unanswered question, however, is how to *find* an inverse matrix. Consider the following two examples.

EXAMPLE 6   Find the inverse of $A = \begin{bmatrix} 1 & 2 \\ 1 & 4 \end{bmatrix}$.

SOLUTION   Find a matrix B (if it exists) so that $AB = I$; since we do not know B, let its entries be variables. That is, let

$$B = \begin{bmatrix} x_1 & x_2 \\ y_1 & y_2 \end{bmatrix}$$

Then

$$AB = \begin{bmatrix} 1 & 2 \\ 1 & 4 \end{bmatrix}\begin{bmatrix} x_1 & x_2 \\ y_1 & y_2 \end{bmatrix} = \begin{bmatrix} x_1 + 2y_1 & x_2 + 2y_2 \\ x_1 + 4y_1 & x_2 + 4y_2 \end{bmatrix} = \begin{bmatrix} 1 & 0 \\ 0 & 1 \end{bmatrix}$$

By the definition of the equality of matrices, we see that

$$\begin{cases} x_1 + 2y_1 = 1 \\ x_1 + 4y_1 = 0 \end{cases} \qquad \begin{cases} x_2 + 2y_2 = 0 \\ x_2 + 4y_2 = 1 \end{cases}$$

Solve these systems by using Gauss–Jordan elimination:

(x_1, y_1): $\begin{bmatrix} 1 & 2 & \vdots & 1 \\ 1 & 4 & \vdots & 0 \end{bmatrix} \rightarrow \begin{bmatrix} 1 & 2 & \vdots & 1 \\ 0 & 2 & \vdots & -1 \end{bmatrix} \rightarrow \begin{bmatrix} 1 & 2 & \vdots & 1 \\ 0 & 1 & \vdots & -\frac{1}{2} \end{bmatrix} \rightarrow \begin{bmatrix} 1 & 0 & \vdots & 2 \\ 0 & 1 & \vdots & -\frac{1}{2} \end{bmatrix}$

(x_2, y_2): $\begin{bmatrix} 1 & 2 & \vdots & 0 \\ 1 & 4 & \vdots & 1 \end{bmatrix} \rightarrow \begin{bmatrix} 1 & 2 & \vdots & 0 \\ 0 & 2 & \vdots & 1 \end{bmatrix} \rightarrow \begin{bmatrix} 1 & 2 & \vdots & 0 \\ 0 & 1 & \vdots & \frac{1}{2} \end{bmatrix} \rightarrow \begin{bmatrix} 1 & 0 & \vdots & -1 \\ 0 & 1 & \vdots & \frac{1}{2} \end{bmatrix}$

$$\begin{cases} x_1 = 2 \\ y_1 = -\frac{1}{2} \end{cases} \qquad \begin{cases} x_2 = -1 \\ y_2 = \frac{1}{2} \end{cases}$$

Therefore the inverse is

$$B = \begin{bmatrix} x_1 & x_2 \\ y_1 & y_2 \end{bmatrix} = \begin{bmatrix} 2 & -1 \\ -\frac{1}{2} & \frac{1}{2} \end{bmatrix}$$ ∎

The solutions of the two systems are shown side by side so you can easily see that the steps are identical since the two rows on the left are the same; the third columns are

$$\begin{bmatrix} 1 \\ 0 \end{bmatrix} \quad \text{and} \quad \begin{bmatrix} 0 \\ 1 \end{bmatrix}$$

respectively. It would seem that we might combine steps, but, before we do, consider a 3×3 example.

EXAMPLE 7 Find the inverse for the matrix

$$\begin{bmatrix} 1 & -1 & 0 \\ 3 & 2 & 2 \\ -1 & -1 & -1 \end{bmatrix}$$

(This is matrix B from Example 5.)

SOLUTION We need to find a matrix

$$\begin{bmatrix} x_1 & x_2 & x_3 \\ y_1 & y_2 & y_3 \\ z_1 & z_2 & z_3 \end{bmatrix}$$

so that

$$\begin{bmatrix} 1 & -1 & 0 \\ 3 & 2 & 2 \\ -1 & -1 & -1 \end{bmatrix}\begin{bmatrix} x_1 & x_2 & x_3 \\ y_1 & y_2 & y_3 \\ z_1 & z_2 & z_3 \end{bmatrix} = \begin{bmatrix} 1 & 0 & 0 \\ 0 & 1 & 0 \\ 0 & 0 & 1 \end{bmatrix}$$

$$\begin{cases} x_1 - y_1 + 0z_1 = 1 \\ 3x_1 + 2y_1 + 2z_1 = 0 \\ -x_1 - y_1 - z_1 = 0 \end{cases} \quad \begin{cases} x_2 - y_2 + 0z_2 = 0 \\ 3x_2 + 2y_2 + 2z_2 = 1 \\ -x_2 - y_2 - z_2 = 0 \end{cases} \quad \begin{cases} x_3 - y_3 + 0z_3 = 0 \\ 3x_3 + 2y_3 + 2z_3 = 0 \\ -x_3 - y_3 - z_3 = 1 \end{cases}$$

We could solve these as three separate systems using Gauss–Jordan elimination; however, all the steps would be identical since the variables on the left of the equal signs are the same in each system. Therefore, suppose we augment the matrix of the coefficients by the *three* columns and do all three at once.

$$\begin{bmatrix} 1 & -1 & 0 & \vdots & 1 & 0 & 0 \\ 3 & 2 & 2 & \vdots & 0 & 1 & 0 \\ -1 & -1 & -1 & \vdots & 0 & 0 & 1 \end{bmatrix} \longrightarrow \begin{bmatrix} 1 & -1 & 0 & \vdots & 1 & 0 & 0 \\ 0 & 5 & 2 & \vdots & -3 & 1 & 0 \\ 0 & -2 & -1 & \vdots & 1 & 0 & 1 \end{bmatrix} \xrightarrow{2R3 + R2}$$

$$\begin{bmatrix} 1 & -1 & 0 & \vdots & 1 & 0 & 0 \\ 0 & 1 & 0 & \vdots & -1 & 1 & 2 \\ 0 & -2 & -1 & \vdots & 1 & 0 & 1 \end{bmatrix} \longrightarrow \begin{bmatrix} 1 & 0 & 0 & \vdots & 0 & 1 & 2 \\ 0 & 1 & 0 & \vdots & -1 & 1 & 2 \\ 0 & 0 & -1 & \vdots & -1 & 2 & 5 \end{bmatrix} \longrightarrow$$

$$\begin{bmatrix} 1 & 0 & 0 & \vdots & 0 & 1 & 2 \\ 0 & 1 & 0 & \vdots & -1 & 1 & 2 \\ 0 & 0 & 1 & \vdots & 1 & -2 & -5 \end{bmatrix}$$

CALCULATOR COMMENT

If you have a calculator that does matrix operations, use the $\boxed{\text{MATRIX}}$ function and then press $[A]^{-1}$ to find the inverse for Examples 7, 8, and 9.

Now, if we relate this back to the original three systems, we see that the inverse is

$$\begin{bmatrix} 0 & 1 & 2 \\ -1 & 1 & 2 \\ 1 & -2 & -5 \end{bmatrix}$$

By studying Examples 6 and 7, we are led to a procedure for finding the inverse of a nonsingular matrix:

1. Augment the given matrix with I; that is, write $[A \vdots I]$, where I is the identity matrix of the same order as the given square matrix A.
2. Perform elementary row operations using Gauss–Jordan elimination in order to change the matrix A into the identity matrix (if possible).
3. If, at any time, you obtain all zeros in a row or column to the left of the dashed line, then there will be no inverse.
4. If steps 1 and 2 can be performed, the result in the augmented part is the inverse of A.

EXAMPLE 8   Find the inverse of the matrix $A = \begin{bmatrix} 1 & 2 \\ 0 & 0 \end{bmatrix}$.

SOLUTION   Write the augmented matrix $[A \vdots I]$:

$$\begin{bmatrix} 1 & 2 & \vdots & 1 & 0 \\ 0 & 0 & \vdots & 0 & 1 \end{bmatrix}$$

We want to make the left-hand side look like the corresponding identity matrix. This is impossible since there are no elementary row operations that will put it into the required form. Thus there is **no inverse.**   ▌

EXAMPLE 9   Find the inverse, if possible, of the matrix $\begin{bmatrix} 1 & 2 \\ 1 & 4 \end{bmatrix}$

SOLUTION $\begin{bmatrix} 1 & 2 & \vdots & 1 & 0 \\ 1 & 4 & \vdots & 0 & 1 \end{bmatrix} \longrightarrow \begin{bmatrix} 1 & 2 & \vdots & 1 & 0 \\ 0 & 2 & \vdots & -1 & 1 \end{bmatrix}$ Multiply row 1 by -1 and add to row 2.

$\longrightarrow \begin{bmatrix} 1 & 2 & \vdots & 1 & 0 \\ 0 & 1 & \vdots & -\frac{1}{2} & \frac{1}{2} \end{bmatrix}$ Multiply row 2 by $\frac{1}{2}$.

$\longrightarrow \begin{bmatrix} 1 & 0 & \vdots & 2 & -1 \\ 0 & 1 & \vdots & -\frac{1}{2} & \frac{1}{2} \end{bmatrix}$ Multiply row 2 by -2 and add to row 1.

The inverse is on the right side of the dotted line:

$$\begin{bmatrix} 2 & -1 \\ -\frac{1}{2} & \frac{1}{2} \end{bmatrix}$$

When a matrix has fractional entries, it is often rewritten with a fractional coefficient and integer entries in order to simplify subsequent arithmetic. For example, we would rewrite the above matrix as

$$\frac{1}{2} \begin{bmatrix} 4 & -2 \\ -1 & 1 \end{bmatrix}$$

The fractions obtained when finding inverses are often ugly.   ▌

EXAMPLE 10 Find the inverse of the matrix $A = \begin{bmatrix} 0 & 1 & 2 \\ 2 & -1 & 1 \\ -1 & 1 & 0 \end{bmatrix}$ if it exists.

SOLUTION Write the augmented matrix $[A \vdots I]$ and make the left-hand side look like the corresponding identity matrix (if possible):

$$
\begin{bmatrix}
0 & 1 & 2 & \vdots & 1 & 0 & 0 \\
2 & -1 & 1 & \vdots & 0 & 1 & 0 \\
-1 & 1 & 0 & \vdots & 0 & 0 & 1
\end{bmatrix}
\xrightarrow[R1 \leftrightarrow R3]{}
\begin{bmatrix}
-1 & 1 & 0 & \vdots & 0 & 0 & 1 \\
2 & -1 & 1 & \vdots & 0 & 1 & 0 \\
0 & 1 & 2 & \vdots & 1 & 0 & 0
\end{bmatrix}
$$

$$
\xrightarrow{}
\begin{bmatrix}
1 & -1 & 0 & \vdots & 0 & 0 & -1 \\
2 & -1 & 1 & \vdots & 0 & 1 & 0 \\
0 & 1 & 2 & \vdots & 1 & 0 & 0
\end{bmatrix}
\xrightarrow{}
\begin{bmatrix}
1 & -1 & 0 & \vdots & 0 & 0 & -1 \\
0 & 1 & 1 & \vdots & 0 & 1 & 2 \\
0 & 1 & 2 & \vdots & 1 & 0 & 0
\end{bmatrix}
$$

$$
\xrightarrow{}
\begin{bmatrix}
1 & 0 & 1 & \vdots & 0 & 1 & 1 \\
0 & 1 & 1 & \vdots & 0 & 1 & 2 \\
0 & 0 & 1 & \vdots & 1 & -1 & -2
\end{bmatrix}
\xrightarrow{}
\begin{bmatrix}
1 & 0 & 0 & \vdots & -1 & 2 & 3 \\
0 & 1 & 0 & \vdots & -1 & 2 & 4 \\
0 & 0 & 1 & \vdots & 1 & -1 & -2
\end{bmatrix}
$$

Thus $A^{-1} = \begin{bmatrix} -1 & 2 & 3 \\ -1 & 2 & 4 \\ 1 & -1 & -2 \end{bmatrix}$. ∎

Systems of Equations

In Example 3 we saw how a system of equations can be written in matrix form. We can now see how to solve a system of linear equations by using the inverse. Consider a system of n linear equations with n unknowns whose matrix of coefficients A has an inverse A^{-1}:

$$AX = B \qquad \text{Given system}$$
$$(A^{-1})AX = A^{-1}B \qquad \text{Multiply both sides by } A^{-1}.$$
$$(A^{-1}A)X = A^{-1}B \qquad \text{Associative property}$$
$$I_n X = A^{-1}B \qquad \text{Inverse property}$$
$$X = A^{-1}B \qquad \text{Identity property}$$

This says that to solve a system, simply multiply A^{-1} and B to find X.

EXAMPLE 11 Solve the system

$$
\begin{cases}
y + 2z = 0 \\
2x - y + z = -1 \\
y - x = 1
\end{cases}
$$

SOLUTION Write in matrix form:

$$
A = \begin{bmatrix} 0 & 1 & 2 \\ 2 & -1 & 1 \\ -1 & 1 & 0 \end{bmatrix}
\qquad
X = \begin{bmatrix} x \\ y \\ z \end{bmatrix}
\qquad
B = \begin{bmatrix} 0 \\ -1 \\ 1 \end{bmatrix}
$$

From Example 10,

$$A^{-1} = \begin{bmatrix} -1 & 2 & 3 \\ -1 & 2 & 4 \\ 1 & -1 & -2 \end{bmatrix}$$

Thus

$$X = A^{-1}B = \begin{bmatrix} -1 & 2 & 3 \\ -1 & 2 & 4 \\ 1 & -1 & -2 \end{bmatrix}\begin{bmatrix} 0 \\ -1 \\ 1 \end{bmatrix} = \begin{bmatrix} 0-2+3 \\ 0-2+4 \\ 0+1-2 \end{bmatrix} = \begin{bmatrix} 1 \\ 2 \\ -1 \end{bmatrix}$$

Therefore **$x = 1$, $y = 2$, and $z = -1$.** ∎

The method of solving a system by using the inverse matrix is very efficient if you know the inverse. Unfortunately, *finding* the inverse for one system is usually more work than using another method to solve the system. However, there are certain applications that yield the same system over and over, and the only thing to change is the constants. In this case the inverse method is worthwhile. And, finally, computers can often find approximations for inverse matrices quite easily, so this method might be appropriate if one is using a calculator or computer.

9.3 Problem Set

A *In Problems 1–12 find the indicated matrices if possible.*

$$A = \begin{bmatrix} 1 & 2 \\ 4 & 0 \\ -1 & 3 \\ 2 & 1 \end{bmatrix} \quad B = \begin{bmatrix} 4 & 2 \\ -1 & 3 \end{bmatrix}$$

$$C = \begin{bmatrix} 1 & 0 & 0 & 0 \\ 0 & 1 & 0 & 0 \\ 0 & 0 & 1 & 0 \\ 0 & 0 & 0 & 1 \end{bmatrix} \quad D = \begin{bmatrix} 4 & 1 & 3 & 6 \\ -1 & 0 & -2 & 3 \end{bmatrix}$$

$$E = \begin{bmatrix} 1 & 0 & 2 \\ 3 & -1 & 2 \\ 4 & 1 & 0 \end{bmatrix} \quad F = \begin{bmatrix} 1 & 4 & 0 \\ 3 & -1 & 2 \\ -2 & 1 & 5 \end{bmatrix}$$

$$G = \begin{bmatrix} 8 & 1 & 6 \\ 3 & 5 & 7 \\ 4 & 9 & 2 \end{bmatrix}$$

1. $E + F$
2. EF
3. EG
4. $EF + EG$
5. $E(F + G)$
6. $2E - G$
7. FG
8. GF
9. $(EF)G$
10. $E(FG)$
11. B^2
12. C^3

Use multiplication in Problems 13–18 to determine whether the matrices are inverses.

13. $\begin{bmatrix} 1 & 2 \\ 2 & 3 \end{bmatrix}, \begin{bmatrix} -3 & 2 \\ 2 & -1 \end{bmatrix}$

14. $\begin{bmatrix} 3 & 5 \\ 4 & 7 \end{bmatrix}, \begin{bmatrix} 7 & -5 \\ -4 & 3 \end{bmatrix}$

15. $\begin{bmatrix} 4 & 3 \\ 2 & 2 \end{bmatrix}, \begin{bmatrix} 1 & -\frac{3}{2} \\ -1 & 2 \end{bmatrix}$

16. $\begin{bmatrix} 0 & 1 & 0 \\ 1 & -1 & 0 \\ -1 & 2 & 1 \end{bmatrix}, \begin{bmatrix} -1 & -1 & 0 \\ -1 & 0 & 0 \\ 1 & -1 & -1 \end{bmatrix}$

17. $\begin{bmatrix} 3 & -2 & 4 \\ 2 & 1 & 2 \\ 5 & 3 & 5 \end{bmatrix}, \begin{bmatrix} -1 & 22 & -8 \\ 0 & -5 & 2 \\ 1 & -19 & 7 \end{bmatrix}$

18. $\begin{bmatrix} 1 & 1 & 0 & 1 \\ 2 & 0 & -5 & 8 \\ 0 & 1 & 3 & 2 \\ 0 & 0 & 3 & 31 \end{bmatrix}, \begin{bmatrix} 1 & 95 & -1 & -29 \\ 0 & -92 & 1 & 28 \\ 0 & 31 & 0 & -10 \\ 0 & -3 & 0 & 1 \end{bmatrix}$

B *Find the inverse of each matrix in Problems 19–27, if it exists.*

19. $\begin{bmatrix} 4 & -7 \\ -1 & 2 \end{bmatrix}$

20. $\begin{bmatrix} -1 & 2 \\ 3 & 4 \end{bmatrix}$

21. $\begin{bmatrix} 1 & 3 \\ 2 & 0 \end{bmatrix}$

22. $\begin{bmatrix} 1 & 0 & 2 \\ 2 & 1 & 0 \\ 0 & -2 & 9 \end{bmatrix}$

23. $\begin{bmatrix} 6 & 1 & 20 \\ 1 & -1 & 0 \\ 0 & 1 & 3 \end{bmatrix}$

24. $\begin{bmatrix} 4 & 1 & 0 \\ 2 & -1 & 4 \\ -3 & 2 & 1 \end{bmatrix}$

25. $\begin{bmatrix} 1 & 0 & 0 & 1 \\ 0 & 2 & 0 & 0 \\ 0 & 0 & 0 & 1 \\ 2 & 0 & 1 & 0 \end{bmatrix}$

26. $\begin{bmatrix} 0 & 1 & 2 & 0 \\ 0 & 0 & 0 & 1 \\ 1 & 1 & 3 & 0 \\ 2 & 4 & 0 & 0 \end{bmatrix}$

27. $\begin{bmatrix} 1 & 2 & 0 & 0 \\ 0 & 0 & 1 & 0 \\ 1 & 3 & 0 & 1 \\ 4 & 0 & 0 & 2 \end{bmatrix}$

Solve the systems in Problems 28–60 by solving the corresponding matrix equation with an inverse if possible. Problems 28–33 use the inverse found in Problem 19.

28. $\begin{cases} 4x - 7y = -2 \\ -x + 2y = 1 \end{cases}$

29. $\begin{cases} 4x - 7y = -65 \\ -x + 2y = 18 \end{cases}$

30. $\begin{cases} 4x - 7y = 48 \\ -x + 2y = -13 \end{cases}$

31. $\begin{cases} 4x - 7y = 2 \\ -x + 2y = 3 \end{cases}$

32. $\begin{cases} 4x - 7y = 5 \\ -x + 2y = 4 \end{cases}$

33. $\begin{cases} 4x - 7y = -3 \\ -x + 2y = 8 \end{cases}$

Problems 34–39 all use the same inverse.

34. $\begin{cases} 8x + 6y = 12 \\ -2x + 4y = -14 \end{cases}$

35. $\begin{cases} 8x + 6y = 16 \\ -2x + 4y = 18 \end{cases}$

36. $\begin{cases} 8x + 6y = -6 \\ -2x + 4y = -26 \end{cases}$

37. $\begin{cases} 8x + 6y = -28 \\ -2x + 4y = 18 \end{cases}$

38. $\begin{cases} 8x + 6y = -26 \\ -2x + 4y = 12 \end{cases}$

39. $\begin{cases} 8x + 6y = -36 \\ -2x + 4y = -2 \end{cases}$

Problems 40–45 all use the same inverse.

40. $\begin{cases} 2x + 3y = 9 \\ x - 6y = -3 \end{cases}$

41. $\begin{cases} 2x + 3y = 2 \\ x - 6y = 16 \end{cases}$

42. $\begin{cases} 2x + 3y = 2 \\ x - 6y = -14 \end{cases}$

43. $\begin{cases} 2x + 3y = 9 \\ x - 6y = 42 \end{cases}$

44. $\begin{cases} 2x + 3y = -22 \\ x - 6y = 49 \end{cases}$

45. $\begin{cases} 2x + 3y = 12 \\ x - 6y = -24 \end{cases}$

Problems 46–51 use the inverse found in Problem 22.

46. $\begin{cases} x + 2z = 7 \\ 2x + y = 16 \\ -2y + 9z = -3 \end{cases}$

47. $\begin{cases} x + 2z = 4 \\ 2x + y = 0 \\ -2y + 9z = 19 \end{cases}$

48. $\begin{cases} x + 2z = 7 \\ 2x + y = 0 \\ -2y + 9z = 31 \end{cases}$

49. $\begin{cases} x + 2z = 7 \\ 2x + y = 1 \\ -2y + 9z = 28 \end{cases}$

50. $\begin{cases} x + 2z = 12 \\ 2x + y = 0 \\ -2y + 9z = 10 \end{cases}$

51. $\begin{cases} x + 2z = 5 \\ 2x + y = 8 \\ -2y + 9z = 9 \end{cases}$

Problems 52–54 use the inverse found in Problem 23.

52. $\begin{cases} 6x + y + 20z = 27 \\ x - y = 0 \\ y + 3z = 4 \end{cases}$

53. $\begin{cases} 6x + y + 20z = 14 \\ x - y = 1 \\ y + 3z = 1 \end{cases}$

54. $\begin{cases} 6x + y + 20z = 11 \\ x - y = 5 \\ y + 3z = -3 \end{cases}$

Problems 55–57 use the inverse found in Problem 24.

55. $\begin{cases} 4x + y = 6 \\ 2x - y + 4z = 12 \\ -3x + 2y + z = 4 \end{cases}$

56. $\begin{cases} 4x + y = 7 \\ 2x - y + 4z = -11 \\ -3x + 2y + z = -12 \end{cases}$

57. $\begin{cases} 4x + y = -10 \\ 2x - y + 4z = 20 \\ -3x + 2y + z = 20 \end{cases}$

Problems 58–60 use the inverse found in Problem 26.

58. $\begin{cases} x + 2y = 5 \\ z = 3 \\ x + 3y + w = 9 \\ 4x + 2w = 8 \end{cases}$

59. $\begin{cases} x + 2y = 0 \\ z = -4 \\ x + 3y + w = 4 \\ 4x + 2w = -2 \end{cases}$

60. $\begin{cases} x + 2y = 7 \\ z = -7 \\ x + 3y + w = 16 \\ 4x + 2w = -4 \end{cases}$

C

61. If M and N are nonsingular, then MN is nonsingular. Show that the inverse of MN is $(MN)^{-1} = N^{-1}M^{-1}$.

62. If M is nonsingular, show that if $MN = MP$ then $N = P$.

63. Prove that if M is nonsingular and $MN = M$, then $N = I_3$ for square matrices of order 3.

9.4 Cramer's Rule

A matrix is an array of numbers, and associated with each square matrix is a real number called its **determinant.** There are many ways to motivate its definition, but the most straightforward is motivated by looking at a *formula* solution to a system of equations. This formula method for solving a system of equations is known as **Cramer's Rule.** Consider

$$\begin{cases} a_{11}x_1 + a_{12}x_2 = b_1 \\ a_{21}x_1 + a_{22}x_2 = b_2 \end{cases}$$

Solve this system by using linear combinations.

$$\begin{aligned} a_{22} & \begin{cases} a_{11}x_1 + a_{12}x_2 = b_1 \\ -a_{12} \begin{cases} a_{21}x_1 + a_{22}x_2 = b_2 \end{cases} \end{cases} \end{aligned}$$

$$+ \begin{cases} a_{11}a_{22}x_1 + a_{12}a_{22}x_2 = b_1a_{22} \\ -a_{12}a_{21}x_1 - a_{12}a_{22}x_2 = -b_2a_{12} \end{cases}$$

$$(a_{11}a_{22} - a_{12}a_{21})x_1 = b_1a_{22} - b_2a_{12}$$

If $a_{11}a_{22} - a_{12}a_{21} \neq 0$, then

$$x_1 = \frac{b_1a_{22} - b_2a_{12}}{a_{11}a_{22} - a_{12}a_{21}}$$

Similarly, solving for x_2,

$$x_2 = \frac{b_2a_{11} - b_1a_{21}}{a_{11}a_{22} - a_{12}a_{21}}$$

The difficulty with using this formula is that it is next to impossible to remember. To help with this matter, the following definition is given.

DETERMINANT OF ORDER 2

The **determinant** of the matrix $A = \begin{bmatrix} a & b \\ c & d \end{bmatrix}$ of order 2 is denoted by $|A|$ or

$$\begin{vmatrix} a & b \\ c & d \end{vmatrix}$$

and is defined to be the real number $ad - bc$.

EXAMPLE 1   Evaluate the given determinants.

a. $\begin{vmatrix} 1 & 2 \\ 3 & 4 \end{vmatrix} = (1)(4) - (2)(3)$

$= -2$

b. $\begin{vmatrix} 3 & -2 \\ 4 & 1 \end{vmatrix} = 3 + 8$

$= 11$

c. $\begin{vmatrix} \sin\theta & \cos\theta \\ -\cos\theta & \sin\theta \end{vmatrix} = \sin^2\theta + \cos^2\theta$ **d.** $\begin{vmatrix} a_{11} & a_{12} \\ a_{21} & a_{22} \end{vmatrix} = a_{11}a_{22} - a_{12}a_{21}$
$= 1$

e. $\begin{vmatrix} b_1 & a_{12} \\ b_2 & a_{22} \end{vmatrix} = b_1 a_{22} - b_2 a_{12}$ **f.** $\begin{vmatrix} a_{11} & b_1 \\ a_{21} & b_2 \end{vmatrix} = b_2 a_{11} - b_1 a_{21}$ ∎

The solution to the general system

$$\begin{cases} a_{11}x_1 + a_{12}x_2 = b_1 \\ a_{21}x_1 + a_{22}x_2 = b_2 \end{cases}$$

can now be stated using determinant notation:

$$x_1 = \frac{\begin{vmatrix} b_1 & a_{12} \\ b_2 & a_{22} \end{vmatrix}}{\begin{vmatrix} a_{11} & a_{12} \\ a_{21} & a_{22} \end{vmatrix}} \qquad x_2 = \frac{\begin{vmatrix} a_{11} & b_1 \\ a_{21} & b_2 \end{vmatrix}}{\begin{vmatrix} a_{11} & a_{12} \\ a_{21} & a_{22} \end{vmatrix}}$$

CALCULATOR COMMENT

Many calculators have a determinant function. For example, on the TI-81 enter a square matrix using the [MATRIX] edit function. To find |A| for [A] press [Det] [[A]] and the value will be displayed

Notice that the denominator of both variables is the same. This determinant is called the **determinant of the coefficients** for the system and is denoted by $|D|$. The numerators are also found by looking at $|D|$. For x_1, replace the first column of $|D|$ by the constant numbers b_1 and b_2, respectively. Denote this by $|D_1|$. Similarly, let $|D_2|$ be the determinant formed by replacing the coefficients of x_2 in $|D|$ by the constant numbers b_1 and b_2, respectively. The solution to the system can now be stated with a result called **Cramer's Rule.**

CRAMER'S RULE (TWO UNKNOWNS)

Let $|D|$ be the determinant of the coefficients ($|D| \neq 0$) of a system of linear equations, and let $|D_1|$ and $|D_2|$ be the determinants where the coefficients of x_1 and x_2 are replaced, respectively, by b_1 and b_2. Then

$$x_1 = \frac{|D_1|}{|D|} \quad \text{and} \quad x_2 = \frac{|D_2|}{|D|}$$

EXAMPLE 2 Solve $\begin{cases} 2x - 3y = -8 \\ x + y = 6 \end{cases}$ using Cramer's Rule.

SOLUTION $x = \dfrac{\begin{vmatrix} -8 & -3 \\ 6 & 1 \end{vmatrix}}{\begin{vmatrix} 2 & -3 \\ 1 & 1 \end{vmatrix}} = \dfrac{-8 + 18}{2 + 3} = \dfrac{10}{5} = 2 \qquad y = \dfrac{\begin{vmatrix} 2 & -8 \\ 1 & 6 \end{vmatrix}}{5} = \dfrac{12 + 8}{5} = \dfrac{20}{5} = 4$

The solution is **(2, 4).**   ∎

One advantage of Cramer's Rule is the ease with which its form can be used when working with more complicated solutions.

EXAMPLE 3   Solve $\begin{cases} 14x - 3y = 1 \\ 5x + 7y = -2 \end{cases}$ using Cramer's Rule.

SOLUTION $x = \dfrac{\begin{vmatrix} 1 & -3 \\ -2 & 7 \end{vmatrix}}{\begin{vmatrix} 14 & -3 \\ 5 & 7 \end{vmatrix}} = \dfrac{7-6}{98+15} = \dfrac{1}{113}$ $y = \dfrac{\begin{vmatrix} 14 & 1 \\ 5 & -2 \end{vmatrix}}{113} = \dfrac{-28-5}{113} = \dfrac{-33}{113}$

The solution is $\left(\frac{1}{113}, \frac{-33}{113} \right)$.   ▌

EXAMPLE 4   Solve $\begin{cases} 2x + 3y = 9 \\ y = -\frac{2}{3}x + 1 \end{cases}$ using Cramer's Rule.

SOLUTION   You must first put both equations into the usual form:

$$\begin{cases} 2x + 3y = 9 \\ 2x + 3y = 3 \end{cases}$$

$$x = \dfrac{\begin{vmatrix} 9 & 3 \\ 3 & 3 \end{vmatrix}}{\begin{vmatrix} 2 & 3 \\ 2 & 3 \end{vmatrix}} = \dfrac{27-9}{6-6} = \dfrac{18}{0}$$

Since division by zero is not defined, **Cramer's Rule fails.**   ▌

Notice, however, from Example 4 that if $a_{11}a_{22} - a_{12}a_{21} = 0$ and the numerators are not zero, the system is inconsistent. To show this, note that

$$a_{11}a_{22} = a_{12}a_{21}$$

since $a_{11}a_{22} - a_{12}a_{21} = 0$. Divide both sides by $a_{12}a_{22}$. (If $a_{12}a_{22} = 0$, then $a_{12}a_{21} = 0$ and the problem would be trivial.) Assume, therefore, that $a_{12}a_{22} \neq 0$:

$$\frac{a_{11}}{a_{12}} = \frac{a_{21}}{a_{22}}$$

Now multiply both sides by -1:

$$-\frac{a_{11}}{a_{12}} = -\frac{a_{21}}{a_{22}}$$

But $-a_{11}/a_{12}$ is the slope of the line represented by the first equation of the system, and $-a_{21}/a_{22}$ is the slope of the line represented by the second equation. This means that if $a_{11}a_{22} - a_{12}a_{21} = 0$, then the slopes of the lines are the same. Thus the lines are either parallel or coincident. Next we show that they are not coincident by considering the numerators:

$$b_1 a_{22} - b_2 a_{12} \neq 0 \qquad\qquad a_{11} b_2 - a_{21} b_1 \neq 0$$

$$b_1 a_{22} \neq b_2 a_{12} \qquad\qquad a_{11} b_2 \neq a_{21} b_1$$

$$\frac{b_1}{b_2} \neq \frac{a_{12}}{a_{22}} \qquad\qquad \frac{b_1}{b_2} \neq \frac{a_{11}}{a_{21}}$$

This shows that the lines differ in at least one point, but since they are parallel or coincident they must be parallel. Therefore, if Cramer's Rule gives a solution of the form where the denominator is zero and the numerators are not zero, the system is *inconsistent*. That is, the lines are parallel. If Cramer's Rule yields a form $\frac{0}{0}$, the system is *dependent*.

9.4 Problem set

A *Evaluate the determinants given in Problems 1–16.*

1. $\begin{vmatrix} 3 & 1 \\ 2 & 4 \end{vmatrix}$

2. $\begin{vmatrix} 6 & 3 \\ 2 & 2 \end{vmatrix}$

3. $\begin{vmatrix} -2 & 4 \\ 1 & 3 \end{vmatrix}$

4. $\begin{vmatrix} -5 & -3 \\ 4 & 2 \end{vmatrix}$

5. $\begin{vmatrix} 6 & -3 \\ 2 & 5 \end{vmatrix}$

6. $\begin{vmatrix} -3 & 2 \\ -4 & 5 \end{vmatrix}$

7. $\begin{vmatrix} 6 & 8 \\ -2 & -1 \end{vmatrix}$

8. $\begin{vmatrix} 6 & -3 \\ 4 & -2 \end{vmatrix}$

9. $\begin{vmatrix} 8 & -3 \\ 4 & -5 \end{vmatrix}$

10. $\begin{vmatrix} \sqrt{3} & \sqrt{5} \\ \sqrt{5} & \sqrt{3} \end{vmatrix}$

11. $\begin{vmatrix} \sqrt{2} & 3 \\ 1 & \sqrt{2} \end{vmatrix}$

12. $\begin{vmatrix} \pi & 3 \\ 0 & 1 \end{vmatrix}$

13. $\begin{vmatrix} \sin\theta & -\cos\theta \\ \cos\theta & \sin\theta \end{vmatrix}$

14. $\begin{vmatrix} \tan\theta & 1 \\ -1 & \tan\theta \end{vmatrix}$

15. $\begin{vmatrix} \csc\theta & 1 \\ 1 & \csc\theta \end{vmatrix}$

16. $\begin{vmatrix} \cos x & \sin x \\ \sin y & \cos y \end{vmatrix}$

Solve the systems in Problems 17–32 by using Cramer's Rule.

17. $\begin{cases} x + y = 1 \\ 3x + y = -5 \end{cases}$

18. $\begin{cases} 2s + t = 7 \\ 3s + t = 12 \end{cases}$

19. $\begin{cases} c + d = 2 \\ 2c - d = 1 \end{cases}$

20. $\begin{cases} 2x + 3y = 1 \\ 3x - 2y = 0 \end{cases}$

21. $\begin{cases} 2s_1 + 3s_2 = 3 \\ 5s_1 - 2s_2 = 1 \end{cases}$

22. $\begin{cases} 3q_1 - 4q_2 = 3 \\ 5q_1 + 3q_2 = 4 \end{cases}$

23. $\begin{cases} 4x + 3y = 5 \\ 3x + 2y = 2 \end{cases}$

24. $\begin{cases} 2x - 3y = 6 \\ 5x + 2y = 3 \end{cases}$

25. $\begin{cases} y = \frac{2}{3}x - 7 \\ 2x + 3y = 3 \end{cases}$

26. $\begin{cases} y = \frac{3}{5}x + 2 \\ 3x - 5y = 10 \end{cases}$

27. $\begin{cases} 2x - 3y = 9 \\ y = \frac{2}{3}x - 5 \end{cases}$

28. $\begin{cases} 2p - 3q = 9 \\ q = \frac{2}{3}p - 8 \end{cases}$

29. $\begin{cases} y = \frac{2}{3}x + 5 \\ x + 7y = 1 \end{cases}$

30. $\begin{cases} 3x + 5y + 6 = 0 \\ y = -\frac{3}{5}x + 2 \end{cases}$

31. $\begin{cases} 2x - 3y - 4 = 0 \\ y = \frac{3}{2}x - 8 \end{cases}$

32. $\begin{cases} 4x - 2y - 4 = 0 \\ y = \frac{5}{2}x - 5 \end{cases}$

Solve the systems in Problems 33–54 for (x, y) by using Cramer's Rule.

33. $\begin{cases} 2x - 3y - 5 = 0 \\ 3x - 5y + 2 = 0 \end{cases}$

34. $\begin{cases} y = \frac{2}{3}x + 3 \\ y = -\frac{3}{4}x - 5 \end{cases}$

35. $\begin{cases} y = \frac{1}{2}x - 4 \\ y = \frac{2}{3}x + 5 \end{cases}$

36. $\begin{cases} x + y = a \\ x - y = b \end{cases}$

37. $\begin{cases} 2x + 3y = a \\ x - 5y = b \end{cases}$

38. $\begin{cases} 2x + y = \alpha \\ x - 3y = \beta \end{cases}$

39. $\begin{cases} ax + by = 1 \\ bx + ay = 0 \end{cases}$

40. $\begin{cases} ax + by = 0 \\ bx + ay = 1 \end{cases}$

41. $\begin{cases} cx + dy = s \\ ex + fy = t \end{cases}$

42. $\begin{cases} y = m_1 x + b_1 \\ y = m_2 x + b_2 \end{cases}$

43. $\begin{cases} ax + by = \alpha \\ cx + dy = \beta \end{cases}$

44. $\begin{cases} ax - by = \gamma \\ cx + dy = \delta \end{cases}$

45. $\begin{cases} .12x + .06y = 108 \\ x + y = 1{,}000 \end{cases}$

46. $\begin{cases} .12x + .06y = 210 \\ x + y = 2{,}000 \end{cases}$

47. $\begin{cases} .12x + .06y = 228 \\ x + y = 2{,}000 \end{cases}$

48. $\begin{cases} .12x + .06y = 1{,}140 \\ x + y = 10{,}000 \end{cases}$

49. $\begin{cases} x + y = 147 \\ .25x + .1y = 24.15 \end{cases}$

50. $\begin{cases} x + y = 10 \\ .4x + .9y = .5 \end{cases}$

51. $\begin{cases} 2x + 3y = \cos^2 45° \\ x - y = -\sin^2 45° \end{cases}$

52. $\begin{cases} 3x - y = \sec^2 30° \\ x - y = \tan^2 30° \end{cases}$

53. $\begin{cases} x + 3y = \csc^2 60° \\ x - 2y = \cot^2 60° \end{cases}$

54. $\begin{cases} x + y = \sec^2 45° \\ 2x - y = \tan^2 45° \end{cases}$

For Problems 55–59 let $|A| = \begin{vmatrix} a & b \\ c & d \end{vmatrix}$.

55. Show that $\begin{vmatrix} c & d \\ a & b \end{vmatrix} = -|A|$.

56. Show that $\begin{vmatrix} a & b \\ a + c & b + d \end{vmatrix} = |A|$.

57. Show that $\begin{vmatrix} a & a + b \\ c & c + d \end{vmatrix} = |A|$.

58. Show that for any constant k, $\begin{vmatrix} ak & bk \\ c & d \end{vmatrix} = k|A|$

59. Show that $\begin{vmatrix} a & b \\ c + ka & d + kb \end{vmatrix} = \begin{vmatrix} a + kc & b + kd \\ c & d \end{vmatrix} = |A|$.

C *Solve the systems in Problems 60–63 by using Cramer's Rule.*

60. $\begin{cases} 3x^2 + 4y^2 = 16 \\ x^2 + y^2 = 5 \end{cases}$

61. $\begin{cases} 13x^2 + 12y^2 = 169 \\ x^2 + y^2 = 12 \end{cases}$

62. $\begin{cases} \dfrac{3}{x} + \dfrac{4}{y} = 3 \\ \dfrac{9}{x} - \dfrac{2}{y} = 2 \end{cases}$

63. $\begin{cases} \dfrac{2}{x - 1} - \dfrac{5}{y + 2} = 12 \\ \dfrac{4}{x - 1} - \dfrac{2}{y + 2} = -12 \end{cases}$

9.5 Properties of Determinants

The strength of Cramer's Rule is that it can be applied to linear systems of n equations with n unknowns. However, to apply Cramer's Rule to these systems, some additional notation and definitions are required. Since the number of rows and columns in a determinant is the same, we call the size of either the **order** or **dimension** of the determinant. The determinants of the previous section were of order 2. Consider a determinant of order 3:

$$\begin{vmatrix} a_{11} & a_{12} & a_{13} \\ a_{21} & a_{22} & a_{23} \\ a_{31} & a_{32} & a_{33} \end{vmatrix} \quad \substack{\longleftarrow \\ \longleftarrow \\ \longleftarrow} \; 3 \text{ rows}$$

3 columns

The entry a_{ij} refers to the entry in the ith row and jth column. That is, a_{23} refers to the entry in the second row and third column. The **minor** of an element in a determinant is the (smaller) determinant that remains after deleting all entries in its row and column.

EXAMPLE 1 Consider

$$\begin{vmatrix} 2 & -4 & 0 \\ 1 & -3 & -1 \\ 6 & 5 & 3 \end{vmatrix}$$

a. The a_{13} entry is 0; the a_{31} entry is 6; and the a_{23} entry is -1.

b. The minor of -3 is
$$\begin{vmatrix} 2 & -4 & 0 \\ 1 & -3 & -1 \\ 6 & 5 & 3 \end{vmatrix} \quad \text{or} \quad \begin{vmatrix} 2 & 0 \\ 6 & 3 \end{vmatrix} = 6$$

Delete shaded portions.

c. The minor of 0 is
$$\begin{vmatrix} 1 & -3 \\ 6 & 5 \end{vmatrix} = 5 + 18 = 23$$

d. The minor of 2 is
$$\begin{vmatrix} -3 & -1 \\ 5 & 3 \end{vmatrix} = -9 + 5 = -4$$

The **cofactor** of an entry a_{ij} is $(-1)^{i+j}$ times the minor of the a_{ij} entry. This says that if the sum of the row and column number is even, the cofactor is the same as the minor. If the sum of the row and column of an entry is odd, the cofactor of that entry is the opposite of its minor.

EXAMPLE 2 Consider
$$\begin{vmatrix} 2 & -4 & 0 \\ 1 & -3 & -1 \\ 6 & 5 & 3 \end{vmatrix}$$

a. The cofactor of 1 is
$$-\begin{vmatrix} -4 & 0 \\ 5 & 3 \end{vmatrix} = -(-12 - 0) = 12$$

minor

Entry is in row 2, column 1, and $2 + 1$ is odd so use the negative sign. Notice that this sign is independent of whether the entry is positive or negative or whether the minor is positive or negative.

b. The cofactor of -3 is
$$+\begin{vmatrix} 2 & 0 \\ 6 & 3 \end{vmatrix} = 6$$

sign   minor

c. The cofactor of -1 is
$$-\begin{vmatrix} 2 & -4 \\ 6 & 5 \end{vmatrix} = -(10 + 24) = -34$$

EXAMPLE 3 Consider
$$\begin{vmatrix} 1 & 32 & -3 & 4 & 5 \\ -6 & 7 & 8 & 9 & 10 \\ 11 & -10 & -9 & -8 & -7 \\ -6 & -5 & -4 & 3 & ② \\ 1 & 0 & 3 & 5 & 12 \end{vmatrix}$$

The a_{45} entry is 2 and its cofactor is

$$-\begin{vmatrix} 1 & 32 & -3 & 4 \\ -6 & 7 & 8 & 9 \\ 11 & -10 & -9 & -8 \\ 1 & 0 & 3 & 5 \end{vmatrix}$$

minor

Sign: Row number is 4; column number is 5; the sum is 9, which is an odd number, so use the negative sign.

A **determinant of order** n is a real number whose value is the sum of the products obtained by multiplying each element of a row (or column) by its cofactor.

EXAMPLE 4 Expand $\begin{vmatrix} 1 & -2 & -5 \\ 2 & -1 & 0 \\ -4 & 5 & 6 \end{vmatrix}$ about the second row.

SOLUTION $\quad -(2)\begin{vmatrix} -2 & -5 \\ 5 & 6 \end{vmatrix} + (-1)\begin{vmatrix} 1 & -5 \\ -4 & 6 \end{vmatrix} - (0)\begin{vmatrix} 1 & -2 \\ -4 & 5 \end{vmatrix}$

signs of cofactors

$$= -2(-12 + 25) - (6 - 20) - 0 = -12$$

EXAMPLE 5 Expand the determinant of Example 4 about the first column.

SOLUTION $\quad +(1)\begin{vmatrix} -1 & 0 \\ 5 & 6 \end{vmatrix} - (2)\begin{vmatrix} -2 & -5 \\ 5 & 6 \end{vmatrix} + (-4)\begin{vmatrix} -2 & -5 \\ -1 & 0 \end{vmatrix}$

$$= (-6 - 0) - 2(-12 + 25) - 4(0 - 5) = -12$$

Notice from Examples 4 and 5 that the same value is obtained regardless of the row or column chosen. You are asked to prove this for order 3 in the problem set.

The method of evaluating determinants by rows or columns as shown in Examples 4 and 5 is not very efficient for higher-order determinants. The following theorem considerably simplifies the work in evaluating determinants.

If $|A'|$ is a determinant obtained from a determinant $|A|$ by multiplying any row by a constant k and adding the result to any other row (entry by entry), then $|A'| = |A|$. The same result holds for columns.

PROOF We will prove this theorem for the determinant of order 2; in the problems you are asked to prove it for order 3. For order 2, let

$$|A| = \begin{vmatrix} a & b \\ c & d \end{vmatrix} = ad - bc$$

Multiply row 1 by k and add it to row 2:

$$|A'| = \begin{vmatrix} a & b \\ c + ak & d + bk \end{vmatrix} = a(d + bk) - b(c + ak)$$

$$= ad + abk - bc - abk$$

$$= ad - bc$$

$$= |A|$$

Thus $|A'| = |A|$. Next multiply row 2 by k and add it to row 1 to obtain the same result.   ⬜⬜⬜

EXAMPLE 6 **Expand**

$$\begin{vmatrix} 1 & -2 & -5 \\ 2 & -1 & 0 \\ -4 & 5 & 6 \end{vmatrix}$$

by using the determinant reduction theorem.

SOLUTION *You want to obtain some row or column with two zeros to simplify the arithmetic.* Add twice the second column to the first column:

$$\begin{vmatrix} 1 & -2 & -5 \\ 2 & -1 & 0 \\ -4 & 5 & 6 \end{vmatrix} = \begin{vmatrix} -3 & -2 & -5 \\ 0 & -1 & 0 \\ 6 & 5 & 6 \end{vmatrix}$$

×2   These columns are unchanged.

Now expand about the second row (do not even bother to write down the products that are zero):

positive understood

$$= (-1) \begin{vmatrix} -3 & -5 \\ 6 & 6 \end{vmatrix}$$

$$= (-1)(-18 + 30)$$

$$= -12$$ ▮

EXAMPLE 7 **Expand**

$$\begin{vmatrix} 2 & -3 & 2 & 5 & 0 \\ 4 & 2 & -1 & 4 & 0 \\ 5 & 1 & 0 & -2 & 0 \\ 6 & 2 & 3 & 6 & 0 \\ 3 & 4 & 6 & 1 & -2 \end{vmatrix}$$

by using the determinant reduction theorem.

SOLUTION First notice that all the entries in the fifth column except one are zeros, so begin by expanding about the fifth column:

$$+ (-2) \begin{vmatrix} 2 & -3 & 2 & 5 \\ 4 & 2 & -1 & 4 \\ 5 & 1 & 0 & -2 \\ 6 & 2 & 3 & 6 \end{vmatrix}$$

Row 5, column 5; 5 + 5 is even, so this sign is positive.

Next obtain a row or column with all entries zero except one; row 3 or column 3 seem to be likely candidates. We will work toward obtaining zeros in column 3:

$$
(-2)\begin{vmatrix} 2 & -3 & 2 & 5 \\ 4 & 2 & -1 & 4 \\ 5 & 1 & 0 & -2 \\ 6 & 2 & 3 & 6 \end{vmatrix} \!\!\!\overset{\times 2}{} = (-2)\begin{vmatrix} 10 & 1 & 0 & 13 \\ 4 & 2 & -1 & 4 \\ 5 & 1 & 0 & -2 \\ 6 & 2 & 3 & 6 \end{vmatrix} \!\!\!\overset{\times 3}{}
$$

$$
= (-2)\begin{vmatrix} 10 & 1 & 0 & 13 \\ 4 & 2 & -1 & 4 \\ 5 & 1 & 0 & -2 \\ 18 & 8 & 0 & 18 \end{vmatrix} = (-2)\left[-(-1)\begin{vmatrix} 10 & 1 & 13 \\ 5 & 1 & -2 \\ 18 & 8 & 18 \end{vmatrix} \right]
$$

Finished when all entries but one are zero; now reduce the order of the determinant.

Row 2, column 3; $2 + 3$ is odd.

$$
\times (-5)
$$

$$
= (-2)\begin{vmatrix} 5 & 1 & 13 \\ 0 & 1 & -2 \\ -22 & 8 & 18 \end{vmatrix} = (-2)\begin{vmatrix} 5 & 1 & 15 \\ 0 & 1 & 0 \\ -22 & 8 & 34 \end{vmatrix}
$$

$$
\times 2
$$

$$
= (-2)\left[+(1)\begin{vmatrix} 5 & 15 \\ -22 & 34 \end{vmatrix} \right] = -2(170 + 330) = -2(500) = \mathbf{-1{,}000}
$$

Row 2, column 2; $2 + 2$ is even.

We can now state Cramer's Rule for n linear equations with n unknowns. Consider the general $n \times n$ system

$$
\begin{cases}
a_{11}x_1 + a_{12}x_2 + a_{13}x_3 + \cdots + a_{1n}x_n = b_1 \\
a_{21}x_1 + a_{22}x_2 + a_{23}x_3 + \cdots + a_{2n}x_n = b_2 \\
a_{31}x_1 + a_{32}x_2 + a_{33}x_3 + \cdots + a_{3n}x_n = b_3 \\
\quad\quad\quad\quad\quad\quad\vdots \\
a_{n1}x_1 + a_{n2}x_2 + a_{n3}x_3 + \cdots + a_{nn}x_n = b_n
\end{cases}
$$

The unknowns are $x_1, x_2, x_3, \ldots, x_n$; a_{ij} are the coefficients and $b_1, b_2, b_3, \ldots, b_n$ are the constants. When written in this form, the system is said to be in **standard form.**

CRAMER'S RULE

Let $|D|$ be the determinants of the coefficients ($|D| \neq 0$) of a system of n linear equations with n unknowns. Let $|D_i|$ be the determinant where the coefficients of x_i have been replaced by the constants $b_1, b_2, b_3, \ldots, b_n$, respectively. Then

$$
x_i = \frac{|D_i|}{|D|}
$$

As before, if $|D_i|$ and $|D|$ are zero, the system is dependent; if at least one of the $|D_i|$ is not zero when $|D|$ is zero, the system is inconsistent.

EXAMPLE 8 Solve

$$\begin{cases} x - 2y - 5z = -12 \\ 2x - y = 7 \\ 5y + 6z = 4x + 1 \end{cases}$$

SOLUTION Rewrite the system in standard form:

$$\begin{cases} x - 2y - 5z = -12 \\ 2x - y \qquad = 7 \\ -4x + 5y + 6z = 1 \end{cases}$$

Be sure to consider "missing terms" as terms with zero coefficient.

$$|D| = \begin{vmatrix} 1 & -2 & -5 \\ 2 & -1 & 0 \\ -4 & 5 & 6 \end{vmatrix} = -12 \qquad \text{From Example 6}$$

$$|D_1| = \begin{vmatrix} -12 & -2 & -5 \\ 7 & -1 & 0 \\ 1 & 5 & 6 \end{vmatrix} = \begin{vmatrix} -26 & -2 & -5 \\ 0 & -1 & 0 \\ 36 & 5 & 6 \end{vmatrix} = (-1)\begin{vmatrix} -26 & -5 \\ 36 & 6 \end{vmatrix}$$

$$\times 7$$

$$= -(-156 + 180)$$

$$= -24$$

$$|D_2| = \begin{vmatrix} 1 & -12 & -5 \\ 2 & 7 & 0 \\ -4 & 1 & 6 \end{vmatrix} \Big) \times 1 = \begin{vmatrix} 1 & -12 & -5 \\ 2 & 7 & 0 \\ -3 & -11 & 1 \end{vmatrix} \Big) \times 5$$

$$= \begin{vmatrix} -14 & -67 & 0 \\ 2 & 7 & 0 \\ -3 & -11 & 1 \end{vmatrix} = \begin{vmatrix} -14 & -67 \\ 2 & 7 \end{vmatrix} = -98 + 134 = 36$$

$$|D_3| = \begin{vmatrix} 1 & -2 & -12 \\ 2 & -1 & 7 \\ -4 & 5 & 1 \end{vmatrix} = \begin{vmatrix} -3 & -2 & -26 \\ 0 & -1 & 0 \\ 6 & 5 & 36 \end{vmatrix} = (-1)\begin{vmatrix} -3 & -26 \\ 6 & 36 \end{vmatrix}$$

$$\times 2 \qquad \times 7$$

$$= -(-108 + 156) = -48$$

Thus

$$x = \frac{|D_1|}{|D|} = \frac{-24}{-12} = 2 \qquad y = \frac{|D_2|}{|D|} = \frac{36}{-12} = -3 \qquad z = \frac{|D_3|}{|D|} = \frac{-48}{-12} = 4$$

The solution is $(x, y, z) = (2, -3, 4)$. ∎

Problem Set

A *Evaluate the determinants in Problems 1–16. If you want to check, evaluate the determinant by using a different row or column.*

1.
$$\begin{vmatrix} 3 & 0 & 0 \\ 2 & 1 & 4 \\ 3 & 6 & -1 \end{vmatrix}$$

2.
$$\begin{vmatrix} 1 & -2 & 3 \\ 0 & 4 & 0 \\ 3 & -1 & -3 \end{vmatrix}$$

3.
$$\begin{vmatrix} 4 & -2 & 6 \\ 3 & 1 & 4 \\ 0 & 0 & -2 \end{vmatrix}$$

4.
$$\begin{vmatrix} 0 & 1 & 1 \\ -3 & 2 & 4 \\ 0 & -2 & -3 \end{vmatrix}$$

5.
$$\begin{vmatrix} 1 & -2 & 3 \\ -2 & 0 & 4 \\ 3 & 0 & 5 \end{vmatrix}$$

6.
$$\begin{vmatrix} 3 & 1 & 0 \\ -2 & 4 & 0 \\ -3 & 5 & -4 \end{vmatrix}$$

7.
$$\begin{vmatrix} 1 & 1 & 1 \\ 1 & 3 & 2 \\ 1 & -2 & 1 \end{vmatrix}$$

8.
$$\begin{vmatrix} 4 & 1 & 1 \\ 4 & 3 & 2 \\ 7 & -2 & 1 \end{vmatrix}$$

9.
$$\begin{vmatrix} 1 & 4 & 1 \\ 1 & 4 & 2 \\ 1 & 7 & 1 \end{vmatrix}$$

10.
$$\begin{vmatrix} -3 & -1 & 1 \\ 14 & 2 & -3 \\ 12 & 3 & -1 \end{vmatrix}$$

11.
$$\begin{vmatrix} 2 & -3 & 1 \\ 1 & 14 & -3 \\ 3 & 12 & -1 \end{vmatrix}$$

12.
$$\begin{vmatrix} 2 & -1 & -3 \\ 1 & 3 & 14 \\ 3 & 3 & 12 \end{vmatrix}$$

13.
$$\begin{vmatrix} 2 & 4 & 3 \\ -2 & 3 & -2 \\ 4 & 3 & 5 \end{vmatrix}$$

14.
$$\begin{vmatrix} 6 & 3 & -3 \\ 2 & 0 & 5 \\ 3 & 5 & -2 \end{vmatrix}$$

15.
$$\begin{vmatrix} 4 & 8 & 5 \\ 3 & 2 & 3 \\ 5 & 5 & 4 \end{vmatrix}$$

16.
$$\begin{vmatrix} 3 & 7 & 1 \\ 2 & 4 & 3 \\ 5 & 6 & 2 \end{vmatrix}$$

B

Evaluate the determinants in Problems 17–32.

17.
$$\begin{vmatrix} 5 & 2 & 6 & -11 \\ -3 & 0 & 3 & 1 \\ 4 & 0 & 0 & 6 \\ 5 & 0 & 0 & -1 \end{vmatrix}$$

18.
$$\begin{vmatrix} 4 & 3 & 2 & 1 \\ -5 & 0 & 0 & 0 \\ 11 & -4 & 0 & 0 \\ 9 & 6 & 3 & -5 \end{vmatrix}$$

19.
$$\begin{vmatrix} 7 & 2 & -5 & 3 \\ 1 & 0 & -3 & 2 \\ 4 & 0 & 1 & -5 \\ 0 & 0 & 3 & 0 \end{vmatrix}$$

20.
$$\begin{vmatrix} 2 & 1 & -1 & 3 \\ 4 & 0 & 0 & 0 \\ 2 & 1 & -2 & 3 \\ 1 & 4 & 3 & 5 \end{vmatrix}$$

21.
$$\begin{vmatrix} 6 & 3 & 0 & -2 \\ 4 & 3 & 4 & -1 \\ 1 & 2 & 0 & 5 \\ 3 & -2 & 0 & 5 \end{vmatrix}$$

22.
$$\begin{vmatrix} 1 & 6 & 2 & -1 \\ 0 & -5 & 0 & 0 \\ 3 & 4 & 5 & -3 \\ 2 & 1 & 4 & 0 \end{vmatrix}$$

23.
$$\begin{vmatrix} 2 & 1 & 2 & 4 \\ 3 & -1 & 2 & 5 \\ -3 & 2 & 3 & -4 \\ -3 & 2 & 8 & -4 \end{vmatrix}$$

24.
$$\begin{vmatrix} 3 & 1 & -1 & 2 \\ 4 & 0 & 3 & 0 \\ 2 & 4 & 3 & -3 \\ 6 & 1 & 4 & 0 \end{vmatrix}$$

25.
$$\begin{vmatrix} 2 & 1 & 3 & 1 \\ 6 & 3 & -3 & 2 \\ 2 & 0 & 5 & 1 \\ 3 & 5 & -2 & -1 \end{vmatrix}$$

26.
$$\begin{vmatrix} 1 & 2 & 1 & 1 \\ 2 & 3 & 7 & 1 \\ 3 & 2 & 4 & 3 \\ 1 & 5 & 6 & 2 \end{vmatrix}$$

27.
$$\begin{vmatrix} 0 & -3 & 5 & 6 & -1 \\ 0 & 0 & 1 & 1 & 3 \\ 5 & -3 & 8 & -5 & 1 \\ 0 & 0 & 2 & 0 & 0 \\ 0 & 0 & 4 & -1 & 2 \end{vmatrix}$$

28.
$$\begin{vmatrix} 3 & 1 & -3 & -2 & 5 \\ -1 & 5 & 1 & 0 & 1 \\ 5 & 0 & 3 & 0 & 0 \\ 4 & 0 & 0 & 0 & 0 \\ 2 & 2 & 4 & 0 & -1 \end{vmatrix}$$

29.
$$\begin{vmatrix} -3 & 4 & 5 & 8 & -9 \\ 0 & 0 & 0 & 0 & 6 \\ 3 & 0 & 0 & -1 & 4 \\ 0 & 0 & 4 & 0 & 9 \\ -3 & 0 & 0 & 1 & 8 \end{vmatrix}$$

30.
$$\begin{vmatrix} 3 & 4 & 1 & -2 & 0 \\ 1 & -2 & 3 & 0 & 1 \\ 0 & 4 & -2 & 1 & 0 \\ 5 & 0 & -3 & 0 & 0 \\ 2 & 1 & -2 & 2 & 0 \end{vmatrix}$$

31.
$$\begin{vmatrix} 2 & -1 & 0 & 1 & -1 \\ 3 & 0 & 0 & 2 & 0 \\ 2 & 1 & 3 & 1 & 3 \\ 0 & 0 & 0 & -2 & 0 \\ 1 & 3 & -1 & 2 & 1 \end{vmatrix}$$

32.
$$\begin{vmatrix} 2 & 1 & 3 & 0 & 0 \\ 6 & 1 & 5 & 2 & -1 \\ 1 & 4 & -5 & 9 & 3 \\ 3 & 2 & 5 & 7 & 2 \\ 2 & -3 & -2 & 4 & 6 \end{vmatrix}$$

Use Cramer's Rule to solve the systems in Problems 33–42.

33. $\begin{cases} x + y + z = 6 \\ 2x - y + z = 3 \\ x - 2y - 3z = 6 \end{cases}$

34. $\begin{cases} 2x - y + z = 3 \\ x - 3y + 2z = 7 \\ x - y - z = -1 \end{cases}$

35. $\begin{cases} x + y + z = 4 \\ x + 3y + 2z = 4 \\ x - 2y + z = 7 \end{cases}$

36. $\begin{cases} x + y + z = 3 \\ 2x + z = -1 \\ y = 5 \end{cases}$

37. $\begin{cases} x + y + z = 3 \\ 3y - z = -11 \\ x = 4 \end{cases}$

38. $\begin{cases} 2x + y + z = -5 \\ 3x - y = -9 \\ z = -4 \end{cases}$

39. $\begin{cases} 2x + 2y + 3z = 1 \\ 2x - z = -11 \\ 3y + 2z = 6 \end{cases}$

40. $\begin{cases} x + 2y + z = 1 \\ x - 3y - 2z = 2 \\ 3x - 2y + z = 3 \end{cases}$

41. $\begin{cases} 2x - y + z = 4 \\ 3x - 2y + 2z = 3 \\ x - y + 3z = 2 \end{cases}$

42. $\begin{cases} 5x - 3y + 2z = 10 \\ 4x + 2y - 3z = 4 \\ 3x + y + 4z = -8 \end{cases}$

Use the determinant equation given in Problem 55 to find the equations of the lines passing through the points given in Problems 43–46.

43. $(-3, -2), (1, -3)$

44. $(4, -5), (7, -8)$

45. $(1, 5), (-2, -3)$

46. $(8, 3), (-5, -2)$

Find the absolute value of the determinant expression given in Problem 56 to find the areas of the triangles with vertices given in Problems 47–50.

47. $(1, 1), (-2, -3), (11, -3)$

48. $(-3, 12), (5, 6), (-3, -9)$

49. $(-8, 0), (12, 10), (4, -5)$

50. $(6, 2), (-1, -3), (2, 5)$

Solve the systems in Problems 51–54.

51. $\begin{cases} w + x - y + z = 7 \\ 2w + y - 3z = 1 \\ 2x - z + w = 4 \\ y - w + z = -4 \end{cases}$

52. $\begin{cases} 3t - u + x = 20 \\ x - 2t - u = 0 \\ 3u - 2x + 5t = 1 \end{cases}$

53. $\begin{cases} s + 3t - 2u = 4 \\ u + 2x - t = -5 \\ x - u - t = 0 \\ s - 2x = -1 \end{cases}$

54. $\begin{cases} 2x - y + w - v = -4 \\ 3x + 2w = 0 \\ x + y + 3z + w + 3v = 5 \\ -2w = -6 \\ x + 3y - z + 2w + v = 10 \end{cases}$

C

55. Prove that

$$\begin{vmatrix} x & y & 1 \\ x_1 & y_1 & 1 \\ x_2 & y_2 & 1 \end{vmatrix} = 0$$

is the equation of a line passing through (x_1, y_1) and (x_2, y_2).

56. Prove that the area of a triangle with vertices at (x_1, y_1), (x_2, y_2), and (x_3, y_3) is the absolute value of

$$\frac{1}{2}\begin{vmatrix} x_1 & y_1 & 1 \\ x_2 & y_2 & 1 \\ x_3 & y_3 & 1 \end{vmatrix}$$

57. If r_1, r_2, r_3, and r_4 are the fourth roots of 1, show that

$$\begin{vmatrix} r_1 & r_2 & r_3 & r_4 \\ r_2 & r_3 & r_4 & r_1 \\ r_3 & r_4 & r_1 & r_2 \\ r_4 & r_1 & r_2 & r_3 \end{vmatrix} = 0$$

For Problems 58–64 let

$$|A| = \begin{vmatrix} a_{11} & a_{12} & a_{13} \\ a_{21} & a_{22} & a_{23} \\ a_{31} & a_{32} & a_{33} \end{vmatrix}$$

58. Show that you obtain the same result if you expand along the first or third row.

59. Show that you obtain the same result if you expand along the second row or second column.

60. Prove the determinant reduction theorem for $|A|$. (Without loss of generality, prove it by multiplying the first row by k and adding it to the second row.)

61. Prove that $|A| = 0$ if two rows (say, the first and second rows) are identical.

62. Prove that $|A| = 0$ if two columns (say, the first and second columns) are identical.

63. Prove that if two rows of $|A|$ are interchanged (say, the first and third), the resulting determinant is $-|A|$.

64. Prove that

$$k|A| = \begin{vmatrix} a_{11} & a_{12} & a_{13} \\ ka_{21} & ka_{22} & ka_{23} \\ a_{31} & a_{32} & a_{33} \end{vmatrix}$$

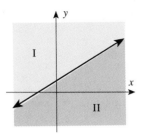

FIGURE 9.5   Half-planes

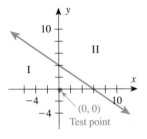

FIGURE 9.6

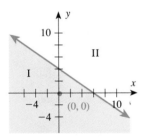

FIGURE 9.7
Graph of $2x + 3y \leq 12$

9.6 Systems of Inequalities

In previous sections we have discussed the simultaneous solution of a system of equations. In this section we will discuss the simultaneous solution of a **system of linear inequalities.** Let us begin by considering some preliminary ideas. The **solution** for an inequality in x and y is defined as an ordered pair that when substituted for (x, y) makes the inequality true. To **solve an inequality** means to find the set of all solutions, which is usually an infinite set. We therefore usually represent the solution graphically so that the **graph of the inequality** is the graph of all solutions of that inequality.

Graphing a linear inequality is similar to graphing a linear equation. A line divides the plane into three regions as shown in Figure 9.5. Regions I and II in Figure 9.5 are called **open half-planes.** Thus the three regions determined by the line are the open half-planes labeled I and II and the set of points on the line. The line is called the **boundary** of each open half-plane. An open half-plane, along with its boundary, is called a **closed half-plane.**

Every linear inequality with one or two variables determines an associated linear equation that is the boundary for the solution set of the inequality. For example, $2x + 3y \leq 12$ has a boundary line $2x + 3y = 12$. To graph the solution set of an inequality, begin by graphing the boundary line as shown in Figure 9.6. Next decide whether the solution is half-plane I or half-plane II. To do this, **choose *any* point not on the boundary.** For example, choose the point $(0, 0)$ in Figure 9.6—this choice, if not on the boundary, is usually the best because of the ease of the arithmetic involved. Notice from Figure 9.6 that the point $(0, 0)$ is in half-plane I. If $(0, 0)$ makes the inequality true, then the solution is the half-plane containing $(0, 0)$; if $(0, 0)$ makes the inequality false, then the solution set is the half-plane not containing $(0, 0)$. Checking by substituting $(0, 0)$ into $2x + 3y \leq 12$, you have

$$2(0) + 3(0) \leq 12 \qquad \text{True}$$

Therefore the solution set is the area shown as half-plane I. This is the shaded portion of Figure 9.7. Notice that the solution set is the closed half-plane I, since it includes the boundary. This is shown on the graph by using a solid line for the boundary. If the boundary is not included (when the inequality symbols are $<$ or $>$), a dashed line is used to indicate the boundary.

EXAMPLE 1   Graph $150x - 75y > 1{,}875$.

SOLUTION   Graph the boundary line $150x - 75y = 1{,}875$:

$$y = 2x - 25$$

Check some point, say $(0, 0)$, not on the boundary and test:

$$150(0) - 75(0) > 1{,}875 \qquad \text{False}$$

The solution is the half-plane not including $(0, 0)$. It is the shaded portion of Figure 9.8.

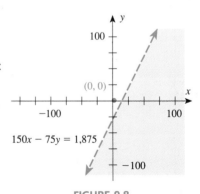

FIGURE 9.8
Graph of $150x - 75y > 1{,}875$

Many curves divide the plane into two regions with the curve serving as a boundary. The test point method illustrated for lines works efficiently in many different settings, as illustrated by Examples 2 and 3.

EXAMPLE 2 Graph $y \geq |x + 3| + 2$.

SOLUTION Begin by graphing the boundary:

$$y = |x + 3| + 2$$

We recognize this as the *form* $y = |x|$ translated to coordinate axes with origin at the point $(-3, 2)$. This is drawn as a solid curve in Figure 9.9. Now this curve divides the plane into two regions, so plot a test point $(0, 0)$ and check the truth or falsity in the given inequality:

$$y \geq |x + 3| + 2$$

$$0 \geq |0 + 3| + 2$$

$$0 \geq 5 \qquad \text{False}$$

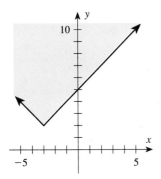

FIGURE 9.9
Graph of $y \geq |x + 3| + 2$

Since it is false, shade the region *not* containing the test point $(0, 0)$, as shown in Figure 9.9. ∎

EXAMPLE 3 Sketch the graph $y > x^2 + 2x + 3$.

SOLUTION Consider the associated equation

$$y = x^2 + 2x + 3$$

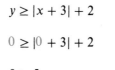

or $y - 2 = (x + 1)^2$, which is a parabola that has vertex at $(-1, 2)$ and opens upward. Sketch this boundary equation as in Figure 9.10. In this case, the boundary is dashed because it is not included. Plot a test point $(0, 0)$ and check the inequality:

$$y > x^2 + 2x + 3$$

$$0 > 0^2 + 2 \cdot 0 + 3 \qquad \text{False}$$

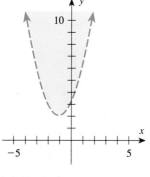

FIGURE 9.10
Graph of $y > x^2 + 2x + 3$

Since it is false, $(0, 0)$ is not in the solution set. Shade the appropriate region on the graph. ∎

The solution of a **system of inequalities** refers to the intersection of the solutions of the individual inequalities in the system. The procedure for graphing the solution of a system of inequalities is to graph the solution of the individual inequalities and then shade in the intersection.

EXAMPLE 4 Graph the solution of the system

$$\begin{cases} y > -20x + 100 \\ x \geq 0 \\ y \geq 0 \\ x \leq 8 \end{cases}$$

SOLUTION STEP 1: Graph $y > -20x + 100$

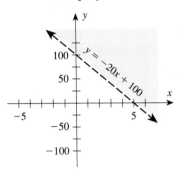

STEP 2: Graph $x \geq 0$.

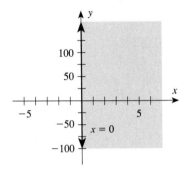

STEP 3: Combine Steps 1 and 2. The intersection of the two graphs is the part that is more darkly shaded.

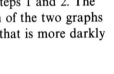

STEP 4: Graph $y \geq 0$.

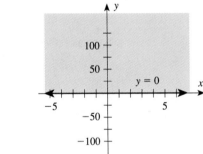

STEP 5: Look at the intersection of Steps 3 and 4.

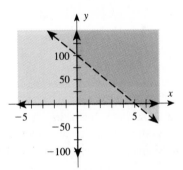

STEP 6: Graph $x \leq 8$.

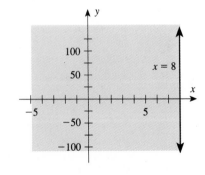

STEP 7: Look at the intersection of Steps 5 and 6.

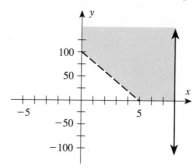

The way we have illustrated the steps of this solution was for explanation only; your work will not look at all like this. In practice, show the individual steps with little arrows on the boundaries and then shade in the intersection only after you have drawn in all of the boundaries. This device replaces the use of a lot of shading, which can be confusing if there are many inequalities in the system. The work for Example 4 is shown in Figure 9.11.

FIGURE 9.11 Graph of the solution of a system of inequalities

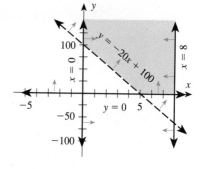

EXAMPLE 5 Graph the solution of the system

$$\begin{cases} 2x + y \leq 3 \\ x - y > 5 \\ x \geq 0 \\ y > -10 \end{cases}$$

SOLUTION The graphs of the individual inequalities and their intersections are shown in Figure 9.12. Notice the use of arrows to show the solutions of the individual inequalities.

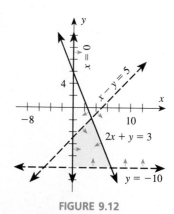

FIGURE 9.12

9.6 Problem Set

A *Graph the solution for each of the linear inequalities given in Problems 1–30.*

1. $y \geq 2x - 3$
2. $y \geq 3x + 2$
3. $y \geq 4x + 5$
4. $y \leq \frac{1}{2}x + 2$
5. $y \leq \frac{2}{3}x - 4$
6. $y \leq \frac{4}{5}x - 3$
7. $x - y > 3$
8. $2x + y < -3$
9. $3x + y < 4$
10. $x + 2y > 4$
11. $x - 3y > 12$
12. $3x - 4y > 8$
13. $y \leq 6$
14. $x \geq 2$
15. $y < -2$
16. $y > 0$
17. $x < 0$
18. $x < 8$
19. $y \geq |x|$
20. $y < |x|$
21. $y < |x + 1|$
22. $y - 3 > |x + 1|$
23. $y + 1 \leq |x - 3|$
24. $y < |x - 1| + 2$
25. $y \leq x^2$
26. $y \geq (x - 3)^2$
27. $y - 1 \geq (x + 1)^2$
28. $y < x^2 - 6x + 7$
29. $x^2 - 4x + y + 5 < 0$
30. $x^2 + 6x + y + 7 > 0$

B *Graph the solution of each system given in Problems 31–56.*

31. $\begin{cases} x \geq 0 \\ y \leq 0 \end{cases}$
32. $\begin{cases} x \geq 0 \\ y \geq 0 \end{cases}$
33. $\begin{cases} x \leq 0 \\ y \leq 0 \end{cases}$

34. $\begin{cases} x \geq 0 \\ y \geq 0 \\ x < 8 \\ y < 5 \end{cases}$
35. $\begin{cases} x \geq 0 \\ y \geq 0 \\ x < 5 \\ y < 6 \end{cases}$
36. $\begin{cases} x \geq 0 \\ y \geq 0 \\ x < 500 \\ y < 1,000 \end{cases}$

37. $\begin{cases} 2x + y > 3 \\ 3x - y < 2 \end{cases}$
38. $\begin{cases} y \leq 3x - 4 \\ y \geq -2x + 5 \end{cases}$

39. $\begin{cases} 3x - 2y \geq 6 \\ 2x + 3y \leq 6 \end{cases}$
40. $\begin{cases} y - 5 \leq 0 \\ y \geq 0 \end{cases}$

41. $\begin{cases} x - 10 \leq 0 \\ x \geq 0 \end{cases}$
42. $\begin{cases} y - 25 \leq 0 \\ y \geq 0 \end{cases}$

43. $\begin{cases} -10 \leq x \\ x \leq 6 \\ 3 < y \\ y < 8 \end{cases}$
44. $\begin{cases} -5 < x \\ 3 \geq x \\ 5 > y \\ -2 \leq y \end{cases}$

45. $\begin{cases} -5 < x \\ x \leq 2 \\ -4 \leq y \\ y < 9 \end{cases}$
46. $\begin{cases} x \geq 0 \\ y \geq 0 \\ x + y \leq 9 \\ 2x - 3y \geq -6 \\ x - y \leq 3 \end{cases}$

47. $\begin{cases} x \geq 0 \\ y \geq 0 \\ x + y \leq 8 \\ y \leq 4 \\ x \leq 6 \end{cases}$
48. $\begin{cases} x \geq 0 \\ y \geq 0 \\ 2x + y \geq 8 \\ y \leq 5 \\ x - y \leq 2 \\ 3x - y \geq 5 \end{cases}$

49. $\begin{cases} 2x - 3y + 30 \geq 0 \\ 3x - 2y + 20 \leq 0 \\ x \leq 0 \\ y \geq 0 \end{cases}$
50. $\begin{cases} 2x + 3y \leq 30 \\ 3x + 2y \geq 20 \\ x \geq 0 \\ y \geq 0 \end{cases}$

51. $\begin{cases} x + y - 10 \leq 0 \\ x + y + 4 \geq 0 \\ x - y \leq 6 \\ y - x \leq 4 \end{cases}$
52. $\begin{cases} 3y = x^2 + 21 \\ y \leq 10 \end{cases}$

53. $\begin{cases} 3y + 16x = x^2 + 85 \\ y \leq 10 \end{cases}$
54. $\begin{cases} 3y + x^2 = 2x + 35 \\ -1 \leq x \leq 2 \end{cases}$

55. $\begin{cases} 4y + 6x + 3 = x^2 \\ |x - 3| \leq 4 \end{cases}$
56. $\begin{cases} 3y + x^2 + 23 = 14x \\ |x - 7| \leq 2 \end{cases}$

C *In Problems 57–62 sketch the systems of inequalities.*

57. $\begin{cases} y \geq x^2 \\ 5x - 4y + 26 \geq 0 \end{cases}$
58. $\begin{cases} y < 4 - x^2 \\ y > \frac{1}{2}x \end{cases}$

59. $\begin{cases} y \geq \frac{1}{2}(x - 2)^2 \\ y - 4 < \frac{1}{4}(x - 2)^2 \end{cases}$
60. $\begin{cases} y - 2 \geq \frac{1}{3}(x - 3)^2 \\ y - 4 \leq -\frac{1}{3}(x - 3)^2 \end{cases}$

61. $\begin{cases} y + 2 \geq \frac{1}{8}x^2 \\ x - \frac{3}{2} \geq \frac{1}{2}(y + 1)^2 \end{cases}$
62. $\begin{cases} x + 1 \geq (y - 5)^2 \\ y - 3 \geq (x + 1)^2 \end{cases}$

Chapter 9 Summary

The material of this chapter is reviewed in the following list of objectives. After each objective there are some practice questions. For a sample test, select the first question of each set and check your answers with the answer section. For a sample test without answers, use the second question of each set. Additional practice is given by the other questions in each set. If you are having trouble with a particular type of problem, look back to that section for extra help.

9.1 Systems of Equations ────────────────────────────

OBJECTIVE 1 *Solve a system of equations by graphing.*

1. $\begin{cases} 2x - y = -8 \\ y = \frac{3}{5}x + 1 \end{cases}$

2. $\begin{cases} y = \frac{1}{3}x - 5 \\ 2x - 4y - 12 = 0 \end{cases}$

3. $\begin{cases} 2x - 3y + 21 = 0 \\ 5x + 4y - 5 = 0 \end{cases}$

4. $\begin{cases} y = x^2 - 6x + 5 \\ x + y - 5 = 0 \end{cases}$

7. $\begin{cases} x + 2y = 26 \\ 5x - 2y = -122 \end{cases}$

8. $\begin{cases} y = x^2 + 4x - 3 \\ x - y + 1 = 0 \end{cases}$

OBJECTIVE 2 *Solve a system of equations by substitution.*

5. $\begin{cases} 4x + 3y = -18 \\ y = -\frac{2}{3}x - 2 \end{cases}$

6. $\begin{cases} 5y - 3x = 0 \\ y = x + 2 \end{cases}$

OBJECTIVE 3 *Solve a system of equations by linear combinations.*

9. $\begin{cases} 2x + 3y = 6 \\ 3x + 2y = -1 \end{cases}$

10. $\begin{cases} 5x - 2y = 30 \\ 3x - 2y = -2 \end{cases}$

11. $\begin{cases} 4x + 3y = -7 \\ 2x - 5y = 55 \end{cases}$

12. $\begin{cases} 3x^2 - 2y^2 = 9 \\ x^2 + y^2 = 8 \end{cases}$

9.2 Matrix Solution of a System of Equations ──────────

OBJECTIVE 4 *Solve a system of equations by using the Gauss–Jordan method.*

13. $\begin{cases} 2x + 3y - z = -2 \\ x + y - 5z = -3 \\ 5x - 7y - 10z = 60 \end{cases}$

14. $\begin{cases} x - 3y + z = 11 \\ 2x + y - z = 0 \\ x - 2y + 4z = 25 \end{cases}$

15. $\begin{cases} 2x - y + z = -3 \\ 3x + y - 2z = 11 \\ 5x - 2y + 3z = -8 \end{cases}$

16. $\begin{cases} w + 3x = -5 \\ x + y = -4 \\ 2x - 3z = -7 \\ z + y = -1 \end{cases}$

9.3 Inverse Matrices ────────────────────────────────

OBJECTIVE 5 *Perform matrix operations.* Let

$$A = \begin{bmatrix} 3 & -2 & 1 \\ -2 & 1 & 4 \\ 1 & -3 & 1 \end{bmatrix} \quad B = \begin{bmatrix} 2 & 1 & 0 \\ -1 & 7 & 3 \\ 2 & -3 & 5 \end{bmatrix}$$

and

$$C = \begin{bmatrix} 2 & 4 & -2 \\ 1 & -1 & 2 \end{bmatrix}$$

Find, if possible:

17. $A + B$

18. $A(B + C)$

19. AB

20. $C(BA)$

OBJECTIVE 6 *Find the inverse of a matrix (if it exists).*

21. $\begin{bmatrix} 1 & -2 \\ -3 & 7 \end{bmatrix}$

22. $\begin{bmatrix} 4 & -3 \\ 5 & 4 \end{bmatrix}$

23. $\begin{bmatrix} 1 & 1 \\ 1 & 1 \end{bmatrix}$

24. $\begin{bmatrix} 1 & 2 & 2 \\ 1 & 3 & 2 \\ 1 & 2 & 3 \end{bmatrix}$

OBJECTIVE 7 *Solve a system of equations by using an inverse matrix.* The matrices

$$\begin{bmatrix} -1 & 1 & -1 \\ 2 & -1 & 2 \\ 2 & -1 & 1 \end{bmatrix} \quad \text{and} \quad \begin{bmatrix} 1 & 0 & 1 \\ 2 & 1 & 0 \\ 0 & 1 & -1 \end{bmatrix}$$

are inverses. Use this information to solve the systems in Problems 25–28.

25. $\begin{cases} -x + y - z = 6 \\ 2x - y + 2z = -5 \\ 2x - y + z = -5 \end{cases}$ **26.** $\begin{cases} y = x + z \\ 2x + 2z = y + 1 \\ 2x + z = y + 3 \end{cases}$ **27.** $\begin{cases} x + z = 11 \\ 2x + y = 14 \\ y - z = -5 \end{cases}$ **28.** $\begin{cases} x + z = -4 \\ 2x + y = -4 \\ y - z = 1 \end{cases}$

9.4 Cramer's Rule

<u>OBJECTIVE 8</u> *Solve a system of two equations in two unknowns by using Cramer's Rule.*

29. $\begin{cases} 3x - y = 2 \\ x + 5y = -3 \end{cases}$ **30.** $\begin{cases} 5x - 3y = 2 \\ 2x + 4y = 5 \end{cases}$ **31.** $\begin{cases} 3x + 4y = 0 \\ 6x + 8y = 1 \end{cases}$ **32.** $\begin{cases} y = m_1 x + b_1 \\ y = m_2 x + b_2 \end{cases}$

9.5 Properties of Determinants

<u>OBJECTIVE 9</u> *Evaluate determinants.*

33. $\begin{vmatrix} 1 & 3 & -2 \\ 4 & 5 & 1 \\ 3 & -2 & 4 \end{vmatrix}$ **34.** $\begin{vmatrix} 3 & 1 & -2 \\ 4 & 4 & 0 \\ 2 & -3 & 1 \end{vmatrix}$

35. $\begin{vmatrix} 3 & -2 & 0 \\ 2 & 5 & 8 \\ 5 & 3 & 5 \end{vmatrix}$ **36.** $\begin{vmatrix} 0 & 3 & 0 & 0 \\ 3 & 4 & 0 & 2 \\ 3 & 8 & -4 & 1 \\ 1 & -5 & 0 & -1 \end{vmatrix}$

<u>OBJECTIVE 10</u> *Solve a system of n unknowns by using Cramer's Rule.*

37. $\begin{cases} 3x - 2y + z = 9 \\ 2x + 5y - 3z = 17 \\ x - 3y + 2z = -2 \end{cases}$ **38.** $\begin{cases} 3x + y - 2z = 2 \\ 4x + 4y = 8 \\ 2x - 3y + z = 0 \end{cases}$

39. $\begin{cases} 2x + z = 6 + y \\ 3x + z = 8 + 2y \\ x + y = 11 - 3z \end{cases}$ **40.** $\begin{cases} w + 2z = 7 \\ x - 3y = 5 \\ z - 4w = -1 \\ 2w + y = 5 \end{cases}$

9.6 Systems of Inequalities

<u>OBJECTIVE 11</u> *Graph a linear inequality in two variables.*

41. $3x - 2y + 16 \geq 9$ **42.** $y < x - 1$

43. $25x - 5y + 120 \geq 0$ **44.** $y \geq x^2 - 8x + 19$

<u>OBJECTIVE 12</u> *Graph the solutions of a system of inequalities.*

45. $\begin{cases} 2x + 7y \geq 420 \\ 2x + 2y \leq 500 \\ x > 50 \\ y < 100 \end{cases}$ **46.** $\begin{cases} y \geq 0 \\ 3x + 2y > -3 \\ x - y < 0 \end{cases}$

47. $\begin{cases} x - y - 8 \leq 0 \\ x - y + 4 \geq 0 \\ x + y \geq -5 \\ y + x \leq 4 \end{cases}$ **48.** $\begin{cases} 2x + 3y \leq 600 \\ x + 2y \leq 360 \\ 3x - 2y \leq 480 \\ x \geq 0 \\ y \geq 0 \end{cases}$

Sir Isaac Newton (1642–1727)

Analysis and natural philosophy owe their most important discoveries to this fruitful means, which is called "induction." Newton was indebted to it for his theorem of the binomial and the principle of universal gravity

PIERRE-SIMON LAPLACE
A Philosophical Essay on Probabilities

I don't know what I may seem to the world, but, as to myself, I seem to have been only as a boy playing on the seashore, and diverting myself in now and then finding a smoother pebble or a prettier shell than ordinary, whilst the great ocean of truth lay all undiscovered before me.

SIR ISAAC NEWTON
Quoted by Rev. J. Spence,
Anecdotes. Observations, and Characters of Books and Men (1858)

Isaac Newton was one of the greatest mathematicians of all time. However, when he was a schoolboy at Grantham he was last in his class until he wanted to beat a bully both physically and mentally. He succeeded on both counts. At 18, he entered Trinity College, Cambridge and remained there until 1696, first as a student and later as a professor. In 1665, the university closed for a year because of bubonic plague, and Newton found himself at home for a year. During this year he interested himself in various physical questions, and among other things he formulated the basis of his theory of gravitation and invented calculus. Leibniz, who was featured at the beginning of Chapter 9, invented calculus at about the same time.

The Binomial Theorem of this chapter was proved for real exponents by Newton in 1665. His statement of this theorem is awkward by today's standards, but remember that he found it by a laborious trial-and-error procedure. The following statement of the Binomial Theorem is attributed to Newton:*

$$\overline{P + PQ} \rceil \frac{m}{n}$$

$$= P\frac{m}{n} + \frac{m}{n} + \frac{m}{n} AQ + \frac{m-n}{2n} BQ$$

$$+ \frac{m-2n}{3n} CQ + \frac{m-3n}{4n} DQ$$

$$+ \frac{m-4n}{5n} EQ + \cdots$$

*From D. E. Smith, *Source Book in Mathematics* (New York: McGraw-Hill), p. 224.

<table>
</table>

10	# ADDITIONAL TOPICS IN ALGEBRA

Contents

Preview

Two of the most far-reaching results in algebra are presented in this chapter, *mathematical induction* and the *Binomial Theorem*. The three sections of this chapter that deal with *sequences* and *series* should be considered together as a unit. There are 9 objectives in this chapter, which are listed on pages 458–459.

Perspective

Mathematical induction is used in calculus in the introduction to the idea of integration. It forms the basis of a number of the results that are needed to formulate some integral theorems. In addition, series are a major topic of consideration in calculus. The material introduced in this chapter forms the starting point for the ideas of convergence and divergence of series in calculus. In the following example taken from a calculus book, we see the so-called *p-series*, which you can find in Section 10.5:

$$\frac{1}{1} + \frac{1}{2^p} + \frac{1}{3^p} + \frac{1}{4^p} + \frac{1}{5^p} + \cdots$$

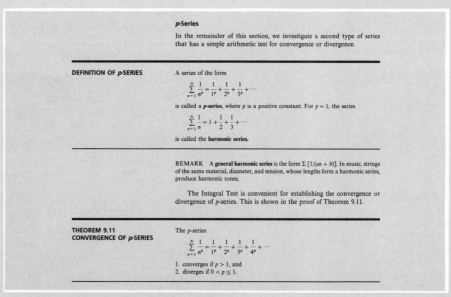

p-Series

In the remainder of this section, we investigate a second type of series that has a simple arithmetic test for convergence or divergence.

DEFINITION OF p-SERIES

A series of the form

$$\sum_{n=1}^{\infty} \frac{1}{n^p} = \frac{1}{1^p} + \frac{1}{2^p} + \frac{1}{3^p} + \cdots$$

is called a **p-series**, where p is a positive constant. For $p = 1$, the series

$$\sum_{n=1}^{\infty} \frac{1}{n} = 1 + \frac{1}{2} + \frac{1}{3} + \cdots$$

is called the **harmonic series.**

REMARK A **general harmonic series** is the form $\sum [1/(an + b)]$. In music, strings of the same material, diameter, and tension, whose lengths form a harmonic series, produce harmonic tones.

The Integral Test is convenient for establishing the convergence or divergence of p-series. This is shown in the proof of Theorem 9.11.

THEOREM 9.11 CONVERGENCE OF p-SERIES

The p-series

$$\sum_{n=1}^{\infty} \frac{1}{n^p} = \frac{1}{1^p} + \frac{1}{2^p} + \frac{1}{3^p} + \frac{1}{4^p} + \cdots$$

1. converges if $p > 1$, and
2. diverges if $0 < p \leq 1$.

From *Calculus*, 4*th* edition, by R. E. Larson and R. P. Hostetler, p. 565. Copyright © 1990 by D. C. Heath and Company. Reprinted by permission.

10.1 Mathematical Induction

Mathematical induction is an important method of proof in mathematics. Let us begin with a simple example. Suppose you want to know the sum of the first n odd integers. You could begin by looking for a pattern.

$$
\begin{array}{ll}
1 & = 1 \\
1 + 3 & = 4 \\
1 + 3 + 5 & = 9 \\
1 + 3 + 5 + 7 & = 16 \\
1 + 3 + 5 + 7 + 9 & = 25
\end{array}
$$

Do you see a pattern here? It appears that the sum of the first n odd numbers is n^2 since $1 = 1^2$, $4 = 2^2$, $9 = 3^2$, $16 = 4^2$, and so on. Now *prove* that

$$1 + 3 + 5 + \cdots + \underbrace{(2n - 1)}_{n\text{th odd number}} = n^2$$

is true for all positive integers n. How can you proceed? Use a method called **mathematical induction,** which is used to prove certain propositions about the positive integers. The proposition is denoted by $P(n)$. In this case, $P(n)$ is the proposition

$$P(n): \quad 1 + 3 + 5 + \cdots + (2n - 1) = n^2$$

Then

$$
\begin{array}{ll}
P(1): & 1 = 1^2 \\
P(2): & 1 + 3 = 2^2 \\
P(3): & 1 + 3 + 5 = 3^2 \\
P(4): & 1 + 3 + 5 + 7 = 4^2 \\
\vdots & \qquad \vdots \\
P(100): & 1 + 3 + 5 + \cdots + 199 = 100^2 \\
P(x - 1): & 1 + 3 + 5 + \cdots + (2x - 3) = (x - 1)^2 \\
P(x): & 1 + 3 + 5 + \cdots + (2x - 1) = x^2 \\
P(x + 1): & 1 + 3 + 5 + \cdots + (2x + 1) = (x + 1)^2
\end{array}
$$

We want to show that $P(n)$ is true for *all* positive integers n. The following statement tells us how to proceed.

PRINCIPLE OF MATHEMATICAL **INDUCTION (PMI)**	If a given proposition $P(n)$ is true for $P(1)$ and if the truth of $P(k)$ implies the truth of $P(k + 1)$, then $P(n)$ is true for all positive integers.

This sets up the following procedure for proof by mathematical induction.

1. Prove $P(1)$ is true.
2. Assume $P(k)$ is true.
3. Prove $P(k + 1)$ is true.
4. Conclude that $P(n)$ is true for all positive integers n.

Students often have a certain uneasiness when they first use the Principle of Mathematical Induction as a method of proof. Suppose this principle is used with a stack of dominoes, as shown in the cartoon. How can the cat in the cartoon be certain of knocking over all the dominoes?

1. The cat would have to be able to knock over the first one.

2. Also, the dominoes would be arranged so that *if* the kth domino falls, then the next one, the $(k + 1)$st, will also fall. That is, each domino is set up so that if it falls, it causes the next one to fall. This is a kind of "chain reaction." The first domino falls; this knocks over the next one (the second domino); the second one knocks over the next one (the third domino); the third one knocks over the next one; and so on. This continues until all the dominoes are knocked over.

EXAMPLE 1 Prove that $1 + 3 + 5 + \cdots + (2n - 1) = n^2$ is true for all positive integers n.

SOLUTION Proof:

You want to prove a theorem, then
To prove it for every integer n,
You prove it first for n equal one,
And then the induction is begun.

STEP 1   Prove $P(1)$ true: $1 = 1^2$ is true.

STEP 2   Assume $P(k)$ true: $1 + 3 + 5 + \cdots + (2k - 1) = k^2$.

The proof goes on in the following way,
You assume it next for n equal k.
If you can show it then for k + 1,
Then the induction is truly done.
          Author unknown

STEP 3   Prove $P(k + 1)$ true.

To prove:

$$1 + 3 + 5 + \cdots + [2(k + 1) - 1] = (k + 1)^2$$

or

$$1 + 3 + 5 + \cdots + (2k + 1) = (k + 1)^2$$ Since $2(k + 1) - 1 = 2k + 2 - 1$
                                                                    $= 2k + 1$

STATEMENTS	REASONS
1. $1 + 3 + 5 + \cdots + (2k - 1) = k^2$	1. By hypothesis (Step 2)
2. $1 + 3 + 5 + \cdots + (2k - 1) + (2k + 1) = k^2 + (2k + 1)$	2. Add $(2k + 1)$ to both sides
3. $1 + 3 + 5 + \cdots + (2k - 1) + (2k + 1) = k^2 + 2k + 1$	3. Associative
4. $1 + 3 + 5 + \cdots + (2k - 1) + (2k + 1) = (k + 1)^2$	4. Factoring (distributive)

STEP 4   The proposition $P(n)$ is true for all positive integers n by PMI (the Principle of Mathematical Induction).

Remember: A single example showing that a proposition is false serves as a counterexample to disprove the proposition.

EXAMPLE 2 Prove or disprove that $2 + 4 + 6 + \cdots + 2n = n(n + 1)$ is true for all positive integers n.

SOLUTION STEP 1 Prove $P(1)$ true: $2 \overset{?}{=} 1(1 + 1)$
$$2 = 2; \text{ it is true}$$

STEP 2 Assume $P(k)$.
Hypothesis: $2 + 4 + 6 + \cdots + 2k = k(k + 1)$

STEP 3 Prove $P(k + 1)$.
To prove: $2 + 4 + 6 + \cdots + 2(k + 1) = (k + 1)(k + 2)$

STATEMENTS	REASONS
1. $2 + 4 + 6 + \cdots + 2k = k(k + 1)$	1. Hypothesis (Step 2)
2. $2 + 4 + 6 + \cdots + 2k + 2(k + 1)$ $= k(k + 1) + 2(k + 1)$	2. Add $2(k + 1)$ to both sides.
3. $\quad = (k + 1)(k + 2)$	3. Factoring

STEP 4 The proposition $P(n)$ is true for all positive integers n by PMI. ∎

Mathematical induction does not apply only to propositions involving sums of terms, as shown by the next examples.

EXAMPLE 3 Prove or disprove that $n^3 + 2n$ is divisible by 3 for all positive integers n.

SOLUTION Proof: An integer is divisible by 3 if it has a factor of 3.

STEP 1 Prove $P(1)$: $1^3 + 2 \cdot 1 = 3$, which is divisible by 3; thus, $P(1)$ is true.

STEP 2 Assume $P(k)$.
Hypothesis: $k^3 + 2k$ is divisible by 3

STEP 3 Prove $P(k + 1)$.
To prove: $(k + 1)^3 + 2(k + 1)$ is divisible by 3

STATEMENTS	REASONS
1. $(k + 1)^3 + 2(k + 1)$ $= k^3 + 3k^2 + 3k + 1 + 2k + 2$	1. Distributive, associative, and commutative properties
2. $= (3k^2 + 3k + 3) + (k^3 + 2k)$	2. Commutative and associative
3. $= 3(k^2 + k + 1) + (k^3 + 2k)$	3. Distributive
4. $3(k^2 + k + 1)$ is divisible by 3	4. Definition of divisibility by 3
5. $k^3 + 2k$ is divisible by 3	5. Hypothesis
6. $(k + 1)^3 + 2(k + 1)$ is divisible by 3	6. Both terms are divisible by 3; therefore the sum is divisible by 3.

STEP 4 The proposition $P(n)$ is true for all positive integers n by PMI. ∎

EXAMPLE 4 Prove or disprove that $n + 1$ is prime for all positive integers n.

SOLUTION Proof:

STEP 1 Prove $P(1)$: $1 + 1 = 2$ is a prime

STEP 2 Assume $P(k)$.
Hypothesis: $k + 1$ is a prime

STEP 3 **Prove $P(k + 1)$.**
To prove: $(k + 1) + 1$ is a prime; it is not possible since $(k + 1) + 1 = k + 2$, which is not prime whenever k is an even positive integer.

STEP 4 **Any conclusion?** You cannot conclude that it is false—only that induction does not work. But it is in fact false, and a counterexample is found by letting $n = 3$; then $n + 1 = 4$ is not prime. ∎

Example 4 shows that even though you made an assumption in Step 2, it is not going to change the conclusion. If the proposition you are trying to prove is not true, making the assumption in Step 2 that it is true will *not* enable you to prove it true. Another common mistake of students working with mathematical induction for the first time is illustrated by Example 5.

EXAMPLE 5 Prove that $1 \cdot 2 \cdot 3 \cdot 4 \cdots n < 0$ for all positive integers n.

SOLUTION Students often slip into the habit of skipping either the first or second step in a proof by mathematical induction. This is dangerous; it is important to check every step. Suppose a careless person did not verify the first step. Then the results could be as follows:

STEP 2 **Assume $P(k)$.**
Hypothesis: $1 \cdot 2 \cdot 3 \cdots k < 0$

STEP 3 **Prove $P(k + 1)$.**
To prove: $1 \cdot 2 \cdot 3 \cdots k \cdot (k + 1) < 0$
Proof: $1 \cdot 2 \cdot 3 \cdots k < 0$ by hypothesis; $k + 1$ is positive since k is a positive integer; therefore

$$\underbrace{1 \cdot 2 \cdot 3 \cdots k}_{\uparrow \atop \text{negative}} \cdot \underbrace{(k + 1)}_{\uparrow \atop \text{positive}} < 0$$

Step 3 is proved.

STEP 4 The proposition is not true for all positive integers, since the first step, $1 < 0$, does not hold. ∎

10.1 Problem Set

A *If $P(n)$ represents each statement given in Problems 1–10, state and prove or disprove $P(1)$.*

1. $5 + 9 + 13 + \cdots + (4n + 1) = n(2n + 3)$

2. $3 + 9 + 15 + \cdots + (6n - 3) = 3n^2$

3. $2^2 + 4^2 + 6^2 + \cdots + (2n)^2 = \dfrac{2n(n + 1)(2n + 1)}{3}$

4. $1^3 + 2^3 + 3^3 + \cdots + n^3 = \dfrac{n^2(n + 1)^2}{4}$

5. $\cos(\theta + n\pi) = (-1)^n \cos \theta$

6. $\sin\left(\dfrac{\pi}{4} + n\pi\right) = (-1)^n \left(\dfrac{\sqrt{2}}{2}\right)$

7. $n^2 + n$ is even

8. $n^3 - n + 3$ is divisible by 3   (Steps in proof may vary.)

9. $\left(\dfrac{2}{3}\right)^{n+1} < \left(\dfrac{2}{3}\right)^n$

10. $1 + 2n \leq 3^n$

If P(n) represents each statement given in Problems 11–20, state P(k) and P(k + 1).

11. $5 + 9 + 13 + \cdots + (4n + 1) = n(2n + 3)$

12. $3 + 9 + 15 + \cdots + (6n - 3) = 3n^2$

13. $2^2 + 4^2 + 6^2 + \cdots + (2n)^2 = \dfrac{2n(n + 1)(2n + 1)}{3}$

14. $1^3 + 2^3 + 3^3 + \cdots + n^3 = \dfrac{n^2(n + 1)^2}{4}$

15. $\cos(\theta + n\pi) = (-1)^n \cos\theta$

16. $\sin\left(\dfrac{\pi}{4} + n\pi\right) = (-1)^n\left(\dfrac{\sqrt{2}}{2}\right)$

17. $n^2 + n$ is even

18. $n^3 - n + 3$ is divisible by 3

19. $\left(\dfrac{2}{3}\right)^{n+1} < \left(\dfrac{2}{3}\right)^{n}$

20. $1 + 2n \le 3^n$

In each of Problems 21–35, prove that the given formula is true for all positive integers n.

21. $1 + 2 + 3 + \cdots + n = \dfrac{n(n + 1)}{2}$

22. $1 + 4 + 7 + \cdots + (3n - 2) = \dfrac{n(3n - 1)}{2}$

23. $5 + 9 + 13 + \cdots + (4n + 1) = n(2n + 3)$

24. $3 + 9 + 15 + \cdots + (6n - 3) = 3n^2$

25. $2 + 7 + 12 + \cdots + (5n - 3) = \dfrac{n(5n - 1)}{2}$

26. $1^2 + 2^2 + 3^2 + \cdots + n^2 = \dfrac{n(n + 1)(2n + 1)}{6}$

27. $1^2 + 3^2 + 5^2 + \cdots + (2n - 1)^2 = \dfrac{n(2n - 1)(2n + 1)}{3}$

28. $1^3 + 2^3 + 3^3 + \cdots + n^3 = \dfrac{n^2(n + 1)^2}{4}$

29. $2^2 + 4^2 + 6^2 + \cdots + (2n)^2 = \dfrac{2n(n + 1)(2n + 1)}{3}$

30. $1 \cdot 2 + 2 \cdot 3 + 3 \cdot 4 + \cdots + n(n + 1) = \dfrac{n(n + 1)(n + 2)}{3}$

31. $1 \cdot 3 + 2 \cdot 4 + 3 \cdot 5 + \cdots + n(n + 2) = \dfrac{n(n + 1)(2n + 7)}{6}$

32. $1 + 3 + 3^2 + \cdots + 3^{n-1} = \dfrac{3^n - 1}{2}$

33. $1 + 5 + 5^2 + \cdots + 5^{n-1} = \dfrac{5^n - 1}{4}$

34. $1 + r + r^2 + \cdots + r^{n-1} = \dfrac{r^n - 1}{r - 1}$

35. $\log(a_1 a_2 \cdots a_n) = \log a_1 + \log a_2 + \cdots + \log a_n$ for $n \ge 2$ (Assume all the a_i are positive real numbers.)

B *Define $b^{n+1} = b^n \cdot b$ and $b^0 = 1$. Use this definition to prove the properties of exponents in Problems 36–39 for all positive integers n.*

36. $b^m \cdot b^n = b^{m+n}$

37. $(b^m)^n = b^{mn}$

38. $(ab)^n = a^n b^n$

39. $\left(\dfrac{a}{b}\right)^n = \dfrac{a^n}{b^n}$

Prove that the statements in Problems 40–53 are true for every positive integer n.

40. $\cos(\theta + n\pi) = (-1)^n(\cos\theta)$

41. $\sin\left(\dfrac{\pi}{4} + n\pi\right) = (-1)^n\left(\dfrac{\sqrt{2}}{2}\right)$

42. $\cos\left(\dfrac{\pi}{3} + n\pi\right) = \dfrac{(-1)^n}{2}$

43. $n^2 + n$ is even

44. $n^5 - n$ is divisible by 5

45. $n(n + 1)(n + 2)$ is divisible by 6

46. $n^3 - n + 3$ is divisible by 3

47. $10^{n+1} + 3 \cdot 10^n + 5$ is divisible by 9

48. $(1 + n)^2 \ge 1 + n^2$

49. $2^n > n$

50. $\left(\dfrac{2}{3}\right)^{n+1} < \left(\dfrac{2}{3}\right)^{n}$

51. $1 + 2n \le 3^n$

52. $\dfrac{1}{2} + \dfrac{1}{3} + \dfrac{1}{4} + \dfrac{1}{5} + \cdots + \dfrac{1}{n + 1} < n$

53. $1 + 2 + 3 + \cdots + n < \dfrac{(2n + 1)^2}{8}$

54. Prove the generalized distributive property:
$$a(b_1 + b_2 + \cdots + b_n) = ab_1 + ab_2 + \cdots + ab_n.$$

C

55. Notice the following:
$$1^3 = 1^2$$
$$1^3 + 2^3 = 3^2$$
$$1^3 + 2^3 + 3^3 = 6^2$$
$$1^3 + 2^3 + 3^3 + 4^3 = 10^2$$

Make a conjecture based on this pattern and then prove your conjecture.

56. Notice the following:

$$1 = 1$$
$$1 + 4 = 5$$
$$1 + 4 + 7 = 12$$
$$1 + 4 + 7 + 10 = 22$$

Make a conjecture based on this pattern and then prove your conjecture.

57. Notice the following:

$$2 = 2$$
$$2 + 2 \cdot 3 = 8$$
$$2 + (2 \cdot 3) + (2 \cdot 3^2) = 26$$
$$2 + (2 \cdot 3) + (2 \cdot 3^2) + (2 \cdot 3^3) = 80$$

Make a conjecture based on this pattern and then prove your conjecture.

58. Notice the following:

$$(-1)^1 = -1$$
$$(-1)^1 + (-1)^2 = 0$$
$$(-1)^1 + (-1)^2 + (-1)^3 = -1$$
$$(-1)^1 + (-1)^2 + (-1)^3 + (-1)^4 = 0$$

Make a conjecture based on this pattern and then prove your conjecture.

59. Prove $\sin x + \sin^2 x + \cdots + \sin^n x = \dfrac{\sin^{n+1} x - \sin x}{\sin x - 1}$ for all positive integers n.

60. Prove $e^x + e^{2x} + \cdots + e^{nx} = \dfrac{e^{(n+1)x} - e^x}{e^x - 1}$ for all positive integers n.

61. Use mathematical induction to prove De Moivre's Theorem: $(r \operatorname{cis} \theta)^n = r^n \operatorname{cis}(n\theta)$ for every positive integer n.

10.2 Binomial Theorem

In mathematics it is frequently necessary to expand $(a + b)^n$. If n is very large, direct calculation is rather tedious so we try to find an easy pattern that will not only help us find $(a + b)^n$ but will also allow us to find any given term in that expansion.

Consider the powers of $(a + b)$, which are found by direct multiplication:

$$(a + b)^0 = \mathbf{1}$$
$$(a + b)^1 = \mathbf{1} \cdot a + \mathbf{1} \cdot b$$
$$(a + b)^2 = \mathbf{1} \cdot a^2 + \mathbf{2} \cdot ab + \mathbf{1} \cdot b^2$$
$$(a + b)^3 = \mathbf{1} \cdot a^3 + \mathbf{3} \cdot a^2 b + \mathbf{3} \cdot ab^2 + \mathbf{1} \cdot b^3$$
$$(a + b)^4 = \mathbf{1} \cdot a^4 + \mathbf{4} \cdot a^3 b + \mathbf{6} \cdot a^2 b^2 + \mathbf{4} \cdot ab^3 + \mathbf{1} \cdot b^4$$
$$(a + b)^5 = \mathbf{1} \cdot a^5 + \mathbf{5} \cdot a^4 b + \mathbf{10} \cdot a^3 b^2 + \mathbf{10} \cdot a^2 b^3 + \mathbf{5} \cdot ab^4 + \mathbf{1} \cdot b^5$$
$$\vdots$$

Ignore the coefficients and focus your attention only on the variables:

$$(a + b)^1: \quad a \quad b$$
$$(a + b)^2: \quad a^2 \quad ab \quad b^2$$
$$(a + b)^3: \quad a^3 \quad a^2 b \quad ab^2 \quad b^3$$
$$(a + b)^4: \quad a^4 \quad a^3 b \quad a^2 b^2 \quad ab^3 \quad b^4$$
$$\vdots$$

Do you see a pattern? As you read from left to right, the powers of a decrease and the powers of b increase. Notice that the sum of the exponents for each term is the same as the original exponent:

$$(a + b)^n: \quad a^n b^0 \quad a^{n-1} b^1 \quad a^{n-2} b^2 \cdots a^{n-r} b^r \cdots a^2 b^{n-2} \quad a^1 b^{n-1} \quad a^0 b^n$$

Next consider the coefficients:

$$
\begin{array}{llccccccc}
(a+b)^0: & & & & & 1 & & & \\
(a+b)^1: & & & & 1 & & 1 & & \\
(a+b)^2: & & & 1 & & 2 & & 1 & \\
(a+b)^3: & & 1 & & 3 & & 3 & & 1 \\
(a+b)^4: & 1 & & 4 & & 6 & & 4 & & 1 \\
(a+b)^5: & 1 & & 5 & & 10 & & 10 & & 5 & & 1 \\
\end{array}
$$

$$\vdots$$

Do you see this pattern? This arrangement of numbers is called **Pascal's triangle**. The rows and columns of Pascal's triangle are usually numbered as shown in Figure 10.1.

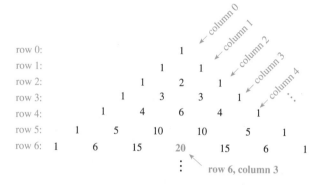

FIGURE 10.1   Pascal's triangle

Notice that the rows of Pascal's triangle are numbered to correspond to the exponent n on $(a - b)^n$. There are many relationships associated with this pattern, but we are concerned with an expression representing the entries in the pattern. Do you see how to generate additional rows of the triangle?

1. Each row begins and ends with a 1.
2. Notice that we began counting the rows with row 0. This is because after row 0, the second entry in the row is the same as the row number. Thus row 7 begins 1   7   . . . .
3. The triangle is symmetric about the middle. This means that the entries of each row are the same at the beginning and the end. Thus row 7 ends with . . .   7   1. (This property is proved in Problem 58.)
4. To find new entries, we can simply add the two entries just above in the preceding row. Thus row 7 is found by looking at row 6:

$$
\begin{array}{lccccccccc}
\text{Row 6:} & 1 & & 6 & & 15 & & 20 & & 15 & & 6 & & 1 \\
\text{Row 7:} & 1 & & 7 & & 21 & & 35 & & 35 & & 21 & & 7 & & 1 \\
\end{array}
$$

(This property is proved in Problem 57.)

Write $\dbinom{n}{r}$ to represent the element in the nth row and rth column of the triangle.

Therefore:

$$\binom{0}{0} = 1$$

$$\binom{1}{0} = 1 \qquad \binom{1}{1} = 1$$

$$\binom{2}{0} = 1 \qquad \binom{2}{1} = 2 \qquad \binom{2}{2} = 1$$

$$\binom{3}{0} = 1 \qquad \binom{3}{1} = 3 \qquad \binom{3}{2} = 3 \qquad \binom{3}{3} = 1$$

$$\vdots$$

Using this notation, we can now state a very important theorem in mathematics called the **Binomial Theorem**:

BINOMIAL THEOREM

For any positive integer n,

$$(a + b)^n = \binom{n}{0}a^n + \binom{n}{1}a^{n-1}b + \binom{n}{2}a^{n-2}b^2 + \cdots + \binom{n}{r}a^{n-r}b^r + \cdots$$

$$+ \binom{n}{n-2}a^2b^{n-2} + \binom{n}{n-1}ab^{n-1} + \binom{n}{n}b^n$$

The proof of the Binomial Theorem is by mathematical induction and the procedure is lengthy, so we leave the proof for the problem set. You are led through the steps and are asked to fill in the details in Problems 60–63.

EXAMPLE 1 Find $(x + y)^8$.

SOLUTION Use Pascal's triangle to obtain the coefficients in the expansion. Thus

$$(x + y)^8 = x^8 + 8x^7y + 28x^6y^2 + 56x^5y^3 + 70x^4y^4$$
$$+ 56x^3y^5 + 28x^2y^6 + 8xy^7 + y^8$$

EXAMPLE 2 Find $(x - 2y)^4$.

SOLUTION In this example, $a = x$ and $b = -2y$ and the coefficients are in Pascal's triangle.

$$(x - 2y)^4 = 1 \cdot x^4 + 4 \cdot x^3(-2y)^1 + 6 \cdot x^2(-2y)^2 + 4 \cdot x(-2y)^3 + 1 \cdot (-2y)^4$$
$$= x^4 - 8x^3y + 24x^2y^2 - 32xy^3 + 16y^4$$

If the power of the binomial is very large, then Pascal's triangle is not efficient, so the next step is to find a formula for $\binom{n}{r}$. This formula is found by using a notation called **factorial notation.**

The symbol

$$n! = n \cdot (n - 1)(n - 2) \cdots \cdot 3 \cdot 2 \cdot 1$$

is called **n factorial** (n a natural number). Also, we define $0! = 1$ and $1! = 1$.

EXAMPLE 3

$$0! = 1$$
$$1! = 1$$
$$2! = 2 \cdot 1 = 2$$
$$3! = 3 \cdot 2 \cdot 1 = 6$$
$$4! = 4 \cdot 3 \cdot 2 \cdot 1 = 24$$
$$5! = 5 \cdot 4 \cdot 3 \cdot 2 \cdot 1 = 120$$
$$6! = 1 \cdot 2 \cdot 3 \cdot 4 \cdot 5 \cdot 6 = 720$$
$$7! = 1 \cdot 2 \cdot 3 \cdot 4 \cdot 5 \cdot 6 \cdot 7 = 5{,}040$$
$$8! = 1 \cdot 2 \cdot 3 \cdots \cdot 7 \cdot 8 = 40{,}320$$
$$9! = 1 \cdot 2 \cdot 3 \cdots \cdot 8 \cdot 9 = 362{,}880$$
$$10! = 1 \cdot 2 \cdot 3 \cdots \cdot 9 \cdot 10 = 3{,}628{,}800$$

In finding the values of Example 3, notice that $3! = 3 \cdot 2!$, $4! = 4 \cdot 3!$, $5! = 5 \cdot 4!, \ldots$; in general, $n! = n(n - 1)!$.

EXAMPLE 4 Evaluate the given expressions.

a. $5! - 4! = 120 - 24$ **b.** $(5 - 4)! = 1!$ **c.** $\dfrac{8!}{4!} = \dfrac{8 \cdot 7 \cdot 6 \cdot 5 \cdot \cancel{4} \cdot \cancel{3} \cdot \cancel{2} \cdot \cancel{1}}{\cancel{4} \cdot \cancel{3} \cdot \cancel{2} \cdot \cancel{1}}$

$\qquad = 96 \qquad\qquad\qquad\qquad = 1 \qquad\qquad\qquad = 8 \cdot 7 \cdot 6 \cdot 5$

$\qquad\qquad\qquad\qquad\qquad\qquad\qquad\qquad\qquad\qquad = 1{,}680$

d. $\left(\dfrac{8}{4}\right)! = 2!$ **e.** $\dfrac{10!}{8!} = \dfrac{10 \cdot 9 \cdot \cancel{8!}}{\cancel{8!}}$ Notice that

$\qquad = 2 \qquad\qquad\qquad\quad = 90 \qquad\qquad\qquad 10! = 10 \cdot 9!$

$\qquad\qquad\qquad\qquad\qquad\qquad\qquad\qquad\qquad\qquad = 10 \cdot 9 \cdot 8!$

$\qquad\qquad\qquad\qquad\qquad\qquad\qquad\qquad\qquad\qquad = 10 \cdot 9 \cdot 8 \cdot 7!$

$\qquad\qquad\qquad\qquad\qquad\qquad\qquad\qquad\qquad$ and so on.

Now we can state a formula for $\dbinom{n}{r}$ using this notation.

The symbol $\dbinom{n}{r}$ is defined for integers r and n such that $0 \le r \le n$.

$$\binom{n}{r} = \frac{n!}{r!(n - r)!}$$

is called the **binomial coefficient n, r.**

EXAMPLE 5 Find $\begin{pmatrix} 3 \\ 2 \end{pmatrix}$ both by Pascal's triangle and by formula.

SOLUTION Row 3, column 2; entry in Pascal's triangle is **3**. By formula,

$$\begin{pmatrix} 3 \\ 2 \end{pmatrix} = \frac{3!}{2!(3-2)!}$$
$$= \frac{3!}{2!1!}$$
$$= 3$$

EXAMPLE 6 Find $\begin{pmatrix} 6 \\ 4 \end{pmatrix}$ both by Pascal's triangle and by formula.

SOLUTION Row 6, column 4; entry is **15**. By formula,

$$\begin{pmatrix} 6 \\ 4 \end{pmatrix} = \frac{6!}{4!(6-4)!}$$
$$= \frac{6!}{4!2!}$$
$$= \frac{6 \cdot 5 \cdot \cancel{4!}}{2 \cdot 1 \cdot \cancel{4!}}$$
$$= 15$$

EXAMPLE 7 Evaluate the given expressions. (You would not use Pascal's triangle for these.)

a. $\begin{pmatrix} 52 \\ 2 \end{pmatrix} = \dfrac{52!}{2!(52-2)!}$
$= \dfrac{52!}{2!50!}$
$= \dfrac{52 \cdot 51 \cdot \cancel{50!}}{2 \cdot 1 \cdot \cancel{50!}}$
$= 1{,}326$

b. $\begin{pmatrix} n \\ n \end{pmatrix} = \dfrac{n!}{n!(n-n)!}$
$= \dfrac{n!}{n!0!}$
$= 1$

c. $\begin{pmatrix} n \\ n-1 \end{pmatrix} = \dfrac{n!}{(n-1)![n-(n-1)]!}$
$= \dfrac{n!}{(n-1)!1!}$
$= \dfrac{n(n-1)!}{(n-1)!}$
$= \boldsymbol{n}$

EXAMPLE 8   Find $(a + b)^{15}$.

SOLUTION   The power is rather large, so use the formula to find the coefficients.

$$(a + b)^{15} = \binom{15}{0}a^{15} + \binom{15}{1}a^{14}b + \binom{15}{2}a^{13}b^2 + \cdots + \binom{15}{14}ab^{14} + \binom{15}{15}b^{15}$$

$$= \frac{15!}{0!15!}a^{15} + \frac{15!}{1!14!}a^{14}b + \frac{15!}{2!13!}a^{13}b^2 + \cdots + \frac{15!}{14!1!}ab^{14} + \frac{15!}{15!0!}b^{15}$$

$$= a^{15} + 15a^{14}b + 105a^{13}b^2 + \cdots + 15ab^{14} + b^{15}$$

EXAMPLE 9   Find the coefficient of the term x^2y^{10} in the expansion of $(x + 2y)^{12}$.

SOLUTION $n = 12, r = 10, a = x$, and $b = 2y$; thus

$$\binom{12}{10}x^2(2y)^{10} = \frac{12!}{10!2!}(2)^{10}x^2y^{10}$$

The coefficient is $66(1,024) = \mathbf{67,584}$.

10.2   Problem Set

A   *Evaluate the expressions in Problems 1–24.*

1. $4! - 2!$ **2.** $5! - 3!$ **3.** $(4 - 2)!$

4. $(6 - 3)!$ **5.** $\dfrac{9!}{7!}$ **6.** $\dfrac{10!}{6!}$

7. $\dfrac{12!}{10!}$ **8.** $\dfrac{10!}{4!6!}$ **9.** $\dfrac{12!}{3!(12 - 3)!}$

10. $\dfrac{15!}{5!(15 - 5)!}$ **11.** $\dfrac{20!}{3!(20 - 3)!}$ **12.** $\dfrac{52!}{3!(52 - 3)!}$

13. $\binom{8}{1}$ **14.** $\binom{5}{4}$ **15.** $\binom{8}{2}$

16. $\binom{52}{3}$ **17.** $\binom{8}{3}$ **18.** $\binom{7}{4}$

19. $\binom{8}{4}$ **20.** $\binom{5}{5}$ **21.** $\binom{52}{2}$

22. $\binom{10}{1}$ **23.** $\binom{1,000}{1}$ **24.** $\binom{574}{2}$

B   *In Problems 25–40, expand by using Pascal's triangle or by formula.*

25. $(a + b)^6$ **26.** $(a + b)^7$ **27.** $(2x + 3)^3$

28. $(2x - 3)^4$ **29.** $(x + y)^5$ **30.** $(x - y)^6$

31. $(3x + 2)^5$ **32.** $(3x - 2)^4$ **33.** $(x + y)^4$

34. $(2x + 3y)^4$ **35.** $(\frac{1}{2}x + y^3)^3$ **36.** $(x^{-2} + y^{-2})^4$

37. $(x^{1/2} + y^{1/2})^4$ **38.** $(1 + x)^{10}$ **39.** $(1 - x)^8$

40. $(1 - 2y)^6$

Find the first four terms in the expressions given in Problems 41–49.

41. $(a + b)^{10}$ **42.** $(a + b)^{12}$ **43.** $(a + b)^{14}$

44. $(x - y)^{15}$ **45.** $(x + 2y)^{16}$ **46.** $(x + \sqrt{2})^8$

47. $(x - 2y)^{12}$ **48.** $(1 - .03)^{13}$ **49.** $(1 - .02)^{12}$

50. Find the coefficient of a^5b^6 in $(a - b)^{11}$.

51. Find the coefficient of a^4b in $(a^2 - 2b)^3$.

52. Find the coefficient of $a^{10}b^4$ in $(a + b)^{14}$.

53. Find the coefficient of $x^{10}y^2$ in $(2x^2 + \sqrt{y})^9$.

C

54. What is the constant term in the expansion of $(9x^{-1} + x^2/3)^6$?

55. What is the constant term in the expansion of $(4y^{-2} + y^3/2)^5$?

56. Show that

$$\binom{n}{0} + \binom{n}{1} + \binom{n}{2} + \cdots + \binom{n}{n-1} + \binom{n}{n} = 2^n$$

This says that the sum of the entries of the nth row of Pascal's triangle is 2^n.

57. Show that

$$\binom{n-1}{r-1} + \binom{n-1}{r} = \binom{n}{r}$$

This says that to find any entry in Pascal's triangle (except the first and last), simply add the entries above.

58. Show that

$$\binom{n}{r} = \binom{n}{n-r}$$

This says that Pascal's triangle is symmetric.

59. Use the formula $\binom{n}{k} = \dfrac{n!}{k!(n-k)!}$ to prove

$$\binom{k}{r} + \binom{k}{r-1} = \binom{k+1}{r}$$

Problems 60–63 will lead you through the induction proof of the Binomial Theorem. Let n be any positive integer. To prove:

$$(a+b)^n = \binom{n}{0}a^n + \binom{n}{1}a^{n-1}b + \binom{n}{2}a^{n-2}b^2 + \cdots$$
$$+ \binom{n}{r}a^{n-r}b^r + \cdots + \binom{n}{n-2}a^2b^{n-2}$$
$$+ \binom{n}{n-1}ab^{n-1} + \binom{n}{n}b^n$$

60. Prove it true for $n = 1$.

61. Assume it true for $n = k$. Fill in the statement of the hypothesis.

62. Prove it true for $n = k + 1$.
To prove:

$$(a+b)^{k+1} =$$
$$a^{k+1} + \cdots + \left[\binom{k}{r} + \binom{k}{r-1}\right]a^{k-r+1}b^r + \cdots + b^{k+1}$$

Hint: To prove this you will need to use the results of Problem 59.

63. Tie together Problems 60–62 to complete the proof of the Binomial Theorem.

64. **WHAT IS WRONG**, with the following "proof," which results in $1 = 2$?

$$(a+b)^n = a^n + na^{n-1}b + \frac{n(n-1)}{2!}a^{n-2}b^2$$
$$+ \cdots + nab^{n-1} + b^n \qquad \text{From the Binomial Theorem}$$

Let $n = 0$. Then

$$(a+b)^0 = a^0 + 0 + 0 + \cdots + 0 + b^0$$

By substitution and zero multiplication

$$1 = 1 + 0 + 0 + \cdots + 1$$

By definition of zero exponent

$$1 = 2$$

Sequences, Series, and Summation Notation

Patterns and proof are two of the cornerstones upon which mathematics is founded. In this chapter these two ideas are tied together. Consider a function whose domain is the set of counting numbers, $N = \{1, 2, 3, 4, \ldots, n, \ldots\}$.

INFINITE SEQUENCE

An **infinite sequence** is a function s with a domain that consists of the set of counting numbers. The number $s(1)$ is called the *first term* of the sequence, $s(2)$ the *second term*, and $s(n)$ the *nth term* or **general term** of the sequence.

For convenience we sometimes refer to an infinite sequence simply as a **sequence** or **progression.** Sometimes we talk of a **finite sequence,** which means that the domain is the finite set $\{1, 2, 3, \ldots, n-1, n\}$ for some natural number n.

Two special types of sequences are studied in this chapter. The first is an **arithmetic sequence.** In an arithmetic sequence, there is a common difference between successive terms. That is, if any term is subtracted from the next term, the result is always the same, and this number is called the **common difference.**

EXAMPLE 1   Show that $1, 4, 7, 10, 13, \underline{\hspace{1cm}}, \dots$ is an arithmetic sequence and find the missing term.

SOLUTION   Look for a common difference by subtracting each term from the succeeding term:

$$4 - 1 = 3$$
$$7 - 4 = 3$$
$$10 - 7 = 3$$
$$13 - 10 = 3$$

The common difference is 3 (be sure to find this for *each* of the *given* terms).

$$x - 13 = 3$$ The common difference is 3; x is the missing term.

Thus $x = 16$ is the next term (which is found by solving the equation). In order to view this as a function whose domain is the set of counting numbers, you will need to wait until we discuss the general term of the sequence in the next section. ▌

The second type of sequence is called a **geometric sequence.** In a geometric sequence, there is a common ratio between successive terms. If any term is divided into the next term, the result is always the same, and this number is called the **common ratio.**

EXAMPLE 2   Show that $2, 4, 8, 16, 32, \underline{\hspace{1cm}}, \dots$ is a geometric sequence and find the missing term.

SOLUTION $\dfrac{4}{2} = 2; \quad \dfrac{8}{4} = 2; \quad \dfrac{16}{8} = 2; \quad \dfrac{32}{16} = 2; \quad \dfrac{x}{32} = 2$ (where x is the next term)

The common ratio is 2.

Thus

$$x = 64 \quad \leftarrow \text{The next term} \qquad\qquad ▌$$

There are other sequences, as shown by Example 3, that are neither arithmetic nor geometric.

EXAMPLE 3   Show that $1, 1, 2, 3, 5, 8, 13, \underline{\hspace{1cm}}, \underline{\hspace{1cm}}, \underline{\hspace{1cm}}, \dots$ is neither arithmetic nor geometric. Find the missing terms.

SOLUTION   First check to see if it is arithmetic: $1 - 1 = 0$; $2 - 1 = 1$; there is no *common* difference. Next check to see if it is geometric: $\frac{1}{1} = 1$; $\frac{2}{1} = 2$; there is no *common* ratio. Look for another pattern:

$$1 + 1 = 2; \quad 1 + 2 = 3; \quad 2 + 3 = 5; \quad 3 + 5 = 8; \quad 5 + 8 = 13$$

It looks like the pattern is obtained by adding the two preceding terms:

$$8 + 13 = 21$$
$$13 + 21 = 34$$
$$21 + 34 = 55$$

The missing terms are 21, 34, and 55. ▌

A new notation is generally used when working with sequences. Remember that the domain is the set of counting numbers, so a sequence could be defined by

$$s(n) = 3n - 2 \quad \text{where } n = 1, 2, 3, \ldots$$

Thus

$$s(1) = 3(1) - 2 = 1$$
$$s(2) = 3(2) - 2 = 4$$
$$s(3) = 3(3) - 2 = 7$$
$$\vdots$$

Instead of writing $s(1)$, however, we use the notation s_1; in place of $s(2)$, s_2; in place of $s(n)$, s_n. Thus s_{15} means the fifteenth term of the sequence. It is found in the same fashion as though the notation $s(15)$ were used:

$$s_{15} = 3(15) - 2$$
$$= 43$$

EXAMPLE 4 Find the first four terms of $s_n = 26 - 6n$.

SOLUTION
$$s_1 = 26 - 6(1) = 20$$
$$s_2 = 26 - 6(2) = 14$$
$$s_3 = 26 - 6(3) = 8$$
$$s_4 = 26 - 6(4) = 2$$

The sequence is **20, 14, 8, 2, ...** . It is an arithmetic sequence.

EXAMPLE 5 Find the first four terms of $s_n = (-2)^n$.

SOLUTION
$$s_1 = (-2)^1 = -2$$
$$s_2 = (-2)^2 = 4$$
$$s_3 = (-2)^3 = -8$$
$$s_4 = (-2)^4 = 16$$

The sequence is **-2, 4, -8, 16, ...** . It is a geometric sequence.

EXAMPLE 6 Find the first four terms of $s_n = s_{n-1} + s_{n-2}$, $n \geq 3$, where $s_1 = 1$ and $s_2 = 2$.

SOLUTION
$$s_1 = 1 \qquad \text{Given}$$
$$s_2 = 2 \qquad \text{Given}$$
$$s_3 = s_2 + s_1$$
$$= 2 + 1 \qquad \text{By substitution}$$
$$= 3$$
$$s_4 = s_3 + s_2$$
$$= 3 + 2$$
$$= 5$$

The sequence is **1, 2, 3, 5, ...** . This sequence is neither arithmetic nor geometric.

EXAMPLE 7 Find the first four terms of $s_n = 2n$.

SOLUTION
$$s_1 = 2, \quad s_2 = 4, \quad s_3 = 6, \quad s_4 = 8$$

The sequence is **2, 4, 6, 8, ...**.

EXAMPLE 8 Find the first four terms of $s_n = 2n + (n-1)(n-2)(n-3)(n-4)$.

SOLUTION
$$s_1 = 2(1) + 0 = 2$$
$$s_2 = 2(2) + 0 = 4$$
$$s_3 = 2(3) + 0 = 6$$
$$s_4 = 2(4) + 0 = 8$$

The sequence is **2, 4, 6, 8, ...**.

If you are given a general term, you can find a unique sequence. However, Examples 7 and 8 show that if only a finite number of successive terms is known and no general term is given, then a *unique* general term cannot be given. That is, if we are given the sequence

$$2, 4, 6, 8, \underline{\quad\quad}$$

the next term is probably 10 (if we are thinking of the general term of Example 7), but it *may* be something different. In Example 8, $s_1 = 2$, $s_2 = 4$, $s_3 = 6$, $s_4 = 8$, and

$$s_5 = 2(5) + (5-1)(5-2)(5-3)(5-4)$$
$$= 10 + (4)(3)(2)(1)$$
$$= 34$$

In general, you are looking for the simplest general term; nevertheless, you must remember that answers are not unique *unless the general term is given.*

It is sometimes necessary to consider the sum of the terms of a sequence. This sum is called a **series.**

FINITE SERIES | The indicated sum of the terms of a finite sequence $s_1, s_2, s_3, \ldots, s_n$ is called a **finite series** and is denoted by
$$S_n = s_1 + s_2 + s_3 + \cdots + s_n$$

EXAMPLE 9 Let $s_n = 26 - 6n$ from Example 4. Find S_4.

SOLUTION
$$S_4 = s_1 + s_2 + s_3 + s_4$$
$$= 20 + 14 + 8 + 2 \qquad \text{From Example 4}$$
$$= \mathbf{44}$$

EXAMPLE 10 Let $s_n = (-1)^n n^2$. Find S_3.

SOLUTION $\quad s_1 = (-1)^1(1)^2 = -1; \qquad s_2 = (-1)^2(2)^2 = 4; \qquad s_3 = (-1)^3(3)^2 = -9.$

Now find S_3:

$$S_3 = s_1 + s_2 + s_3$$
$$= (-1) + 4 + (-9)$$
$$= -6$$

The terms of the sequence in Example 10 alternate in sign:

$$-1, \quad 4, \quad -9, \quad 16, \quad \ldots$$

A factor of $(-1)^n$ or $(-1)^{n+1}$ in the general term will cause the sign of the terms to alternate, creating a series called an **alternating series.**

The next two sections discuss more efficient ways of finding the general term of a sequence, as well as more efficient ways of finding the sum of n terms of these sequences. We will use the following **summation notation** to simplify the way we express sums.

Consider the function $s_4 = 2k$ with the domain $N = \{1, 2, 3, 4\}$. The sum of the terms of this finite arithmetic sequence is the sum of the series

$$2 + 4 + 6 + 8$$

as shown in the table in the margin. Now denote this sum by using the symbol Σ (called **sigma**) as follows:

This is the last natural number in the domain.

k	$s_k = 2k$
1	2
2	4
3	6
4	8

$$\sum_{k=1}^{4} 2k = 2 + 4 + 6 + 8$$

This is the function being evaluated; it is the general term of the sequence.

This is the first natural number in the domain.

This variable is called the **index of summation.**

This symbol means to evaluate the function for each number in the domain and *add* the resulting terms.

Thus

$$\sum_{k=1}^{4} 2k = 2 + 4 + 6 + 8 = 20$$

EXAMPLE 11 Let $s_k = 2k + 1$ and $N = \{3, 4, 5, 6\}$. Evaluate $\sum_{k=3}^{6} (2k + 1)$.

SOLUTION

k	$s_k = 2k + 1$
First natural number in domain; $k = 3 \to$ 3	7
4	9
5	11
Last natural number in domain; $k = 6 \to$ 6	13

Σ means to add these values

Thus

$$\sum_{k=3}^{6} (2k + 1) = 7 + 9 + 11 + 13 = 40$$

$k = 4 \quad k = 5 \quad k = 6$

This is obtained by letting $k = 3$ and evaluating $(2k + 1)$.

EXAMPLE 12 Expand $\sum\limits_{k=1}^{n} \dfrac{1}{2^k}$.

SOLUTION

$$\sum_{k=3}^{n} \frac{1}{2^k} = \underset{\underset{k=3}{\uparrow}}{\frac{1}{8}} + \underset{\underset{k=4}{\uparrow}}{\frac{1}{16}} + \underset{\underset{k=5}{\uparrow}}{\frac{1}{32}} + \cdots + \underset{\underset{k=n}{\uparrow}}{\frac{1}{2^n}}$$

EXAMPLE 13 Write the sum of the arithmetic series

$$S_n = a + (a + d) + (a + 2d) + \cdots + (a + nd)$$

using sigma notation.

SOLUTION

$$S_n = \sum_{k=1}^{n+1} [a + (k-1)d]$$

EXAMPLE 14 Write the sum of the finite geometric series $S_n = a + ar + ar^2 + \cdots + ar^{n-1}$ using sigma notation.

SOLUTION

$$S_n = \sum_{k=1}^{n} ar^{k-1}$$

EXAMPLE 15 Write the Binomial Theorem using summation notation.

SOLUTION

$$(a + b)^n = \sum_{k=0}^{n} \binom{n}{k} a^{n-k} b^k$$

10.3 Problem Set

A *For the sequences in Problems 1–16, answer the following questions. The answers are not necessarily unique.*
a. *Classify each as arithmetic, geometric, or neither.*
b. *If arithmetic, state the common difference d; if geometric, state the common ratio r: if neither, state a pattern in your own words.*
c. *Supply the missing term.*

1. 2, 5, 8, 11, 14, ____
2. 1, 2, 1, 1, 2, 1, 1, 1, 2, 1, 1, 1, 1, ____
3. 3, 6, 12, 24, 48, ____
4. 5, −15, 45, −135, 405, ____
5. 100, 99, 97, 94, 90, ____   6. 1, 1, 2, 3, 5, 8, 13, ____
7. p, pq, pq^2, pq^3, pq^4, ____   8. 97, 86, 75, 64, ____
9. 8, 12, 18, 26, ____   10. $5^5, 5^4, 5^3, 5^2$, ____
11. 2, 5, 2, 5, 5, 2, 5, 5, 5, ____
12. 5, −5, −15, −25, −35, ____
13. $1, \frac{1}{2}, \frac{1}{3}, \frac{2}{3}, \frac{1}{4}, \frac{3}{4}, \frac{1}{5}, \frac{2}{5}, \frac{3}{5}, \frac{4}{5}, \frac{1}{6}$, ____

14. $\frac{4}{3}$, 2, 3, $4\frac{1}{2}$, ____
15. 1, 8, 27, 64, 125, ____
16. 2, 8, 18, 32, ____

Find the first three terms of the sequence with the nth term given in Problems 17–28.

17. $s_n = 4n - 3$ 18. $s_n = a + nd$

19. $s_n = ar^{n-1}$ 20. $s_n = \dfrac{n-1}{n+1}$

21. $s_n = (-1)^n$ 22. $s_n = (-1)^n(n+1)$

23. $s_n = 1 + \dfrac{1}{n}$ 24. $s_n = \dfrac{n+1}{n}$

25. $s_n = 2$ 26. $s_n = -5$

27. $\cos nx$ 28. $\dfrac{\sin n}{n^2}$

Evaluate the expressions in Problems 29–40.

29. $\sum\limits_{k=2}^{6} k$ **30.** $\sum\limits_{m=1}^{4} m^2$ **31.** $\sum\limits_{n=0}^{6} (2n+1)$

32. $\sum\limits_{p=1}^{6} 2p$ **33.** $\sum\limits_{k=2}^{5} (10-2k)$ **34.** $\sum\limits_{k=2}^{5} (100-5k)$

35. $\sum\limits_{k=1}^{5} (-2)^{k-1}$ **36.** $\sum\limits_{k=0}^{4} 3(-2)^k$ **37.** $\sum\limits_{k=0}^{3} 2(3^k)$

38. $\sum\limits_{k=1}^{3} (-1)^k(k^2+1)$ **39.** $\sum\limits_{k=1}^{10} [1^k + (-1)^k]$

40. $\sum\limits_{k=0}^{5} [2^k + (-2)^k]$

B

41. Find the fifteenth term of the sequence in Problem 17.

42. Find the 102nd term of the sequence in Problem 20.

43. Find the tenth term of the sequence in Problem 21.

44. Find the twentieth term of the sequence in Problem 22.

45. Find the third term of the sequence $(-1)^{n+1}5^{n+1}$.

46. Find the second term of the sequence $(-1)^{n-1}7^{n-1}$.

47. Find the first five terms of the sequence where $s_1 = 2$ and $s_n = 3s_{n-1}$, $n \ge 2$.

48. Find the first five terms of the sequence where $s_1 = 3$ and $s_n = \frac{1}{3}s_{n-1}$, $n \ge 2$.

49. Find the first five terms of the sequence where $s_1 = 1$, $s_2 = 1$, and $s_n = s_{n-1} + s_{n-2}$, $n \ge 3$.

50. Find the first five terms of the sequence where $s_1 = 1$, $s_2 = 2$, and $s_n = s_{n-1} + s_{n-2}$, $n \ge 3$.

51. Write $\dfrac{1}{2} + \dfrac{1}{4} + \dfrac{1}{8} + \cdots + \dfrac{1}{2^r} + \cdots + \dfrac{1}{128}$ using summation notation.

52. Write $2 + 4 + 6 + \cdots + 2n + \cdots + 100$ using summation notation.

53. Write $1 + 6 + 36 + 216 + 1{,}296$ using summation notation. The rth term is 6^{r-1}.

54. Write $5 + 15 + 45 + 135 + 405$ using summation notation. The rth term is $5 \cdot 3^{r-1}$.

Classify the situations in Problems 55–59 as arithmetic or geometric. Do NOT answer the questions.

55. Suppose that a teacher obtains a job with a starting salary of $15,000 and receives a $500 raise every year thereafter. What will the teacher's salary be in 10 years?

56. Suppose that an autoworker obtains a job with a starting salary of $20,000 and receives a 8% raise every year thereafter. What will the autoworker's salary be in 10 years?

57. Suppose that a teacher obtains a job with a starting salary of $15,000 and receives a $3\frac{1}{3}$% raise every year thereafter. What will the teacher's salary be in 10 years?

58. A grocery clerk must stack 30 cases of canned fruit, each containing 24 cans. He decides to display the cans by stacking them in a pyramid where each row after the bottom row contains one less can. It is possible to use all the cans and end up with a top row of only one can?

59. Suppose that a chain letter asks you to send copies to 10 of your friends. If everyone carries out the directions and sends the chain letter, and if nobody receives more than one chain letter, how many people will be involved in 10 mailings?

60. Compare Problems 55 and 57. Since $3\frac{1}{3}$% of $15,000 is $500 do you think that the total amount earned in 10 years will be the same for both of these problems? If not, which one do you think will be the larger, and why?

C

61. a. Write out $\sum\limits_{j=1}^{r} a_j b_j$ without summation notation.

     b. Let $b_j = m$ and show that $\sum\limits_{j=1}^{r} ma_j = m \sum\limits_{j=1}^{r} a_j$.

     c. Let $a_j = 1$ and show that $\sum\limits_{j=1}^{r} m = mr$.

62. Show that $\sum\limits_{k=1}^{n} (a_k + b_k) = \sum\limits_{k=1}^{n} a_k + \sum\limits_{k=1}^{n} b_k$.

Find the next term for the sequences in Problems 63–65. Use any rule you can defend.

63. 1, 3, 4, 7, 11, 18, 29, _____

64. 225, 625, 1,225, 2,025, _____

65. 8, 5, 4, 9, 1, _____

10.4 Arithmetic Sequences and Series

This section focuses on arithmetic sequences and series.

GENERAL TERM OF AN ARITHMETIC SEQUENCE

> The **general term of an arithmetic sequence** $a_1, a_2, a_3, \ldots, a_n, \ldots$ with common difference d is
>
> $$a_n = a_1 + (n-1)d$$
>
> for $n \geq 1$.

Even though this formula can be proved using mathematical induction, let us consider the first few terms to see where it came from.

Remember that an arithmetic sequence is one in which each succeeding term follows from the previous term by adding the common difference d. Thus,

for $n = 2$, then $a_2 = a_1 + d$;
for $n = 3$, then $a_3 = a_2 + d = a_1 + d + d = a_1 + 2d$;
for $n = 4$, then $a_4 = a_3 + d = a_1 + 2d + d = a_1 + 3d$;
$$\vdots$$

Do you see how each step is substituted into the next step? Look for a pattern:

$$a_5 = a_1 + 4d$$

$\uparrow \qquad \uparrow$
$\,n \qquad$ 1 less than n

Thus, it is reasonable to write

$$a_n = a_1 + (n-1)d$$

EXAMPLE 1 Find the general term of the arithmetic sequence $18, 14, 10, 6, \ldots$.

SOLUTION $a_1 = 18; \quad d = 14 - 18 = -4; \quad$ thus

$$
\begin{aligned}
a_n &= a_1 + (n-1)d \\
&= 18 + (n-1)(-4) \\
&= 18 - 4n + 4 \\
&= \mathbf{22 - 4n}
\end{aligned}
$$

EXAMPLE 2 If $a_5 = 14$, $a_{10} = 34$, find d.

SOLUTION Use the formula $a_n = a_1 + (n-1)d$:

$$a_5 = a_1 + 4d \qquad \text{or, since } a_5 = 14, \qquad 14 = a_1 + 4d$$
$$a_{10} = a_1 + 9d \qquad \text{or, since } a_{10} = 34, \qquad 34 = a_1 + 9d$$

This can be written as the system

$$\begin{cases} a_1 + 4d = 14 \\ a_1 + 9d = 34 \end{cases}$$

Multiply the first equation by -1 and add:

$$5d = 20$$
$$\mathbf{d = 4}$$

What is the total number of blocks shown in Figure 10.2?

FIGURE 10.2 How many blocks?

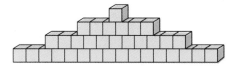

The numbers of blocks in the successive rows form an arithmetic sequence: 1, 6, 11, 16. The total number of blocks is the sum

$$\sum_{j=1}^{4} (-4 + 5j)$$

The indicated sum of an arithmetic sequence is called an **arithmetic series.** Let A_n denote the arithmetic series

$$\sum_{k=1}^{n} a_k = a_1 + a_2 + a_3 + \cdots + a_n$$

EXAMPLE 3 **a.** How many blocks are in the stack shown in Figure 10.2?
b. How many blocks are in 10 rows of a stack of blocks similar to the one shown in Figure 10.2?

SOLUTION **a.**

$$A_1 = a_1 = 1$$
$$A_2 = a_1 + a_2 = 1 + 6 = 7$$
$$A_3 = a_1 + a_2 + a_3 = 1 + 6 + 11 = 18$$
$$A_4 = a_1 + a_2 + a_3 + a_4 = 1 + 6 + 11 + 16 = \mathbf{34}$$

This is the number of blocks in Figure 10.2.

b.

$$A_{10} = a_1 + a_2 + \cdots + a_9 + a_{10}$$
$$= 1 + 6 + 11 + 16 + 21 + 26 + 31 + 36 + 41 + 46$$
$$= \mathbf{235 \text{ blocks in a stack 10 rows high}}$$

In Example 3, we found A_{10} by brute-force addition. This method would not be practical for large n, however, so it is desirable to find a general formula for A_n as we did for a_n. Consider the following method for finding A_{10} of Example 3:

$$A_{10} = 1 + 6 + 11 + 16 + 21 + 26 + 31 + 36 + 41 + 46$$

and

$$A_{10} = 46 + 41 + 36 + 31 + 26 + 21 + 16 + 11 + 6 + 1$$

Add these equations term by term:

$$A_{10} + A_{10} = (1 + 46) + (6 + 41) + (11 + 36) + \cdots + (41 + 6) + (46 + 1)$$
$$2A_{10} = 47 + 47 + 47 + \cdots + 47 + 47$$

Notice that the sums of all the numbers within parentheses are equal to the sum of the first and last terms. Instead of doing a lot of addition, we can find the result by multiplication. In this example $n = 10$, so the number of terms (without directly counting them) is 10. Thus

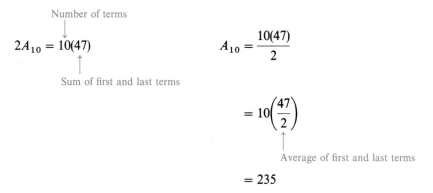

$$= 235$$

ARITHMETIC SERIES

For an arithmetic sequence $a_1, a_2, a_3, \ldots, a_n$, with common difference d, the sum of the arithmetic series

$$\sum_{k=1}^{n} a_k = a_1 + a_2 + a_3 + \cdots + a_n$$

is

$$A_n = n\left(\frac{a_1 + a_n}{2}\right) \qquad \text{or, equivalently,} \qquad A_n = \frac{n}{2}[2a_1 + (n-1)d]$$

PROOF

$$A_n = a_1 + a_2 + a_3 + \cdots + a_{n-1} + a_n$$
$$A_n = a_n + a_{n-1} + \cdots + a_3 + a_2 + a_1$$

Add these equations term by term:

$$2A_n = (a_1 + a_n) + (a_2 + a_{n-1}) + \cdots + (a_{n-1} + a_2) + (a_n + a_1)$$

All the quantities enclosed by parentheses are equal because

$$a_1 + a_n = a_1 + [a_1 + (n-1)d]$$
$$= 2a_1 + (n-1)d$$
$$a_2 + a_{n-1} = (a_1 + d) + [a_1 + (n-2)d]$$
$$= 2a_1 + (n-1)d$$
$$a_3 + a_{n-2} = (a_1 + 2d) + [a_1 + (n-3)d]$$
$$= 2a_1 + (n-1)d$$
$$\vdots$$

Since there is a total of n such sums,

$$2A_n = n[2a_1 + (n-1)d]$$

$$A_n = \frac{n}{2}[2a_1 + (n-1)d]$$

For the other part of the formula, replace $2a_1 + (n-1)d$ with $a_1 + a_n$ because

$$a_1 + a_n = a_1 + [a_1 + (n-1)d]$$
$$= 2a_1 + (n-1)d$$

Therefore $A_n = n\left(\dfrac{a_1 + a_n}{2}\right)$

□□□

EXAMPLE 4   Find A_{100} for the sequence of Example 3.

SOLUTION $a_1 = 1;\quad n = 100;\quad d = 6 - 1 = 5;\quad$ thus

$$A_{100} = \frac{100}{2}[2(1) + 99(5)] = \mathbf{24{,}850}$$

∎

EXAMPLE 5   Find A_{10} where $a_1 = 6$ and $a_7 = -18$.

SOLUTION   First find d:

$$a_n = a_1 + (n-1)d$$
$$a_7 = a_1 + (7-1)d \qquad \text{Replace } n \text{ by } 7.$$
$$-18 = 6 + 6d \qquad\qquad \text{Substitute the given values of } a_7 \text{ and } a_1.$$
$$-24 = 6d$$
$$-4 = d$$

Then find A_{10}:

$$A_n = \frac{n}{2}[2a_1 + (n-1)d]$$

$$A_{10} = \frac{10}{2}[2(6) + (10-1)(-4)] = \mathbf{-120}$$

∎

10.4   Problem Set

A   *Write out the first four terms of the arithmetic sequences in Problems 1–12.*

1. $a_1 = 5, d = 4$

2. $a_1 = -5, d = 3$

3. $a_1 = 85, d = 3$

4. $a_1 = 50, d = -10$

5. $a_1 = 100, d = -5$

6. $a_1 = 20, d = -4$

7. $a_1 = -\frac{5}{2}, d = -\frac{1}{2}$

8. $a_1 = \frac{2}{3}, d = -\frac{5}{3}$

9. $a_1 = \sqrt{12}, d = \sqrt{3}$

10. $a_1 = 5, d = x$

11. $a_1 = x, d = y$

12. $a_1 = m, d = 2m$

Find a_1 and d for the arithmetic sequences in Problems 13–21.

13. $5, 8, 11, \ldots$

14. $5, 5, 5, \ldots$

15. $6, 11, 16, \ldots$

16. $35, 46, 57, \ldots$

17. $-8, -1, 6, \ldots$

18. $-1, 1, 3, \ldots$

19. $x, 2x, 3x, \ldots$

20. $x + \sqrt{3}, x + \sqrt{12}, x + \sqrt{27}, \ldots$

21. $x - 5b, x - 3b, x - b, \ldots$

B *Find an expression for the general term of each arithmetic sequence in Problems 22–30.*

22. 5, 8, 11, …

23. 5, 5, 5, …

24. 6, 11, 16, …

25. 35, 46, 57, …

26. $-8, -1, 6, …$

27. $-1, 1, 3, …$

28. $x, 2x, 3x, …$

29. $x + \sqrt{3}, x + \sqrt{12}, x + \sqrt{27}, …$

30. $x - 5b, x - 3b, x - b, …$

Find the indicated quantity for each arithmetic sequence in Problems 31–45.

31. $a_1 = 6, d = 5; a_{20}$

32. $a_1 = 35, d = 11; a_{10}$

33. $a_1 = -20, d = 5; a_{10}$

34. $a_1 = 35, d = 11; A_{10}$

35. $a_1 = -7, d = -2; A_{100}$

36. $a_1 = 15, d = -4; A_{50}$

37. $a_1 = -5, a_{30} = -63; d$

38. $a_1 = 4, a_6 = 24; d$

39. $a_1 = -13, a_{10} = 5; d$

40. $a_1 = 4, a_6 = 24; A_{15}$

41. $a_1 = -5, a_{30} = -63; A_{10}$

42. $a_1 = 110, a_{11} = 0; d$

43. $a_5 = 27, a_{10} = 47; d$

44. $a_4 = 36, a_5 = 60; d$

45. $a_3 = 36, a_5 = 60; d$

46. a. How many blocks are shown in Figure 10.3?
 b. How many blocks would there be in 100 rows of a stack of blocks like the one shown in Figure 10.3?

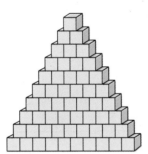

FIGURE 10.3
How many blocks?

47. Find the sum of the first 20 terms of the arithmetic sequence with first term 100 and common difference 50.

48. Find the sum of the first 50 terms of the arithmetic sequence with first term -15 and common difference 5.

49. Find the sum of the even integers between 41 and 99.

50. Find the sum of the odd integers between 100 and 80.

51. Find the sum of the first n odd integers.

52. Find the sum of the first n even integers.

53. A sequence $s_1, s_2, …, s_n$ is a **harmonic sequence** if its reciprocals form an arithmetic sequence. Which of the following are harmonic sequences?
 a. $1, \frac{1}{2}, \frac{1}{3}, \frac{1}{4}, \frac{1}{5}, …$
 b. $\frac{1}{2}, \frac{1}{5}, \frac{1}{8}, \frac{1}{11}, \frac{1}{14}, …$
 c. $2, \frac{2}{3}, \frac{2}{5}, \frac{2}{7}, …$
 d. $\frac{1}{5}, -\frac{1}{5}, -\frac{1}{15}, -\frac{1}{25}, …$
 e. $\frac{3}{4}, \frac{1}{2}, \frac{1}{3}, \frac{2}{9}, …$

54. Consider the arithmetic sequence a_1, x, a_3. The number x can be found as follows:

$$+\begin{cases} x = a_1 + d \\ x = a_3 - d \end{cases}$$
$$\overline{\quad 2x = a_1 + a_3 \quad} \quad \text{By adding}$$
$$x = \frac{a_1 + a_3}{2}$$

Here x is called the **arithmetic mean** between a_1 and a_3. Find the arithmetic mean between each of the given pairs of numbers.
 a. 1, 8 **b.** 1, 7 **c.** $-5, 3$ **d.** 80, 88 **e.** 40, 56

55. Find the arithmetic mean (see Problem 54) between each of the given pairs of numbers.
 a. 4, 20 **b.** 4, 15 **c.** $\frac{1}{2}, \frac{1}{3}$ **d.** $-10, -2$ **e.** $-\frac{2}{3}, \frac{4}{5}$

C

56. Suppose you were hired for a job paying $21,000 per year and were given the following options:

 Option A: annual salary increase of $1,440
 Option B: semiannual salary increase of $360

 Which is the better option?

57. Repeat Problem 56 for the following options:

 Option C: quarterly salary increase of $90
 Option D: monthly salary increase of $10

58. What are the differences in the amounts earned in the first year from options A to D in Problems 56 and 57?

59. Repeat Problem 58 for the first 2 years.

60. Write the arithmetic series for the total amount of money earned in 10 years under the following options described in Problem 56.
 a. Option A **b.** Option B

10.5 Geometric Sequences and Series

As with arithmetic sequences, we will denote the terms of a geometric sequence by using a special notation. Let $g_1, g_2, g_3, \ldots, g_n$ be the terms of a geometric sequence. To find the general term of a geometric sequence, use r to represent the common ratio. Thus

$$g_2 = rg_1$$
$$g_3 = rg_2 = r(rg_1) = r^2g_1$$
$$g_4 = rg_3 = r(r^2g_1) = r^3g_1$$
$$\vdots$$
$$g_n = g_1 r^{n-1}$$

GENERAL TERM OF A GEOMETRIC SEQUENCE

For a geometric sequence $g_1, g_2, g_3, \ldots, g_n$ with a common ratio r,

$$g_n = g_1 r^{n-1}$$

for every $n \geq 1$.

EXAMPLE 1 Find the general term of the geometric sequence $50, 100, 200, \ldots$.

SOLUTION
$$g_1 = 50 \qquad r = \frac{100}{50} = 2$$
$$g_n = 50(2)^{n-1} = 2 \cdot 5^2 \cdot 2^{n-1} = \mathbf{2^n \cdot 5^2}$$

EXAMPLE 2 Find r when $g_1 = 20$ and $g_5 = 200$.

SOLUTION Use $g_n = g_1 r^{n-1}$:

$$g_5 = 20r^4$$
$$200 = 20r^4$$
$$10 = r^4$$
$$r = \sqrt[4]{10}$$

Dear Friend,

This is a chain letter …

Copy this letter six times and send it to six of your friends. In twenty days, you will have good luck.

If you break this chain, you will have bad luck! …

Suppose you receive the letter shown in the margin. Consider the number of people that could become involved with this chain letter if we assume that everyone carries out their task and does not break the chain. The first mailing would consist of six letters with seven people involved:

$$1 + 6 = 7$$

The second mailing would involve 43 people since the second mailing of 36 letters (each of the six people receiving a letter sends out six more letters) is added to the total:

$$1 + 6 + 36 = 43$$

The number of letters in each successive mailing is a number of a geometric sequence:

1st mailing: $6 = 6$
2nd mailing: $6^2 = 36$
3rd mailing: $6^3 = 216$
4th mailing: $6^4 = 1{,}296$

$\vdots$

10th mailing: $6^{10} = 60{,}466{,}176$
11th mailing: $6^{11} = 362{,}797{,}056$

By the eleventh mailing, more letters would have to be sent than there are people in the United States! The number of letters in only two more mailings would exceed the number of men, women, and children in the whole world.

How many people are involved in 11 mailings assuming that no person receives a letter more than once? To answer this question, consider the series associated with the geometric sequence. Then G_n represents the sum of the first n terms of the geometric series. In this problem, then, you need to find G_{11}:

$$G_{11} = 1 + 6 + 36 + 216 + \cdots + 60{,}466{,}176 + 362{,}797{,}056$$

You could add these on your calculator, but that would take a long time; instead write these numbers by using exponents:

$$G_{11} = 1 + 6 + 6^2 + 6^3 + \cdots + 6^{10} + 6^{11}$$

Next multiply both sides by 6:

$$6G_{11} = 6 + 6^2 + 6^3 + 6^4 + \cdots + 6^{11} + 6^{12}$$

Finally, consider $G_{11} - 6G_{11}$:

$$G_{11} - 6G_{11} = 1 - 6^{12} \qquad \text{Subtract (see two preceding equations).}$$
$$-5G_{11} = 1 - 6^{12}$$
$$G_{11} = \frac{1 - 6^{12}}{-5} \quad \text{or} \quad \frac{1}{5}(6^{12} - 1)$$

This number is easy to find on your calculator. But, more important, it leads to a procedure for finding G_n in general, where G_n represents the geometric series.

$$\sum_{k=1}^{n} g_k = g_1 + g_2 + g_3 + \cdots + g_n$$

The next step is to find a formula for G_n:

$$G_n = g_1 + g_1r + g_1r^2 + g_1r^3 + \cdots + g_1r^{n-1}$$

Multiply both sides by r:

$$rG_n = g_1r + g_1r^2 + g_1r^3 + g_1r^4 + \cdots + g_1r^n$$

Notice that, except for the first and last terms, all the terms in the expressions for

G_n and rG_n are the same, so that

$$G_n - rG_n = g_1 - g_1 r^n$$

Now solve for G_n:

$$(1 - r)G_n = g_1(1 - r^n)$$
$$G_n = \frac{g_1(1 - r^n)}{1 - r} \quad (r \neq 1)$$

GEOMETRIC SERIES

For a geometric series $\displaystyle\sum_{k=1}^{n} g_k = g_1 + g_2 + g_3 + \cdots + g_n$ with common ratio $r, r \neq 1$,

$$G_n = \frac{g_1(1 - r^n)}{1 - r}$$

EXAMPLE 3   Find the sum of the first five terms of a geometric series with $g_1 = -15$ and $r = 2$.

SOLUTION
$$G_n = \frac{g_1(1 - r^n)}{1 - r} = \frac{-15(1 - 2^n)}{1 - 2}$$

$$= \frac{-15}{-1}(1 - 2^n) = 15(1 - 2^n)$$

$$G_5 = 15(1 - 2^5) = 15(1 - 32) = \mathbf{-465}$$ ∎

EXAMPLE 4   Find

$$\sum_{k=1}^{10} \left(\frac{1}{2}\right)^k = \left(\frac{1}{2}\right)^1 + \left(\frac{1}{2}\right)^2 + \left(\frac{1}{2}\right)^3 + \left(\frac{1}{2}\right)^4 + \cdots + \left(\frac{1}{2}\right)^{10}$$

$$= \frac{1}{2} + \frac{1}{4} + \frac{1}{8} + \frac{1}{16} + \cdots + \frac{1}{1,024}$$

This is a geometric series with $g_1 = \frac{1}{2}$ and $r = \frac{1}{2}$.

$$\sum_{k=1}^{10} \left(\frac{1}{2}\right)^k = G_{10} = \frac{\frac{1}{2}[1 - (\frac{1}{2})^{10}]}{1 - \frac{1}{2}} = 1 - \left(\frac{1}{2}\right)^{10} = \frac{\mathbf{1,023}}{\mathbf{1,024}}$$ ∎

The geometric series presented above is finite. Consider now an infinite geometric series. Remember:

1. $s_1, s_2, s_3, \ldots,$ is a **sequence.**
2. $s_1 + s_2 + s_3 + \cdots + s_n$ is a **series** denoted by S_n.
3. If we consider $s_1 + s_2 + s_3 + \cdots + s_n$, then S_n is called the **nth partial sum.**
4. Now consider $S_1, S_2, S_3, \ldots$. This is a **sequence of partial sums.**

Consider the sequence of partial sums for Example 4:

$$G_1 = \frac{1}{2}$$

$$G_2 = \frac{1}{2} + \frac{1}{4} = \frac{3}{4}$$

$$G_3 = \frac{1}{2} + \frac{1}{4} + \frac{1}{8} = \frac{7}{8}$$

$$\vdots$$

$$G_{10} = \frac{1,023}{1,024} \qquad \text{This result was found in Example 4.}$$

It appears that the partial sums are getting closer to 1 as n becomes larger. We *can* find the sum of an infinite geometric sequence. Consider

$$G_n = \frac{g_1(1 - r^n)}{1 - r} = \frac{g_1 - g_1 r^n}{1 - r} = \frac{g_1}{1 - r} - \frac{g_1}{1 - r} r^n$$

Now g_1, r, and $1 - r$ are fixed numbers. If $|r| < 1$, then $r^n \to 0$ as $n \to \infty$ and thus

$$G_n \to \frac{g_1}{1 - r} \quad \text{as } n \to \infty$$

because the second term is approaching zero. This is not a proof, of course, but it does lead to the following result, which can be proved in a calculus course.

SUM OF AN INFINITE GEOMETRIC SERIES

If $g_1, g_2, g_3, \ldots, g_n, \ldots$ is an infinite geometric sequence with a common ratio r such that $|r| < 1$, then its sum is denoted by G and is found by

$$G = \frac{g_1}{1 - r}$$

If $|r| \geq 1$, the infinite geometric series has no sum.

EXAMPLE 5   Find the sum of the series $100 + 50 + 25 + \cdots$ if possible.

SOLUTION   Since $g_1 = 100$ and $r = \frac{1}{2}$, then $G = \dfrac{100}{(1 - \frac{1}{2})} = \mathbf{200}$.   ▮

EXAMPLE 6   Find the sum of the series $-5 + 10 - 20 + \cdots$ if possible.

SOLUTION $g_1 = -5$ and $r = -2$; since $|r| \geq 1$, this infinite series does not have a sum.   ▮

EXAMPLE 7   The repeating decimal $.\overline{72} = .72727272 \ldots$ is a rational number and can therefore be written as the quotient of two integers. Find this representation by using a geometric series.

$\text{SOLUTION}\quad .72727272\ldots = .72 + .0072 + .000072 + \cdots$

Notice that two digits are repeating, so form a series using the repeating digits and leading zeros as placeholders.

$$= .72 + .72(.01) + .72(.0001) + \cdots$$

Factor out the common (repeating) digits.

$$= .72 + .72(.01) + .72(.01)^2 + \cdots$$

Now you can see the common ratio.

Now $g_1 = .72$ and $r = .01$; then

$$G = \frac{.72}{1 - .01} = \frac{.72}{.99} = \frac{\frac{72}{100}}{\frac{99}{100}} = \frac{72}{99} = \frac{8}{11}$$

10.5 Problem Set

A *Write out the first three terms of the geometric sequences in Problems 1–8 with the first term g_1 and common ratio r.*

1. $g_1 = 5, r = 3$

2. $g_1 = -12, r = 3$

3. $g_1 = 1, r = -2$

4. $g_1 = 1, r = 2$

5. $g_1 = -15, r = \frac{1}{5}$

6. $g_1 = 625, r = -\frac{1}{5}$

7. $g_1 = 8, r = x$

8. $g_1 = a, r = \frac{1}{2}$

Find g_1 and r for the geometric sequences in Problems 9–14.

9. $3, 6, 12, \ldots$

10. $7, 14, 28, \ldots$

11. $1, \frac{1}{2}, \frac{1}{4}, \ldots$

12. $100, 50, 25, \ldots$

13. $x, x^2, x^3, \ldots$

14. $xyz, xy, \ldots$

Find the general term for each of the geometric sequences in Problems 15–20.

15. $3, 6, 12, \ldots$

16. $7, 14, 28, \ldots$

17. $1, \frac{1}{2}, \frac{1}{4}, \ldots$

18. $100, 50, 25, \ldots$

19. $x, x^2, x^3, \ldots$

20. $xyz, xy, \ldots$

Find the sum, if possible, of the infinite geometric series in Problems 21–26.

21. $1 + \frac{1}{2} + \frac{1}{4} + \cdots$

22. $1,000 + 500 + 250 + \cdots$

23. $100 + 50 + 25 + \cdots$

24. $-20 + 10 - 5 + \cdots$

25. $-45 - 15 - 5 - \cdots$

26. $-216 - 36 - 6 - \cdots$

B *Find the indicated quantities in Problems 27–38 for the given geometric sequences.*

27. $g_1 = 6, r = 3; g_5$

28. $g_1 = 100, r = \frac{1}{10}; g_{10}$

29. $g_1 = 6, r = 3; G_5$

30. $g_1 = 7, g_8 = 896; r$

31. $g_1 = 1, r = 10; G_{10}$

32. $g_1 = 3, r = \frac{1}{5}; G_3$

33. $g_1 = \frac{1}{3}, r = \frac{1}{3}; G$

34. $g_1 = \frac{1}{4}, r = \frac{1}{4}; G$

35. $g_1 = 1, r = .08; G$

36. $\sum\limits_{k=1}^{4} \left(\frac{1}{10}\right)^k$

37. $\sum\limits_{k=1}^{4} \left(\frac{1}{3}\right)^k$

38. $\sum\limits_{k=1}^{4} \left(\frac{1}{4}\right)^k$

Represent each repeating decimal in Problems 39–50 as the quotient of two integers by considering an infinite geometric series.

39. $.\overline{4}$

40. $.\overline{5}$

41. $.\overline{9}$

42. $.\overline{27}$

43. $.\overline{18}$

44. $.\overline{45}$

45. $.\overline{418}$

46. $.\overline{218}$

47. $.\overline{123}$

48. $2.\overline{45}$

49. $5.03\overline{1}$

50. $2.25\overline{34}$

51. SOCIAL SCIENCE Suppose that a chain letter asks you to send copies to 10 of your friends. If everyone carries out the directions and sends the chain letter, and if nobody receives more than one chain letter, how many people will be involved in five mailings? Count yourself, the 10 letters you mail, the letters they mail, and so forth.

52. SOCIAL SCIENCE According to the 1990 census, the U.S. population is 249,632,692. If everyone follows the directions in the chain letter mentioned in Problem 51, how many mailings would be necessary to include the *entire* U.S. population?

53. A new type of Superball advertises that it will rebound to nine-tenths of its original height. If it is dropped from a height of 10 ft, how far will the ball travel before coming to rest?

54. Repeat Problem 53 for a ball that rebounds to two-thirds of its original height.

55. BUSINESS Suppose that a piece of machinery costing $10,000 depreciates 20% of its present value each year. That is, the first year $10,000(.20) = $2,000 is depreciated. The second year's depreciation is

$$\$8,000(.20) = \$1,600$$

since the value for the second year is $10,000 − $2,000 = $8,000. The third year's depreciation is

$$6,400(.20) = \$1,280$$

If the depreciation is calculated this way indefinitely, what is the total depreciation?

56. BUSINESS   Winnie Winner wins $100 in a pie-baking contest run by the Hi-Do Pie Co. The company gives Winnie the $100. However, the tax collector wants 20% of the $100. Winnie pays the tax. But then she realizes that she didn't really win a $100 prize and tells her story to the Hi-Do Co. The friendly Hi-Do Co. gives Winnie the $20 she paid in taxes. Unfortunately, the tax collector now wants 20% of the $20. She pays the tax again and then goes back to the Hi-Do Co. with her story. Assume that this can go on indefinitely. How much money does the Hi-Do Co. have to give Winnie so that she will really win $100? How much does she pay in taxes?

57. Consider the geometric sequence g_1, x, g_3. The number x can be found by considering

$$\frac{x}{g_1} = r \quad \text{and} \quad \frac{g_3}{x} = r$$

Thus $\dfrac{x}{g_1} = \dfrac{g_3}{x}$ so $x^2 = g_1 g_3$

This equation has two solutions:

$$x = \sqrt{g_1 g_3} \quad \text{and} \quad x = -\sqrt{g_1 g_2}$$

If g_1 and g_3 are both positive, then $\sqrt{g_1 g_3}$ is called the **geometric mean** of g_1 and g_3. If g_1 and g_3 are both negative, then $-\sqrt{g_1 g_3}$ is called the **geometric mean.** Find the geometric mean of each of the given pairs of numbers:

a. 1, 8   **b.** 2, 8,   **c.** −5, −3
d. −10, −2   **e.** 4, 20

C

Find the sum of the infinite geometric series in Problems 58–61.

58. $2 + \sqrt{2} + 1 + \cdots$ **59.** $3 + \sqrt{3} + 1 + \cdots$

60. $(1 + \sqrt{2}) + 1 + (-1 + \sqrt{2}) + \cdots$

61. $(\sqrt{2} - 1) + 1 + (\sqrt{2} + 1) + \cdots$

62. Find three distinct numbers with a sum equal to 9 so that these numbers form an arithmetic sequence and their squares form a geometric sequence.

63. The infinite series

$$\lim_{n \to \infty} \sum_{k=1}^{n} \frac{1}{k^p} = \frac{1}{1^p} + \frac{1}{2^p} + \frac{1}{3^p} + \cdots$$

is called the **p-series.** Look at the sequence of partial sums $s_1, s_2, s_3, s_4, \ldots$, and state whether you think the given infinite p-series has a limit.
a. $p = 1$ **b.** $p = 2$ **c.** $p = 3$ **d.** $p = .5$

64. Can you make a conjecture (guess) about the values of p in Problem 63 for which the infinite p-series has a limit?

65. Square $ABCD$ has sides of length 1. Square $EFGH$ is formed by connecting the midpoints of the sides of the first square, as shown in Figure 10.4. Assume that the pattern of shaded regions in the square is continued indefinitely. What is the total area of the shaded regions?

FIGURE 10.4   Find the area of the shaded regions.

66. Repeat Problem 65 for a square whose sides have length a.

10.6   Chapter 10 Summary

The material of this chapter is reviewed in the following list of objectives. After each objective there are some practice questions. For a sample test, select the first question of each set and check your answers with the answer section. For a sample test without answers, use the second question of each set. Additional practice is given by the other questions in each set. If you are having trouble with a particular type of problem, look back to that section for extra help.

10.1   Mathematical Induction

OBJECTIVE 1   *State the Principle of Mathematical Induction and prove propositions using this principle.*

1. State the Principle of Mathematical Induction.

2. If $4 + 8 + 12 + \cdots + 4n = 2n(n + 1)$, then state the proposition $P(1)$ and prove it is true.

3. State the propositions $P(k)$ and $P(k + 1)$ for the proposition given in Problem 2.

4. Prove that $4 + 8 + 12 + \cdots + 4n = 2n(n + 1)$ for all positive integers n.

10.2 Binomial Theorem

OBJECTIVE 2 *Use Pascal's triangle to expand binomials.*
Expand:

5. $(a + b)^5$

6. $(x - y)^5$

7. $(2x + y)^5$

8. $(3x^2 - y^3)^5$

OBJECTIVE 3 *Simplify expressions containing factorial and binomial coefficient notations.*

9. $\dfrac{52!}{5!47!}$

10. $\left(\dfrac{8}{4}\right)!$

11. $\dbinom{8}{4}$

12. $\dbinom{p}{q}$

OBJECTIVE 4 *State the Binomial Theorem, expand binomials by formula, and find particular terms in a binomial expansion.*

13. State the binomial formula with and without summation notation.

14. Write out the first four terms in the expansion of $(a + b)^{18}$.

15. Write the term that includes y^r in the expansion of $(x - y)^{15}$.

16. What is the coefficient of $x^8 y^4$ in the expansion of $(x + 2y)^{12}$?

10.3 Sequences, Series, and Summation Notation

OBJECTIVE 5 *Classify a sequence as arithmetic, geometric, or neither. If it is arithmetic or geometric find d or r and give the general term; if it is neither, find the pattern* and give the next two terms.

17. 1, 11, 21, 31, ...

18. 1, 11, 121, 1331, ...

19. 1, 11, 111, 1111, ...

20. 54, 18, 6, 2, ...

10.4, 10.5 Arithmetic and Geometric Sequences and Series

OBJECTIVE 6 *Use summation notation with series.* Evaluate:

21. $\displaystyle\sum_{k=0}^{3} 3^k$

22. $\displaystyle\sum_{k=1}^{4} 5$

23. $\displaystyle\sum_{k=1}^{10} 2(3)^{k-1}$

24. $\displaystyle\sum_{k=1}^{100} [5 + (k - 1)4]$

Find the indicated quantities for the given sequences:

29. $a_1 = 2$, $a_{10} = 20$; d, A_{10}

30. $a_2 = 5$, $d = 13$; a_n

31. $g_1 = 5$, $r = 2$; g_{10}, G_5

32. $a_1 = 50$, $d = -5$; a_{10}, A_5

OBJECTIVE 7 *Find the sum of n terms of an arithmetic or geometric sequence.* Find the sum of the first 10 terms of the following sequences.

25. 1, 11, 21, 31, ...

26. 1, 11, 121, 1331, ...

27. 54, 18, 6, 2, ...

28. 5, 5, 5, 5, ...

OBJECTIVE 8 *Work with the following formulas for sequences and series:*

Arithmetic Sequence	Geometric Sequence
$a_n = a_1 + (n - 1)d$	$g_n = g_1 r^{n-1}$

Arithmetic Series	Geometric Series
$A_n = n\left(\dfrac{a_1 + a_n}{2}\right)$	$G_n = \dfrac{g_1(1 - r^n)}{1 - r}$
$A_n = \dfrac{n}{2}[2a_1 + (n - 1)d]$	$G = \dfrac{g_1}{1 - r}, \quad |r| < 1$

OBJECTIVE 9 *Find the sum of an infinite geometric series including the fractional representation of a repeating decimal.*

33. Find the sum of the infinite series $1{,}000 + 500 + 250 + \cdots$.

34. Find the sum of the infinite series

$$\frac{1}{8} + \frac{1}{16} + \frac{1}{32} + \cdots.$$

35. Find $2.\overline{18}$ as the quotient of two integers by considering an infinite geometric series.

36. Find $3.1\overline{6}$ as the quotient of two integers by considering an infinite geometric series.

**Charlotte Angas Scott
(1858–1931)**

Indeed, mathematics, the
indispensable tool of the sciences,
defying the senses to follow its
splendid flights, is demonstrating
today, as it never has been
demonstrated before, the
supremacy of the pure reason.

NICHOLAS BUTLER
*The Meaning of Education and
Other Essays and Addresses*

The first prominent female mathematician in America was Charlotte Angas Scott. She is best known because of her lifelong work as an educator. She studied in England, but soon after receiving her Ph.D., she came to America and developed the mathematics program at Bryn Mawr College. Today, it is difficult for us to understand the problems encountered by a woman in Scott's day who wished to receive a college education. Scott, for example, was permitted to take her final undergraduate examinations at Cambridge only informally. According to Karen Rappaport,*

"Scott tied for eighth place in the mathematics exam. Mathematics was an unprecedented area for a woman to excel in and the achievement attracted public attention, especially when her name was not mentioned at the official ceremony. When the official name was read, loud shouts of 'Scott of Girton' could be heard in the gallery. The public also responded to the slight. The February 7, 1880 issue of Punch stated: 'But when the academy doors are reopened to the Ladies let them be opened to their full worth. Let us not hear of any restrictions or exclusions from this or that function or privilege. . . .' As a result of this support and a public petition, women were formally admitted to the Tripos Exams the following year but they were still not permitted Cambridge degrees. (That event did not occur until 1948.)"

Mathematically, Scott was concerned with the study of specific algebraic curves of degree higher than 2. She published thirty papers in the field of algebraic geometry, as well as a text, *Modern Analytic Geometry*, in 1894. In 1922, members of the American Mathematical Society and many former students organized a dinner to honor Scott. Alfred North Whitehead came from England to give the main address, in which he stated: "A life's work such as that of Professor Charlotte Angas Scott is worth more to the world than many anxious efforts of diplomatists."

---

*Karen D. Rappaport, "Two American Women Mathematicians," *MATYC Journal*, Fall (1980), p. 203.

11

ANALYTIC GEOMETRY

Contents

Preview

The conic sections consist of the circles (already considered), parabolas, ellipses, and hyperbolas. In this chapter we consider the standard and general forms of each of these curves as well as the translated and rotated forms. (Section 11.4 on rotations may easily be omitted.) Next, we look at vectors and vector operations, and conclude this chapter with polar coordinates, polar-form curves, and parametric equations. There are 24 objectives for this chapter, which are listed on pages 524–526.

Perspective

Although many calculus books include analytic geometry, they often assume that students have a rudimentary knowledge of the fundamentals, in particular of conic sections. In the following page taken from a calculus book, we see hyperbolas being used in a navigational system.

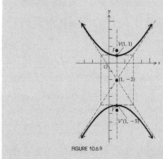

$$9(x^2 - 2x + 1) - 4(y^2 + 4y + 4) = -29 + 9 - 16$$

$$9(x - 1)^2 - 4(y + 2)^2 = -36$$

$$\frac{(y + 2)^2}{9} - \frac{(x - 1)^2}{4} = 1 \qquad (14)$$

Equation (14) has the form of (11); so the graph is a hyperbola whose principal axis is parallel to the y axis and whose center is at $(1, -2)$.

EXAMPLE 5
Find the vertices and foci of the hyperbola of Example 4. Draw a sketch showing the hyperbola, its asymptotes, and the foci.

SOLUTION
From (14) we observe that $a = 3$ and $b = 2$. Because the principal axis is vertical, the center is at $(1, -2)$, and $a = 3$, it follows that the vertices are at the points $V(1, 1)$ and $V'(1, -5)$. For a hyperbola, $c^2 = a^2 + b^2$; therefore $c^2 = 9 + 4$ and $c = \sqrt{13}$. Therefore the foci are at $F(1, -2 + \sqrt{13})$ and $F'(1, -2 - \sqrt{13})$. Figure 10.6.9 shows the hyperbola, its asymptotes, and the foci.

FIGURE 10.6.9

The property of the hyperbola given in Definition 10.6.1 forms the basis of several important navigational systems. These systems involve a network of pairs of radio transmitters at fixed positions at a known distance from one another. The transmitters send out radio signals that are received by a navigator. The difference in arrival time of the two signals determines the difference $2a$ of the distances from the navigator. Thus the navigator's position is known to be somewhere along one arc of a hyperbola having foci at the locations of the two transmitters. One arc, rather than both, is determined because of the signal delay between the two transmitters that is built into the system. The procedure is then repeated for a different pair of radio transmitters, and another arc of a hyperbola that contains the navigator's position is determined. The point of intersection of the two hyperbolic arcs is the actual position. For example, in Fig. 10.6.10 suppose a pair of transmitters is located at points T_1 and S_1 and the signals from this pair determine the hyperbolic arc A_1. Another pair of transmitters is located at points T_2 and S_2 and hyperbolic arc A_2 is determined from their signals. Then the intersection of A_1 and A_2 is the position of the navigator.

FIGURE 10.6.10

11.1 Parabolas

In Chapter 3 we looked at quadratic functions of the form

$$y = ax^2 + bx + c \quad (a \neq 0)$$

The graph of this quadratic equation is a parabola, but not all parabolas can be represented by this equation, because not all parabolas are graphs of functions. Consider the general second-degree equation

$$Ax^2 + Bxy + Cy^2 + Dx + Ey + F = 0$$

for any constants A, B, C, D, E, and F. If $A = B = C = 0$, then the equation is not quadratic but linear (first degree); and if at least one of A, B, or C is not zero, then the equation is quadratic.

Historically, second-degree equations in two variables were first considered in a geometric context and were called **conic sections** because the curves they represent can be described as the intersections of a double-napped right circular cone and a plane. There are three general ways a plane can intersect a cone, as shown in Figure 11.1. (Several special cases are discussed later.)

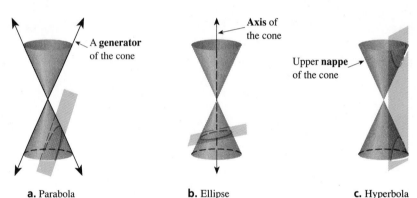

FIGURE 11.1 Conic sections

a. Parabola

The plane is parallel to one of the generators of the cone.

b. Ellipse

The plane intersects only one nappe but is not parallel to a generator. A circle is a special ellipse in which the plane is perpendicular to the axis of the cone.

c. Hyperbola

The plane interesects both nappes of the cone.

Reconsider the parabola; this time take a different geometric viewpoint:

PARABOLA

> A **parabola** is a set of all points in the plane equidistant from a given point (called the **focus**) and a given line (called the **directrix**).

To obtain the graph of a parabola from this definition, you can use the special type of graph paper shown in Figure 11.2, where F is the focus and L is the directrix.

To sketch a parabola using the definition, let F be any point and let L be any line, as shown in Figure 11.2. Plot points in the plane equidistant from the focus, F, and the directrix, L. Draw a line through the focus and perpendicular to the directrix. This line is called the **axis** of the parabola. Let V be the point on this line halfway between the focus and the directrix. This is the point of the parabola

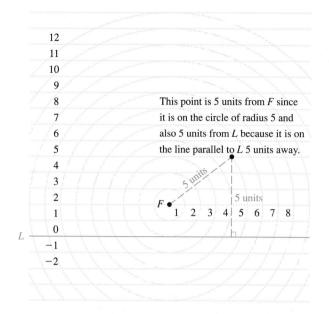

This point is 5 units from F since it is on the circle of radius 5 and also 5 units from L because it is on the line parallel to L 5 units away.

FIGURE 11.2 Parabola graph paper

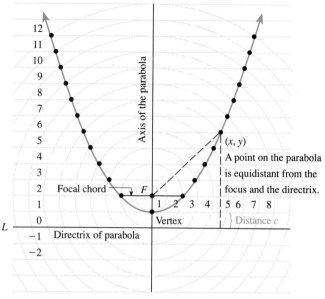

FIGURE 11.3 Parabola graphed from definition

nearest to both the focus and the directrix. It is called the **vertex** of the parabola. Plot other points equidistant from F and L as shown in Figure 11.3.

In Figure 11.3, let c be the distance from the vertex to the focus. Notice that the distance from the vertex to the directrix is also c. Consider the segment that passes through the focus perpendicular to the axis and with endpoints on the parabola. This segment has length $4c$ and is called the **focal chord.**

To obtain the equation of a parabola, first consider a special case—a parabola with focus $F(0, c)$ and directrix $y = -c$, where c is any positive number. This parabola must have its vertex at the origin (remember that the vertex is halfway between the focus and the directrix) and must open upward, as shown in Figure 11.4.

Let (x, y) be any point on the parabola. Then, from the definition of a parabola,

Distance from (x, y) to $(0, c)$ = distance from (x, y) to directrix

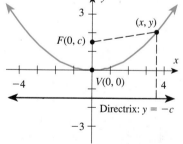

FIGURE 11.4
Graph of the parabola $x^2 = 4cy$

or

$$\sqrt{(x - 0)^2 + (y - c)^2} = y + c$$

Square both sides:

$$x^2 + (y - c)^2 = (y + c)^2$$
$$x^2 + y^2 - 2cy + c^2 = y^2 + 2cy + c^2$$
$$x^2 = 4cy$$

This is the equation of the parabola with vertex $(0, 0)$ and directrix $y = -c$.

You can repeat this argument for parabolas that have vertex at the origin and open downward, to the left, and to the right to obtain the results summarized below. A positive number c, the distance from the focus to the vertex, is assumed given. These are called the **standard-form parabola equations** with vertex $(0, 0)$.

STANDARD-FORM EQUATIONS FOR PARABOLAS WITH VERTEX (0, 0)

PARABOLA	FOCUS	DIRECTRIX	VERTEX	EQUATION
Opens *upward*	$(0, c)$	$y = -c$	$(0, 0)$	$x^2 = 4cy$
Opens *downward*	$(0, -c)$	$y = c$	$(0, 0)$	$x^2 = -4cy$
Opens *right*	$(c, 0)$	$x = -c$	$(0, 0)$	$y^2 = 4cx$
Opens *left*	$(-c, 0)$	$x = c$	$(0, 0)$	$y^2 = -4cx$

EXAMPLE 1 Graph $x^2 = 8y$.

SOLUTION This equation represents a parabola that opens upward. The vertex is $(0, 0)$, and notice by inspection that

$$4c = 8$$
$$c = 2$$

Thus the focus is $(0, 2)$. After the vertex $V(0, 0)$ and the focus $F(0, 2)$ are plotted, the only question is the width of the parabola. In Chapter 3 you plotted a couple of points or found the y-intercept. The method in these examples is more useful and efficient for determining the graph of a parabola. In this chapter we will determine the width of the parabola by using the focal chord. Remember that the **length of the focal chord is 4c**, so that in this case it is 8. Do you see that in each case $4c$ is the absolute value of the coefficient of the first-degree term in the standard-form equation? Since a parabola is symmetric with respect to its axis, draw a segment of length 8 with the midpoint at F. Using these three points (the vertex and the endpoints of the focal chord), sketch the parabola as shown in Figure 11.5.

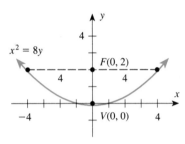

FIGURE 11.5
Graph of the parabola $x^2 = 8y$

EXAMPLE 2 Graph $y^2 = -12x$.

SOLUTION This equation represents a parabola that opens left. The vertex is $(0, 0)$ and

$$4c = 12$$
$$c = 3$$

(recall that c is positive), so the focus is $(-3, 0)$. The length of the focal chord is 12 and the parabola is drawn as in Figure 11.6.

FIGURE 11.6 Graph of the parabola $y^2 = -12x$

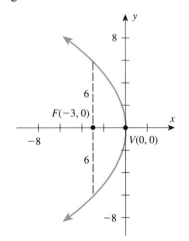

EXAMPLE 3 Graph $2y^2 - 5x = 0$.

SOLUTION You might first put the equation into standard form by solving for the second-degree term:

$$y^2 = \tfrac{5}{2}x$$

The vertex is $(0, 0)$ and

$$4c = \tfrac{5}{2}$$
$$c = \tfrac{5}{8}$$

Thus the parabola opens to the right, the focus is $(\tfrac{5}{8}, 0)$ and the length of the focal chord is $\tfrac{5}{2}$, as shown in Figure 11.7. ▮

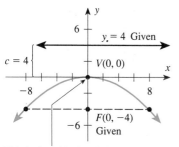

This is half the length of the focal chord, or $2c$.

$V(0, 0)$

$F(\tfrac{5}{8}, 0)$

FIGURE 11.7 Graph of the parabola $2y^2 - 5x = 0$

There are two basic types of problems you will need to solve concerning each curve in analytic geometry:

1. Given the equation, draw the graph; this is what you did in Examples 1–3.
2. Given the graph (or information about the graph), write the equation. Example 4 is a problem of this type.

EXAMPLE 4 Find the equation of the parabola with directrix $y = 4$ and focus at $F(0, -4)$.

SOLUTION This curve is a parabola that opens downward with vertex at the origin, as shown in Figure 11.8. The value for c is found by inspection: $c = 4$. Thus $4c = 16$. Since the equation is of the form $x^2 = -4cy$, the desired equation is (by substitution)

$$x^2 = -16y$$ ▮

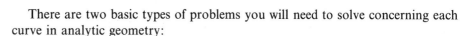

$y = 4$ Given

$c = 4$

$V(0, 0)$

$F(0, -4)$
Given

This is the midpoint of that segment of the axis between the focus and the directrix.

FIGURE 11.8 Graph of the parabola with focus at $(0, -4)$ and directrix $y = 4$

The types of parabolas we have been considering are quite limited, since we have assumed that the vertex is at the origin and the directrix is parallel to one of the coordinate axes. Suppose, however, that you are given a parabola with vertex at (h, k) and a directrix parallel to one of the coordinate axes. In Section 2.3 we showed that you can translate the axes to (h, k) by a substitution:

$$x' = x - h$$
$$y' = y - k$$

Therefore the **standard-form parabola equations** with vertex (h, k) can be summarized by the following table.

STANDARD-FORM EQUATIONS FOR TRANSLATED PARABOLAS

PARABOLA	FOCUS	DIRECTRIX	VERTEX	EQUATION
Opens *upward*	$(h, k + c)$	$y = k - c$	(h, k)	$(x - h)^2 = 4c(y - k)$
Opens *downward*	$(h, k - c)$	$y = k + c$	(h, k)	$(x - h)^2 = -4c(y - k)$
Opens *right*	$(h + c, k)$	$x = h - c$	(h, k)	$(y - k)^2 = 4c(x - h)$
Opens *left*	$(h - c, k)$	$x = h + c$	(h, k)	$(y - k)^2 = -4c(x - h)$

For the rest of the material in this chapter you will need to remember the procedure for **completing the square.** Examples 5 and 6 should provide enough detail to refresh your memory; if you need additional review, see Section 3.3, pages 134–135.

EXAMPLE 5   Sketch $x^2 + 4y + 8x + 4 = 0$.

SOLUTION   STEP 1   Associate together the terms involving the variable that is squared:

$$x^2 + 8x = -4y - 4$$

STEP 2   Complete the square for the variable that is squared:

Coefficient is 1; if it is not, divide both sides by this coefficient.

$$x^2 + 8x + \left(\frac{1}{2} \cdot 8\right)^2 = -4y - 4 + \left(\frac{1}{2} \cdot 8\right)^2$$

Take one-half of this coefficient, square it, and add it to both sides.

$$x^2 + 8x + 16 = -4y - 4 + 16$$
$$(x + 4)^2 = -4y + 12$$

STEP 3   Factor out the coefficient of the first-degree term:

$$(x + 4)^2 = -4(y - 3)$$

STEP 4   Determine the vertex by inspection. Plot (h, k); in this example, the vertex is $(-4, 3)$. (See Figure 11.9.)

STEP 5   Determine the focus. By inspection, $4c = 4$, $c = 1$, and the parabola opens downward from the vertex as shown in Figure 11.9.

STEP 6   Plot the endpoints of the focal chord; $4c = 4$. Draw the parabola as shown in Figure 11.9.

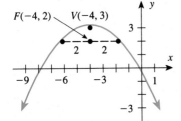

FIGURE 11.9
Graph of $x^2 + 4y + 8x + 4 = 0$

EXAMPLE 6   Sketch $2y^2 + 6y + 5x + 10 = 0$.

SOLUTION

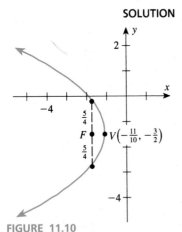

FIGURE 11.10
Graph of $2y^2 + 6y + 5x + 10 = 0$

$$2y^2 + 6y = -5x - 10$$

$$y^2 + 3y = -\tfrac{5}{2}x - 5$$

Divide both sides by 2 so the leading coefficient is 1; then complete the square.

$$y^2 + 3y + \tfrac{9}{4} = -\tfrac{5}{2}x - 5 + \tfrac{9}{4}$$

$$(y + \tfrac{3}{2})^2 = -\tfrac{5}{2}x - \tfrac{11}{4}$$

$$(y + \tfrac{3}{2})^2 = -\tfrac{5}{2}(x + \tfrac{11}{10})$$

The last step is to factor out the coefficient of the first-degree term.

The vertex is $(-\tfrac{11}{10}, -\tfrac{3}{2})$ and $4c = \tfrac{5}{2}$, $c = \tfrac{5}{8}$. Sketch the curve as shown in Figure 11.10.

EXAMPLE 7 Find the equation of the parabola with focus at $(4, -3)$ and directrix the line $x + 2 = 0$.

SOLUTION Sketch the given information as shown in Figure 11.11. The vertex is $(1, -3)$ since it must be equidistant from F and the directrix. Note that $c = 3$. Thus substitute into the equation

$$(y - k)^2 = 4c(x - h)$$

since the parabola opens to the right. The desired equation is

$$(y + 3)^2 = 12(x - 1)$$

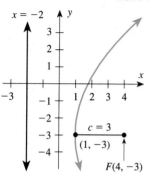

FIGURE 11.11

By sketching the focus and directrix, it is easy to find c and the vertex by inspection. The sketch of the curve is not necessary in order to find this equation.

EXAMPLE 8 Find the equation of the parabola with vertex at $(2, -1)$, axis parallel to the y-axis, and passing through $(3, 2)$.

SOLUTION Sketch the given information as shown in Figure 11.12. The parabola opens upward and thus has the form

$$(x - h)^2 = 4c(y - k)$$

Since the vertex is $(2, -1)$,

$$(x - 2)^2 = 4c(y + 1)$$

Also, since it passes through $(3, 2)$, this point must satisfy the equation

$$(3 - 2)^2 = 4c(2 + 1)$$

Solving for c,

$$c = \tfrac{1}{12}$$

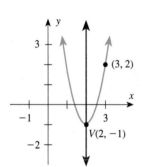

FIGURE 11.12

Therefore the desired equation is

$$(x - 2)^2 = \tfrac{1}{3}(y + 1)$$

11.1 **Problem Set**

A *In Problems 1–6 use the definition of a parabola to sketch a parabola such that the distance between the focus and the directrix is the given number. (See Figure 11.2.)*

1. 4   **2.** 6   **3.** 8

4. 10   **5.** 5   **6.** 7

Sketch the curve satisfying the conditions given in Problems 7–16.

7. Directrix $x = 0$; focus at $(5, 0)$

8. Directrix $y = 0$; focus at $(0, -3)$

9. Directrix $x - 3 = 0$; vertex at $(-1, 2)$

10. Directrix $y + 4 = 0$; vertex at $(4, -1)$

11. Vertex at $(-2, -3)$; focus at $(-2, 3)$

12. Vertex at $(-3, 4)$; focus at $(1, 4)$

13. Vertex at $(-3, 2)$ and passing through $(-2, -1)$; axis parallel to the y-axis

14. Vertex at $(4, 2)$ and passing through $(-3, -4)$; axis parallel to the x-axis

15. The set of all points with distances from $(4, 3)$ that equal their distances from $(0, 3)$

16. The set of all points with distances from $(4, 3)$ that equal their distances from $(-2, 1)$

Sketch the curves given by the equations in Problems 17–40.
Label the focus F and the vertex V in Problems 17–30.

17. $y^2 = 8x$
18. $y^2 = -12x$
19. $y^2 = -20x$
20. $4x^2 = 10y$
21. $3x^2 = -12y$
22. $2x^2 = -4y$
23. $2x^2 + 5y = 0$
24. $5y^2 + 15x = 0$
25. $3y^2 - 15x = 0$
26. $4y^2 + 3x = 12$
27. $5x^2 + 4y = 20$
28. $4x^2 + 3y = 12$
29. $(y - 1)^2 = 2(x + 2)$
30. $(y + 3)^2 = 3(x - 1)$
31. $(x + 2)^2 = 2(y - 1)$
32. $(x - 1)^2 = 3(y + 3)$
33. $(x - 1) = -2(y + 2)$
34. $(x + 3) = -3(y - 1)$
35. $y^2 + 4x - 3y + 1 = 0$
36. $y^2 - 4x + 10y + 13 = 0$
37. $2y^2 + 8y - 20x + 148 = 0$
38. $x^2 + 9y - 6x + 18 = 0$
39. $9x^2 + 6x + 18y - 23 = 0$
40. $9x^2 + 6y + 18x - 23 = 0$

B *Find the equation of each curve in Problems 41–50. You sketched these curves in Problems 7–16.*

41. Directrix $x = 0$; focus at $(5, 0)$
42. Directrix $y = 0$; focus at $(0, -3)$
43. Directrix $x - 3 = 0$; vertex at $(-1, 2)$
44. Directrix $y + 4 = 0$; vertex at $(4, -1)$
45. Vertex at $(-2, -3)$; focus at $(-2, 3)$
46. Vertex at $(-3, 4)$; focus at $(1, 4)$
47. Vertex at $(-3, 2)$ and passing through $(-2, -1)$; axis parallel to the y-axis
48. Vertex at $(4, 2)$ and passing through $(-3, -4)$; axis parallel to the x-axis

49. The set of all points with distances from $(4, 3)$ that equal their distances from $(0, 3)$
50. The set of all points with distances from $(4, 3)$ that equal their distances from $(-2, 1)$
51. PHYSICS If the path of a baseball is parabolic and is 200 ft wide at the base and 50 ft high in the vertex, write the equation that gives the path of the baseball if the origin is the point of departure for the ball.
52. ENGINEERING A parabolic archway has the dimensions shown in Figure 11.13. Find the equation of the parabolic portion.

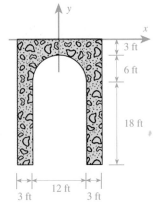

FIGURE 11.13
A parabolic archway

53. ENGINEERING A radar antenna is constructed so that a cross section along its axis is a parabola with the receiver at the focus. Find the focus if the antenna is 12 m across and its depth is 4 m. See Figure 11.14.
54. ENGINEERING If the diameter of the parabolic reflector in Problem 53 is 16 cm and the depth is 8 cm, find the focus. (See Figure 11.14.)

a. Radar antenna

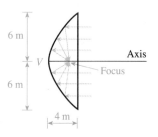

b. Dimensions for radar antenna in Problem 53

FIGURE 11.14 A three-dimensional model of a parabola is a **parabolic reflector** (or **parabolic mirror**). A radar antenna serves as an example. If a source of light is placed at the focus of the mirror, the light rays will reflect from the mirror as rays parallel to the axis (as in an automobile headlamp). The radar antenna works in reverse—parallel incoming rays are focused at a single location. (Photo courtesy of Wide World Photos, Inc.)

C *Find the points of intersection (if any) for each line and parabola in Problems 55–58. Show the result both algebraically and geometrically.*

55. $\begin{cases} y = 2x + 10 \\ y = x^2 + 4x + 7 \end{cases}$

56. $\begin{cases} y = -2x + 4 \\ y = x^2 - 12x + 25 \end{cases}$

57. $\begin{cases} 2x + y - 7 = 0 \\ y^2 - 6y - 4x + 17 = 0 \end{cases}$

58. $\begin{cases} 3x + 2y - 5 = 0 \\ y^2 + 4y + 3x - 4 = 0 \end{cases}$

59. SPORTS Phil Lee, a physical education expert, has made a study and determined that a woman who runs the 100-m dash reaches her peak at age 20. Her time T, for a 100-m dash at a particular age A, is

$$T = c_1(A - 20)^2 + c_2$$

for constants c_1 and c_2. If Wyomia Tyus ran the race when she was 16 years old in 11.4 sec and when she was 20 in 11.0 sec, predict her time for running the race when she is 40 years old. Graph this function.

60. SPORTS According to Jim Kintzi, another physical education expert, the peak age for a woman runner is 24. Using this information, answer the questions posed in Problem 59.

61. Derive the equation of a parabola with $F(0, -c)$, where c is a positive number and the directrix is the line $y = c$.

62. Derive the equation of a parabola with $F(c, 0)$, where c is a positive number and the directrix is the line $x = -c$.

63. Derive the equation of a parabola with $F(-c, 0)$, where c is a positive number and the directrix is the line $x = c$.

64. Show that the length of the focal chord for the parabola $y^2 = 4cx$ is $4c$ $(c > 0)$.

65. Let $L: Ax + By + C = 0$ be any nonvertical line and let $P(x_0, y_0)$ be any point not on the line.
   a. Find the slope of L.
   b. Let L' be a line through P perpendicular to L. Find the slope of L'.
   c. Find the equation of L'.
   d. Let Q be the point of intersection of L and L'. Find Q.
   e. Find the distance between P and Q to show that the distance from a point to a line is

$$d = \frac{|Ax_0 + By_0 + C|}{\sqrt{A^2 + B^2}}$$

66. Find the equation of the parabola with focus at $(4, -3)$ and directrix $x - y + 3 = 0$. (*Hint:* Use the definition of a parabola and the formula derived in Problem 65.)

67. Find the equation of the parabola with focus at $(3, -5)$ and directrix $12x - 5y + 4 = 0$. (*Hint:* Use the definition of a parabola and the formula derived in Problem 65.)

11.2 Ellipses

The second conic section considered in this chapter is called an *ellipse*.

ELLIPSE | An **ellipse** is the set of all points in the plane such that, for each point on the ellipse, the sum of its distances from two fixed points is a constant.

The fixed points are called the **foci** (plural of **focus**). To see what an ellipse looks like, we will use the special type of graph paper shown in Figure 11.15a (page 470), where F_1 and F_2 are the foci.

Let the given constant be 12. Plot all the points in the plane so that the sum of their distances from the foci is 12. If a point is 8 units from F_1, for example, then it is 4 units from F_2 and you can plot the points P_1 and P_2. The completed graph of this ellipse is shown in Figure 11.15b.

The line passing through F_1 and F_2 is called the **major axis.** The **center** is the midpoint of the segment $\overline{F_1F_2}$. The line passing through the center perpendicular

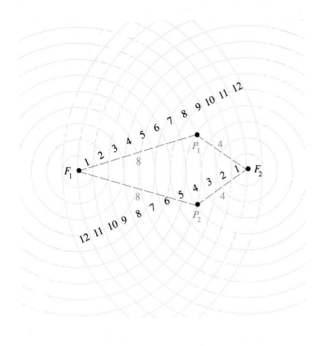

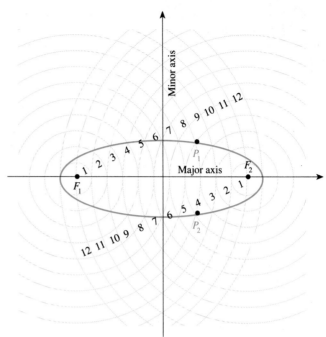

a. Ellipse graph paper

b. Graphing an ellipse

FIGURE 11.15

to the major axis is called the **minor axis.** The ellipse is symmetric with respect to both the major and minor axes.

To find the equation of an ellipse, first consider a special case where the center is at the origin. Let the distance from the center to a focus be the positive number c; that is, let $F_1(-c, 0)$ and $F_2(c, 0)$ be the foci and let the constant distance be $2a$, as shown in Figure 11.16.

FIGURE 11.16 Developing the equation of an ellipse by using the definition

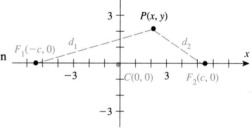

Notice that the center of the ellipse in Figure 11.16 is $(0, 0)$. Let $P(x, y)$ be any point on the ellipse, and use the distance formula and the definition of an ellipse to derive the equation of this ellipse:

$$d_1 + d_2 = 2a$$

or

$$\sqrt{(x + c)^2 + (y - 0)^2} + \sqrt{(x - c)^2 + (y - 0)^2} = 2a$$

Simplifying,

$$\sqrt{(x + c)^2 + y^2} = 2a - \sqrt{(x - c)^2 + y^2} \qquad \text{Isolate one radical.}$$

$$(x + c)^2 + y^2 = 4a^2 - 4a\sqrt{(x - c)^2 + y^2} + (x - c)^2 + y^2 \qquad \text{Square both sides.}$$

$$x^2 + 2cx + c^2 + y^2 = 4a^2 - 4a\sqrt{(x - c)^2 + y^2} + x^2 - 2cx + c^2 + y^2$$

$$4a\sqrt{(x - c)^2 + y^2} = 4a^2 - 4cx$$

$$\sqrt{(x - c)^2 + y^2} = a - \frac{c}{a}x \qquad \text{Since } a \neq 0, \text{ divide by } 4a.$$

$$(x - c)^2 + y^2 = \left(a - \frac{c}{a}x\right)^2 \qquad \text{Square both sides again.}$$

$$x^2 - 2cx + c^2 + y^2 = a^2 - 2cx + \frac{c^2}{a^2}x^2$$

$$x^2 + y^2 = a^2 - c^2 + \frac{c^2}{a^2}x^2$$

$$x^2 - \frac{c^2}{a^2}x^2 + y^2 = a^2 - c^2$$

$$\left(1 - \frac{c^2}{a^2}\right)x^2 + y^2 = a^2 - c^2$$

$$\frac{a^2 - c^2}{a^2}x^2 + y^2 = a^2 - c^2$$

$$\frac{x^2}{a^2} + \frac{y^2}{a^2 - c^2} = 1 \qquad \text{Divide both sides by } a^2 - c^2.$$

Let $b^2 = a^2 - c^2$; then

$$\frac{x^2}{a^2} + \frac{y^2}{b^2} = 1$$

This substitution is a notational change that will simplify the equation.

If $x = 0$, the y-intercepts are obtained:

$$\frac{y^2}{b^2} = 1$$

$$y = \pm b$$

If $y = 0$, the x-intercepts are obtained: $x = \pm a$. The intercepts on the major axis are called the **vertices** of the ellipse.

The equation of the ellipse with major axis vertical, $F_1(0, c)$, $F_2(0, -c)$, and constant distance $2a$ is found in a similar fashion. Simplifying the equation as before,

$$\frac{y^2}{a^2} + \frac{x^2}{b^2} = 1$$

where $b^2 = a^2 - c^2$.

Notice that in both cases a^2 must be larger than both c^2 and b^2. If it were not, a square number would be equal to a negative number, which is a contradiction in the set of real numbers.

STANDARD-FORM EQUATIONS FOR ELLIPSES WITH CENTER (0, 0)

| ELLIPSE | CENTER | FOCI | INTERCEPTS | | CONSTANT DISTANCE | EQUATION |
			MAJOR AXIS VERTICES	MINOR AXIS VERTICES		
Horizontal	$(0, 0)$	$(-c, 0), (c, 0)$	$(a, 0), (-a, 0),$	$(0, b), (0, -b)$	$2a$	$\dfrac{x^2}{a^2} + \dfrac{y^2}{b^2} = 1$
Vertical	$(0, 0)$	$(0, c), (0, -c)$	$(0, a), (0, -a)$	$(b, 0), (-b, 0)$	$2a$	$\dfrac{y^2}{a^2} + \dfrac{x^2}{b^2} = 1$
						where $b^2 = a^2 - c^2$ or $c^2 = a^2 - b^2$

EXAMPLE 1   Sketch $\dfrac{x^2}{9} + \dfrac{y^2}{4} = 1$.

SOLUTION   The center of the ellipse is $(0, 0)$. The x-intercepts are ± 3 (these are the vertices) and the y-intercepts are ± 2. Sketch the ellipse as shown in Figure 11.17.

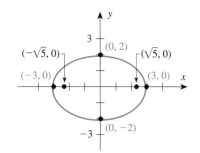

FIGURE 11.17   Graph of $\dfrac{x^2}{9} + \dfrac{y^2}{4} = 1$

The foci can also be found, since

$$c^2 = a^2 - b^2$$
$$c^2 = 9 - 4$$
$$c = \pm\sqrt{5}$$

EXAMPLE 2   Sketch $\dfrac{x^2}{4} + \dfrac{y^2}{9} = 1$.

SOLUTION   Here $a^2 = 9$ and $b^2 = 4$, which is an ellipse with major axis vertical. The x-intercepts are ± 2 and the y-intercepts are ± 3 (these are the vertices). The sketch is shown in Figure 11.18.

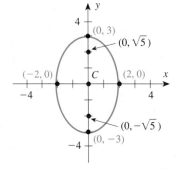

FIGURE 11.18   Graph of $\dfrac{x^2}{4} + \dfrac{y^2}{9} = 1$

The foci are found by

$$c^2 = a^2 - b^2$$
$$c^2 = 9 - 4$$
$$c = \pm\sqrt{5}$$

Remember: The foci are always on the major axis, so plot $(0, \sqrt{5})$ and $(0, -\sqrt{5})$ for the foci

If the center of the ellipse is at (h, k), the equations can be written in terms of a translation as shown by the following table.

STANDARD-FORM EQUATIONS FOR TRANSLATED ELLIPSES

ELLIPSE	CENTER	FOCI	INTERCEPTS MAJOR AXIS VERTICES	INTERCEPTS MINOR AXIS VERTICES	EQUATION
Horizontal	(h, k)	$(h - c, k)$ $(h + c, k)$	$(h - a, k)$ $(h + a, k)$	$(h, k + b)$ $(h, k - b)$	$\dfrac{(x - h)^2}{a^2} + \dfrac{(y - k)^2}{b^2} = 1$
Vertical	(h, k)	$(h, k + c)$ $(h, k - c)$	$(h, k + a)$ $(h, k - a)$	$(h + b, k)$ $(h - b, k)$	$\dfrac{(y - k)^2}{a^2} + \dfrac{(x - h)^2}{b^2} = 1$

The segment from the center to a vertex on the major axis is called a **semimajor axis** and has length a; the segment from the center to an intercept on the minor axis is called a **semiminor axis** and has length b. Use a and b to plot four points, which are in turn used to draw the curve.

EXAMPLE 3   Graph $\dfrac{(x - 3)^2}{25} + \dfrac{(y - 1)^2}{16} = 1$.

SOLUTION   STEP 1   Plot the center (h, k). By inspection, the center of this ellipse is $(3, 1)$. This becomes the center of a new translated coordinate system. The vertices and foci are now measured with reference to the new origin at $(3, 1)$.

STEP 2   Plot the x'- and y'-intercepts. These are ± 5 and ± 4, respectively. Remember to measure these distances from $(3, 1)$, as shown in Figure 11.19.

FIGURE 11.19   Graph of $\dfrac{(x - 3)^2}{25} + \dfrac{(y - 1)^2}{16} = 1$

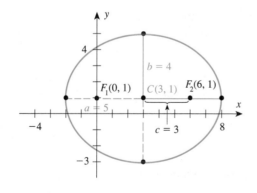

The foci are found by

$c^2 = a^2 - b^2$

$= 25 - 16$

$= \pm 9$

The distance from the center to each focus is 3, so the coordinates of the foci are $(6, 1)$ and $(0, 1)$.

EXAMPLE 4   Graph $3x^2 + 4y^2 + 24x - 16y + 52 = 0$.

SOLUTION   STEP 1   Associate together the x and the y terms:

$$(3x^2 + 24x) + (4y^2 - 16y) = -52$$

STEP 2   Complete the squares in both x and y. This requires that the coefficients of the squared terms be 1. You can accomplish this by factoring:

$$3(x^2 + 8x \quad) + 4(y^2 - 4y \quad) = -52$$

Next complete the square for both x and y, being sure to add the same number to both sides:

Add 16 to both sides.

$$3(x^2 + 8x + 16) + 4(y^2 - 4y + 4) = -52 + 48 + 16$$

Add 48 to both sides.

STEP 3 Factor:

$$3(x + 4)^2 + 4(y - 2)^2 = 12$$

STEP 4 Divide both sides by 12:

$$\frac{(x + 4)^2}{4} + \frac{(y - 2)^2}{3} = 1$$

STEP 5 Plot the center (h, k). By inspection, you can see the center is $(-4, 2)$. The vertices are at ± 2, and the length of the semiminor axis is $\sqrt{3}$, as shown in Figure 11.20.

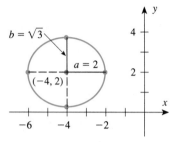

FIGURE 11.20 Graph of $3x^2 + 4y^2 + 24x - 16y + 52 = 0$

We have seen that some ellipses are more circular and some more flat than others. A measure of the amount of flatness of an ellipse is called its **eccentricity**, which is defined as

$$\varepsilon = \frac{c}{a}$$

Notice that

$$\varepsilon = \frac{c}{a} = \frac{\sqrt{a^2 - b^2}}{a} = \sqrt{\frac{a^2 - b^2}{a^2}} = \sqrt{1 - \left(\frac{b}{a}\right)^2}$$

Since $c < a$, ε is between 0 and 1. If $a = b$, then $\varepsilon = 0$ and the conic is a **circle**. If the ratio b/a is small, then the ellipse is very flat. Thus, for an ellipse,

$$0 \leq \varepsilon < 1$$

and ε measures the amount of roundness of the ellipse.

Consider a circle; that is, suppose $a = b$. In this case, let $r = a = b$ and call this distance the **radius.** You can see that a circle is a special case of an ellipse, as given in the following box:

STANDARD-FORM EQUATION OF A CIRCLE

The equation of a **circle** with center at (h, k) and radius r is

$$(x - h)^2 + (y - k)^2 = r^2$$

EXAMPLE 5 Graph $x^2 + y^2 + 6x - 14y + 22 = 0$.

SOLUTION Complete the square in x and y:

$$(x^2 + 6x \quad) + (y^2 - 14y \quad) = -22$$
$$(x^2 + 6x + 9) + (y^2 - 14y + 49) = -22 + 9 + 49$$
$$(x + 3)^2 + (y - 7)^2 = 36$$

This is a circle with center at $(-3, 7)$ and radius 6, as shown in Figure 11.21.

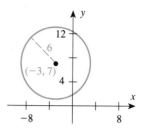

FIGURE 11.21 Graph of $x^2 + y^2 + 6x - 14y + 22 = 0$

EXAMPLE 6 Find the equation of the circle passing through the points $(-3, 4)$, $(4, 5)$, and $(1, -4)$. Sketch the graph.

SOLUTION Now $(x - h)^2 + (y - k)^2 = r^2$ can be written as $x^2 + y^2 + C_1 x + C_2 y + C_3 = 0$ for some constants C_1, C_2, and C_3. Since the points lie on the circle, they satisfy the equation. Thus

$$
\begin{aligned}
(-3, 4): && -3C_1 + 4C_2 + C_3 &= -25 \\
(4, 5): && 4C_1 + 5C_2 + C_3 &= -41 \\
(1, -4): && C_1 - 4C_2 + C_3 &= -17
\end{aligned}
$$

This is a system of three linear equations in three unknowns that can be solved simultaneously by using the methods given in Chapter 9. (We will not show the details of this solution here.) After several steps you will find that $C_1 = -2$, $C_2 = -2$, and $C_3 = -23$. Now substitute these values into the equation of the circle:

$$
\begin{aligned}
x^2 + y^2 - 2x - 2y - 23 &= 0 \\
(x^2 - 2x + 1) + (y^2 - 2y + 1) &= 23 + 1 + 1 \\
(x - 1)^2 + (y - 1)^2 &= 25
\end{aligned}
$$

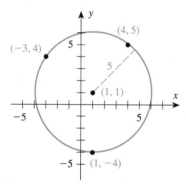

FIGURE 11.22 Graph of the circle passing through $(-3, 4)$, $(4, 5)$, and $(1, -4)$

The graph is shown in Figure 11.22. Note that the center is $(1, 1)$ and the radius is $r = 5$.

EXAMPLE 7 Find the equation of the ellipse with vertices at $(3, 2)$ and $(3, -4)$ and foci at $(3, \sqrt{5} - 1)$ and $(3, -\sqrt{5} - 1)$.

SOLUTION By inspection, the ellipse is vertical, and it is centered at $(3, -1)$, where $a = 3$ and $c = \sqrt{5}$ (see Figure 11.23). Thus

$$
\begin{aligned}
c^2 &= a^2 - b^2 \\
5 &= 9 - b^2 \\
b^2 &= 4
\end{aligned}
$$

The equation is

$$
\frac{(y + 1)^2}{9} + \frac{(x - 3)^2}{4} = 1
$$

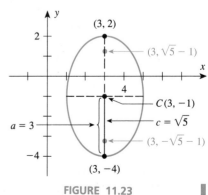

FIGURE 11.23

EXAMPLE 8   Find the equation of the set of all points with the sum of distances from $(-3, 2)$ and $(5, 2)$ equal to 16.

SOLUTION   By inspection, the ellipse is horizontal and is centered at $(1, 2)$. You are given

$$2a = 16 \qquad \text{This is the sum of the distances.}$$
$$a = 8$$

and $c = 4$—the distance from the center $(1, 2)$ to a focus $(5, 2)$. Thus

$$c^2 = a^2 - b^2$$
$$16 = 64 - b^2$$
$$b^2 = 48$$

The equation is

$$\frac{(x-1)^2}{64} + \frac{(y-2)^2}{48} = 1$$

EXAMPLE 9   Find the equation of the ellipse with foci at $(-3, 6)$ and $(-3, 2)$ with $\varepsilon = \frac{1}{5}$.

SOLUTION   By inspection, the ellipse is vertical and is centered at $(-3, 4)$ with $c = 2$. Since

$$\varepsilon = \frac{c}{a} = \frac{1}{5}$$

and $c = 2$, then $\dfrac{2}{a} = \dfrac{1}{5}$, which implies $a = 10$. Just because $\dfrac{c}{a} = \dfrac{1}{5}$, you cannot assume that $c = 1$ and $a = 5$; all you know is that the reduced *ratio* of c to a is $\frac{1}{5}$. Also,

$$c^2 = a^2 - b^2$$
$$4 = 100 - b^2$$
$$b^2 = 96$$

Thus the equation is

$$\frac{(y-4)^2}{100} + \frac{(x+3)^2}{96} = 1$$

11.2 Problem Set

A *Sketch the curves in Problems 1–18.*

1. $\dfrac{x^2}{4} + \dfrac{y^2}{9} = 1$

2. $\dfrac{x^2}{25} + \dfrac{y^2}{36} = 1$

3. $x^2 + \dfrac{y^2}{9} = 1$

4. $4x^2 + 9y^2 = 36$

5. $25x^2 + 16y^2 = 400$

6. $36x^2 + 25y^2 = 900$

7. $3x^2 + 2y^2 = 6$

8. $4x^2 + 3y^2 = 12$

9. $5x^2 + 10y^2 = 7$

10. $(x - 2)^2 + (y + 3)^2 = 25$

11. $(x + 4)^2 + (y - 2)^2 = 49$

12. $(x - 1)^2 + (y - 1)^2 = \frac{1}{4}$

13. $\dfrac{(x+3)^2}{81} + \dfrac{(y-1)^2}{49} = 1$ **14.** $\dfrac{(x-3)^2}{16} + \dfrac{(y-2)^2}{9} = 1$

15. $\dfrac{(x+2)^2}{25} + \dfrac{(y+4)^2}{9} = 1$ **16.** $3(x + 1)^2 + 4(y - 1)^2 = 12$

17. $10(x - 5)^2 + 6(y + 2)^2 = 60$

18. $5(x + 2)^2 + 3(y + 4)^2 = 60$

Sketch the curves in Problems 19–30.

19. The set of points 6 units from the point $(4, 5)$

20. The set of points 3 units from the point $(-2, 3)$

21. The set of points 6 units from $(-1, -4)$

22. The set of points such that the sum of the distances from $(-6, 0)$ and $(6, 0)$ is 20

23. The set of points such that the sum of the distances from $(4, 0)$ and $(-4, 0)$ is 10

24. The set of points such that the sum of the distances from $(-4, 1)$ and $(2, 1)$ is 10

25. The ellipse with vertices at $(0, 7)$ and $(0, -7)$ and foci at $(0, 5)$ and $(0, -5)$

26. The ellipse with vertices at $(4, 3)$ and $(4, -5)$ and foci at $(4, 2)$ and $(4, -4)$

27. The ellipse with vertices $(-6, 3)$ and $(4, 3)$ and foci at $(-4, 3)$ and $(2, 3)$

28. The ellipse with foci at $(-4, -3)$ and $(2, -3)$ with eccentricity $\frac{4}{5}$

29. The circle passing through $(2, 2)$, $(-2, -6)$, and $(5, 1)$

30. The ellipse passing through $(5, 2)$ and $(3, \sqrt{5})$ with axes along the coordinate axes

   B Find the equations of the curves in Problems 31–42. These are the curves you sketched in Problems 19–30.

31. The set of points 6 units from the point $(4, 5)$

32. The set of points 3 units from the point $(-2, 3)$

33. The set of points 6 units from $(-1, -4)$

34. The set of points such that the sum of the distances from $(-6, 0)$ and $(6, 0)$ is 20

35. The set of points such that the sum of the distances from $(4, 0)$ and $(-4, 0)$ is 10

36. The set of points such that the sum of the distances from $(-4, 1)$ and $(2, 1)$ is 10

37. The ellipse with vertices at $(0, 7)$ and $(0, -7)$ and foci at $(0, 5)$ and $(0, -5)$

38. The ellipse with vertices at $(4, 3)$ and $(4, -5)$ and foci at $(4, 2)$ and $(4, -4)$

39. The ellipse with vertices $(-6, 3)$ and $(4, 3)$ and foci at $(-4, 3)$ and $(2, 3)$

40. The ellipse with foci at $(-4, -3)$ and $(2, -3)$ and eccentricity $\frac{4}{5}$

41. The circle passing through $(2, 2)$, $(-2, -6)$, and $(5, 1)$

42. The ellipse passing through $(5, 2)$ and $(3, \sqrt{5})$ with axes along the coordinate axes

Sketch the curves in Problems 43–54.

43. $x^2 + 4x + y^2 + 6y - 12 = 0$
44. $9x^2 + 4y^2 - 18x + 16y - 11 = 0$
45. $16x^2 + 9y^2 + 96x - 36y + 36 = 0$
46. $y^2 + 6y + 25x + 159 = 0$
47. $3x^2 + 4y^2 + 2x - 8y + 4 = 0$

48. $144x^2 + 72y^2 - 72x + 48y - 7 = 0$
49. $y^2 + 4x^2 + 2y - 8x + 1 = 0$
50. $4y^2 + x^2 - 16y + 4x - 8 = 0$
51. $x^2 + y^2 - 10x - 14y - 70 = 0$
52. $x^2 + y^2 - 4x + 10y + 15 = 0$
53. $4x^2 + y^2 + 24x + 4y + 16 = 0$
54. $x^2 + 9y^2 - 4x - 18y - 14 = 0$

55. Derive the equation of the ellipse with foci at $(0, c)$ and $(0, -c)$ and constant distance $2a$. Let $b^2 = a^2 - c^2$. Show all your work.

56. Derive the equation of a circle with center (h, k) and radius r by using the distance formula.

The following is needed for Problems 57–59. The orbit of a planet can be described by

$$\frac{x^2}{a^2} + \frac{y^2}{b^2} = 1$$

with the sun at one focus. The orbit is commonly identified by its major axis 2a and its eccentricity ε.

$$\varepsilon = \frac{c}{a} = \sqrt{1 - \frac{b^2}{a^2}} \quad (a \geq b > 0)$$

where c is the distance between the center and a focus.

57. The orbits of the planets can be described as standard-form ellipses with the sun at one focus (see Figure 11.24). The point at which a planet is farthest from the sun is called **aphelion,** and the point at which it is closest is called **perihelion.** (For a more detailed discussion of planetary orbits, see the Extended Application at the end of this chapter.) If the major axis of the earth's orbit is 186,000,000 miles and its eccentricity is $\frac{1}{62}$, how far is the earth from the sun at aphelion and at perihelion?

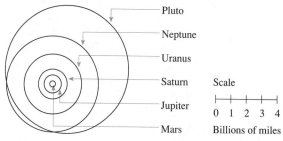

FIGURE 11.24 Planetary orbits: The orbits of planets or satellites serve as an example of ellipses.

58. If the planet Mercury is 28 million miles from the sun at perihelion, and the eccentricity of its orbit is $\frac{1}{5}$, how long is the major axis of Mercury's orbit? (See Problem 57.)

59. The moon's orbit is elliptical with the earth at one focus. The point at which the moon is farthest from the earth is called **apogee,** and the point at which it is closest is called **perigee.** If the moon is 199,000 miles from the earth at apogee and the major axis of its orbit is 378,000 miles, what is the eccentricity of the moon's orbit?

C

60. ENGINEERING A stone tunnel is to be constructed such that the opening is a semielliptic arch as shown in Figure 11.25. It is necessary to know the height at 4-ft intervals from the center. That is, how high is the tunnel at 4, 8, 12, 16, and 20 ft from the center? (Answer to the nearest tenth of a foot.)

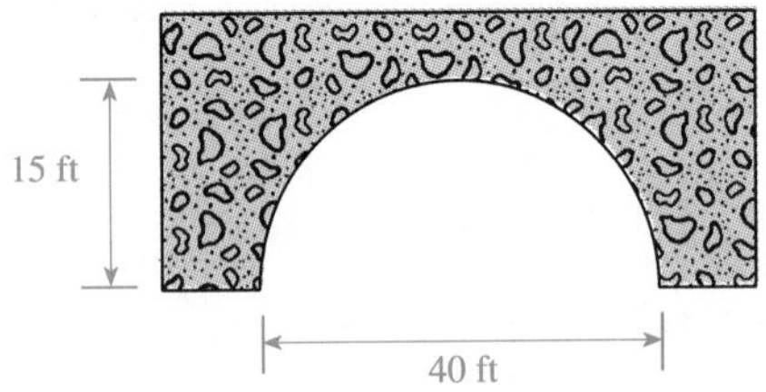

15 ft

40 ft

FIGURE 11.25 Semielliptic arch

61. Find the points of intersection (if any) of the ellipse given by $16(x - 2)^2 + 9(y + 1)^2 = 144$ and the line $4x - 3y - 23 = 0$. Graph both equations.

62. Find the points of intersection (if any) of the circle given by $x^2 + y^2 + 4x + 6y + 4 = 0$ and the parabola $x^2 + 4x + 8y + 4 = 0$. Graph both equations.

63. If we are given an ellipse with foci at $(-c, 0)$ and $(c, 0)$ and vertices at $(-a, 0)$ and $(a, 0)$, we define the *directrices* of the ellipse as the lines $x = a/\varepsilon$ and $x = -a/\varepsilon$. Show that an ellipse is the set of all points with distances from $F(c, 0)$ equal to ε times their distances from the line $x = a/\varepsilon$ $(a > 0, c > 0)$.

64. A line segment through a focus parallel to a directrix (see Problem 63) and cut off by the ellipse is called the *focal chord.* Show that the length of the focal chord of the following ellipse is $2b^2/a$:

$$\frac{x^2}{a^2} + \frac{y^2}{b^2} = 1$$

11.3 Hyperbolas

The last of the conic sections to be considered has a definition similar to that of the ellipse.

HYPERBOLA

> A **hyperbola** is the set of all points in the plane such that, for each point on the hyperbola, the difference of its distances from the two fixed points is a constant.

The fixed points are called the **foci.** A hyperbola with foci at F_1 and F_2, where the given constant is 6, is shown in Figure 11.26.

The line passing through the foci is called the **transverse axis.** The **center** is the midpoint of the segment connecting the foci. The line passing through the center perpendicular to the transverse axis is called the **conjugate axis.** The hyperbola is symmetric with respect to both the transverse and the conjugate axes.

If you use the definition, you can derive the equation for a hyperbola with foci at $(-c, 0)$ and $(c, 0)$ with constant distance $2a$. If (x, y) is any point on the curve, then

$$\left| \sqrt{(x + c)^2 + (y - 0)^2} - \sqrt{(x - c)^2 + (y - 0)^2} \right| = 2a$$

The procedure for simplifying this expression is the same as that shown for the ellipse, so the details are left as a problem (see Problems 59 and 60). After several

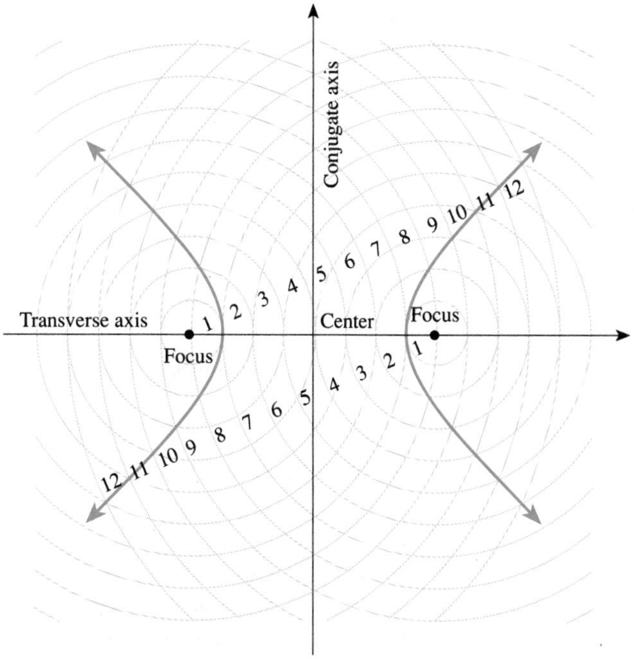

FIGURE 11.26 Graph of a hyperbola from the definition

steps, you should obtain

$$\frac{x^2}{a^2} - \frac{y^2}{c^2 - a^2} = 1$$

If $b^2 = c^2 - a^2$, then

$$\frac{x^2}{a^2} - \frac{y^2}{b^2} = 1$$

which is the standard-form equation. Notice that $c^2 = a^2 - b^2$ for the ellipse and that $c^2 = a^2 + b^2$ for the hyperbola. For the ellipse it is necessary that $a^2 > b^2$, but for the hyperbola there is no restriction on the relative sizes for a and b.

Repeat the argument for a hyperbola with foci $(0, c)$ and $(0, -c)$, and you will obtain the other standard-form equation for a hyperbola with a vertical transverse axis.

STANDARD-FORM EQUATIONS FOR THE HYPERBOLA WITH CENTER (0, 0)

HYPERBOLA	CENTER	FOCI	VERTICES	PSEUDOVERTICES	CONSTANT DISTANCE	EQUATION
Horizontal	$(0, 0)$	$(-c, 0), (c, 0)$	$(a, 0), (-a, 0)$	$(0, b), (0, -b)$	$2a$	$\dfrac{x^2}{a^2} - \dfrac{y^2}{b^2} = 1$
Vertical	$(0, 0)$	$(0, c), (0, -c)$	$(0, a), (0, -a)$	$(b, 0), (-b, 0)$	$2a$	$\dfrac{y^2}{a^2} - \dfrac{x^2}{b^2} = 1$

where $b^2 = c^2 - a^2$ or $c^2 = a^2 + b^2$

As with the other conic sections, we will sketch a hyperbola by determining some information about the curve directly from the equation by inspection. The points of intersection of the hyperbola with the transverse axis are called the **vertices.** For

$$\frac{x^2}{a^2} - \frac{y^2}{b^2} = 1 \qquad \text{and} \qquad \frac{y^2}{a^2} - \frac{x^2}{b^2} = 1$$

notice that the vertices occur at $(a, 0)$, $(-a, 0)$ and $(0, a)$, $(0, -a)$, respectively. The number $2a$ is the **length of the transverse axis.** The hyperbola does not intersect the conjugate axis, but if you plot the points $(0, b)$, $(0, -b)$ and $(-b, 0)$, $(b, 0)$, respectively, you determine a segment on the conjugate axis called the **length of the conjugate axis.**

EXAMPLE 1   Sketch $\dfrac{x^2}{4} - \dfrac{y^2}{9} = 1$.

SOLUTION   The center of the hyperbola is $(0, 0)$, $a = 2$, and $b = 3$. Plot the vertices at ± 2, as shown in Figure 11.27. The transverse axis is along the x-axis and the conjugate axis is along the y-axis. Plot the length of the conjugate axis at ± 3. We call the points the **pseudovertices,** since the curve does not actually pass through these points.

FIGURE 11.27

Next draw lines through the vertices and pseudovertices parallel to the axes of the hyperbola. These lines form what we will call the **central rectangle.** The diagonal lines passing through the corners of the central rectangle are **slant asymptotes** for the hyperbola, as shown in Figure 11.28; they aid in sketching the hyperbola.

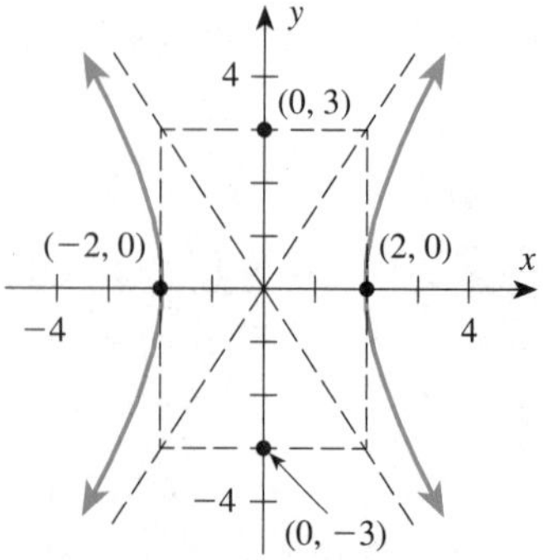

FIGURE 11.28   Graph of $\dfrac{x^2}{4} - \dfrac{y^2}{9} = 1$

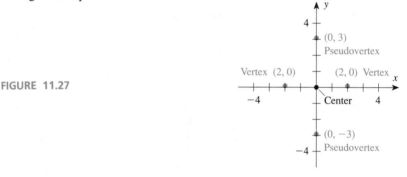

For the general hyperbola given by the equation

$$\frac{x^2}{a^2} - \frac{y^2}{b^2} = 1$$

the equations of the slant asymptotes described in Example 1 are

$$y = \frac{b}{a}x \quad \text{and} \quad y = -\frac{b}{a}x$$

To justify this result, solve the equation of the hyperbola for y:

$$\frac{y^2}{b^2} = \frac{x^2}{a^2} - 1$$

$$y^2 = b^2 \left(\frac{x^2 - a^2}{a^2} \right)$$

$$y = \pm \frac{b}{a} \sqrt{x^2 - a^2}$$

Now write

$$y = \pm \frac{b}{a} \sqrt{x^2 \left(1 - \frac{a^2}{x^2} \right)} = \pm \frac{bx}{a} \sqrt{1 - \frac{a^2}{x^2}}$$

in order to see that, as $|x| \to \infty$,

$$\sqrt{1 - \frac{a^2}{x^2}} \to 1$$

So, as $|x|$ becomes large, $\quad y \to \pm \frac{b}{a}x.$

The same result is obtained for the hyperbola

$$\frac{y^2}{a^2} - \frac{x^2}{b^2} = 1$$

If the center of the hyperbola is (h, k), the following equations for a hyperbola are obtained:

STANDARD-FORM EQUATIONS FOR TRANSLATED HYPERBOLAS

HYPERBOLA	CENTER	FOCI	VERTICES	PSEUDOVERTICES	EQUATION
Horizontal	(h, k)	$(h - c, k)$ $(h + c, k)$	$(h - a, k)$ $(h + a, k)$	$(h, k + b)$ $(h, k - b)$	$\dfrac{(x - h)^2}{a^2} - \dfrac{(y - k)^2}{b^2} = 1$
Vertical	(h, k)	$(h, k + c)$ $(h, k - c)$	$(h, k + a)$ $(h, k - a)$	$(h - b, k)$ $(h + b, k)$	$\dfrac{(y - k)^2}{a^2} - \dfrac{(x - h)^2}{b^2} = 1$

EXAMPLE 2 Sketch $16x^2 - 9y^2 - 128x - 18y + 103 = 0$.

SOLUTION Complete the square in both x and y:

$$(16x^2 - 128x) + (-9y^2 - 18y) = -103$$

$$16(x^2 - 8x \quad\ ) - 9(y^2 + 2y \quad\ ) = -103$$

$$16(x^2 - 8x + 16) - 9(y^2 + 2y + 1) = -103 + 256 - 9$$

$$16(x - 4)^2 - 9(y + 1)^2 = 144$$

$$\frac{(x - 4)^2}{9} - \frac{(y + 1)^2}{16} = 1$$

The graph is shown in Figure 11.29.

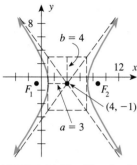

FIGURE 11.29 Sketch of
$16x^2 - 9y^2 - 128x - 18y + 103 = 0$

EXAMPLE 3 Find the equation of the hyperbola with vertices at $(2, 4)$ and $(2, -2)$ and foci at $(2, 6)$ and $(2, -4)$.

SOLUTION Plot the given points as shown in Figure 11.30. Notice that the center of the hyperbola is $(2, 1)$ since it is the midpoint of the segment connecting the foci. Also, $c = 5$ and $a = 3$. Since

$$c^2 = a^2 + b^2$$

we have

$$25 = 9 + b^2$$

$$b^2 = 16$$

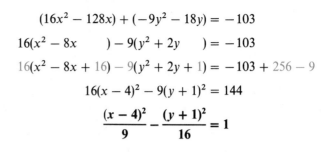

FIGURE 11.30

and the equation is

$$\frac{(y - 1)^2}{9} - \frac{(x - 2)^2}{16} = 1$$

The eccentricity of the hyperbola and parabola is defined by the same equation that was used for the ellipse, namely

$$\varepsilon = \frac{c}{a}$$

Remember that for the ellipse, $0 \le \varepsilon < 1$; however, for the hyperbola $c > a$ so $\varepsilon > 1$, and for the parabola $c = a$ so $\varepsilon = 1$.

EXAMPLE 4 Find the equation of the hyperbola with foci at $(-3, 2)$ and $(5, 2)$ and with eccentricity $\frac{3}{2}$.

SOLUTION The center of the hyperbola is $(1, 2)$ and $c = 4$. Also, since $\varepsilon = \dfrac{c}{a} = \dfrac{3}{2}$ we have

$$\frac{4}{a} = \frac{3}{2} \quad \text{so} \quad a = \frac{8}{3}$$

Since $c^2 = a^2 + b^2$,

$$16 = \frac{64}{9} + b^2$$

$$b^2 = \frac{80}{9}$$

Thus the equation is

$$\frac{(x - 1)^2}{\dfrac{64}{9}} - \frac{(y - 2)^2}{\dfrac{80}{9}} = 1 \quad \text{or} \quad \frac{9(x - 1)^2}{64} - \frac{9(y - 2)^2}{80} = 1$$

EXAMPLE 5 Find the set of points such that the difference of their distances from $(6, 2)$ and $(6, -5)$ is always 3.

SOLUTION This is a hyperbola with center $(6, -\frac{3}{2})$ and $c = \frac{7}{2}$. Also $2a = 3$, so $a = \frac{3}{2}$. Since

$$c^2 = a^2 + b^2$$

we have

$$\frac{49}{4} = \frac{9}{4} + b^2$$

$$b^2 = 10$$

The equation is

$$\frac{\left(y + \dfrac{3}{2}\right)^2}{\dfrac{9}{4}} - \frac{(x - 6)^2}{10} = 1 \quad \text{or} \quad \frac{4\left(y + \dfrac{3}{2}\right)^2}{9} - \frac{(x - 6)^2}{10} = 1$$

Conic Section Summary

We have now considered the graphs of equations of the form

$$Ax^2 + Bxy + Cy^2 + Dx + Ey + F = 0$$

Geometrically they represent the intersection of a plane and a cone, usually resulting in a line, parabola, ellipse, or hyperbola. However, there are certain positions of the plane that result in what are called **degenerate conics.** To visualize some of these degenerate conics, first consider Figure 11.31 on page 484.

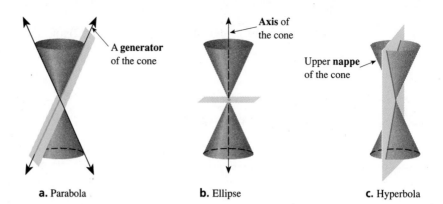

FIGURE 11.31 Conic sections

a. Parabola   **b.** Ellipse   **c.** Hyperbola

For a **degenerate parabola,** visualize the cone (see Figure 11.31a) situated so that one of its generators lies in the plane; a line results. For a **degenerate ellipse,** visualize the plane intersecting at the vertex of the upper and lower nappes (see Figure 11.31b); a point results. And finally, for a **degenerate hyperbola,** visualize the plane situated so that the axis of the cone lies in the plane (see Figure 11.31c); a pair of intersecting lines results.

EXAMPLE 6   Sketch $\dfrac{(x-2)^2}{4} + \dfrac{(y+3)^2}{9} = 0$

SOLUTION   There is only one point that satisfies this equation—namely $(2, -3)$. This is an example of a *degenerate ellipse*. Notice that, except for the zero, the equation has the "form of an ellipse." See Figure 11.32.

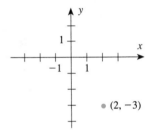

FIGURE 11.32 Graph of $\dfrac{(x-2)^2}{4} + \dfrac{(y+3)^2}{9} = 0$

EXAMPLE 7   Sketch $\dfrac{x^2}{4} - \dfrac{y^2}{9} = 0$.

SOLUTION   This equation has the "form of a hyperbola," but because of the zero it cannot be put into standard form. You can, however, treat this as a factored form:

$$\left(\frac{x}{2} - \frac{y}{3}\right)\left(\frac{x}{2} + \frac{y}{3}\right) = 0$$

$$\frac{x}{2} - \frac{y}{3} = 0 \quad \text{or} \quad \frac{x}{2} + \frac{y}{3} = 0$$

The graph (Figure 11.33) is a *degenerate hyperbola*.

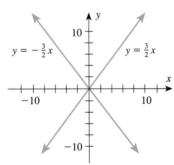

FIGURE 11.33 Graph of $\dfrac{x^2}{4} - \dfrac{y^2}{9} = 0$

It is important to be able to recognize the curve by inspection of the equation before you begin. The first thing to notice is whether there is an xy term.

Up to now we have considered conics for which $B = 0$ (that is, no xy term.) If $B = 0$, then:

TYPE OF CURVE	DEGREE OF EQUATION	DEGREE IN x	DEGREE IN y	RELATIONSHIP TO THE GENERAL EQUATION $Ax^2 + Bxy + Cy^2 + Dx + Ey + F = 0$
1. Line	First	First	First	$A = C = 0$
2. Parabola	Second	First	Second	$A = 0$ and $C \neq 0$
		Second	First	$A \neq 0$ and $C = 0$
3. Ellipse	Second	Second	Second	A and C have same sign (not zero)
4. Circle	Second	Second	Second	$A = C$ (not zero)
5. Hyperbola	Second	Second	Second	A and C have opposite signs

In the next section, we will graph conics for which $B \neq 0$ (that is, there is an xy term). However, we will begin in this section by identifying this type of conic. This identification consists of calculating $B^2 - 4AC$.

These tests do not distinguish degenerate cases. This means that the test may tell you the curve is an ellipse, but it may turn out to be a single point. Remember too that a circle is a special case of the ellipse. The expression $B^2 - 4AC$ is called the **discriminant.**

TYPE OF CURVE	RELATIONSHIP TO GENERAL EQUATION
1. Ellipse	$B^2 - 4AC < 0$
2. Parabola	$B^2 - 4AC = 0$
3. Hyperbola	$B^2 - 4AC > 0$

EXAMPLE 8 Identify each curve.

SOLUTION **a.** $x^2 + 4xy + 4y^2 = 9$

$B^2 - 4AC = 16 - 4(1)(4) = 0$; **parabola**

b. $2x^2 + 3xy + y^2 = 25$

$B^2 - 4AC = 9 - 4(2)(1) > 0$; **hyperbola**

c. $x^2 + xy + y^2 - 8x + 8y = 0$

$B^2 - 4AC = 1 - 4(1)(1) < 0$; **ellipse**

d. $xy = 5$

$B^2 - 4AC = 1 - 4(0)(0) > 0$; **hyperbola**

It is important to remember the standard-form equations of the conics and some basic information about these curves. The important ideas are summarized in the following table.

		PARABOLA	ELLIPSE	HYPERBOLA
Definition		All points equidistant from a given point and a given line	All points with the sum of distances from two fixed points constant	All points with the difference of distances from two fixed points constant
Equations		Up: $x^2 = 4cy$ Down: $x^2 = -4cy$ Right: $y^2 = 4cx$ Left: $y^2 = -4cx$ c is a positive number and the distance from the vertex to the focus; $4c$ is the length of the focal chord	$c^2 = a^2 - b^2$ Horizontal axis: $\dfrac{x^2}{a^2} + \dfrac{y^2}{b^2} = 1$ Vertical axis: $\dfrac{y^2}{a^2} + \dfrac{x^2}{b^2} = 1$	$c^2 = a^2 + b^2$ Horizontal axis: $\dfrac{x^2}{a^2} - \dfrac{y^2}{b^2} = 1$ Vertical axis: $\dfrac{y^2}{a^2} - \dfrac{x^2}{b^2} = 1$
Recognition		Second-degree equation; linear in one variable, quadratic in the other variable	Second-degree equation; coefficients of x^2 and y^2 have same sign	Second-degree equation; coefficients of x^2 and y^2 have different signs
Eccentricity		$\varepsilon = 1$	$0 \le \varepsilon < 1$	$\varepsilon > 1$
Directrix		Perpendicular to axis c units from the vertex (one directrix)	Perpendicular to major axis $\pm a/\varepsilon$ units from center (two directrices); see Problem Set 11.2, Problem 63	Perpendicular to transverse axis $\pm a/\varepsilon$ units from center (two directrices); see Problem 64 below
Translations to the point (h, k): $x' = x - h$ and $y' = y - k$				

11.3 Problem Set

A *Sketch the curves in Problems 1–21.*

1. $x^2 - y^2 = 1$

2. $x^2 - y^2 = 4$

3. $y^2 - x^2 = 1$

4. $\dfrac{x^2}{4} - \dfrac{y^2}{9} = 1$

5. $\dfrac{x^2}{9} - \dfrac{y^2}{4} = 1$

6. $\dfrac{y^2}{9} - \dfrac{x^2}{4} = 1$

7. $\dfrac{x^2}{16} - \dfrac{y^2}{25} = 1$

8. $\dfrac{y^2}{16} - \dfrac{x^2}{25} = 1$

9. $\dfrac{x^2}{36} - \dfrac{y^2}{9} = 1$

10. $36y^2 - 25x^2 = 900$

11. $3x^2 - 4y^2 = 12$

12. $3y^2 = 4x^2 + 12$

13. $3x^2 - 4y^2 = 5$

14. $4y^2 - 4x^2 = 5$

15. $4y^2 - x^2 = 9$

16. $\dfrac{(x-2)^2}{4} - \dfrac{(y+3)^2}{16} = 1$

17. $\dfrac{(x+3)^2}{8} - \dfrac{(y-1)^2}{5} = 1$

18. $\dfrac{(y-1)^2}{6} - \dfrac{(x+2)^2}{8} = 1$

19. $\dfrac{(x-2)^2}{16} - \dfrac{(y+1)^2}{9} = 1$

20. $\dfrac{(y+2)^2}{25} - \dfrac{(x+1)^2}{16} = 1$

21. $\dfrac{(y-1)^2}{\frac{1}{9}} - \dfrac{(x+2)^2}{\frac{1}{4}} = 1$

In Problems 22–33, *identify and sketch the curve.*

22. $2x - y - 8 = 0$

23. $2x + y - 10 = 0$

24. $4x^2 - 16y = 0$

25. $\dfrac{(x-3)^2}{4} - \dfrac{(y+2)}{6} = 1$

26. $\dfrac{(x-3)^2}{9} - \dfrac{(y+2)^2}{25} = 1$

27. $\dfrac{x-3}{9} + \dfrac{y-2}{25} = 1$

28. $(x+3)^2 + (y-2)^2 = 0$

29. $9(x+3)^2 + 4(y-2) = 0$

30. $9(x+3)^2 - 4(y-2)^2 = 0$ **31.** $x^2 + 8(y-12)^2 = 16$

32. $x^2 + 64(y+4)^2 = 16$

33. $x^2 + y^2 - 3y = 0$

B *Find the equations of the curves in Problems 34–39.*

34. The hyperbola with vertices at $(0, 5)$ and $(0, -5)$ and foci at $(0, 7)$ and $(0, -7)$

35. The set of points such that the difference of their distances from $(-6, 0)$ and $(6, 0)$ is 10

36. The hyperbola with foci at $(5, 0)$ and $(-5, 0)$ and eccentricity 5

37. The hyperbola with vertices at $(4, 4)$ and $(4, 8)$ and foci at $(4, 3)$ and $(4, 9)$

38. The set of points such that the difference of their distances from $(4, -3)$ and $(-4, -3)$ is 6

39. The hyperbola with vertices at $(-2, 0)$ and $(6, 0)$ passing through $(10, 3)$

Sketch the curves in Problems 40–47.

40. $5(x-2)^2 - 2(y+3)^2 = 10$

41. $4(x+4)^2 - 3(y+3)^2 = -12$

42. $3x^2 - 4y^2 = 12x + 80y + 88$

43. $9x^2 - 18x - 11 = 4y^2 + 16y$

44. $4y^2 - 8y + 4 = 3x^2 - 2x$

45. $x^2 - 4x + y^2 + 6y - 12 = 0$

46. $x^2 - y^2 = 2x + 4y - 3$

47. $3x^2 - 5y^2 + 18x + 10y - 8 = 0$

In Problems 48–57, *identify and sketch the curve.*

48. $4x^2 - 3y^2 - 24y - 112 = 0$

49. $y^2 - 4x + 2y + 21 = 0$

50. $9x^2 + 2y^2 - 48y + 270 = 0$

51. $x^2 + 4x + 12y + 64 = 0$

52. $y^2 - 6y - 4x + 5 = 0$

53. $100x^2 - 7y^2 + 98y - 368 = 0$

54. $x^2 + y^2 + 2x - 4y - 20 = 0$

55. $4x^2 + 12x + 4y^2 + 4y + 1 = 0$

56. $x^2 - 4y^2 - 6x - 8y - 11 = 0$

57. $9x^2 + 25y^2 - 54x - 200y + 256 = 0$

C

58. Consider a person A who fires a rifle at a distant gong B. Assuming that the ground is flat, where must you stand to hear the sound of the gun and the sound of the gong simultaneously? *Hint:* To answer this question, let d be the distance sound travels in the length of time it takes the bullet to travel from the gun to the gong. Show that the person who hears the sounds simultaneously must stand on a branch of a hyperbola (the one nearest the target) so that the difference of the distances from A to B is d.

59. Derive the equation of the hyperbola with foci at $(-c, 0)$ and $(c, 0)$ and constant distance $2a$. Let $b^2 = c^2 - a^2$. Show all your work.

60. Derive the equation of the hyperbola with foci at $(0, c)$ and $(0, -c)$ and constant distance $2a$. Let $b^2 = c^2 - a^2$. Show all your work.

61. Let d represent the vertical distance between the hyperbola in the first quadrant

$$y = \frac{b}{a}\sqrt{x^2 - a^2} \quad \text{and the line} \quad y = \frac{b}{a}x$$

in the first quadrant. Show that as $|x| \to \infty$, then $d \to 0$.

62. Refer to Problem 61. Show what happens to d as $|x| \to \infty$ if (x, y) is in Quadrant IV.

63. Given the hyperbola

$$\frac{x^2}{a^2} - \frac{y^2}{b^2} = 1$$

show that the length of the diagonal of the central rectangle of the hyperbola is $2c$.

64. Given the hyperbola

$$\frac{x^2}{a^2} - \frac{y^2}{b^2} = 1$$

we define the *directrices* of the hyperbola as the lines

$$x = \frac{a}{\varepsilon} \quad \text{and} \quad x = -\frac{a}{\varepsilon}$$

Show that the hyperbola is the set of all points with distances from $F(c, 0)$ that are equal to ε times their distances from the line $x = a/\varepsilon$.

65. A line through a focus parallel to a directrix and cut off by the hyperbola is called the *focal chord.* Show that the length of the focal chord of the following hyperbola is $2b^2/a$:

$$\frac{x^2}{a^2} - \frac{y^2}{b^2} = 1$$

11.4 Rotations

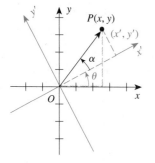

FIGURE 11.34 Rotation of axes

In Section 2.3 we introduced the idea of a **translation** in order to simplify the equation of a curve by writing it relative to a new translated coordinate system. In this section we will consider the idea of a **rotation** in which the equation of a curve can be simplified by writing it in terms of a rotated coordinate system. Suppose the coordinate axes are rotated through an angle θ ($0 < \theta < 90°$). The relationship between the old coordinates (x, y) and the new coordinates (x', y') can be found by considering Figure 11.34.

Let O be the origin and P be a point with coordinates (x, y) relative to the old coordinate system and (x', y') relative to the new rotated coordinate system. Let θ be the amount of rotation and let α be the angle between the x' axis and $|OP|$. Then, using the definition of sine and cosine,

$$x = |OP|\cos(\theta + \alpha) \qquad x' = |OP|\cos\alpha$$
$$y = |OP|\sin(\theta + \alpha) \qquad y' = |OP|\sin\alpha$$

Now

$$
\begin{aligned}
x = |OP|\cos(\theta + \alpha) &= |OP|[\cos\theta\cos\alpha - \sin\theta\sin\alpha] \\
&= |OP|\cos\theta\cos\alpha - |OP|\sin\theta\sin\alpha \\
&= (|OP|\cos\alpha)\cos\theta - (|OP|\sin\alpha)\sin\theta \\
&= x'\cos\theta - y'\sin\theta
\end{aligned}
$$

Also,

$$
\begin{aligned}
y = |OP|\sin(\theta + \alpha) &= |OP|\sin\theta\cos\alpha + |OP|\cos\theta\sin\alpha \\
&= x'\sin\theta + y'\cos\theta
\end{aligned}
$$

Therefore:

ROTATION OF AXES FORMULAS

$$x = x'\cos\theta - y'\sin\theta \qquad y = x'\sin\theta + y'\cos\theta$$

All the curves considered in this chapter can be characterized by the general second-degree equation

$$Ax^2 + Bxy + Cy^2 + Dx + Ey + F = 0$$

Notice that the xy term has not appeared before. The presence of this term indicates that the conic has been rotated. Thus in this section we assume that $B \neq 0$ and now need to determine the amount this conic has been rotated from standard position. That is, the new axes should be rotated the same amount as the given conic so that it will be in standard position after the rotation. To find out how much to rotate the axes, substitute

$$x = x'\cos\theta - y'\sin\theta \qquad y = x'\sin\theta + y'\cos\theta$$

into

$$Ax^2 + Bxy + Cy^2 + Dx + Ey + F = 0 \quad (B \neq 0)$$

After a lot of simplifying you will obtain

$$(A \cos^2\theta + B \cos\theta \sin\theta + C \sin^2\theta)x'^2$$
$$+ [B(\cos^2\theta - \sin^2\theta) + 2(C - A) \sin\theta \cos\theta]x'y'$$
$$+ (A \sin^2\theta - B \sin\theta \cos\theta + C \cos^2\theta)y'^2 + (D \cos\theta + E \sin\theta)x'$$
$$+ (-D \sin\theta + E \cos\theta)y' + F = 0$$

This looks terrible, but it is still in the form

$$A'x'^2 + B'x'y' + C'y'^2 + D'x' + E'y' + F = 0$$

You want to choose θ so that $B' = 0$. This will give you a standard position relative to the new coordinate axes. That is,

$$B(\cos^2\theta - \sin^2\theta) + 2(C - A) \sin\theta \cos\theta = 0$$

$$B \cos 2\theta + (C - A) \sin 2\theta = 0 \qquad \text{Using double-}$$
$$\qquad\qquad\qquad\qquad\qquad\qquad\qquad \text{angle identities,}$$
$$B \cos 2\theta = (A - C) \sin 2\theta \qquad B \neq 0, \theta \neq 0.$$

$$\frac{\cos 2\theta}{\sin 2\theta} = \frac{A - C}{B}$$

Simplifying, you obtain the following result:

AMOUNT OF ROTATION FORMULA

$$\cot 2\theta = \frac{A - C}{B}$$

Notice that we required $0 < \theta < 90°$, so 2θ is in Quadrant I or Quadrant II. This means that if $\cot 2\theta$ is positive, then 2θ must be in Quadrant I; if $\cot 2\theta$ is negative, then 2θ is in Quadrant II.

In Examples 1–4, find the appropriate rotation so that the given curve will be in standard position relative to the rotated axes. Also find the x and y values in the new coordinate system.

EXAMPLE 1 $xy = 6$

SOLUTION $\cot 2\theta = \dfrac{A - C}{B}$ $\qquad x = x' \cos\theta - y' \sin\theta \qquad y = x' \sin\theta + y' \cos\theta$

$\qquad\qquad = \dfrac{0 - 0}{1}$ $\qquad\qquad = x' \cos 45° - y' \sin 45° \qquad = x' \sin 45° + y' \cos 45°$

$\qquad\qquad = \dfrac{0 - 0}{1}$ $\qquad\qquad = x'\left(\dfrac{1}{\sqrt{2}}\right) - y'\left(\dfrac{1}{\sqrt{2}}\right) \qquad = x'\left(\dfrac{1}{\sqrt{2}}\right) + y'\left(\dfrac{1}{\sqrt{2}}\right)$

$\qquad\qquad = 0$ $\qquad\qquad\qquad = \dfrac{1}{\sqrt{2}} (x' - y') \qquad\qquad = \dfrac{1}{\sqrt{2}} (x' + y')$

Thus $2\theta = 90°$
and $\theta = 45°$

EXAMPLE 2 $7x^2 - 6\sqrt{3}xy + 13y^2 - 16 = 0$

SOLUTION $\cot 2\theta = \dfrac{A - C}{B}$ $\qquad x = x' \cos\theta - y' \sin\theta$ $\qquad y = x' \sin\theta + y' \cos\theta$

$\qquad\qquad = \dfrac{7 - 13}{-6\sqrt{3}}$ $\qquad\qquad = x' \cos 30° - y' \sin 30°$ $\qquad = x' \sin 30° + y' \cos 30°$

$\qquad\qquad = \dfrac{1}{\sqrt{3}}$ $\qquad\qquad = x'\left(\dfrac{\sqrt{3}}{2}\right) - y'\left(\dfrac{1}{2}\right)$ $\qquad = x'\left(\dfrac{1}{2}\right) + y'\left(\dfrac{\sqrt{3}}{2}\right)$

Thus $2\theta = 60°$ $\qquad\qquad = \dfrac{1}{2}\left(\sqrt{3}x' - y'\right)$ $\qquad = \dfrac{1}{2}\left(x' + \sqrt{3}y'\right)$
and $\theta = 30°$ ∎

EXAMPLE 3 $x^2 - 4xy + 4y^2 + 5\sqrt{5}y - 10 = 0$

SOLUTION $\qquad\qquad\qquad\qquad\qquad\qquad\qquad \cot 2\theta = \dfrac{1 - 4}{-4} = \dfrac{3}{4}$

Since this is not an exact value for θ (as were those found in Examples 1 and 2), you will need to use some trigonometric identities to find $\cos\theta$ and $\sin\theta$. If $\cot 2\theta = \frac{3}{4}$, then $\tan 2\theta = \frac{4}{3}$ and $\sec 2\theta = \sqrt{1 + (\frac{4}{3})^2} = \frac{5}{3}$. Then $\cos 2\theta = \frac{3}{5}$ and you can now apply the half-angle identities:

$\qquad \cos\theta = \pm\sqrt{\dfrac{1 + \cos 2\theta}{2}} \qquad \sin\theta = \pm\sqrt{\dfrac{1 - \cos 2\theta}{2}}$

$\qquad \cos\theta = \sqrt{\dfrac{1 + (\frac{3}{5})}{2}} \qquad\qquad \sin\theta = \sqrt{\dfrac{1 - (\frac{3}{5})}{2}} \qquad$ Positive because θ is in Quadrant I

$\qquad\qquad = \dfrac{2}{\sqrt{5}} \qquad\qquad\qquad\quad = \dfrac{1}{\sqrt{5}}$

To find the amount of rotation, use a calculator and one of the preceding equations to find $\theta \approx 26.6°$. Finally, the rotation of axes formulas provide

$\qquad x = x' \cos\theta - y' \sin\theta \qquad\qquad\qquad y = x' \sin\theta + y' \cos\theta$

$\qquad\quad = x'\left(\dfrac{2}{\sqrt{5}}\right) - y'\left(\dfrac{1}{\sqrt{5}}\right) \qquad\qquad = x'\left(\dfrac{1}{\sqrt{5}}\right) + y'\left(\dfrac{2}{\sqrt{5}}\right)$

$\qquad\quad = \dfrac{1}{\sqrt{5}}\left(2x' - y'\right) \qquad\qquad\qquad = \dfrac{1}{\sqrt{5}}\left(x' + 2y'\right)$ ∎

EXAMPLE 4 $10x^2 + 24xy + 17y^2 - 9 = 0$

SOLUTION $\qquad\qquad\qquad\qquad\qquad\qquad\qquad \cot 2\theta = \dfrac{10 - 17}{24} = -\dfrac{7}{24}$

Since this is negative, 2θ must be in Quadrant II; this means that $\sec 2\theta$ is negative in the following sequence of identities: Since $\cot 2\theta = -\frac{7}{24}$, then

$\tan 2\theta = -\frac{24}{7}$ and $\sec 2\theta = -\sqrt{1 + (-\frac{24}{7})^2} = -\frac{25}{7}$. Thus $\cos 2\theta = -\frac{7}{25}$, which gives

$$\cos\theta = \sqrt{\frac{1 + (-\frac{7}{25})}{2}} \qquad\qquad \sin\theta = \sqrt{\frac{1 - (-\frac{7}{25})}{2}}$$

$$= \frac{3}{5} \qquad\qquad\qquad\qquad = \frac{4}{5}$$

Using either of these equations and a calculator, you find that the rotation is $\theta \approx 53.1°$. The rotation of axes formulas provide

$$x = x'\cos\theta - y'\sin\theta \qquad\qquad y = x'\sin\theta + y'\cos\theta$$

$$= x'\left(\frac{3}{5}\right) - y'\left(\frac{4}{5}\right) \qquad\qquad = x'\left(\frac{4}{5}\right) + y'\left(\frac{3}{5}\right)$$

$$= \frac{1}{5}(3x' - 4y') \qquad\qquad\qquad = \frac{1}{5}(4x' + 3y')$$

▮

PROCEDURE FOR SKETCHING A ROTATED CONIC

1. Identify curve.
2. a. Find the angle of rotation.
   b. Find x and y in the new coordinate system.
   c. Substitute the values found in Step 2b into the given equation and simplify.
3. Translate axes, if necessary.
4. Sketch the resulting equation relative to the new x' and y' axes.

EXAMPLE 5 Sketch $xy = 6$.

SOLUTION From Example 1, the rotation is $\theta = 45°$ and

$$x = \frac{1}{\sqrt{2}}(x' - y') \qquad y = \frac{1}{\sqrt{2}}(x' + y')$$

Substitute these values into the original equation $xy = 6$:

$$\left[\frac{1}{\sqrt{2}}(x' - y')\right]\left[\frac{1}{\sqrt{2}}(x' + y')\right] = 6$$

Simplify (see Problem 1 for the details) to obtain

$$x'^2 - y'^2 = 12$$

$$\frac{x'^2}{12} - \frac{y'^2}{12} = 1$$

This curve is a hyperbola that has been rotated 45°. Draw the rotated axis and sketch this equation relative to the rotated axes. The result is shown in Figure 11.35.

▮

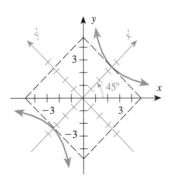

FIGURE 11.35 Graph of $xy = 6$

EXAMPLE 6   Sketch $7x^2 - 6\sqrt{3}xy + 13y^2 - 16 = 0$.

SOLUTION   From Example 2, $\theta = 30°$ and

$$x = \frac{1}{2}(\sqrt{3}x' - y') \qquad y = \frac{1}{2}(x' + \sqrt{3}y')$$

Substitute into the original equation:

$$7\left(\frac{1}{2}\right)^2(\sqrt{3}x' - y')^2$$

$$- 6\sqrt{3}\left(\frac{1}{2}\right)(\sqrt{3}x' - y')\left(\frac{1}{2}\right)(x' + \sqrt{3}y')$$

$$+ 13\left(\frac{1}{2}\right)^2(x' + \sqrt{3}y')^2 - 16 = 0$$

Simplify (see Problem 2 for the details) to obtain

$$\frac{x'^2}{4} + \frac{y'^2}{1} = 1$$

CALCULATOR COMMENT

A curve such as the one shown in Example 5 can be sketched much more easily as $y = 6/x$. However, to use a calculator to sketch one like Example 6 would require that you solve for y:

$$Y1 = (6X\sqrt{3} + \sqrt{(36 * 3X^2 -}$$
$$4 * 13(7X^2 - 16)))/26$$

$$Y2 = (6X\sqrt{3} - \sqrt{(36 * 3X^2 -}$$
$$4 * 13(7X^2 - 16)))/26$$

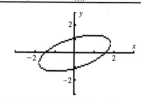

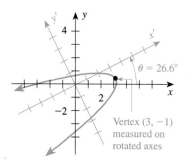

FIGURE 11.36   Graph of
$7x^2 - 6\sqrt{3}xy + 13y^2 - 16 = 0$

This curve is an ellipse with a 30° rotation. The sketch is shown in Figure 11.36.

EXAMPLE 7   Sketch $x^2 - 4xy + 4y^2 + 5\sqrt{5}y - 10 = 0$.

SOLUTION   From Example 3, the rotation is $\theta \approx 26.6°$ and

$$x = \frac{1}{\sqrt{5}}(2x' - y') \qquad y = \frac{1}{\sqrt{5}}(x' + 2y')$$

Substitute

$$\frac{1}{5}(2x' - y')^2 - 4\left(\frac{1}{5}\right)(2x' - y')(x' + 2y') + 4\left(\frac{1}{5}\right)(x' + 2y')^2$$

$$+ 5\sqrt{5}\left(\frac{1}{\sqrt{5}}\right)(x' + 2y') - 10 = 0$$

Simplify (see Problem 3 for the details) to obtain

$$y'^2 + 2y' = -x' + 2$$

This curve is a parabola with a rotation of about 26.6°. Next complete the square to obtain

$$(y' + 1)^2 = -(x' - 3)$$

This sketch is shown in Figure 11.37.

FIGURE 11.37   Graph of
$x^2 - 4xy + 4y^2 + 5\sqrt{5}y - 10 = 0$

Vertex $(3, -1)$ measured on rotated axes

To graph the general second-degree equation

$$Ax^2 + Bxy + Cy^2 + Dx + Ey + F = 0$$

follow the steps shown in Figure 11.38.

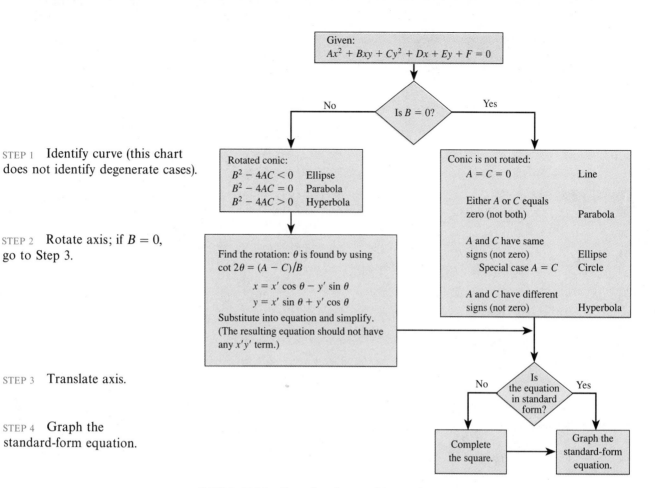

FIGURE 11.38 Procedure for graphing conics

11.4 Problem Set

A *Because of the amount of arithmetic and algebra involved, many algebraic steps were left out of the examples of this section. In Problems 1–3, fill in the details left out of the indicated example.*

1. Example 5   **2.** Example 6

3. Example 7

Identify the curves whose equations are given in Problems 4–21.

4. $xy = 10$ **5.** $xy = -1$ **6.** $xy = -4$

7. $13x^2 - 10xy + 13y^2 - 72 = 0$

8. $5x^2 - 26xy + 5y^2 + 72 = 0$

9. $x^2 + 4xy + 4y^2 + 10\sqrt{5}x = 9$

10. $5x^2 - 4xy + 8y^2 = 36$

11. $23x^2 + 26\sqrt{3}xy - 3y^2 - 144 = 0$

12. $3x^2 + 2\sqrt{3}xy + y^2 + 16x - 16\sqrt{3}y = 0$

13. $24x^2 + 16\sqrt{3}xy + 8y^2 - x + \sqrt{3}y - 8 = 0$

14. $3x^2 - 2\sqrt{3}xy + y^2 + 24x + 24\sqrt{3}y = 0$

15. $13x^2 - 6\sqrt{3}xy + 7y^2 + (16\sqrt{3} - 8)x$
$+ (-16 - 8\sqrt{3})y + 16 = 0$

16. $3x^2 - 10xy + 3y^2 - 32 = 0$

17. $5x^2 - 3xy + y^2 + 65x - 25y + 203 = 0$

18. $5x^2 - 6xy + y^2 + 4x - 3y + 10 = 0$

19. $4x^2 + 3xy + 2y^2 - 5x - 4y - 18 = 0$

20. $4x^2 + 2\sqrt{3}xy + 3y^2 - 6x + 12y - 15 = 0$

21. $2x^2 - \sqrt{56}xy + 7y^2 + 9x + 7y + 112 = 0$

B *Find the appropriate rotation in Problems 22–35 so that the given curve will be in standard position relative to the rotated axes. Also find x and y values in the new coordinate system by using the rotation of axes formulas.*

22. $xy = 10$ 23. $xy = -1$

24. $xy = -4$

25. $13x^2 - 10xy + 13y^2 - 72 = 0$

26. $5x^2 - 26xy + 5y^2 + 72 = 0$

27. $x^2 + 4xy + 4y^2 + 10\sqrt{5}x = 9$

28. $5x^2 - 4xy + 8y^2 = 36$

29. $23x^2 + 26\sqrt{3}xy - 3y^2 - 144 = 0$

30. $3x^2 + 2\sqrt{3}xy + y^2 + 16x - 16\sqrt{3}y = 0$

31. $24x^2 + 16\sqrt{3}xy + 8y^2 - x + \sqrt{3}y - 8 = 0$

32. $3x^2 - 2\sqrt{3}xy + y^2 + 24x + 24\sqrt{3}y = 0$

33. $13x^2 - 6\sqrt{3}xy + 7y^2 + (16\sqrt{3} - 8)x$
    $+ (-16 - 8\sqrt{3})y + 16 = 0$

34. $3x^2 - 10xy + 3y^2 - 32 = 0$

35. $5x^2 - 3xy + y^2 + 65x - 25y + 203 = 0$

C *Sketch the curves in Problems 36–59.*

36. $xy = 10$ 37. $xy = -1$

38. $xy = -4$ 39. $xy = 8$

40. $8x^2 - 4xy + 5y^2 = 36$

41. $13x^2 - 10xy + 13y^2 - 72 = 0$

42. $5x^2 - 26xy + 5y^2 + 72 = 0$

43. $x^2 + 4xy + 4y^2 + 10\sqrt{5}x = 9$

44. $5x^2 - 4xy + 8y^2 = 36$

45. $23x^2 + 26\sqrt{3}xy - 3y^2 - 144 = 0$

46. $3x^2 + 2\sqrt{3}xy + y^2 + 16x - 16\sqrt{3}y = 0$

47. $24x^2 + 16\sqrt{3}xy + 8y^2 - x + \sqrt{3}y - 8 = 0$

48. $3x^2 - 2\sqrt{3}xy + y^2 + 24x + 24\sqrt{3}y = 0$

49. $3x^2 - 10xy + 3y^2 - 32 = 0$

50. $x^2 + 2xy + y^2 + 12\sqrt{2}x - 6 = 0$

51. $10x^2 + 24xy + 17y^2 - 9 = 0$

52. $5x^2 - 3xy + y^2 + 65x - 25y + 203 = 0$

53. $3xy - 4y^2 + 18 = 0$

54. $17x^2 - 12xy + 8y^2 - 80 = 0$

55. $5x^2 - 8xy + 5y^2 - 9 = 0$

56. $16x^2 - 24xy + 9y^2 - 60x - 80y + 100 = 0$

57. $13x^2 - 6\sqrt{3}xy + 7y^2 + (16\sqrt{3} - 8)x$
    $+ (-16 - 8\sqrt{3})y + 16 = 0$

58. $x^2 + 2\sqrt{3}xy + 3y^2 + 2\sqrt{3}x + 2y - 16 = 0$

59. $21x^2 + 10\sqrt{3}xy + 31y^2 - 72x - 16\sqrt{3}x$
    $- 72\sqrt{3}y + 16y + 16 = 0$

60. Let $Ax^2 + Bxy + Cy^2 + Dx + Ey + F = 0$. Show that $B^2 - 4AC = B'^2 - 4A'C'$ for any angle θ through which the axes may be rotated and that A', B', and C' are the values given on pages 488–489. Use this fact to prove that (if the graph exists)
    If $B^2 - 4AC = 0$, the graph is a parabola.
    If $B^2 - 4AC < 0$, the graph is an ellipse.
    If $B^2 - 4AC > 0$, the graph is a hyperbola.

11.5 Vectors

Many applications of mathematics involve quantities that have *both* magnitude and direction, such as forces, velocities, accelerations, and displacements. Vectors are used to describe such quantities. A **vector** is a directed line segment specifying both a magnitude and a direction. The length of the vector represents the **magnitude** of the quantity being represented; the **direction** of the vector represents the direction of the quantity. Two vectors are **equal** if they have the same magnitude and direction.

Suppose we choose a point O in the plane and call it the origin. A vector is a directed line segment from O to a point $P(x, y)$ in the plane. This vector is denoted by $\overrightarrow{OP}$ or **v**. In the text we use **v**; in your work you will write $\vec{v}$. The magnitude of $\overrightarrow{OP}$ is denoted by $|\overrightarrow{OP}|$ or $|\mathbf{v}|$. The vector from O to O is called the **zero vector 0**.

If **v** and **w** represent any two vectors having different (but not opposite) directions, then the **sum** or **resultant** is the vector drawn as the diagonal of the parallelogram having **v** and **w** as the adjacent sides, as shown in Figure 11.39. The vectors **v** and **w** are called **components**.

There are basically two types of vector problems dealing with addition of

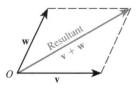

FIGURE 11.39
Resultant of vectors **v** and **w**

vectors. The first is to find the *resultant vector*. To do this you use a right triangle, the Law of Sines, or the Law of Cosines. The second problem is to *resolve* a vector into two component vectors. If the two vectors form a right angle, they are called **rectangular components.** You will usually resolve a vector into rectangular components.

EXAMPLE 1   Consider two forces, one with magnitude 3.0 in a N20°W direction and the other with magnitude 7.0 in a S50°W direction. Find the resultant vector.

SOLUTION   Sketch the given vectors and draw the parallelogram formed by these vectors. The diagonal is the resultant vector as shown in Figure 11.40. You can easily find $\theta = 110°$, but you really need an angle inside the shaded triangle. Use the property from geometry that adjacent angles in a parallelogram are supplementary (they add up to 180°). This tells you that $\phi = 70°$. Thus you know SAS, so you should use the Law of Cosines to find the magnitude $|\mathbf{v}|$:

$$|\mathbf{v}|^2 = 3^2 + 7^2 - 2(3)(7) \cos 70°$$
$$|\mathbf{v}| \approx 6.60569103$$

The direction of $\mathbf{v}$ can be found by using the Law of Sines to derive α:

$$\frac{\sin 70°}{|\mathbf{v}|} = \frac{\sin \alpha}{7}$$

$$\sin \alpha = \frac{7}{|\mathbf{v}|} \sin 70° \approx .99578505$$

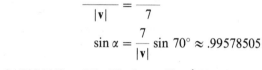

Thus $\alpha \approx 84.737565° \approx 85°$. (Find $\alpha = \text{Sin}^{-1}.99578505$; do not forget significant digits.)

Since $20° + 85° = 105°$, you can see that the direction of $\mathbf{v}$ should be measured from the south (since the measurement from the north is greater than 90°; it is $180° - 105° = 75°$). Thus the magnitude of $\mathbf{v}$ is **6.6** and the direction is **S75°W**.

In the figure:

$\phi = 70°$

$\theta = 180° - 20° - 50°$
$\quad = 110°$
$\phi = 180° - \theta$
$\quad = 70°$

FIGURE 11.40

EXAMPLE 2   Suppose a vector $\mathbf{v}$ has a magnitude of 5.00 and a direction given by $\theta = 30.0°$, where θ is the angle the vector makes with the positive x-axis. Resolve this vector into horizontal and vertical components. (Do not forget significant digits—see Appendix A.)

SOLUTION   Let $\mathbf{v}_x$ be the horizontal component and $\mathbf{v}_y$ be the vertical component as shown in Figure 11.41. Then

$$\cos \theta = \frac{|\mathbf{v}_x|}{|\mathbf{v}|} \qquad\qquad \sin \theta = \frac{|\mathbf{v}_y|}{|\mathbf{v}|}$$

$$|\mathbf{v}_x| = |\mathbf{v}| \cos \theta \qquad\qquad |\mathbf{v}_y| = |\mathbf{v}| \sin \theta$$

$$= 5 \cos 30° \qquad\qquad\qquad = 5 \sin 30°$$

$$= \frac{5}{2}\sqrt{3} \qquad\qquad\qquad\quad = \frac{5}{2}$$

$$\approx 4.33 \qquad\qquad\qquad\quad = 2.50$$

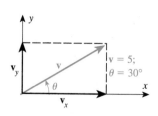

FIGURE 11.41   Resolving a vector

Thus $\mathbf{v}_x$ is a horizontal vector with magnitude 4.33 and $\mathbf{v}_y$ is a vertical vector with magnitude 2.50.

The process of resolving a vector can be simplified by using scalar multiplication and two special vectors. **Scalar multiplication** is the multiplication of a vector by a real number. It is called scalar multiplication because real numbers are sometimes called scalars. If c is a positive real number and $\mathbf{v}$ is a vector, then the vector $c\mathbf{v}$ is a vector representing the scalar multiplication of c and $\mathbf{v}$. It is defined geometrically as a vector in the same direction as $\mathbf{v}$ but with a magnitude c times the original magnitude of $\mathbf{v}$, as shown in Figure 11.42. If c is a negative real number, then the scalar multiplication results in a vector in exactly the opposite direction as $\mathbf{v}$ with length c times as long, as shown in Figure 11.42. If $c = 0$, then $c\mathbf{v}$ is the zero vector.

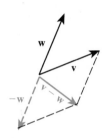

FIGURE 11.43
Vector subtraction

FIGURE 11.42
Examples of scalar multiplication

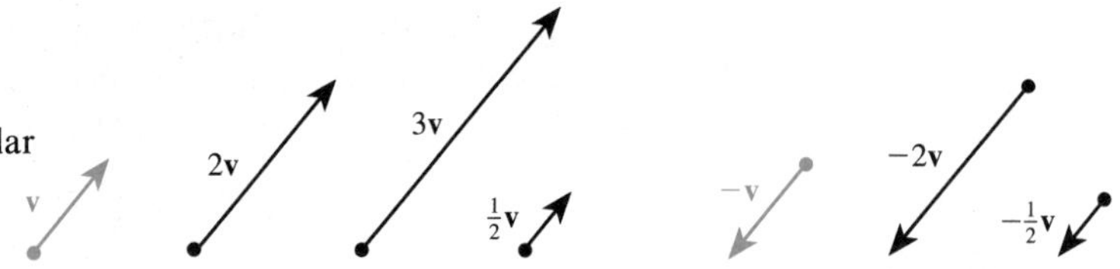

Scalar multiplication allows us to define **subtraction** for vectors:

$$\mathbf{v} - \mathbf{w} = \mathbf{v} + (-\mathbf{w})$$

Geometrically, subtraction is shown in Figure 11.43.

Two special vectors help us to treat vectors algebraically as well as geometrically:

DEFINITION OF i AND j VECTORS

$\mathbf{i}$ is the vector of unit length in the direction of the positive x-axis.

$\mathbf{j}$ is the vector of unit length in the direction of the positive y-axis.

Example 2 shows how the vector $\mathbf{v}$ with magnitude 5.00 and $\theta = 30.0°$ can be resolved into rectangular components:

1. $\mathbf{v}_x$ (horizontal component) with magnitude 4.33
2. $\mathbf{v}_y$ (vertical component) with magnitude 2.50

However, since $\mathbf{i}$ has unit length and is horizontal and $\mathbf{j}$ has unit length and is vertical, we can write

$$\mathbf{v}_x = 4.33\mathbf{i} \quad \text{and} \quad \mathbf{v}_y = 2.50\mathbf{j}$$

Consequently,

$$\mathbf{v} = \mathbf{v}_x + \mathbf{v}_y$$
$$= 4.33\mathbf{i} + 2.50\mathbf{j}$$

In general, any vector $\mathbf{v}$ can be written as

$$\mathbf{v} = a\mathbf{i} + b\mathbf{j}$$

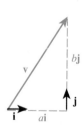

$v = ai + bj$

FIGURE 11.44 Algebraic representation of a vector

where a and b are the magnitude of the horizontal and vertical components, respectively. This is called the **algebraic representation of a vector** and is shown in Figure 11.44.

EXAMPLE 3 Find the algebraic representation for a vector **v** with magnitude 10 making an angle of 60° with the positive *x*-axis.

SOLUTION Figure 11.45 shows the general procedure for writing the algebraic representation of a vector when given the magnitude and direction of that vector.

$$|b\mathbf{j}| = b$$
$$|a\mathbf{i}| = a$$

$$\cos\theta = \frac{a}{|\mathbf{v}|} \qquad a = |\mathbf{v}|\cos\theta$$

$$\sin\theta = \frac{b}{|\mathbf{v}|} \qquad b = |\mathbf{v}|\sin\theta$$

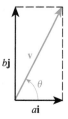

FIGURE 11.45
$$\mathbf{v} = |\mathbf{v}|\cos\theta\,\mathbf{i} + |\mathbf{v}|\sin\theta\,\mathbf{j}$$

From Figure 11.45,

$a = |\mathbf{v}|\cos\theta$	$b = |\mathbf{v}|\sin\theta$
$= 10\cos 60°$	$= 10\sin 60°$
$= 5.0$	≈ 8.7

Therefore $\mathbf{v} = 5.0\mathbf{i} + 8.7\mathbf{j}$.

EXAMPLE 4 Find the algebraic representation for a vector **v** with initial point $(4, -3)$ and endpoint $(-2, 4)$.

SOLUTION Figure 11.46 shows the general procedure for writing the algebraic representation of a vector when given the endpoints of that vector:

$$a = x_2 - x_1 \quad\text{and}\quad b = y_2 - y_1$$

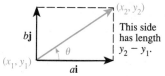

This side has length $x_2 - x_1$.

FIGURE 11.46
$$\mathbf{v} = (x_2 - x_1)\mathbf{i} + (y_2 - y_1)\mathbf{j}$$

From Figure 11.46,

$a = x_2 - x_1$	$b = y_2 - y_1$
$= -2 - 4$	$= 4 - (-3)$
$= -6$	$= 7$

Thus $\mathbf{v} = -6\mathbf{i} + 7\mathbf{j}$.

EXAMPLE 5 Find the magnitude of the vector in Example 4.

SOLUTION $\mathbf{v} = -6\mathbf{i} + 7\mathbf{j}$. Thus

$$|\mathbf{v}| = \sqrt{(-6)^2 + (7)^2} = \sqrt{85}$$

Example 5 leads to the following general result.

MAGNITUDE OF A VECTOR

The **magnitude** of a vector $\mathbf{v} = a\mathbf{i} + b\mathbf{j}$ is given by

$$|\mathbf{v}| = \sqrt{a^2 + b^2}$$

The operations of addition, subtraction, and scalar multiplication can also be stated algebraically. Let $\mathbf{v} = a\mathbf{i} + b\mathbf{j}$ and $\mathbf{w} = c\mathbf{i} + d\mathbf{j}$. Then

$$\mathbf{v} + \mathbf{w} = (a + c)\mathbf{i} + (b + d)\mathbf{j}$$
$$\mathbf{v} - \mathbf{w} = (a - c)\mathbf{i} + (b - d)\mathbf{j}$$
$$c\mathbf{v} = ca\mathbf{i} + cb\mathbf{j}$$

EXAMPLE 6   Let $\mathbf{v} = 6\mathbf{i} + 4\mathbf{j}$ and $\mathbf{w} = -2\mathbf{i} + 3\mathbf{j}$.

SOLUTION
a. $|\mathbf{v}| = \sqrt{6^2 + 4^2}$
$= \sqrt{36 + 16}$
$= 2\sqrt{13}$

b. $|\mathbf{w}| = \sqrt{(-2)^2 + 3^2}$
$= \sqrt{4 + 9}$
$= \sqrt{13}$

c. $\mathbf{v} + \mathbf{w} = (6 - 2)\mathbf{i} + (4 + 3)\mathbf{j}$
$= 4\mathbf{i} + 7\mathbf{j}$

d. $\mathbf{v} - \mathbf{w} = (6 + 2)\mathbf{i} + (4 - 3)\mathbf{j}$
$= 8\mathbf{i} + \mathbf{j}$

e. $-\mathbf{v} = (-1)6\mathbf{i} + (-1)4\mathbf{j}$
$= -6\mathbf{i} - 4\mathbf{j}$

f. $-2\mathbf{w} = (-2)(-2)\mathbf{i} + (-2)(3)\mathbf{j}$
$= 4\mathbf{i} - 6\mathbf{j}$

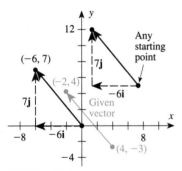

FIGURE 11.47   Some vectors represented by $\mathbf{v} = -6\mathbf{i} + 7\mathbf{j}$

Note the usage of the words *product* and *multiplication*.

You can see from Example 6 that the algebraic representation of a vector makes it easy to handle vectors and their operations. Another advantage of the algebraic representation is that it specifies a direction and a magnitude, but not a particular location. The directed line segment in Example 4 defined a given vector. However, there are infinitely many other vectors represented by the form $-6\mathbf{i} + 7\mathbf{j}$. Two of these are shown in Figure 11.47.

In arithmetic, the words *multiplication* and *product* are used to mean the same thing. When working with vectors, however, these words are used to denote different ideas. Recall that scalar multiplication does not tell how to multiply two vectors; instead it tells how to multiply a scalar and a vector. Now we define an operation called **scalar product** in order to multiply two vectors to obtain a number. It is called *scalar* product because a number or scalar is obtained as an answer. In more advanced courses another vector multiplication, called **vector product,** is defined in which a vector is obtained as an answer.

DEFINITION OF SCALAR PRODUCT

Let $\mathbf{v} = a\mathbf{i} + b\mathbf{j}$ and $\mathbf{w} = c\mathbf{i} + d\mathbf{j}$. Then the **scalar product,** written $\mathbf{v} \cdot \mathbf{w}$, is defined by

$$\mathbf{v} \cdot \mathbf{w} = ac + bd$$

Sometimes this product is called the **dot product,** from the form in which it is written.

EXAMPLE 7   Find the scalar product of the given vectors.

SOLUTION   **a.** If $\mathbf{v} = 2\mathbf{i} + 5\mathbf{j}$ and $\mathbf{w} = 6\mathbf{i} - 3\mathbf{j}$, then

$$\mathbf{v} \cdot \mathbf{w} = 2(6) + 5(-3)$$
$$= 12 - 15$$
$$= -3$$

b. If $\mathbf{v} = \cos 30°\mathbf{i} + \sin 30°\mathbf{j}$ and $\mathbf{w} = \cos 60°\mathbf{i} - \sin 60°\mathbf{j}$, then

$$\begin{aligned}
\mathbf{v} \cdot \mathbf{w} &= \cos 30° \cos 60° - \sin 30° \sin 60° \\
&= \cos(30° + 60°) \\
&= 0
\end{aligned}$$

c. If $\mathbf{v} = -\sqrt{3}\mathbf{i} + \sqrt{2}\mathbf{j}$ and $\mathbf{w} = 3\sqrt{3}\mathbf{i} + 5\sqrt{2}\mathbf{j}$, then

$$\begin{aligned}
\mathbf{v} \cdot \mathbf{w} &= (-\sqrt{3})(3\sqrt{3}) + \sqrt{2}(5\sqrt{2}) \\
&= -9 + 10 \\
&= 1
\end{aligned}$$

d. If $\mathbf{v} = 2\mathbf{i} - 3\mathbf{j}$ and $\mathbf{w} = 4\mathbf{i} + a\mathbf{j}$, then

$$\mathbf{v} \cdot \mathbf{w} = 8 - 3a$$ ▮

A very useful geometric property for scalar product is apparent if we find an expression for the angle between two vectors:

ANGLE BETWEEN VECTORS	The angle θ between vectors $\mathbf{v}$ and $\mathbf{w}$ is found by $$\cos \theta = \frac{\mathbf{v} \cdot \mathbf{w}}{|\mathbf{v}||\mathbf{w}|}$$

PROOF Let $\mathbf{v} = a\mathbf{i} + b\mathbf{j}$ and $\mathbf{w} = c\mathbf{i} + d\mathbf{j}$ be drawn with their bases at the origin, as shown in Figure 11.48. Let x be the distance between the endpoints of the vectors. Then, by the Law of Cosines,

$$\begin{aligned}
\cos \theta &= \frac{|\mathbf{v}|^2 + |\mathbf{w}|^2 - x^2}{2|\mathbf{v}||\mathbf{w}|} \\
&= \frac{(\sqrt{a^2 + b^2})^2 + (\sqrt{c^2 + d^2})^2 - (\sqrt{(a - c)^2 + (b - d)^2})^2}{2|\mathbf{v}||\mathbf{w}|} \\
&= \frac{a^2 + b^2 + c^2 + d^2 - (a^2 - 2ac + c^2 + b^2 - 2bd + d^2)}{2|\mathbf{v}||\mathbf{w}|} \\
&= \frac{2ac + 2bd}{2|\mathbf{v}||\mathbf{w}|} \\
&= \frac{ac + bd}{|\mathbf{v}||\mathbf{w}|} \\
&= \frac{\mathbf{v} \cdot \mathbf{w}}{|\mathbf{v}||\mathbf{w}|}
\end{aligned}$$ ▯▯▯

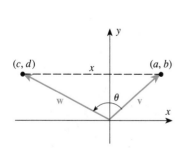

FIGURE 11.48
Finding the angle between two vectors

There is a useful geometric property of vectors whose directions differ by 90°. If you are dealing with lines that meet at a 90° angle, they are called perpendicular lines, but if vectors form a 90° angle, they are called **orthogonal vectors**. Notice that if $\theta = 90°$, then $\cos 90° = 0$; therefore, from the formula for the angle between vectors,

$$0 = \frac{\mathbf{v} \cdot \mathbf{w}}{|\mathbf{v}||\mathbf{w}|}$$

If you multiply both sides by $|\mathbf{v}||\mathbf{w}|$, the result is the following important condition.

ORTHOGONAL VECTOR THEOREM

Vectors $\mathbf{v}$ and $\mathbf{w}$ are orthogonal if and only if $\mathbf{v} \cdot \mathbf{w} = 0$.

EXAMPLE 8 Show that $\mathbf{v} = 3\mathbf{i} - 2\mathbf{j}$ and $\mathbf{w} = 6\mathbf{i} + 9\mathbf{j}$ are orthogonal.

SOLUTION
$$\mathbf{v} \cdot \mathbf{w} = 3(6) + (-2)(9)$$
$$= 18 - 18$$
$$= 0$$

Since the scalar product is zero, the vectors are orthogonal. ∎

EXAMPLE 9 Find a so that $\mathbf{v} = 3\mathbf{i} + a\mathbf{j}$ and $\mathbf{w} = \mathbf{i} - 2\mathbf{j}$ are orthogonal.

SOLUTION
$$\mathbf{v} \cdot \mathbf{w} = 3 - 2a$$

If they are orthogonal, then

$$3 - 2a = 0$$
$$a = \frac{3}{2}$$ ∎

11.5 Problem Set

A *Find the algebraic representation for each vector given in Problems 1–18.* $|\mathbf{v}|$ *is the magnitude of the vector* $\mathbf{v}$, *and* θ *is the angle the vector makes with the positive x-axis. A and B are the endpoints of the vector* $\mathbf{v}$, *and A is the base point. Draw each vector.*

1. $|\mathbf{v}| = 12$, $\theta = 60°$
2. $|\mathbf{v}| = 8$, $\theta = 30°$
3. $|\mathbf{v}| = \sqrt{2}$, $\theta = 45°$
4. $|\mathbf{v}| = 9$, $\theta = 45°$
5. $|\mathbf{v}| = 7$, $\theta = 23°$
6. $|\mathbf{v}| = 5$, $\theta = 72°$
7. $|\mathbf{v}| = 4$, $\theta = 112°$
8. $|\mathbf{v}| = 10$, $\theta = 214°$
9. $A(4, 1)$, $B(2, 3)$
10. $A(-1, -3)$, $B(4, 5)$
11. $A(1, -2)$, $B(-5, -7)$
12. $A(6, -8)$, $B(5, -2)$
13. $A(-3, 2)$, $B(5, -8)$
14. $A(0, 0)$, $B(-3, -4)$
15. $A(7, 1)$, $B(0, 0)$
16. $A(2, 9)$, $B(-5, 8)$
17. $A(6, -1)$; $B(-3, -7)$
18. $A(-2, -8)$, $B(-4, 7)$

Find the magnitude of each of the vectors given in Problems 19–27.

19. $\mathbf{v} = 3\mathbf{i} + 4\mathbf{j}$
20. $\mathbf{v} = 5\mathbf{i} - 12\mathbf{j}$
21. $\mathbf{v} = 6\mathbf{i} - 7\mathbf{j}$
22. $\mathbf{v} = -3\mathbf{j} + 5\mathbf{j}$
23. $\mathbf{v} = -2\mathbf{i} + 2\mathbf{j}$
24. $\mathbf{v} = 5\mathbf{i} - 8\mathbf{j}$
25. $\mathbf{v} = \mathbf{i} - 3\mathbf{j}$
26. $\mathbf{v} = 2\mathbf{i} - \mathbf{j}$
27. $\mathbf{v} = 4\mathbf{i} + 5\mathbf{j}$

State whether the given pairs of vectors in Problems 28–33 are orthogonal.

28. $\mathbf{v} = 3\mathbf{i} - 2\mathbf{j}$; $\mathbf{w} = 6\mathbf{i} + 9\mathbf{j}$
29. $\mathbf{v} = 2\mathbf{i} + 3\mathbf{j}$; $\mathbf{w} = 6\mathbf{i} - 9\mathbf{j}$
30. $\mathbf{v} = 4\mathbf{i} - 5\mathbf{j}$; $\mathbf{w} = 8\mathbf{i} + 10\mathbf{j}$
31. $\mathbf{v} = 5\mathbf{i} + 4\mathbf{j}$; $\mathbf{w} = 8\mathbf{i} - 10\mathbf{j}$
32. $\mathbf{v} = 2\mathbf{i} - 3\mathbf{j}$; $\mathbf{w} = 3\mathbf{i} + 2\mathbf{j}$
33. $\mathbf{v} = \mathbf{i}$; $\mathbf{w} = \mathbf{j}$

B

34. NAVIGATION A woman sets out in a rowboat heading due west and rows at 4.8 mph. The current is carrying the boat due south at 12 mph. What is the true course of the rowboat, and how fast is the boat traveling relative to the ground?

35. AVIATION An airplane is headed due west at 240 mph. The wind is blowing due south at 43 mph. What is the true course of the plane, and how fast is it traveling across the ground?

36. AVIATION An airplane is heading S35°W with a velocity of 723 mph. How far south has it traveled in 1 hr?

37. AVIATION An airplane is heading N43.0°E with a velocity of 248 mph. How far east has it traveled in 2 hr?

In Problems 38–49, find $\mathbf{v} \cdot \mathbf{w}$*,* $|\mathbf{v}|$*,* $|\mathbf{w}|$*, and* $\cos \theta$*, where* θ *is the angle between* $\mathbf{v}$ *and* $\mathbf{w}$*.*

38. $\mathbf{v} = 3\mathbf{i} + 4\mathbf{j}$
$\mathbf{w} = 5\mathbf{i} + 12\mathbf{j}$

39. $\mathbf{v} = 8\mathbf{i} - 6\mathbf{j}$
$\mathbf{w} = -5\mathbf{i} + 12\mathbf{j}$

40. $\mathbf{v} = 2\mathbf{i} + \sqrt{5}\mathbf{j}$
$\mathbf{w} = 3\sqrt{5}\mathbf{i} - 3\mathbf{j}$

41. $\mathbf{v} = 7\mathbf{i} - \sqrt{15}\mathbf{j}$
$\mathbf{w} = 2\sqrt{15}\mathbf{i} + 14\mathbf{j}$

42. $\mathbf{v} = -2\mathbf{i} + 3\mathbf{j}$
$\mathbf{w} = 6\mathbf{i} + 5\mathbf{j}$

43. $\mathbf{v} = 3\mathbf{i} + 9\mathbf{j}$
$\mathbf{w} = 2\mathbf{i} - 5\mathbf{j}$

44. $\mathbf{v} = \mathbf{i}$
$\mathbf{w} = \mathbf{i}$

45. $\mathbf{v} = \mathbf{j}$
$\mathbf{w} = \mathbf{j}$

46. $\mathbf{v} = \mathbf{i}$
$\mathbf{w} = -\mathbf{j}$

47. $\mathbf{v} = 5\mathbf{i} - \mathbf{j}$
$\mathbf{w} = 2\mathbf{i} + 3\mathbf{j}$

48. $\mathbf{v} = 4\mathbf{i} + 2\mathbf{j}$
$\mathbf{w} = 3\mathbf{i} - \mathbf{j}$

49. $\mathbf{v} = \mathbf{i} + \mathbf{j}$
$\mathbf{w} = \mathbf{i}$

In Problems 50–55, find the angle θ *to the nearest degree,* $0° \leq \theta \leq 180°$*, between the vectors* $\mathbf{v}$ *and* $\mathbf{w}$*.*

50. $\mathbf{v} = \dfrac{1}{2}\mathbf{i} + \dfrac{\sqrt{3}}{2}\mathbf{j}$

$\mathbf{w} = \dfrac{1}{2}\mathbf{i} + \dfrac{1}{2}\mathbf{j}$

51. $\mathbf{v} = \sqrt{2}\mathbf{i} - \sqrt{2}\mathbf{j}$

$\mathbf{w} = \dfrac{\sqrt{3}}{2}\mathbf{i} + \dfrac{1}{2}\mathbf{j}$

52. $\mathbf{v} = \mathbf{j}$

$\mathbf{w} = \dfrac{1}{2}\mathbf{i} - \dfrac{\sqrt{3}}{2}\mathbf{j}$

53. $\mathbf{v} = -\mathbf{i}$

$\mathbf{w} = -2\sqrt{2}\mathbf{i} + 2\sqrt{2}\mathbf{j}$

54. $\mathbf{v} = 2\mathbf{i} + 3\mathbf{j}$
$\mathbf{w} = -\mathbf{i} + 4\mathbf{j}$

55. $\mathbf{v} = -3\mathbf{i} + 2\mathbf{j}$
$\mathbf{w} = 6\mathbf{i} + 9\mathbf{j}$

In Problems 56–58, find a number a so that the given vectors are orthogonal.

56. $\mathbf{v} = 2\mathbf{i} + 3\mathbf{j}$
$\mathbf{w} = 5\mathbf{i} + a\mathbf{j}$

57. $\mathbf{y} = 4\mathbf{i} - a\mathbf{j}$
$\mathbf{w} = -2\mathbf{i} + 5\mathbf{j}$

58. $\mathbf{v} = a\mathbf{i} + 5\mathbf{j}$
$\mathbf{w} = a\mathbf{i} - 15\mathbf{j}$

C

59. AVIATION A pilot is flying at an airspeed of 241 mph in a wind blowing 20.4 mph from the east. In what direction must the pilot head in order to fly due north? What is the pilot's speed relative to the ground?

60. AVIATION Answer the questions posed in Problem 59 with the pilot wishing to fly due south.

61. SPACE SCIENCE The weight of astronauts on the moon is about one-sixth of their weight on earth. This fact has a marked effect on such simple acts as walking, running, and jumping. To study these effects and to train astronauts for working under lunar-gravity conditions, scientists at NASA's Langley Research Center have designed an inclined-plane apparatus to simulate reduced gravity. The apparatus consists of a sling that holds the astronaut in a position perpendicular to the inclined plane (see Figure 11.49). The sling is attached to one end of a long cable that runs parallel to the inclined plane. The other end of the cable is attached to a trolley that runs along an overhead track. This device allows the astronaut to move freely in a plane perpendicular to the inclined plane. Let *W* be the astronaut's mass and θ be the angle between the inclined plane and the ground. Make a vector diagram showing the tension in the cable and the force exerted by the inclined plane against the feet of the astronaut.

FIGURE 11.49

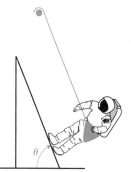

62. SPACE SCIENCE From the point of view of the astronaut in Problem 61, the inclined plane is the ground and the astronaut's simulated mass (that is, the downward force against the inclined plane) is $W \cos \theta$. What value of θ is required in order to simulate lunar gravity?

11.6 Properties of Vectors

In this section we will look at several properties of vectors that will be useful in more advanced mathematics courses. Suppose you are given a line $ax + by + c = 0$. A vector perpendicular to a line is called a **normal** to the line. We will show that

$$\mathbf{N} = a\mathbf{i} + b\mathbf{j}$$

is normal to

$$L: ax + by + c = 0$$

Notice that, once L is given, $\mathbf{N}$ can be found by inspection. For example,

$$\text{If} \quad L: 3x + 4y + 5 = 0 \quad \text{then} \quad \mathbf{N} = 3\mathbf{i} + 4\mathbf{j}.$$
$$\text{If} \quad L: 5x - 3y + 7 = 0 \quad \text{then} \quad \mathbf{N} = 5\mathbf{i} - 3\mathbf{j}.$$

In general, if $P_1(x_1, y_1)$ and $P_2(x_2, y_2)$ are any two points on L, then $\overrightarrow{P_1P_2}$ is a representative of the vector determined by the line. That is,

$$\overrightarrow{P_1P_2} = (x_2 - x_1)\mathbf{i} + (y_2 - y_1)\mathbf{j}$$

We must show that $\overrightarrow{P_1P_2}$ and $\mathbf{N} = a\mathbf{i} + b\mathbf{j}$ are orthogonal. That is, we must show that

$$\overrightarrow{P_1P_2} \cdot \mathbf{N} = 0$$

Now,

$$\overrightarrow{P_1P_2} \cdot \mathbf{N} = a(x_2 - x_1) + b(y_2 - y_1)$$

Since $P_1(x_1, y_1)$ and $P_2(x_2, y_2)$ are on the line, they make the equation true:

$$ax_1 + by_1 + c = 0$$
$$ax_2 + by_2 + c = 0$$

Subtracting,

$$a(x_2 - x_1) + b(y_2 - y_1) = 0$$

Thus,

$$\overrightarrow{P_1P_2} \cdot \mathbf{N} = a(x_2 - x_1) + b(y_2 - y_1) = 0$$

EXAMPLE 1 Find a vector determined by the line $5x - 3y - 15 = 0$. Also find a normal vector. Show that they are orthogonal.

SOLUTION A normal vector is $5\mathbf{i} - 3\mathbf{j}$.

A vector determined by the line is found by first obtaining two points on the line. If $x = 0$, then $y = -5$; if $y = 0$, then $x = 3$. Thus, two points on the line are $(0, -5)$ and $(3, 0)$. Hence $3\mathbf{i} + 5\mathbf{j}$ is a vector determined by the line.

Finally, $(5\mathbf{i} - 3\mathbf{j}) \cdot (3\mathbf{i} + 5\mathbf{j}) = 15 - 15 = 0$, so they are orthogonal (see Figure 11.50).

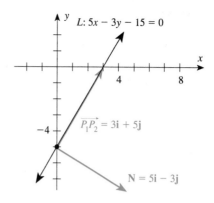

FIGURE 11.50

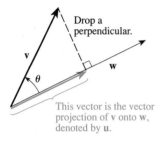

Drop a perpendicular.

This vector is the vector projection of **v** onto **w**, denoted by **u**.

FIGURE 11.51
Projection of **v** onto **w**

Let **v** and **w** be two vectors whose representatives have a common base point. If we drop a perpendicular from the head of **v** to the line determined by **w**, we determine a vector called *the vector projection of* **v** *onto* **w**, denoted by **u**, as shown in Figure 11.51.

The *scalar projection* is the *length* of the vector projection. Let θ be an acute angle between **v** and **w**. Then

$$\cos \theta = \frac{|\mathbf{u}|}{|\mathbf{v}|} \qquad \text{By definition of cosine}$$

Thus,

$$|\mathbf{u}| = |\mathbf{v}| \cos \theta = |\mathbf{v}| \frac{\mathbf{v} \cdot \mathbf{w}}{|\mathbf{v}||\mathbf{w}|} = \frac{\mathbf{v} \cdot \mathbf{w}}{|\mathbf{w}|}$$

If $90° < \theta < 180°$, then $\cos \theta \leq 0$ so $|\mathbf{v}| \cos \theta$ is a negative number. Since we want $|\mathbf{u}|$ to be nonnegative (since it is a length), we introduce an absolute value:

$$|\mathbf{u}| = \left| \frac{\mathbf{v} \cdot \mathbf{w}}{|\mathbf{w}|} \right|$$

In order to find the vector projection, we notice

$$\mathbf{u} = s\mathbf{w}$$

for some scalar s. Since we do not know **u** and do not know s, this equation is not much help. Therefore we use

$$|\mathbf{u}| = s|\mathbf{w}|$$

since we know $|\mathbf{u}|$ (it is the scalar projection we found above). Thus,

$$\frac{\mathbf{v} \cdot \mathbf{w}}{|\mathbf{w}|} = s|\mathbf{w}|$$

$$s = \frac{\mathbf{v} \cdot \mathbf{w}}{|\mathbf{w}|^2} = \frac{\mathbf{v} \cdot \mathbf{w}}{\mathbf{w} \cdot \mathbf{w}}$$

We can now find **u**:

$$\mathbf{u} = s\mathbf{w} = \left(\frac{\mathbf{v} \cdot \mathbf{w}}{\mathbf{w} \cdot \mathbf{w}} \right) \mathbf{w}$$

In summary:

PROJECTIONS OF v ONTO w

Scalar projection (a number): $\left| \dfrac{\mathbf{v} \cdot \mathbf{w}}{|\mathbf{w}|} \right|$

Vector projection (a vector): $\left(\dfrac{\mathbf{v} \cdot \mathbf{w}}{\mathbf{w} \cdot \mathbf{w}} \right) \mathbf{w}$

EXAMPLE 2 Find the scalar and vector projections of $\mathbf{v} = 5\mathbf{i} - 3\mathbf{j}$ onto $\mathbf{w} = 7\mathbf{i} + 4\mathbf{j}$.

SOLUTION *Vector projection*:

$$\left(\frac{\mathbf{v} \cdot \mathbf{w}}{\mathbf{w} \cdot \mathbf{w}} \right) \mathbf{w} = \frac{35 - 12}{49 + 16} \mathbf{w} = \frac{23}{65} (7\mathbf{i} + 4\mathbf{j}) = \frac{161}{65} \mathbf{i} + \frac{92}{65} \mathbf{j}$$

Scalar projection: We could find the length of the vector projection by calculating

$$\sqrt{\left(\frac{161}{65} \right)^2 + \left(\frac{92}{65} \right)^2}$$

or we can use

$$\left| \frac{\mathbf{v} \cdot \mathbf{w}}{|\mathbf{w}|} \right|$$

Since the latter is easier to calculate, we find

$$\left| \frac{23}{\sqrt{49 + 16}} \right| = \frac{23}{\sqrt{65}}$$

▌

EXAMPLE 3 Find the scalar projection of $\mathbf{v} = 3\mathbf{i} - 2\mathbf{j}$ onto $\mathbf{w} = 2\mathbf{i} + 4\mathbf{j}$.

SOLUTION The scalar projection of $\mathbf{v}$ onto $\mathbf{w}$ is

$$\left| \frac{\mathbf{v} \cdot \mathbf{w}}{|\mathbf{w}|} \right| = \left| \frac{6 - 8}{\sqrt{4 + 16}} \right| = \left| \frac{-2}{2\sqrt{5}} \right| = \frac{1}{\sqrt{5}}$$

We can check this result by finding the length of the vector projection of $\mathbf{v}$ onto $\mathbf{w}$:

$$\mathbf{u} = \left(\frac{\mathbf{v} \cdot \mathbf{w}}{\mathbf{w} \cdot \mathbf{w}} \right) \mathbf{w} \qquad |\mathbf{u}| = \sqrt{\left(-\frac{1}{5} \right)^2 + \left(-\frac{2}{5} \right)^2}$$

$$= \left(\frac{-2}{4 + 16} \right) \mathbf{w} \qquad = \sqrt{\frac{1 + 4}{25}}$$

$$= \frac{-1}{10} (2\mathbf{i} + 4\mathbf{j}) \qquad = \frac{1}{5} \sqrt{5}$$

$$= -\frac{1}{5} \mathbf{i} - \frac{2}{5} \mathbf{j}$$

▌

The final property of vectors we will consider in this section enables us to derive a formula for finding the distance from a point to a line. In Section 11.1 (Problem 65) this formula was derived with a great deal of effort without using vectors. We can now derive the same formula quite easily by using vector ideas.

Let L be any given line and P any given point not on L. By the distance from P to L we mean the perpendicular distance d as shown in Figure 11.52a. We wish to find this distance.

FIGURE 11.52

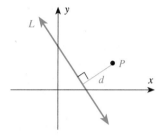

a. Distance from P to L **b.** Procedure for finding the distance from a point to a line

If L is a vertical line, then the distance from P to L is easy to find. (Why?) If L is not vertical, then we let B be the y-intercept and $\mathbf{N}$ a normal to L. Actually, B can be any point on L, but it is convenient to use the y-intercept. Let the base of $\mathbf{N}$ be drawn at B, as shown in Figure 11.52b.

The distance we seek is seen to be the scalar projection of $\overrightarrow{BP}$ onto $\mathbf{N}$. Thus

VECTOR FORMULA FOR THE DISTANCE FROM A POINT TO A LINE

$$d = \left| \frac{\overrightarrow{BP} \cdot \mathbf{N}}{|\mathbf{N}|} \right|$$

EXAMPLE 4   Find the distance from the point $(5, -3)$ to the line $4x + 3y - 15 = 0$.

SOLUTION P is $(5, -3)$ and B is $(0, 5)$. Then

$$\overrightarrow{BP} = 5\mathbf{i} - 8\mathbf{j}$$
$$\mathbf{N} = 4\mathbf{i} + 3\mathbf{j}$$
$$|\mathbf{N}| = \sqrt{4^2 + 3^2} = 5$$
$$\overrightarrow{BP} \cdot \mathbf{N} = 20 - 24 = -4$$

Thus,

$$d = \left| \frac{\overrightarrow{BP} \cdot \mathbf{N}}{|\mathbf{N}|} \right|$$
$$= \left| \frac{-4}{5} \right|$$
$$= \frac{4}{5}$$

11.6 Problem Set

A *Find a vector normal to each line given in Problems 1–12.*

1. $2x - 3y + 4 = 0$

2. $x + y - 1 = 0$

3. $x - y + 3 = 0$

4. $4x + 5y - 3 = 0$

5. $3x - 2y + 1 = 0$

6. $5x + y - 3 = 0$

7. $9x + 7y - 5 = 0$

8. $6x - 3y + 2 = 0$

9. $4x - y - 12 = 0$

10. $y = \frac{2}{3}x - 5$

11. $y = -\frac{1}{2}x - 10$

12. $y = -\frac{5}{8}x + 4$

Find a vector determined by each line given in Problems 13–24.

13. $2x - 3y + 4 = 0$

14. $x + y - 1 = 0$

15. $x - y + 3 = 0$

16. $4x + 5y - 3 = 0$

17. $3x - 2y + 1 = 0$

18. $5x + y - 3 = 0$

19. $9x + 7y - 5 = 0$

20. $6x - 3y + 2 = 0$

21. $4x - y - 12 = 0$

22. $y = \frac{2}{3}x - 5$

23. $y = -\frac{1}{2}x - 10$

24. $y = -\frac{5}{8}x + 4$

In Problems 25–30 find the scalar projection of **v** *onto* **w.**

25. $\mathbf{v} = 3\mathbf{i} + 4\mathbf{j}$
$\mathbf{w} = 5\mathbf{i} + 12\mathbf{j}$

26. $\mathbf{v} = 8\mathbf{i} - 6\mathbf{j}$
$\mathbf{w} = -5\mathbf{i} + 12\mathbf{j}$

27. $\mathbf{v} = 7\mathbf{i} - \sqrt{15}\mathbf{j}$
$\mathbf{w} = 2\sqrt{15}\mathbf{i} + 14\mathbf{j}$

28. $\mathbf{v} = 2\mathbf{i} - \sqrt{5}\mathbf{j}$
$\mathbf{w} = 3\sqrt{5}\mathbf{i} - 3\mathbf{j}$

29. $\mathbf{v} = -2\mathbf{i} + 3\mathbf{j}$
$\mathbf{w} = 6\mathbf{i} + 5\mathbf{j}$

30. $\mathbf{v} = 3\mathbf{i} + 9\mathbf{j}$
$\mathbf{w} = 2\mathbf{i} - 5\mathbf{j}$

In Problems 31–36 find the vector projection of **v** *onto* **w.**

31. $\mathbf{v} = 3\mathbf{i} + 4\mathbf{j}$
$\mathbf{w} = 5\mathbf{i} + 12\mathbf{j}$

32. $\mathbf{v} = 8\mathbf{i} - 6\mathbf{j}$
$\mathbf{w} = -5\mathbf{i} + 12\mathbf{j}$

33. $\mathbf{v} = 7\mathbf{i} - \sqrt{15}\mathbf{j}$
$\mathbf{w} = 2\sqrt{15}\mathbf{i} + 14\mathbf{j}$

34. $\mathbf{v} = 2\mathbf{i} + \sqrt{5}\mathbf{j}$
$\mathbf{w} = 3\sqrt{5}\mathbf{i} - 3\mathbf{j}$

35. $\mathbf{v} = -2\mathbf{i} + 3\mathbf{j}$
$\mathbf{w} = 6\mathbf{i} + 5\mathbf{j}$

36. $\mathbf{v} = 3\mathbf{i} + 9\mathbf{j}$
$\mathbf{w} = 2\mathbf{i} - 5\mathbf{j}$

B *Find the distance from the given point to the given line in Problems 37–50.*

37. $3x - 4y + 8 = 0$; $(4, 5)$

38. $5x - 12y + 15 = 0$; $(6, -3)$

39. $3x - 4y + 8 = 0$; $(9, -3)$

40. $5x - 12y + 15 = 0$; $(-2, 6)$

41. $4x + 3y - 5 = 0$; $(-1, -1)$

42. $12x + 5y - 2 = 0$; $(3, 5)$

43. $4x + 3y - 5 = 0$; $(6, 1)$

44. $12x + 5y - 2 = 0$; $(4, -3)$

45. $x - 3y + 15 = 0$; $(1, -6)$

46. $6x - y - 10 = 0$; $(-5, 6)$

47. $x - 3y + 15 = 0$; $(8, 14)$

48. $6x - y - 10 = 0$; $(8, 10)$

49. $2x - 5y = 0$; $(4, 5)$

50. $4x + 7y = 0$; $(5, 10)$

Find the area of the triangle determined by the given points in Problems 51–56.

51. $(1, 2)$, $(4, 5)$, $(-5, 3)$

52. $(-1, 1)$, $(4, 3)$, $(1, -1)$

53. $(0, 0)$, $(5, -3)$, $(-2, -7)$

54. $(5, 6)$, $(-3, 5)$, $(0, 0)$

55. $(3, 0)$, $(0, 8)$, $(-4, 6)$

56. $(0, 6)$, $(-5, 2)$, $(-3, -6)$

C

57. Prove the commutative law $\mathbf{u} \cdot \mathbf{v} = \mathbf{v} \cdot \mathbf{u}$ for any vectors **u** and **v.**

58. Prove the distributive law $\mathbf{u} \cdot (\mathbf{v} + \mathbf{w}) = \mathbf{u} \cdot \mathbf{v} + \mathbf{u} \cdot \mathbf{w}$ for any vectors **u, v,** and **w.**

59. Let **u** be the projection of **v** onto **w.** Show that $\mathbf{v} - \mathbf{u}$ is orthogonal to **w.**

60. Let $\mathbf{v} = a\mathbf{i} + b\mathbf{j}$ and $\mathbf{w} = c\mathbf{i} + d\mathbf{j}$ be two vectors. Use the Law of Cosines to show that

$$\cos \theta = \frac{ac + bd}{|\mathbf{v}||\mathbf{w}|}$$

where θ is the angle between the vectors. *Hint:* The Law of Cosines states that $c^2 = a^2 + b^2 - 2ab \cos \gamma$, where a, b, and c are sides of a triangle and γ is the angle opposite side c.

---

*This section does not require Sections 11.5–11.6.

11.7 Polar Coordinates*

Plotting Points

Up to this point in the book, we have used a rectangular coordinate system. Now, consider a different system, called the **polar coordinate system.** In this system, fix a point O, called the *origin*, or **pole**, and represent a point in the plane by an ordered pair $P(r, \theta)$, where θ measures the angle from the positive x-axis and r represents the directed distance from the pole to the point P. Both r and θ can be any real numbers. When plotting points, consider a **polar axis,** fixed at the pole and coinciding with the x-axis. Now rotate the polar axis through an angle θ as shown in Figure 11.53. You might find it helpful to rotate your pencil as the axis—the tip points in the positive direction. If θ is positive, the angle is measured in a counterclockwise direction, and if θ is negative, it is measured in a clockwise direction.

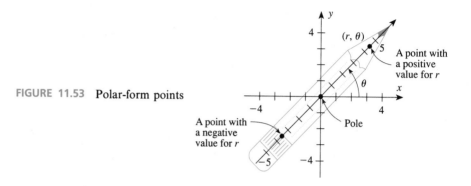

FIGURE 11.53   Polar-form points

Next, plot r on the polar axis (pencil). Notice that any real number can be plotted on this real number line. Plotting points seems easy, but it is necessary that you completely understand this process. Study each part of Example 1 and make sure you understand how each point is plotted.

EXAMPLE 1   Plot each of the following polar-form points:

$$A(4, \tfrac{\pi}{3}), \ B(-4, \tfrac{\pi}{3}), \ C(3, \ -\tfrac{\pi}{6}), \ D(-3, \ -\tfrac{\pi}{6}), \ E(-3, 3),$$

$$F(-3, \ -3), \ G(-4, \ -2), \ H(5, \tfrac{3\pi}{2}), \ I(-5, \tfrac{\pi}{2}), \ J(5, \ -\tfrac{\pi}{2}).$$

SOLUTION   Points A–D illustrate the basic ideas of plotting polar-form points.

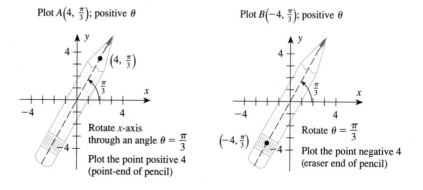

Plot $C\left(3, -\frac{\pi}{6}\right)$; negative θ

Rotate $\theta = -\dfrac{\pi}{6}$

Plot the point positive 3
(point end)

Plot $D\left(-3, -\frac{\pi}{6}\right)$; negative θ

Rotate $\theta = -\dfrac{\pi}{6}$

Plot the point positive 3
(eraser end)

Points E–H illustrate common situations that can sometimes be confusing. Make sure you understand each example.

Plot $E(-3, 3)$.

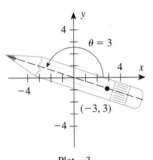

Plot -3
(eraser end)

Plot $F(-3, -3)$.

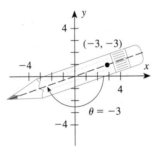

Plot -3
(eraser end)

Plot $G(-4, -2)$.

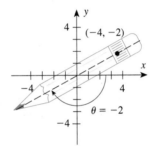

Plot -4
(eraser end)

Plot $H\left(5, \frac{3\pi}{2}\right)$.

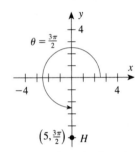

Coordinates $I\left(-5, \frac{\pi}{2}\right)$, $J\left(5, -\frac{\pi}{2}\right)$, and $H\left(5, \frac{3\pi}{2}\right)$ all represent the same point.

Primary Representations of a Point

⟲ Remember to simplify so the second component represents a nonnegative angle less than one revolution. ⟳

One thing you will notice from Example 1 is that ordered pairs in polar form are not associated in a one-to-one fashion with points in the plane. Indeed, given any point in the plane, there are infinitely many ordered pairs of polar coordinates associated with that point in polar form. If you are given a point (r, θ) other than the pole in polar coordinates, then $(-r, \theta + \pi)$ also represents the same point. In addition, there are also infinitely many others, all of which have the same first component as one of these, and have second components that are multiples of 2π added to these angles. We call (r, θ) and $(-r, \theta + \pi)$ the **primary representations of the point** if the angles θ and $\theta + \pi$ are between zero and 2π.

PRIMARY REPRESENTATIONS OF A POINT IN POLAR FORM

Every point in polar form has two primary representations:

(r, θ), where $0 \le \theta < 2\pi$ and $(-r, \pi + \theta)$, where $0 \le \pi + \theta < 2\pi$

508   CHAPTER 11   ANALYTIC GEOMETRY

EXAMPLE 2 Give both primary representations for each of the given points:

a. $(3, \frac{\pi}{4})$ has primary representations $\mathbf{(3, \frac{\pi}{4})}$ and $\mathbf{(-3, \frac{5\pi}{4})}$.

$$\frac{\pi}{4} + \pi = \frac{5\pi}{4}$$

b. $(5, \frac{5\pi}{4})$ has primary representations $\mathbf{(5, \frac{5\pi}{4})}$ and $\mathbf{(-5, \frac{\pi}{4})}$.

$\frac{5\pi}{4} + \pi = \frac{9\pi}{4}$, but $(-5, \frac{9\pi}{4})$ is not a primary representation of the point $(5, \frac{5\pi}{4})$ since $\frac{9\pi}{4} > 2\pi$.
Use $\frac{\pi}{4}$ since it is coterminal with $\frac{9\pi}{4}$ and satisfies $0 \le \frac{\pi}{4} < 2\pi$.

c. $(-6, -\frac{2\pi}{3})$ has primary representations $\mathbf{(-6, \frac{4\pi}{3})}$ and $\mathbf{(6, \frac{\pi}{3})}$.

d. $(9, 5)$ has primary representations $\mathbf{(9, 5)}$ and $\mathbf{(-9, 5 - \pi)}$; a point like $(-9, 5 - \pi)$ is usually approximated by writing $(-9, 1.86)$.

Notice that $(-9, 5 + \pi)$ is not a primary representation, since $5 + \pi > 2\pi$.

e. $(9, 7)$ has primary representations $(9, 7 - 2\pi)$ or $\mathbf{(9, .72)}$ and $(-9, 7 - \pi)$ or $\mathbf{(-9, 3.86)}$.

The relationship between rectangular and polar coordinates can easily be found by using the definition of the trigonometric functions (see Figure 11.54).

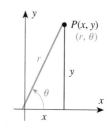

FIGURE 11.54 Relationship between rectangular and polar coordinates

RELATIONSHIP BETWEEN RECTANGULAR AND POLAR COORDINATES

1. To change **from polar to rectangular coordinates**:

$$x = r \cos \theta$$
$$y = r \sin \theta$$

2. To change **from rectangular to polar coordinates**:

$$r = \sqrt{x^2 + y^2} \qquad \theta' = \tan^{-1}\left|\frac{y}{x}\right|, \quad x \ne 0$$

where θ' is the reference angle for θ. Place θ in the proper quadrant by noting the signs of x and y. If $x = 0$, then $\theta' = \frac{\pi}{2}$.

EXAMPLE 3 Change the polar coordinates $(-3, \frac{5\pi}{4})$ to rectangular coodinates.

SOLUTION
$$x = -3 \cos \frac{5\pi}{4} = -3\left(-\frac{\sqrt{2}}{2}\right) = \frac{3\sqrt{2}}{2}$$

$$y = -3 \sin \frac{5\pi}{4} = -3\left(-\frac{\sqrt{2}}{2}\right) = \frac{3\sqrt{2}}{2}$$

$$\underbrace{\left(-3, \frac{5\pi}{4}\right)}_{\text{polar form}} = \underbrace{\left(\frac{3\sqrt{2}}{2}, \frac{3\sqrt{2}}{2}\right)}_{\text{rectangular form}}$$

EXAMPLE 4 Write both primary representations of the polar-form coordinates for the point whose rectangular coordinates are

$$\left(\frac{5\sqrt{3}}{2}, -\frac{5}{2}\right)$$

SOLUTION
$$r = \sqrt{\left(\frac{5\sqrt{3}}{2}\right)^2 + \left(-\frac{5}{2}\right)^2} \qquad \theta' = \tan^{-1}\left|\frac{-\dfrac{5}{2}}{\dfrac{5\sqrt{3}}{2}}\right|$$

$$= \sqrt{\frac{75}{4} + \frac{25}{4}}$$

$$= 5 \qquad\qquad\qquad\qquad = \tan^{-1}\left(\frac{1}{\sqrt{3}}\right)$$

$$= \frac{\pi}{6} \qquad\qquad \theta = \frac{11\pi}{6} \text{ (Quadrant IV)}$$

$$\underbrace{\left(\frac{5\sqrt{3}}{2}, -\frac{5}{2}\right)}_{\substack{\text{rectangular} \\ \text{form}}} = \underbrace{\left(5, \frac{11\pi}{6}\right) = \left(-5, \frac{5\pi}{6}\right)}_{\text{polar form}} \qquad \frac{11\pi}{6} + \pi = \frac{17\pi}{6}, \text{ and } \frac{17\pi}{6} \text{ is} \\ \text{coterminal with } \frac{5\pi}{6}.$$

Polar-Form Graphing

We now turn our attention to polar-form graphing. The basic procedure is to plot points to determine a curve's general shape and then to make some generalizations that will simplify the graphing of similar curves. Because the representation of points in polar form by ordered pairs of real numbers is not unique, we need the following definition of what it means for a polar-form point to **satisfy** an equation.

A POINT SATISFYING AN EQUATION

> An ordered pair representing a polar-form point (other than the pole) **satisfies an equation** involving a trigonometric function of $n\theta$ (n an integer) if and only if at least one of its primary representations satisfies the given equation.

The easiest graphs to consider are those for which either r or θ is a constant. For example, $r = 5$ is the equation of the set of all ordered pairs (r, θ) for which the first component (r) is 5. This means that points $(5, 0), (5, 1), (5, 2), (5, 3), \ldots$ all satisfy this equation. The graph shown in Figure 11.55a is seen to be a circle with center at the pole and radius 5. Similarly, $\theta = 1$ is the set of all ordered pairs for which the second component (θ) is 1. This means that the points $(0, 1), (1, 1), (2, 1), (3, 1), (-1, 1), \ldots$ all satisfy the equation $\theta = 1$. The graph shown in Figure 11.55b is seen to be a line.

The following examples show some nontrivial polar graphs. Remember, the basic ideas of polar-form graphing are identical to those of rectangular-form graphing. You are looking for ordered pairs satisfying the given equation. Each of these ordered pairs is then plotted on a Cartesian coordinate system; the only difference with polar-form graphing is that the first component is a distance from the pole and the second component represents an angle of rotation.

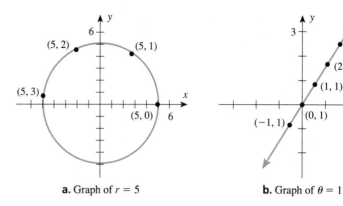

FIGURE 11.55 Polar-form graphs for which one of the components is a constant

a. Graph of $r = 5$

b. Graph of $\theta = 1$

EXAMPLE 5 Graph $r = 2(1 - \cos \theta)$.

SOLUTION First construct a table of values by choosing values for θ and approximating the corresponding values for r:

θ	0	$\dfrac{\pi}{6}$	$\dfrac{\pi}{3}$	$\dfrac{\pi}{2}$	$\dfrac{2\pi}{3}$	$\dfrac{5\pi}{6}$	π	$\dfrac{7\pi}{6}$	$\dfrac{4\pi}{3}$	$\dfrac{3\pi}{2}$	$\dfrac{5\pi}{3}$	$\dfrac{11\pi}{6}$
r (approx. value)	0	.27	1	2	3	3.7	4	3.7	3	2	1	.27

These points are plotted and then connected as in Figure 11.56.

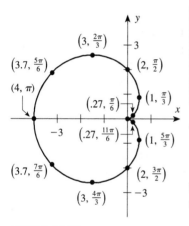

FIGURE 11.56
Graph of $r = 2(1 - \cos \theta)$

The curve in Example 5 is called a **cardioid** because it is heart-shaped. Compare the curve graphed in Example 5 with the general curve $r = a(1 - \cos \theta)$, which is called the **standard-position cardioid**. Consider the following table of values:

θ	0	$\dfrac{\pi}{2}$	π	$\dfrac{3\pi}{2}$
r	0	a	$2a$	a

These values for θ should be included whenever you are making a graph in polar coordinates and using the method of plotting points.

These reference points are the only ones to plot for future standard-position cardioids, because they will all have the same shape as the one shown in Figure 11.56.

What about cardioids that are not in standard position? In Chapter 2 translations were considered, but the translation of a polar-form curve is rather difficult since points are not labeled in a rectangular fashion. You can, however, easily rotate a polar-form curve.

ROTATION OF POLAR-FORM GRAPHS

The polar graph $r = f(\theta - \alpha)$ is the same as the polar graph of $r = f(\theta)$ that has been rotated through an angle α.

EXAMPLE 6   Graph $r = 3 - 3\cos(\theta - \frac{\pi}{6})$.

SOLUTION   Recognize this as a cardioid with $a = 3$ and a rotation of $\frac{\pi}{6}$. Plot the four points shown in Figure 11.57 and draw the cardioid.

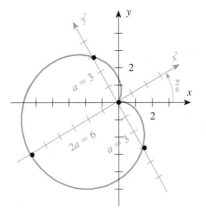

FIGURE 11.57
Graph of $r = 3 - 3\cos(\theta - \frac{\pi}{6})$

Notice that the $x'y'$-axis is drawn by rotating the xy-axis through an angle of $\frac{\pi}{6}$. The standard cardioid is then drawn on the rotated axis.

EXAMPLE 7   Graph $r = 4(1 - \sin\theta)$.

SOLUTION   Notice that

$$\sin\theta = \cos\left(\frac{\pi}{2} - \theta\right)$$

$$= \cos\left(\theta - \frac{\pi}{2}\right)$$

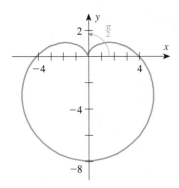

FIGURE 11.58
Graph of $r = 4(1 - \sin\theta)$

Thus, $r = 4[1 - \cos(\theta - \frac{\pi}{2})]$. This is a cardioid with $a = 4$ that has been rotated $\frac{\pi}{2}$. The graph is shown in Figure 11.58. A curve of the form $r = a(1 - \sin\theta)$ is a cardioid that has been rotated $\frac{\pi}{2}$.

The cardioid is only one of the interesting polar-form curves. It is a special case of a curve called a **limaçon**, which is developed in the problem set (see Problem 55). The next example illustrates a curve called a **rose curve**.

EXAMPLE 8   Graph $r = 4\cos 2\theta$.

SOLUTION   When presented with a polar-form curve that you do not recognize, graph the curve by plotting points. You can use tables, exact values, or a calculator. A calculator was used to find the values in the table.

θ	0	$\dfrac{\pi}{12}$	$\dfrac{\pi}{6}$	$\dfrac{\pi}{4}$	$\dfrac{\pi}{3}$	$\dfrac{5\pi}{12}$	$\dfrac{\pi}{2}$	$\dfrac{7\pi}{12}$	$\dfrac{2\pi}{3}$	$\dfrac{3\pi}{4}$	$\dfrac{5\pi}{6}$	$\dfrac{11\pi}{12}$
r (approx. value)	4	3.5	2	0	-2	-3.5	-4	-3.5	-2	0	2	3.5

θ	π	$\dfrac{13\pi}{12}$	$\dfrac{7\pi}{6}$	$\dfrac{5\pi}{4}$	$\dfrac{4\pi}{3}$	$\dfrac{17\pi}{12}$	$\dfrac{3\pi}{2}$	$\dfrac{19\pi}{12}$	$\dfrac{5\pi}{3}$	$\dfrac{7\pi}{4}$	$\dfrac{11\pi}{6}$	$\dfrac{23\pi}{12}$
r (approx. value)	4	3.5	2	0	-2	-3.5	-4	-3.5	-2	0	2	3.5

The graph is shown in Figure 11.59. Notice that as θ increases from the starting point 0 to $\frac{\pi}{4}$, r decreases from 4 to 0. As θ increases from $\frac{\pi}{4}$ to $\frac{\pi}{2}$, you can see that r is negative and decreases from 0 to -4. Study the table at the right as it relates to the graph in Figure 11.59.

FIGURE 11.59
Graph of $r = 4\cos 2\theta$

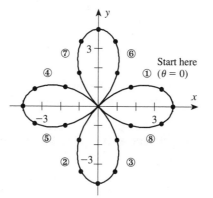

θ	r	SECTION OF CURVE
0 to $\frac{\pi}{4}$	4 to 0	1
$\frac{\pi}{4}$ to $\frac{\pi}{2}$	0 to -4	2
$\frac{\pi}{2}$ to $\frac{3\pi}{4}$	-4 to 0	3
$\frac{3\pi}{4}$ to π	0 to 4	4
π to $\frac{5\pi}{4}$	4 to 0	5
$\frac{5\pi}{4}$ to $\frac{3\pi}{2}$	0 to -4	6
$\frac{3\pi}{2}$ to $\frac{7\pi}{4}$	-4 to 0	7
$\frac{7\pi}{4}$ to 2π	0 to 4	8

The equation

$$r = a\cos n\theta$$

is a four-leaved rose if $n = 2$, and the length of the leaves is a. In general, if n is an even number, the curve has $2n$ leaves; if n is odd, the number of leaves is n. These leaves are equally spaced on a radius of a.

EXAMPLE 9 Graph $r = 4\cos 2(\theta - \frac{\pi}{4})$.

SOLUTION This is a rose curve with four leaves of length 4 equally spaced on a circle. However, this curve has been rotated $\frac{\pi}{4}$, as shown in Figure 11.60.

FIGURE 11.60 Graph of $r = 4\cos 2(\theta - \frac{\pi}{4})$

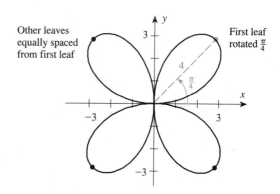

Other leaves equally spaced from first leaf

First leaf rotated $\frac{\pi}{4}$

Notice that Example 9 can be rewritten as a sine curve:

$$r = 4 \cos 2\left(\theta - \frac{\pi}{4}\right) = 4 \cos\left(2\theta - \frac{\pi}{2}\right) = 4 \cos\left(\frac{\pi}{2} - 2\theta\right) = 4 \sin 2\theta$$

These steps can be reversed to graph a rose curve written in terms of a sine function.

EXAMPLE 10 Graph $r = 5 \sin 4\theta$.

SOLUTION

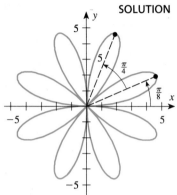

FIGURE 11.61
Graph of $r = 5 \sin 4\theta$

$$r = 5 \sin 4\theta$$
$$= 5 \cos\left(\frac{\pi}{2} - 4\theta\right)$$
$$= 5 \cos\left(4\theta - \frac{\pi}{2}\right)$$
$$= 5 \cos 4\left(\theta - \frac{\pi}{8}\right)$$

Recognize this as a rose curve rotated $\pi/8$. There are eight leaves of length 5. The leaves are a distance $\frac{\pi}{4}$ apart, as shown in Figure 11.61.

The third and last general type of polar-form curve we will consider is called a **lemniscate** and has the general form

$$r^2 = a^2 \cos 2\theta$$

EXAMPLE 11 Graph $r^2 = 16 \cos 2\theta$.

SOLUTION As before, when graphing a curve for the first time, begin by plotting points. For this example be sure to obtain two values for r when solving this quadratic equation. For example, if $\theta = 0$, then $\cos 2\theta = 1$ and $r^2 = 16$, so $r = 4$ or -4.

θ	0	$\dfrac{\pi}{12}$	$\dfrac{\pi}{6}$	$\dfrac{\pi}{4}$	$\dfrac{\pi}{4}$ to $\dfrac{3\pi}{4}$	$\dfrac{5\pi}{6}$	$\dfrac{11\pi}{12}$	π
r (approx. value)	± 4	± 3.7	± 2.8	0	undefined	± 2.8	± 3.7	± 4

Notice that for $\frac{\pi}{4} < \theta < \frac{3\pi}{4}$ there are no values for r since $\cos 2\theta$ is negative. For $\pi \le \theta \le 2\pi$, the values repeat the sequence given above, so these points are plotted and then connected, as shown in Figure 11.62.

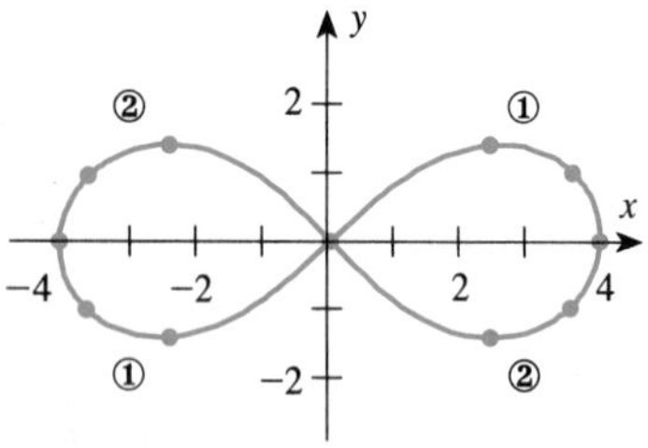

FIGURE 11.62
Graph of $r^2 = 9 \cos 2\theta$

θ	r	SECTION OF CURVE	
0 to $\frac{\pi}{4}$	4 to 0	1 (two values)	Quadrants I and III
$\frac{\pi}{4}$ to $\frac{3\pi}{4}$	No values		
$\frac{3\pi}{4}$ to π	0 to 4	2 (two values)	Quadrants II and IV
π to 2π	Repeat of values from 0 to π		

The graph of $r^2 = a^2 \sin 2\theta$ is also a lemniscate. There are always two leaves to a lemniscate and the length of the leaves is a. The sine function can be considered as a rotation of the cosine function.

EXAMPLE 12 Graph $r^2 = 16 \sin 2\theta$.

SOLUTION
$$r^2 = 16 \sin 2\theta$$

$$= 16 \cos\left(\frac{\pi}{2} - 2\theta\right)$$

$$= 16 \cos\left(2\theta - \frac{\pi}{2}\right)$$

$$= 16 \cos 2\left(\theta - \frac{\pi}{4}\right)$$

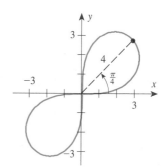

FIGURE 11.63
Graph of $r^2 = 16 \sin 2\theta$

This is a lemniscate whose leaf has length $\sqrt{16} = 4$ and is rotated $\frac{\pi}{4}$, as shown in Figure 11.63.

Summary of Polar Curves

We conclude this section by summarizing the special types of polar-form curves we have examined. There are many others, some of which are presented in the problems, but these three special types are the most common.

CARDIOID $r = a(1 \pm \cos \theta)$ or $r = a(1 \pm \sin \theta)$

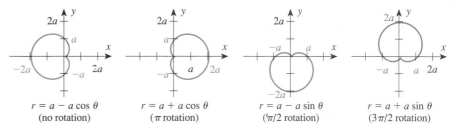

$r = a - a \cos \theta$
(no rotation)

$r = a + a \cos \theta$
(π rotation)

$r = a - a \sin \theta$
($\pi/2$ rotation)

$r = a + a \sin \theta$
($3\pi/2$ rotation)

ROSE CURVE $r = a \cos n\theta$ or $r = a \sin n\theta$ (n is a positive integer). Length of each leaf is a.

1. If n is odd, the rose is n-leaved.

   One leaf: If $n = 1$, the rose is a curve with one petal and is circular.

2. If n is even, the rose is $2n$-leaved.

   Two leaves: See the lemniscate described below.

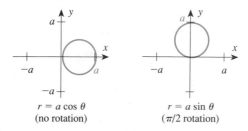

$r = a \cos \theta$
(no rotation)

$r = a \sin \theta$
($\pi/2$ rotation)

Three leaves: $n = 3$

Four leaves: $n = 2$

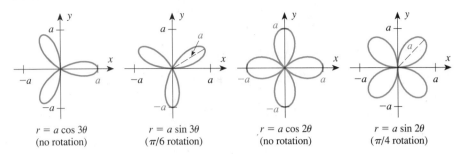

$r = a \cos 3\theta$
(no rotation)

$r = a \sin 3\theta$
($\pi/6$ rotation)

$r = a \cos 2\theta$
(no rotation)

$r = a \sin 2\theta$
($\pi/4$ rotation)

LEMNISCATE $r^2 = a^2 \cos 2\theta$ or $r^2 = a^2 \sin 2\theta$

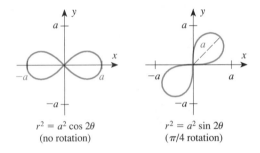

$r^2 = a^2 \cos 2\theta$
(no rotation)

$r^2 = a^2 \sin 2\theta$
($\pi/4$ rotation)

11.7 Problem Set

A In Problems 1–10, plot each of the given polar-form points. Give both primary representations, and give the rectangular coordinates of the point. In Problems 7–8 approximate values to two decimal places. Otherwise use exact values.

1. $(4, \frac{\pi}{4})$
2. $(6, \frac{\pi}{3})$
3. $(5, \frac{2\pi}{3})$
4. $(3, -\frac{\pi}{6})$
5. $(\frac{3}{2}, -\frac{5\pi}{6})$
6. $(5, -\frac{\pi}{2})$
7. $(-4, 4)$
8. $(4, 10)$
9. $(-4, 5\pi)$
10. $(-3, -\frac{7\pi}{3})$

In Problems 11–20 plot the given rectangular-form points and give both primary representations in polar form. In Problems 18–20 approximate values to the nearest hundredth.

11. $(5, 5)$
12. $(-1, \sqrt{3})$
13. $(2, -2\sqrt{3})$
14. $(-2, -2)$
15. $(3, -3)$
16. $(-6, 6)$
17. $(-\sqrt{3}, 1)$
18. $(4, 3)$
19. $(-12, 5)$
20. $(3, 7)$

Identify each of the curves in Problems 21–36 as a cardioid, a rose curve (state number of leaves), a lemniscate, or none of these.

21. $r^2 = 9 \cos 2\theta$
22. $r = 2 \sin 2\theta$
23. $r = 3 \sin 3\theta$
24. $r^2 = 2 \cos 2\theta$
25. $r = 2 - 2 \cos \theta$
26. $r = 3 + 3 \sin \theta$
27. $r^2 = \sin 3\theta$
28. $r = 4 \sin 30°$
29. $r = 5 \cos 60°$
30. $r = 5 \sin 8\theta$
31. $r = 3\theta$
32. $r\theta = 3$
33. $\theta = \tan \frac{\pi}{4}$
34. $r = 5(1 - \sin \theta)$
35. $\cos \theta = 1 - r$
36. $\sin \theta = 1 - r$

B Sketch each of the curves given in Problems 37–50. If it is not one of the types discussed in the text, graph it by plotting points.

37. $r = 2(1 + \cos \theta)$
38. $r = 3(1 - \sin \theta)$
39. $r = 4(1 + \sin \theta)$
40. $r = 4 \cos 2\theta$
41. $r = 5 \sin 3\theta$
42. $r = 3 \cos 3\theta$
43. $r = 2 \cos \theta$
44. $r^2 = 9 \cos 2\theta$
45. $r^2 = 16 \cos 2\theta$
46. $r^2 = 16 \sin 2\theta$
47. $r = 5 \sin \frac{\pi}{6}$
48. $r = 9 \cos \frac{\pi}{3}$
49. $r = \theta$
50. $r = 3\theta$
51. Derive the equations for changing from polar coordinates to rectangular coordinates.

52. Derive the equations for changing from rectangular coordinates to polar coordinates.

C

53. What is the distance between the polar-form points $(3, \frac{\pi}{3})$ and $(7, \frac{\pi}{4})$?

54. What is the distance between the polar-form points (r, θ) and (c, α)?

55. The **limaçon** is a curve of the form $r = b \pm a \cos \theta$ or $r = b \pm a \sin \theta$, where $a > 0$, $b > 0$. There are four types of limaçons.
a. $b/a < 1$ (limaçon with inner loop): graph $r = 2 - 3 \cos \theta$ by plotting points.
b. $b/a = 1$ (cardioid): graph $r = 2 - 2 \cos \theta$.
c. $1 < b/a < 2$ (limaçon with a dimple): graph $r = 3 - 2 \cos \theta$ by plotting points.
d. $b/a \geq 2$ (convex limaçon): graph $r = 3 - \cos \theta$ by plotting points.

56. Graph the following limaçons (see Problem 55):
a. $r = 1 - 2 \cos \theta$ **b.** $r = 2 - \cos \theta$
c. $r = 2 + 3 \cos \theta$ **d.** $r = 2 - 3 \sin \theta$
e. $r = 2 + 3 \sin \theta$ **f.** $r = 3 - 2 \sin \theta$

57. *Spirals* are interesting mathematical curves. There are three general types of spirals:
a. A spiral of Archimedes has the form $r = a\theta$; graph $r = 2\theta$ $(\theta > 0)$ by plotting points
b. A hyperbolic spiral has the form $r\theta = a$; graph $r\theta = 2$ $(\theta > 0)$ by plotting points.
c. A logarithmic spiral has the form $r = a^{k\theta}$; graph $r = 2^\theta$ $(\theta > 0)$ by plotting points.

58. Identify and graph the following spirals (see Problem 57). Assume $\theta > 0$.
a. $r = \theta$ **b.** $r = -\theta$ **c.** $r\theta = 1$
d. $r\theta = -1$ **e.** $r = 2^{2\theta}$ **f.** $r = 3^\theta$

59. The *strophoid* is a curve of the form $r = a \cos 2\theta \sec \theta$; graph this curve where $a = 2$ by plotting points.

60. The *bifolium* has the form $r = a \sin \theta \cos^2\theta$; graph this curve where $a = 1$ by plotting points.

61. The *folium of Descartes* has the form

$$r = \frac{3a \sin \theta \cos \theta}{\sin^3\theta + \cos^3\theta}$$

Graph this curve where $a = 2$ by plotting points.

11.8 Parametric Equations

Up to now, the curves we have discussed have been represented by a single equation or, in the case of piecewise-defined curves, a single equation for a single piece of the curve. In Chapter 9 we introduced the concept of a **parameter**, which is an arbitrary constant used to distinguish various specific cases in a system of equations. In this section, we shall see how we can use a parameter to represent various curves. This new **parametric representation** defines the x and y in (x, y) so that *each* is a function of some parameter, say t. That is, let

$$x = g(t) \quad \text{and} \quad y = h(t)$$

for functions g and h, where the domain of these functions is some interval I. For example, let

$$g(t) = 1 + 3t \quad \text{and} \quad h(t) = 2t \quad \text{(for } 0 \leq t \leq 5)$$

Then, if $t = 1$

$$x = g(1) = 1 + 3(1) = (4) \quad \text{and} \quad y = h(1) = 2(1) = 2$$

Then the point $(x, y) = (4, 2)$ when $t = 1$. Other values are shown in the following table and plotted in Figure 11.64 on page 518.

t	0	1	2	3	4	5
x	1	4	7	10	13	16
y	0	2	4	6	8	10

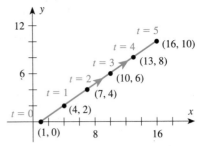

FIGURE 11.64   Graph of $x = 1 + 3t$, $y = 2t$

The variable t is called a **parameter** and the equations $x = 1 + 3t$ and $y = 2t$ are called **parametric equations** for the line segment shown in Figure 11.64.

PARAMETER AND PARAMETRIC EQUATIONS

Let t be a number in an interval I. Consider the curve defined by the set of ordered pairs (x, y), where

$$x = f(t) \quad \text{and} \quad y = g(t)$$

for functions f and g defined on I. Then the variable t is called a **parameter** and the equations $x = f(t)$ and $y = g(t)$ are called **parametric equations** for the curve defined by (x, y). The direction on the curve of increasing values of t is called the **orientation** of the curve.

EXAMPLE 1   Plot the curve represented by the parametric equations

$$x = \cos \theta$$
$$y = \sin \theta$$

SOLUTION   The parameter is θ and you can generate a table of values:

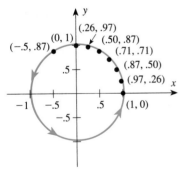

FIGURE 11.65
Graph of $x = \cos \theta$, $y = \sin \theta$

θ	0°	15°	30°	45°	60°	75°	90°	120°	$\cdots$
x	1.00	.97	.87	.71	.50	.26	.00	−.50	$\cdots$
y	0.00	.26	.50	.71	.87	.97	1.00	.87	$\cdots$

These points are plotted in Figure 11.65. If the plotted points are connected, you can see that the curve is a circle. The orientation is shown by arrows on the curve.

It is possible to recognize the parametric equations in Example 1 as a unit circle if you square both sides of the equation and add:

$$x^2 = \cos^2\theta$$
$$y^2 = \sin^2\theta$$
$$x^2 + y^2 = \cos^2\theta + \sin^2\theta$$

So

$$x^2 + y^2 = 1$$

This process is called **eliminating the parameter.**

EXAMPLE 2 Eliminate the parameter for the parametric equations

$$x = t + 2, \, y = t^2 + 2t - 1.$$

SOLUTION Solve the first equation for t: $t = x - 2$. Substitute into the second equation:

$$
\begin{aligned}
y &= (x - 2)^2 + 2(x - 2) - 1 \\
&= x^2 - 4x + 4 + 2x - 4 - 1 \\
&= x^2 - 2x - 1
\end{aligned}
$$

You can recognize $y = x^2 - 2x - 1$ as a parabola. It can now be graphed by completing the square or by using the parametric equations and plotting points as illustrated by Example 1.

EXAMPLE 3 Eliminate the parameter for the parametric equations

$$x = t^2 - 3t + 1, \quad y = -t^2 + 2t + 3$$

SOLUTION It is not as easy to solve one of these equations for t as it was in Example 2. You can, however, add one equation to the other:

$$x + y = -t + 4 \quad \text{or} \quad t = 4 - x - y$$

This can be substituted into either equation to give

$$
\begin{aligned}
x &= (4 - x - y)^2 - 3(4 - x - y) + 1 \\
&= 16 - 4x - 4y - 4x + x^2 + xy - 4y + xy + y^2 - 12 + 3x + 3y + 1 \\
&= x^2 + 2xy + y^2 - 5x - 5y + 5
\end{aligned}
$$

The curve whose equation is $x^2 + 2xy + y^2 - 6x - 5y + 5 = 0$ is a rotated parabola since $B^2 - 4AC = 4 - 4(1)(1) = 0$.

EXAMPLE 4 Eliminate the parameter for the parametric equations $x = 2^t, \, y = 2^{t+1}$.

SOLUTION

$$\frac{y}{x} = \frac{2^{t+1}}{2^t} = 2^{t+1-t} = 2$$

$$y = 2x \qquad \text{⊗ This answer is not complete; see discussion below. ⊗}$$

Consider the parametric equations given in this example a little more closely. You can plot a curve by using the parametric equations or by eliminating the parameter. Plot the equations both ways:

1. **Parametric form:**

t	0	1	2	3
x	1	2	4	8
y	2	4	8	16

The values on this table are found by substitution. For example, if $t = 0$ then

$$x = 2^0 \quad \text{and} \quad y = 2^{0+1}$$
$$= 1 \qquad\qquad = 2$$

Can $x = 3$? Solve

$$3 = 2^t$$
$$\log 3 = t \log 2$$
$$t = \frac{\log 3}{\log 2} \approx 1.5850$$

and then

$$y = 2^{1.5850+1} \approx 6$$

Can $x = 0$? No, since $0 = 2^t$ has no solution.
Can $x = -1$? No, since $2^t > 0$.
The graph is shown in Figure 11.66.

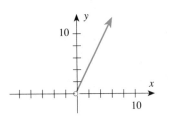

FIGURE 11.66
Graph of $x = 2^t$, $y = 2^{t+1}$

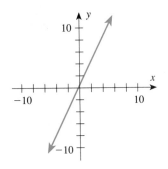

FIGURE 11.67
Graph of $y = 2x$

2. Eliminate the parameter as shown above:

$$y = 2x$$

Can $x = 3$? Solve $y = 2(3) = 6$.
Can $x = 0$? Solve $y = 2(0) = 0$.
Can $x = -1$? Solve $y = 2(-1) = -2$.
The graph is shown in Figure 11.67.

Of course, what you notice by comparing Figures 11.66 and 11.67 is that **you must be very careful about the domain for x when eliminating the parameter.** Thus when you eliminate the parameter t and write $y = 2x$, you must include the condition $x > 0$ (from $x = 2^t$, x is nonnegative). ∎

Note also that sometimes it is impossible to eliminate the parameter in any simple way, as in the equations

$$x = 3t^5 - 4t^2 + 7t + 11$$
$$y = 4t^{13} + 12t^5 + 6t^4 + 16$$

You must therefore be able to graph parametric equations using *both* methods since one or the other might not be appropriate for a particular graph.

Parametrization of a Curve

In calculus it is sometimes necessary to find a parametric representation for a given curve whose rectangular equation, domain, and orientation are known. This process is called **parametrization** of a curve.

EXAMPLE 5   Find a parametrization of the curve C that is the portion of the parabola $y = x^2$ traversed from $(0, 0)$ to $(2, 4)$.

SOLUTION   If $x = t$, then $y = t^2$. If let $t = 0$ we have the point $(0, 0)$, and if we let $t = 2$, we find $(2, 4)$, so we see this curve has the proper orientation if we let $0 \le t \le 2$.

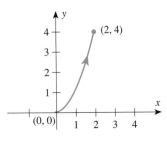

EXAMPLE 6 Find a parametrization of the curve C that is the portion of the line $y = -2x + 3$ traversed from $(2, -1)$ to $(0, 3)$.

SOLUTION If $x = t$, then $y = -2t + 3$. If we let $t = 0$ we have the point $(0, 3)$, and if we let $t = 2$ we find $(2, -1)$, but we see that the orientation is in the wrong direction if we let $0 \le t \le 2$. To reverse the direction, we write

$$x = 2 - 2t$$
$$y = -1 + 4t$$

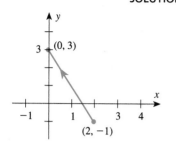

so that if $t = 0$ we obtain the point $(2, -1)$ and if $t = 1$ we find $(0, 3)$ so $0 \le t \le 1$. This gives the parametric representation

$$x = 2 - 2t, \quad y = -1 + 4t, \quad 0 \le t \le 1$$

EXAMPLE 7 Find a parametrization of the curve C that is the portion of the circle $x^2 + y^2 = 9$, oriented so that the interior of the circle is on the left as you travel the path of the curve,

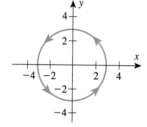

SOLUTION Let $x = 3 \cos t$ and $y = 3 \sin t$. We see that the proper orientation is obtained if we let $0 \le t \le 2\pi$.

EXAMPLE 8 Find a parametrization of the curve C that is a three-leaved rose with polar equation $r = 3 \cos 3\theta$ with the orientation shown in the graph in Figure 11.68.

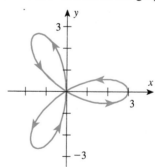

FIGURE 11.68

SOLUTION We can parametrize a polar equation of the form $r = f(\theta)$ by remembering the formulas for converting from polar form to rectangular form:

$$x = r \cos \theta \quad \text{and} \quad y = r \sin \theta$$

Since $r = f(\theta)$, we see that to convert to a parameter θ we need to substitute $x = f(\theta) \cos \theta$ and $y = f(\theta) \sin \theta$. Thus, for this example, we have a parametrization

$$x = 3 \cos 3\theta \cos \theta, \quad y = 3 \cos 3\theta \sin \theta$$

for $0 \le \theta \le 2\pi$.

We now summarize in Table 11.1 the common parametrizations you will need for calculus.

Table 11.1 Parametrizations of Common Curves

CLASSIFICATION	EQUATION	CURVE	PARAMETRIZATION
Function	$y = f(x)$		$x = t$ $y = f(t)$
Line segment	$P_1(x_1, y_1)$ to $P_2(x_2, y_2)$		$x = x_1 + (x_2 - x_1)t$ $y = y_1 + (y_2 - y_1)t$ $0 \le t \le 1$
Circle	$x^2 + y^2 = r^2$		$x = r \cos t$ $y = r \sin t$ $0 \le t \le 2\pi$
Ellipse	$\dfrac{x^2}{a^2} + \dfrac{y^2}{b^2} = 1$		$x = a \cos t$ $y = b \sin t$ $0 \le t < 2\pi$
Polar equation	$r = f(\theta)$		$x = f(t) \cos t$ $y = f(t) \sin t$ $0 \le t < 2\pi$

11.8 Problem Set

A *Plot the curves in Problems 1–14 by plotting points.*

1. $x = t, y = 2 + \frac{2}{3}(t - 1)$ **2.** $x = t, y = 3 - \frac{3}{5}(t + 2)$

3. $x = 2t, y = t^2 + t + 1$ **4.** $x = 3t, y = t^2 - t + 6$

5. $x = 3 \cos \theta, y = 3 \sin \theta$

6. $x = 2 \cos \theta, y = 2 \sin \theta$

7. $x = 4 \cos \theta, y = 3 \sin \theta$

8. $x = 5 \cos \theta, y = 2 \sin \theta$

9. $x = t^2 + 2t + 3, y = t^2 + t - 4$

10. $x = t^2 - 2t + 3, y = t^2 - t + 4$

11. $x = 3^t, y = 3^{t+1}$

12. $x = 2^t, y = 2^{1-t}$

13. $x = e^t, y = e^{t+1}$

14. $x = e^t, y = e^{1-t}$

Eliminate the parameter in Problems 15–28 and plot the resulting equations.

15. $x = t, y = 2 + \frac{2}{3}(t - 1)$ **16.** $x = t, y = 3 - \frac{3}{5}(t + 2)$

17. $x = 2t, y = t^2 + t + 1$ **18.** $x = 3t, y = t^2 - t + 6$

19. $x = 3 \cos \theta, y = 3 \sin \theta$ **20.** $x = 2 \cos \theta, y = 2 \sin \theta$

21. $x = 4 \cos \theta, y = 3 \sin \theta$ **22.** $x = 5 \cos \theta, y = 2 \sin \theta$

23. $x = t^2 + 2t + 3, y = t^2 + t - 4$

24. $x = t^2 - 2t + 3, y = t^2 - t + 4$

25. $x = 3^t, y = 3^{t+1}$ **26.** $x = 2^t, y = 2^{1-t}$

27. $x = e^t, y = e^{t+1}$ **28.** $x = e^t, y = e^{1-t}$

Find an appropriate parametrization for the curves described in Problems 29–40. Be sure to consider the orientation and give the domain for the parameter. Answers are not unique.

29. The line segment from $(0, 0)$ to $(4, 9)$

30. The line segment from $(4, 9)$ to $(0, 0)$

31. The parabola $y = x^2$ from $(4, 9)$ to $(0, 0)$

32. The parabola $y = x^2$ from $(0, 0)$ to $(3, 9)$

33. The arc of the circle $x^2 + y^2 = 16$ in the first quadrant from $(4, 0)$ to $(0, 4)$

34. The arc of the semicircle $x^2 + y^2 = 9$ from $(-3, 0)$ to $(3, 0)$ oriented counterclockwise

35. The ellipse $4x^2 + 9y^2 = 36$ oriented counterclockwise

36. The semiellipse $9x^2 + 4y^2 = 36$ from $(2, 0)$ to $(-2, 0)$ oriented counterclockwise

37. The graph of $y = \sqrt{x}$ from $(1, 1)$ to $(9, 3)$

38. The graph of $y = \sqrt[3]{x}$ from $(27, 3)$ to $(0, 0)$

39. The graph of $r = 2 \sin 5\theta$ for $0 \le \theta \le 2\pi$

40. The graph of $r = 4 \cos \theta$ oriented counterclockwise

B *Plot the curves in Problems 41–60 by any convenient method.*

41. $x = 60t, y = 80t - 16t^2$ **42.** $x = 30t, y = 60t - 9t^2$

43. $x = 10 \cos t, y = 10 \sin t$ **44.** $x = 8 \sin t, y = 8 \cos t$

45. $x = 5 \cos \theta, y = 3 \sin \theta$ **46.** $x = 4 \cos \theta, y = 2 \sin \theta$

47. $x = t^2, y = t^3$ **48.** $x = t^3 + 1, y = t^2 - 1$

49. $x = e^t, y = e^{t+2}$ **50.** $x = e^t, y = e^{t-2}$

51. $x = \theta - \sin \theta, y = 1 - \cos \theta$

52. $x = \theta + \sin \theta, y = 1 - \cos \theta$

53. $x = 4 \tan 2t, y = 3 \sec 2t$

54. $x = 2 \tan 2t, y = 4 \sec 2t$

55. $x = 1 + \cos t, y = 3 - \sin t$

56. $x = 2 - \sin t, y = -3 + \cos t$

57. $x = 3 \cos \theta + \cos 3\theta, y = 3 \sin \theta - \sin 3\theta$

58. $x = \cos t + t \sin t, y = \sin t - t \cos t, t > 0$

59. $x = \tan t, y = \cot t$ **60.** $x = \cos \theta, y = \sec \theta$

C

61. PHYSICS Suppose a light is attached to the edge of a bike wheel. The path of the light is shown in Figure 11.69.

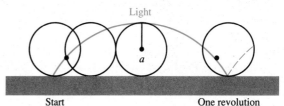

Light

Start   One revolution

FIGURE 11.69   Graph of a cycloid

If the radius of the wheel is a, find the equation for the path of the light. Such a curve is called a *cycloid*.
Hint: Consider Figure 11.70, and find the coordinates of $P(x, y)$. Notice that

$$x = |OA| \qquad y = |PA|$$

Find x and y in terms of θ, the amount of rotation in radians.

FIGURE 11.70

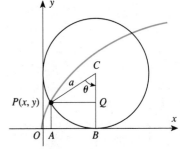

62. Suppose a string is wound around a circle of radius a. The string is then unwound in the plane of the circle while it is held tight, as shown in Figure 11.71. Find the equation for this curve, called the *involute of a circle*.

Hint: Consider Figure 11.72, and find the coordinates of $P(x, y)$. Notice that

$$x = |OB| \qquad y = |PB|$$

Find x and y in terms of θ, the amount of rotation in radians.

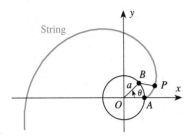

FIGURE 11.71 Graph of the involute of a circle

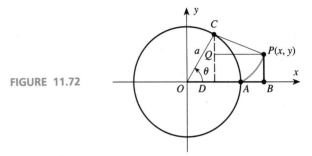

FIGURE 11.72

11.9 Chapter 11 Summary

The material of this chapter is reviewed in the following list of objectives. After each objective there are some practice questions. For a sample test, select the first question of each set and check your answers with the answer section. For a sample test without answers, use the second question of each set. Additional practice is given by the other questions in each set. If you are having trouble with a particular type of problem, look back to that section for extra help.

11.1 Parabolas

OBJECTIVE 1 *Graph parabolas.*

1. $x^2 = y$
2. $(y - 1)^2 = 8(x + 2)$
3. $8y^2 - x - 32y + 31 = 0$
4. $y^2 + 4x + 4y = 0$

OBJECTIVE 2 *Find the equations of parabolas given certain information about the graph.*

5. Vertex at $(6, 3)$; directrix $x = 1$

6. Directrix $y - 3 = 0$; focus $(-3, -2)$
7. Vertex $(-3, 5)$; focus $(-3, -1)$
8. Vertex at $(4, 2)$ and passing through $(-3, -4)$; axis parallel to the y-axis

11.2 Ellipses

OBJECTIVE 3 *Graph ellipses.*

9. $25x^2 + 16y^2 = 400$
10. $5(x + 3)^2 + 9(y - 2)^2 = 45$
11. $x^2 + y^2 = 4x + 2y - 3$
12. $9x^2 + 16y^2 - 90x - 32y + 97 = 0$

OBJECTIVE 4 *Find the equations of ellipses given certain information about the graph.*

13. The ellipse with the center at $(4, 1)$, a focus at $(5, 1)$, and a semimajor axis of length 2

14. The set of points such that the sum of the distances from $(-3, 4)$ and $(-7, 4)$ is 12
15. The set of points 8 units from the point $(-1, -2)$
16. The ellipse with foci at $(2, 3)$ and $(-1, 3)$ with eccentricity $\frac{3}{5}$

11.3 Hyperbolas

OBJECTIVE 5 *Graph hyperbolas.*

17. $x^2 - y^2 + x - y = 3$

18. $x(x - y) = y(y - x) - 1$

19. $5(x + 2)^2 - 3(y + 4)^2 = 60$

20. $12x^2 - 4y^2 + 24x - 8y + 4 = 0$

OBJECTIVE 6 *Find the equations of hyperbolas given certain information about the graph.*

21. The set of points with the difference of distances from $(-3, 4)$ and $(-7, 4)$ equal to 2

22. The hyperbola with vertices at $(-3, 0)$ and $(3, 0)$ and foci at $(5, 0)$ and $(-5, 0)$

23. The hyperbola with vertices at $(0, -3)$ and $(0, 3)$ and eccentricity $\frac{5}{3}$

24. The hyperbola with vertices at $(-3, 1)$ and $(-5, 1)$ and foci at $(-4 - \sqrt{6}, 1)$ and $(-4 + \sqrt{6}, 1)$

OBJECTIVE 7 *Know the definition and standard-form equations for the conic sections. State the appropriate standard-form equation.*

25. Horizontal ellipse

26. Vertical hyperbola

27. Parabola opening right

28. Circle

OBJECTIVE 8 *Graph conic sections. Name the type of curve (by inspection) and then graph the curve.*

29. $3x - 2y^2 - 4y + 7 = 0$

30. $\dfrac{x}{16} + \dfrac{y}{4} = 1$

31. $25x^2 + 9y^2 = 225$

32. $25(x - 2)^2 + 25(y + 1)^2 = 400$

11.4 Rotations

OBJECTIVE 9 *Use the rotation of axes formulas and the amount of rotation formula.*

33. What is the rotation for the curve whose equation is $xy - 7 = 0$?

34. What is the rotation for the curve whose equation is
$$5x^2 + 4xy + 5y^2 + 3x - 2y + 5 = 0?$$

35. What is the rotation for the curve whose equation is $4x^2 + 4xy + y^2 + 3x - 2y + 7 = 0$?

36. Use the rotation formulas to rewrite the equation in Problem 33 so that there is no xy term.

OBJECTIVE 10 *Identify the conic by looking at its equation.*

37. $xy + x^2 - 3x = 5$

38. $x^2 + y^2 + xy + 3x - y = 3$

39. $x^2 + 2xy + y^2 = 10$

40. $(x - 1)(y + 1) = 7$

11.5 Vectors

OBJECTIVE 11 *Find the resultant vector of two given forces.*

41. If $\mathbf{v} = 5\mathbf{i} - 2\mathbf{j}$ and $\mathbf{w} = -7\mathbf{i} + 3\mathbf{j}$, find the resultant vector.

42. If $\mathbf{v} = -3\mathbf{i} + 2\mathbf{j}$ and $\mathbf{w} = 8\mathbf{i} - 3\mathbf{j}$, find the resultant vector.

43. Consider two forces, one with magnitude 8.0 in a N20°E direction and the other with a magnitude of 9.0 in a S50°E direction. What is the resultant vector?

44. An object is hurled from a catapult due east with a velocity of 38 feet per second (fps). If the wind is blowing due south at 15 mph (22 fps), what are the true bearing and the velocity of the object?

OBJECTIVE 12 *Resolve a given vector and find its algebraic representation.*

45. Resolve the vector with magnitude 4.5 and $\theta = 51°$ into horizontal and vertical components.

46. Resolve the vector with magnitude 12 and $\theta = 30°$ into horizontal and vertical components.

47. Find the algebraic representation of a vector determined by $\overrightarrow{AB}$ with $A(0, 0)$ and $B(5, 12)$.

48. Find the algebraic representation of a vector determined by $\overrightarrow{AB}$ with $A(-3, -5)$, $B(-6, -10)$.

OBJECTIVE 13 *Find the magnitude of a vector.*

49. $\mathbf{v} = 5\mathbf{i} - 2\mathbf{j}$ **50.** $\mathbf{w} = -7\mathbf{i} + 3\mathbf{j}$

51. $\mathbf{v} = 6\mathbf{i}$ **52.** $\mathbf{w} = \cos 42°\mathbf{i} + \sin 42°\mathbf{j}$

OBJECTIVE 14 *Know the definition of scalar product and be able to find the scalar product of two vectors.*

53. Define scalar product.

54. Find the scalar product of $\mathbf{v}$ and $\mathbf{w}$ where $\mathbf{v} = 4\mathbf{i} - 2\mathbf{j}$ and $\mathbf{w} = -7\mathbf{i} + 3\mathbf{j}$.

55. Find $(3\mathbf{i} + 5\mathbf{j}) \cdot (-2\mathbf{i} + 3\mathbf{j})$

56. Find $(e\mathbf{i} - e^{-1}\mathbf{j}) \cdot (e^2\mathbf{i} - e\mathbf{j})$

OBJECTIVE 15 *Find the cosine of the angle between two vectors, or use the inverse cosine function to find that angle.*

57. Find the cosine of the angle between the vectors $3\mathbf{i} - 2\mathbf{j}$ and $5\mathbf{i} + 12\mathbf{j}$.

58. Find the angle between the vectors $5\mathbf{i} - 2\mathbf{j}$ and $-7\mathbf{i} + 3\mathbf{j}$.

59. Find the angle between the vectors $\cos 15°\mathbf{i} + \sin 15°\mathbf{j}$ and $\cos 20°\mathbf{i} + \sin 20°\mathbf{j}$.

60. Find the cosine of the angle between the vectors $5\mathbf{i}$ and $-3\mathbf{i} - \mathbf{j}$.

11.6 Properties of Vectors

OBJECTIVE 16 *Find a vector normal to a given line and a vector determined by a line.* After you have found the vector normal to the given line and the vector determined by the line, check your answer by showing that the dot product of these vectors is 0.

61. $5x - 12y + 3 = 0$

62. $2x + 3y + 12 = 0$

63. $x - 5y + 4 = 0$

64. $x + y = 0$

OBJECTIVE 17 *Find scalar and vector projections.* For the given vectors, find the vector and scalar projections of $\mathbf{v}$ onto $\mathbf{w}$.

65. $\mathbf{v} = 2\mathbf{i} + \sqrt{5}\mathbf{j}$ $\mathbf{w} = 3\sqrt{5}\mathbf{i} + 3\mathbf{j}$

66. $\mathbf{v} = 5\mathbf{i} - 12\mathbf{j}$ $\mathbf{w} = 3\mathbf{i} + 4\mathbf{j}$

67. $\mathbf{v} = 9\mathbf{i} + \mathbf{j}$ $\mathbf{w} = \mathbf{i} - 9\mathbf{j}$

68. $\mathbf{v} = -2\mathbf{j} - 7\mathbf{j}$ $\mathbf{w} = 4\mathbf{i} + 3\mathbf{j}$

OBJECTIVE 18 *Find the distance from a given point to a given line.*

69. $5x + 12y + 8 = 0$; $(1, 5)$

70. $6x - 2y + 5 = 0$; $(-5, -10)$

71. $2x - 3y - 5 = 0$; $(-10, 2)$

72. $5x + 3y = 0$; $(8, 10)$

11.7 Polar Coordinates

OBJECTIVE 19 *Plot points in polar form and give both primary representations. Approximate to four decimal places.*

73. $(5, \sqrt{75})$

74. $(3, -\frac{2\pi}{3})$

75. $(-2, 2)$

76. $(-5, 9.4248)$

OBJECTIVE 20 *Change from rectangular to polar form and from polar form to rectangular form. Use exact values if possible; otherwise approximate to four decimal places.*

77. Change the polar-form point $(3, -\frac{2\pi}{3})$ to rectangular form.

78. Change the polar-form point $(5, \sqrt{75})$ to rectangular form.

79. Change the rectangular-form point $(3, -3)$ to polar form.

80. Change the rectangular-form point $(3, \sqrt{3})$ to polar form.

OBJECTIVE 21 *Sketch polar-form curves.*

81. $r = 2 \cos 2\theta$

82. $r = 2 + 2 \cos \theta$

83. $r^2 = 25 \cos 2\theta$

84. $r = \tan 45°$

OBJECTIVE 22 *Identify cardioids, rose curves, and lemniscates by looking at the equation.*

85. a. $r^2 = 5 \cos 2\theta$ **b.** $r = 5 \cos 2\theta$

86. a. $r = 5\theta$ **b.** $r = 5 \cos 3\theta$

87. a. $r = 3 + 5 \cos \theta$ **b.** $r = 5 + 5 \cos \theta$

88. a. $\theta = \frac{\pi}{2}$ **b.** $\frac{\pi}{2} \cos \theta = \frac{\pi}{2} - r$

11.8 Parametric Equations

OBJECTIVE 23 *Sketch a curve represented by parametric equations by plotting points.*

89. $x = 2 + 5t, y = -1 - 3t$

90. $x = 3 \cos \theta, y = 5 \sin \theta$

91. $x = e^{4t}, y = e^{4t-2}$

92. $x = t^2 + 3t - 1, y = t^2 + 2t + 5$

OBJECTIVE 24 *Eliminate the parameter, if possible, to sketch a curve defined by parametric equations.*

93. $x = 2 + 5t, y = -1 - 3t$

94. $x = 3 \cos \theta, y = 5 \sin \theta$

95. $x = e^{4t}, y = e^{4t-2}$

96. $x = t^2 + 3t - 1, y = t^2 + 2t + 5$

Cumulative Review III

Suggestions for study of Chapters 9–11:

MAKE A LIST OF IMPORTANT IDEAS FROM CHAPTERS 9–11. Study this list. Use the objectives at the end of each chapter to help you make up this list.

WORK SOME PRACTICE PROBLEMS. A good source of problems is the set of chapter objectives at the end of each chapter. You should try to work at least one problem from each objective:

Chapter 9, pp. 425–426; work 12 problems

Chapter 10, pp. 458–459; work 9 problems

Chapter 11, pp. 524–526; work 24 problems

Check the answers for the practice problems you worked. The first and third problems of each objective have their answers listed in the back of the book.

ADDITIONAL PROBLEMS. Work additional odd-numbered problems (answers in the back of the book) from the problem sets as needed. Focus on the problems you missed in the chapter summaries.

WORK THE PROBLEMS IN THE FOLLOWING CUMULATIVE REVIEW. These problems should be done after you have studied the material. They should take you about 1 hour and will serve as a sample test for Chapters 9–11. Assume that all variables are restricted so that each expression is defined. All the answers for these questions are provided in the back of the book for self-checking.

Practice Test for Chapters 9–11

1. Find the next term of each given sequence, and classify it as arithmetic, geometric, or neither. If arithmetic, give the common difference; if geometric, give the common ratio; and if neither, explain the pattern.
   a. 45, 30, 20   b. 45, 30, 15   c. 45, 30, 25, 30, 5

2. Solve each system by the indicated method.

   a. By addition: $\begin{cases} 5x + 3y = 5 \\ 3x + 2y = 4 \end{cases}$

   b. By substitution: $\begin{cases} x + y = 1 \\ y = x^2 + 4x + 5 \end{cases}$

   c. By graphing $\begin{cases} 2x - y - 5 = 0 \\ 2x + y + 1 = 0 \end{cases}$

   d. Represent the system $\begin{cases} x + z = 1 \\ x + y + z = 0 \\ 2x + y = 3 \end{cases}$

   as an augmented matrix and solve using the Gauss–Jordan method.

3. Find the indicated matrices, if possible, where:

   $$A = \begin{bmatrix} 1 & 0 \\ 2 & -1 \end{bmatrix} \qquad B = \begin{bmatrix} 0 & 1 & 0 \\ 1 & 0 & -1 \end{bmatrix}$$

   $$C = \begin{bmatrix} 1 & 0 \\ -2 & 1 \\ 0 & 1 \end{bmatrix}$$

   a. $A + BC$ b. $AB - AC$
   c. BCA

4. Identify the curve represented by the given equation.
   a. $3x - 2y^2 - 4y + 7 = 0$ b. $y^2 = x^2 - 1$
   c. $xy + y^2 - 3x = 5$ d. $x^2 + y^2 + x - y = 3$
   e. $\dfrac{x}{4} + \dfrac{y}{9} = 1$ f. $(x - 1)(y + 1) = 7$
   g. $r = 3 \sin 2\theta$ h. $r^2 = 3 \sin 2\theta$
   i. $r - 5 = 5 \cos \theta$ j. $r = 2 \sin \frac{\pi}{2}$

5. Write the equations for the given curves.
   a. The ellipse with center (2, 1), focus (1, 1), and semimajor axis of length 2
   b. The set of points so that the difference of the distance from (3, 0) and (9, 0) is always 4
   c. The parabola whose vertex is (4, 3) and directrix is $x = 1$

6. Sketch the given curves.
   a. $3x - 2y^2 - 4y + 7 = 0$ b. $25x^2 - 16y^2 = 400$

7. Sketch the given curves.
   a. $x = 4 \cos \theta$, $y = 3 \sin \theta$ b. $r = 3 \sin 3\theta$
   c. $r = 5 - 5 \cos \theta$

8. a. State the Binomial Theorem using summation notation.
   b. State the Binomial Theorem without using summation notation.
   c. Use the Binomial Theorem to find $(x - 3y)^4$.

9. Prove that $1^3 + 2^3 + 3^3 + \cdots + n^3 = \dfrac{n^2(n + 1)^2}{4}$.

10. Let $\mathbf{v} = 5\mathbf{i} - \sqrt{5}\mathbf{j}$ and $\mathbf{w} = -2\mathbf{i} + 3\sqrt{5}\mathbf{j}$. Find the requested numbers or vectors, if possible.
    a. $\mathbf{v} - 2\mathbf{w}$ b. $\mathbf{v} \cdot \mathbf{w}$ c. $|\mathbf{w}|$

▌*Planetary Orbits*

A 'Planet X' Way Out There?

LIVERMORE, Calif. (UPI)—A scientist at the Lawrence Livermore Laboratory suggested Friday the existence of a "Planet X" three times as massive as Saturn and nearly six billion miles from earth.

The planet, far beyond Pluto, which is currently the outermost of the nine known planets of the solar system, was predicted on sophisticated mathematical computations of the movements of Halley's Comet.

Joseph L. Brady, a Lawrence mathematician and an authority on the comet, reported the calculations in the Journal of the Astronomical Society of the Pacific.

Brady said he and his colleagues, Edna M. Carpenter and Francis H. McMahon, used a computer to process mathematical observations of the strange deviations in Halley's Comet going back to before Christ.

Lawrence officials said the existence of a 10th planet has been predicted

before, but Brady is the first to predict its orbit, mass and position.

Brady said the planet was about 65 times as far from the sun as earth, which is about 93 million miles from the sun. From earth "Planet X" would be located in the constellation Casseiopeia on the border of the Milky Way.

The size and location of "Planet X" were proposed to account for mysterious deviations in the orbit of Halley's Comet. But the calculations subsequently were found to account for deviations in the orbits of two other reappearing comets, Olbers and Pons-Brooks, Brady said.

No contradiction between the proposed planet and the known orbits of comets and other planets has been found.

The prediction of unseen planets is not new. The location of Neptune was predicted in 1846 on the basis of deviations in the orbit of Uranus. Deviations in Neptune's orbit led to a prediction of Pluto's location in 1915.

Although no such deviations of Pluto have been found, Brady pointed out that since its discovery in 1930, Pluto has been observed through less than one-fourth of its revolution around the sun and a complete picture of its orbit is not available.

Brady said "Planet X" may be as elusive as Pluto was a half century ago. It took 15 years to find it from the time of its prediction.

"The proposed planet is located in the densely populated Milky Way where even a tiny area encompasses thousands of stars, many of which are brighter than we expect this planet to be," he said. "If it exists, it will be extremely difficult to find."

"Planet X," in its huge orbit, takes 600 years to complete a revolution around the sun, he said.

SOURCE: Courtesy of *The Press Democrat*, Santa Rosa, California, April 30, 1972. Reprinted by permission.

In 1986 Halley's Comet returned for our once-every-76-year view. By using mathematics and a knowledge of the conic sections, we can predict the comet's path as well as its exact arrival time (closest to earth on April 10, 1986).

After the planet Uranus was discovered in 1781, its motion revealed gravitational perturbations caused by an unknown planet. Independent mathematical calculations by Urbain Leverrier and John Couch Adams predicted the position of this unknown planet—the discovery of Neptune in 1846 is one of the greatest triumphs of celestial mechanics and mathematics in the history of astronomy. A similar search led to the discovery of Pluto in 1930. The article reproduced here is dated April 30, 1972, but at the time of this printing no planet has yet been found. Remember, though, that it took 15 years to find Pluto after the time of its prediction.

The orbits of the planets are elliptical in shape. If the sun is placed at one of the foci of a giant ellipse, the orbit of the earth is elliptical. The **perihelion** is the point where the planet comes closest to the sun; the **aphelion** is the farthest distance the planet travels from the sun. The eccentricity of a planet tells us the amount of roundness of that planet's orbit. The eccentricity of a circle is 0 and that of a parabola is 1. The eccentricity for each planet in our solar system is given here:

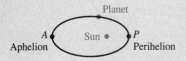

PLANET	ECCENTRICITY	PLANET	ECCENTRICITY
Mercury	.194	Saturn	.056
Venus	.007	Uranus	.047
Earth	.017	Neptune	.009
Mars	.093	Pluto	.249
Jupiter	.048		

EXAMPLE 1 The orbit of the earth around the sun is elliptical with the sun at one focus. If the semimajor axis of this orbit is 9.3×10^7 mi and the eccentricity is about .017, determine the greatest and least distance of the earth from the sun (correct to two significant digits).

SOLUTION We are given $a = 9.3 \times 10^7$ and $\varepsilon = .017$. Now

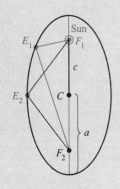

$$\varepsilon = \frac{c}{a}$$

$$.017 = \frac{c}{9.3 \times 10^7}$$

$$c \approx 1.581 \times 10^6$$

The greatest distance is $a + c \approx \mathbf{9.5 \times 10^7}$.
The least distance is $a - c \approx \mathbf{9.1 \times 10^7}$.

The orbit of a satellite can be calculated from Kepler's third law:

$$\frac{\text{Mass of (planet + satellite)}}{\text{Mass of (sun + planet + satellite)}} = \frac{(\text{Semimajor axis of satellite orbit})^3}{(\text{Semimajor axis of planet orbit})^3} \times \frac{(\text{Period of planet})^2}{(\text{Period of satellite})^2}$$

EXAMPLE 2 If the mass of the earth is 6.58×10^{21} tons, the sun 2.2×10^{27} tons, and the moon 8.1×10^{19} tons, calculate the orbit of the moon. Assume that the semimajor axis of the earth is 9.3×10^7 mi, the period of the earth 365.25 days, and the period of the moon 27.3 days.

SOLUTION First solve the equation given as Kepler's third law for the unknown—the semimajor axis of the satellite orbit:

$$(\text{Semimajor axis of satellite orbit})^3$$
$$= \frac{\text{Mass of (planet + satellite)}}{\text{Mass of (sun + planet + satellite)}} \times \frac{(\text{Period of satellite})^2}{(\text{Period of planet})^2} \times (\text{Semimajor axis of planet orbit})^3$$

Thus

(Semimajor axis of satellite orbit)^3

$$= \frac{(6.58 \cdot 10^{21}) + (8.1 \cdot 10^{19})}{(2.2 \cdot 10^{27}) + (6.58 \cdot 10^{21}) + (8.1 \cdot 10^{19})} \times \frac{(27.3)^2}{(365.25)^2} \times (9.3 \cdot 10^7)^3$$

$$\approx \frac{3.993131141 \cdot 10^{48}}{2.934975261 \cdot 10^{32}}$$

$$\approx 1.360533151 \cdot 10^{16}$$

Semimajor axis of satellite orbit $\approx 2.387278258 \cdot 10^5$

The moon's orbit has a semimajor axis of about 239,000 mi.

Extended Application Problems—Planetary Orbits

1. The orbit of Mars about the sun is elliptical with the sun at one focus. If the semimajor axis of this orbit is 1.4×10^8 mi and the eccentricity is about .093, determine the greatest and least distance of Mars from the sun, correct to two significant digits.

2. The orbit of Venus about the sun is elliptical with the sun at one focus. If the semimajor axis of this orbit is 6.7×10^7 mi and the eccentricity is about .007, determine the greatest and least distance of Venus from the sun, correct to two significant digits.

3. The orbit of Neptune about the sun is elliptical with the sun at one focus. If the semimajor axis of this orbit is 3.66×10^9 mi and the eccentricity is about .009, determine the greatest and least distance of Neptune from the sun, correct to two significant digits.

4. If an unknown Planet X has an elliptical orbit about the sun with a semimajor axis of 6.89×10^{10} mi and eccentricity of about .35, what is the least distance between Planet X and the sun?

5. If the mass of Mars is 7.05×10^{20} tons, its satellite Phobos 4.16×10^{12} tons, and the sun 2.2×10^{27} tons, calculate the length of the semimajor axis of Phobos. Assume that the semimajor axis of Mars is 1.4×10^8 mi, the period of Mars 693.5 days, and the period of Phobos .3 day.

6. Use the information in Problem 5 to find the approximate mass of the Martian satellite Deimos if its elliptical orbit has a semimajor axis of 15,000 mi and a period of 1.26 days.

7. If the perihelion distance of Mercury is 2.9×10^7 mi and the aphelion distance is 4.3×10^7 mi, write the equation for the orbit of Mercury.

8. If the perihelion distance of Venus is 6.7234×10^7 mi and the aphelion distance is 6.8174×10^7 mi, write the equation for the orbit of Venus.

9. If the perihelion distance of earth is 9.225×10^7 mi and the aphelion distance is 9.542×10^7 mi, write the equation for the orbit of earth.

10. If the perihelion distance of Mars is 1.29×10^8 mi and the aphelion distance is 1.55×10^8 mi, write the equation for the orbit of Mars.

APPENDIXES

SIGNIFICANT DIGITS

When you work with numbers that arise from counting, the numbers are **exact.** If you work with measurements, however, the quantities are necessarily **approximations.** The digits known to be correct in a number obtained by a measurement are called **significant digits.** The digits, 1, 2, 3, 4, 5, 6, 7, 8, and 9 are always significant, whereas the digit 0 may or may not be significant, as described by the following two rules:

1. Zeros that come between two other digits are significant, as in 203 or 10.04.

2. If the only function of the zero is to place the decimal point, it is not significant, as in

$$.000023 \quad \text{or} \quad 23,000$$

placeholders   placeholders

If it does more than fix the decimal point, it is significant, as in

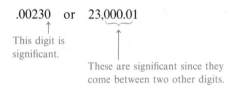

This second rule can, of course, result in certain ambiguities, as in 23,000 (measured to the *exact* unit). To avoid confusion, we use scientific notation in this case:

EXAMPLE 1   Two significant digits: 46,   0.00083, 4.0×10^1,   .050

EXAMPLE 2   Three significant digits: 523,   403, 4.00×10^2,   .000800

EXAMPLE 3   Four significant digits: 600.1, 4.000×10^1,   .0002345

In this book, we do not wish to resort to scientific notation to resolve this ambiguity. We will agree to violate this rule and assume the maximum degree of accuracy possible. This means for calculation purposes, we will assume that 23,000 has five significant digits.

When we are doing calculations with approximate numbers (particularly when using a calculator), it is often necessary to round off results.

To round off numbers:
1. Increase the last retained digit by 1 if the residue is 5 or greater.
2. Retain the last digit unchanged if the residue is less than 5.

Elaborate rules for calculating approximate data can be developed (when it is necessary for some applications, as in chemistry), but there are two simple rules that will work satisfactorily for the material in this book.

SIGNIFICANT-DIGIT OPERATIONS

ADDITION–SUBTRACTION: Add or subtract in the usual fashion, and then round off the result so that the last digit retained is in the column farthest to the right in which both given numbers have significant digits.

MULTIPLICATION–DIVISION: Multiply or divide in the usual fashion, and then round off the result to the smaller number of significant digits found in either of the given numbers.

These results are particularly important when we are using a calculator, since the results obtained will look much more accurate on the calculator than they actually are.

In solving triangles in this text, we will assume a certain relationship in the accuracy of the measurement between the sides and the angles.

SIGNIFICANT DIGITS IN SOLVING TRIANGLES

ACCURACY IN SIDES	ACCURACY IN ANGLES
Two significant digits	Nearest degree
Three significant digits	Nearest tenth of a degree
Four significant digits	Nearest hundredth of a degree

This chart means that if the data include one side given with two significant digits and another with three significant digits, the angle would be computed to the nearest degree. If one side is given to four significant digits and an angle to the nearest tenth of a degree, the other sides would be given to three significant digits and the angles computed to the nearest tenth of a degree. In general, results should not be more accurate than the least accurate item of the given data.

B

ANSWERS

1.1 Problem Set, Pages 9–10

1. a. $\mathbb{Z}, \mathbb{Q}, \mathbb{R}$ **b.** $\mathbb{Q}, \mathbb{R}$ **c.** $\mathbb{W}, \mathbb{N}, \mathbb{Z}, \mathbb{Q}, \mathbb{R}$ **d.** $\mathbb{Q}', \mathbb{R}$
   e. $\mathbb{Q}', \mathbb{R}$ **f.** $\mathbb{Q}', \mathbb{R}$
3. a. $\mathbb{Q}, \mathbb{R}$ **b.** $\mathbb{Q}, \mathbb{R}$ **c.** $\mathbb{Q}', \mathbb{R}$ **d.** $\mathbb{Q}, \mathbb{R}$ **e.** $\mathbb{Q}', \mathbb{R}$
   f. $\mathbb{Q}, \mathbb{R}$
5. $\neq$; π is irrational, 3.141592654 is rational
7.
9.
11.
13. a. $<$ **b.** $<$
15. a. $<$ **b.** $=$ **17. a.** $=$ **b.** $<$
19. a. $\pi - 2$ **b.** $5 - \pi$ **c.** $2\pi - 6$ **d.** $7 - 2\pi$
21. a. $\sqrt{2} - 1$ **b.** $2 - \sqrt{2}$ **c.** $1 - \frac{\pi}{6}$ **d.** $\frac{2\pi}{3} - 1$
23. a. $\pi - 3$ **b.** 119   **c.** $3 - \sqrt{5}$ **25.** Reflexive
27. Distributive or distributive and commutative   **29.** Closure
31. Transitive   **33.** Substitution   **35.** Closure
37. Multiplicative inverse   **39.** Distributive   **41.** Identity
43. No   **45.** No   **47.** Yes   **49.** No
51. $12 \div 4 \neq 4 \div 12$ **53.** No   **55.** $-\frac{\pi}{3} - 1$ **57.** $\dfrac{3}{\pi + 3}$
59. Addition: closure, commutative, associative; multiplication: closure, commutative, associative, identity; distributive for multiplication over addition
61. Distributive for multiplication over addition; addition: closure, commutative, associative, identity, inverse; multiplication: closure, commutative, associative, identity
63. Not closed, so the following properties hold only for results in the set: commutative and associative for both addition and multiplication; distributive for multiplication over addition
65. Not closed for addition; multiplication: closure, commutative, associative, identity, inverse

1.2 Problem Set, Pages 15–16

1. a. $x^2 + 3x + 2$ **b.** $y^2 + y - 6$ **c.** $x^2 - x - 2$
   d. $y^2 - y - 6$
3. a. $2x^2 - x - 1$ **b.** $2x^2 - 5x + 3$
   c. $3x^2 + 4x + 1$ **d.** $3x^2 + 5x + 2$
5. a. $a^2 + 4a + 4$ **b.** $b^2 - 4b + 4$
   c. $x^2 + 8x + 16$ **d.** $y^2 - 6y + 9$
7. a. $m(e + i + y)$ **b.** $(a - b)(a + b)$ **c.** Irreducible
   d. $(a - b)(a^2 + ab + b^2)$
9. a. $(a + b)^3$ **b.** $(p - q)^3$ **c.** $(d - c)^3$ **d.** $xy(x + y)$
11. a. $(3x + 1)(x - 2)$ **b.** $(3y - 2)(2y - 1)$
   c. $b(4a - 1)(2a + 3)$ **d.** $2(s - 8)(s + 3)$

13. a. $3x^3 + 8x^2 - 9x + 2$ **b.** $2x^3 + 5x^2 - 8x - 5$
15. a. $2x^3 - 3x^2 - 8x - 3$ **b.** $6x^3 + 17x^2 - 4x - 3$
17. a. $x^3 - 3x^2 + 4$ **b.** $x^3 - 3x - 2$
19. $(x - y - 1)(x - y + 1)$ **21.** $5(a - 1)(5a + 1)$
23. $-\frac{1}{25}(3x + 10)(7x + 10)$ **25.** $\dfrac{1}{y^8}(x^3 - 13y^4)(x^3 + 13y^4)$
27. $(a + b - x - y)(a + b + x + y)$ **29.** $(2x - 3)(x + 2)$
31. $(6x - 1)(x + 8)$ **33.** $(6x + 1)(x + 8)$
35. $(4x - 3)(x + 4)$ **37.** $(9x - 2)(x - 6)$
39. $(2x - 1)(2x + 1)(x + 2)(x - 2)$
41. $(x + 2)(x + 1)(x^2 - 2x + 4)(x^2 - x + 1)$
43. $\frac{1}{36}(2x + 1)(2x - 1)(3x - 1)(3x + 1)$
45. $(x + y)^3 = x^3 + 3x^2y + 3xy^2 + y^3$
47. $\sqrt{x^2 + y^2} = (x^2 + y^2)^{1/2}$; exponent does not distribute
49. True, since $\sqrt{x}$ is defined only for $x \geq 0$ in $\mathbb{R}$
51. $-2x^{-3}(2x - 5)^{-3}(4x - 5)$
53. $(2x + 1)(3x + 2)^2(30x + 17)$
55. $\dfrac{(7x + 11)^2(7x^2 - 22x + 63)}{(x^2 + 3)^2}$
57. $12x^6 - 20x^5 + 23x^4 - 34x^3 + 71x^2 - 53x + 12$
59. $20x^6 - 4x^5 - 31x^4 - 2x^3 - 7x^2 - 16x - 5$
61. $8x^9 + 36x^8 + 30x^7 + 3x^6 + 114x^5 + 48x^4$
   $- 56x^3 + 192x^2 - 96x + 64$
63. $(x^n - y^n)(x^n + y^n)$ **65.** $(x^n + y^n)(x^{2n} - x^ny^n + y^{2n})$
67. $x(x - 5)$ **69.** $(z - 2)^2(z + 2)(z^2 + 2z + 4)$
71. $(x - y - a - b)(x - y + a + b)$
73. $(x - y - a + b)(x - y + a - b)$ **75.** $4y(x + 2z)$
77. $(2x + 2y + a + b)(x + y - 3a - 3b)$
79. $(x + 1)^2(x - 2)^3(7x - 2)$
81. $6(2x - 1)^2(3x + 2)(5x + 1)$
83. $2(x + 5)^3(x^2 - 2)^2(5x^2 + 15x - 4)$

1.3 Problem Set, Pages 25–26

1. a. 5   **b.** 13   **3. a.** $\sqrt{37}$ **b.** $\sqrt{37}$ **5.** $-5x$
7. a. $(\frac{13}{2}, 3)$ **b.** $(7, \frac{13}{2})$ **9. a.** $(\frac{7}{2}, 2)$ **b.** $(-\frac{3}{2}, -2)$
11. $(-x, \frac{7x}{2})$
13.    **15.**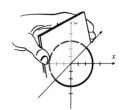

APPENDIX B • Answers   535

17.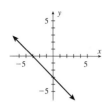

19.

21.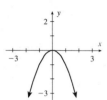

23.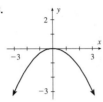

25.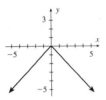

27.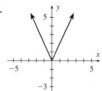

29.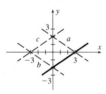

31.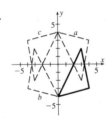

33.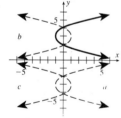

35. y-axis **37.** None **39.** None **41.** All **43.** Origin
45. Origin **47.** $(x-5)^2+(y+1)^2=16$
49. $(x-2)^2+(y-1)^2=26$ **51.** $(x-a)^2+(y-b)^2=b^2$
53. No **55.** No **57.** Yes **59.** $(x-2)^2+(y-3)^2=49$
61. $(x+4)^2+(y-1)^2=9$ **63.** $(0, 4+2\sqrt{15}), (0, 4-2\sqrt{15})$
65. 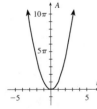 **67.**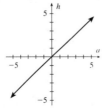

69. Answers vary. **71.** Answers vary.

73. $9x^2+25y^2=225$ **75.** $9y^2-16x^2=144$

1.4 Problem Set, Pages 36–37

1. a. $(3,7)$ **b.** $(-4,-1)$
c. $[-2,6]$ **d.** $(-3,0]$
3. a. $(-\infty,-3]$ **b.** $[-2,\infty)$
c. $(-\infty,0)$ **d.** $(2,\infty)$
5. a. **b.**
c. **d.**
7. a. **b.**
c. **d.**
9. a. $-4\le x\le 2$ **b.** $-1\le x\le 2$
c. $0<x<8$ **d.** $-5<x\le 3$
11. a. $x<2$ **b.** $x>6$ **c.** $x>-1$ **d.** $x\le 3$
13. False; $5>x>1$ is equivalent to $1<x<5$
15. Not true if $y<0$ **17.** No; x is more than 4 units from 3.
19. $\{5, -5\}$ **21.** $\{\ \}$ or $\varnothing$ **23.** $\{7, -1\}$ **25.** $\{24, -6\}$
27. $\{\ \}$ or $\varnothing$ **29.** $\{\frac{2}{5}, -2\}$ **31.** $[7,\infty)$ **33.** $(-\infty,-41]$
35. $(-\infty,2)$ **37.** $(-\frac{8}{5},0)$ **39.** $(1,3]$ **41.** $[-4,2)$
43. $(-\frac{15}{2},4]$ **45.** $[-2,8]$ **47.** $(2.999, 3.001)$
49. $\{\ \}$ or $\varnothing$ **51.** $(\frac{1}{3},\frac{7}{3})$ **53.** $(-\infty,-6)\cup(-1,\infty)$
55. Under 200 mi per day **57.** $68°<F<86°$
59. Values are in thousands of dollars. **a.** $s<10$ or $s>20$
b. $|s-15|>5$
61. $(-\infty,-3)\cup(9,\infty)$ **63.** $[-2,-1)\cup(-1,0]$
65. $\{1,-\frac{3}{2}\}$ **67.** Answers vary. **69.** Answers vary.

1.5 Problem Set, Pages 39–40

1. $6i$ **3.** $7i$ **5.** $2i\sqrt{5}$ **7.** $8+7i$ **9.** $-5i$
11. $1-6i$ **13.** $3+3i$ **15.** 10 **17.** $2+5i$
19. $7+i$ **21.** 29 **23.** 34 **25.** 1 **27.** $-i$ **29.** -1
31. 1 **33.** $-i$ **35.** -1 **37.** $32-24i$ **39.** $-9+40i$
41. $-198-10i$ **43.** $-\frac{3}{2}+\frac{3}{2}i$ **45.** $1+i$ **47.** $-2i$
49. $-\frac{1}{5}-\frac{3}{5}i$ **51.** $\frac{1}{5}-\frac{2}{5}i$ **53.** $1-i$ **55.** $-1+i$
57. $-\frac{45}{53}+\frac{28}{53}i$ **59.** $.3131+2.2281i$
61. $(-1-\sqrt{3})+2i$ **63.** $\dfrac{5+2\sqrt{3}}{13}+\dfrac{1+3\sqrt{3}}{13}i$
65. $(13+8\sqrt{3})+(12+4\sqrt{3})i$ **67.** Answers vary.

1.6 Problem Set, Page 46

1. $\{3,-5\}$ **3.** $\{2,-9\}$ **5.** $\{0,\frac{5}{6}\}$
7. $\{\frac{4}{5},-\frac{1}{2}\}$ **9.** $\{4,-\frac{2}{9}\}$ **11.** $\{0,-1\}$
13. $\{1,-5\}$ **15.** $\{2,-4\}$ **17.** $\{-3,-4\}$
19. $\{5\pm 3\sqrt{3}\}$ **21.** $\left\{\dfrac{3\pm\sqrt{13}}{2}\right\}$ **23.** $\{\frac{2}{3},-\frac{1}{2}\}$

25. $\{1, -6\}$ **27.** $\{5\}$ **29.** $\{\frac{1}{4}, -\frac{2}{3}\}$

31. $\emptyset$; complex: $\{\frac{2}{5} \pm \frac{1}{5}i\}$ **33.** $\left\{\pm\dfrac{\sqrt{5}}{2}\right\}$ **35.** $\{0, \frac{7}{3}\}$

37. $\{2, -\frac{1}{3}\}$ **39.** $\left\{\dfrac{-5 \pm \sqrt{73}}{8}\right\}$ **41.** $\left\{\dfrac{-3 \pm \sqrt{17}}{4}\right\}$

43. $\left\{\dfrac{-3 \pm \sqrt{9 + 16\sqrt{5}}}{8}\right\}$ **45.** $\left\{\dfrac{2 \pm \sqrt{4 + 3\sqrt{5}}}{3}\right\}$

47. $\{\pm\sqrt{\sqrt{34} - 3}\}$; complex: $\{\pm\sqrt{-3 \pm \sqrt{34}}\}$

49. $\left\{\dfrac{-1 \pm \sqrt{1 + 8w}}{4}\right\}$ **51.** $\left\{\dfrac{-1 \pm \sqrt{-3y - 5}}{3}\right\}$

53. $\left\{\dfrac{1 \pm |t|}{2}\right\}$ **55.** $\left\{\dfrac{-3 \pm \sqrt{8y - 23}}{4}\right\}$ **57.** $\left\{2, \dfrac{3t + 2}{4}\right\}$

59. $\{-2 \pm \sqrt{9 + (y + 1)^2}\}$ **61.** Answers vary.

1.7   **Problem Set, Page 51**

1. False; use a number line for correct solution.
3. False; do not know that $x \neq 0$
5. True only if $ax^2 + bx + c = 0$ **7.** $(-3, 0)$
9. $(-\infty, 2) \cup [6, \infty)$ **11.** $(-7, 8)$ **13.** $[-2, \frac{1}{2}]$
15. $(-\infty, -\frac{2}{3}) \cup (3, \infty)$ **17.** $(-\infty, -2] \cup [8, \infty)$
19. $(-\infty, \frac{1}{3}) \cup (4, \infty)$ **21.** $(-\infty, -4] \cup [0, 3]$
23. $[-3, 2] \cup [4, \infty)$ **25.** $(-\infty, -\frac{5}{2}) \cup (-1, \frac{7}{3})$
27. $(-2, 0)$ **29.** $(-\infty, 0) \cup (8, \infty)$ **31.** $(-5, 2]$
33. $(-\infty, 0) \cup (\frac{1}{2}, 5)$ **35.** $(-\infty, -2) \cup (0, 3)$
37. $(-\infty, -3] \cup [3, \infty)$ **39.** $(-\infty, \infty)$
41. $(-\infty, -2) \cup (3, \infty)$ **43.** $[2, 3]$
45. $[-5, 1]$ **47.** $(1 - \sqrt{3}, 1 + \sqrt{3})$ **49.** $(-\infty, \infty)$
51. $\left(-\infty, \dfrac{-3 - \sqrt{37}}{2}\right] \cup \left[\dfrac{-3 + \sqrt{37}}{2}, \infty\right)$
53. $[-5, -3) \cup [0, 3] \cup [4, \infty)$ **55.** $(-3, 2) \cup [12, \infty)$
57. $[-13, -\frac{1}{2}) \cup [0, \frac{1}{3})$ **59.** $(-\infty, 2) \cup (3, 4]$

1.8   **Problem Set, Pages 59–61**

1. 62, 63   **3.** 31   **5.** 15   **7.** 5 ft $\times$ 11 ft
9. Height = 3 cm; base = 4 cm
11. True for any such rectangle   **13.** $600   **15.** $2,500
17. $600 at $9\frac{1}{2}\%$; $900 at 14%   **19.** $8,500   **21.** 288 mi
23. 4 hr   **25.** $2\frac{1}{2}$ mi   **27.** 6 mph; 8 mph   **29.** 24 mph
31. 50 gal milk with 100 gal cream
33. 2.5 liters of 72% with 2 liters of 45%
35. 40 cc of water with 60 cc of 35%
37. 12 oz of water
39. Add 20 oz of 15% alcohol.
41. All numbers except those between -17 and 20
43. All numbers except those between -3 and 0, inclusive
45. The width must be greater than 3.
47. Produce either 2 or 10 radios.   **49.** 14 sec ($1 \leq t \leq 15$)
51. $2\frac{2}{3}$ qt drained and replaced   **53.** 2,240 kwh
55. 2 hr; 7 mi and 8 mi   **57.** 1 hr
59. **a.** 15 in.   **b.** 21 in.   **c.** 29 in.   **d.** 19 in.

Chapter 1 Summary, Pages 62–63

1. $\frac{14}{7}$: $\mathbb{N}, \mathbb{W}, \mathbb{Z}, \mathbb{Q}, \mathbb{R}$; $\sqrt{144}$: $\mathbb{N}, \mathbb{W}, \mathbb{Z}, \mathbb{Q}, \mathbb{R}$; $6.\overline{2}$: $\mathbb{Q}, \mathbb{R}$; π: $\mathbb{Q}', \mathbb{R}$
3. $3.\overline{1}$: $\mathbb{Q}, \mathbb{R}$; $\frac{5\pi}{6}$: $\mathbb{Q}', \mathbb{R}$; $\frac{22}{7}$: $\mathbb{Q}, \mathbb{R}$; $\sqrt{10}$: $\mathbb{Q}', \mathbb{R}$
5. ⟨number line⟩   **7.** ⟨number line⟩
9. $=$ **11.** $<$ **13.** $\sqrt{11}$ **15.** $\sqrt{11} - 3$ **17.** 8
19. $\pi + 2$ **21.** $a(b + c)$ **23.** $a(b + c) = 5$
25. $(b + c)a$ **27.** $ab + ac$
29. $a_n x^n + a_{n-1} x^{n-1} + \cdots + a_2 x^2 + a_1 x + a_0, a_n \neq 0$ **31.** b^{m+n}
33. $9x^4 + 15x^3 - 101x^2 - 71x - 12$
35. $-3x^3 - 4x^2 + 38x + 13$
37. $\dfrac{1}{y^2}(2x - 2xy - y^2)(2x + 2xy + y^2)$
39. $(2x - 1)(x + 1)(4x^2 + 2x + 1)(x^2 - x + 1)$
41, 43.

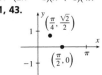

45. $\sqrt{(\gamma - \alpha)^2 + (\delta - \beta)^2}$ **47.** $-5x$ **49.** $\left(\dfrac{\alpha + \gamma}{2}, \dfrac{\beta + \delta}{2}\right)$
51. $\left(3x, \dfrac{5x}{2}\right)$ **53.** a set of ordered pairs   **55.** satisfies
57. ⟨graph⟩   **59.** ⟨graph⟩
61. ⟨graph⟩   **63.** ⟨graph⟩
65. $(-4, 2)$ **67.** $(-3, \infty)$ **69.** ⟨number line⟩
71. ⟨number line⟩
73. $-8 \leq x < -5$ **75.** $x < 3$ **77.** $(-\infty, 4]$
79. $(-8, -4]$ **81.** $\{-8, 8\}$ **83.** $\{-\frac{11}{2}, \frac{5}{2}\}$ **85.** $(-1, 9)$
87. $[-10, 15]$ **89.** i **91.** 29   **93.** $\{-3, 4\}$
95. $\{3, -\frac{5}{2}\}$ **97.** $\{-3, 5\}$ **99.** $\{-4, -5\}$
101. $\left\{\dfrac{5 \pm \sqrt{13}}{2}\right\}$ **103.** $\{-1 \pm \sqrt{6}\}$ **105.** $(-\frac{1}{3}, 1)$
107. $(-\infty, \infty)$ **109.** $(-1, 0) \cup (3, \infty)$
111. $(-\infty, -1) \cup (0, 2) \cup (2, \infty)$ **113.** 12 ft by 20 ft
115. 5 gal of milk with 10% butterfat mixed with 10 gal of milk with 70% butterfat

2.1 Problem Set, Pages 74–76

1. One-to-one function   **3.** Not a function
5. One-to-one function   **7.** One-to-one function
9. Not a function   **11.** Function
13. This is not an equation; cannot solve
15. No; it is equal to $\dfrac{(x + h)^2 - x^2}{h}$.

17. a. 1   **b.** 5   **c.** -5 **d.** $2\sqrt{5} + 1$ **e.** $2\pi + 1$
19. a. $2w + 1$ **b.** $2w^2 - 1$ **c.** $2t^2 - 1$ **d.** $2v^2 - 1$
    e. $2m + 1$

21. a. $3 + 2\sqrt{2}$ **b.** $5 + 4\sqrt{2}$ **c.** $2t^2 + 12t + 17$
    d. $2t^2 + 4t + 3$ **e.** $2m^2 - 4m + 1$
23. 2   **25.** 2   **27.** $4t + 2h$
29. a. $w^2 - 1$ **b.** $h^2 - 1$ **c.** $w^2 + 2wh + h^2 - 1$
    d. $w^2 + h^2 - 2$
31. a. $x^4 - 1$ **b.** $x - 1$ **c.** $x^2 + 2xh + h^2 - 1$ **d.** $x^2 - 1$
33. 2   **35.** 9   **37.** $\dfrac{|x + h| - |x|}{h}$ **39.** $10x + 5h$

41. $4x + 3 + 2h$ **43.** $\dfrac{-2}{(x + h - 1)(x - 1)}$

45. a. \$.92   **b.** \$.45   **47.** \$1.15
49. a. \$.51   **b.** $e(1984) - e(1944)$
51. a. \$.02225
    b. Average annual change of price of gasoline from 1944 to 1984

53. a. \$.018; $\dfrac{s(1954) - s(1944)}{10}$ **b.** \$.0125; $\dfrac{s(1964) - s(1944)}{20}$
    c. \$.058; $\dfrac{s(1974) - s(1944)}{30}$ **d.** \$.02875; $\dfrac{s(1984) - s(1944)}{40}$
    e. $\dfrac{s(1944 + h) - s(1944)}{h}$

55. a. $-14,800$
    b. The average annual change in the number of marriages from 1982 to the year h years after 1982
57. $C(50) = 650$; $C(100) = 700$ **59.** \$1.98
61. a. 128   **b.** 96   **c.** 80   **d.** $64 + 16h$ **e.** $32x + 16h$
63. Answers vary. Let $f(x) = x^2 + 2$.

2.2 Problem Set, Pages 86–87

1. $(-\infty, \infty)$ **3.** $(-\infty, -2) \cup (-2, \infty)$ **5.** $[-\frac{1}{2}, \infty)$
7. $[-2, 1]$ **9.** Not equal   **11.** Not equal
13. Not equal   **15.** Even   **17.** Neither   **19.** Even
21. $R: [x_0, G(x_0)]$; $S: [x_0 + h, G(x_0 + h)]$
23. $D: [-4, 7]$; $R: [-4, 5]$; intercepts: $(0, 5)$, $(3, 0)$, $(6, 0)$; increasing: $(3, 5)$; decreasing: $(1, 3)$; constant: $(-4, 1)$; turning points: $(3, 0)$, $(5, 5)$
25. $D: [-5, 3) \cup (3, \infty)$; $R: [-3, 6) \cup (6, \infty)$; intercepts: $(-\frac{7}{4}, 0)$, $(0, -3)$, $(\frac{7}{4}, 0)$; increasing: $(0, 6) \cup (6, \infty)$; decreasing: $(-2, 0)$; constant: $(-5, -2)$; turning point: $(0, -3)$
27. $D: [-6, 6]$; $R: [-5, 5]$; intercepts: $(-3, 0)$, $(0, 5)$, $(3, 0)$; increasing: $(-6, 0)$; decreasing: $(0, 6)$; turning point: $(0, 5)$
29. $(0, 5)$, $(-5, 0)$ **31.** $(0, -4)$, $(2, 0)$, $(-2, 0)$ **33.** $(0, 0)$
35. $(0, 3)$ **37.** $(0, \frac{9}{2})$, $(3, 0)$, $(-3, 0)$ **39.** $(0, -2)$, $(2, 0)$
41. $(0, 1)$ **43.** $(0, \sqrt{3})$, $(-3, 0)$ **45.** $(\frac{5}{2}, 0)$, $(-\frac{5}{2}, 0)$

47. None   **49.** $(0, 2)$, $(0, -2)$, $(2, 0)$, $(-2, 0)$
Also show excluded regions for Problems 51–62.
51. $D: [0, \infty)$; $R: [0, \infty)$ **53.** $D: [2, \infty)$; $R: [0, \infty)$
55. $D: [\frac{1}{2}, \infty)$; $R: [0, \infty)$ **57.** $D: [0, \infty)$; $R: [0, \infty)$
59. $D: [0, \infty)$; $R: [0, \infty)$ **61.** $D: [0, \infty)$; $R: (-\infty, \infty)$
63. $D: (-\infty, -3) \cup [3, \infty)$; $R: (-\infty, \infty)$
65. $D: (-\infty, -2] \cup [2, \infty)$; $R: [0, \infty)$
67. $D: (-\infty, -4] \cup [3, \infty)$; $R: [0, \infty)$ **69.** $(-\frac{1}{5}, 3)$, $(3, 3)$
71. $(-4, -16)$ **73.** None
75. Answers vary (y has exactly 5 elements)   **77.** Answers vary.

2.3 Problem Set, Pages 95–96

1. a. $(6, 3)$ **b.** $(-3, 5)$ **3. a.** $(0, \sqrt{2})$ **b.** $(0, 6)$
5. a. $(0, 0)$ **b.** $(-\sqrt{2}, 0)$ **7.** $y = x^2 - 3$
9. $y + 4 = (x + 4)^2$ **11.** $y - 4 = |x|$ **13.** $y = \sqrt{x} + 3$
15.

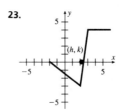

17.

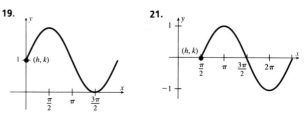

19.

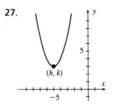

21.

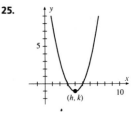

23.

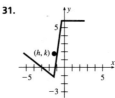

25.

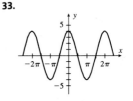

27.

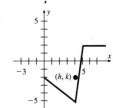

29.

31.

33.

35.

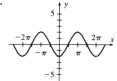

37.

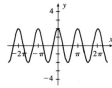

39.

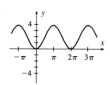

41.

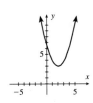

43.

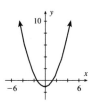

45.

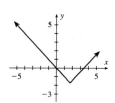

47.

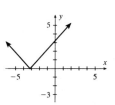

49.

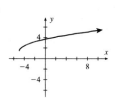

51.

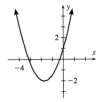

53.

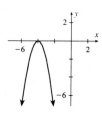

55.

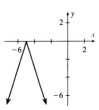

57.

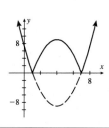

59.

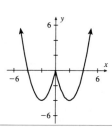

61.

63.

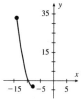

65.

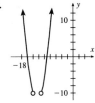

67.

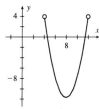

2.4   Problem Set, Pages 105–107

1. 33   **3.** 5   **5.** 7   **7.** $-\frac{35}{2}$ **9.** $\dfrac{1}{(99+1)^2} = .0001$

11. -6 **13.** 980 (factor first)   **15.** -8

17. 7   **19.** $\frac{1}{3}$

21. $(f+g)(x) = x^2 + 2x - 2$; $(f-g)(x) = -x^2 + 2x - 4$

23. $(f+g)(x) = \dfrac{x^3 - x^2 - x + 1}{x-2}$;

$(f-g)(x) = \dfrac{-x^3 + 5x^2 - x - 7}{x-2}$

25. $(fg)(x) = 2x^3 - 3x^2 + 2x - 3$; $(f/g)(x) = \dfrac{2x-3}{x^2+1}$

27. $(fg)(x) = (2x-3)(x+1)^2$; $(f/g)(x) = \dfrac{2x-3}{(x-2)^2}$

29. $(f \circ g)(x) = 2x^2 - 1$; $(g \circ f)(x) = 4x^2 - 12x + 10$

31. $(f \circ g)(x) = 2x^2 - 2x - 3$, $x \neq 2$;

$(g \circ f)(x) = 4x^2 + 2x - 2$, $x \neq 2$

33.   **35.**   **37.**

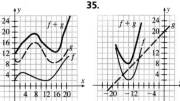

39. $f[u(x)] = \sqrt[3]{2x^2 + 1}$

41. $f[u(x)] = 3(x^3)^2 + 5(x^3) + 1 = 3x^6 + 5x^3 + 1$

43. $(f \circ g)(x) = 4x^2 - 4x + 1$; $(g \circ h)(x) = 6x + 3$

45. $36x^2 + 36x + 9$ **47.** $9x^4 + 6x^2 + 1$

49. $(f \circ g)(x) = \sqrt{x^2 - 2}$; $(g \circ h)(x) = (x+2)^2 - 2$

51. $\sqrt{x^2 + 4x + 2}$ **53.** x **55.** $u(x) = 2x^2 - 1$; $g(x) = x^4$

57. $u(x) = 5x - 1$; $g(x) = \sqrt{x}$

59. $u(x) = x^2 - 1$; $g(x) = x^3 + \sqrt{x} + 5$

61. a. $\frac{16\pi}{3}$ **b.** $\frac{2\pi}{3}t^3$ **c.** $(0, 3]$

63. Answers vary. $f(x) = 2x + 1$

65. a. $\dfrac{2x+1}{x+1}$ **b.** $\dfrac{3x+2}{2x+1}$ **c.** $\dfrac{5x+3}{3x+2}$ **d.** Answers vary.

67. It is close to 4.   **69.** It is close to k^2.

2.5 Problem Set, Pages 113–114

1. Inverses　**3.** Inverses　**5.** Not inverses
7. Not inverses　　**9.** {(5, 4), (3, 6), (1, 7), (4, 2)}

11. $x - 3$　**13.** $\dfrac{x}{5}$　**15.** Does not exist　**17.** x

19. Does not exist　**21.** $\dfrac{1 + 2x}{x},\ x \neq 0$

23. $\dfrac{-3(x + 2)}{3x - 2},\ x \neq \frac{2}{3}$

25. a. -4　**b.** -2　**c.** 3　**d.** 5　**e.** 8
27. a. -5　**b.** 5　**c.** 3　**d.** -4.5　**e.** 6.5
29. a. 9　**b.** 3.5　**c.** 3.5　**d.** 4　**e.** 13

31.
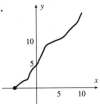

33. Inverses　**35.** Not inverses　**37.** Inverses

39.
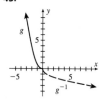
41.

43.
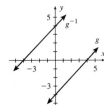
45.

47.
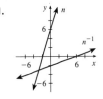

49.
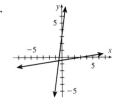
51.

53. $-\sqrt{x}$ on $[0, \infty)$　**55.** $\sqrt{x - 1}$ on $[1, \infty)$

57. $-\frac{1}{2}\sqrt{2x}$ on $[2, 200]$　**59.** $\dfrac{1 + x}{2x}$ on $(0, \infty)$

61. $x - 1$ on $[0, \infty)$

Chapter 2 Summary, Pages 114–115

1. One-to-one function　**3.** Not a function
5. a. 11　**b.** -11　**7. a.** $3w - 1$　**b.** $5 - w^2 - 2wh - h^2$

9. $-2x - h$　**11.** 0　**13.** $D: (-\infty, \infty);\ R = [5, \infty)$
15. $D: [-5, 1);\ R = [0, \infty)$　**17.** $(0, -20), (2, 0), (-\frac{5}{3}, 0), (-2, 0)$
19. $(\frac{1}{2}\sqrt{2}, 0), (-\frac{1}{2}\sqrt{2}, 0)$　**21.** Not equal
23. Not equal　**25.** Neither　**27.** Neither　**29.** $(-\pi, 6)$
31. $(4, 0)$　**33.** $y - 3 = 9(x + \sqrt{2})^2$　**35.** $y = -2(x - \pi)^2$
37. $y' = 3x'^2$　**39.** $y' = 5x'^2$

41.
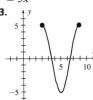
43.

45.

47.
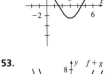
49. $x^2 + 5x - 2$
51. $5x^3 - 2x^2$

53.

55.
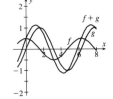

57. $14 - 3x^2$　**59.** 17　**61.** No　**63.** No　**65.** $\dfrac{x + 1}{3}$
67. $2x - 10$

3.1 Problem Set, Pages 123–124

1. a. 3　**b.** -9　**c.** -6　**d.** $3x^2 - 3x + 5$
3. a. -2　**b.** -4　**c.** -10　**d.** 16　**e.** 12
　f. -12　**g.** $2x^2 - 8x + 6$
5. a. -4　**b.** 1　**c.** 3　**d.** 8　**e.** -4　**f.** -20
　g. -4　**h.** $x^2 + 3x - 4 + \dfrac{-4}{x + 4}$
7. a. 1　**b.** 1　**c.** 2　**d.** 0　**e.** -5　**f.** -2
　g. 2　**h.** 0　**i.** $x^3 + 3x^2 + 3x - 2$
9. a. -2　**b.** 1　**c.** 0　**d.** 0　**e.** 0　**f.** -8
　g. 0　**h.** 16　**i.** 16　**j.** -32　**k.** -32　**l.** -64
　m. $x^4 - 2x^3 + 4x^2 - 8x + 16 + \dfrac{-64}{x + 2}$
11. $3x^2 + 7x + 25$　**13.** $x^3 - 7x^2 + 8x - 8$
15. $2x^3 - 6x^2 + 3x - 1$　**17.** $4x^4 + x^3 - 4x^2 - 4x - 4$
19. $3x^2 + 4x + 12$　**21.** $x^2 + x - 6$　**23.** $x^2 + x - 12$
25. $4x^3 + 8x^2 - 7x - 7$　**27.** $x^3 - 2x^2 + 3x - 4$
29. $3x^3 - 2x^2 - 5$　**31.** $Q(x) = 2x^3 - 6x^2 + 2x - 2;\ R(x) = 0$
33. $Q(x) = x^3 - x^2 + 1;\ R(x) = 1 - x$
35. $Q(x) = 3x^2 - 7x + 5;\ R(x) = 0$　**37.** $3x^2 + 2x + 1$
39. $x^3 + x^2 + 2$　**41.** $5x^3 - 5x^2 - 5x + 3$
43. $x^4 - 5x^3 + 10x^2 - 18x + 36$　**45.** $4x^2 - 5x + 7$
47. $5x^4 - 10x^3 + 20x^2 - 40x + 78$　**49.** $x^4 + 3x^2 - 7x$
51. $x^2 + 2x - 5$　**53.** $2x - 1$　**55.** $2x^2 + 3x - 4$

57. Factoring the denominator to $(x + 5)(x - 1)$ helps to find the quotient: $x^2 + 3x - 2$

59. $K = 7$

The lines in Problems 1–21 should also be drawn.

1. 1   **3.** $\frac{13}{5}$ **5.** $-\frac{7}{3}$ **7.** Undefined   **9.** 0

11. $m = -4; b = -1$ **13.** $m = \frac{1}{5}; b = -\frac{6}{5}$

15. $m = 300; b = 0$ **17.** $m = 0; b = -2$ **19.** $m = \frac{1}{3}; b = \frac{2}{3}$

21. $m = \frac{2}{5}; b = -240$

23. 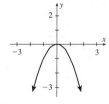   **25.**   **27.**

29. Yes   **31.** No   **33.** Yes   **35.** Yes

37. $5x - y + 6 = 0$ **39.** $y = 0$ **41.** $3x - y - 3 = 0$

43. $x - 2y + 3 = 0$ **45.** $x - 2y + 2 = 0$

47. $2x - y - 4 = 0$ **49.** $2x + 3y - 16 = 0$

51. $2x + y + 4 = 0$

53. a. $A(x_0, f(x_0))$ and $B(x_0 + \Delta x, f(x_0 + \Delta x))$

   b. $m = \dfrac{f(x_0 + \Delta x) - f(x_0)}{\Delta x}$

55. a. $A(x_0, H(x_0))$ and $B(x_0 + h, H(x_0 + h))$

   b. $m = \dfrac{H(x_0 + h) - H(x_0)}{h}$

57. $(1, 10)$, $(2, 20)$; $10x - y = 0$

59. $(0, 14.2)$, $(10, 16.8)$; $13x - 50y + 710 = 0$; 19.7 million

61. $(50, 60)$, $(260, 60)$; $y - 60 = 0$; $60   **63.** Answers vary.

65. Answers vary.   **67.** Answers vary.

1. 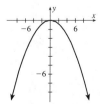   **3.**

5.   **7.**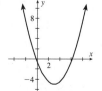

9.   **11.**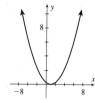

13. 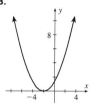   **15.** 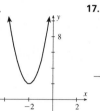   **17.**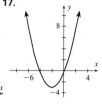

19. Simplified, $f(x) = 5x + 2$ is linear.

21. $y = 5(x^2 + 2x)$: Add $5 \cdot 1^2 = 5$ to both sides.

23. The parabola opens up: $y = 3$ is a minimum value.

25.   **27.**

29. 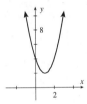   **31.**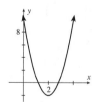

33.   **35.**

37.   **39.**

41. 3   **43.** -15 **45.** $\frac{2}{3}$

47. a. 375   **b.** $250,000 loss   **c.** $1,156,250

49. $650   **51. a.** 16 ft   **b.** 10 ft   **53.** 25 ft by 25 ft; 625 ft²

55. 3,456 ft (rounded)

57.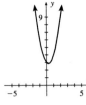

59.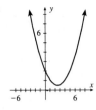

49.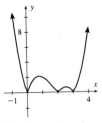

51.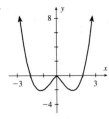

61.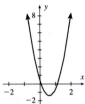

53. Must be continuous
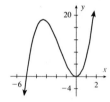

55. Must be continuous
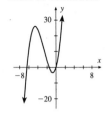

3.4 **Problem Set, Pages 147–150**

1. Quadratic **3.** Cubic **5.** Quartic **7.** Cubic
9. Cubic **11.** Quartic **13.** I **15.** G **17.** B
19. H **21.** L **23.** M **25.** Q **27.** T **29.** E
31. U

33.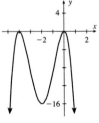

35.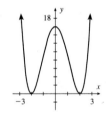

57. Must be smooth (no corners)
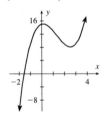

59. $a < 0$, $n = 3$, direction is
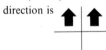

61. $a > 0$, $n = 4$, direction is

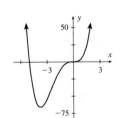

63. $a > 0$, $n = 4$, direction is
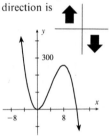

37.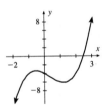

39.

41.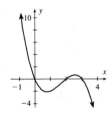

43.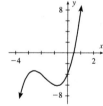

45.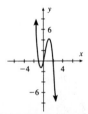

47.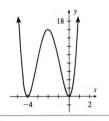

3.5 **Problem Set, Pages 153–155**

1. a. -3 **b.** -19 **c.** -4 **d.** 842 **e.** -448
3. a. -824 **b.** -758 **c.** -812 **d.** -890 **e.** $-1,022$
5. a. -10 **b.** -8 **c.** 68 **d.** $2,140$ **e.** 380
7. a. -3 **b.** 8 **c.** 0 **d.** $-\frac{5}{2}$ **e.** 840
9. a. 0 **b.** 42 **c.** 0 **d.** 0 **e.** 0
11. a. 0 **b.** 0 **c.** 0 **d.** -54 **e.** -10

13.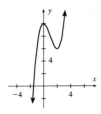

15.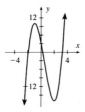

37.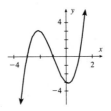

39.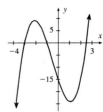

17.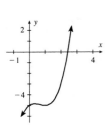

19. *n* odd, *a* > 0, not smooth

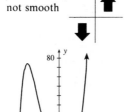

41.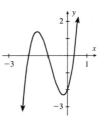

43.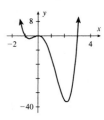

45.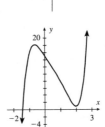

47.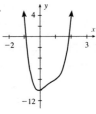

21. Not continuous

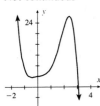

23. Not smooth, not continuous

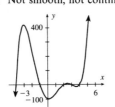

49.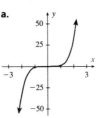

25.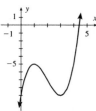

27.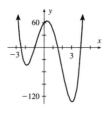

51. a. 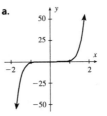 **b.**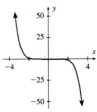

29.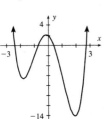

31.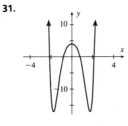

53. a. **b.**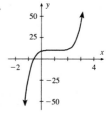

33.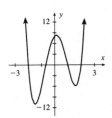

35.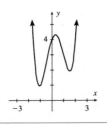

55.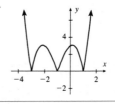

57.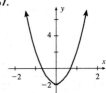

59.

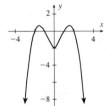

3.6 Problem Set, Pages 163–164

1. 2, -3 (multiplicity 2)
3. 0 (multiplicity 3), $\frac{3}{2}$ (multiplicity 2)
5. -2, -1 (multiplicity 2), 1 (multiplicity 2)
7. -5 (multiplicity 2), 3 (multiplicity 2)
9. 0 (multiplicity 2), 4 (multiplicity 2)
11. 0 (multiplicity 2), 3 (multiplicity 2), -3 (multiplicity 2)
13. Positive: 4, 2, or 0; negative: 0
15. Positive: 2 or 0; negative: 1   **17.** Positive: 1; negative: 2 or 0
19. Positive: 1; negative: 2 or 0
21. Positive: 2 or 0; negative: 2 or 0
23. Positive: 2 or 0; negative: 1   **25.** $\pm 1, \pm 3, \pm 5, \pm 15$
27. $\pm 1, \pm 2, \pm 3, \pm 4, \pm 6, \pm 12, \pm\frac{1}{2}, \pm\frac{3}{2}$
29. $\pm 1, \pm 2, \pm 3, \pm 4, \pm 6, \pm 12$
31. $\pm 1, \pm 3, \pm 5, \pm 15, \pm\frac{1}{2}, \pm\frac{3}{2}, \pm\frac{5}{2}, \pm\frac{15}{2}$
33. $\pm 1, \pm 2, \pm 3, \pm 6, \pm 9, \pm 18$ **35.** $\pm 1, \pm\frac{1}{2}, \pm\frac{1}{3}, \pm\frac{1}{6}$
37. $\{1, 2, -2\}$ **39.** $\{2, 3, -3\}$ **41.** $\{2, -2, -3\}$
43. $\{-3, -\frac{1}{2}, 5\}$ **45.** $\{1, -1, -6, 3\}$ **47.** $\{-3, -5, -7\}$
49. $\{\frac{1}{2}, -\frac{5}{2}, \frac{7}{2}\}$ **51.** $\{-1, 3, -3, -4\}$ **53.** $\{0, 1, -1, -3\}$
55. $\{-2, 2\}$
57. $x - 1 = x^3$; no rational roots. By Descartes' Rule of Signs there are 2 or 0 positive roots and 1 negative root. Use synthetic division to show that there exists one real root between 0 and -2. Graphical solution:

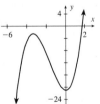

59. $x^2(x - 2) = 384$; side length is 8 cm
61. $x^3 + 5x^2 - 21 = 0$; one real root ≈ 1.8 (1.76224 ...)
Graphical solution:

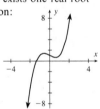

63. $\{-.2, 2.6, .4, -4.8\}$
65. a. Answers vary (3, for example).   **b.** Answers vary.
  c. Varies   **d.** Answers vary.   **e.** Answers vary.

3.7 Problem Set, Pages 171–172

1. $-24 + 4i$ **3.** $24 - 14\sqrt{2}$ **5.** $12 + 16i$ **7.** $-48i$
9. 0   **11.** 0   **13.** 5   **15.** 0   **17.** 0   **19.** -3
21. 0   **23.** $30 + 6i$ **25.** Not a root   **27.** Yes; $1 + 2i$

---

29. Not a root   **31.** $\{2, -1 \pm i\sqrt{3}\}$

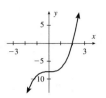

33. $\{5, -\frac{5}{2} \pm \frac{5}{2}\sqrt{3}i\}$

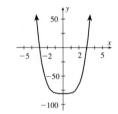

35. $\{\pm 3, \pm 3i\}$

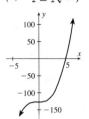

37. $\{\pm 2i, \pm i\sqrt{5}\}$

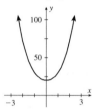

39. $\{\pm 3i, \pm 2i\}$

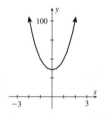

41. $\{3 \pm i, 4 \pm i\}$

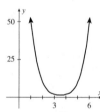

43. $\{2 \pm \sqrt{5}, \frac{3}{2} \pm \frac{1}{2}\sqrt{11}i\}$

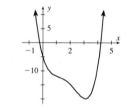

45. $\{-\frac{1}{2}, 1 \pm i\sqrt{2}\}$

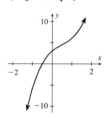

47. $\{\pm 1, 1 \pm 2i\}$

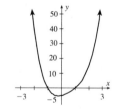

49. $\left\{5, -\frac{1}{3}, \dfrac{3 \pm \sqrt{7}}{3}\right\}$

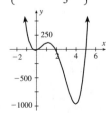

51. Answers vary.
$y = (x - 3)^3(x + 2)^2$

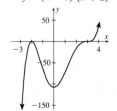

53. Answers vary.
$y = (x - 1)(x - 2)^2(x + 1)(x + 2)$

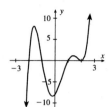

55. Answers vary.
$y = (x - 1)^2(x - 2)^2(x + 1)^2$

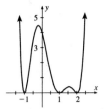

41. Polynomial functions are continuous; this is not.

57. $\{\pm 3i, \pm 2i\}$ **59.** $\{2 \pm i, \pm \sqrt{2}\}$ **61.** $\{-1, \frac{3}{2}, \pm \sqrt{5}\}$
63. $\{\frac{3}{2}, -3 \pm \sqrt{2}, \pm i\sqrt{2}\}$ **65.** $\{\pm \sqrt{2}, -\frac{1}{2} \pm \frac{1}{2}\sqrt{3}i\}$
67. Answers vary.

Chapter 3 Summary, Pages 172–175

1. $x^3 - 2x^2 + 6x - 13$ **3.** $3x^2 + 8x + 4$
5.

7.

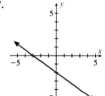

9. $-\frac{5}{4}$ **11.** $\frac{7}{5}$ **13.**

15.

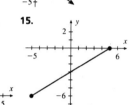

17. Not a right triangle **19.** $-\frac{2}{3}$
21. $Ax + By + C = 0$, where (x, y) is any point on the line and A, B, C are constants (not all 0)
23. $y = mx + b$, where m is the slope and b is the y-intercept
25. $2x - 3y - 27 = 0$ **27.** $5x + 8y + 23 = 0$
29.

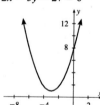

31.

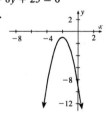

33.

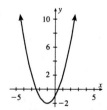

35.

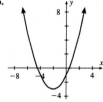

37. Maximum of 250 at $x = -6$
39. Maximum of -140 at $x = -5$

43. Leading coefficient is positive; even power graph has the pattern

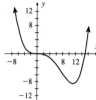

45.

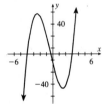

47.

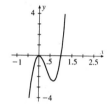

49. -42 **51.** 0
53.

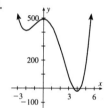

55.

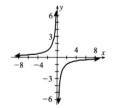

57. 1 (multiplicity 3), -2 (multiplicity 3)
59. 1 (multiplicity 2), -1 (multiplicity 5)
61. $\pm 1, \pm 2, \pm 3, \pm 4, \pm 6, \pm 12, \pm \frac{1}{3}, \pm \frac{2}{3}, \pm \frac{4}{3}$
63. $\pm 1, \pm 2, \pm \frac{1}{2}, \pm \frac{1}{3}, \pm \frac{1}{6}, \pm \frac{2}{3}$
65. Positive: 1; negative: 2 or 0
67. Positive: 3 or 1; negative: 1 **69.** $\{-\frac{1}{3}, -4, 3\}$
71. $\{1, -\frac{5}{6}\}$ **73.** 0 **75.** 0 **77.** $\{2, -3, 2 \pm \sqrt{3}\}$
79. $\{0, -1, \pm \sqrt{14}\}$ **81.** $\{1 \pm i, \pm i\sqrt{3}\}$
83. $\{1 \pm i\sqrt{3}, \pm \sqrt{5}\}$

4.1 **Problem Set, Pages 181–182**

1.

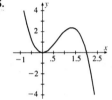

3.

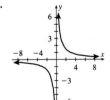

5.

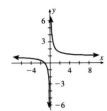

7.

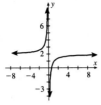

29.

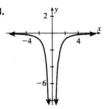

31.

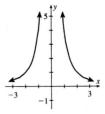

9.

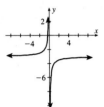

11.

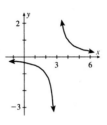

33.

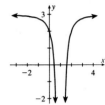

35.

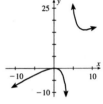

13.

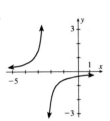

15.

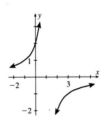

37.

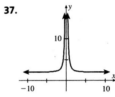

39.

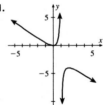

17.

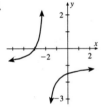

19.

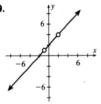

41.

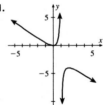

43.

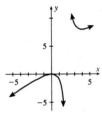

21.

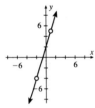

23.

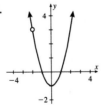

45.

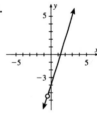

47.

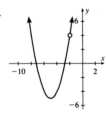

25.

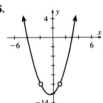

27.

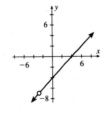

49.

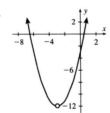

51.

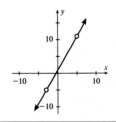

53.

55.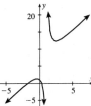

57.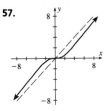

59.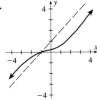

61.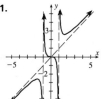

63. Answers vary.

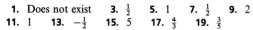

4.2 Problem Set, Pages 188–189

1. Does not exist **3.** $\frac{1}{2}$ **5.** 1 **7.** $\frac{1}{2}$ **9.** 2
11. 1 **13.** $-\frac{1}{2}$ **15.** 5 **17.** $\frac{4}{3}$ **19.** $\frac{3}{5}$
21. $x = 0, y = 0$ **23.** $x = 0, y = 1$ **25.** $x = 0, y = 2$
27. $x = -3, y = 0$ **29.** $x = 4, y = x + 4$
31. $x = 1, y = -x - 1$ **33.** None **35.** None
37. None **39.** $x = 2, x = -4, y = 1$

41.

43.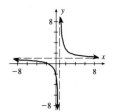

45.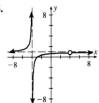

47.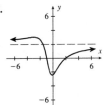

49.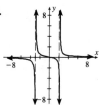

51.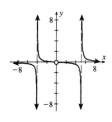

53.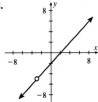

55.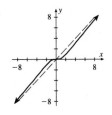

4.3 Problem Set, Page 196

1.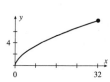

3.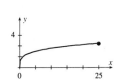

5.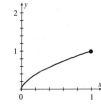

7.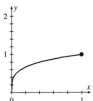

9.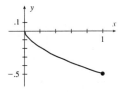

11.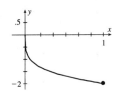

13.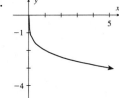

15.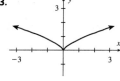

17.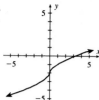

19.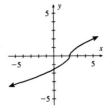

41.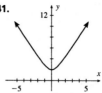

43.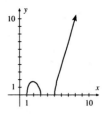

21.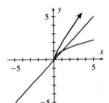

23.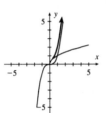

45.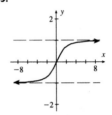

47.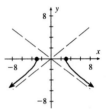

25.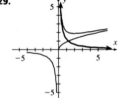

27.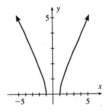

49.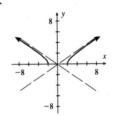

51.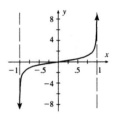

29.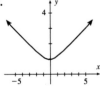

31.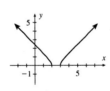

53.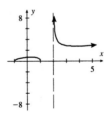

55.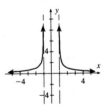

33.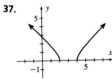

35.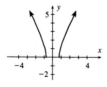

57.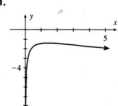

59.

61.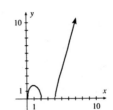

63.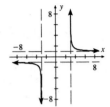

Concave up

65.

Concave up

67.

Concave down / Concave up

29. Not a function; $D = (-\infty, -3) \cup (-3, 3] \cup (5, \infty)$;
$R = (-\infty, -1) \cup (-1, 1) \cup (1, \infty)$ with deleted points at
$\left(-3, \dfrac{\sqrt{3}}{2}\right)$ and $\left(-3, -\dfrac{\sqrt{3}}{2}\right)$; asymptotes: $x = 5$, $y = 1$,
$y = -1$; symmetric with respect to the x-axis; intercepts:
$\left(0, \dfrac{\sqrt{15}}{5}\right), \left(0, -\dfrac{\sqrt{15}}{5}\right), (3, 0)$

4.4 **Problem Set, Pages 200–201**

1. $\{2 \pm \sqrt{13}\}$ **3.** $\{\frac{4}{3}\}$ **5.** $\{4\}$ **7.** $\{1, 6\}$ **9.** $\{-8\}$
11. $\{2\}$ **13.** $\{5, -\frac{1}{5}\}$ **15.** $\{1, -\frac{9}{5}\}$ **17.** $\{1\}$ **19.** $\{7\}$
21. $\{1\}$ **23.** $\{4\}$ **25.** $\{3\}$ **27.** $\varnothing$ **29.** $\{1, -7\}$
31. $\{\frac{3}{2}\}$ **33.** $\{1, 5\}$ **35.** $\left\{\dfrac{-1 \pm \sqrt{3}}{2}\right\}$ **37.** $\{\frac{5}{3}, -2\}$
39. $\{4, 7\}$ **41.** $\{1\}$ **43.** $\{12\}$ **45.** $\varnothing$ **47.** $\{0, -1\}$
49. $\{3\}$ **51.** $\{1, -2\}$ **53.** $\{\pm\frac{1}{3}\}$ **55.** $\{-1, -2, -3\}$
57. $\{2, 17\}$ **59.** $\{4, 9\}$ **61.** $\{4\}$ **63.** $\{1, \frac{3}{2}\}$
65. Let $u = \sqrt{x^2 - 2x + 2}$. Solution is $\{1 \pm 2\sqrt{6}\}$
67. $\{\frac{1}{10}, \frac{9}{10}\}$

4.5 **Problem Set, Page 208**

1. A function; $D = (-\infty, 0) \cup (0, \infty)$; $R = (-\infty, 0) \cup (0, \infty)$;
symmetric with respect to the origin; asymptotes: $x = 0$,
$y = 0$; no intercepts
3. A function; $D = (-\infty, 0) \cup (0, \infty)$; $R = (-\infty, 1) \cup (1, \infty)$;
asymptotes: $x = 0$, $y = 1$; intercept: $(-1, 0)$
5. A function; $D = (-\infty, 4]$; $R = [0, \infty)$; intercepts: $(0, 2)$, $(4, 0)$
7. A function; $D = [6, \infty)$; $R = (-\infty, 0]$; intercept: $(6, 0)$
9. A function; $D = (-\infty, -2) \cup (-2, 2) \cup (2, \infty)$;
$R = (-\infty, -\frac{1}{4}] \cup (0, \infty)$; symmetric with respect to the
y-axis; asymptotes: $x = 2$, $x = -2$, $y = 0$; intercept:
$(0, -\frac{1}{4})$
11. A function; $D = (-\infty, -2) \cup (-2, \infty)$;
$R = (-\infty, 1) \cup (1, \infty)$; intercepts: $(0, 5)$, $(-\frac{5}{2}, 0)$; it is a line
with a deleted point.
13. A function; $D = (-\infty, -\frac{1}{2}) \cup (-\frac{1}{2}, \infty)$; $R = [-1, \frac{5}{4}) \cup (\frac{5}{4}, \infty)$;
intercepts: $(0, 0)$, $(2, 0)$. It is the parabola $y = x^2 - 2x$ with a
deleted point at $x = -\frac{1}{2}$.
15. A function; $D = [0, \infty)$; $R = (-\infty, \frac{1}{4}]$; intercepts: $(0, 0)$, $(1, 0)$
17. A function; $D = (-\infty, -3] \cup [1, \infty)$; $R = [0, \infty)$; intercepts:
$(-3, 0)$, $(1, 0)$
19. A function; $D = (-2, 2)$; $R = (-\infty, \infty)$; asymptotes: $x = 2$,
$x = -2$; intercept: $(0, 0)$
21. Not a function; $D = [-5, 5]$; $R = [-5, 5]$; symmetric with
respect to the x-axis, y-axis, and origin; intercepts: $(5, 0)$,
$(-5, 0)$, $(0, 5)$, $(0, -5)$
23. Not a function; $D = [-4, 4]$; $R = [-3, 3]$; symmetric with
respect to the x-axis, y-axis, and origin; intercepts: $(0, -3)$,
$(0, 3)$, $(4, 0)$, $(-4, 0)$
25. A function; $D = (-\infty, 0) \cup (0, \infty)$; $R = (-\infty, -2] \cup [2, \infty)$;
asymptotes: $x = 0$, $y = x$; no intercepts; symmetric with
respect to the origin
27. Not a function; $D = [-1, 1] \cup (4, \infty)$;
$R = (-\infty, -\sqrt{8 + 2\sqrt{15}}] \cup (-\sqrt{8 - 2\sqrt{15}},$
$\sqrt{8 - 2\sqrt{15}}) \cup [\sqrt{8 + 2\sqrt{15}}, \infty)$; asymptote: $x = 4$;
intercepts: $(0, \frac{1}{2})$, $(0, -\frac{1}{2})$, $(1, 0)$, $(-1, 0)$

31.

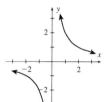

33.

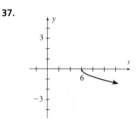

35.

37.

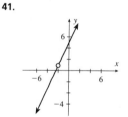

39.

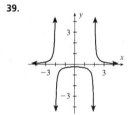

41.

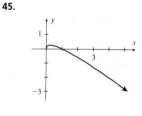

43.

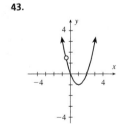

45.

47.

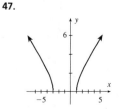

49.

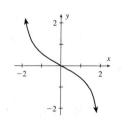

51.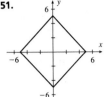

53.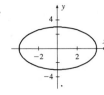

11. $(x - 5), (x - 5)^2$ **13.** $x, (x - 1), (x - 1)^2$

15. $(x - 2), (x - 2)^2, (x - 2)^3$ **17.** $(x - 3), (x + 1)$

19. $x, (x + 1), (x - 2)$ **21.** $x, (x - 4), (x - 1)$

23. $(1 - x), (1 + x), (1 + x^2)$ **25.** $\dfrac{1}{x} + \dfrac{2}{x^2} + \dfrac{5}{x^3}$

27. $\dfrac{2}{x} - \dfrac{5}{x^2} + \dfrac{4}{x^3}$ **29.** $\dfrac{1}{x + 4} - \dfrac{1}{x + 5}$ **31.** $\dfrac{4}{x - 2} + \dfrac{3}{x - 1}$

33. $\dfrac{2}{x + 2} + \dfrac{5}{x - 4}$ **35.** $\dfrac{4}{x + 3} - \dfrac{2}{x - 2}$ **37.** $\dfrac{1}{x - 2} + \dfrac{3}{x + 2}$

39. $\dfrac{3}{x - 4} - \dfrac{2}{x - 5}$ **41.** $\dfrac{3}{x} - \dfrac{3}{x - 1} + \dfrac{4}{x + 1}$

43. $\dfrac{2}{x - 2} + \dfrac{3}{(x - 2)^2}$ **45.** $\dfrac{1}{x} + \dfrac{3}{(x + 1)^2}$

47. $\dfrac{2}{x + 1} + \dfrac{4}{(x + 1)^2} - \dfrac{3}{(x + 1)^3}$ **49.** $\dfrac{5}{6(x + 5)} + \dfrac{1}{6(x - 1)}$

51. $\dfrac{13}{3(x - 2)} + \dfrac{8}{3(x + 1)}$ **53.** $\dfrac{1}{x} + \dfrac{3}{x + \cdot 3} - \dfrac{4}{x - 2}$

55. $\dfrac{3}{x - 1} + \dfrac{2x - 4}{x^2 + 1}$ **57.** $x + 2 + \dfrac{3}{x - 1} + \dfrac{1}{(x - 1)^2}$

59. $\dfrac{1}{1 - x} - \dfrac{3}{1 + x} + \dfrac{2x + 1}{1 + x^2}$

55.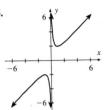

57.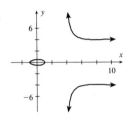

59.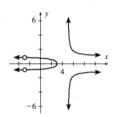

61. Not a function; $D = [-8, 8]$; $R = [-8, 8]$; symmetric with respect to the x-axis, y-axis, and origin; intercepts: $(0, 8)$, $(0, -8), (8, 0), (-8, 0)$

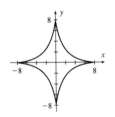

63. Not a function; $D = (-\infty, -3) \cup [-2, 2] \cup (3, \infty)$; $R = (-\infty, -1) \cup [-\frac{2}{3}, \frac{2}{3}] \cup (1, \infty)$; asymptotes: $x = 3$, $x = -3, y = 1, y = -1$; symmetric with respect to the x-axis, y-axis, and origin; intercepts: $(0, \frac{2}{3}), (0, -\frac{2}{3}), (2, 0), (-2, 0)$

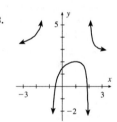

65. Not a function; $D = [-2, 2]$; $R = [-2, 2]$; symmetric with respect to the x-axis, y-axis, and origin; intercepts: $(0, 0)$, $(2, 0), (-2, 0)$

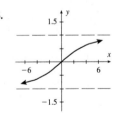

4.6 **Problem Set, Pages 212–213**

1. $(x + 3), (x + 4)$ **3.** $(x - 3), (2x + 1)$ **5.** $(x - 8), (x - 6)$

7. $x, (x - 1), (x + 1)$ **9.** $x, (x - 2), (x + 2)$

Chapter 4 Summary, Pages 213–215

1.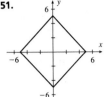

3.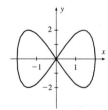

5. $\frac{3}{4}$ **7.** 0 **9.** $x = \frac{1}{2}; y = 3x - 4$ **11.** $x = \pm 3$

13. **15.**

17. $\{2\}; x \neq -3, -\frac{1}{2}$ **19.** $\{\frac{5}{3}\}; x \neq 0, 3, 6$

21. $\{3\}; 2$ is extraneous **23.** $\{4\}$

25. Not a function; $y = \pm \sqrt{\dfrac{x^2 + 25}{x^2 + 1}}$;

$x = \pm \sqrt{\dfrac{y^2 - 25}{1 - y^2}}; D = (-\infty, \infty)$;

$R = [-5, -1) \cup (1, 5]$;

symmetric with respect to the x-axis, y-axis, and origin; intercepts: $(0, 5), (0, -5)$

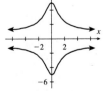

27. Not a function; $D = [-1, 1]$; $R = [-1, 1]$; symmetric with respect to the x-axis, y-axis, and origin; intercepts: $(1, 0)$, $(-1, 0)$, $(0, -1)$, $(0, 1)$

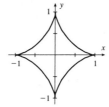

29. $\dfrac{3}{x-1} + \dfrac{2}{x-2} + \dfrac{-1}{(x-2)^2}$ **31.** $2 + \dfrac{3}{x+5} + \dfrac{6}{x-3}$

5.1   Problem Set, Pages 225–226

1.
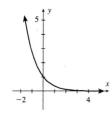

3.
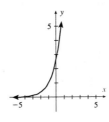

5.
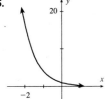

7.
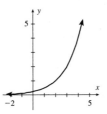

9.
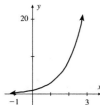

11.
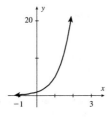

13.
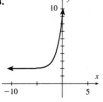

15.
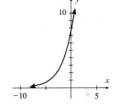

17.
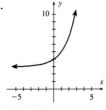

19.
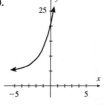

21.
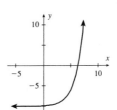

23.
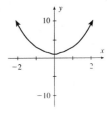

25.
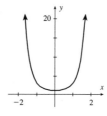

27.
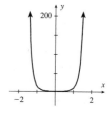

29.
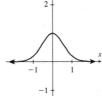

31. 20.08553692   **33.** 1.127496852   **35.** 2.117000017
37. 1.172887932   **39.** 1.127474616   **41.** 2.716923932
43. 2.7   **45.** 26.0
47. a. If $b = 1$, then $b^x = 1$, a constant function. It is algebraic.
     b. If $b = 0$, then $b^x = 0$, a constant function. It is algebraic.
49. $5,427.43   **51.** $54,598.15   **53.** $14,527.69
55. $10,724.94   **57.** $1,131.47   **59.** $1,530.69
61. a.

   b.
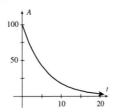

63.
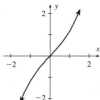

65.
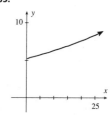

5.2   Problem Set, Pages 231–232

1. $\log_2 64 = 6$ **3.** $\log 1{,}000 = 3$ **5.** $\log_5 125 = 3$
7. $\log_n m = p$ **9.** $\log_{1/3} 9 = -2$ **11.** $\log_9 \frac{1}{3} = -\frac{1}{2}$

13. $10^4 = 10,000$ **15.** $10^0 = 1$ **17.** $e^2 = e^2$
19. $e^5 = x$ **21.** $2^{-3} = \frac{1}{8}$ **23.** $4^{1/2} = 2$ **25.** $m^p = n$
27. $x^3 = 8$ **29.** 2 **31.** 4 **33.** -1 **35.** 2 **37.** .630
39. .926 **41.** 4.8549 **43.** .1209 **45.** .820
47. .8883 **49.** 2.6 **51.** 2.0 **53.** 3.55 **55.** -1.3
57. 4.023 **59.** 2.8
61. $\log_b N$ is undefined when N is negative.
63. $\log_b N$ is the exponent. **65.** The base is 10.
67. True; nothing wrong
69. a. 3,572 **b.** 4,746 **c.** $116,619
71. a. $M \approx 3.89$ **b.** $M \approx 8.8$

5.3 Problem Set, Pages 238–240

1. a. 2 **b.** 7 **3. a.** -1 **b.** 4
5. a. $2\sqrt{7}$ **b.** 3 **7. a.** $\sqrt{e}$ **b.** e
9. a. e^4 **b.** 14 **11. a.** $\pm 5\sqrt{5}$ **b.** $\pm 2\sqrt{3}$
13. a. $\frac{7}{2}$ **b.** All positive x **15. a.** 2^5 **b.** 10^5
17. True **19.** Concave down
21. $\log_b AB = \log_b A + \log_b B$ **23.** $b > 1$; concave down
25. $0 < b < 1$; concave up
27.
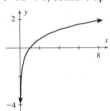
29.
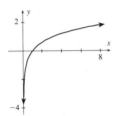
31.
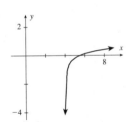
33.
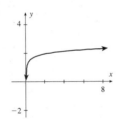
35.
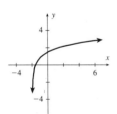
37.
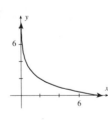
39.
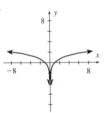
41.
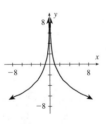

43.
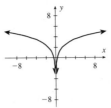

45. 0 **47.** $1,250\sqrt{3}$ **49.** 100 **51.** 2, 6 **53.** 25
55. 3, 6 **57.** 243 **59.** $5e^{-2}$ **61.** 3
63. a. About 29 days **b.** No **c.** $N = 80(1 - e^{-t/62.5})$
65. a. $E = 10^{1.5M + 11.8}$ **b.** 1.12202×10^{23} ergs
67. Answers vary. **69.** 2

5.4 Problem Set, Pages 243–244

1. 7 **3.** $\frac{5}{3}$ **5.** $\frac{2}{3}$ **7.** $\frac{3}{7}$ **9.** -2 **11.** -3
13. -2 **15.** $\frac{1}{9}$ **17.** $-\frac{2}{3}$ **19.** $-\frac{1}{3}$ **21.** $-\frac{1}{6}$
23. 1.62324929 **25.** -2.48811664 **27.** $-.66958623$
29. .78746047 **31.** 2.12907205 **33.** 1.15129255
35. $-.57564627$ **37.** $-.049306144$ **39.** 2.7429396
41. 1.4306766 **43.** -2.321928 **45.** -1.2920297
47. 11.89566 periods; 6 years
49. 30.99891276; 7 years and 3 quarters **51.** .023104906
53. 10.813529 **55.** $t \approx 11,400$ **57.** About 17,400 years
59. $\frac{1}{2}$ mile above sea level
61. a. 108° **b.** .08 **c.** $t = -\frac{1}{k}\ln\left(\frac{T - A}{B - A}\right)$
63. $-.0208493$ **65.** 2.2924317, -2.2924317
67. $\log_2 \frac{4}{3} \approx .415$

Chapter 5 Summary, Pages 245–246

1.
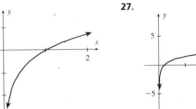

3.
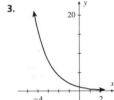

5. $\log\sqrt{10} = .5$
7. $\log_9 729 = 3$ **9.** $10^0 = 1$ **11.** $2^6 = 64$
13. .47712121547 **15.** -2.67780705 **17.** 1.098612289
19. -4.342805922 **21.** 1.953445298 **23.** 164.6194501
25.

27.

29. $\log_b A + \log_b B$ **31.** $p \log_b A$ **33.** 2 **35.** 243
37. .096910013 **39.** $\pm.5888866497$ **41.** 1.304007668
43. -2.768345799 **45.** 3 **47.** 1.771243749
49. $k \approx .026$ **51.** $131,894.24 **53.** $36,018.87

Cumulative Review I, Page 247

1. a. 14 **b.** $2t - 1$ **c.** $3t^2 + 6th + 3h^2 + 2$
 d. $12x^2 - 12x + 5$ **e.** $6x^2 + 3$ **f.** $6x + 3h$

2. a. None (not one-to-one) **b.** $y = \dfrac{x+1}{2}$

3. a. 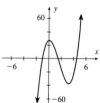 **b.**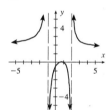

4. a. 0 **b.** 27 **c.** -36 **d.** 0 **e.** $-1, \frac{3}{2}, \frac{9}{2}$

5.   **6.** $\frac{1}{2}, -\frac{3}{2}, 3 \pm \sqrt{7}$

7. a. $(-\infty, -1) \cup (-1, 2) \cup (2, \infty)$ **b.** No symmetry
 c. Yes; $x = 2, x = -1, y = 2$; no slant asymptote
 d. $(\frac{1}{2}, 0), (1, 0), (0, -\frac{1}{2})$

e. 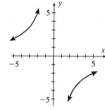 **8. a.**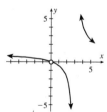

b. (graph) **c.** (graph)

9. a. $\frac{5}{3}$ **b.** 92.4 **c.** 1 **d.** 81

10. a. $k = \dfrac{1}{5,700}$ **b.** 2,700 yr **c.** $t = -\dfrac{1}{k} \log_2 \dfrac{A}{A_0}$

Extended Application: Population Growth, Pages 250–251

1. .866% **3.** 1.918% **5.** 1994 **7.** 2007
9. a. $-.5\%$ **b.** 2,627,416 **11.** See page 226
13. Answers vary. **15.** Answers vary.

6.1 Problem Set, Pages 261–262

1. a. $\frac{\pi}{6}$ **b.** $\frac{\pi}{2}$ **c.** $\frac{3\pi}{2}$ **d.** $\frac{\pi}{4}$
3. a. $180°$ **b.** $45°$ **c.** $60°$ **d.** $360°$

5. a.   **b.**

c.   **d.**

7. a.   **b.**

c.   **d.**

9. a.   **b.**

c.   **d.**

11. a. $40°$ **b.** $180°$ **c.** $30°$ **d.** $330°$
13. a. $240°$ **b.** $140°$ **c.** $180°$ **d.** $280°$
15. a. $\frac{7\pi}{4}$ **b.** $\frac{\pi}{4}$ **c.** $\frac{5\pi}{3}$ **d.** $2\pi - 2$
17. a. 2.7168 **b.** 1.2832 **c.** .7879 **d.** .2832
19. a. .5236 **b.** 5.4978 **c.** 5.4867 **d.** 3.1415
21. a. $60°$ **b.** $60°$ **c.** $60°$ **d.** $45°$
23. a. $\frac{\pi}{12}$ **b.** $\frac{\pi}{3}$ **c.** $\frac{\pi}{6}$ **d.** $\frac{\pi}{4}$
25. a. $\frac{\pi}{4}$ **b.** 0 **c.** $\frac{\pi}{6}$ **d.** $\frac{\pi}{3}$
27. a. .7879 **b.** .4250 **c.** .5236 **d.** .7854
29. $18.00°$ **31.** $300.00°$ **33.** $30.00°$ **35.** $-171.89°$
37. $-143.24°$ **39.** $29.22°$ **41.** $\frac{\pi}{9}$ **43.** $-\frac{11\pi}{9}$
45. $\frac{17\pi}{36}$ **47.** 5.48 **49.** 6.11 **51.** 17.19 **53.** 14.04 cm
55. 4.19 m **57.** 4.89 ft **59.** 14.07 cm **61.** 87.27 cm
63. 440 km **65.** 5,200 km

1. .6   3. −.6   5. −.4   7. 2.9
9. a. .64278761   b. .34202014
11. a. .36397023   b. −1.4142136
13. a. 3.0776835   b. 1.1923633
15. a. .84147098   b. .0174524
17. a. −1.3386481   b. −14.136833
19. a. −1.2281621   b. −.49996641
21. a. .2876553   b. .5455785
23. a. 2.1164985   b. 1.8899139
25. a. 2.6327477   b. −3.4184818
27. a. −5.4914632   b. −1.6389075
29. Positive   31. Negative
33. Negative   35. Negative   37. Negative   39. Negative
41. II, III   43. III   45. III
Note: Also draw a picture for Problems 47–53.
47. $\cos\theta = -\frac{3}{5}$, $\sin\theta = \frac{4}{5}$, $\tan\theta = -\frac{4}{3}$, $\sec\theta = -\frac{5}{3}$, $\csc\theta = \frac{5}{4}$, $\cot\theta = -\frac{3}{4}$
49. $\cos\theta = \frac{5}{13}$, $\sin\theta = \frac{12}{13}$, $\tan\theta = \frac{12}{5}$, $\sec\theta = \frac{13}{5}$, $\csc\theta = \frac{13}{12}$, $\cot\theta = \frac{5}{12}$
51. $\cos\theta = \frac{5}{13}$, $\sin\theta = -\frac{12}{13}$, $\tan\theta = -\frac{12}{5}$, $\sec\theta = \frac{13}{5}$, $\csc\theta = -\frac{13}{12}$, $\cot\theta = -\frac{5}{12}$
53. $\cos\theta = -\dfrac{6}{\sqrt{37}}$, $\sin\theta = \dfrac{1}{\sqrt{37}}$, $\tan\theta = -\frac{1}{6}$, $\sec\theta = -\dfrac{\sqrt{37}}{6}$, $\csc\theta = \sqrt{37}$, $\cot\theta = -6$
55. $|x|\sqrt{2}$ 57. $-x\sqrt{2}$ 59. $3x$ 61. $\frac{1}{2}$
63. −4   65. a. Answers vary.   b. Answers vary.
67. .5403

1. a. 1   b. 1   c. $\dfrac{\sqrt{3}}{2}$ d. $\dfrac{\sqrt{3}}{2}$
3. a. 0   b. 1   c. 0   d. $\dfrac{\sqrt{2}}{2}$
5. a. $\sqrt{2}$ b. Undefined   c. 2   d. $\sqrt{3}$
7. a. 1   b. −1   c. −1   d. 0
9. a. $\frac{1}{2}$ b. $\frac{1}{2}$ c. $\dfrac{\sqrt{2}}{2}$ d. 1   11. a. .5650   b. .5850
13. a. .6428   b. .1763   15. a. −.3640   b. −.1736
17. a. .7337   b. −.4791   19. a. .9320   b. .7961
21. a. 1.5574   b. .0709   23. a. −.7470   b. .1411
25. a. .9004   b. .3555   27. Answers vary.   29. $-\dfrac{\sqrt{2}}{2}$
31. $-\dfrac{\sqrt{2}}{2}$ 33. $-\dfrac{\sqrt{3}}{2}$ 35. 1   37. −1   39. $\sqrt{2}$
41. $\frac{1}{2}$ 43. 1   45. $\frac{1}{2}$ 47. 1   49. $\dfrac{\sqrt{2}}{2}$ 51. $\dfrac{2\sqrt{3}}{3}$
53. $\dfrac{\sqrt{3}}{3}$ 55. $\dfrac{\sqrt{3}}{2}$ 57. $-\sqrt{3}$ 59. $\frac{1}{2}$
61. a–d. Answers vary.   63. Answers vary.

1.

x = angle	$\frac{2\pi}{3}$	$\frac{3\pi}{4}$	$\frac{5\pi}{6}$	$\frac{7\pi}{6}$
Quadrant	II; −	II; −	II; −	III; −
$y = \cos x$	$-\dfrac{1}{2}$	$-\dfrac{\sqrt{2}}{2}$	$-\dfrac{\sqrt{3}}{2}$	$-\dfrac{\sqrt{3}}{2}$
y (approximate)	−.50	−.71	−.87	−.87

x = angle	$\frac{5\pi}{4}$	$\frac{4\pi}{3}$	$\frac{7\pi}{4}$	$\frac{11\pi}{6}$
Quadrant	III; −	III; −	IV; +	IV; +
$y = \cos x$	$-\dfrac{\sqrt{2}}{2}$	$-\dfrac{1}{2}$	$\dfrac{\sqrt{2}}{2}$	$\dfrac{\sqrt{3}}{2}$
y (approximate)	−.71	−.50	.71	.87

See Figure 6.17.
3. See graph in text
5.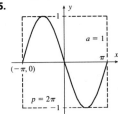

7.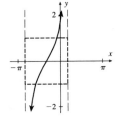

9.

11.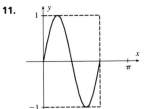

13.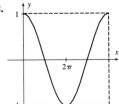

15.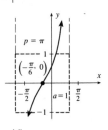

17.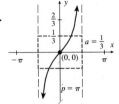

19.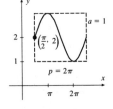

21.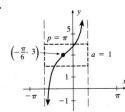

23.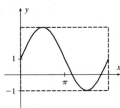

45.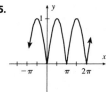

47.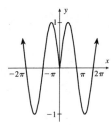

25.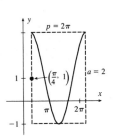

27.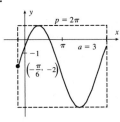

49.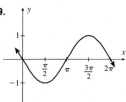

51.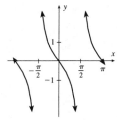

29.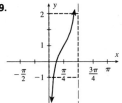

31.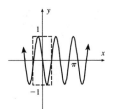

53.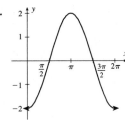

55.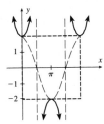

33.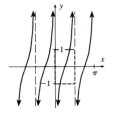

35.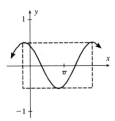

57.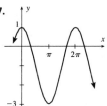

59.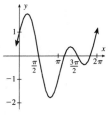

37.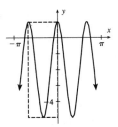

39.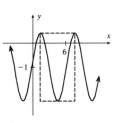

61.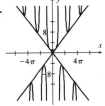

63.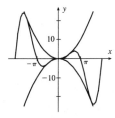

41.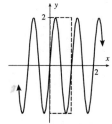

43.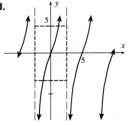

65.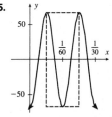

67. a.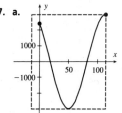

b. 3,000 km **c.** 120 min

6.5 Problem Set, Page 292

1. a. 0 **b.** $\frac{\pi}{6}$ **c.** $\frac{\pi}{6}$ **d.** 0
3. a. $\frac{\pi}{3}$ **b.** $\frac{\pi}{4}$ **c.** $\frac{\pi}{4}$ **d.** $\frac{\pi}{4}$
5. a. $\frac{5\pi}{6}$ **b.** $-\frac{\pi}{4}$ **c.** $-\frac{\pi}{4}$ **d.** $\frac{2\pi}{3}$
7. a. 0 **b.** $\frac{\pi}{3}$ **c.** $\frac{\pi}{3}$ **d.** $\frac{\pi}{3}$ **9.** .58 **11.** 1.65
13. 2.81 **15.** 1.32 **17.** .98 **19.** 1.34 **21.** 69°
23. $-28°$ **25.** 46° **27.** 85° **29.** 135° **31.** 153°
33. $-72°$ **35.** True **37.** Quadrant II
39. True **41.** True only for $x > 0$

43. $\operatorname{Csc}^{-1}x = \operatorname{Sin}^{-1}\left(\dfrac{1}{x}\right)$ **45.** $\frac{\pi}{6}$ **47.** $\frac{\pi}{15}$

49. $\frac{2\pi}{15}$ **51.** .4163 **53.** 1.28 **55.** .2836

57. $-.64$ **59.** $\frac{1}{5}\sqrt{21} \approx .917$

61.

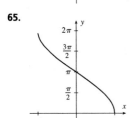

63.

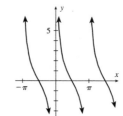

65.

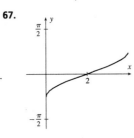

67.

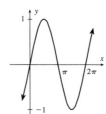

Chapter 6 Summary, Pages 293–295

1. a. lambda **b.** theta **c.** phi **d.** alpha
e. beta
3. $x^2 + y^2 = 1$ **5.** 145° **7.** $\frac{7\pi}{6}$

9.

11.

13.

15.

17. 270° **19.** $-315°$ **21.** $\frac{5\pi}{3}$ **23.** $\frac{3\pi}{10}$ or .9425
25. $s = r\theta$; radius; angle measured in radians **27.** $5\pi \approx 15.7$ cm
29. 60° **31.** $4 - \pi$ **33.** an angle in standard position

34–36. (*Answer in any order*) $\cos\theta = a$; $\sec\theta = \dfrac{1}{a}$, $a \neq 0$;

$\sin\theta = b$; $\csc\theta = \dfrac{1}{b}$, $b \neq 0$; $\tan\theta = \dfrac{b}{a}$, $a \neq 0$; $\cot\theta = \dfrac{a}{b}$, $b \neq 0$

37. All **39.** tangent and cotangent **41.** $\frac{4}{3}$ **43.** secant
45. 1.0896 **47.** 1.4587
49. θ is an angle in standard position with a point $P(x, y)$ on the terminal side of θ a nonzero distance of r from the origin

50–52. (*Answer in any order*) $\sin\theta = \dfrac{y}{r}$; $\cos\theta = \dfrac{x}{r}$; $\tan\theta = \dfrac{y}{x}$,

$x \neq 0$; $\csc\theta = \dfrac{r}{y}$, $y \neq 0$; $\sec\theta = \dfrac{r}{x}$, $x \neq 0$; $\cot\theta = \dfrac{x}{y}$, $y \neq 0$

53. $\cos\theta = \frac{5}{13}$, $\sin\theta = -\frac{12}{13}$, $\tan\theta = -\frac{12}{5}$, $\sec\theta = \frac{13}{5}$,
$\csc\theta = -\frac{13}{12}$, $\cot\theta = -\frac{5}{12}$

55. $\cos\theta = -\dfrac{5}{\sqrt{29}}$, $\sin\theta = \dfrac{2}{\sqrt{29}}$, $\tan\theta = -\frac{2}{5}$, $\sec\theta = -\dfrac{\sqrt{29}}{5}$,

$\csc\theta = \dfrac{\sqrt{29}}{2}$, $\cot\theta = -\frac{5}{2}$

57. a. 1 **b.** $\dfrac{\sqrt{3}}{2}$ **c.** $\dfrac{\sqrt{2}}{2}$ **d.** $\frac{1}{2}$ **e.** 0 **f.** -1 **g.** 0

59. a. 0 **b.** $\dfrac{\sqrt{3}}{3}$ **c.** 1 **d.** $\sqrt{3}$ **e.** Undefined
f. 0 **g.** Undefined
61. $\frac{1}{2}$ **63.** -1 **65.** 1.4587 **67.** 1.0896
69.

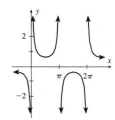

71.

73.

75.

77.

79.

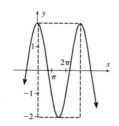

81. $-\frac{\pi}{2} < y < \frac{\pi}{2}$ **83.** $0 < y < \pi$ **85.** $\frac{\pi}{6}$
87. $\frac{\pi}{6}$ **89.** .31940 **91.** 1.2741

93.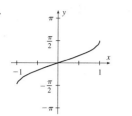

95.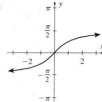

97. 1   **99.** −.4292

Problem Set, Pages 304–305

1. $\frac{\pi}{6}$ **3.** $-\frac{\pi}{6}$ **5.** $\frac{\pi}{6} + 2n\pi, \frac{5\pi}{6} + 2n\pi$ **7.** $\frac{\pi}{3}$ **9.** 120°
11. $\frac{\pi}{3} + 2n\pi, \frac{5\pi}{3} + 2n\pi$ **13.** $\frac{\pi}{6}, \frac{5\pi}{6}, \frac{7\pi}{6}, \frac{11\pi}{6}$
15. $\frac{\pi}{3}, \frac{2\pi}{3}, \frac{4\pi}{3}, \frac{5\pi}{3}$ **17.** $\frac{2\pi}{3}, \frac{5\pi}{6}, \frac{5\pi}{3}, \frac{11\pi}{6}$
19. $\frac{\pi}{12}, \frac{5\pi}{12}, \frac{3\pi}{4}, \frac{13\pi}{12}, \frac{17\pi}{12}, \frac{7\pi}{4}$ **21.** $\frac{5\pi}{12}, \frac{7\pi}{12}, \frac{17\pi}{12}, \frac{19\pi}{12}$
23. $0, \pi$ **25.** $\frac{\pi}{2}, \frac{3\pi}{2}$ **27.** $\frac{\pi}{6}, \frac{\pi}{3}, \frac{5\pi}{6}, \frac{5\pi}{3}$ **29.** $0, \pi, \frac{\pi}{4}, \frac{5\pi}{4}$
31. $\frac{\pi}{4}, \frac{3\pi}{4}, \frac{5\pi}{4}, \frac{7\pi}{4}$ **33.** $0, \pi, \frac{\pi}{3}, \frac{5\pi}{3}$ **35.** 2.2370, 4.0461
37. .3649, 1.2059, 3.5065, 4.3475   **39.** .6662, 2.4754
41. .8213, 2.3203   **43.** .00, 1.57, 2.09, 3.14, 4.19, 4.71
45. .41, 1.16, 3.55, 4.30   **47.** .32, 1.89, 3.46, 5.03
49. .00, 2.09, 4.19   **51.** .00, .52, 2.62, 3.14
53. .00. 1.05, 1.22, 1.92, 2.09, 3.14, 3.32, 4.01, 4.19, 5.24, 5.41, 6.11
55. $\frac{5\pi}{6} + 2n\pi, \frac{7\pi}{6} + 2n\pi$ **57.** .4014 + 2n\pi, 2.7402 + 2n\pi
59. .9423 + n\pi   **61.** .185   **63.** .00, 2.36, 3.14, 5.50
65. No solution   **67.** 1.86

Problem Set, Pages 307–308

1. See text.   **3.** III, IV   **5.** I, IV   **7.** II   **9.** I
11. $\cot(A + B)$ **13.** $\tan \frac{\pi}{15}$ **15.** $\cos \frac{\pi}{8}$ **17.** $\cos 127°$
19. 1   **21.** 1   **23.** −1   **25.** Answers vary.
27. Answers vary.   **29.** Answers vary.

31. $\cot \theta = \cot \theta; \tan \theta = \dfrac{1}{\cot \theta}; \csc \theta = \pm\sqrt{1 + \cot^2 \theta};$

$\sin \theta = \dfrac{\pm 1}{\sqrt{1 + \cot^2 \theta}}; \cos \theta = \dfrac{\pm \cot \theta}{\sqrt{1 + \cot^2 \theta}};$

$\sec \theta = \dfrac{\pm\sqrt{1 + \cot^2 \theta}}{\cot \theta}$

33. $\csc \theta = \csc \theta; \sin \theta = \dfrac{1}{\csc \theta}; \cot \theta = \pm\sqrt{\csc^2 \theta - 1};$

$\tan \theta = \dfrac{\pm 1}{\sqrt{\csc^2 \theta - 1}}; \cos \theta = \dfrac{\pm\sqrt{\csc^2 \theta - 1}}{\csc \theta};$

$\sec \theta = \dfrac{\pm \csc \theta}{\sqrt{\csc^2 \theta - 1}}$

35. $\sin \theta = -\frac{4}{5}, \tan \theta = -\frac{4}{3}, \sec \theta = \frac{5}{3}, \csc \theta = -\frac{5}{4}, \cot \theta = -\frac{3}{4}$
37. $\sin \theta = \frac{12}{13}, \tan \theta = \frac{12}{5}, \sec \theta = \frac{13}{5}, \csc \theta = \frac{13}{12}, \cot \theta = \frac{5}{12}$
39. $\cos \theta = -\frac{12}{13}, \sin \theta = -\frac{5}{13}, \sec \theta = -\frac{13}{12}, \csc \theta = -\frac{13}{5}, \cot \theta = \frac{12}{5}$

41. $\cos \theta = -\dfrac{\sqrt{5}}{3}, \tan \theta = -\dfrac{2}{\sqrt{5}}, \sec \theta = -\dfrac{3}{\sqrt{5}},$

$\csc \theta = \frac{3}{2}, \cot \theta = -\dfrac{\sqrt{5}}{2}$

43. $\cos \theta = \dfrac{5}{\sqrt{34}}, \sin \theta = \dfrac{3}{\sqrt{34}}, \tan \theta = \frac{3}{5}, \csc \theta = \dfrac{\sqrt{34}}{3},$

$\cot \theta = \frac{5}{3}$

45. $\cos \theta = -\dfrac{1}{\sqrt{10}}, \sin \theta = -\dfrac{3}{\sqrt{10}}, \tan \theta = 3,$

$\sec \theta = -\sqrt{10}, \cot \theta = \frac{1}{3}$

47. $\sin \theta$ **49.** $\dfrac{2}{1 - \cos^2 \theta}$ or $\dfrac{2}{\sin^2 \theta}$ **51.** $\dfrac{1}{\sin \theta}$

53. $\dfrac{\sin^2 \theta - \cos^2 \theta}{\sin \theta \cos \theta}$ **55.** $\dfrac{1}{\cos \theta}$ **57.** $\dfrac{1 + \sin \theta}{\cos \theta}$

59. $\dfrac{\sin \theta + \cos \theta}{\cos \theta \sin \theta}$ **61.** $\dfrac{1 + \cos^2 \theta}{\sin^2 \theta}$ **63.** $\sin^2\theta - \cos \theta$

Problem Set, Pages 312–313

1–71. All answers vary. Individual hints (or first steps) are provided.
1. Factor $\sin^3\theta + \cos^2\theta \sin \theta$.   **3.** Factor $\cot \theta \tan^2\theta$.
5. $\tan^2\theta \sin^2\theta = \tan^2(1 - \cos^2\theta)$
7. Change to sines and cosines.
9. $\dfrac{\sec x + \csc x}{\csc x \sec x} = \dfrac{\sec x}{\csc x \sec x} + \dfrac{\csc x}{\csc x \sec x}$
11. Break up the fraction.   **13.** Multiply out $(\sec \theta - \cos \theta)^2$.
15. Factor $\dfrac{1 - \sin^2 2\theta}{1 + \sin 2\theta}$.   **17.** Factor $\dfrac{\sin^2 \lambda + \sin \lambda \cos \lambda + \sin \lambda}{\sin \lambda + \cos \lambda + 1}$.
19. Change to sines and cosines.
21. Change to sines and cosines.
23. Factor $\dfrac{\sin^2 B - \cos^2 B}{\sin B + \cos B}$.
25. Use fundamental identities to simplify $\tan^2 2\gamma + \sin^2 2\gamma + \cos^2 2\gamma$.
27. Change to sines and cosines.
29. Use fundamental identities to simplify $1 + \sin^2\lambda$.
31. Change to sines and cosines.
33. Change to sines and cosines.
35. Change to sines and cosines.
37. Change to secants and tangents.
39. Multiply by $\dfrac{1 + \tan C}{1 + \tan C}$.
41. Factor $\dfrac{\sin^3 x - \cos^3 x}{\sin x - \cos x}$.   **43.** Multiply out $\left(\dfrac{1 - \cos \theta}{\sin \theta}\right)^2$.
45. Factor $\dfrac{(\cos^2 \gamma - \sin^2 \gamma)^2}{\cos^4 \gamma - \sin^4 \gamma}$.   **47.** Multiply by $\dfrac{\sec \theta - \tan \theta}{\sec \theta - \tan \theta}$.
49. Change to sines and cosines.   **51.** Factor $\dfrac{1 - \sec^3 \theta}{1 - \sec \theta}$.
53. Factor $\dfrac{\tan^2\theta - 2 \tan \theta}{2 \tan \theta - 4}$.
55. Multiply out $\sqrt{(3 \cos \theta - 4 \sin \theta)^2 + (3 \sin \theta + 4 \cos \theta)^2}$.
57. Change to sines and cosines.   **59.** Change to sines and cosines.
61. Multiply out $(\cos \alpha - \cos \beta)^2 + (\sin \alpha - \sin \beta)^2$.
63. Change to sines and cosines.

65. Change to sines and cosines.
67. Multiply $\sin\theta + \cos\theta + 1$ by $\dfrac{\sin\theta + \cos\theta - 1}{\sin\theta + \cos\theta - 1}$.
69. Change to sines and cosines.
71. Change to sines and cosines.

7.4 Problem Set, Pages 319–320

1. $\dfrac{\sqrt{3}\cos\theta - \sin\theta}{2}$ 3. $\dfrac{\sqrt{2}}{2}(\cos\theta + \sin\theta)$ 5. $\dfrac{1 + \tan\theta}{1 - \tan\theta}$
7. $\dfrac{2\tan\theta}{1 - \tan^2\theta}$ 9. $\cos^2\theta - \sin^2\theta$ 11. $\cos 52°$ 13. $\cos\frac{\pi}{3}$
15. $-\tan\frac{\pi}{6}$ 17. $-\tan 49°$ 19. $-\sin 31°$ 21. $-\tan 24°$
23. .8746   25. .6561   27. .6745
29. $\cos\theta = \dfrac{\sqrt{6} + \sqrt{2}}{4}$, $\sin\theta = \dfrac{\sqrt{2} - \sqrt{6}}{4}$, $\tan\theta = -2 + \sqrt{3}$
31. $\cos\theta = \dfrac{\sqrt{6} - \sqrt{2}}{4}$, $\sin\theta = \dfrac{\sqrt{6} + \sqrt{2}}{4}$, $\tan\theta = 2 + \sqrt{3}$
33. $\cos\theta = \dfrac{\sqrt{2} - \sqrt{6}}{4}$, $\sin\theta = \dfrac{\sqrt{6} + \sqrt{2}}{4}$, $\tan\theta = -2 - \sqrt{3}$
35. a. $\cos(\theta - \frac{2\pi}{3})$ b. $-\sin(\theta - \frac{2\pi}{3})$ c. $-\tan(\theta - \frac{2\pi}{3})$
37.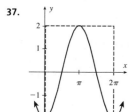
39.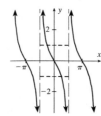
41.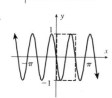
43.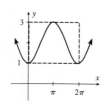
45.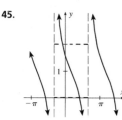
47. a. Odd   b. Even   49. a. Even   b. Odd
51–71. Answers vary for proofs. Individual hints on first steps are given.
51. Use Identity 17 and replace β by $-\beta$.
53. Same as proof of Identity 19   55. Subtract the fractions.
57. Same as proof of Identity 10
59. Same as proof of Identity 10
61. Same as proof of Identity 12
63. Multiply out $\sin(\alpha + \beta)\cos\beta - \cos(\alpha + \beta)\sin\beta$.

65. Use $\alpha = [(\alpha + \beta) - \beta]$ in $\tan\alpha$.   67. Use Identity 15.
69. Associate: $(\alpha + \beta + \gamma) = [(\alpha + \beta) + \gamma]$
71. Associate: $(\alpha + \beta + \gamma) = [(\alpha + \beta) + \gamma]$

7.5 Problem Set, Pages 327–328

1. $\dfrac{\sqrt{2}}{2}$ 3. $\frac{1}{2}$ 5. $\frac{1}{2}\sqrt{2 + \sqrt{2}}$
7. $\cos(\theta + \phi) = \cos\theta\cos\phi - \sin\theta\sin\phi$
9. $\dfrac{\sin 2\theta}{2} = \dfrac{2\sin\theta\cos\theta}{2} = \sin\theta\cos\theta$
11. $\pm$ missing; also sinus is minus
13. $\cos 2\theta = \frac{7}{25}$, $\sin 2\theta = \frac{24}{25}$, $\tan 2\theta = \frac{24}{7}$
15. $\cos 2\theta = \frac{119}{169}$, $\sin 2\theta = -\frac{120}{169}$, $\tan 2\theta = -\frac{120}{119}$
17. $\cos 2\theta = -\frac{31}{81}$, $\sin 2\theta = \dfrac{20\sqrt{14}}{81}$, $\tan 2\theta = -\dfrac{20\sqrt{14}}{31}$
19. $\cos\frac{1}{2}\theta = \dfrac{3}{\sqrt{10}}$, $\sin\frac{1}{2}\theta = \dfrac{1}{\sqrt{10}}$, $\tan\frac{1}{2}\theta = \frac{1}{3}$
21. $\cos\frac{1}{2}\theta = \dfrac{-5}{\sqrt{26}}$, $\sin\frac{1}{2}\theta = \dfrac{1}{\sqrt{26}}$, $\tan\frac{1}{2}\theta = -\frac{1}{5}$
23. $\cos\frac{1}{2}\theta = \dfrac{\sqrt{7}}{3}$, $\sin\frac{1}{2}\theta = \dfrac{\sqrt{2}}{3}$, $\tan\frac{1}{2}\theta = \dfrac{\sqrt{14}}{7}$
25. $\cos 11° - \cos 59°$ 27. $\frac{1}{2}\cos 75° - \frac{1}{2}\cos 165°$
29. $\frac{1}{2}\cos 3\theta - \frac{1}{2}\cos 7\theta$ 31. $2\sin 53°\cos 10°$
33. $2\sin 257.5°\cos 42.5°$ 35. $2\cos 4x\cos 2x$
37. $0, \pi, \frac{\pi}{4}, \frac{3\pi}{4}, \frac{5\pi}{4}, \frac{7\pi}{4}$
39. $\frac{\pi}{8}, \frac{3\pi}{8}, \frac{\pi}{2}, \frac{5\pi}{8}, \frac{7\pi}{8}, \frac{9\pi}{8}, \frac{11\pi}{8}, \frac{3\pi}{2}, \frac{13\pi}{8}, \frac{15\pi}{8}$
41. $0, \frac{\pi}{4}, \frac{\pi}{2}, \frac{3\pi}{4}, \pi, \frac{5\pi}{4}, \frac{3\pi}{2}, \frac{7\pi}{4}$ 43. $0, \frac{\pi}{8}, \frac{3\pi}{8}, \frac{5\pi}{8}, \frac{7\pi}{8}, \pi, \frac{9\pi}{8}, \frac{11\pi}{8}, \frac{13\pi}{8}, \frac{15\pi}{8}$
45. $\cos\theta = \dfrac{1}{\sqrt{5}}$, $\sin\theta = \dfrac{2}{\sqrt{5}}$, $\tan\theta = 2$
47. $\cos\theta = \dfrac{1}{\sqrt{10}}$, $\sin\theta = \dfrac{3}{\sqrt{10}}$, $\tan\theta = 3$
49. $\cos\theta = \frac{1}{2}$, $\sin\theta = \dfrac{\sqrt{3}}{2}$, $\tan\theta = \sqrt{3}$
51. a. 3.9   b. $\dfrac{2}{\sqrt{2 - \sqrt{3}}}$ or $\sqrt{2}(\sqrt{3} + 1)$
53–65. Answers vary. Side of equation to start with is shown and individual hints are given.
53. Left side; replace 2θ with θ.   55. Right side; let $\theta = \dfrac{3\beta}{4}$.
57. Left side; use Problem 56 result and conjugate.
59. Left side; use sum identities.   61. Left side; $\sin(2\theta + \theta)$.
63. Left side; use double-angle identity (twice).
65. Right side; use double-angle identity.

7.6 Problem Set, Pages 333–334

1. cis 90° is on positive imaginary axis.
3. True   5. Quadrant IV
7. a. $\sqrt{10}$ b. $5\sqrt{2}$ c. $\sqrt{13}$ d. $\sqrt{13}$
9. a. $\sqrt{29}$ b. $\sqrt{41}$ c. $\sqrt{17}$ d. $\sqrt{2}$
11. a. 2 cis 330°   b. 2 cis 30°

13. a. cis 0° **b.** 2 cis 0° **c.** 3 cis 0° **d.** 4 cis 0°
15. a. 5 cis 270° **b.** 6 cis 270° **c.** 7 cis 270° **d.** 8 cis 270°

17. 6 cis 345° **19.** 4 cis 100° **21.** $\dfrac{3}{2} + \dfrac{3\sqrt{3}}{2} i$

23. $-5i$ **25.** $-8.8633 - 1.5628i$ **27.** 15 cis 140°
29. $\frac{5}{2}$ cis 267° **31.** 3 cis 130° **33.** 81 cis 240°

35. 64 cis 180° or -64 **37.** 256 cis 120° or $-128 + 128\sqrt{3}i$
39. 3 cis 22°, 3 cis 112°, 3 cis 202°, 3 cis 292°
41. cis 60°, cis 180°, cis 300°
43. $\sqrt[8]{2}$ cis 11.25°, $\sqrt[8]{2}$ cis 101.25°, $\sqrt[8]{2}$ cis 191.25°, $\sqrt[8]{2}$ cis 281.25°
45. 2 cis 40°, 2 cis 112°, 2 cis 184°, 2 cis 256°, 2 cis 328°
47. 3 cis 0°, 3 cis 120°, 3 cis 240°
49. $\sqrt[8]{2}$ cis 56.25°, $\sqrt[8]{2}$ cis 146.25°, $\sqrt[8]{2}$ cis 236.25°, $\sqrt[8]{2}$ cis 326.25°
51. $2^{1/18}$cis 15°, $2^{1/18}$cis 55°, $2^{1/18}$cis 95°, $2^{1/18}$cis 135°, $2^{1/18}$cis 175°, $2^{1/18}$cis 215°, $2^{1/18}$cis 255°, $2^{1/18}$cis 295°, $2^{1/18}$cis 335°
53. 2 cis 30°, 2 cis 90°, 2 cis 150°, 2 cis 210°, 2 cis 270°, 2 cis 330°
55. cis 9°, cis 45°, cis 81°, cis 117°, cis 153°, cis 189°, cis 225°, cis 261°, cis 297°, cis 333°

57. $-2, 1 + \sqrt{3}i, 1 - \sqrt{3}i$
59. $-.6840 + 1.8794i, -1.2856 - 1.5321i, 1.9696 - .3473i$
61. $1.9696 + .3473i, -.3473 + 1.9696i, -1.9696 - .3473i,$ $.3473 - 1.9696i$
63. 8 cis 0°, 8 cis 72°, 8 cis 144°, 8 cis 216°, 8 cis 288°
65. Answers vary. **67.** Answers vary.

Chapter 7 Summary, Pages 335–337

1. $\frac{\pi}{12}, \frac{7\pi}{12}, \frac{13\pi}{12}, \frac{19\pi}{12}$ **3.** $\frac{\pi}{12}, \frac{\pi}{4}, \frac{3\pi}{4}, \frac{11\pi}{12}, \frac{17\pi}{12}, \frac{19\pi}{12}$ **5.** $\frac{\pi}{3}, \frac{2\pi}{3}, \frac{4\pi}{3}, \frac{5\pi}{3}$
7. No solution **9.** See text. **11.** See text.
13. $\cos \delta = -\frac{4}{5}, \tan \delta = -\frac{3}{4}, \sec \delta = -\frac{5}{4}, \csc \delta = \frac{5}{3}, \cot \delta = -\frac{4}{3}$
15. $\cos \omega = -\frac{4}{5}, \tan \omega = \frac{3}{4}, \sec \omega = -\frac{5}{4}, \csc \omega = -\frac{5}{3}, \cot \omega = \frac{4}{3}$
17. $\dfrac{\sin^2\theta + \cos\theta}{\sin\theta\cos\theta}$ **19.** $\dfrac{2\sin^2\theta + \cos^2\theta}{\cos^2\theta}$ or $\dfrac{\sin^2\theta + 1}{\cos^2\theta}$
21–27. Answers vary. Side to start with and individual hints are given.
21. Left side; use fundamental identity.
23. Right side; change to sines and cosines.
25. Left side; factor. **27.** Left side; change to sines and cosines.
29. $\cos 52°$ **31.** $\cot \frac{3\pi}{8}$ **33.** $\cos(\theta - \frac{\pi}{6})$
35.

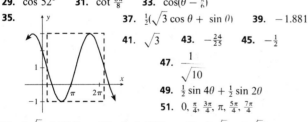

37. $\frac{1}{2}(\sqrt{3} \cos \theta + \sin \theta)$ **39.** -1.881
41. $\sqrt{3}$ **43.** $-\frac{24}{25}$ **45.** $-\frac{1}{2}$
47. $\dfrac{1}{\sqrt{10}}$
49. $\frac{1}{2} \sin 4\theta + \frac{1}{2} \sin 2\theta$
51. $0, \frac{\pi}{4}, \frac{3\pi}{4}, \pi, \frac{5\pi}{4}, \frac{7\pi}{4}$
53. $7\sqrt{2}$ cis 315° **55.** 7 cis 330° **57.** $2\sqrt{2} - 2\sqrt{2}i$
59. $-5i$ **61.** $-128 - 128\sqrt{3}i$ **63.** cis 154° **65.** See text.
67. $-2.5556 + .6848i, 2.5556 - .6848i$; also plot points.

8.1 Problem Set, Pages 343–344

1. $\alpha = 30°, \beta = 60°, \gamma = 90°, a = 80, b = 140, c = 160$
3. $\alpha = 45°, \beta = 45°, \gamma = 90°, a = 9.0, b = 9.0, c = 13$
5. $\alpha = 37°, \beta = 53°, \gamma = 90°, a = 11, b = 15, c = 19$
7. $\alpha = 71°, \beta = 19°, \gamma = 90°, a = 69, b = 24, c = 73$

9. $\alpha = 54°, \beta = 36°, \gamma = 90°, a = 18, b = 13, c = 22$
11. $\alpha = 40°, \beta = 50°, \gamma = 90°, a = 24, b = 29, c = 38$
13. $\alpha = 76°, \beta = 14°, \gamma = 90°, a = 29, b = 7.2, c = 30$
15. $\alpha = 45°, \beta = 45°, \gamma = 90°, a = 49, b = 49, c = 69$
17. $\alpha = 77°, \beta = 13°, \gamma = 90°, a = 390, b = 90, c = 400$
19. $\alpha = 50°, \beta = 40°, \gamma = 90°, a = 98, b = 82, c = 130$
21. $\alpha = 6°, \beta = 84°, \gamma = 90°, a = 3.8, b = 36, c = 36$
23. $\alpha = 56.00°, \beta = 34.00°, \gamma = 90.00°, a = 3,484, b = 2,350,$ $c = 4,202$
25. $\alpha = 27.66°, \beta = 62.34°, \gamma = 90.00°, a = 1,625, b = 3,100,$ $c = 3,500$
27. $\alpha = 42°, \beta = 48°, \gamma = 90°, a = 320, b = 350, c = 470$
29. $\alpha = 17.54°, \beta = 72.46°, \gamma = 90.00°, a = 1,296, b = 4,100,$ $c = 4,300$
31. 23 m tall **33.** 200 m **35.** 340 ft **37.** 350 ft
39. 13 ft **41.** 60° **43.** 1,251 ft **45.** 263 m
47. 170 m **49.** 222.0 ft **51.** 14.6 ft **53.** 6.34×10^7 mi
55. 633.9 ft **57.** Answers vary. **59.** About 12 ft (12.1667)
61. Answers vary.

8.2 Problem Set, Pages 347–348

1. 54° **3.** 80° **5.** 21° **7.** 13 **9.** 10 **11.** 17
13. 14 **15.** 46.6° **17.** 604 **19.** 80°
21. 22.2° **23.** 78°
25. $\alpha = 54°, \beta = 85°, \gamma = 41°, a = 4.2, b = 5.2, c = 3.4$
27. $\alpha = 25.8°, \beta = 139.4°, \gamma = 14.8°, a = 214, b = 320, c = 126$
29. $\alpha = 94.3°, \beta = 37.3°, \gamma = 48.4°, a = 140, b = 85.0, c = 105$
31. No solution since no side is given.
33. $\alpha = 110.5°, \beta = 49.4°, \gamma = 20.1°, a = 641, b = 520, c = 235$
35. $\alpha = 78.7°, \beta = 77.9°, \gamma = 23.4°, a = 341, b = 340, c = 138$
37. 537 ft **39.** 362 ft **41.** 293 ft
43. Proof varies. **45.** Proof varies. **47.** Proof varies.
49. 700 mi **51.** 484 mi
53. $d = 2.25, q = 1.88, p = 2.13, \angle D = 68.1°, \angle P = 61.2°,$ $\angle Q = 50.7°$
55. 16 in. **57.** 450 in.^{2} **59.** Proof varies.
61. Proof varies. **63.** Proof varies.

8.3 Problem Set, Pages 351–353

1. $\alpha = 48°, \beta = 62°, \gamma = 70°, a = 10, b = 12, c = 13$
3. $\alpha = 30°, \beta = 50°, \gamma = 100°, a = 30, b = 46, c = 59$
5. $\alpha = 50°, \beta = 70°, \gamma = 60°, a = 33, b = 40, c = 37$
7. $\alpha = 120°, \beta = 53°, \gamma = 7°, a = 310, b = 280, c = 43$
9. $\alpha = 18.3°, \beta = 54.0°, \gamma = 107.7°, a = 107, b = 276, c = 325$
11. $\alpha = 49°, \beta = 59°, \gamma = 72°, a = 85, b = 97, c = 110$
13. $\alpha = 48.5°, \beta = 62.7°, \gamma = 68.8°, a = 105, b = 125, c = 131$
15. No solution; $\alpha + \beta + \gamma > 180°$
17. $\alpha = 9.5°, \beta = 148.0°, \gamma = 22.5°, a = 45.3, b = 145, c = 105$
19. $\alpha = 45.2°, \beta = 21.5°, \gamma = 113.3°, a = 41.0, b = 21.2, c = 53.1$
21. No solution; $\alpha + \beta + \gamma > 180°$
23. $\alpha = 14°, \beta = 42°, \gamma = 123°, a = 26, b = 71, c = 88$
25. No solution; one side must be given.
27. $\alpha = 74.7°, \beta = 16.1°, \gamma = 89.2°, a = 80.6, b = 23.2, c = 83.6$
29. $\alpha = 35.60°, \beta = 42.10°, \gamma = 102.30°, a = 16.90, b = 19.46,$ $c = 28.36$
31. $\alpha = 11°, \beta = 61°, \gamma = 108°, a = 10, b = 48, c = 52$
33. $\alpha = 4.8°, \beta = 82.0°, \gamma = 93.2°, a = 48.1, b = 569, c = 574$

35. No solution; $\alpha + \beta + \gamma > 180°$
37. $\alpha = 29°$, $\beta = 62°$, $\gamma = 90°$, $a = 8.1$, $b = 15$, $c = 17$
39. $\alpha = 16.2°$, $\beta = 153.2°$, $\gamma = 10.6°$, $a = 6.74$, $b = 10.9$, $c = 4.45$
41. 260 ft　　**43.** 280 ft　　**45.** 970.4 m and 1,532 m
47. The height of the building is 31 ft and it is 60 ft from Mr. T to the top.
49. 8 ft　　**51.** Proof varies.　　**53.** Proof varies.
55. Proof varies.　　**57.** Proof varies.　　**59.** Proof varies.
61. A complete discussion of at least five different solutions is found in the October 1977 issue of *Popular Science*. The problem is to find r where r is the rate of the car.

Step 1: $|AC| = \text{rate} \cdot \text{time} = r\left(\frac{1}{2}\right)$

Step 2: $|AB| + |AC| = \text{rate} \cdot \text{time}$

$|AC| = \text{rate} \cdot \text{time} - |AC|$

$= r\left(\frac{35}{60}\right) - 10$

$= \dfrac{7r}{12} - 10$

Step 3: $|AD| + |DC| = \text{rate} \cdot \text{time}$

$|AD| = \text{rate} \cdot \text{time} - |DC|$

$= r\left(\frac{40}{60}\right) - 10$

$= \dfrac{2r}{3} - 10$

Step 4: Let $R = \dfrac{r}{12}$; then $|AB| = 7R - 10$; $|BC| = 10$;

$|AC| = 6R$; $|AD| = 8R - 10$, and $|DC| = 10$.

Let $\angle ACD = \theta$; then $\angle ACB = 90° - \theta$ since $\angle DCB$ is a right angle.

Step 5: $\cos\theta = \dfrac{10^2 + (6R)^2 - (8R - 10)^2}{2(10)(6R)}$ from the Law of Cosines

for $\triangle ADC$

$= \dfrac{160 - 28R}{120}$ Do not reduce this; keep the common denominator for Step 7.

Step 6: $\cos(90° - \theta) = \dfrac{10^2 + (6R)^2 - (7R - 10)^2}{2(10)(6R)} = \dfrac{140 - 13R}{120}$

Step 7: Since $\cos(90° - \theta) = \sin\theta$ and $\cos^2\theta + \sin^2\theta = 1$,

$\left(\dfrac{160 - 28R}{120}\right)^2 + \left(\dfrac{140 - 13R}{120}\right)^2 = 1$

Simplify to: $953R^2 - 12{,}600R + 30{,}800 = 0$

Solve for R using the quadratic formula:

$R = \dfrac{12{,}600 \pm \sqrt{12{,}600^2 - 4(953)(30{,}800)}}{2(953)}$

$= \dfrac{6{,}300 \pm 20\sqrt{25{,}844}}{953}$

$\approx 3.2369,\ 9.9845$

Thus, the rate of the car is 38.8428 mph or 119.8128 mph. Since the latter solution is highly unlikely, the published solution of 38.843 is correct to the nearest thousandth; a better approximation is 38.8430577.

1. $\alpha > 90°$, OPP $\le$ ADJ, no triangle formed
3. $\alpha < 90°$, OPP $< h <$ ADJ, no triangle formed
5. $\alpha = 125°$, $\beta = 41°$, $\gamma = 14°$, $a = 5.0$, $b = 4.0$, $c = 1.5$
7. $\alpha = 75°$, $\beta = 49°$, $\gamma = 56°$, $a = 9.0$, $b = 7.0$, $c = 7.8$
9. $\alpha < 90°$, OPP $< h <$ ADJ, no triangle formed
11. $\alpha = 47.0°$, $\beta = 90.0°$, $\gamma = 43.0°$, $a = 8.63$, $b = 11.8$, $c = 8.05$
13. $\alpha = 52.1°$, $\beta = 115.0°$, $\gamma = 12.9°$, $a = 14.2$, $b = 16.3$, $c = 4.00$
15. $\alpha = 147.0°$, $\beta = 15.0°$, $\gamma = 18.0°$, $a = 49.5$, $b = 23.5$, $c = 28.1$
17. Ambiguous case; $\alpha = 90.8°$, $\beta = 57.1°$, $\gamma = 32.1°$, $a = 98.2$, $b = 82.5$, $c = 52.2$; or $\alpha = 25.0°$, $\beta = 122.9°$, $\gamma = 32.1°$, $a = 41.6$, $b = 82.5$, $c = 52.2$
19. No solution; $\alpha + \beta + \gamma > 180°$
21. No solution, since the sum of the two smaller sides must be larger than the third side.
23. $\alpha = 54.5°$, $\beta = 62.5°$, $\gamma = 63.0°$, $a = 179$, $b = 195$, $c = 196$
25. 37 square units　　**27.** 83 square units
29. 680 square units　　**31.** 560 square units
33. 6.4 square units　　**35.** 35 square units
37. 54.0 square units
39. Ambiguous case; 58.2 or 11.3 square units
41. Ambiguous case; 2,150 or 911 square units
43. 180 square units　　**45.** No triangle formed
47. 20,100 square units　　**49.** 5.39 mi
51. Altitude is 1,850 ft and the distances are 1.926 mi (10,170 ft) and .5363 mi (2,832 ft).
53. 19.0°　　**55.** 7.83 cm　　**57.** 18,000 m²
59. 21,000 ft³ or 760 yd³　　**61.** 985 ft　　**63.** Answers vary.
65. Answers vary.

Chapter 8 Summary, Pages 363–364

1. See text.
3. $\alpha = 35.0°$, $\beta = 55.0°$, $\gamma = 90.0°$, $a = 475$, $b = 678$, $c = 828$
5. See text.　　**7.** Answers vary.
9. Procedure: $\cos\alpha = \dfrac{b^2 + c^2 - a^2}{2bc}$;

solution: $\alpha = \cos^{-1}\left(\dfrac{b^2 + c^2 - a^2}{2bc}\right)$

11. Procedure: $\alpha + \beta + \gamma = 180°$; solution: $\alpha = 180° - \beta - \gamma$
13. $\alpha = 23°$, $\beta = 58°$, $\gamma = 99°$, $a = 24$, $b = 52$, $c = 61$
15. Ambiguous case; $\alpha = 137°$, $\beta = 27°$, $\gamma = 16°$, $a = 51$, $b = 34$, $c = 21$; or $\alpha = 11°$, $\beta = 153°$, $\gamma = 16°$, $a = 14$, $b = 34$, $c = 21$
17. 460 ft²　　**19.** Ambiguous case; 122 in.² or 17.4 in.²
21. 430 ft; angle of elevation, 48.9°　　**23.** 278 ft

Cumulative Review II, Page 365

1. See text.
2. a. $-\dfrac{\sqrt{3}}{2}$　**b.** $-\dfrac{\sqrt{3}}{2}$　**c.** 0　**d.** $-\dfrac{\pi}{3}$　**e.** $\dfrac{\pi}{3}$
f. No values　**g.** 1.162253　**h.** $2\sqrt{2}$

3. a.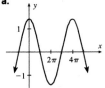

b.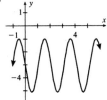

c.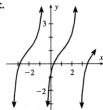

4. Proofs vary.

5. a. $\frac{\pi}{18}, \frac{7\pi}{18}, \frac{13\pi}{18}, \frac{19\pi}{18}, \frac{25\pi}{18}, \frac{31\pi}{18}$

b. .1183, 3.0238

6. a. Start with left side; combine fractions.

b. Start with left side; change to sines and cosines.

c. Start with left side. *Prove* double-angle identity (don't use it); write $\cos 2\theta = \cos(\theta + \theta)$.

7. a. $\dfrac{\sin \psi}{s} = \dfrac{\sin \theta}{t} = \dfrac{\sin \phi}{u}$ **b.** $u^2 = s^2 + t^2 - 2st \cos \phi$;

$s^2 = u^2 + t^2 - 2ut \cos \psi$; $t^2 = s^2 + u^2 - 2us \cos \theta$

8. a. $\alpha = 29°, \beta = 109°, \gamma = 42°, a = 14, b = 27, c = 19$

b. $\alpha = 113°, \beta = 20°, \gamma = 47°, a = 19, b = 7.2, c = 15$

c. $\alpha = 26°, \beta = 35°, \gamma = 119°, a = 35, b = 45, c = 69$

d. $\alpha = 50.8°, \beta = 83.5°, \gamma = 45.7°, a = 92.8, b = 119, c = 85.7$

9. a. 320 ft^2 **b.** 51.7 in.^2 **c.** $1,900 \text{ cm}^2$

10. a. Ambiguous case; 22 mi or 67 mi **b.** 4,620 ft

Extended Application: Solar Power, Page 369

1. 16 hr 50 min, 4:50 P.M. **3.** 19 hr 1 min, 7:01 P.M.

5. 6 hr 18 min, 6:18 A.M. **7.**

9.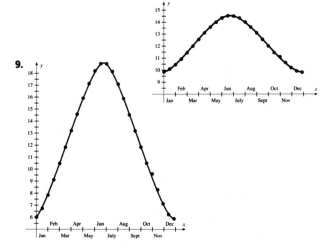

11. Answers vary.

---

1. $(3, 2)$ **3.** $(6, -3)$ **5.** Dependent system

7. $(-5, 9), (-2, 6)$ **9.** Inconsistent system

11. $(s, t) = (-3, 4)$ **13.** Inconsistent system

15. $(t_1, t_2) = (57, 274)$ **17.** $(\gamma, \delta) = (-1, 1)$

19. $(c, d) = (1, 1)$ **21.** $(q_1, q_2) = (1, 0)$ **23.** $(\frac{2}{7}, 3)$

25. $(\alpha, \beta) = (\frac{20}{3}, \frac{16}{3})$ **27.** $(\theta, \phi) = (-4, 3)$

29. Dependent system **31.** $(3, 15)$ **33.** $(6, -1)$

35. $(2\gamma - \delta, \gamma - \delta)$ **37.** $\left(\dfrac{a + b}{2}, \dfrac{a - b}{2}\right)$

39. $\left(\dfrac{c + d}{5}, \dfrac{2d - 3c}{5}\right)$ **41.** $(-2, 2), (-5, -4)$

43. $(1, 2), (1, -2), (-1, 2), (-1, -2)$

45. $(\sqrt{13}, 1), (-\sqrt{13}, 1), (5, 2), (-5, 2)$ **47.** Answers vary.

49. Proof varies. **51.** 1 and -3 or 3 and -1

53. 12 ft and 5 ft **55.** $(\frac{\pi}{3}, \frac{7\pi}{6}), (\frac{\pi}{3}, \frac{11\pi}{6}), (\frac{5\pi}{3}, \frac{7\pi}{6}), (\frac{5\pi}{3}, \frac{11\pi}{6})$

57. Inconsistent system **59.** $(2\sqrt{2}, 0), (-2\sqrt{2}, 0)$

61. $(\frac{17}{21}, -\frac{20}{9})$

63. $\left(\dfrac{-9 + \sqrt{113}}{2}, \dfrac{-15 + \sqrt{113}}{8}\right), \left(\dfrac{-9 - \sqrt{113}}{2}, \dfrac{-15 - \sqrt{113}}{8}\right)$

65. $(h, k) = (1, 8)$

67. $\left(\dfrac{a + \sqrt{a^2 - 4}}{2}, \dfrac{a - \sqrt{a^2 - 4}}{2}\right), \left(\dfrac{a - \sqrt{a^2 - 4}}{2}, \dfrac{a + \sqrt{a^2 - 4}}{2}\right)$

---

1. a. $\begin{bmatrix} 4 & 5 & | & -16 \\ 3 & 2 & | & 5 \end{bmatrix}$ **b.** $\begin{bmatrix} 1 & 1 & 1 & | & 4 \\ 3 & 2 & 1 & | & 7 \\ 1 & -3 & 2 & | & 0 \end{bmatrix}$

c. $\begin{bmatrix} 1 & 3 & 1 & 1 & | & 3 \\ 1 & 0 & -2 & 2 & | & 0 \\ 0 & 0 & 1 & 5 & | & -14 \\ 0 & 1 & -3 & -1 & | & 2 \end{bmatrix}$

3. $R1 \leftrightarrow R3$; $\begin{bmatrix} 1 & 3 & -4 & | & 9 \\ 0 & 2 & 4 & | & 5 \\ 3 & 1 & 2 & | & 1 \end{bmatrix}$

5. $\frac{1}{2}R1$; $\begin{bmatrix} 1 & 2 & 5 & | & -6 \\ 6 & 3 & 4 & | & 6 \\ 10 & -1 & 0 & | & 1 \end{bmatrix}$

7. $-R2 + R1$; $\begin{bmatrix} 1 & 5 & -12 & | & 2 \\ 4 & 1 & 9 & | & 2 \\ 7 & 6 & 1 & | & 3 \end{bmatrix}$

9. $\begin{bmatrix} 1 & 2 & -3 & | & 0 \\ 0 & 3 & 1 & | & 4 \\ 0 & 1 & 7 & | & 6 \end{bmatrix}$ **11.** $\begin{bmatrix} 1 & 2 & 4 & | & 1 \\ 0 & 9 & 8 & | & 4 \\ 0 & 13 & 17 & | & 7 \end{bmatrix}$

13. $\begin{bmatrix} 1 & 4 & -1 & 3 & | & 3 \\ 0 & 16 & 3 & 13 & | & 9 \\ 0 & -19 & 14 & -14 & | & -17 \\ 0 & -4 & 3 & -3 & | & -3 \end{bmatrix}$

15. $\frac{1}{2}R2;$ $\begin{bmatrix} 1 & 3 & 5 & \vdots & 2 \\ 0 & 1 & 3 & \vdots & -4 \\ 0 & 3 & 4 & \vdots & 1 \end{bmatrix}$ **17.** $\frac{1}{5}R2;$ $\begin{bmatrix} 1 & 3 & -2 & \vdots & 4 \\ 0 & 1 & \frac{1}{5} & \vdots & \frac{3}{5} \\ 0 & 7 & 9 & \vdots & 2 \end{bmatrix}$

19. $-R3 + R2;$ $\begin{bmatrix} 1 & 3 & -2 & \vdots & 0 \\ 0 & 1 & -4 & \vdots & 8 \\ 0 & 3 & 6 & \vdots & 1 \end{bmatrix}$

21. $\begin{bmatrix} 1 & 5 & -3 & \vdots & 2 \\ 0 & 1 & 4 & \vdots & 5 \\ 0 & 0 & -8 & \vdots & -13 \end{bmatrix}$ **23.** $\begin{bmatrix} 1 & 7 & 6 & 6 & \vdots & 2 \\ 0 & 1 & 9 & 2 & \vdots & 1 \\ 0 & 0 & 42 & 9 & \vdots & 5 \\ 0 & 0 & -37 & 0 & \vdots & -2 \end{bmatrix}$

25. $\begin{bmatrix} 1 & -2 & 5 & \vdots & 6 \\ 0 & 1 & 9 & \vdots & 1 \\ 0 & 0 & 1 & \vdots & -\frac{5}{3} \end{bmatrix}$ **27.** $\begin{bmatrix} 1 & 3 & 0 & \vdots & 9 \\ 0 & 1 & 0 & \vdots & -2 \\ 0 & 0 & 1 & \vdots & 4 \end{bmatrix}$

29. $\begin{bmatrix} 1 & 6 & 0 & 0 & \vdots & -2 \\ 0 & 1 & 0 & 0 & \vdots & 1 \\ 0 & 0 & 1 & 0 & \vdots & 0 \\ 0 & 0 & 0 & 1 & \vdots & -6 \end{bmatrix}$

31. $\begin{bmatrix} 1 & 0 & 0 & \vdots & -\frac{23}{3} \\ 0 & 1 & 0 & \vdots & -\frac{8}{3} \\ 0 & 0 & 1 & \vdots & \frac{3}{5} \end{bmatrix}$ $\begin{array}{l} x = -\frac{23}{3} \\ y = -\frac{8}{3} \\ z = \frac{3}{5} \end{array}$

33. $(3, 2)$ **35.** $(1, \frac{3}{2})$ **37.** $(3, 1)$
39. $(1, 2, 3)$ **41.** $(3, -1, 2)$ **43.** $(5, 6, 1)$ **45.** $(1, 1, 1)$
47. $(-4, 0, 3)$ **49.** $(5, \frac{15}{2}, \frac{3}{2})$ **51.** Inconsistent system
53. $(w, x, y, z) = (-1, 1, 2, 3)$ **55.** $y = x^2 + 4x + 5$
57. $y = x^2 - 10x + 20$
59. 7 bars of alloy I; 3 bars of alloy II
61. 3 units of I; 7 units of II; 5 units of III

9.3 Problem Set, Pages 406–407

1. $\begin{bmatrix} 2 & 4 & 2 \\ 6 & -2 & 4 \\ 2 & 2 & 5 \end{bmatrix}$ **3.** $\begin{bmatrix} 16 & 19 & 10 \\ 29 & 16 & 15 \\ 35 & 9 & 31 \end{bmatrix}$

5. $\begin{bmatrix} 13 & 25 & 20 \\ 25 & 31 & 23 \\ 42 & 24 & 33 \end{bmatrix}$ **7.** $\begin{bmatrix} 20 & 21 & 34 \\ 29 & 16 & 15 \\ 7 & 48 & 5 \end{bmatrix}$

9. $\begin{bmatrix} 34 & 117 & 44 \\ 45 & 143 & 97 \\ 109 & 100 & 151 \end{bmatrix}$ **11.** $\begin{bmatrix} 14 & 14 \\ -7 & 7 \end{bmatrix}$ **13.** Yes

15. Yes **17.** Yes **19.** $\begin{bmatrix} 2 & 7 \\ 1 & 4 \end{bmatrix}$ **21.** $\begin{bmatrix} 0 & \frac{1}{2} \\ \frac{1}{3} & -\frac{1}{6} \end{bmatrix}$

23. $\begin{bmatrix} 3 & -17 & -20 \\ 3 & -18 & -20 \\ -1 & 6 & 7 \end{bmatrix}$ **25.** $\begin{bmatrix} 1 & 0 & -1 & 0 \\ 0 & \frac{1}{2} & 0 & 0 \\ -2 & 0 & 2 & 1 \\ 0 & 0 & 1 & 0 \end{bmatrix}$

27. $\begin{bmatrix} .6 & 0 & -.4 & .2 \\ .2 & 0 & .2 & -.1 \\ 0 & 1 & 0 & 0 \\ -1.2 & 0 & .8 & .1 \end{bmatrix}$ **29.** $(-4, 7)$ **31.** $(25, 14)$

33. $(50, 29)$ **35.** $(-1, 4)$ **37.** $(-5, 2)$ **39.** $(-3, -2)$
41. $(4, -2)$ **43.** $(12, -5)$ **45.** $(0, 4)$ **47.** $(-2, 4, 3)$
49. $(3, -5, 2)$ **51.** $(-5, 18, 5)$ **53.** $(5, 4, -1)$ **55.** $(1, 2, 3)$

57. $(-3, 2, 7)$ **59.** $(w, x, y, z) = (3, -2, 1, -4)$
63. 1 Proof varies. **63.** Proof varies.

9.4 Problem Set, Pages 411–412

1. 10 **3.** -10 **5.** 36 **7.** 10 **9.** -28 **11.** -1
13. 1 **15.** $\cot^2\theta$ **17.** $(-3, 4)$ **19.** $(c, d) = (1, 1)$
21. $(s_1, s_2) = (\frac{9}{19}, \frac{13}{19})$ **23.** $(-4, 7)$ **25.** $(6, -3)$
27. Inconsistent system **29.** $(-6, 1)$ **31.** $(8, 4)$
33. $(31, 19)$ **35.** $(-54, -31)$ **37.** $\left(\dfrac{5a + 3b}{13}, \dfrac{a - 2b}{13}\right)$
39. $\left(\dfrac{a}{a^2 - b^2}, \dfrac{-b}{a^2 - b^2}\right)$ **41.** $\left(\dfrac{sf - td}{cf - ed}, \dfrac{ct - es}{cf - ed}\right)$
43. $\left(\dfrac{\alpha d - b\beta}{ad - bc}, \dfrac{a\beta - c\alpha}{ad - bc}\right)$ **45.** $(800, 200)$ **47.** $(1800, 200)$
49. $(63, 84)$ **51.** $(-\frac{1}{5}, \frac{3}{10})$ **53.** $(\frac{11}{15}, \frac{1}{5})$
55. Answers vary. **57.** Answers vary. **59.** Answers vary.
61. Inconsistent system **63.** $(\frac{17}{21}, -\frac{20}{9})$

9.5 Problem Set, Pages 418–419

1. -75 **3.** -20 **5.** -44 **7.** 3 **9.** -3 **11.** 38
13. -4 **15.** 21 **17.** -204 **19.** -78 **21.** -664
23. 15 **25.** 61 **27.** 150 **29.** 0 **31.** 24
33. $(5, 4, -3)$ **35.** $(3, -1, 2)$ **37.** $(4, -3, 2)$
39. $(-4, 0, 3)$ **41.** $(5, \frac{15}{2}, \frac{3}{2})$ **43.** $x + 4y + 11 = 0$
45. $8x - 3y + 7 = 0$ **47.** 26 **49.** 110
51. $(w, x, y, z) = (3, 1, -2, 1)$
53. $(s, t, u, x) = (-3, 1, -2, -1)$ **55.** Answers vary.
57. Answers vary. **59.** Proof varies. **61.** Proof varies.
63. Proof varies.

9.6 Problem Set, Page 424

1. 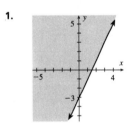 **3.**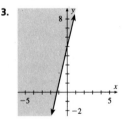

5. 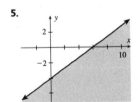 **7.**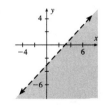

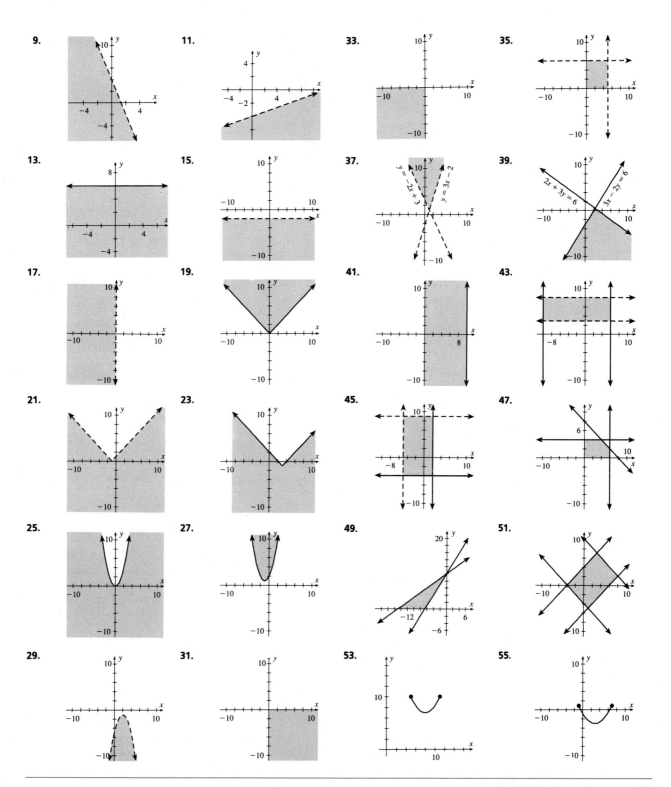

57.

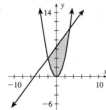

59.

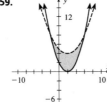

61.

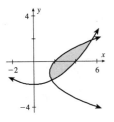

11–19. $P(k)$ is the same as $P(n)$ with n replaced by k. The $P(k + 1)$ for each problem follows.

11. $P(k + 1)$; $5 + 9 + \cdots + (4k + 5) = (k + 1)(2k + 5)$

13. $P(k + 1)$: $2^2 + 4^2 + \cdots + (2k + 2)^2 = \dfrac{(2k + 2)(k + 2)(2k + 3)}{3}$

15. $P(k + 1)$: $\cos(\theta + k\pi + \pi) = (-1)^{k+1} \cos \theta$

17. $P(k + 1)$: $(k + 1)^2 + (k + 1)$ is even.

19. $P(k + 1)$: $\left(\frac{2}{3}\right)^{k+2} < \left(\frac{2}{3}\right)^{k+1}$ **21–53.** Proofs vary.

55. $1^3 + 2^3 + 3^3 + \cdots + n^3 = \dfrac{n^2(n + 1)^2}{4}$

57. $2 + 6 + 18 + \cdots + (2 \cdot 3^{n-1}) = 3^n - 1$ **59.** Proof varies.

61. Proof varies.

10.2 Problem Set, Pages 440–441

1. 22   **3.** 2   **5.** 72   **7.** 132   **9.** 220   **11.** 1,140

13. 8   **15.** 28   **17.** 56   **19.** 70   **21.** 1,326

23. 1,000

25. $(a + b)^6 = a^6 + 6a^5b + 15a^4b^2 + 20a^3b^3 + 15a^2b^4$
$\qquad + 6ab^5 + b^6$

27. $(2x + 3)^3 = 8x^3 + 36x^2 + 54x + 27$

29. $(x + y)^5 = x^5 + 5x^4y + 10x^3y^2 + 10x^2y^3 + 5xy^4 + y^5$

31. $(3x + 2)^5 = 243x^5 + 810x^4 + 1,080x^3 + 720x^2 + 240x + 32$

33. $(x + y)^4 = x^4 + 4x^3y + 6x^2y^2 + 4xy^3 + y^4$

35. $(\frac{1}{2}x + y^3)^3 = \frac{1}{8}x^3 + \frac{3}{4}x^2y^3 + \frac{3}{2}xy^6 + y^9$ or
$\qquad \frac{1}{8}(x^3 + 6x^2y^3 + 12xy^6 + 8y^9)$

37. $(x^{1/2} + y^{1/2})^4 = x^2 + 4x^{3/2}y^{1/2} + 6xy + 4x^{1/2}y^{3/2} + y^2$

39. $(1 - x)^8 = 1 - 8x + 28x^2 - 56x^3 + 70x^4 - 56x^5$
$\qquad + 28x^6 - 8x^7 + x^8$

41. $a^{10} + 10a^9b + 45a^8b^2 + 120a^7b^3$

43. $a^{14} + 14a^{13}b + 91a^{12}b^2 + 364a^{11}b^3$

45. $x^{16} + 32x^{15}y + 480x^{14}y^2 + 4,480x^{13}y^3$

47. $x^{12} - 24x^{11}y + 264x^{10}y^2 - 1,760x^9y^3$

49. $1 - .24 + .0264 - .00176$ **51.** -6 **53.** 4,032

55. 160   **57–63.** Answers vary.

10.3 Problem Set, Pages 446–447

1. a. Arithmetic   **b.** $d = 3$ **c.** 17

3. a. Geometric   **b.** $r = 2$ **c.** 96

5. a. Neither   **b.** The difference increases by 1 each term.
c. 85

7. a. Geometric   **b.** $r = q$ **c.** pq^5

9. a. Neither   **b.** The difference increases by 2 each term
c. 36

11. a. Neither   **b.** The number of 5's between 2's increases by 1.
c. 2

13. a. Neither   **b.** Listing of fractions   **c.** $\frac{5}{6}$

15. a. Neither   **b.** Listing of perfect cubes   **c.** 216

17. 1, 5, 9   **19.** a, ar, ar^2

21. $-1, 1, -1$ **23.** $2, \frac{3}{2}, \frac{4}{3}$ **25.** 2, 2, 2

27. $\cos x, \cos 2x, \cos 3x$ **29.** 20   **31.** 49   **33.** 12

35. 11   **37.** 80   **39.** 10   **41.** 57   **43.** 1   **45.** 5^4

47. 2, 6, 18, 54, 162   **49.** 1, 1, 2, 3, 5   **51.** $\displaystyle\sum_{k=1}^{7} \frac{1}{2^k}$

53. $\displaystyle\sum_{k=1}^{5} 6^{k-1}$ **55.** Arithmetic   **57.** Geometric

---

■ **Chapter 9 Summary, Pages 425–426**

1. $(-5, -2)$ **3.** $(-3, 5)$ **5.** $(-6, 2)$ **7.** $(-16, 21)$

9. $(-3, 4)$ **11.** $(5, -9)$ **13.** $(7, -5, 1)$ **15.** $(1, 2, -3)$

17. $\begin{bmatrix} 5 & -1 & 1 \\ -3 & 8 & 7 \\ 3 & -6 & 6 \end{bmatrix}$ **19.** $\begin{bmatrix} 10 & -14 & -1 \\ 3 & -7 & 23 \\ 7 & -23 & -4 \end{bmatrix}$

21. $\begin{bmatrix} 7 & 2 \\ 3 & 1 \end{bmatrix}$ **23.** Does not exist   **25.** $(1, 7, 0)$

27. $(8, -2, 3)$ **29.** $(\frac{7}{16}, -\frac{11}{16})$ **31.** Inconsistent system

33. 29   **35.** -57 **37.** $(4, 0, -3)$ **39.** $(-1, -3, 5)$

41.

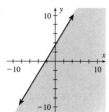

43.

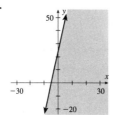

45.

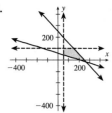

47.

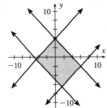

10.1 Problem Set, Pages 433–435

1. True; $5 = 1(2 \cdot 1 + 3)$

3. True: $2^2 = \dfrac{2 \cdot 1(1 + 1)(2 \cdot 1 + 1)}{3}$

5. True; $\cos(\theta + \pi) = (-1)^1 \cos \theta$ **7.** True; $1^2 + 1$ is even.

9. True; $(\frac{2}{3})^2 < (\frac{2}{3})^1$

---

59. Geometric

61. a. $a_1b_1 + a_2b_2 + a_3b_3 + \cdots + a_rb_r$ **b.** Answers vary.
c. Answers vary.

63. 47; add preceding terms. **65.** 7 (alphabetical)

10.4 Problem Set, Pages 451–452

1. 5, 9, 13, 17 **3.** 85, 88, 91, 94 **5.** 100, 95, 90, 85
7. $-\frac{5}{2}, -2, -\frac{3}{2}, -1$ **9.** $2\sqrt{3}, 3\sqrt{3}, 4\sqrt{3}, 5\sqrt{3}$
11. $x, x + y, x + 2y, x + 3y$ **13.** $a_1 = 5, d = 3$
15. $a_1 = 6, d = 5$ **17.** $a_1 = -8, d = 7$
19. $a_1 = x, d = x$ **21.** $a_1 = x - 5b, d = 2b$ **23.** $a_n = 5$
25. $a_n = 24 + 11n$ **27.** $a_n = -3 + 2n$ **29.** $a_n = x + n\sqrt{3}$
31. 101 **33.** 25 **35.** $-10,600$ **37.** -2 **39.** 2
41. -140 **43.** 4 **45.** 12 **47.** 11,500 **49.** 2,030
51. n^2
53. a. Harmonic **b.** Harmonic **c.** Harmonic
d. Harmonic **e.** Not harmonic
55. a. 12 **b.** $\frac{19}{2}$ **c.** $\frac{5}{12}$ **d.** -6 **e.** $\frac{1}{15}$
57. Option D **59.** A: \$43,440; B: \$44,160 (\$720 more than A);
C: \$44,520 (\$1,080 more than A and \$360 more than B); D:
\$44,760 (\$1,320 more than A, \$600 more than B, and \$240 more
than C)

10.5 Problem Set, Pages 457–458

1. 5, 15, 45 **3.** 1, -2, 4 **5.** $-15, -3, -\frac{3}{5}$
7. 8, 8x, $8x^2$ **9.** $g_1 = 3, r = 2$ **11.** $g_1 = 1, r = \frac{1}{2}$
13. $g_1 = x, r = x$ **15.** $g_n = 3 \cdot 2^{n-1}$ **17.** $g_n = 2^{1-n}$
19. $g_n = x^n$ **21.** 2 **23.** 200 **25.** $-\frac{135}{2}$ **27.** 486
29. 726 **31.** 1,111,111,111 **33.** $\frac{1}{2}$ **35.** $\frac{25}{23}$ **37.** $\frac{40}{81}$
39. $\frac{4}{9}$ **41.** 1 **43.** $\frac{2}{11}$ **45.** $\frac{418}{999}$ **47.** $\frac{41}{333}$ **49.** $\frac{1,132}{225}$
51. 111,111 **53.** 190 ft **55.** \$10,000
57. a. $2\sqrt{2}$ **b.** 4 **c.** $-\sqrt{15}$ **d.** $-2\sqrt{5}$ **e.** $4\sqrt{5}$
59. $\dfrac{9 + 3\sqrt{3}}{2}$ **61.** No sum since $|r| > 1$
63. a. No limit **b.** Limit **c.** Limit **d.** No limit **65.** $\frac{1}{4}$

Chapter 10 Summary, Pages 458–459

1. See text.
3. $P(k): 4 + 8 + 12 + \cdots + 4k = 2k(k + 1): P(k + 1):$
$4 + 8 + 12 + \cdots + 4(k + 1) = 2(k + 1)(k + 2)$
5. $a^5 + 5a^4b + 10a^3b^2 + 10a^2b^3 + 5ab^4 + b^5$
7. $32x^5 + 80x^4y + 80x^3y^2 + 40x^2y^3 + 10xy^4 + y^5$
9. 2,598,960 **11.** 70 **13.** See text.
15. $\binom{15}{r}x^{15-r}(-y)^r$ **17.** Arithmetic; $d = 10; a_n = -9 + 10n$
19. Neither; 11111, 111111 **21.** 40 **23.** $3^{10} - 1$ or 59,048
25. 460 **27.** $81 - 3^{-6} \approx 80.99862826$
29. $d = 2; A_{10} = 110$ **31.** $g_{10} = 2,560; G_5 = 155$
33. 2,000 **35.** $\frac{24}{11}$

1.
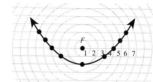

3.
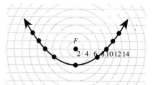

5.
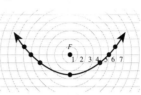

7.
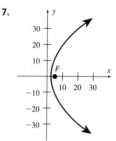

9.
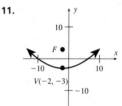

11.
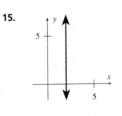

13.

15.
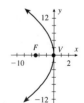

17.
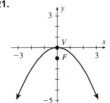

19.
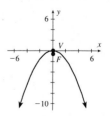

21.

23.

25.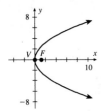

27.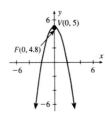

5.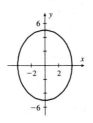

7.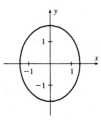

29.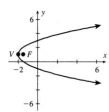

31.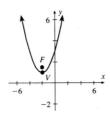

9.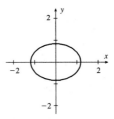

11.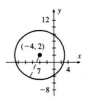

33.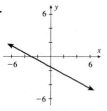

35.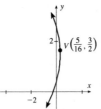

13.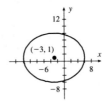

15.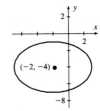

37.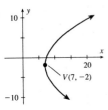

39.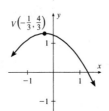

17.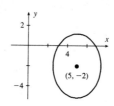

19.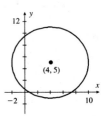

41. $y^2 = 10(x - \frac{5}{2})$ **43.** $(y - 2)^2 = -16(x + 1)$
45. $(x + 2)^2 = 24(y + 3)$ **47.** $(x + 3)^2 = -\frac{1}{3}(y - 2)$
49. $x = 2$ **51.** $(x - 100)^2 = -200(y - 50)$; D: $[0, 200]$
53. The focus is 2.25 m from the vertex on the axis of the parabola.
55. $(-3, 4)$, $(1, 12)$ **57.** $(2, 3)$, $(3, 1)$
59. 21.0 sec **61.** Answers vary. **63.** Answers vary.
65. a. $m = -\dfrac{A}{B}$ **b.** $m' = \dfrac{B}{A}$ **c.** $Bx - Ay - Bx_0 + Ay_0 = 0$

 d. $\left(\dfrac{B^2x_0 - ABy_0 - AC}{A^2 + B^2}, \dfrac{-ABx_0 + A^2y_0 - BC}{A^2 + B^2} \right)$

 e. Answers vary.
67. $25x^2 + 120xy + 144y^2 - 1{,}110x + 1{,}730y + 5{,}730 = 0$

21.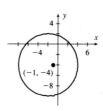

23.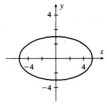

25.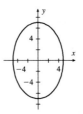

27.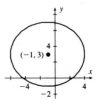

29.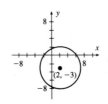

31. $(x - 4)^2 + (y - 5)^2 = 36$
33. $(x + 1)^2 + (y + 4)^2 = 36$
35. $\dfrac{x^2}{25} + \dfrac{y^2}{9} = 1$
37. $\dfrac{x^2}{24} + \dfrac{y^2}{49} = 1$

11.2 **Problem Set, Pages 476–478**

1.

3.

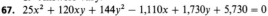

39. $\dfrac{(x+1)^2}{25} + \dfrac{(y-3)^2}{16} = 1$

41. $x^2 + y^2 - 4x + 6y - 12 = 0$ or $(x-2)^2 + (y+3)^2 = 25$

43. $(x+2)^2 + (y+3)^2 = 25$ **45.** $\dfrac{(x+3)^2}{9} + \dfrac{(y-2)^2}{16} = 1$

47. $\dfrac{(x+\frac{1}{3})^2}{\frac{1}{9}} + \dfrac{(y-1)^2}{\frac{1}{12}} = 1$ **49.** $\dfrac{(x-1)^2}{1} + \dfrac{(y+1)^2}{4} = 1$

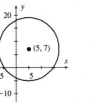

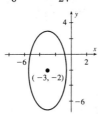

51. $(x-5)^2 + (y-7)^2 = 144$ **53.** $\dfrac{(x+3)^2}{6} + \dfrac{(y+2)^2}{24} = 1$

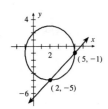

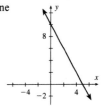

55. Answers vary.

57. Aphelion: 94,500,000 mi; perihelion: 91,500,000 mi

59. $\varepsilon = \dfrac{10}{189} \approx .053$

61. $(5, -1), (2, -5)$ **63.** Answers vary.

1.   **3.**

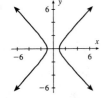

5.

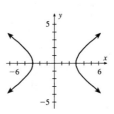

7.

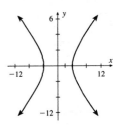

9.   **11.**

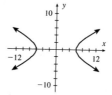

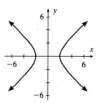

13.   **15.**

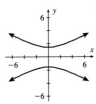

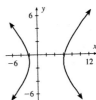

17.   **19.**

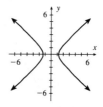

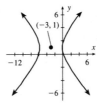

21.   **23.** Line

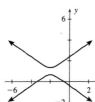

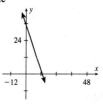

25. Parabola   **27.** Line

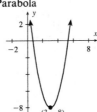

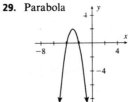

29. Parabola

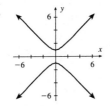

31. Ellipse

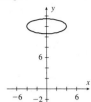

33. Circle

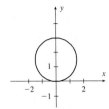

57. Ellipse

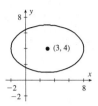

59–65. Answers vary.

35. $\dfrac{x^2}{25} - \dfrac{y^2}{11} = 1$ **37.** $\dfrac{(y-6)^2}{4} - \dfrac{(x-4)^2}{5} = 1$

39. $\dfrac{(x-2)^2}{16} - \dfrac{y^2}{3} = 1$

1. $xy = 6$

$$\left[\frac{1}{\sqrt{2}}(x' - y')\right]\left[\frac{1}{\sqrt{2}}(x' + y')\right] = 6$$

$$\tfrac{1}{2}(x' - y')(x' + y') = 6$$

$$x'^2 - y'^2 = 12$$

41. $\dfrac{(y+3)^2}{4} - \dfrac{(x+4)^2}{3} = 1$ **43.** $\dfrac{(x-1)^2}{\frac{4}{9}} - \dfrac{(y+2)^2}{1} = 1$

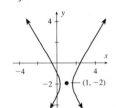

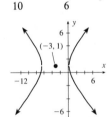

3. $x^2 - 4xy + 4y^2 + 5\sqrt{5}\,y - 10 = 0$

$$\tfrac{1}{5}(2x' - y')^2 - 4(\tfrac{1}{5})(2x' - y')(x' + 2y') + 4(\tfrac{1}{5})(x' + 2y')^2$$

$$+ 5\sqrt{5}\left(\frac{1}{\sqrt{5}}\right)(x' + 2y') - 10 = 0$$

$$4x'^2 - 4x'y' + y'^2 - 8x'^2 - 12x'y' + 8y'^2 + 4x'^2$$

$$+ 16x'y' + 16y'^2 + 25x' + 50y' - 50 = 0$$

$$25y'^2 + 50y' = -25x' + 50$$

$$y'^2 + 2y' = -x' + 2$$

$$y'^2 + 2y' + 1 = -x' + 3$$

$$(y' + 1)^2 = -(x' - 3)$$

45. $(x-2)^2 + (y+3)^2 = 25$ **47.** $\dfrac{(x+3)^2}{10} - \dfrac{(y-1)^2}{6} = 1$

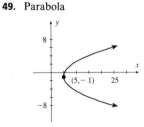

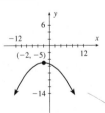

5. Hyperbola **7.** Ellipse **9.** Parabola
11. Hyperbola **13.** Parabola **15.** Ellipse
17. Ellipse **19.** Ellipse **21.** Parabola

23–25. $\theta = 45°;\ x = \dfrac{1}{\sqrt{2}}(x' - y');\ y = \dfrac{1}{\sqrt{2}}(x' + y')$

27. $\tan \theta = 2;\ \theta \approx 63.4°;\ x = \dfrac{1}{\sqrt{5}}(x' - 2y');\ y = \dfrac{1}{\sqrt{5}}(2x' + y')$

29–31. $\theta = 30°;\ x = \tfrac{1}{2}(\sqrt{3}x' - y');\ y = \tfrac{1}{2}(x' + \sqrt{3}y')$

33. $\theta = 60°;\ x = \tfrac{1}{2}(x' - \sqrt{3}y');\ y = \tfrac{1}{2}(\sqrt{3}x' + y')$

35. $\tan \theta = 3;\ \theta \approx 71.6°;\ x = \dfrac{1}{\sqrt{10}}(x' - 3y');\ y = \dfrac{1}{\sqrt{10}}(3x' + y')$

49. Parabola

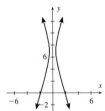

51. Parabola

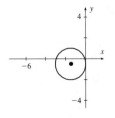

37. $\dfrac{y'^2}{2} - \dfrac{x'^2}{2} = 1$ **39.** $\dfrac{x'^2}{16} + \dfrac{y'^2}{16} = 1$

53. Hyperbola

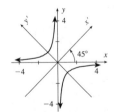

55. Circle

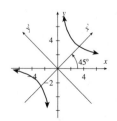

41. $\dfrac{x'^2}{9} + \dfrac{y'^2}{4} = 1$

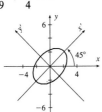

43. $(x'+1)^2 = 4\left(y' + \tfrac{7}{10}\right)$

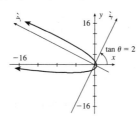

45. $\dfrac{x'^2}{4} - \dfrac{y'^2}{9} = 1$

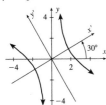

47. $x'^2 = -\tfrac{1}{16}(y'-4)$

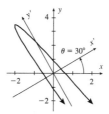

49. $\dfrac{y'^2}{4} - \dfrac{x'^2}{16} = 1$

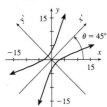

51. $\dfrac{x'^2}{\frac{9}{26}} + \dfrac{y'^2}{9} = 1$

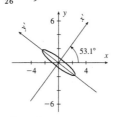

53. $\dfrac{y'^2}{4} - \dfrac{x'^2}{36} = 1$

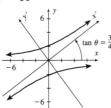

55. $\dfrac{x'^2}{9} + \dfrac{y'^2}{1} = 1$

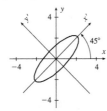

57. $\dfrac{(x'-2)^2}{4} + \dfrac{(y'-1)^2}{1} = 1$

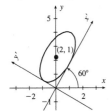

59. $\dfrac{(x'-2)^2}{4} + \dfrac{(y'+1)^2}{9} = 1$

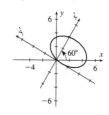

Also draw vector in Problems 1–17.
1. $6\mathbf{i} + 6\sqrt{3}\mathbf{j}$ **3.** $\mathbf{i} + \mathbf{j}$

5. $7\cos 23°\mathbf{i} + 7\sin 23°\mathbf{j}$ or $6.4435\mathbf{i} + 2.7351\mathbf{j}$
7. $4\cos 112°\mathbf{i} + 4\sin 112°\mathbf{j}$ or $-1.4989\mathbf{i} + 3.7087\mathbf{j}$
9. $-2\mathbf{i} + 2\mathbf{j}$ **11.** $-6\mathbf{i} - 5\mathbf{j}$ **13.** $8\mathbf{i} - 10\mathbf{j}$ **15.** $-7\mathbf{i} - \mathbf{j}$
17. $-9\mathbf{i} - 6\mathbf{j}$ **19.** 5 **21.** $\sqrt{85} \approx 9.2195$
23. $2\sqrt{2} \approx 2.8284$ **25.** $\sqrt{10} \approx 3.1623$ **27.** $\sqrt{41} \approx 6.4031$
29. No   **31.** Yes   **33.** Yes
35. 240 mph; S80°W (heading 260°)   **37.** 338 mi
39. $\mathbf{v} \cdot \mathbf{w} = -112; |\mathbf{v}| = 10; |\mathbf{w}| = 13; \cos\theta = -\tfrac{56}{65}$
41. $\mathbf{v} \cdot \mathbf{w} = 0; |\mathbf{v}| = 8; |\mathbf{w}| = 16; \cos\theta = 0$
43. $\mathbf{v} \cdot \mathbf{w} = -39; |\mathbf{v}| = 3\sqrt{10}; |\mathbf{w}| = \sqrt{29}; \cos\theta = -\dfrac{13\sqrt{290}}{290}$
45. $\mathbf{v} \cdot \mathbf{w} = 1; |\mathbf{v}| = 1; |\mathbf{w}| = 1; \cos\theta = 1$
47. $\mathbf{v} \cdot \mathbf{w} = 7; |\mathbf{v}| = \sqrt{26}; |\mathbf{w}| = \sqrt{13}; \cos\theta = \dfrac{7\sqrt{2}}{26}$
49. $\mathbf{v} \cdot \mathbf{w} = 1; |\mathbf{v}| = \sqrt{2}; |\mathbf{w}| = 1; \cos\theta = \dfrac{\sqrt{2}}{2}$
51. $75°$ **53.** $45°$ **55.** $90°$ **57.** $-\tfrac{8}{5}$
59. N4.9°E; 240 mph
61. The weight of the astronaut is resolved into two components, one parallel to the inclined plane with length y and the other perpendicular to it with length x. The weight of the astronaut is $|\mathbf{x}|$.

1. $2\mathbf{i} - 3\mathbf{j}$ **3.** $\mathbf{i} - \mathbf{j}$ **5.** $3\mathbf{i} - 2\mathbf{j}$ **7.** $9\mathbf{i} + 7\mathbf{j}$
9. $4\mathbf{i} - \mathbf{j}$ **11.** $\mathbf{i} + 2\mathbf{j}$ **13.** $3\mathbf{i} + 2\mathbf{j}$ **15.** $\mathbf{i} + \mathbf{j}$
17. $2\mathbf{i} + 3\mathbf{j}$ **19.** $7\mathbf{i} - 9\mathbf{j}$ **21.** $\mathbf{i} + 4\mathbf{j}$ **23.** $2\mathbf{i} - \mathbf{j}$
25. $\tfrac{63}{13}$ **27.** 0 **29.** $\dfrac{3\sqrt{61}}{61}$ **31.** $\tfrac{315}{169}\mathbf{i} + \tfrac{756}{169}\mathbf{j}$
33. 0 **35.** $\tfrac{18}{61}\mathbf{i} + \tfrac{15}{61}\mathbf{j}$ **37.** 0 **39.** $\tfrac{47}{5}$ **41.** $\tfrac{12}{5}$
43. $\tfrac{22}{5}$ **45.** $\tfrac{17}{5}\sqrt{10}$ **47.** $\tfrac{19}{10}\sqrt{10}$ **49.** $\tfrac{17}{29}\sqrt{29}$ **51.** $\tfrac{21}{2}$
53. $\tfrac{41}{2}$ **55.** 19 **57.** Answers vary.   **59.** Answers vary.

Also plot points in Problems 1–19.
1. $P: (4, \tfrac{\pi}{4}), (-4, \tfrac{5\pi}{4}); R: (2\sqrt{2}, 2\sqrt{2})$
3. $P: (5, \tfrac{2\pi}{3}), (-5, \tfrac{5\pi}{3}); R: (-\tfrac{5}{2}, \tfrac{5}{2}\sqrt{3})$
5. $P: (\tfrac{3}{2}, \tfrac{7\pi}{6}), (-\tfrac{3}{2}, \tfrac{\pi}{6}); R: (-\tfrac{3}{4}\sqrt{3}, -\tfrac{3}{4})$
7. $P: (-4, 4), (4, .86); R: (2.61, 3.03)$
9. $P: (-4, \pi), (4, 0); R: (4, 0)$ **11.** $(5\sqrt{2}, \tfrac{\pi}{4}); (-5\sqrt{2}, \tfrac{5\pi}{4})$
13. $(4, \tfrac{5\pi}{3}); (-4, \tfrac{2\pi}{3})$ **15.** $(3\sqrt{2}, \tfrac{7\pi}{4}); (-3\sqrt{2}, \tfrac{3\pi}{4})$
17. $(2, \tfrac{5\pi}{6}); (-2, \tfrac{11\pi}{6})$ **19.** $(13, 2.75); (-13, 5.89)$
21. Lemniscate   **23.** 3-leaved rose   **25.** Cardioid
27. None   **29.** None   **31.** None   **33.** None
35. Cardioid
37.

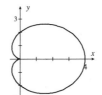

39.

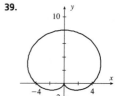

41.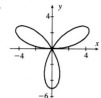

43.

59.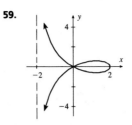

61.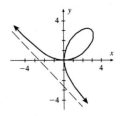

45.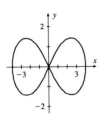

47.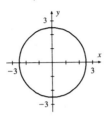

1.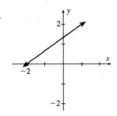

3.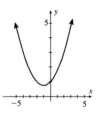

49.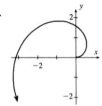

51. Answers vary.

53. 4.1751 (Use Law of Cosines.)

5.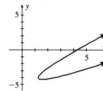

7.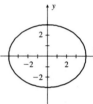

9.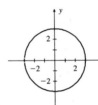

11.

55. a.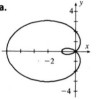

b.

13.

c.

d.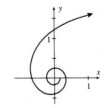

57. a.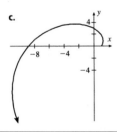

b.

c.

15. $y = \frac{2}{3}x + \frac{4}{3}$; same graph as in Problem 1.
17. $(x + 1)^2 = 4(y - \frac{3}{4})$; same graph as in Problem 3.
19. $x^2 + y^2 = 9$; same graph as in Problem 5.
21. $\dfrac{x^2}{16} + \dfrac{y^2}{9} = 1$; same graph as in Problem 7.
23. $x^2 - 2xy + y^2 - 13x + 12y + 38 = 0$; same graph as in Problem 9.
25. $y = 3x$, $x > 0$; same graph as in Problem 11.
27. $y = ex$, $x > 0$; same graph as in Problem 13.
29. $x = 4t$, $y = 9t$, $0 \le t \le 1$
31. $x = 3 - t$, $y = (3 - t)^2$, $0 \le t \le 3$
33. $x = 4 \cos t$, $y = 4 \sin t$, $0 \le t \le \frac{\pi}{2}$
35. $x = 3 \cos t$, $y = 2 \sin t$, $0 \le t \le 2\pi$
37. $x = t^2$, $y = t$, $1 \le t \le 3$

39. $x = 2 \sin 5\theta \cos \theta$, $y = 2 \sin 5\theta \sin \theta$

41.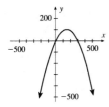

43.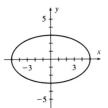

45.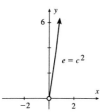

47.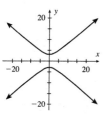

49.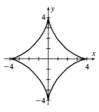

51.

53.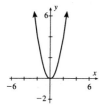

55.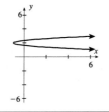

57.

5. $(y - 3)^2 = 20(x - 6)$

7. $(x + 3)^2 = -24(y - 5)$

9. $\dfrac{x^2}{16} + \dfrac{y^2}{25} = 1$

11. $(x - 2)^2 + (y - 1)^2 = 2$

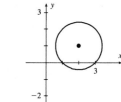

13. $\dfrac{(x - 4)^2}{4} + \dfrac{(y - 1)^2}{3} = 1$

15. $(x + 1)^2 + (y + 2)^2 = 64$

17. $\dfrac{(x + \frac{1}{2})^2}{3} - \dfrac{(y + \frac{1}{2})^2}{3} = 1$

19. $\dfrac{(x + 2)^2}{12} - \dfrac{(y + 4)^2}{20} = 1$

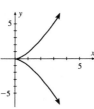

21. $\dfrac{(x + 5)^2}{1} - \dfrac{(y - 4)^2}{3} = 1$

23. $\dfrac{y^2}{9} - \dfrac{x^2}{16} = 1$

25. $\dfrac{(x - h)^2}{a^2} + \dfrac{(y - h)^2}{b^2} = 1$

27. $(y - k)^2 = 4c(x - h)^2$, $c > 0$

29. Parabola; $(y + 1)^2 = \frac{3}{2}(x + 3)$

31. Ellipse; $\dfrac{x^2}{9} + \dfrac{y^2}{25} = 1$

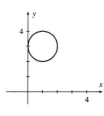

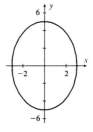

61. $x = |OA| = |OB| - |AB| = \text{arc } PB - |PQ| = a\theta - a \sin \theta$;
$y = |PA| = |QB| = |CB| - |CQ| = a - a \cos \theta$

Chapter 11 Summary, Pages 524–526

1.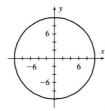

3.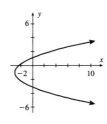

33. $45°$ **35.** $26.6°$ **37.** Hyperbola **39.** Parabola

41. $-2i + j$ **43.** 9.8; N80°E **45.** $v_x \approx 2.8i$; $v_y \approx 3.5j$

47. $5i + 12j$ **49.** $\sqrt{29}$ **51.** 6

53. Let $v = ai + bj$ and $w = ci + dj$; $v \cdot w = ac + bd$

55. 9 **57.** $-\frac{9}{169}\sqrt{13}$ **59.** $5°$

61. $N = 5i - 12j$; $v = 12i + 5j$ **63.** $N = i - 5j$; $v = 5i + j$

65. $\dfrac{5}{2}i + \dfrac{\sqrt{5}}{2}j$; $\dfrac{\sqrt{30}}{2}$ **67.** $0; 0$ **69.** $\frac{73}{13}$

71. $\frac{31}{13}\sqrt{13}$ **73.** $(5, 2.3771)$; $(-5, 5.5187)$; also plot point

75. $(-2, 2)$; $(2, 5.1416)$; also plot point

77. $(-1.5000, -2.5981)$ **79.** $(4.2426, 5.4978)$

81.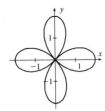

83.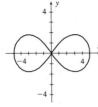

6. a. 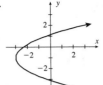 **b.**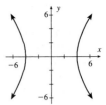

85. a. Lemniscate **b.** 4-leaved rose
87. a. None **b.** Cardioid

7. a. 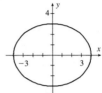 **b.**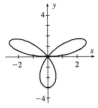

89.

91.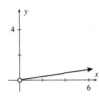

c.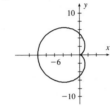

93. $3x + 5y - 1 = 0$; same graph as in Problem 89.
95. $y = e^{-2}x$, $x > 0$; same graph as in Problem 91.

Cumulative Review III, Page 527

1. a. $\frac{40}{3}$; geometric; $r = \frac{2}{3}$ **b.** 0; arithmetic; $d = -15$
  c. 30; neither
2. a. $(-2, 5)$ **b.** $(-1, 2), (-4, 5)$ **c.** $(1, -3)$
  d. $(2, -1, -1)$
3. a. $\begin{bmatrix} -1 & 1 \\ 3 & -2 \end{bmatrix}$ **b.** Not conformable **c.** $\begin{bmatrix} 0 & -1 \\ -1 & 1 \end{bmatrix}$
4. a. Parabola **b.** Hyperbola **c.** Hyperbola
  d. Circle **e.** Line **f.** Hyperbola **g.** 4-leaved rose
  h. Lemniscate **i.** Cardioid **j.** Circle
5. a. $\dfrac{(x - 2)^2}{4} + \dfrac{(y - 1)^2}{3} = 1$
  b. $\dfrac{(x - 6)^2}{4} - \dfrac{y^2}{5} = 1$
  c. $(y - 3)^2 = 12(x - 4)$

8. a. $(a + b)^n = \sum\limits_{k=0}^{n} \binom{n}{k} a^{n-k} b^k$
  b. $(a + b)^n = \binom{n}{0}a^n + \binom{n}{1}a^{n-1}b + \binom{n}{2}a^{n-2}b^2$
    $+ \cdots + \binom{n}{n-1}ab^{n-1} + \binom{n}{n}b^n$
  c. $x^4 - 12x^3y + 54x^2y^2 - 108xy^3 + 81y^4$
9. Answers vary. **10. a.** $9i - 7\sqrt{5}j$ **b.** -25 **c.** 7

Extended Application: Planetary Orbits, Page 530

1. Perihelion 1.5×10^8; aphelion 1.3×10^8
3. Perihelion 3.7×10^9; aphelion 3.6×10^9
5. 5,500 mi **7.** $1.2x^2 + 1.3y^2 = 1.6 \times 10^{15}$
9. $8.802x^2 + 8.805y^2 = 7.751 \times 10^{16}$

INDEX

TO THE OWNER OF THIS BOOK

We hope that you have found *Precalculus with Graphing and Problem Solving, Fifth Edition* useful. So that this book can be improved in a future edition, would you take the time to complete this sheet and return it to the publisher?

Thank you.

School and address: _____

_____

_____

Department: _____

Instructor's name: _____

1. What I like most about this book is: _____

_____

2. What I like least about this book is: _____

_____

3. My general reaction to this book is: _____

_____

_____

4. The name of the course in which I used this book is:_____

_____

5. Were all the chapters of the book assigned for you to read?   Yes   No

   If not, which ones weren't?_____

6. What specific suggestions do you have for improving this book?

_____

_____

_____

_____

_____

_____

OPTIONAL   Your name _____   Date _____

May Brooks/Cole Publishing Company quote you, either in promotion for
Precalculus with Graphing and Problem Solving, Fifth Edition, or in future
publishing ventures?

Yes _____   No _____

Sincerely,
Karl J. Smith

- FOLD HERE -

‖‖‖

┌───┐
│ BUSINESS REPLY MAIL   │
│ FIRST CLASS   PERMIT NO. 358   PACIFIC GROVE, CA │
└───┘

POSTAGE WILL BE PAID BY ADDRESSEE

ATT: *Karl J. Smith* _____

Brooks/Cole Publishing Company
511 Forest Lodge Road
Pacific Grove, California 93950-9968

‖‖·‖‖·‖‖‖·‖‖·‖‖‖·‖‖·‖·‖‖·‖‖·‖‖

- FOLD HERE -

▌ FORMULAS

Distance between (x_1, y_1) and (x_2, y_2): $d = \sqrt{(\Delta x)^2 + (\Delta y)^2}$, where $\Delta x = x_2 - x_1$ and $\Delta y = y_2 - y_1$.

Slope of the line passing through (x_1, y_1) and (x_2, y_2): $m = \Delta y / \Delta x$.

Pythagorean Theorem: The sum of the squares of the lengths of the legs of a right triangle is equal to the square of the length of the hypotenuse.

Quadratic Formula: If $ax^2 + bx + c = 0$, $a \neq 0$, then $x = \dfrac{-b \pm \sqrt{b^2 - 4ac}}{2a}$.

Discriminant: $b^2 - 4ac$
$b^2 - 4ac < 0 \rightarrow$ *no real* solutions
$b^2 - 4ac = 0 \rightarrow$ *one real* solution
$b^2 - 4ac > 0 \rightarrow$ *two real* solutions

Arithmetic Sequence: $a_n = a_1 + (n - 1)d$

Geometric Sequence: $g_n = g_1 r^{n-1}$

Arithmetic Series: $A_n = \dfrac{n}{2}[2a_1 + (n - 1)d]$ or $n\left(\dfrac{a_1 + a_n}{2}\right)$

Geometric Series: $G_n = \dfrac{g_1(1 - r^n)}{1 - r}$; $G = \dfrac{g_1}{1 - r}$ if $|r| < 1$

Binomial Expansion: $(a + b)^n = \displaystyle\sum_{k=0}^{n} \binom{n}{k} a^{n-k} b^k$, where $\binom{n}{k} = \dfrac{n!}{k!(n - k)!}$

Forms of a Complex Number z:

Rectangular form: $z = a + bi$

Polar form: $z = r \operatorname{cis} \theta$
$\qquad = r(\cos \theta + i \sin \theta)$

Conversions: $a = r \cos \theta \qquad r = \sqrt{a^2 + b^2}$
$\qquad\qquad b = r \sin \theta \qquad \theta' = \tan^{-1}\left|\dfrac{b}{a}\right|$, where θ' is the reference angle for θ

De Moivre's Theorem: If n is a natural number, $(r \operatorname{cis} \theta)^n = r^n \operatorname{cis} n\theta$.

nth Root Theorem: If n is a positive integer, $(r \operatorname{cis} \theta)^{1/n} = \sqrt[n]{r} \operatorname{cis} \dfrac{1}{n}(\theta + 360°k)$, where $k = 0, 1, 2, \ldots, n - 1$.

Cramer's Rule:

If $\begin{cases} a_{11}x + a_{12}y + a_{13}z = b_1 \\ a_{21}x + a_{22}y + a_{23}z = b_2 \\ a_{31}x + a_{32}y + a_{33}z = b_3 \end{cases}$ then $x = \dfrac{D_x}{D}, \qquad y = \dfrac{D_y}{D}, \qquad z = \dfrac{D_z}{D} \qquad (D \neq 0)$

D is the determinant of the coefficients and the constants b_1, b_2, b_3 replace the coefficients x, y, z for D_x, D_y, and D_z, respectively.

Matrices

$A + B = [a_{ij}]_{m \times n} + [b_{ij}]_{m \times n} = [a_{ij} + b_{ij}]_{m \times n} \qquad cA = c[a_{ij}]_{m \times n} = [ca_{ij}]_{m \times n}$

$I = [a_{ij}]_{n \times n}$, where $a_{ij} = \begin{cases} 1 & \text{if } i = j \\ 0 & \text{if } i \neq j \end{cases} \qquad AB = [a_{ij}]_{m \times n}[b_{ij}]_{n \times p} = [a_{i1}b_{1j} + a_{i2}b_{2j} + \cdots + a_{in} + b_{nj}]_{m \times p}$

$IA = AI = A \qquad\qquad\qquad A^{-1}A = AA^{-1} = I$

Principle of Mathematical Induction: If a given proposition $P(n)$ is true for $P(1)$ and if the truth of $P(k)$ implies its truth for $P(k + 1)$, then $P(n)$ is true for all positive integers n.

FUNCTIONS

Definition: A function f of a real number x is a rule that assigns a single real number $f(x)$ to each x in the domain of the function. This definition can be characterized as a mapping or as a set of ordered pairs.

Constant Function: $f(x) = a$

Linear Function: $f(x) = mx + b$

Standard form:	$Ax + By + C = 0$
Point–slope form:	$y - k = m(x - h)$
Slope–intercept form:	$y = mx + b$

Two–point form: $\quad y - y_1 = \left(\dfrac{y_2 - y_1}{x_2 - x_1}\right)(x - x_1)$ or $\begin{vmatrix} x & y & 1 \\ x_1 & y_1 & 1 \\ x_2 & y_2 & 1 \end{vmatrix} = 0$

Intercept form: $\quad \dfrac{x}{a} + \dfrac{y}{b} = 1$

Horizontal line: $\quad y = k$

Vertical line: $\quad x = h$

Factoring Rules:

$x^2 - y^2 = (x - y)(x + y)$

$x^3 - y^3 = (x - y)(x^2 + xy + y^2)$

$x^3 + y^3 = (x + y)(x^2 - xy + y^2)$

See also Table 1.2, page 13

Quadratic Function: $f(x) = ax^2 + bx + c, a \neq 0$

Absolute Value Function: $f(x) = |x| = \begin{cases} x & \text{if } x \geq 0 \\ -x & \text{if } x < 0 \end{cases}$

Polynomial Function: $f(x) = a_n x^n + a_{n-1} x^{n-1} + a_{n-2} x^{n-2} + \cdots + a_2 x^2 + a_1 x + a_0, a_n \neq 0$

Rational Function: $f(x) = P(x)/D(x)$, where $P(x)$ and $D(x)$ are polynomial functions and $D(x) \neq 0$

Asymptotes (from optional section):

Let $f(x) = P(x)/D(x)$, where $P(x)$ and $D(x)$ are polynomial functions and $P(x)$ and $D(x)$ have no common factors. Let $P(x)$ have degree m with leading coefficient p and $D(x)$ have degree n with leading coefficient d.

Vertical asymptote: $\quad x = r$, where $D(r) = 0$

Horizontal asymptote: $\quad y = 0$ if $m < n$; $y = p/d$ if $m = n$; none if $m > n$

Slant asymptote: $\quad y = mx + b$ if $m = n + 1$ and $f(x) = mx + b + \dfrac{R(x)}{D(x)}$ by long division

Exponential Function: $f(x) = b^x$ $(b > 0, b \neq 1)$

Laws of exponents: Let a and b be nonzero real numbers and let m and n be any real numbers, except that the form 0^0 and division by zero are excluded.

FIRST LAW: $b^m \cdot b^n = b^{m+n}$

SECOND LAW: $\dfrac{b^m}{b^n} = b^{m-n}$

THIRD LAW: $(b^n)^m = b^{mn}$

FOURTH LAW: $(ab)^m = a^m b^m$

FIFTH LAW: $\left(\dfrac{a}{b}\right)^m = \dfrac{a^m}{b^m}$

Logarithmic Function: $f(x) = \log_b x$ $(b > 0, b \neq 1)$

Laws of logarithms:

FIRST LAW: $\log_b AB = \log_b A + \log_b B$

SECOND LAW: $\log_b \dfrac{A}{B} = \log_b A - \log_b B$

THIRD LAW: $\log_b A^p = p \log_b A$

Log of Both Sides Theorem: $\log_b A = \log_b B \Leftrightarrow A = B$

Change of Base Theorem: $\log_a x = \dfrac{\log_b x}{\log_b a}$

Trigonometric Functions: See the definitions on the inside back cover.